VICTIMOLOGY

SAGE Text/Reader Series in Criminology and Criminal Justice

Craig Hemmens, Series Editor

1. *Introduction to Criminology: A Text/Reader 2nd Edition* by Anthony Walsh/Craig Hemmens
2. *Juvenile Justice: A Text/Reader* by Richard Lawrence/Craig Hemmens
3. *Corrections: A Text/Reader 2nd Edition* by Mary Stohr/Anthony Walsh/Craig Hemmens
4. *Courts: A Text/Reader 2nd Edition* by Cassia Spohn/Craig Hemmens
5. *Policing: A Text/Reader (forthcoming)* by Carol Archbold
6. *Community Corrections : A Text/Reader* by Shannon Barton-Belessa/Robert Hanser
7. *Race and Crime: A Text/Reader* by Helen Greene/Shaun Gabbidon
8. *Criminological Theory: A Text/Reader* by Stephen Tibbetts/Craig Hemmens
9. *Victimology: A Text/Reader* by Leah Daigle
10. *Women and Crime: A Text/Reader* by Stacy Mallicoat
11. *White Collar Crime: A Text/Reader* by Brian Payne

Other Titles of Related Interest

Victims of Crime, 3rd edition by Robert Davis/Arthur Lorigio/Susan Herman
Unsafe in the Ivory Tower by Bonnie Fisher/Leah Daigle/Frank Cullen
Responding to Domestic Violence, 4th edition by Eve Buzawa/Carl Buzawa/Evan Stark
White Collar Crime: The Essentials by Brian Payne
Deviance and Social Control by Michelle Inderbitzen/ Kristin Bates/Randy Gainey
Criminal Justice Ethics, 3rd edition by Cyndi Banks
Introduction to Policing by Gene Scaramella/ Steven Cox/ William McCamey
Corrections: The Essentials by Mary Stohr/Anthony Walsh
Community Corrections by Robert Hanser
Correctional Theory by Frank Cullen/Cheryl Lero Jonson
Addicted to Incarceration by Travis Pratt
Introduction to Criminology, 7th edition by Frank Hagan
Criminological Theory, 5th edition by J. Robert Lilly/Frank Cullen/ Richard Ball
Criminological Theory: The Essentials by Stephen Tibbetts
Key Ideas in Criminology and Criminal Justice by Travis Pratt/Jacinta Gau/ Travis Franklin
Crime and Everyday Life, 5th edition by Marcus Felson/Rachel Boba
Race and Crime, 3rd edition by Shaun Gabbidon, Helen Greene
Juvenile Justice, 7th edition by Steven Cox, Jennifer Allen/Robert Hanser/John Conrad
Juvenile Justice: The Essentials by Richard Lawrence/Mario Hesse
Understanding Terrorism 4th edition by Gus Martin
Gangs in America's Communities by James Howell
Critical Issues in Criminology and Criminal Justice by Mary Maguire/Dan Okada
Criminal Courts by Craig Hemmens, David Brody/Cassia Spohn
Criminal Procedure by Matthew Lippman
Contemporary Criminal Law, 2nd edition by Matthew Lippman
Crime Analysis with Crime Mapping, 3rd edition by Rachel Boba Santos
The Practice of Research in Criminology and Criminal Justice, 4th edition by Ronet Bachman/Russell Schutt
Fundamentals of Research in Criminology and Criminal Justice, 2nd edition by Ronet Bachman/Russell Schutt
Statistics for Criminal Justice by Jacinta Gau
The Mismeasure of Crime, 2nd edition by Clay Mosher/Terence Miethe/Timothy Hart

VICTIMOLOGY
A Text/Reader

Leah E. Daigle

Georgia State University

Los Angeles | London | New Delhi
Singapore | Washington DC

Los Angeles | London | New Delhi
Singapore | Washington DC

FOR INFORMATION:

SAGE Publications, Inc.
2455 Teller Road
Thousand Oaks, California 91320
E-mail: order@sagepub.com

SAGE Publications Ltd.
1 Oliver's Yard
55 City Road
London EC1Y 1SP
United Kingdom

SAGE Publications India Pvt. Ltd.
B 1/I 1 Mohan Cooperative Industrial Area
Mathura Road, New Delhi 110 044
India

SAGE Publications Asia-Pacific Pte. Ltd.
33 Pekin Street #02-01
Far East Square
Singapore 048763

Acquisitions Editor: Jerry Westby
Production Editor: Karen Wiley
Copy Editor: Megan Granger
Typesetter: C&M Digitals (P) Ltd.
Proofreader: Emily Bakely
Indexer: Sheila Bodell
Cover Designer: Janet Kiesel
Marketing Manager: Erica DeLuca
Permissions Editor: Karen Ehrmann

Printed in the United States of America

Library of Congress Cataloging-in-Publication Data

Daigle, Leah E.
Victimology : a text/reader / Leah E. Daigle.

p. cm. — (SAGE text/reader series in criminology and criminal justice)

Includes bibliographical references and index.

ISBN 978-1-4129-8732-5 (pbk. : alk. paper)

1. Victims of crimes. I. Title.

HV6250.25.D336 2012
362.88—dc23   2011041241

This book is printed on acid-free paper.

12 13 14 15 10 9 8 7 6 5 4 3 2

Brief Contents

Detailed Contents

READINGS

READINGS

Foreword

You hold in your hands a book that is part of a series we have created at Sage, and that we think represents an innovative approach to criminology and criminal justice pedagogy. It is a "text/reader." What that means is that we have attempted to blend the two most commonly used types of books, the textbook and the reader, in a way that will appeal to both students and faculty.

Our experience as teachers and scholars has been that textbooks for the core classes in criminal justice and criminology (or any other social science discipline) leave many students and professors cold. The textbooks are huge, crammed with photographs, charts, highlighted material, and all sorts of pedagogical devices intended to increase student interest. Too often, though, these books end up creating a sort of sensory overload for students; they suffer from a focus on "bells and whistles," such as fancy graphics, at the expense of coverage of the most current research on the subject matter.

Readers, on the other hand, are typically comprised of recent and classic research articles on the subject matter. They generally suffer, however, from an absence of meaningful textual material. Articles are simply lined up and presented to the students, with little or no context or explanation. Students, particularly undergraduate students, are often confused and overwhelmed by the jargon and detailed statistical analysis presented in the articles.

This text/reader represents our attempt to take the best of both the textbook and reader approaches. It is comprised of research articles on victimology and is intended to serve either as a supplement to a core textbook or as a stand-alone text. The book includes a combination of previously published articles and textual material that introduces and provides some structure and context for the selected readings. The book is divided into a number of sections. The sections follow the typical content and structure of a textbook on the subject. Each section of the book has an overview of the topic that serves to introduce, explain, and provide context for the readings that follow. The readings are a selection of the best recent research that has appeared in academic journals, as well as some essential older readings. The articles are edited as necessary. This variety of research and perspectives provides students with an understanding of both the development of research and the current status of research on victimology. This approach gives the student the opportunity to learn the basics (in the text portion of each section) and to read some of the most interesting research on the subject.

There is also a preface and an introductory chapter on how to read a research article. The preface explains the organization and content of the book. The introductory chapter provides a framework for the text and articles that follow and introduces relevant themes, issues, and concepts. This will assist the student in understanding the articles.

Each section also includes a summary of the material covered and a selection of discussion questions. These summaries and discussion questions should facilitate student thought and class discussion of the material.

We acknowledge that this approach may be viewed by some as more challenging than the traditional textbook. To that we say, "Yes! It is!" But we believe that if we raise the bar, our students will rise to the challenge. Research shows that students and faculty often find textbooks boring to read. It is our belief that many criminology and criminal justice instructors welcome the opportunity to teach without having to rely on a standard textbook that covers only the most basic information and that lacks both depth of coverage and an attention to current research. This book provides an alternative for instructors who want to get more out of the basic criminal justice courses or curriculum than one can get from a basic textbook that is filled with flashy but often useless features that merely serve to drive up the cost of the textbook. This book is intended for instructors who want to go beyond the ordinary, basic coverage provided in textbooks.

We also believe students will find this approach more interesting. They are given the opportunity to read current, cutting-edge research on the subject, while also being provided with background and context for this research. We hope that this unconventional approach will make learning and teaching more fun. Crime and criminal justice are fascinating subjects, and they deserve to be presented in an interesting manner. We hope you will agree.

Craig Hemmens, JD, PhD, Series Editor
Department of Criminology & Criminal Justice
Missouri State University

Preface

While offender behavior and the impacts of crime long have been studied, how crime victimization shapes the lives of victims was not similarly studied until recently. Now, policymakers, practitioners, academics, and activists alike have recognized the importance of studying the other half of the crime-victim dyad. Indeed, it is an exciting time to study victimology—an academic field that is growing rapidly. Hence, this text fills a void in what is currently available in the market. As noted below, it is a hybrid text/reader that incorporates up-to-date original textual information with empirical research. In addition, it uses a consistent framework throughout to orient the reader. Finally, although victimology is relatively "new," the latest topics are discussed.

I have attempted to incorporate a general framework in each section—one that examines the causes and consequences of specific types of victimization and the responses to them. My intent was to create a comprehensive work that examines many types of victimization from a common framework so that similarities and differences can be easily identified.

Within this framework, I pay particular attention to identifying the characteristics of victims and incidents so that theory can be applied to understanding why some people are victims while others remain unscathed. Although the earliest forays into the study of victimology were focused on identifying victim typologies, theory development in this field has lagged behind that in criminology. Aside from routine activities and lifestyles theory, there are few theories that explicitly identify causes of victimization. This is not to say that the field of victimology is devoid of theory—it is just that the theories that have been applied to victimization are largely derived from other fields of study. I have included a section that discusses these theories, and in each section about a specific type of victimization, I have identified the causes and how theory may apply. Knowing this is a critical first step in preventing victimization and revictimization.

I also wanted to include throughout the text emerging issues in the field of victimology. To this end, each section discusses current issues germane to its particular topic and the latest research. For example, same-sex intimate partner violence is covered in depth, as are cyberbullying, identity theft victimization, and the offender-victim overlap. Other sections wholly address contemporary issues. Specifically, there are sections devoted to victims of terrorism, hate crime, and human trafficking. I believe that the inclusion of the latest issues within the field of victimology will expose the reader to the topics likely to garner the most attention in the years to come.

This text covers these topics with textual material and supplements this material with scholarly journal articles, an additional unique contribution of the book. The readings cover issues related to the textual material but provide a more in-depth treatment of the concepts and ideas presented. As such, the book is appropriate for undergraduate students and can be used as a primary text but can also be used as a supplemental text for classes using a more traditional textbook. Given the book's inclusion of original research, instructors may also find it appropriate to use in graduate classes. The book is appropriate for classes within criminal justice and criminology programs (e.g., victimology, crime victims, gender and crime) but is also relevant for women's studies, social work, psychology, and sociology courses.

The book contains 11 sections that were selected because they address the topics typically covered in victimology courses. These sections include the following:

- Introduction to Victimology
- Extent, Theories, and Factors of Victimization
- Consequences of Victimization
- Victims' Rights and Remedies
- Sexual Victimization
- Intimate Partner Violence
- Victimization at the Beginning and End of Life: Child and Elder Abuse
- Victimization of Special Populations
- Victimization at School and Work
- Property and Identity Theft Victimization
- Contemporary Issues in Victimology: Victims of Hate Crimes, Human Trafficking, and Terrorism

The text also includes a range of features to aid both professors and students:

- Each section is summarized in bullet points.
- Discussion questions are included at the end of each section.
- A list of key terms is included at the end of each section.
- Internet resources relevant for each section are provided.
- After each section, several articles that address the topics covered are included. Original empirical research and review pieces are included, and additional readings are available on SAGE's website.
- Discussion questions following each article orient the student to important points addressed in the reading. The questions will also be used as a basis to ensure a critical assessment of the material.
- Each section includes graphics pertinent to the topic presented.
- The book has a glossary of key terms.

Acknowledgments

As excited as I am for this book to be published, it was a bit of an unplanned project. I am in Craig Hemmens's debt, actually, for it was he who called me to pitch the idea of writing this text/reader. I owe Brian Payne, my department chair, a thank you for convincing me to say "yes" to Craig and Jerry Westby at SAGE Publications. Brian and I were working on text/readers at the same time—without sharing this experience with him, I do not think I would have been able to finish mine in a timely manner. He kept me laughing and motivated when I was overwhelmed—it was harder than I thought it would be (TWSS)!

I also would like to thank the editorial and production staff at SAGE Publications for their assistance. Jerry Westby has been wonderfully supportive throughout this process. I appreciate your continued confidence in my work. Erim Sarbuland and Karen Wiley, Jerry's editors, have provided valuable assistance. Thanks also to Megan Granger for her copyediting work and to Erica Deluca for her work marketing the book.

I owe a great debt to a number of students at Georgia State who assisted me in various ways—especially on those tasks I did not want to do myself. Sarah Keller and Samantha Role read articles and gave me feedback on ones that undergraduates may enjoy. Erin Marsh located references for me. Susannah Tapp was a great help with the glossary and references during my final push to get the book finished. Andia Azimi helped collect research. Sadie Mummert helped with editing. Candace Johnson created tables, worked on references, and collected Internet resources—a special thanks goes out to her.

Many special people have provided me with much-needed encouragement and feedback. I cannot think of a better place to start the work day than in the coffee shop at Georgia State University with Wendy Guastaferro, Dean Dabney, and Brent Teasdale. Thank you for listening to me talk about this book almost every day for the past year. Wendy, your friendship has truly seen me through the ups and downs of this project and others. Bonnie Fisher and Frank Cullen at the University of Cincinnati were great supports, as they have been since I was a student there. Finally, thank you, Zack. With you, I have laughed more and worried less than I ever thought possible.

Although the review process was always daunting, a number of scholars provided wonderful feedback that improved the final book. I cannot thank them enough for their time and effort for providing such detailed and invaluable reviews. Along with SAGE Publications, I wish to thank these reviewers:

Kathryn A. Branch, University of Tampa

Karen Bune, Marymount University/George Mason University

Julia Campbell, University of Northern Kentucky

Alison Cares, University of Massachusetts at Lowell

Ellen G. Cohn, Florida International University

Andrew Davies, State University of New York at Albany

Amanda Gendon, University of Missouri, St. Louis

Lee Michael Johnson, University of West Georgia

Lynn Jones, Northern Arizona University

Howard Kurtz, Oklahoma City University

Stephanie P. Manzi, Roger Williams University

Elizabeth Quinn, Fayetteville State University

Shannon A. Santana, University of North Carolina, Wilmington

Cheryl Swanson, University of West Florida

Jeffrey Walsh, Illinois State University

Janet Wilson, University of Central Arkansas

How to Read a Research Article

L et's use one of the articles in Section 2, "Specifying the Influence of Family and Peers on Violent Victimization: Extending Routine Activities and Lifestyles Theories" by Christopher J. Schreck and Bonnie S. Fisher, as an example for the following steps.

1. What is the thesis or main idea of this article?

The thesis or main idea is found in the introduction of this article. Although it is not directly stated, in the final paragraph of the introduction, Schreck and Fisher note what they are testing. That is, they are examining whether the family and associating with delinquent peers impacts juveniles' level of violent victimization.

2. What is the hypothesis?

Schreck and Fisher identify two hypotheses, although these hypotheses are not specifically stated. You can identify the general hypotheses by reading the introduction. The first hypothesis is located in the second paragraph. The authors suggest that a strong family will reduce the likelihood that an adolescent will experience violent victimization. The second hypothesis is discussed in the following paragraph. The authors hypothesize that membership in delinquent peer groups will lead to increased risk of violent victimization.

3. Is there any prior literature related to the hypothesis?

Literature linking victimization to offending is cited throughout the introduction. In addition, Schreck and Fisher use routine activities/lifestyles theory to justify why the family and peer group should be considered when trying to understand violent victimization risk among adolescents. They link the elements of routine activities theory to family and peers and explain how these activities impact violent victimization risk.

4. What methods are used to test the hypothesis?

Schreck and Fisher use survey research to test their hypotheses. They used data from the National Longitudinal Study of Adolescent Health, which had been collected by other researchers. This is known as secondary data analysis. They used a type of regression analysis called overdispersed Poisson regression as their analytic tool to examine factors related to higher levels of violent victimization.

5. Is this a qualitative or a quantitative study?

Qualitative studies are those that use nonnumerical data or methods, while quantitative studies use numbers to tests hypotheses. Schreck and Fisher use statistics to test their hypotheses; therefore, their study is quantitative.

6. *What are the results, and how does the author present them?*

The results of the study are presented in the "Results" section. The authors present their results in two forms—in a table (Table 1) and in the write-up of the results section. If the statistical technique is one with which you are unfamiliar, the table can be examined by finding those variables that have asterisks by their coefficients or values. The asterisks indicate that the variable is significant in the model. Those variables' coefficients or values that do not have asterisks beside them are not significant. The write-up also will tell you what variables are significant and how to interpret the values. Schreck and Fisher found that "teenagers whose friends engaged in higher levels of delinquency and risky activity tended to experience more victimization." Spending time with those friends also increased risk of experiencing violence. Two family context variables were related to violent victimization. A warm, accepting family climate and parents' positive feelings toward the adolescent reduced risk of violent victimization.

7. *Do you believe that the authors provided a persuasive argument? Why or why not?*

The answer to this question is subjective; however, the authors do make a strong case for examining both the peer and family context in risk for violent victimization. A close reading of the front end of the paper provides the reader the basis for the argument. The analytic test shows support for their hypotheses—variables measuring both peer and family context do "matter" in predicting risk of violent victimization. It should be noted, though, that variables operationalizing the routine activities/lifestyles perspective were also shown to be relevant (they were significant). This finding suggests that both the routine activities/lifestyle perspective and the peer and family context should be further examined.

8. *Who is the intended audience of this article?*

As you read the article, you should consider to whom the authors are speaking. Most academic articles have professors as their audience, of course, but they are also often written for others. Schreck and Fisher are writing for students, practitioners, and policymakers as well as for professors. When determining who the audience is, you should ask yourself who would benefit from reading the article. Also, consider for whom the findings would be beneficial.

9. *What does the article add to your knowledge of the subject?*

This question is obviously specific to the reader. One way to answer this question may be as follows: This article helps the reader understand how the peer and family context influence risk for violent victimization. It also helps the reader understand how the impact of peer and family variables on the risk of violent victimization can be explained through a routine activities/lifestyles theory approach.

10. *What are the implications for criminal justice policy that can be derived from this article?*

Implications for criminal justice policy are likely to be found in the discussion or conclusion sections of the article. In the conclusion, Schreck and Fisher discuss the implications of their findings mainly in terms of how future research and theory may be impacted. They do, however, question policies intended to reduce victimization by educating potential victims about ways to reduce their risk, since it is possible that victimization risk is tied to family attachment, which they link to socialization. Socialization through the family via its effect on the development of self-control has been previously linked to victimization risk. Schreck and Fisher suggest that victimization may be linked to something that cannot be easily changed (i.e., low self-control) rather than something that can be easily manipulated (i.e., a person's routine activities).

In this way, policy implications are discussed. You will notice that some articles have more criminal justice policy implications.

Now that we have examined the elements of a research article, it is your turn to answer these questions as you read through the various articles in the text. Some of the articles may be easier for you to read than others, and not all the articles will have all the elements you just read about: introduction, literature review, methods, results, and discussion. You should refer to this introduction on how to read a research article for guidance if you have any problems.

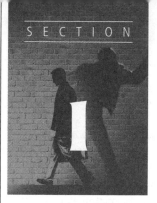

Introduction to Victimology

What Is Victimology?

The term *victimology* is not new. In fact, Benjamin Mendelsohn first used it in 1947 to describe the scientific study of crime victims. Victimology is often considered a subfield of criminology, and the two fields do share much in common. Just as criminology is the study of criminals—what they do, why they do it, and how the criminal justice system responds to them—victimology is the study of victims. **Victimology**, then, is the study of the etiology (or causes) of victimization, its consequences, how the criminal justice system accommodates and assists victims, and how other elements of society, such as the media, deal with crime victims. Victimology is a science; victimologists use the scientific method to answer questions about victims. For example, instead of simply wondering or hypothesizing why younger people are more likely to be victims than are older people, victimologists conduct research to attempt to identify the reasons why younger people seem more vulnerable.

The History of Victimology: Before the Victims' Rights Movement

As previously mentioned, the term *victimology* was coined in the mid-1900s. Crime was, of course, occurring prior to this time; thus, people were being victimized long before the scientific study of crime victims began. Even though they were not scientifically studied, victims were recognized as being harmed by crime and their role in the criminal justice process has evolved over time.

Before and throughout the Middle Ages (about the 5th through the 16th century), the burden of the justice system, informal as it was, fell on the victim. When a person or property was harmed, it was up to the victim and the victim's family to seek justice. This was typically achieved via retaliation. The justice system operated under the principle of **lex talionis,** an eye for an eye. A criminal would be punished because he or she deserved it, and the punishment would be equal to the harm caused. Punishment based on these notions is consistent with **retribution.** During this time period, a crime was considered a harm against the victim, not the state. The concepts of restitution and retribution governed action against criminals. Criminals were expected to pay back the victim through **restitution.** During this time period, a criminal who stole a person's cow likely would have to compensate the owner (the victim) by returning the stolen cow and also giving him or her another one.

Early criminal codes incorporated these principles. The **Code of Hammurabi** was the basis for order and certainty in Babylon. In the code, restoration of equity between the offender and victim was stressed. Notice that the early response to crime centered on the victim, not the state. This focus on the victim continued until the Industrial Revolution, when criminal law shifted to considering crimes violations against the state rather than the victim. Once the victim ceased to be seen as the entity harmed by the crime, the victim became secondary. Although this shift most certainly benefited the state—by allowing it to collect fines and monies from these newly defined harms—the victim did not fare as well. Instead of being the focus, the crime victim was effectively excluded from the formal aspects of the justice system.

Since then, this state-centered system has largely remained in place, but attention—at least from researchers and activists—returned to the crime victim during the 1940s. Beginning in this time period, concern was shown for the crime victim, but this concern was not entirely sympathetic. Instead, scholars and others became preoccupied with how the crime victim contributes to his or her own victimization. Scholarly work during this time period focused not on the needs of crime victims but on identifying to what extent victims could be held responsible for being victimized. In this way, the damage that offenders cause was ignored. Instead, the ideas of victim precipitation, victim facilitation, and victim provocation emerged.

⊠ The Role of the Victim in Crime: Victim Precipitation, Victim Facilitation, and Victim Provocation

Although the field of victimology has largely moved away from simply investigating how much a victim contributes to his or her own victimization, the first forays into the study of crime victims were centered on such investigations. In this way, the first studies of crime victims did not portray victims as innocents who were wronged at the hands of an offender. Rather, concepts such as victim precipitation, victim facilitation, and victim provocation developed from these investigations. **Victim precipitation** is defined as the extent to which a victim is responsible for his or her own victimization. The concept of victim precipitation is rooted in the notion that, although some victims are not at all responsible for their victimization, other victims are. In this way, victim precipitation acknowledges that crime victimization involves least two people—an offender and a victim—and that both parties are acting and often reacting before, during, and after the incident. Identifying victim precipitation does not necessarily lead to negative outcomes. It is problematic, however, when it is used to blame the victim while ignoring the offender's role.

Similar to victim precipitation is the concept of victim facilitation. **Victim facilitation** occurs when a

▲ **Photo 1.1** A bar fight breaks out after a man yells an insult at the other. By yelling an insult, the victim is precipitating the victimization.

Photo credit: © Ilona Habben/Corbis

victim unintentionally makes it easier for an offender to commit a crime. A victim may, in this way, be a catalyst for victimization. A woman who accidentally left her purse in plain view in her office while she went to the restroom and then had it stolen would be a victim who facilitated her own victimization. This woman is not blameworthy—the offender should not steal, regardless of whether the purse is in plain view or not. But the victim's actions certainly made her a likely target and made it easy for the offender to steal her purse. Unlike precipitation, facilitation helps understand why one person may be victimized over another but does not connote blame and responsibility.

Contrast victim facilitation with victim provocation. **Victim provocation** occurs when a person does something that incites another person to commit an illegal act. Provocation suggests that without the victim, the crime would not have occurred. Provocation, then, most certainly connotes blame. In fact, the offender is not at all responsible. An example of victim provocation would be if a person attempted to mug a man who was walking home from work and the man, instead of willingly giving the offender his wallet, instead pulled out a gun and shot the mugger. The offender in this scenario ultimately is a victim, but he would not have been shot if not for attempting to mug the shooter. The distinctions between victim precipitation, facilitation, and provocation, as you probably noticed, are not always clear-cut. These terms were developed, described, studied, and used in somewhat different ways in the mid-1900s by several scholars.

Hans von Hentig

In his book *The Criminal and His Victim: Studies in the Sociobiology of Crime,* **Hans von Hentig** (1948) recognized the importance of investigating what factors underpin why certain people are victims, just as criminology attempts to identify those factors that produce criminality. He determined that some of the same characteristics that produce crime also produce victimization. We will return to this link between victims and offenders in Section 2, but for now, recognize that one of the first discussions of criminal victimization connected it to offending.

In studying victimization, then, von Hentig looked at the criminal-victim dyad, thus recognizing the importance of considering the victim and the criminal not in isolation but together. He attempted to identify the characteristics of a victim that may effectively serve to increase victimization risk. He considered that victims may provoke victimization—acting as agent provocateurs—based on their characteristics. He argued that crime victims could be placed into one of 13 categories based on their propensity for victimization: (1) young; (2) females; (3) old; (4) immigrants; (5) depressed; (6) mentally defective/deranged; (7) the acquisitive; (8) dull normals; (9) minorities; (10) wanton; (11) the lonesome and heartbroken; (12) tormentor; and (13) the blocked, exempted, and fighting. All these victims are targeted and contribute to their own victimization because of their characteristics. For example, the young, old, and females may be victimized because of their ignorance or risk taking, or may be taken advantage of, such as when women are sexually assaulted. Immigrants, minorities, and dull normals are likely to be victimized due to their social status and inability to activate assistance in the community. The mentally defective or deranged may be victimized because they do not recognize or appropriately respond to threats in the environment. Those who are depressed, acquisitive, wanton, lonesome, or heartbroken may place themselves in situations in which they do not recognize danger because of their mental state, their sadness over a lost relationship, their desire for companionship, or their greed. Tormentors are people who provoke their own victimization via violence and aggression toward others. Finally, the blocked, exempted, and fighting victims are those who are enmeshed in poor decisions and unable to defend themselves or seek assistance if victimized. An example of such a victim is a person who is blackmailed because of his behavior, which places him in a precarious situation if he reports the blackmail to the police (Dupont-Morales, 2009).

Benjamin Mendelsohn

Known as the "father of victimology," **Benjamin Mendelsohn** coined the term for this area of study in the mid-1940s. As an attorney, he became interested in the relationship between the victim and the criminal as he conducted interviews with victims and witnesses and realized that victims and offenders often knew each other and had some kind of existing relationship. He then created a classification of victims based on their culpability, or the degree of the victim's blame. His classification entailed the following:

1. *Completely innocent victim:* a victim who bears no responsibility at all for victimization; victimized simply because of his or her nature, such as being a child

2. *Victim with minor guilt:* a victim who is victimized due to ignorance; a victim who inadvertently places him- or herself in harm's way

3. *Victim as guilty as offender/voluntary victim:* a victim who bears as much responsibility as the offender; a person who, for example, enters into a suicide pact

4. *Victim more guilty than offender:* a victim who instigates or provokes his or her own victimization

5. *Most guilty victim:* a victim who is victimized during the perpetration of a crime or as a result of crime

6. *Simulating or imaginary victim:* a victim who is not victimized at all but, instead, fabricates a victimization event

Mendelsohn's classification emphasized degrees of culpability, recognizing that some victims bear no responsibility for their victimization, while others, based on their behaviors or actions, do.

Stephen Schafer

One of the earliest victimologists, **Stephen Schafer** (1968) wrote *The Victim and His Criminal: A Study in Functional Responsibility.* Much like von Hentig and Mendelsohn, Schafer also proposed a victim typology. Using both social characteristics and behaviors, his typology places victims in groups based on how responsible they are for their own victimization. In this way, it includes facets of von Hentig's typology based on personal characteristics and Mendolsohn's typology rooted in behavior. He argued that victims have a functional responsibility not to provoke others into victimizing or harming them and that they also should actively attempt to prevent that from occurring. He identified seven categories and labeled their levels of responsibility as follows:

1. Unrelated victims—no responsibility

2. Provocative victims—share responsibility

3. Precipitative victims—some degree of responsibility

4. Biologically weak victims—no responsibility

5. Socially weak victims—no responsibility

6. Self-victimizing—total responsibility

7. Political victims—no responsibility

Marvin Wolfgang

The first person to empirically investigate victim precipitation was **Marvin Wolfgang** (1957) in his classic study of homicides occurring in Philadelphia from 1948 to 1952. He examined some 558 homicides to see to what extent victims precipitated their own deaths. In those instances in which the victim was the direct, positive precipitator in the homicide, Wolfgang labeled the incident as victim precipitated. For example, the victim in such an incident would be the first to brandish or use a weapon, the first to strike a blow, and the first to initiate physical violence. He found that 26% of all homicides in Philadelphia during this time period were victim precipitated.

Beyond simply identifying the extent to which homicides were victim precipitated, he also identified those factors that were common in such homicides. He determined that often in this kind of homicide, the victim and the offender knew each other. He also found that most victim-precipitated homicides involved male offenders and male victims and that the victim was likely to have a history of violent offending himself. Alcohol was also likely to play a role in victim-precipitated homicides, which makes sense, especially considering that Wolfgang determined these homicides often started as minor altercations that escalated to murder.

▲ **Photo 1.2**   Marvin Wolfgang studied homicides in Philadelphia and found that about a quarter were victim precipitated. He has been recognized as one of the most influential criminologists in the English-speaking world (Kaufman, 1998).

Photo credit: University of Pennsylvania

Since Wolfgang's study of victim-precipitated homicide, others have expanded his definition to include felony-related homicide and subintentional homicide. **Subintentional homicide** occurs when the victim facilitates her or his own demise by using poor judgment, placing him- or herself at risk, living a risky lifestyle, or using alcohol or drugs. Perhaps not surprising, a study of subintentional homicide found that as many as three fourths of victims were subintentional (Allen, 1980).

Menachem Amir

The crime of rape is not immune from victim blaming today, and it certainly has not been in the past either. **Menachem Amir**, a student of Wolfgang's, conducted an empirical investigation into rape incidents that were reported to the police. Like Wolfgang, he conducted his study using data from Philadelphia, although he examined rapes that occurred from 1958 to 1960. He examined the extent to which victims precipitated their own rapes and also identified common attributes of victim-precipitated rape. Amir labeled almost 1 in 5 rapes as victim precipitated. He found that these rapes were likely to involve alcohol, the victim was likely to engage in seductive behavior, likely to wear revealing clothing, likely to use risqué language, and she likely had a bad reputation. What Amir also determined was that it is the offender's interpretation of actions that is important, rather than what the victim actually does. The offender may view the victim—her actions, words, and clothing—as going against what he considers appropriate female behavior. In this way, the victim may be viewed as being "bad" in terms of how women should behave sexually. He may then choose to rape her because of his misguided view of how women should act, because he thinks she deserves it, or because he thinks she has it coming to her. Amir's study was quite controversial—it was attacked for blaming victims, namely women, for their own victimization. As you will learn in Section 5, rape and sexual assault victims today still must overcome this view that women (since such victims are usually female) are largely responsible for their own victimization.

⊠ The History of Victimology: The Victims' Rights Movement

Beyond the attention victims began to get based on how much they contributed to their own victimization, researchers and social organizations started to pay attention to victims and their plight during the mid-1900s. This marked a shift in how victims were viewed not only by the public but also by the criminal justice system. As noted, scholars began to examine the role of the victim in criminal events, but more sympathetic attention was also given to crime victims, largely as an outgrowth of other social movements.

During the 1960s, concern about crime was growing. This time period saw a large increase in the amount of crime occurring in the United States. As crime rates soared, so too did the number of people directly and indirectly harmed by crime. In 1966, in response to the growing crime problem, the President's Commission on Law Enforcement and the Administration of Justice was formed. One of the commission's responsibilities was to conduct the first ever government-sponsored victimization survey, called the **National Crime Survey** (which later became the National Crime Victimization Survey). This survey is discussed in depth in Section 2. Importantly, it showed that although official crime rates were on the rise, they paled in comparison with the amount of victimization uncovered. This discrepancy was found because official data sources of crime rates are based on those crimes reported or otherwise made known to the police, whereas the National Crime Survey relied on victims to recall their own experiences. Further, victims were asked in the survey whether they reported their victimization to the police and, if not, why they chose not to report. For the first time, a picture of victimization emerged, and this picture was far different than previously depicted. Victimization was more extensive than originally thought, and the reluctance of victims to report was discovered. This initial data collection effort did not occur in a vacuum. Instead, several social movements were underway that further moved crime victims into the collective American consciousness.

The Women's Movement

One of the most influential movements for victims was the **women's movement**. In recognition that victimization such as sexual assault and domestic violence was a byproduct of sexism, traditional sex roles, emphasis on traditional family values, and economic subjugation of women, the women's movement took on as part of its mission helping female victims of crime. Feminists were, in part, concerned with how female victims were treated by the criminal justice system and pushed for victims of rape and domestic violence to receive special care and services. As a result, domestic violence shelters and rape crisis centers started appearing in the 1970s. Closely connected to the women's movement was the push toward giving children rights. Not before viewed as crime victims, children were also identified as being in need of services, as they could be victims of child abuse, could become runaways, and could be victimized in much the same ways as older people. The effects of victimization on children were, at this time, of particular concern.

Three critical developments arose from the recognition of women and children as victims and from the opening of victims' services devoted specifically to them. First, the movement brought awareness that victimization often entails emotional and mental harm, even in the absence of physical injury. To address this harm, counseling for victims was advocated. Second, the criminal justice system was no longer relied on to provide victims with assistance in rebuilding their lives, thus additional victimization by the criminal justice system could be lessened or avoided altogether. Third, because these shelters and centers relied largely on volunteers, services were able to run and stay open even without significant budgetary support (Young & Stein, 2004).

The Civil Rights Movement

Also integral to the development of victims' rights was the **civil rights movement**. This movement advocated against racism and discrimination, noting that all Americans have rights protected by the U.S. Constitution. The civil rights movement, as it created awareness of the mistreatment of minorities, served as a backdrop for the **victims' rights movement** in that it identified how minorities were mistreated by the criminal justice system, both as offenders and victims. The ideologies of the women's movement and the civil rights movement merged to create a victims' rights movement largely supported by females, minorities, and young persons who pushed forward a victims' agenda that concentrated on making procedural changes in the operation of the criminal justice system (Smith, Sloan, & Ward, 1990).

✉ Contributions of the Victims' Rights Movement

We will discuss the particulars of programs and services available for crime victims today in Section 4, but to understand the importance of the victims' rights movement, its contributions should be outlined.

Early Programs for Crime Victims

In the United States, the first crime victims' compensation program was started in California in 1965. Victim compensation programs allow for victims to be financially compensated for uncovered costs resulting from their victimization. Not long after, in 1972, the first three victim assistance programs in the nation, two of which were rape crisis centers, were founded by volunteers. The first prototypes for what today are victim/witness assistance programs housed in district attorneys' offices were funded in 1974 by the Federal Law Enforcement Assistance Administration. These programs were designed to notify victims of critical dates in their cases and to create separate waiting areas for victims. Some programs began to make social services referrals for victims, providing them with input on criminal justice decisions that involved them, such as bail and plea bargains, notifying them about critical points in their cases—not just court dates—and going to court with them. Victim/witness assistance programs continue to provide similar services today.

Development of Victim Organizations

With women and children victims and their needs at the forefront of the victims' rights movement, other crime victims found that special services were not readily available to them. One group of victims whose voices emerged during the 1970s was persons whose loved ones had been murdered—called secondary victims. After having a loved one become a victim of homicide, many survivors found that people around them did not know how to act or how to help them. As one woman whose son was murdered remarked, "I soon found that murder is a taboo subject in our society. I found, to my surprise, that nice people apparently just don't get killed" (quoted in Young & Stein, 2004, p. 5). In response to the particular needs of homicide survivors, Families and Friends of Missing Persons was organized in 1974 and Parents of Murdered Children was formed in 1978. Mothers Against Drunk Driving was formed in 1980. These groups provide support for their members and others but also advocate for laws and policy changes that reflect the groups' missions. The National Organization for Victim Assistance was developed in 1975 to consolidate the purposes of the victims' movement and eventually to hold national conferences and provide training for persons working with crime victims.

Legislation and Policy

In 1980, Wisconsin became the first state to pass a Victims' Bill of Rights. Also in 1980, the National Organization for Victim Assistance created a new policy platform that included the initiation of a National Campaign for Victim Rights, which included a National Victims' Rights Week, implemented by then-President Ronald Reagan.

The attorney general at the time, William French Smith, created a Task Force on Violent Crime, which recommended that a President's Task Force on Victims of Crime be commissioned. President Reagan followed the recommendation. The President's Task Force held six hearings across the country from which 68 recommendations on how crime victims could be better assisted were made. Major initiatives were generated from these recommendations.

1. Federal legislation to fund state victim compensation programs and local victim assistance programs

2. Recommendations to criminal justice professionals and other professionals about how to better treat crime victims

3. Creation of a task force on violence within families

4. An amendment to the U.S. Constitution to provide crime victims' rights (yet to be passed)

As part of the first initiative, the Victims of Crime Act (1984) was passed and created the Office for Victims of Crime in the Department of Justice and established the Crime Victims Fund, which provides money to state victim compensation and local victim assistance programs. The Crime Victims Fund and victim compensation are discussed in detail in Section 4. The Victims of Crime Act was amended in 1988 to require victim compensation eligibility to include victims of domestic violence and drunk-driving accidents. It also expanded victim compensation coverage to nonresident commuters and visitors.

Legislation and policy continued to be implemented through the 1980s and 1990s. The Violent Crime Control and Law Enforcement Act, passed in 1994 by Congress, included the Violence Against Women Act. This law provides funding for research and for the development of professional partnerships to address the issues of violence against women. Annually, the attorney general reports to Congress the status of monies awarded under the act, including the amount of money awarded and the number of grants funded. The act also mandates that federal agencies engage in research specifically addressing violence against women.

In 1998, a publication called *New Directions from the Field: Victims' Rights and Services for the 21st Century* was released by then-Attorney General Janet Reno and the Office for Victims of Crime. This publication reviewed the status of the recommendations and initiatives put forth by President Reagan's task force. It also identified some 250 new recommendations for victims' rights, victim advocacy, and services. Also integral, during the 1990s, the federal government and many states implemented victims' rights legislation that enumerated specific rights to be guaranteed to crime victims. These rights will be discussed in detail in Section 4, but some basic rights typically afforded to victims include the right to be present at trial, to be provided a waiting area separate from the offender and people associated with the offender during stages of the criminal justice process, to be notified of key events in the criminal justice process, to testify at parole hearings, to be informed of rights, to be informed of compensation programs, and to be treated with dignity and respect. These rights continue to be implemented and expanded through various pieces of legislation, such as the Crime Victims' Rights Act, which is part of the Justice for All Act of 2004 signed into law by then-President George W. Bush. Despite this push among the various legislatures, a federal victims' rights constitutional amendment has not been passed. Some states have been successful in amending their constitutions to ensure that the rights of crime victims are protected, but the U.S. Constitution has not been similarly amended. Various rights afforded to crime victims through these amendments are outlined in Section 4.

◪ Victimology Today

Today, the field of victimology covers a wide range of topics, including crime victims, causes of victimization, consequences of victimization, interaction of victims with the criminal justice system, interaction of victims with other social service agencies and programs, and prevention of victimization. Each of these topics is discussed throughout the text. As a prelude to the text, a brief treatment of the contents is provided in the following subsections.

The Crime Victim

To study victimization, one of the first things victimologists needed to know was who was victimized by crime. In order to determine who victims were, victimologists looked at official data sources—namely, the Uniform Crime Reports—but found them to be imperfect sources for victim information because they do not include detailed information on crime victims. As a result, victimization surveys were developed to determine the extent to which people were victimized, the typical characteristics of victims, and the characteristics of victimization incidents. The most widely cited and used victimization survey is the National Crime Victimization Survey (NCVS), which is discussed in detail in Section 2.

From the NCVS and other victimization surveys, victimologists discovered that victimization is more prevalent than originally thought. Also, the "typical" victim was identified—a young male who lives in urban areas. This is not to say that other people are not victimized. In fact, children, women, and older people are all prone to victimization. These groups are discussed in detail in later sections. In addition, victimologists have uncovered other vulnerable groups. Homeless individuals, persons with mental illness, disabled persons, and prisoners all have been recognized as deserving of special attention given their victimization rates. Special populations vulnerable to victimization are discussed in Section 8.

The Causes of Victimization

It is difficult to know why a person is singled out and victimized by crime. Is it something he did? Did an offender choose a particular individual because she seemed like an easy target? Or does victimization occur because somebody is simply in the wrong place at the wrong time? Perhaps there is an element of "bad luck" or chance involved, but victimologists have developed some theories to explain victimization. Theories are sets of propositions that explain phenomena. In relation to victimology, victimization theories explain why some people are more likely than others to be victimized. As you will read in Section 2, the most widely utilized theory of victimization is routine activities theory. In the past two decades, however, victimologists and criminologists alike have developed additional theories and identified other correlates of victimization both generally and to explain why particular types of victimization, such as child abuse, occur.

Costs of Crime

Victimologists are particularly interested in studying victims of crime because of the mass costs they often incur. These **costs of crime** can be tangible, such as the cost of stolen or damaged property or the costs of receiving treatment at the emergency room, but can also be harder to quantify. Crime victims may experience mental anguish, fear, or other more serious mental health issues such as posttraumatic stress disorder. Costs also include monies spent by the criminal justice system preventing and responding to crime and monies spent to assist crime victims. One other significant cost of victimization is the real risk that many victims face of being victimized again. Unfortunately, some victims do not suffer only a single victimization event but, rather, are victimized again, and sometimes again and again. In this way, a certain subset of victims appears to be particularly vulnerable to revictimization. Revictimization and the other costs of crime are discussed in Section 3.

The Crime Victim and the Criminal Justice System

Another experience of crime victims that is important to understand is how they interact with the criminal justice system. As is discussed in detail in Section 3, many persons who are victimized by crime do not report their experiences to the police. The reasons victims choose to remain silent, at least in terms of not calling the police, are varied but often include an element of suspicion and distrust of the police. Some victims worry that police will not take them seriously or will not think what happened to them is worth the police's time. Others may be worried that calling the police will effectively invoke a system response that cannot be erased or stopped, even when the victim wishes not to have the system move forward. An example of such a victim is one who does not want to call the police after being hit by her partner because she fears the police will automatically and mandatorily arrest him. Whatever the reason, without a report, the victim will not activate the formal criminal justice system, which will preclude an arrest and also may preclude the victim from receiving victim services explicitly tied to reporting.

When victims do report, they then enter the world of criminal justice, a world in which they are often seen as witnesses rather than victims, given that the U.S. criminal justice system recognizes crimes as harms against the state. This being the case, victims do not always find they are treated with dignity and respect, even though the victims' rights movement stresses the importance of doing so. The police are not the only ones with whom victims must contend. If an offender is apprehended and charged with a crime, the victim will also interact with the prosecutor and perhaps a judge. Fortunately, many police departments and prosecutors' offices offer victim assistance programs through which victims can receive information about available services. These programs also offer personal assistance and support, such as attending court sessions with the victim or helping submit a victim impact statement. The experience of the crime victim after the system is put into motion is an area of research ripe for study by victimologists. It is important to understand how victims view their interactions with the criminal justice system so that victim satisfaction can be maximized and any additional harm caused to the victim can be minimized. The criminal justice response will be discussed throughout this text, especially since different victim types have unique experiences with the police.

The Crime Victim and Social Services

The criminal justice system is not the only organization with which crime victims may come into contact. After being victimized, victims may need medical attention. As a result, emergency medical technicians, hospital and doctor's office staff, nurses, doctors, and clinicians may all be persons with whom victims interact. Although some of these professionals will have training or specialize in dealing with victims, others may not treat victims with the care and sensitivity they need. To combat this, sometimes victims will have persons from the police department or prosecutor's office with them at the hospital to serve as mediators and provide counsel. Also to aid victims, many hospitals and clinics now have sexual assault nurse examiners, who are specially trained in completing forensic and health exams for sexual assault victims.

In addition to medical professionals, mental health clinicians also often serve victims, as large number of victims seek mental health services after being victimized. Beyond mental health care, victims may use the services of social workers or other social service workers. But not all persons with whom victims interact as a consequence of being victimized are part of social service agencies accustomed to serving victims. Crime victims may seek assistance from insurance agents and repair and maintenance workers. Crime victims may need special accommodations from their employers or schools. In short, being victimized may touch multiple aspects of a person's life, and agencies, businesses, and organizations alike may find themselves in the position of dealing with the aftermath, one to which they may not be particularly attuned. The more knowledge people have about crime victimization and its impact on victims, the more likely victims will be satisfactorily treated.

Prevention

Knowing the extent to which people are victimized, who is likely targeted, and the reasons why people are victimized can help in the development of prevention efforts. To be effective, prevention programs and policies need to target the known causes of victimization. Although the offender is ultimately responsible for crime victimization, it is difficult to change offender behavior. Reliance on doing so limits complete prevention, since victimization involves at least two elements—the offender and the victim—that both need to be addressed to stop crime victimization. In addition, as noted by scholars, it is easier to reduce the opportunity than the motivation to offend (Clarke, 1980, 1982). Nonetheless, offenders should be discouraged from committing crimes, likely through informal mechanisms of social control. For example, colleges could provide crime awareness seminars directed at teaching leaders of student organizations how to dissuade their members from committing acts of aggression, using drugs or alcohol, or engaging in other conduct that could lead to victimization.

In addition to discouraging offenders, potential victims also play a key role in preventing victimization. Factors that place victims at risk need to be addressed to the extent that victims can change them. For example, since routine activities and lifestyles theories identify daily routines and risky lifestyles as being key risk factors for victimization, people should attempt to reduce their risk by making changes they are able to make. Other theories and risk factors related to victimization should also be targeted (these are discussed in Section 2). Because different types of victimization have different risk factors—and, therefore, different risk-reduction strategies—prevention will be discussed in each section that deals with a specific victim type.

As victimology today focuses on the victim, the causes of victimization, the consequences associated with victimization, and how the victim is treated within and outside the criminal justice system, this text will address these issues for the various types of crime victims. In this way, each section that deals with specific types of victimization—such as sexual victimization and intimate partner violence—will include an overview of who is victimized, why they are victimized, the outcomes of being victimized, and the services provided to and challenges faced by victims. The specific remedies in place for crime victims are discussed in each section and also in a stand-alone section.

SUMMARY

- The field of victimology originated in the early to mid-1900s, with the first victimologists attempting to identify how victims contribute to their own victimization. To this end, the concepts of victim precipitation, victim facilitation, and victim provocation were examined.
- Hans von Hentig, Benjamin Mendelsohn, and Stephen Schafer each proposed victim typologies that were used to classify victims in terms of their responsibility or role in their own victimization.
- Marvin Wolfgang and Menachem Amir conducted the first empirical examinations of victim precipitation. Wolfgang studied homicides in Philadelphia, and Amir focused on forcible rapes. Wolfgang found that

26% of homicides were victim precipitated. Amir concluded that 19% of forcible rapes were precipitated by the victim.

- The victims' rights movement gained momentum during the 1960s. It was spurred by the civil rights and women's movements. This time period saw the recognition of children and women as victims of violence. The first victim services agencies were developed in the early 1970s.
- The victims' rights movement influenced the development of multiple advocacy groups, such as Mothers Against Drunk Driving, Families and Friends of Missing Persons, and Parents of Murdered Children.
- Important pieces of legislation came out of the victims' rights movement, including the Victims of Crime Act, the Violence Against Women Act, and the Crime Victims' Rights Act. Many states have victims' rights amendments and/ or legislation that guarantees victim protections.
- Victimology today is concerned with the extent to which people are victimized, the different types of victimization they experience, the causes of victimization, the consequences associated with victimization, the criminal justice system's response to victims, and the response of other agencies and people. Victimology is a science—victimologists use the scientific method to study these areas.
- As victimologists become aware of who is likely to be victimized and the reasons for this, risk-reduction and prevention strategies can be developed. These should target not only offender behavior but also opportunity. In this way, victims can play an important role in reducing their likelihood of being victimized.

DISCUSSION QUESTIONS

1. Compare and contrast victim precipitation, victim facilitation, and victim provocation.

2. Why do you think the first explorations into victimization in terms of explaining why people are victimized centered not on offender behavior but on victim behavior?

3. What are the reasons behind labeling crimes as acts against the state rather than against victims?

4. How does the victims' rights movement correspond to the treatment of offenders and rights afforded to offenders?

5. Does examining victim behavior when attempting to identify causes of victimization lead to victim blaming? Is it wrong to consider the role of the victim?

KEY TERMS

victimology	victim provocation	National Crime Survey
lex talionis	Hans von Hentig	women's movement
retribution	Benjamin Mendelsohn	civil rights movement
restitution	Stephen Schafer	victims' rights movement
Code of Hammurabi	Marvin Wolfgang	costs of crime
victim precipitation	subintentional homicide	
victim facilitation	Menachem Amir	

INTERNET RESOURCES

The American Society of Victimology (http://www.american-society-victimology.us)

This organization advances the discipline of victimology by promoting evidence-based practices and providing leadership in research and education. The website contains information about victimology and victimologists. This organization looks at advancements in victimology through research, practice, and teaching.

An Oral History of the Crime Victim Assistance Field Video and Audio Archive (http://vroh.uakron.edu/index.php)

This website contains information from the Victim Oral History Project, intended to capture the development and evolution of the crime victims' movement. You will find video clips of interviews with more than 50 persons critical to this movement, in which they discuss their contributions to and perspectives of the field.

Crime in the United States (http://www2.fbi.gov/ucr/cius2008/index.html)

The Federal Bureau of Investigation compiles all the information for both the Uniform Crime Reports and National Incident-Based Reporting System. The information is then put into several annual publications, such as Crime in the United States and Hate Crime Statistics. The data for these statistics are provided by nearly 17,000 law enforcement agencies across the United States. This website provides the crime information for 2008.

Crime Prevention Tips (http://www.crimepreventiontips.org)

This website provides many tips on how to reduce your chances of becoming a crime victim. There is also a section to help you determine whether you already are a crime victim. Some of the prevention tips specifically address how to be safer when you use public transportation and on college campuses.

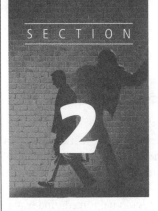

SECTION

2

Extent, Theories, and Factors of Victimization

I t was not exactly a typical night for Polly. Instead of studying at the library as she normally did during the week, she decided to meet two of her friends at a local bar. They spent the evening catching up and drinking a few beers before they decided to head home. Since Polly lived within walking distance of the bar, she bid her friends goodnight and started on her journey home. It was dark out, but since she had never confronted trouble in the neighborhood before—even though it was in a fairly crime-ridden part of a large city—she felt relatively safe.

As Polly walked by an alley, two young men whom she had never seen before stepped out, and one of them grabbed her arm and demanded that she give them her school bag, in which she had her wallet, computer, keys, and phone. Since Polly refused, the other man shoved her while the first man grabbed her bag. Despite holding on as tightly as she could, the men were able to take her bag before running off into the night. Slightly stunned, Polly stood there trying to calm down. Without her bag, which held her phone and keys, she felt there was little she could do other than continue to walk home and hope her roommates were there to let her in. As she walked home, she wondered why she had such bad luck. Why was she targeted? Was she simply in "the wrong place at the wrong time," or did she do something to place herself in harm's way? Although it is hard to know why Polly was victimized, we can compare her to other victims to see how similar she is to them. To this end, a description of the "typical" crime victim is presented in this section. But what about why she was targeted? Fortunately, we can use the theories presented in this section to understand why Polly fell victim on that particular night.

▲ **Photo 2.1**   Polly, on her way home from the bar.

Photo credit: © iStockphoto.com/ Rasmus Rasmussen

14

⬚ Measuring Victimization

Before we can begin to understand *why* some people are the victims of crime and others are not, we must first know how often victimization occurs. Also important is knowing who the typical crime victim is. Luckily, these characteristics of victimization can be readily gleaned from existing data sources.

Uniform Crime Reports

Begun in 1929, the **Uniform Crime Reports (UCR)** show the amount of crime known to the police in a year. Police departments around the country submit to the Federal Bureau of Investigation (FBI) monthly law enforcement reports on crimes that are reported to them or that they otherwise know about. The FBI then compiles this data and each year publishes a report called *Crime in the United States,* which details the crime that occurred in the United States for the year. This report includes information on eight offenses, known as the Part I index offenses: murder and nonnegligent manslaughter, forcible rape, robbery, aggravated assault, burglary, larceny-theft, motor vehicle theft, and arson. Arrest data are also included in the report on Part II offenses, which include an additional 21 crime categories.

Advantages and Disadvantages

The UCR is a valuable data source for learning about crime and victimization. Because more than 90% of the population is represented by agencies participating in the UCR program, it provides an approximation of the total amount of crime experienced by almost all Americans (FBI, 2006). It presents the number of crimes for regions, states, cities, towns, areas under tribal law enforcement, and colleges and universities. It does so annually so that crime trends can be determined for the country and for these geographical units. Another benefit of the UCR is that crime characteristics are also reported. It includes demographic information (age, sex, and race) on people who are arrested and some information on the crimes, such as location and time of occurrence.

Despite these advantages, it does not provide detailed information on crime victims. Also important to consider, the UCR includes information only on crimes that are reported to the police or of which the police are aware. In this way, all crimes that occur are not represented, especially since, as discussed below, crime victims often do not report their victimization to the police. Another limitation of the UCR as a crime data source is that the Part I index offenses do not cover the wide range of crimes that occur, such as simple assault and sexual assaults other than forcible rape, and federal crimes are not counted. Furthermore, the UCR uses the **hierarchy rule**. If more than one Part I offense occurs within the same incident report, the law enforcement agency counts only the highest offense in the reporting process (FBI, 2009). These exclusions also contribute to the UCR's underestimation of the extent of crime. Accuracy of the UCR data are also affected by law enforcement's willingness to participate in the program and to do so by reporting to the FBI all offenses of which they are aware.

Crime as Measured by the UCR

Nonetheless, the UCR can be used to paint a picture of crime in the United States. In 2008, the police became aware of 1,382,012 violent crimes and 9,767,915 property crimes. According to the UCR data shown in Figure 2.1 later in this section, the most common offense is larceny-theft. Aggravated assaults are the most common violent crime, although they are outnumbered by larceny-thefts. The typical criminal is a young (less than 30 years old), white male (FBI, 2009).

National Incident-Based Reporting System

As noted, the UCR includes little information about the characteristics of criminal incidents. To overcome this deficiency, the FBI began the National Incident-Based Reporting System (NIBRS), an expanded data collection effort that includes detailed information about crimes. Agencies participating in the NIBRS collect information on each crime incident and arrest in 22 different offense categories (Group A offenses) that encompass 46 specific crimes. Arrest data are reported for an additional 11 offenses (Group B offenses). Information about the offender, the victim, injury, location, property loss, and weapons is included (FBI, n.d.).

Although the NIBRS represents an advancement of the UCR program, not all law enforcement agencies participate in the system. As such, crime trends similar to those based on national data produced by the UCR are not yet available. As more agencies come online, the NIBRS data will likely be an even more valuable tool for understanding patterns and trends of crime victimization.

The National Crime Victimization Survey

As noted, the UCR and NIBRS have some limitations as crime data sources, particularly when information on victimization is of interest. To provide a picture of the extent to which individuals experience a range of crime victimizations, the Bureau of Justice Statistics (BJS) began, in 1973, a national survey of U.S. households. Originally called the National Crime Survey, it provides a picture of crime incidents and victims. In 1993, the BJS redesigned the survey, making extensive methodological changes, and renamed it the **National Crime Victimization Survey (NCVS)**.

The NCVS is administered by the U.S. Census Bureau to a nationally representative sample of about 43,000 households. Each member of participating households who is 12 years old or older completes the survey, resulting in about 76,000 persons being interviewed (FBI, 2006). Each household selected remains in the study for 3 years and completes seven interviews 6 months apart. Each interview serves a **bounding** purpose by giving respondents a concrete event to reference (i.e., since the last interview) when answering questions in the next interview. Bounding is used to improve recall.

The NCVS is conducted in two stages. In the first stage, individuals are asked if they experienced any of seven types of victimization during the previous 6 months. The victimizations that respondents are asked about are rape and sexual assault, robbery, aggravated and simple assault, personal theft, household burglary, motor vehicle theft, and theft. The initial questions asked in the first stage are known as **screen questions,** which are used to cue respondents or jog their memories as to whether they experienced any of these criminal victimizations in the previous 6 months. An example of a screen question is shown in Table 2.1. In the second stage, if the respondent answers affirmatively to any of the screen questions, the respondent then completes an **incident report** for each victimization experienced. In this way, if an individual stated that he had experienced one theft and one aggravated assault, he would fill out two incident reports—one for the theft and a separate one for the aggravated assault. In the incident report, detailed questions are asked about the incident, such as where it happened, whether it was reported to the police and why the victim did or did not report it, who the offender was, and whether the victim did anything to protect himself during the incident. Table 2.2 shows an example of a question from the incident report. As you can see, responses to the questions from the incident report can help reveal the context of victimization.

Another advantage to using this two-stage procedure is that the incident report is used to determine what, if any, incident occurred. The incident report, as discussed, includes detailed questions about what happened, including questions that are used to classify an incident into its appropriate crime victimization type. For example, in order for a rape to be counted as such, the questions in the incident report that concern the elements of rape, which are discussed in Section 5 (force, penetration), must be answered affirmatively for the incident to be counted as rape in the NCVS. This process is fairly conservative in that all elements of the criminal victimization must have occurred for it to be included in the estimates of that type of crime victimization.

The NCVS has several advantages as a measure of crime victimization. First, it includes in its estimates of victimization several offenses that are not included in Part I of the UCR; for example, simple assault and sexual assault are both included in NCVS estimates of victimization. Second, the NCVS does not measure only crimes reported to the police as does the UCR. Third, the NCVS asks individuals to recall incidents that occurred only during the previous 6 months, which is a relatively short recall period. In addition, its two-stage measurement process allows for a more conservative way of estimating the amount of victimization that occurs each year in that incidents are counted only if they meet the criteria for inclusion.

Despite these advantages, the NCVS is not without its limitations. Estimates of crime victimization depend on the ability of respondents

Table 2.1 Example of Screen Question From NCVS

(Other than any incidents already mentioned,) has anyone attacked or threatened you in any of these ways (exclude telephone threats)?

   (a) With any weapon, for instance, a gun or knife
   (b) With anything like a baseball bat, frying pan, scissors, or stick
   (c) By something thrown, such as a rock or bottle
   (d) Include any grabbing, punching, or choking
   (e) Any rape, attempted rape, or other type of sexual attack
   (f) Any face-to-face threats

OR

   (g) Any attack or threat or use of force by anyone at all? Please mention it even if you are not certain it was a crime.

Table 2.2 Example of Question From Incident Report in NCVS

Did the offender have a weapon such as a gun or knife, or something to use as a weapon, such as a bottle or wrench?

to accurately recall what occurred to them during the previous 6 months. Even though the NCVS attempts to aid in recall by spanning a short period (6 months) and by providing bounding via the previous survey administration, it is still possible that individuals will not be completely accurate in recounting the particulars of an incident. Bounding and using a short recall period also do not combat against someone intentionally being misleading or lying or answering in a way meant to please the interviewer. In addition, murder and "victimless" crimes such as prostitution and drug use are not included in NCVS estimates of crime victimization. Another limitation is that crime that occurs to commercial establishments is not included. Beyond recall issues, the NCVS sample is selected from U.S. households. This sample may not be truly representative, as it excludes individuals who are institutionalized, such as persons in prison, and does not include homeless people. Remember, too, that only those persons over the age of 12 are included. As a result, estimates about victimization of children cannot be determined.

Extent of Crime Victimization

Each year, the BJS publishes *Criminal Victimization in the United States,* which is a report about crime victimization as measured by the NCVS. From this report, we can see what the most typical victimizations are and who is most likely to be victimized. In 2008, more than 21,000,000 victimizations were experienced among the nation's households (Rand, 2009). Property crimes were much more likely to be experienced compared with violent crimes; 4.9 million violent crime victimizations were experienced compared with 16.3 million property crime victimizations. The most common type of property crime reported was theft, while simple assault was the most commonly occurring violent crime (see Figure 2.1).

The Typical Victimization and Victim

The typical crime victim can also be identified from the NCVS. For all violent victimizations except for rape and sexual assaults, males are more likely to be victimized than females. Persons who are Black and those under the age of 24 also have higher victimization rates than others. Characteristics of victimization incidents are also evident. Less than half of all victimizations experienced by individuals in the NCVS are reported to the police. This may be related in part to the fact that most victims of violent crime know their offender; most often, victims identified their attacker as a friend or acquaintance. Strangers accounted for only about one third of violent victimizations in the NCVS. Females were more likely than males to be victimized by an intimate partner. In only 1 in 5 incidents did the

Figure 2.1 Number of Crimes Occurring in 2008, Comparison for UCR and NCVS

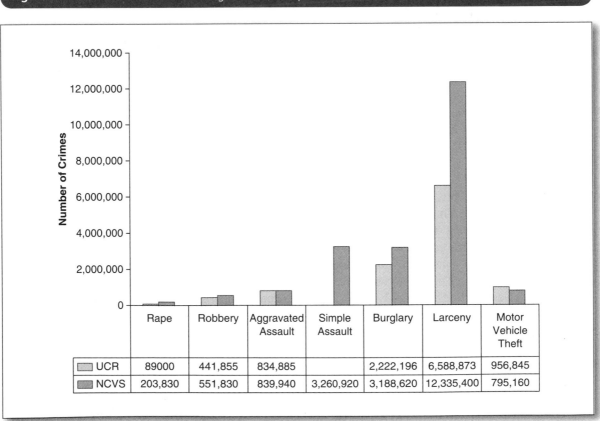

	Rape	Robbery	Aggravated Assault	Simple Assault	Burglary	Larceny	Motor Vehicle Theft
UCR	89000	441,855	834,885		2,222,196	6,588,873	956,845
NCVS	203,830	551,830	839,940	3,260,920	3,188,620	12,335,400	795,160

NOTES: The UCR includes only forcible rape, while the NCVS includes both rape and sexual assault.

The UCR measures only aggravated assault, while the NCVS includes both aggravated and simple assault.

offender have a weapon (Rand, 2009). Now that you know the characteristics of the typical victimization and the typical crime victim, how do Polly and her victimization compare?

The International Crime Victims Survey

As you may imagine, there are many other self-report victimization surveys that are used to understand more specific forms of victimization, such as sexual victimization and those that occur outside the United States. Many of these will be discussed in later sections. One oft-cited survey of international victimization is the International Crime Victims Survey (ICVS), which was created to provide a standardized survey to compare crime victims' experiences across countries (van Dijk, van Kesteren, & Smit, 2008). The first round of the survey was conducted in 1989 and was repeated in 1992, 1996, 2000, and 2004/2005. Collectively, more than 340,000 persons have been surveyed in more than 78 countries as part of the ICVS program (van Dijk et al., 2008). Respondents are asked about 10 different types of victimization that they could have experienced: car theft, theft from or out of a car, motorcycle theft, bicycle theft, attempted or completed burglary, sexual victimization (rapes and sexual assault), threats, assaults, robbery, and theft of personal property (van Dijk et al., 2008). If a person has experienced any of these offenses, he or she then answers follow-up questions about the incident. This survey has provided estimates of the extent of crime victimization in many countries and regions of the world. In addition, characteristics of crime victims and incidents have been produced from these surveys.

⊠ Theories and Explanations of Victimization

Now that you have an idea about who the typical crime victim is, you are probably wondering *why* some people are more likely than others to find themselves victims of crime. Is it because those people provoke the victimization, as von Hentig and his contemporaries thought? Is it because crime victims are perceived by offenders to be more vulnerable than others? Is there some personality trait that influences victimization risk? All these factors may play at least some role in why victimization occurs to particular people. The following sections address these possibilities.

The Link Between Victimization and Offending

One facet about victimization that cannot be ignored is the link between offending and victimization and offenders and victims. As mentioned in Section 1, the first forays into the study of victims included a close look at how victims contribute to their own victimization. In this way, victims were not always assumed to be innocents; rather, some victims were seen as being at least partly responsible for bringing on their victimization—for instance, by being an offender who is victimized when the victim fights back. Although the field of victimology has moved from trying to place blame on victims, the recognition that offenders and victims are often linked—and often the same person— has aided in the understanding of why people are victimized.

Victim and Offender Characteristics

The typical victim and the typical offender have many commonalities. As mentioned before in our discussion of the NCVS, the groups with the highest rates of violent victimization are young, Black males. The UCR also provides information on offenders. The groups with the highest rates of violent offending are also young, Black males. The typical victim and the typical offender, then, share common demographics. In addition, both victims and offenders are likely to live in urban areas. Thus, individuals who spend time with people who have the characteristics of offenders are more likely to be victimized than others.

Explaining the Link Between Victimization and Offending

Some even argue that victims and offenders are often one and the same, with offenders being more likely to be victimized and vice versa. It is not hard to understand why this may be the case. Offending can be viewed as part of a risky lifestyle. Individuals who engage in offending are exposed more frequently to people and contexts in which victimization is likely to occur (Lauritsen, Laub, & Sampson, 1992).

There also may be a link between victimization and offending that is part of a broader cultural belief in the acceptability and sometimes necessity of violence, known as the subculture of violence theory. This theory proposes that for certain subgroups of the population and in certain areas, violence is part of a value system that supports the use of violence, in response to disrespect in particular (Wolfgang & Ferracuti, 1967). In this way, when a subculture exists that supports violence, victims will be likely to respond by retaliating. Offenders may initiate violence that leads to their being victimized—for example, getting into a physical fight to resolve a dispute.

Being victimized may be related to offending in ways that are not directly tied to retaliation. In fact, being victimized at one point in life may increase the likelihood that a person will engage in delinquency and crime later in life. This link has been found especially in individuals who are abused during childhood. As discussed in Section 7 on victimization at the beginning and end of life, those who are victimized as children are significantly more likely than those who do not experience child abuse to be arrested in adulthood (Widom, 2000) or to engage in violence and property offending (Menard, 2002).

Insomuch as victimization and offending are linked, it makes sense then, as you will see in the following sections, that the same influences on offending may also affect victimization. The reasons why victimization may lead to participation in crime are not fully understood, but it may be that being victimized carries psychological consequences, such as depression, anxiety, or posttraumatic stress disorder, that can lead to coping through the use of alcohol or drugs. Victimization may also carry physical consequences, such as brain damage, that can further impede success later in life. Cognitive ability may also be tempered by maltreatment, particularly in childhood, which can hinder school performance. Behavior may also change as a result of being victimized. People may experience problems in their interpersonal relationships or become violent or aggressive. Whatever the reason, it is evident that victimization and offending are intimately intertwined. This is not to say that the only explanations of victimization should be tied to or be an extension of explanations of offending—just remember that when you read about the research that has used criminological theories to explain victimization, it is largely because of the connection between victimization and offending.

Routine Activities and Lifestyles Theory

In the 1970s, two theoretical perspectives—**routine activities and lifestyles theory**—were put forth that both linked crime victimization risk to the fact that victims had to come into contact with a potential offender. Before discussing these theories in detail, first it is important to understand what a **victimization theory** is. A victimization theory is generally a set of testable propositions designed to explain why a person is victimized. Both routine activities and lifestyles theories propose that a person's victimization risk can best be understood by the extent to which the victim's routine activities or lifestyle creates opportunities for a motivated offender to commit crime.

In developing routine activities theory, Lawrence Cohen and Marcus Felson (1979) argued that a person's routine activities, or daily routine patterns, impact risk of being a crime victim. Insomuch as a person's routine activities bring him or her into contact with **motivated offenders**, crime victimization risk abounds. Cohen and Felson thought that motivated offenders were plentiful and that their motivation to offend did not need to be explained. Rather, their selection of particular victims was more interesting. Cohen and Felson noted that there must be something about particular targets, both individuals and places, that encouraged selection by these motivated offenders. In fact, those individuals deemed to be **suitable targets** based on their attractiveness would be chosen by offenders. Attractiveness relates to qualities about the target, such as ease of transport, which is why a burglar may break into a home and leave with an iPod or laptop computer rather than a couch. Attractiveness is further evident when the target does not have **capable guardianship.** Capable guardianship is conceived as means by which a person or target can be effectively guarded to prevent a victimization from occurring. Guardianship is typically considered to be *social,* when the presence of another person makes someone less attractive as a target. Guardianship can also be provided through *physical* means, such as a home with a burglar alarm or a person who carries a weapon for self-protection. A home with a burglar alarm and a person who carries a weapon are certainly less attractive crime targets! When these three elements—motivated offenders, suitable targets, and lack of capable guardianship—coalesce in time and space, victimization is likely to occur.

When Cohen and Felson (1979) originally developed their theory, they focused on predatory crimes—those that involve a target and offender making contact. They originally were interested in explaining changes in rates of these types of crime over time. In doing so, they argued that people's routines had shifted since World War II, taking them away from home and making their homes attractive targets. People began spending more time outside the home, in leisure activities and going to and from work and school. As people spent more time interacting with others, they were more likely to come into contact with motivated offenders. Capable guardianship was unlikely to be present; thus, the risk of criminal victimization increased. Cohen and Felson also linked the increase in crime to the production of durable goods. Electronics began to be produced in portable sizes, making them easier to steal. Similarly, cars and other expensive items that could be stolen, reused, and resold became targets. As Cohen and Felson saw it, prosperity of society could produce an increase in criminal victimization rather than a decline! Also important, they linked victimization to everyday activities rather than to social ills, such as poverty.

Michael Hindelang, Michael Gottfredson, and James Garofalo's (1978) lifestyles theory is a close relative of routine activities theory. Hindelang and colleagues posited that certain lifestyles or behaviors place people in situations in which victimization is likely to occur. Your lifestyle, such as going to bars or working late at night in relative seclusion, places you at more risk of being a crime victim than others. Although the authors of lifestyles theory did not specify how opportunity structures risk as clearly as did the authors of routine activities theory, at its heart, lifestyles theory closely resembles routine activities theory and its propositions. As a person comes into contact—via lifestyle and behavior—with potential offenders, he or she is likely creating opportunities for crime victimization to occur. The lifestyle factors identified by Hindelang and his colleagues that create opportunities for victimization are the people with whom one associates, working outside the home, and engaging in leisure activities. In this way, a person who associates with criminals, works outside the home, and participates in activities—particularly at night, away from home, and with nonfamily members—is a more likely target for personal victimization than others.

Hindelang et al. (1978) further delineated why victimization risk is higher for some people than others using the **principle of homogamy**. According to this principle, the more frequently a person comes into contact with persons in demographic groups with likely offenders, the more likely it is the person will be victimized. This frequency may be a function of demographics or lifestyle. For example, males are more likely to be criminal offenders than females. Males, then, are at greater risk for victimization because they are more likely to spend time with other males. Now that you know about routine activities theory, do you think Polly's routines or lifestyle placed her at risk for being victimized? Today, researchers largely treat routine activities theory and lifestyles theory interchangeably and often refer to them as the routine activities and lifestyles theory perspectives.

One of the reasons that routine activities and lifestyles theories have been the prevailing theories of victimization for more than 30 years is the wide empirical support researchers have found when testing them. It has been shown that a person's routine activities and lifestyles impact risk of being sexually victimized (Cass, 2007; Fisher,

Daigle, & Cullen, 2010; Mustaine & Tewksbury, 1999, 2007; Schwartz & Pitts, 1995). This perspective also has been used to explain auto theft (Rice & Smith, 2002), stalking (Mustaine & Tewksbury, 1999), cybercrime victimization (Holt & Bossler, 2009), adolescent violent victimization (Lauritsen, Laub, & Sampson, 1992), theft (Mustaine & Tewksbury, 1998), victimization at work (Lynch, 1997), and street robbery (Groff, 2007).

Structural and Social Process Factors

In addition to routine activities and lifestyles theory, other factors also increase a person's risk of being victimized. Key components of life—such as **neighborhood context**, family, friends, and personal interaction—also play a role in victimization.

Neighborhood Context

We have already discussed how certain individuals are more at risk of becoming victims of crime than others. So far, we have tied this risk to factors related to the person's lifestyle. Where that person lives and spends time, however, may also place him or her at risk of victimization. Indeed, you are probably not surprised to learn that certain areas have higher rates of victimization than others. Some areas are so crime-prone that they are considered to be **hot spots** for crime. Highlighted by Lawrence Sherman, hot spots are areas that have a concentrated amount of crime. He found through examining police call data in Minneapolis that only 3% of all locations made up most calls to the police. A person living in or frequenting a hot spot will be putting him- or herself in danger. The features of these hot spots and other high-risk areas may create opportunities for victimization that, independent of a person's lifestyle or demographic characteristics, enhance chances of being victimized.

What is it about certain areas that relates them to victimization? A body of recent research has identified many features, particularly of neighborhoods (notice we are not discussing hot spots specifically). One factor related to victimization is **family structure**. Robert Sampson (1985), in his seminal piece on neighborhoods and crime, found that neighborhoods that have a large percentage of female-headed households have higher rates of theft and violent victimization. He also found that **structural density**, as measured by the percentage of units in structures of five or more units, is positively related to victimization. **Residential mobility**, or the percentage of persons 5 years and older living in a different house from 5 years before, also predicted victimization.

Beyond finding that the structure of a neighborhood influences victimization rates for that area, it also has been shown that neighborhood features influence personal risk. In this way, living in a neighborhood that is disadvantaged places individuals at risk of being victimized, even if they do not have risky lifestyles or other characteristics related to victimization (Browning & Erickson, 2009). For example, neighborhood disadvantage and neighborhood residential instability are related to experiencing violent victimization at the hands of an intimate partner (Benson, Fox, DeMaris, & Van Wyk, 2003). Using the notions of collective efficacy, it makes sense that neighborhoods that are disadvantaged are less able to mobilize effective sources of informal social

▲ **Photo 2.2** This area may be a "hot spot" due to lots of people milling about at night.

Photo credit: ©Stockbyte/Thinkstock

control (Sampson, Raudenbush, & Earls, 1997). Informal social controls are often used as mechanisms to maintain order, stability, and safety in neighborhoods. When communities do not have strong informal mechanisms in place, violence and other deviancy is likely to abound. Such communities are less safe; hence, their residents are more likely to be victimized than residents of more socially organized areas.

Exposure to Delinquent Peers

The neighborhood context is but one factor related to risk of victimization. Social process factors, such as peers and family, are also important in understanding crime victimization. Generally, one of the strongest influences on youth is their peers. Peer pressure can lead people, especially juveniles, to act in ways they normally would not and to engage in behavior they otherwise would not. Having **delinquent peers** places youth not only at risk of engaging in delinquent behavior—juvenile delinquency does, after all, often take place in groups—but also of being victimized. Spending time with delinquent peers places people at risk of being victimized because, as lifestyles and routine activities theory suggests, spending time in the presence of motivated offenders increases risk. Never mind that these would-be offenders are your friends! Another reason having delinquent peers may be related to victimization is that a person may find him- or herself in risky situations (such as being present for a fight) in which being harmed is not unlikely. In this situation, it may not be your friends per se who harm you, but others involved in the fight may attack you, or you may feel the need to come to the aid of your friends. As discussed in the reading by Terrance Taylor, Dana Peterson, Finn-Aage Esbensen, and Adrienne Freng (2007), in a specific case of having delinquent peers, being a member of a gang increases a young person's risk of experiencing violence.

Family

Especially during adolescence, the family also plays an important role in individual experiences. Having strong attachments to family members, particularly parents, is likely to insulate a person from many negative events, including being victimized. Not surprisingly, research has found that weak emotional attachment between family members is a strong predictor of victimization (Esbensen, Huizinga, & Menard, 1999; Lauritsen et al., 1992). This may be due to parents being unable and unwilling to exert control over the behavior of their children, such that they are more likely to end up in risky situations. Family units may also spend more time together when there is strong attachment, thus reducing exposure to motivated offenders. Youth may also be less likely to place themselves in risky situations because they do not want to disappoint their parents, as they place high value on the relationships they have with them. In these ways, emotional attachment to family members serves to reduce risky behavior. At this point, you may be noting that familial attachment may be related to lifestyles and routine activities theory—and you would be right! This relationship is explored, along with how delinquent peers increase victimization risk, by Christopher Schreck and Bonnie Fisher (2004) in their article included in this section.

Control-Balance Theory

A general theory of deviancy, **control-balance theory**, may also apply to victimization. Developed by Charles Tittle (1995, 1997), this theory proposes that the amount of control that people possess over others and the amount of control to which one is subject factor into risk of engaging in deviancy. When considered together, a **control ratio** can be determined for individuals. Control-balance theory posits that when the control a person has exceeds the amount of control he or she is subject to, that person has a **control surplus**. When the amount of control a person exercises is outweighed by the control he or she is subject to, that person has a **control deficit**. When a person has a control surplus or deficit, he or she is likely to be predisposed toward deviant behavior. The type of deviant behavior to which a person will be predisposed depends on the control ratio. A control surplus is linked to autonomous forms of deviance such as exploitation of others. Control deficits, on the other hand, are linked to repressive forms of deviance such as defiance.

While not expressly a theory of victimization, Alex Piquero and Matthew Hickman (2003) used control-balance theory to explain victimization. They proposed that having a control surplus or control deficit would increase victimization risk as compared with having a control balance. Individuals with a control surplus are used to having their needs and desires met and have a desire to extend their control. In short, they engage in risky behaviors (in terms of victimization) because there is little to restrain their actions. They may treat others who have control deficits with disrespect in such a way that those individuals act out and victimize them. Those with control deficits are at risk for victimization for different reasons. So used to having little control at their disposal, they lack the confidence or belief that they can protect themselves and are, thus, vulnerable targets. They may also try to overcome their control deficits by lashing out or victimizing those who exercise control over them. Piquero and Hickman tested control-balance's ability to predict victimization and found that both control deficits and control surpluses predicted general and theft victimization.

Social Interactionist Perspective

Richard Felson (1992) posited that distress may be related to victimization. When experiencing stress, peoples' behavior and demeanor are impacted. They are more likely to break rules and to be generally irritating to others. Distressed individuals, thus, may entice a certain measure of aggression from others given their poor attitudes and rule-breaking behavior. Consider a student who goes to class having just learned that he failed a test in his previous class that effectively ruined his chances of passing that class. This student is likely experiencing a level of stress that will negatively impact his behavior in class. While in class, then, he may explode after a fellow student makes a comment that he finds unreasonable. The student who is the "victim" of the outburst may find the other student's behavior unacceptable and offensive. The attacked student then may, as a result, respond aggressively, effectively starting an aggressive exchange. This distress-and-reaction sequence is at the heart of the **social interactionist perspective**.

Stated more formally, Felson (1992) argues that aggressive encounters occur when distressed individuals break social rules and those who are aggrieved by the breaking of rules respond aggressively. The distressed individual is then placed in a situation in which he or she has to respond to aggression. If this person does so unsatisfactorily, the original aggrieved person is likely to implement punishment—in other words, victimization. The distressed individual then may retaliate, thus continuing the cycle of aggression. In this way, distress is a cause of victimization.

The Life-Course Perspective

Emerging in the 1990s in the field of criminology, the **life-course perspective** considers the development of offending over time. In doing so, it uses elements from biology, sociology, and psychology to explain why persons initiate into, continue with, and desist or move out of a life of crime. Contributing to the growth of this field, in large part due to the overlap between victims and offenders discussed in a section below, victimologists have recently begun applying and testing the principles of life-course criminology to victimization. A summary of these theories is presented in Table 2.3.

The General Theory of Crime

In 1990, Michael Gottfredson and Travis Hirschi published a book titled *The General Theory of Crime.* In this seminal work, they proposed their **general theory of crime**, proposing that criminal behavior is caused by a single

| | Table 2.3 | Summary Table of Life-Course Criminological Theories Relevant for Victimization |

Theory	Author(s)	Key Factor Related to Outcome	Research Support for Link to Victimization
General theory of crime	Gottfredson and Hirschi (1990)	Low self-control	Kerley, Hochstetler, and Copes, 2009; Kerley, Xu, and Sirisunyaluck, 2008; Piquero, MacDonald, Dobrin, Daigle, and Cullen, 2005; Schreck (1999); Schreck, Stewart, and Fisher, 2006; Stewart, Elifson, and Sterk, 2004
Age-graded theory of adult social bonds	Sampson and Laub (1993)	Social bonds: Marriage and employment; marriage related to desistance	Daigle, Beaver, and Hartman (2008)

factor—namely, low self-control. They argued that a person with low self-control, when presented with opportunity, will engage in criminal and other analogous behaviors, such as excessive drinking. When examining the characteristics of persons with low self-control, the reasons this trait might lead to criminal behavior are clear. A person with low self-control will exhibit six elements, the first being inability to delay gratification; a person with low self-control will be impulsive and unable or unwilling to delay gratification. Second, the person will be a risk taker who engages in thrill-seeking behavior without thought of consequence. Third, an individual with low self-control will be shortsighted, without any clear long-term goals. Fourth, low self-control is indicated by a preference for physical as compared with mental activity. This preference may lead an individual to respond to disrespect with violence rather than having a discussion about the finer points of being respectful. Fifth, low self-control is evidenced by low frustration tolerance, which results in a person being quick to anger. Sixth, insensitivity and self-centeredness are hallmarks of low self-control. A person with low self-control will be unlikely to exhibit empathy toward others.

Gottfredson and Hirschi (1990) argue that low self-control is fairly immutable once developed, which occurs during early childhood. They believe that, although self-control is an individual-level characteristic, it is not inherent; rather, it is developed through parental socialization. Once the level is set (around age 8), people will be hard-pressed to develop greater abilities to moderate their behavior. Without self-control, a person will act on impulses and seek personal gratification—often engaging in crime. Importantly, as noted, low self-control will lead individuals to engage in other behaviors that are similar to crime.

In 1999, Christopher Schreck applied the general theory of crime to victimization. He was one of the first researchers to apply to victimization what had been conceived as a theory of crime. This innovative approach was rooted in his recognition that persons who engage in crime are also likely to be victimized, a point we return to later. He also noted that since crime and victimization may be closely related, often with the same people engaging in both, the same factors that explain crime participation may also explain crime victimization. He tested his theory and found that low self-control increased the likelihood that a person would experience both personal and property victimization, even when controlling for participation in criminal behavior. This finding suggests that solely being involved in crime does not increase risk of victimization but that low self-control has significant, independent effects on victimization. Others have since applied the theory of low self-control to various types of victimization, as shown in the included reading by Tammatha Clodfelter, Michael Turner, Jennifer Hartman, and Joseph Kuhns (2010), who find that low self-control does not impact sexual harassment.

Age-Graded Theory of Adult Social Bonds

Not all criminologists agree that there is a single cause of crime (or victimization) called low self-control. Others noted that people do indeed move in and out of criminal activity, a phenomenon that is difficult to explain with a persistent personality trait, low self-control. Robert Sampson and John Laub (1993) instead believed that a person's social bonds could serve to insulate him or her from criminal activity. In their **age-graded theory of adult social bonds**, Sampson and Laub identified two key social bonds—marriage and employment—that can aid people in moving out of a life of delinquency and crime as they emerge into young adulthood. If a person enters into marriage and has gainful employment, he or she is developing valuable social capital. In other words, a person who has these two social bonds will have much to lose by engaging in crime, which will promote crime desistance if he or she was previously involved in crime. If that person was not involved in crime, social capital would enable him or her to continue living a crime-free life.

Although this obviously is not a victimization theory, because of the link between victimization and offending, researchers have attempted to connect the attainment of adult social bonds with victimization in that individuals who are married and working will be less likely to be crime victims than those with little to lose. Leah Daigle, Kevin Beaver, and Jennifer Hartman (2008) found that entering into marriage did in fact predict desistance from victimization as individuals moved into early adulthood. They found that employment was not similarly protective; instead, employment reduced the chances that a person would desist from victimization. Looking at routine activities and lifestyles theory, however, this finding is none too surprising! The more time a person spends outside the home, at work or in other activities, the greater the chances of being victimized.

Genes and Victimization

The life-course perspective in criminology has also centered on individual factors, such as genetics, that promote offending. This body of research has found a link between different genetic polymorphisms and behaviors relevant to criminology, such as criminal involvement and alcohol and drug use. A genetic polymorphism is a variant on a gene. Research has shown that sometimes these variations impact the likelihood of engaging in certain behaviors, such as violence, aggression, and delinquency. The genes that have been identified as linked to criminality are those that code for neurotransmitters, such as monoamine oxidase, serotonin, and dopamine. Neurotransmitters are chemical messengers responsible for information transmission. In terms of criminal behavior, relevant neurotransmitters are those linked to behavioral inhibition, mood, reward, and attention deficits. One important aspect of the link between genetics and crime is that possessing a variant for a gene, or having a certain polymorphism for a particular neurotransmitter, appears to "matter" only in certain environments. This is known as a **gene x environment interaction**. Genes tend to be important not for everyone in every circumstance but for particular individuals in particular contexts. For example, a person who is genetically predisposed toward alcoholism will express these alcoholic tendencies only if first exposed to alcohol.

As noted with other life-course perspective approaches, the applicability of genetic factors to the study of victimization has been explored. In fact, a gene x environment interaction for one gene in particular, dopamine, has been found to increase victimization risk. Dopamine is a neurotransmitter linked to the reward and punishment systems of the brain. Dopamine is released when we engage in pleasurable activities, thus reinforcing such behavior. Too much dopamine, however, can be a bad thing. High levels of dopamine are linked to enhanced problem solving and attentiveness, but overproduction of dopamine can be problematic. In fact, it has been linked to violence and aggression. One gene that codes for dopamine is the DRD2 gene, a dopamine receptor gene. Research has found evidence for a gene x environment interaction between DRD2 and having delinquent peers. White males who have delinquent peers and have a certain genetic polymorphism for DRD2 are more likely than others to be violently victimized (Beaver et al., 2007). This is an emerging area of research, so additional research is certainly needed to understand fully how genes impact victimization.

The Role of Alcohol in Victimization

One of the common elements present in victimization is alcohol. According to data from the NCVS, 27.5% of victims perceived their offender to be under the influence of alcohol or drugs at the time of the incident (Bureau of Justice Statistics, 2008). Alcohol use is commonplace among crime offenders, but many crime victims also report that just prior to their victimizations, they had consumed alcohol. Patricia Tjaden and Nancy Thoennes (2006) found in their National Violence Against Women study that 20% of women and 38% of men who experienced rape in adulthood had consumed alcohol or drugs prior to being victimized. Alcohol use is associated with other forms of victimization as well, such as physical assault. This fact should not be too surprising given the effects of alcohol on individuals. Generally, alcohol is linked to victimization since it reduces inhibition and also impedes people's ability to recognize or respond effectively to dangerous situations. Offenders may also see intoxicated persons as particularly vulnerable targets for these reasons. Where a person consumes alcohol is also important. A person who drinks at home alone, or with family is less likely to be victimized than a person who drinks in a bar at night. The latter person is likely interacting with motivated offenders without capable guardianship and may be perceived as a suitable target.

Alcohol use may place a person at risk of being victimized but also may impact how the victim responds to the incident. Research by Barry Ruback, Kim Menard, Maureen Outlaw, and Jennifer Shaffer (1999) shows how alcohol use may be relevant to understanding why victims often do not report their experiences to police. In their study, college students evaluated various hypothetical scenarios that depicted victimization. Study participants were asked whether they would advise a victimized friend to report to the police based on a given scenario. When the friend in the scenario had been drinking, college students were less likely to advise that the police be contacted, and this relationship was particularly strong for victims depicted as being underage and drinking.

As you can see, the explanations of victimization are many. The "hallmark" victimization theory is routine activities and lifestyles theory, which is based on the notion that a person's routines and lifestyle, not social conditions, place him or her at risk. As you have read, however, explanations of victimization have expanded beyond this to include social process and structural factors. The explanations you are drawn to may be tied to the data you are examining, which you now know are impacted by methodology. To understand the causes of victimization, you must first know who the "typical" victim is and what characterizes the "typical" victimization. In some of the following sections, specific types of victims will be examined. Think about what theories can be used to explain their victimizations.

SUMMARY

- The Uniform Crime Reports are an official measure of the amount of crime known to the police. According to these reports, which are published annually by the Federal Bureau of Investigation, the most common crime type is larceny/theft. The most common type of violent crime is aggravated assault. Criminal offending rates are highest for young, Black males.
- The National Crime Victimization Survey (NCVS) is a nationally representative sample of U.S. households. Individuals over the age of 12 in selected households are asked questions about victimization experiences they faced during the previous 6 months. According to the NCVS, the typical victim is a young, White male—although Blacks have higher victimization rates than other racial or ethnic groups and females experience higher rates of rape and sexual assault than males.

- The typical victimization incident is perpetrated by someone known to the victim, is not reported to the police, and does not involve a weapon.
- There is a clear link between victimization and offending, as well as between victims and offenders. Persons who live risky lifestyles are more likely to engage in criminal or delinquent activity and also to be victims of crime. Victims and offenders also share similar demographic profiles.
- Routine activities theory suggests that crime victimization is likely to occur when motivated offenders, lack of capable guardianship, and suitable targets coalesce in time and space. Lifestyles theory is closely linked with routine activities theory in proposing that a person who leads a risky lifestyle is at risk of being victimized.
- Neighborhoods are not equally safe. The risk of being victimized, then, differs across geographical areas, and even when controlling for individual-level factors such as risky lifestyle, neighborhood disadvantage predicts victimization.
- Spending time with friends who participate in delinquent activities places a person at risk of being victimized. These "friends" may victimize their nondelinquent peers and also encourage them to participate in risky behaviors that may lead to victimization.
- Strong attachments to family may serve to protect individuals from victimization, while weak attachments may increase victimization risk.
- According to control-balance theory, individuals with an unequal control-balance ratio—either having a control deficit or a control surplus—are more prone to victimization than those with a balanced ratio. Those with control deficits may be seen as easy targets. They also may get tired of being targeted and lash out, thus increasing their involvement in situations associated with violent victimization. Those with control surpluses may engage in risky behavior with impunity, which could set them up for being victimized or retaliated against.
- Research on the general theory of crime suggests that those individuals who have low self-control are more likely to be victimized than those with higher levels of self-control.
- Adult social bonds may explain why people who were once victimized are not victimized again as they age into young adulthood. Marriage appears to protect individuals from victimization.
- Genetic factors may also play a role in victimization. One specific genetic polymorphism of the DRD2 gene has been found to increase risk for White males who have delinquent peers. A genetic effect that occurs only under certain environmental conditions is known as a gene x environment interaction.
- Alcohol and victimization appear to go hand-in-hand. Alcohol impacts cognitive ability, and persons who are drinking are less likely to assess and recognize situations as being risky even when they are. In addition, alcohol is linked to behavioral inhibition, such that people may act in ways they otherwise would not, which may incite aggression in others. Alcohol is also linked to victimization when offenders purposefully select intoxicated victims because they are seen as easy targets.

DISCUSSION QUESTIONS

1. Compare and contrast the UCR and the NCVS. What are the advantages and disadvantages of each? Which is the best measure of victimization?

2. Apply the concepts of routine activities and lifestyles theory to evaluate your own risk of being victimized. What could you change to reduce your risk?

3. What are the individual-level factors that place people at risk of being crime victims? What are the structural factors and social process factors that place individuals at risk of being crime victims?

4. How should the criminal justice system handle victimized individuals who may also have been involved in a crime? Do you believe in the idea of truly innocent victims? If so, should they be treated differently?

5. Given what you have read about the theories and factors that influence crime victimization, how can victimization be prevented? Be sure to tie your prevention ideas to what is thought to cause victimization.

KEY TERMS

Uniform Crime Reports (UCR)

National Crime Victimization Survey (NCVS)

bounding

screen questions

hierarchy rule

incident report

routine activities and lifestyles theory

victimization theory

motivated offenders

suitable targets

capable guardianship

principle of homogamy

neighborhood context

hot spots

family structure

structural density

residential mobility

delinquent peers

control-balance theory

control ratio

control surplus

control deficit

social interactionist perspective

life-course perspective

general theory of crime

age-graded theory of adult social bonds

gene x environment interaction

INTERNET RESOURCES

Crime in the United States: The Nation's Two Crime Measures (http://www2.fbi.gov/ucr/cius2006/about/crime_measures.html)

This website is part of the FBI's research on various crimes. This one specifically examines the differences, advantages, and disadvantages of the UCR and the NCVS. Both forms of research are important to the study of crime.

Bureau of Justice Statistics: Victim Characteristics (http://bjs.ojp.usdoj.gov/index.cfm?ty=tp&tid=92)

The NCVS provides information on characteristics of victims, including age, race, ethnicity, gender, marital status, and household income. For violent crimes (rape, sexual assault, assault, and robbery) the characteristics are based on the victim who experienced the crime. For property crimes (household burglary, motor vehicle theft, and property theft) the characteristics are based on the household of the respondent who provided information about these crimes. Property crimes are defined as affecting the entire household.

"Opportunity Makes the Thief: Practical Theory for Crime Prevention" (http://www.homeoffice.gov.uk/rds/prgpdfs/fprs98.pdf)

This article combines several theories that focus on the "opportunity" of crimes. This includes the routine activities approach, the rational choice perspective, and crime pattern theory. This publication argues that the root cause of crime is opportunity. This allows for prevention techniques to focus on how to lessen the opportunity for crime to occur.

Crime Times (http://www.crimetimes.org/)

Crime Times is a quarterly publication of the Wacker Foundation concentrating on the links between brain dysfunction and disordered/criminal/psychopathic behavior. Instead of focusing just on sociological problems, this website looks at brain malfunctions that prevent criminals from benefiting from sociological or psychological interventions. This website addresses several topics, including attention-deficit hyperactivity disorder, aggression, antisocial behavior, food and chemical sensitivities, hormonal imbalances, maternal smoking or alcohol abuse, and new medical and nutritional interventions.

"Alcohol and Crime" (http://bjs.ojp.usdoj.gov/content/pub/pdf/ac.pdf)

This report by the Bureau of Justice Statistics, in connection with the U.S. Department of Justice, looks at the link between alcohol and crime. It includes several graphs and figures that show the link between crime, specifically violent crime, and alcohol. These statistics also show that alcohol-related crime is generally decreasing.

Project on Human Development in Chicago Neighborhoods (http://www.icpsr.umich.edu/PHDCN/about.html)

The Project on Human Development in Chicago Neighborhoods is an interdisciplinary study of how families, schools, and neighborhoods affect child and adolescent development. It was designed to advance understanding of the developmental pathways of both positive and negative human social behaviors. In particular, the project examined the pathways to juvenile delinquency, adult crime, substance abuse, and violence. At the same time, the project also provided a detailed look at the environments in which these social behaviors take place by collecting substantial amounts of data about urban Chicago, including its people, institutions, and resources.

Introduction to Reading 1

Two theories that have been used to explain victimization—low self-control and routine activities theory—are used in this article to examine sexual harassment victimization of a sample of 164 college students enrolled in a southeastern, urban university. Students in the sample completed an online survey. Sexual harassment was conceived of as physical sexual harassment, nonverbal sexual harassment, and verbal sexual harassment. Physical sexual harassment includes patting, pinching, tickling, or other unwanted touching; brushing against the body; attempted or actual kissing; attempted or actual fondling; or coerced sexual intercourse. Nonverbal sexual harassment includes public display or sharing of pictures or photographs of a sexual nature; e-mails or websites of a sexual nature; love notes or letters; suggestive comments on memos; graffiti; or gag gifts such as sex toys or games. Finally, verbal sexual harassment includes sexual innuendoes and comments; whistling in a suggestive manner; jokes about sex or gender in general; spreading rumors about sexual activity or performance of a student or employee; sexual propositions, invitations, or other pressures for sex; or implied or overt threats. The authors found that routine activities theory applies to sexual harassment; however, low self-control was not found to be a significant predictor of sexual harassment victimization.

Sexual Harassment Victimization During Emerging Adulthood

A Test of Routine Activities Theory and a General Theory of Crime

Tammatha A. Clodfelter, Michael G. Turner, Jennifer L. Hartman, and Joseph B. Kuhns

The developmental period known as "emerging adulthood" has recently attracted the interest of scholars (Arnett, 2000, 2004). Defined as the developmental period between the ages of 18 and 25, emerging adulthood is a period of identity exploration, instability, self-focus, transition, and possibilities. Unfortunately, it is also a period when the probability of offending and being victimized peaks for a number of different crimes (Piquero, Brame, Mazerolle, & Hapaanen, 2002; see also Fisher, Cullen, & Turner, 2000, for increased risk of being sexually victimized).

Over the past two decades, understanding the nature and extent of the sexual victimization experiences of college students within the period of emerging adulthood has captured the interests of several researchers (DeKeseredy & Kelly, 1993; Fisher et al., 2000; Koss, Gidycz, & Wisniewski, 1987). Recognizing that college campuses are not immune to crime and that the true levels of sexual victimization are unknown, Koss and her colleagues (1987) initiated the first national effort to shed light on the nature and extent of sexual victimization of female college students. More than a decade later,

SOURCE: Clodfelter et al. (2010). Reprinted with permission of Sage Publications.

and following a similar methodology, Fisher and her colleagues (2000) collected sexual victimization data from a second national sample of female college students. Despite minor differences in methodological approaches, each of these national studies documented the measurement issues surrounding the nature and extent of sexual victimizations that occur on college campuses. In fact, these studies indicated that sexual victimizations on college campuses are much more prevalent than official statistics suggest. The ubiquitous impact of this research inevitably has initiated a number of other campus-specific studies investigating the victimization experiences of college students (see Combs-Lane & Smith, 2002; Kalof, 2000; Larimer, Lydum, Anderson, & Turner, 1999; Parks & Fals-Stewart, 2004).

Once type of sexual victimization experienced by individuals during emerging adulthood that has received significant attention within the research community has been sexual harassment (Cortina, Swan, Fitzgerald, & Waldo, 1998; Menard, Hall, Phung, Ghebrial, & Martin, 2003). In a review of more than 120 studies, Spitzberg (1999) found that sexual harassment has a higher prevalence rate than more physically violent sexual assaults. Despite the scholarly attention that sexual harassment has drawn, few studies have investigated the impact that theoretical models possess in empirically explaining this behavior (for an exception, see Fisher et al., 2000). Moreover, even fewer studies have examined the impact of competing theories on the explanation of sexual harassment incidents.

The current study seeks to fill this void in the campus victimization literature. Using a randomly selected sample of students from a medium-sized southeastern university, we address a number of questions related to the sexual harassment experiences of college students. To guide our analyses, we use tenets of routine activities theory and the general theory of crime to test a number of hypotheses (see Cohen & Felson, 1979; Gottfredson & Hirschi, 1990). First, we examine the nature and extent of three types of sexual harassment (physical, verbal, and nonverbal) on a college campus. Second, we investigate the prevalence rate of reporting and the potential reasons students fail to report. Third, we investigate whether measures of routine activities

theory and the general theory of crime effectively explain sexual harassment. Before embarking on these analyses, we present the literature related to sexual harassment and the theories guiding our research.

⬚ Literature Overview

Sexual harassment is a form of sexual victimization historically associated with the workplace (LaRocca & Kromrey, 1999; Rubin, 1995), and its definition has been the topic of continual modification since its inception (Birdeau, Somers, & Lenihan, 2005; Fitzgerald, Swan, & O'Donohue, 1997). However, sexual harassment occurs in many different venues, including academic environments, and has been documented as a pervasive problem (Bradenburg, 1997; Dansky & Kilpatrick, 1997; Dziech, 2003; Dziech & Hawkins, 1998; Hall, Graham, & Hoover, 2004; Paludi, 1997; Williams, Lam, & Shively, 1992). Approximately 50% of all women who were ever employed, and about 30% of all women who attended 4-year colleges, report being sexually harassed (Dansky & Kilpatrick, 1997; Paludi, 1997). Dziech (2003) reported that university faculty members have victimized approximately 30% of undergraduate females and 40% of female graduate students, whereas 90% of undergraduate females reported unwanted behavior from their male peers. However, these estimates may not accurately represent the true rate of victimization because of the general consensus that individuals of differing races and genders perceive, label, and report sexual harassment differently (Birdeau et al., 2005; Fitzgerald & Ormerod, 1991; Ivy & Hamlet, 1996; Kalof, Eby, Matheson, & Kroska, 2001; Kelley & Parsons, 2000; Magley & Shupe, 2005; Shelton & Chavous, 1999). For example, research suggests that sexual harassment is more likely to occur among certain populations including females, graduate students, women in nontraditional fields, minority females, disabled persons, divorced women, young and naïve females, persons who were sexually abused, and homosexuals (Cortina et al., 1998; Dziech, 2003; Kalof et al., 2001; Paludi, 1997; Russell & Oswald, 2001). Related research observes that particular interactions may enhance a student's risk of sexual harassment victimization such as being in an environment that includes alcohol (Collins & Messerschmidt, 1993; Gover, 2004), having a one-on-one relationship with a person of direct authority, or participation in group activities (Bradenburg, 1997). Consequently, when a victim disengages from his or her academic endeavors, the harassing behavior of the offender can be reinforced (Benson & Thomson, 1982; Schneider, 1987).

Routine Activities Theory

Cohen and Felson (1979) contended that the convergence in time and space of a motivated offender, suitable target, and lack of capable guardianship is conducive to criminal offending (also see Clarke & Felson, 1993; Felson & Clarke, 1998). Routine activities explanations were traditionally applied to crimes such as property, but a review of victimization literature suggested that routine activities theory might also be used to account for personal victimization. Mustaine and Tewksbury (1999) reported that activities that placed college women in closer proximity to offenders and actions that reduced their guardianship led to a greater risk of stalking victimization. Fisher, Cullen, and Turner (2002) also explored stalking among college students and determined that factors such as frequenting a place where alcohol was served, living alone, being involved in dating relationships, and prior victimization were significantly related to further victimization. Other research has shown that risky behaviors including drug and alcohol abuse were predictive of dating violence and sexual victimization (Felson & Burchfield, 2004; Gover, 2004; Wilson, Calhoun, & McNair, 2002). That is, substance abuse may decrease a victim's personal guardianship, particularly while in proximity to a motivated offender, thus increasing the likelihood of becoming a suitable target.

Sexual Harassment and Routine Activities Theory

Routine activities theory can be specifically applied to the study of sexual harassment in an academic setting for several reasons. The certain types of individuals at risk for victimization, females in particular, represent the majority of undergraduate and graduate student enrollment (Dziech, 2003; Paludi, 1997; U.S. Department of Education, 2007). Meanwhile, individuals tend to reach their criminal-offending peak in their late teens and begin to desist in their early 20s (Moffitt, 2003), and offenders are predominantly male (Cohen & Vila, 1996; Menard et al., 2003). Therefore, one can reasonably conclude that combining males, who are most prone to committing offenses during their college-age years, with females, who dominate campus student populations, has the potential to increase opportunities for victimization.

Cohen and Felson (1979) suggested that the victim (target) must be perceived as suitable by the offender, which may be because of the proximity of the victim to the offender or the lack of guardianship of the victim. Both of these situations occur regularly on college campuses. Males and females share space such as classrooms, dormitories, and cafeterias. Many classes are held at night, and consequently students walk (sometimes alone) to their cars, dorms, parking garages, or residences after dark.

Alcohol consumption is a significant factor that places individuals at risk of sexual harassment victimization (Abbey, Zawacki, Buck, Clinton, & McAuslan, 2001; Rothman & Silverman, 2007). Campus research indicates that high rates of male and female college students engage in binge drinking and other serious levels of alcohol consumption (Leppel, 2006; Weitzman & Nelson, 2004). From a perspective of routine activities, intoxication can lower the ability to protect oneself from victimization and accurately perceive dangerousness. In fact, Menard et al. (2003) found alcohol to be a significant predictor of male-perpetrated harassment and coercion and female-perpetrated harassment.

The social and structural natures of college campuses shape the routine activities of individuals, which may increase the risk for sexual harassment victimization. Students regularly interact with potential offenders in settings that increase the opportunities for offending. Also, college students are likely to engage in behaviors that lower their ability to safeguard themselves against potential offenders. Finally, students may lack guardianship when engaging in normal campus behaviors such as traveling alone or studying in private areas.

General Theory of Crime

Self-control theory, or the general theory of crime, attempts to explain all individual differences (sex, culture, age, and circumstance) in the propensity to commit or to refrain from committing crime and analogous behaviors (Gottfredson & Hirschi, 1990) and has been found to be one of the strongest predictors of such behaviors (see Pratt & Cullen, 2000). The premise of the theory is that individuals with low self-control (LSC) will be at an increased risk for involvement in criminal and analogous behavior. These behaviors may be defined among similar groups of people and may remain stable throughout the life course (Gottfredson & Hirschi, 2003).

More recent studies have applied the theory to victimization, and significant findings have emerged. Schreck (1999) suggested that when an individual lacks the attributes of self-control, a vulnerable situation emerges in which the individual can place himself or herself at risk. Schreck found that males were more likely than females to be victimized, and females with LSC were at a greater risk of victimization than females with high self-control. Also, victims who reported lower levels of self-control and risky lifestyles also reported a higher number of criminal offenses. Stewart, Elifson, and Sterk (2004) furthered Schreck's study to determine whether LSC could account for variations in violent victimization and whether risky lifestyles mediated the effect of self-control as an explanation of victimization among female offenders. The findings of Stewart et al. supported the hypothesis that women with LSC were more likely to be violently victimized, regardless of demographics or routine activities.

Sexual Harassment and the General Theory of Crime

Using the general theory of crime to explain sexual harassment in an academic setting is appropriate for a variety of reasons. Many attributes of self-control (see Schreck, 1999) are expected to be present among individuals who are considered to have high levels of self-control. However, not all individuals who are enrolled in higher education institutions can be presumed to possess these qualities, thus creating an expectation of varied levels of self-control.

Although it is theorized that individuals with LSC are more likely to engage in risky behaviors, these behaviors may not directly contribute to one's risk of victimization. It can be argued, however, that engaging in risky behaviors with others who are also engaging in risky activities may increase one's vulnerability. Thus, an overlap may exist among victims and offenders. For example, victimization frequently occurs while students are participating in socially acceptable behaviors including risky activities and events involving alcohol (Bradenburg, 1997). Binge drinking is a major concern for college campuses (Gibson, Schreck, & Miller, 2004) because individuals who engage in such activities may be at a greater risk of victimization because of the vulnerabilities created by these environments. Students may not possess sufficient self-controls that might protect them from engaging in behaviors that may lead to victimization. This theoretical perspective is not intended to place blame on victims; rather, its purpose is to highlight factors that are associated with increased risk of victimization.

It is important to remember, however, that sexual harassment may occur without the overlap of victim and offender characteristics. For example, being a member of a previously defined vulnerable population does not imply that a person inevitably possesses LSC. Also, a person may be a victim of sexual harassment without having any type of relationship with the offender. Ultimately, one primary purpose of this research is to determine whether the level of self-control contributes to the risk of sexual harassment victimization.

Summary and Current Focus

Although sexual harassment has often been the focus of research in the workplace, understanding and preventing offending and victimization patterns among emerging adults are particularly important on college campuses because these experiences can result in adverse academic outcomes and high rates of physical and psychological injury. Sexual harassment on college campuses is currently understudied and underreported but also likely to be prevalent. College campuses naturally bring motivated offenders into close proximity with suitable targets, often in the absence of capable guardians. Typical emerging adult behaviors that peak during college years include frequent and increased alcohol consumption, illicit drug experimentation and use (U.S. Department of Health and Human Services, 2006), and periods of freedom mixed with solitude, all of which may increase risky behavior and affect individual levels of self-control. As a result, continued study of sexual harassment prevalence, variations, and theoretical explanations on campus remains important. Using an online survey design, this research contributes to the scholarly literature regarding the nature and prevalence of sexual harassment, the characteristics of victims and offenders, and two theoretical models for explaining the circumstances under which sexual harassment is most likely to occur.

✖ Methods

Sample

Participants comprised a random sample of 750 college students. Participant names were generated from the eligible campus population of a southeastern urban university. Eligibility included being enrolled during the fall semester of 2004. The target population either had assigned e-mail accounts that were provided by the university or had alternate e-mail addresses stored within the contact information provided to the school. E-mail addresses were used to initiate correspondence. All students first received a letter via electronic mail to introduce them to the study. A second letter was sent 2 weeks later with a link to a Web site that would host the survey. Two follow-up request were e-mailed within 3-week intervals for those who did not initially respond. Once a student agreed to participate, he or she was directly routed to the Web address that housed the self-administered questionnaire. The survey was available from March 2005 through April 2005. Completion time ranged from 5 to 30 minutes. Survey directions prompted respondents to recall events between the 2004 fall semester and the end of March of the 2005 spring semester.

In the present study, 216 students (29%) started and completed most of the survey, although some intrasurvey attrition and missing data resulted in 164 completed surveys (22%). Furthermore, the sample fairly closely approximated the university population of interest with some minor exceptions (see Table 1).

Table 1 Descriptive Statistics of Sample Population and Victim Population

Variable	Sample %	Campus %	Victim, n (%)	Nonvictim, n (%)
Sex				
Female	65	55.3	18 (85.7)	88 (62.0)
Male	35	44.7	3 (14.3)	54 (38.0)
Race				
White	83.5	74.8	19 (90.4)	118 (83.0)
Non-White	16.5	25.2	2 (9.6)	25 (17.0)
Class				
Freshman	6.7	19.6	NA	NA
Sophomore	15.9	17.1	NA	NA

(Continued)

Table 1 (Continued)

Variable	Sample %	Campus %	Victim, n (%)	Nonvictim, n (%)
Junior	25.6	19.0	NA	NA
Senior	26.8	22.3	NA	NA
Undergraduate special	NA	2.0	NA	NA
Graduate/doctoral	25.0	20.0	NA	NA
Age, years				
Average age	25.5	25.2	NA	NA
18–25	NA	NA	19 (90.4)	95 (66.9)
26–35	NA	NA	1 (4.8)	28 (19.7)
36+	NA	NA	1 (4.8)	19 (13.4)
Enrollment status				
Full time	74.7	69.3	NA	NA
Part time	25.3	30.7	NA	NA
Location of residence				
On campus	22.2	21.3	NA	NA
Off campus	77.8	78.7	NA	NA
Sexual orientation				
Heterosexual	NA	NA	18 (85.7)	139 (97.9)
Homosexual	NA	NA	3 (14.3)	3 (2.1)
Prior campus sexual harassment				
Yes	NA	NA	10 (47.6)	6 (4.2)
No	NA	NA	11 (52.4)	136 (95.8)

Measures

Independent variables. The primary goals of the research were to estimate the prevalence of campus-based sexual harassment, define and describe the victim and offender characteristics of those who reported victimization, and identify theoretical predictors of sexual harassment in this context. Within the study, routine activities of students were explored in an effort to measure whether the students' behaviors or actions placed them at an increased risk for sexual harassment victimization and whether their own levels of self-control increased harassment victimization and whether their own levels of self-control increased the probability of being victimized. The specific theoretical concepts measured were (a) lack of capable guardianship, (b) proximity to offender, and (c) target suitability (routine activities) and self-control.

Capable guardianship is an index consisting of two variables: (a) whether a person carries self-protection such as pepper spray and (b) how often the person was escorted to his or her car after dark. Self-protection was coded as 1 = *carries self-protection*. Escort was labeled as 1 = *never escorted* to 4 = *escorted all the time*. Each variable was standardized and summed to create a measure of capable guardianship, with higher scores indicating more capable guardianship. These measures are related to routine activities theory because the presence of self-protection or an escort may deter an offender from selecting that particular individual to victimize. Although carrying self-protection may not deter verbal and nonverbal sexual harassment incidents, it may prevent an incident of physical sexual harassment if the offender is aware that the potential victim has protection.

Proximity to offenders is a five-item index that captures the enrollment status of students, whether they lived or worked on campus, whether they spent their weekends on campus, and whether they participated in campus activities (alpha = .52). Enrollment status was coded as 0 = *part time* and 1 = *full time*. The other variables were coded as 1 = yes. The answers for these five questions were summed to create the proximity measure. Each of these items was chosen to indicate whether the normal activities of a student that placed him or her among potential offenders increased his or her risk of being victimized.

An individual perceived to be more vulnerable to victimization by a motivated offender was classified as a suitable target. *Target suitability* is a two-item measure that documents how likely a person was to find a private place to eat and study. The options were labeled as 1 = *least likely to find a private place*, 2 = *less likely to find a private place*, 3 = *likely to find a private place*, 4 = *more likely to find a private place*, and 5 = *most likely to find a private place*. Higher scores are indicative of being a more suitable target to potential offenders. Seeking privacy may isolate a victim from capable guardianship and make him or her more attractive as a target. The reliability of the measure was moderate (alpha = .62).

Self-control was measured by using a scale containing six questions regarding one's perception of his or her own self-control (see Turner & Piquero, 2002). The questions assessed (a) whether a person gets into

a jam because he or she does things without thinking, (b) whether he or she thinks planning takes the fun out of things, (c) whether he or she enjoys taking risks, (d) whether he or she enjoys a new and exciting experience, even if it is a little frightening or unusual, and (e) whether life with no danger would be too dull for him or her (alpha = .71). Response sets were labeled as 1 = *strongly disagree* to 4 = *strongly agree*. The scale scores ranged from 6 to 24, with higher scores being indicative of lower levels of self-control.

Control variables. Prior research suggests that individuals of different races and genders perceive and label sexual harassment differently (Birdeau et al., 2005; Fitzgerald & Ormerod, 1991; Magley & Shupe, 2005; Shelton & Chavous, 1999) and that sexual harassment victimization might be more likely to occur among certain subpopulations (Cortina et al., 1998; Dziech, 2003; Paludi, 1997; Russell & Oswald, 2001). Therefore, several control variables were included, such as age (measured continuously in years), sex (1 = *female*), race (1 = *nonwhite*), sexual orientation (1 = *heterosexual*), and prior sexual harassment victimization on campus (1 = *yes*).

Dependent variables. The dependent variables were physical sexual harassment, nonverbal sexual harassment, and verbal sexual harassment. The definitions used for each type of sexual harassment were selected from the *Prevention Brochure of Sexual Harassment*, previously published by the university under analysis. Physical sexual harassment is defined as patting, pinching, tickling, or other unwanted touching; brushing against the body; attempted or actual kissing; attempted or actual fondling; or coerced sexual intercourse. Nonverbal sexual harassment is defined as public display or sharing of pictures or photographs of a sexual nature; e-mails or Web sites of a sexual nature; love notes or letters; suggestive comments on memos; graffiti; or gag gifts such as sex toys or games. Verbal sexual harassment included sexual innuendoes and comments; whistling in a suggestive manner; jokes about sex or gender in general; spreading rumors about sexual activity or performance of a student or employee; sexual propositions, invitations, or other pressures for sex; or implied or overt threats. Each of these variables were coded as 1 = *experienced sexual harassment.* An overall indicator of sexual harassment was also tested by asking whether a participant answered yes to any of the specific harassment questions.

Findings

The majority of the victims and nonvictims (see Table 1) were within the age range of 18 to 25 years. Both groups were overwhelmingly female, white, and heterosexual. However, among victims, nearly half had previously been sexually harassed on campus.[1]

We begin by reporting the prevalence rate of sexual harassment victimization. The data indicated that 22.7% of the participants were victims of sexual harassment during the study time. More specifically, 7.0% experienced a physical sexual harassment, 7.0% experienced a nonverbal sexual harassment, and 14.5% experienced a verbal sexual harassment.[2] As shown in Table 2, whereas more than nine tenths (92.9%) of the perceived offenders were other students, half (50.0%) were also strangers. As such, unlike more serious forms of sexual assaults on emerging adults where victims and offenders are typically acquaintances, offenders who sexually harass on campus were more likely to be strangers.

Table 2 also indicates that offenders were generally male (70.8%), were usually White (62.5%), and typically offended in isolation when they harassed their

Table 2 Characteristics of Sexual Harassment Offender, Incident, and Reporting

Characteristic	%
Type of offender	
Student	92.9
Both student and employee	7.1
Association with offender	
Stranger	50.0
Acquaintance	25.0
Classmate	10.7
Friend	14.3
Number of offenders	
One	62.5
More than one	37.5
Sex of offender	
All female	4.2
All male	70.8
Both	25.0
Perceived race/ethnicity of offender	
White	62.5
Non-White	37.5

	%
Perceived age of offender, years	
18–24	91.4
25–35	4.3
36 and above	4.3
Location of offense	
Inside your campus living quarters	4.2
Outside, but near your campus living quarters	20.8
Specifically designated parking lot, parking area	12.5
Another campus living quarter	4.2
Classroom, building, or lab	25.0
Gym	4.2
Dining area	4.2
General parking lot or area	4.2
Outside or near classroom building, library, gym	16.7
Other	4.2
Alone during offense	
Yes	66.7
No	33.3
Consumed alcohol during offense	
Yes	4.2
No	95.8

victims (62.5%). In terms of the location of the incident, most offenders chose to victimize students within a classroom building or lab (25.0%), outside but near campus living quarters (20.8%), or outside or near a classroom building or gym (16.7%) but did not consume alcohol during more serious types of sexual victimization of this population (Fisher et al., 2000; Testa & Parks, 1996). In fact, the data from the present study suggest the opposite in that almost all of the sexual harassment incidents on campus occurred without the victim consuming alcohol.

Investigation of the Sources of Sexual Harassment

The final interest of this study is to investigate the predictors of sexual harassment. Table 3 reports the results of three logistic regressions predicting sexual harassment. Specifically, in Model 1 a set of demographic predictors were regressed on the measure of sexual harassment; in Model 2 the measure of self-control was added; and in Model 3 the routine activities measures were added.

Beginning with Model 1, the results indicate that experiencing prior sexual harassment (b = 3.04; p = .00) significantly increased the odds of experiencing subsequent sexual harassment. Inclusion of the self-control measure in Model 2 did not change the significance of the control variables. That is, a prior sexual harassment (b = 3.07; p = .00) still significantly increases the odds of experiencing a subsequent sexual harassment. The addition of the self-control measure, however, did not emerge as a significant predictor of sexual harassment. Finally, in Model 3, age (b = .33; p = .05) emerged as a significant predictor, suggesting that older students were less likely to be victimized. Prior sexual harassment (b = 5.28; p = .00) continued to increase the chances of additional victimization. Sexual orientation (b = 8.37; p = .05) and race (b = 2.72; p = .05) emerged as significant predictors in the final model. Although the measure of self-control was unrelated to sexual harassment, the guardianship (b = 0.95; p = .01) and proximity measures (b = 1.20; p = .00) each independently emerged as significant predictors.

Specifically, individuals with increased levels of guardianship were less likely to experience sexual harassment, whereas individuals who spent more time on campus and were more proximate to potential offenders were more likely to be victimized.

⊠ Discussion

Emerging adulthood has been documented as a period of substantial transition and exploration (Arnett, 2000, 2004). Understanding the nature and extent of victimization experiences within this period has attracted the attention of scholars. Much of this attention has focused on sexually related or sexually explicit offenses (Fisher et al., 2000). The present study contributes to this literature by documenting the nature and extent of sexual harassment victimization of a sample of predominantly emerging adults, providing an in-depth exploration of the reporting characteristics of sexual harassment victims, and investigating the theoretical factors related to the probability of being sexually harassed. Two key findings emerged from these efforts.

First, just less than one fourth (22.7%) of our sample reported experiencing some type of sexual harassment (i.e., verbal, nonverbal, or physical) since school began during that academic year. This suggests that in any given academic year, 2,270 students on a campus with a population of 10,000 students might expect to experience sexual harassment. If prevalence rates in our study are similar to those occurring at other college campuses, these figures would imply that there would be approximately 700 incidents of physical sexual harassment, 700 incidents of nonverbal sexual harassment, and 1,450 incidents of verbal sexual harassment. Depending on the nature of the incident, these figures are cause for increased concern considering how sexual harassment affects the well-being of the student throughout his or her academic career (Fisher et al., 2002).

Second, the results of the multivariate models indicated that two measures of routine activities (proximity and guardianship) were significantly related to experiencing sexual harassment whereas the measure of self-control was not significant.

Table 3 Logistic Regression Predicting Overall Sexual Harassment

Variables	Model 1 (n = 160)				Model 2 (n = 160)				Model 3 (n = 158)			
	B	SE	Wald	Exp(b)	B	SE	Wald	Exp (b)	B	SE	Wald	Exp(b)
Age	−0.26	0.12	4.84	0.77	−0.25	0.12	4.41	0.79	−0.33*	0.16*	4.40*	0.72*
Sex	0.87	0.73	1.43	2.39	1.05	0.75	1.95	2.86	1.36	0.86	2.51	3.90
Race	−0.93	0.97	0.92	0.40	−0.93	0.97	0.93	0.39	−2.72*	1.34*	4.16*	0.07*
Orientation	−5.80	3.12	3.46	0.00	−5.93	3.31	3.21	0.00	−8.37**	3.56**	5.53	0.00*
Prior sexual harassment	3.04**	0.73**	17.63**	21.00**	3.07**	0.73**	17.60**	21.52**	5.28**	1.48**	12.79**	197.03**
Sum of low self-control					0.08	0.10	0.80	1.09	−0.05	0.12	0.14	0.96
Guardianship									−0.95*	0.35*	7.12*	0.39*
Target suitability									0.32	0.19	2.78	1.38
Proximity									1.20**	0.39**	9.49**	3.31**
Constant	8.42	5.05	2.79	4,555.44	7.06	5.40	1.71	1,168.44	15.56	6.82	5.21	5,727,861.6
Nagelkerke R²	.45				.45				.63			

*p < .05. **p < .01.

Specifically, respondents who were more proximate to motivated offenders by having a greater presence on campus were less likely to be sexually harassed. Together, these findings confirm prior research efforts that sought to understand sexually related victimization from a routine activities perspective (Fisher et al., 2000).

It was somewhat surprising that the self-control measure failed to emerge as a significant predictor of sexual harassment given that this variable has been identified as one of the strongest predictors of crime and victimization (see Pratt & Cullen, 2000). A few potential explanations are offered. First, it might be the case that college students possess relatively high levels of self-control. In fact, a comparison of identical self-control scales within our present study (mean = 13.10; SD = 3.261) and the National Longitudinal Survey of Youth (mean = 14.92; SD = 2.74) from which the measures were derived (see Turner & Piquero, 2002) supported these assertions. Second, and related, this study only used a six-item measure of self-control. Although this measure has been used in previous research, it still may have not captured enough variation within a population of college students. Third, it might be the case that an individual's self-control has little to do with his or her likelihood of being sexually harassed. Again, although prior research suggests that self-control was related to victimization (see Schreck, 1999; Unnever & Cornell, 2003), it is possible that an individual's self-control has little to do with this specific type of victimization.

Combined, the results of the present study suggest an approach for policies seeking to reduce the probability of sexual harassment experiences for college students. Specifically, policies should focus on educating potential offenders and victims regarding what it means to sexually harass or to be sexually harassed. It is important that both males and females receive such education to ensure that all students are equally aware of the definition of sexual harassment and situational contexts in which these behaviors are likely to occur.

Although the results of this research provide insight into the nature and extent of sexual harassment, the results must be taken within the context of the limitations of the research.

First, we investigated the effects of what we believed were two important contributors (i.e., measures of routine activities and self-control) to the explanation of sexual harassment. Other theories might be of equal or greater importance in the explanation of sexual harassment. For example, a fruitful line of research might investigate the role peers play in either modeling sexual harassment or insulating individuals from sexual harassment over the course of their lives.

Second, we examined the predictors of sexual harassment in a sample that was primarily within the developmental period of emerging adulthood. We did not, however, investigate whether the predictors of sexual harassment varied across developmental periods. For example, future research may seek to investigate whether the same predictors of sexual harassment were significant in emerging adulthood versus adulthood. It is possible that the stability brought forth with the onset of adulthood (i.e., marriage and childbearing) also adjusts the routine activities of individuals, which correspond with changes in the importance of those predictors in the explanation of sexual harassment. To provide some initial insight into this question, we examined the prevalence rates of sexual harassment across developmental periods within our sample and found that emerging adults (ages 18–25 years) experienced sexual harassment (27% vs. 6%; p < .04) at far greater rates than did adults (ages 26 and above). Similarly, emerging adults appeared to have significantly lower levels of self-control (13.44 vs. 12.20; p < .03) and significantly higher levels of guardianship (5.35 vs. 4.02; p < .00). As such, there appears to be evidence that emerging adults are qualitatively different from adults in terms of some of the predictors of sexual harassment.

Emerging adulthood is an important developmental period because it is a time when individuals experience substantial instability, become highly self-focused, experience significant identity explorations, and embark on substantial transition as they approach adulthood (Arnett, 2002). Unfortunately, emerging adulthood is also a developmental period when prevalence rates of a number of risky behaviors peak, including criminal offending, victimization, alcohol use, and substance use. Sexual harassment is one of those risky behaviors that can affect subsequent routine (e.g., going to class or to work) and could also be a precursor to more serious types of victimization (e.g., sexual assault). We hope that the findings from the current study will initiate more in-depth investigations regarding the causes of risky behaviors and how the causes vary across developmental periods. Comparison of effect sizes across developmental periods would provide insight into how emerging adulthood is similar to, or qualitatively different from, the preceding and subsequent developmental periods.

⊠ Notes

1. Prior campus sexual harassment was significantly correlated with the present measure of sexual harassment ($r = .49$).

2. These figures do not sum to the total victimization rate because individuals could have been victimized via more than one method. The total number of sexual harassment incidents was 49 (verbal = 25; physical = 12; nonverbal = 12).

⊠ References

Abbey, A., Zawacki, T., Buck, P. O., Clinton, A. M., & McAuslan, P. (2001). Alcohol and sexual assault. *Alcohol Research and Health, 25*, 43–51.

Arnett, J. (2000). High hopes in a grim world. *Youth and Society, 31*, 267–286.

Arnett, J. (2002). *Readings on adolescence and emerging adulthood.* Upper Saddle River, NJ: Prentice Hall.

Arnett, J. (2004). Emerging adulthood: The winding road from the late teens through the twenties. New York: Oxford University Press.

Benson, D. J., & Thomson, G. E. (1982). Sexual harassment on a university campus: The confluence of authority relations, sexual interest, and gender stratification. *Social Problems, 29*, 236–251.

Birdeau, D. R., Somers, C. L., & Lenihan, G. O. (2005). Effects of educational strategies on college students' identification of sexual harassment. *Education, 125*, 496–510.

Bradenburg, J. B. (1997). Confronting sexual harassment: What schools and colleges can do. New York: Teachers College Press.

Clarke, R. V., & Felson, M. (Eds.). (1993). *Routine activity and rational choice: Advances in criminological theory* (Vol. 5). New Brunswick, NJ: Transaction Books.

Cohen, L., & Felson, M. (1979). Social change and crime rate trends: A routine activity approach. *American Sociological Review, 44*, 588–608.

Cohen, L. E., & Vila, B. J. (1996). Self-control and social control: An exposition of the Gottfredson-Hirschi/Sampson-Laub debate. *Studies on Crime and Crime Prevention, 5*, 125–150.

Collins, J. J., & Messerschmidt, P. M. (1993). Epidemiology of alcohol-related violence. *Alcohol Health & Research World, 17*, 93–100.

Combs-Lane, A. M., & Smith, D. W. (2002). Risk of sexual victimization in college women: The role of behavioral intention and risk-taking behaviors. *Journal of Interpersonal Violence, 17*, 165–183.

Cortina, L. M., Swan, S., Fitzgerald, L. F., & Waldo, C. (1998). Sexual harassment and assault: Chilling the climate for women in academia. *Psychology of Women Quarterly, 22*, 419–441.

Dansky, B., & Kilpatrick, P. (1997). Effects of sexual harassment. In W. O'Donohue (Ed.), *Sexual harassment: Theory, research, and treatment* (pp. 152–174). Boston: Allyn & Bacon.

DeKeseredy, W., & Kelly, K. (1993). The incidence and prevalence in woman abuse in Canadian university and college dating relationships. *Canadian Journal of Sociology, 18*, 137–159.

Dziech, B. W. (2003). Sexual harassment of college campuses. In M. Paludi & C. Paludi (Eds.), *Academic and workplace sexual harassment: A handbook of cultural, social science, management, and legal perspectives* (pp. 147–172). Westport, CT: Praeger.

Dziech, B. W., & Hawkins, M. W. (1998). Sexual harassment in higher education: Reflections and new perspective. New York: Garland.

Felson, M., & Clarke, R. V. (1998). *Opportunity makes the thief* (Crime Detection and Prevention Series, Paper 98, Policy Research Group). London: Home Office.

Felson, R. B., & Burchfield, K. B. (2004). Alcohol and the risk of physical and sexual assault victimization. *Criminology, 42*, 837–859.

Fisher, B. S., Cullen, F. T., & Turner, M. G. (2000). *The sexual victimization of college women.* Washington, DC: U.S. Department of Justice, National Institute of Justice.

Fisher, B. S., Cullen, F. T., & Turner, M. G. (2002). Being pursued: Stalking victimization in a national study of college women. *Criminology and Public Policy, 1*, 257–308.

Fitzgerald, L. F., & Ormerod, A. J. (1991). Perceptions of sexual harassment: The influence of gender in academic context. *Psychology of Women Quarterly, 15*, 281–294.

Fitzgerald, L., Swan, S., & O'Donohue, W. (1997). But was it really sexual harassment? Legal, behavioral, and psychological definitions of the workplace victimization of women. In W. O'Donohue (Ed.), *Sexual harassment: Theory, research, and treatment* (pp. 5–28). Boston: Allyn & Bacon.

Gibson, C., Schreck, C. J., & Miller, M. (2004). Binge drinking and negative alcohol-related behaviors: A test of self-control theory. *Journal of Criminal Justice, 32*, 411–420.

Gottfredson, M., & Hirschi, T. (1990). *A general theory of crime.* Palo Alto, CA: Stanford University Press.

Gottfredson, M., & Hirschi, T. (2003). A general theory of crime. In F. Cullen & R. Agnew (Eds.), *Criminological theory: Past to present* (pp. 240–252). Los Angeles: Roxbury.

Gover, A. R. (2004). Risky lifestyles and dating violence: A theoretical test of violent victimization. *Journal of Criminal Justice, 32*, 171–180.

Hall, R. F., Graham, R. D., & Hoover, G. A. (2004). Sexual harassment in higher education: A victim's remedies and a private university liability. *Education and the Law, 16*, 33–45.

Ivy, D. K., & Hamlet, S. (1996). College students and sexual dynamics: Two studies of peer sexual harassment. *Communication Education, 45*, 149–166.

Kalof, L. (2000). Vulnerability to sexual coercion among college women: A longitudinal study. *Gender Issues, 18*, 47–58.

Kalof, L., Eby, K. K., Matheson, J. L., & Kroska, R. J. (2001). The influence of race and gender on student self-reports of sexual harassment by college professors. *Gender and Society, 15*, 2828–2302.

Kelley, M. L., & Parsons, B. (2000). Sexual harassment in the 1990s: A university-wide survey of female faculty, administrators, staff, and students. *Journal of Higher Education, 71*, 548–568.

Koss, M. P., Gidycz, C. A., & Wisniewski, N. (1987). The scope of rape· Incidence and prevalence of sexual aggression and victimization in a national sample of higher education students. *Journal of Consulting and Clinical Psychology, 55*, 162–170.

Larimer, M. E., Lydum, A. R., Anderson, B. K., & Turner, A. P. (1999). Male and female recipients of unwanted sexual contact in a college student sample: Prevalence rates, alcohol use, and depression symptoms. *Sex Roles, 40*, 295–308.

LaRocca, M. A., & Kromrey, J. (1999). The perception of sexual harassment in higher education: Impact of gender and attractiveness—Statistical data included. *Sex Roles: A Journal of Research.* Retrieved April 14, 2005, from http\\www.findarticles.com/p/articles/mi_m2294/is_11_40/ai_ai57533232.

Leppel, K. (2006). College binge drinking: Deviant versus mainstream behavior. *American Journal of Drug and Alcohol Abuse, 32*, 519–525.

Magley, V. J., & Shupe, E. I. (2005). Self-labeling sexual harassment. *Sex Roles, 53*, 173–189.

Menard, K., Hall, G., Phung, A., Ghebrial, M. L. (2003). Gender differences in sexual harassment and coercion in college students: Developmental, individual, and situational determinants. *Journal of Interpersonal Violence, 18*, 1222–1239.

Moffitt, T. E. (2003). Pathways in the life course to crime. In F. T. Cullen & R. Agnew (Eds.), *Criminological theory: Past to present* (pp. 450–469). Los Angeles: Roxbury.

Mustaine, E. E., & Tewksbury, R. (1999). A routine activity theory explanation for women's stalking victimization. *Violence Against Women, 5*, 43–62.

Paludi, M. (1997). Sexual harassment in schools. In W. O'Donohue (Ed.), *Sexual harassment: Theory, research, and treatment* (pp. 225–249). Boston: Allyn & Bacon.

Parks, K. A., & Fals-Stewart, W. (2004). The temporal relationship between college women's alcohol consumption and victimization experiences. *Alcohol: Clinical and Experimental Research, 28*, 625–629.

Piquero, A. R., Brame, R., Mazerolle, P., & Hapaanen, R. (2002). Crime in emerging adulthood. *Criminology, 40*, 137–170.

Pratt, T. P., & Cullen, F. T. (2000). The empirical status of Gottfredson and Hirschi's general theory of crime: A meta-analysis. *Criminology, 38*, 964.

Rothman, E., & Silverman, J. (2007). The effect of a college sexual assault prevention program on first-year students' victimization rates. *Journal of American College Health, 55*, 283–290.

Rubin, P. N. (1995). *Civil rights and criminal justice: Primer on sexual harassment.* Research in action (NCJ Publication No. 15663). Washington, DC: Department of Justice.

Russell, B. L., & Oswald, D. L. (2001). Strategies and dispositional correlates of sexual coercion perpetrated by women: An exploratory investigation. *Sex Roles, 45*, 103–115.

Schneider, B. E. (1987). Graduate women, sexual harassment, and university policy. *Journal of Higher Education, 58*, 46–65.

Schreck, C. J. (1999). Criminal victimization and low self-control: An extension and test of a general theory of crime. *Justice Quarterly, 16*, 633–65.

Shelton, J. N., & Chavous, T. M. (1999). Black and white college women's perceptions of sexual harassment. *Sex Roles, 40*, 593–615.

Spitzberg, B. H. (1999). An analysis of empirical estimates of sexual aggression victimization and perpetration. *Violence and Victims, 14*, 241–260.

Stewart, E. A., Elifson, K. W., & Sterk, C. E. (2004). Integrating the general theory of crime into an explanation of violent victimization among female offenders. *Justice Quarterly, 21*, 159–180.

Testa, M., & Parks, K. A. (1996). The role of women's alcohol consumption in sexual victimization. *Aggression and Violent Behavior, 1*, 217–234.

Turner, M. G., & Piquero, A. R. (2002). The stability of self-control. *Journal of Criminal Justice, 30*, 457–471.

Unnever, J. D., & Cornell, D. G. (2003). Bullying, self-control, and ADHD. *Journal of Interpersonal Violence, 18*, 129–147.

U.S. Department of Education, National Center for Education Statistics. (2007). *The Condition of Education 2007* (NCES 2007-064). Washington, DC: U.S. Government Printing Office.

U.S. Department of Health and Human Services. (2006). *Results from the 2006 National Survey on Drug Use and Health: National findings.* Rockville, MD: Substance Abuse and Mental Health Services Administration, Office of Applied Studies.

Weitzman, E. R., & Nelson, T. F. (2004). College student binge drinking and the "Prevention paradox": Implications for prevention and harm reduction. *Journal of Drug Education, 34*, 247–266.

Williams, E. A., Lam, J. A., & Shively, M. (1992). The impact of a university policy on the sexual harassment of female students. *Journal of Higher Education, 63*, 50–64.

Wilson, A. E., Calhoun, K. S., & McNair, L. D. (2002). Alcohol consumption and expectancies among sexually coercive college men. *Journal of Interpersonal Violence, 17*, 1145–1159.

DISCUSSION QUESTIONS

1. Why do you think low self-control was not related to sexual harassment victimization in this study?

2. What are the policy implications for college students, given the finding that routine activities do, in fact, place persons at risk of being sexually harassed?

3. Why do you think sexual harassment is so prevalent among college students?

4. Given the use of the Internet and social networking sites, how do you think routine activities theory and risky lifestyles should be conceptualized in future research on sexual harassment?

---◈---

Introduction to Reading 2

In this article, Christopher Schreck and Bonnie Fisher examined how both the family and peers impact a person's likelihood of experiencing a violent victimization. Four types of violent victimization were examined: being threatened with a knife or gun, being shot, being stabbed, or being "jumped." They also examined variables designed to measure risky lifestyles. In doing so, they use data from the National Longitudinal Study of Adolescent Health on more than 3,500 respondents. The authors found that the lifestyle factors of sneaking out of the house, driving around in a car, and exercising increased the chances of a person being victimized. They also found that general feelings the respondent had for his or her family were negatively related to victimization—the better a person felt about his or her family, the less likely he or she was to be victimized. Finally, they found that peer delinquency and spending time with peers both increased victimization risk. Taken together, they found support for three of the theories or risk factors identified in this chapter: routine activities/lifestyle, family, and delinquent peers.

Specifying the Influence of Family and Peers on Violent Victimization

Extending Routine Activities and Lifestyles Theories

Christopher J. Schreck and Bonnie S. Fisher

Criminologists have long been interested in family conditions as an antecedent of youthful delinquent behavior (Farrington, 1987; Glueck & Glueck, 1950; Gottfredson & Hirschi, 1990; Loeber & Stouthamer-Loeber, 1986; Rebellon, 2002; Sampson & Laub, 1993). As important as family context is in the delinquency research, the family may have broader significance in view of the finding that delinquents

SOURCE: Schreck and Fisher (2004). Reprinted with permission.

tend to experience high levels of violent victimization (Lauritsen, Sampson, & Laub, 1991). This connection between involvement in crime and the experience of victimization has led some criminologists to propose that established correlates of crime—which might include family characteristics—are also relevant for understanding victimization (Piquero & Hickman, 2003; Schreck, 1999; Schreck, Wright, & Miller, 2002). In addition, it appears reasonable to think that families have an interest in preventing their children from being threatened, beaten up, or robbed. Consequently, family context may be particularly important among teenagers, who bear a disproportionately high level of victimization relative to every other age group (Bureau of Justice Statistics, 2003).

In contrast to the longstanding tradition of family and delinquency research, there have been few attempts by researchers to explore a connection between family context and the experience of general violent victimization. The few studies to consider the relationship between family context and general victimization found that level of emotional attachment between family members is perhaps the strongest family-relevant predictor of victimization (Esbensen, Huizinga, & Menard, 1999; Lauritsen, Laub, & Sampson, 1992). Consistent with these studies, we argue that a strong family—one that possesses durable ties of emotional warmth and support—might reduce the chances that adolescents will experience violent victimization.

The primary competitor to the influence of the family is most likely the adolescent peer group, which criminologists and victimologists find to be a risk factor for criminal behavior as well as victimization (Akers, 1985; Lauritsen et al., 1992; Schreck et al., 2002; Warr, 2002). The peer group represents one of the primary social contexts in the lives of adolescents, aside from the family (Coleman, 1980; Corsaro & Eder, 1990). At the same time, the presence of criminal associates in a teenager's life is traditionally one of the strongest and most consistent correlates of individual delinquency (Akers, 1994; Hindelang, Hirschi, & Weis, 1981; Warr & Stafford, 1991; c.f. Haynie, 2001). Membership in peer groups, however, also corresponds

with a higher risk of victimization. Researchers have theorized that membership in a delinquent group can invite retaliation from the group's victims or else group norms might favor the use of violence as an accepted means of settling within-group disputes (Baron, Forde, & Kennedy, 2001; Baron, Kennedy, & Forde, 2001; Decker & Van Winkle, 1996; Singer, 1981). In addition, proximity to other participants in crime leads members of the peer group to use each other as convenient targets (Jensen & Brownfield, 1986; Lauritsen et al., 1991; Schreck et al., 2002). Even research looking exclusively at criminal youth gang members found that membership is a source of risk rather than protection (Miller & Decker, 2001; Sanders, 1994). Taken together, attempts at understanding the sources of adolescent victimization cannot afford to ignore risk factors originating in the peer context. Nevertheless, the literature has not given much attention to the effectiveness of family context as a predictor of victimization after controlling for delinquent peer group characteristics (see Esbensen et al., 1999; Lauritsen et al., 1992).

In view of the importance of the family and peer group for delinquency causation, both appear to have significance beyond simply being precursors of juvenile delinquency. This article explores family and peer contexts and examines how these contexts link to routine activities/lifestyles theories of victimization.

Using data from the National Longitudinal Study of Adolescent Health (Add Health), we then empirically tested whether the combined influence of family and delinquent peer-group associations significantly affects adolescents' level of violent victimization. One important methodological advantage of the Add Health is its national coverage of adolescents. The Add Health also offers a significant improvement over other major data sets (such as the National Youth Survey or the Monitoring the Future study) in that the staff collected data from not only the respondent and his or her parents (which would be relevant for measuring family attachment) but also from the respondent's peers. This data set thus provides a unique opportunity to measure the influence of peer context in conjunction with family characteristics.

◪ Theoretical Framework

Our general framework for explaining how family and peer contexts might influence victimization draws from the routine activities/lifestyles theories (Cohen & Felson, 1979; Hindelang, Gottfredson, & Garofalo, 1978). Together, these theories stress how situations or contexts carry their own level of risk for victimization. The routine activities theory maintains that the convergence in time and space of motivated offenders, attractive targets, and ineffective guardianship determines the risk of victimization. People who spend a lot of time in settings where they are exposed to offenders and cannot adequately protect themselves will tend to have a higher risk of victimization. Social structure and demographic characteristics generally affect daily routines and, thus, exposure and vulnerability to likely offenders (Hindelang et al., 1978; Miethe, Stafford, & Long, 1987; Sampson & Wooldredge, 1987). Family and peer contexts could also be important mechanisms for the meeting of motivated offenders and worthwhile, vulnerable targets; that is, social bonds with family and peers might structure or determine routine daily activities that have relevance for victimization risk (Felson, 1986; Horney, Osgood, & Marshall, 1995; Schreck et al., 2002).

Attachment to the Family and Violent Victimization

The bond of attachment attends to the degree of sensitivity and emotional closeness that people have for others (see Hirschi, 1969). Although the level of emotional closeness with others might directly influence risk, the connection between bonds, similar to attachment, and routine activity is better documented (Felson, 1986; Schreck et al., 2002); that is, attachment could serve to constrain the types of activities we routinely engage in and bring individuals into closer proximity to those with an interest or social obligation to act as protectors. The following discussion thus shows how attachment to the family has relevance for each of the three necessary elements for crime: effective guardianship, exposure to motivated offenders, and target attractiveness.

Family context and exposure to motivated offenders. Bonds to family members should remove would-be victims from the proximity of likely offenders. Strong attachment for parents should tend to keep children closer to home and away from the company of strangers. Felson (1998) noted that although some people have family members who are unquestionably dangerous to be with, it is typically much more dangerous to spend time around strangers and outside the home (see also, Bureau of Justice Statistics, 2003; Hindelang et al., 1978). In addition, strong feelings of attachment might compel parents to regulate their children's friendship groups (which includes getting to know the friends, including them in family activities, and getting to know their families). As noted earlier, the delinquent peer group may well represent a pool of unusually motivated offenders. In short, attachment keeps children closer to home, makes parents interested in the friends of their child, and thus keeps children away from motivated offenders.

Family context and guardianship. Strong bonds represent a catalyst for the meeting, in time and space, of potential victims with effective guardians. Gottfredson and Hirschi (1990) maintained that strong bonds of parental attachment to the child are a prerequisite for effective supervision and, presumably, protection; that is, one might expect that parental presence would make criminal activity against their offspring inconvenient. One may also assume that children who are attached to their parents will choose to spend more time with them enjoying their protection than children who dislike or distrust their parents. Attachment can also lead to improved guardianship in ways other than direct supervision. Given the scattered nature of informal control in modern American society (see Felson, 1998), with parents who work and students who go to school and who may work themselves, parents can ensure guardianship by arranging for surrogate parents, such as child care providers or coaches. Thus, although strong attachment cannot guarantee that parents will always be able to immediately supervise their children, attachment appears more likely to facilitate the direct presence of alternative guardians.

Family context and target attractiveness. Bonds might create proximate so-called handlers, or people who are socially obligated to prevent likely victims from making themselves into attractive targets (for more on handlers, see Felson, 1986); that is, effective handlers will strive to prevent those they feel responsible for from behaving provocatively and risking retaliation, which could happen if the child is behaving belligerently or insolently to others or is victimizing them. Handlers may also be in a position to require precautions intended specifically to promote safety (e.g., "Come straight home after school" or "Make sure you take the bus; don't walk home by yourself").

Social bonds aside, we would expect that deviant parents will be less capable of protecting their children from crime. Gottfredson and Hirschi (1990) suggested that such parents are little inclined to exert themselves monitoring their children. Beyond the effort and time that effective supervision entails, parents who impair their judgment by consuming illegal drugs or drinking alcohol to excess will tend to be less effective as protectors. Deviant parents may also engage in illegal behavior against their own children and represent proximate likely offenders in their own right. The presence of two married adults in the household may also be important, as researchers have speculated that family disruption makes supervision more difficult as well as interferes with the maintenance of bonds with children (Hirschi, 1991; Rutter & Giller, 1983).

Peer Context and Violent Victimization

Although adolescent life typically revolves around the family, the peer group represents a second major context. We have already noted the fact that teenagers spend considerable amounts of time engaged in social activity with their friends. As is the case with the family, the routine activities/lifestyles approaches can offer insights about why contact with the peer group, most notably the delinquent peer group, also suggests exposure to motivated offenders, weakened guardianship, and increased attractiveness as a target.

Delinquent Peer Association and Exposure to Motivated Offenders

Although routine activities theory does not attempt to explain why people are motivated to commit crime,

research reveals that there are individual differences in how easy it is to tempt someone into crime (Felson, 1998; Gottfredson & Hirschi, 1990). To illustrate, exposure to persons with a record of delinquency—particularly when one is vulnerable—would therefore be risky. The principle of homogamy in lifestyle theory (Hindelang et al., 1978, pp. 256–257) addresses this issue, positing that individuals who share daily living conditions and social activities with high-risk offenders (because they possess similar demographic backgrounds) will have a disproportionately high risk of victimization (see also, Miethe & Meier, 1994). Ties of friendship to a delinquent would therefore make it more likely that someone will routinely be in proximity to a motivated offender and vulnerable. If most crimes do not require significant departures from the normal routines of the offender, the closeness of a delinquent's friends would make them accessible targets (see Felson, 1998).

Delinquent peer association and guardianship. In light of the amount of time teens spend with their peer group, other members of that group are potentially well situated to act as guardians and protect their peers. This role for the peer group presupposes that one's associates are interested in serving as guardians. Guardianship, however, potentially necessitates time, effort, and risk. Felson (1986, 1998) suggested that strong social bonds create a stake in protecting others. These bonds influence the amount of time people spend in risky activity away from those who can protect them from the criminal inclinations of others; that is, caring about others may make one spend more time with them, which promotes personal enjoyment and safety from victimization.

The delinquent peer group, on the other hand, undermines guardianship in multiple ways. First, Gottfredson and Hirschi (1990) characterized delinquents as lacking diligence and being impulsive and selfish, thus suggesting that they are unlikely to exert themselves to protect their closest associates when their associates are vulnerable. Second, as Sparks (1982) indicated, someone in a delinquent group would be unlikely to report victimization to the police and can therefore count on having less police protection. Part of this may reflect the delinquent's understandable lack of comfort in dealing with legal authorities. For instance, delinquents might assume that the police will be quick to dismiss their claims because of their connection to other troublemakers, or else might run a higher risk of implicating themselves merely by talking to the police.

Delinquent peer association and target attractiveness. As noted in criminological research, children who participate in youth gangs and who claim the friendship of other delinquent youth are much more likely to participate in crime themselves. In fact, juvenile delinquency frequently occurs as a group activity (Erickson & Jensen, 1977; Warr, 1996). A number of researchers have suggested that participation in crime—or even of merely identifying with a delinquent group—can make a youth a worthwhile target for others. Singer (1981) proposed that retaliation is part of the normative characteristics of a violent subculture (see also, Baron, Forde, et al., 2001). Someone thus committing a crime or who is a member of a group responsible for committing a crime might thus alternate from offender to victim when the former victim attempts to retaliate. Youth groups that encourage aggressive behavior may likewise make themselves attractive targets by precipitating violence through their confrontational behavior.

Research Questions

There is good reason to think that families with conventional parents and where strong bonds are present between family members will be more effective at promoting the safety of offspring. At the same time, the prominence of the peer group in the lives of adolescents represents a plausible source of risk. More specifically, do stronger bonds of attachment between family members—especially between parents and children—promote greater safety from violent victimization? Does spending time with friends, and having delinquent friends, increase risk?

Most existing data sources survey the victimization experiences and traits of only the adolescent

respondent, and not their parents or members of their peer network. The National Longitudinal Study of Adolescent Health (Add Health), however, integrated information from each of these three sources—the adolescent, parents, and friends—which provides more valid measures than would be the case if one relied on data from only the teenagers. This analysis below, which uses these data, considers the relative influences of family context and peer group factors on the level of violent victimization.

Data and Methods

The current research used the first wave of the public-use version of the Add Health study, which was conducted between September 1994 and December 1995. This wave consisted of in-school and in-home administrations. The goal of the Add Health study was to create a sample that is nationally representative of students attending U.S. schools in Grades 7 through 12. Our analyses excluded cases with any missing data, leaving slightly more than 3,500 respondents who provided sufficient data for analysis.

Dependent Variable—Violent Victimization

To measure the occurrences of violent victimization for each individual, we created an additive index. Respondents were asked about having been the victim of four types of serious violence. These items each measure serious forms of violence: (a) threat with a knife or gun, (b) shooting, (c) stabbing, and (d) being "jumped." The reference period was the 12 months preceding the survey. The Add Health survey recorded responses using ordinal categories (*never* = 0, *once* = 1, *more than once* = 2). As expected, the distribution of violent victimization is skewed; nearly 80% of the respondents were not victims of any of the violent crimes noted above during the reference period, approximately 10% reported one instance of victimization, and the remainder accounted for two or more incidences of victimization.

Independent Measures

Family context. There were six family-related measures included in the analysis. The first measure, welfare receipt (coded 1 = *yes*, 0 = *no*), measures the economic condition of the family. Less than 10% of the sample reported receiving any public assistance. The second measure, parental drinking, measures how often parents consumed five or more drinks in the previous month (coded 1 = *never* through 6 = *five or more times*). We used this variable as a proxy for deviant parents. The median parent reported never consuming five or more drinks. The third item, positive parental feelings toward child, is a composite index of four items measuring the parent's opinions about the child, with a higher positive score indicating warmer feelings toward the child. Because the component items in this index had different metrics, we converted the raw item score indicating more positive feelings toward the child. The fourth measure is a two-item index tapping the child's attachment toward the mother, with a higher score indicating stronger attachment (coded 1 = *not at all* through 5 = *very much*). The median score was 5, indicating that the typical child is strongly attached to his or her mother. The fifth item is also a two-item index made up of similar questions but measuring the child's attachment toward the father (median = 4.5). The final item, family climate, is a four-item index measuring the general feelings of the respondent for the family (as opposed to the parents specifically). Descriptive statistics indicate that the average respondent lived in a family characterized by closeness between the members (median = 4).

Peer delinquency. The friends located in the respondent's peer network reported participation in five minor types of delinquency: (a) cigarette smoking, (b) drinking alcohol, (c) getting drunk, (d) skipping school, and (e) doing risky things on a dare. All items had the following response categories: 0 = *not at all*, 1 = *once or twice*, 2 = *3 or 4 times*, and 3 = *5 or more times*. The Add Health survey designers then computed the average level of delinquency among those peers nominated by the respondent, or who nominated the

respondent, and recorded them on the respondent's record. We summed the averages for the five items to create a peer delinquency index. The average respondent had a peer network that engaged in these delinquent or risky activities approximately 4 times during the reference period. Most members of the sample had friends who reported relatively low levels of delinquency, averaging four different acts of delinquency committed once or twice in the previous 12 months.

Peer context. Two variables measure additional dimensions of peer context. The first variable, friends care, measures the feeling that friends have about the respondent. Scores range from 1 (*not at all*) to 5 (*very much*). The second variable, activities with friends, reflects the amount of time the respondent spends simply hanging out with friends (coding ranges from 0 = *not at all* to 3 = *5 or more times per week*). The average respondent felt that friends generally cared (mean = 4.29) and reported going out with friends 3 or 4 times per week.

Lifestyles. We also employed several activities that bivariate analysis indicated as being correlated with victimization: sneaking out of the house at night, driving a car, and exercising. Each of these items represents activity outside the home, which should elevate victimization risk. One might reasonably interpret each of these activities as events that are likely to occur away from parental guardians and home, and therefore likely to incur additional risk.

Controls. The multivariate data analyses include several demographic control variables. Gender represents the effect of being a male respondent (with female respondents being the reference category). Males constituted 47% of the sample. Racial identity is divided between Black and non-Black (reference), with 25% of the sample being Black. The variable for respondent age measures age in discrete years, ranging from 11 to 20. Each of these variables is a correlate of violent victimization (see, e.g., Bureau of Justice Statistics, 2003; Cohen, Kluegel, & Land, 1981; Hindelang et al., 1978).

⊠ Results

We were interested in learning whether family and peer-related variables predict violent victimization and significantly improve prediction beyond baseline demographic variables. In view of the highly skewed distribution of the dependent variable, we use an overdispersed Poisson regression model to estimate the effect coefficients of the independent variables. The first column of coefficient in Table 1 (i.e., Model 1) represents the baseline model. As expected, minorities tend to have significantly higher levels of

Table 1 Poisson Regression Estimates for Violent Victimization

	Model 1		Model 2		Model 3	
	b	β	b	β	b	β
Intercept	−1.72	—	−.47	—	−.40	—
Demographic						
Male	.90***	.25	.98***	.27	.95***	.26
Black	.42***	.10	.35***	.08	.41***	.10
Age	−.05	−.41	−.06*	−.49	−.08*	−.65
Lifestyles						
Driving around	.35*	.08	.32*	.08	.30*	.07
Sneaks out	1.03***	.18	.78***	.13	74***	.13
Exercises	.15***	.09	.15***	.09	.15***	.09
Family context						
Welfare receipt	—	—	.32	.05	.31	.04
Climate	—	—	−.26**	−.10	−.25**	−.09
Parent drinks	—	—	.04	.02	.03	.01
Parents' feelings for child	—	—	−.06**	−.09	−.05*	−.07
Attached to mother	—	—	−.02	−.01	−.02	−.01
Attached to father	—	—	−.03	−.03	−.02	−.02
Peers						
Peer delinquency	—	—	—	—	.03**	.06
Friends care	—	—	—	—	−.04	−.02
Activities w/ friends	—	—	—	—	.10*	.05
−2 Log Likelihood	2097.76	—	2091.52	—	2083.46	—
Generalized R^2	.08	—	.10	—	.11	—

*$p < .05$

**$p < .01$

***$p < .001$

victimization, as do male and younger members of the sample. Model 1 also includes the three indicators of unsupervised and unstructured activities. Consistent with the finding of Schreck and colleagues (2002), the data show that teenagers who engage in leisure activities away from home (and, in many cases, away from supervision) also tend to have higher levels of violent victimization.

Model 2 incorporates the family-related variables into the analysis. The strongest family-related predictors of violent victimization are family climate and parental feelings for their children. Adolescents living in households where there is a warmer, more accepting climate experience less violence, while those living in families where parents dislike them have a greater risk for becoming victims. This supports our explanation for children as well as reduced exposure to motivated offenders. Some types of family closeness do not appear to matter, however. The attachment of child to mother or father did not significantly affect victimization risk. Because attachment from the parent to the child is significant, this finding suggests that parents have the more central role when it comes to protecting teenagers from violence. More interesting, drinking activity by parents failed to significantly influence victimization risk. In addition, incorporating the family-relevant variables did not substantially alter the baseline variable coefficients, thus indicating that demographic and lifestyle effects are independent of the family context variables used in the analysis.

Model 3 adds the peer context variable. Peer delinquency and activities with friends were the significant peer-context predictors. Teenagers whose friends engaged in higher levels of delinquency and risky activity tended to experience more victimization. Readers should be aware, however, that the peer delinquency measures tap minor forms of delinquency, and that a measure for more serious forms could yield a larger influence. This finding supports our explanation that those who associate with delinquent peers risk enhanced exposure to motivated offenders, ineffective guardianship, and suitability as a target for violence. In addition, spending time with friends—whether or not they are delinquent—also increases the risk of experiencing violence. Having friends who care about the respondent, on the other hand, did not significantly influence level of violent victimization. Peer context, however, predicted victimization independently of demographic, lifestyle, or family context variables—the effect coefficients of these variables, in fact, did not seem much affected by the inclusion of the peer context variables. That is, the effects of peer context do not seem to detract from the influence of family variables; each appears to predict violent victimization independently. In addition, the continued strength of demographic predictors even in the full model indicates that researchers looking to explain demographic variation in violent victimization would have to look at risk factors other than the ones considered in the analysis.

Conclusion

What relevance do family and peer variables have for general victimization risk among adolescents? Building from the similarities between the correlates of crime and victimization, the importance of the family and peer contexts in the etiology of delinquency indicates that victimization research might benefit from exploring these contexts further. Moreover, the centrality of family and peers in the lives of teenagers and the likely influence these contexts have on the convergence of motivated offenders, attractive targets, and guardianship further recommend research on family and peer environments.

In the current study, we linked family and peer context to routine activity/lifestyles theories. In the case of the family, we hypothesized that strong bonds of attachment decrease the likelihood that meetings between motivated offenders and attractive and vulnerable targets will occur. To test, we measured attachment in several ways: Attachment included (a) general feelings of closeness with the family, (b) feelings of warmth toward parents, and (c) parental feelings toward the child. Our results indicated that teenagers in households where the family context is characterized by closeness and understanding tend to be safe from violent crime. Likewise, children whose parents are emotionally alienated from them tend to experience a higher risk of victimization. Given the preliminary nature of research linking family and peer contexts to victimization, readers should be aware that although the results are consistent with the expectations of routine activity/lifestyle theories (that family climate and positive parental feelings for offspring appear to promote better guardianship, increase distance from motivated offenders, and prevent children from becoming attractive targets), the interpretation we offer might not be the only one. For instance, strong bonds of attachment might also create unwillingness to take risks that might lead to victimization (e.g., by behaving belligerently or insultingly toward others) because victimization jeopardizes the happiness of loved ones; that is, attachment might show how much more one has to lose through victimization and thus deter provocative behavior. Further research about the nature of the attachment-victimization link might clarify this issue.

The presence of delinquent peers appears to be a risk factor, in addition to family-related correlates. The results support the interpretation that spending time around delinquent peers is a risky venture. First, group delinquency, or identification with delinquents, can invite retaliation (Baron, Forde, et al., 2001; Singer, 1981). Affiliation with delinquent peers also suggests exposure to motivated offenders (Lauritsen et al., 1991; Schreck et al., 2002). Second, delinquents may exacerbate risk because they are not the most effective of guardians for their friends. We should note that the magnitude coefficients in our estimated model appear somewhat weaker than those coefficients reported in Lauritsen et al. (1992). This may be a reflection of the fact that the National Youth Survey asked respondents to report on the behavior of their peers, whereas the Add Health asked the peers themselves. Haynie (2001) found a similar reduction in effect magnitude, although in the case of delinquent behavior. As is the case with the family, the relation between peer group affiliation and general risk of victimization deserves further investigation.

The findings also revealed that demographic variables remain important predictors net of the routine activities/lifestyles, family, and peer variables. Lifestyles theory was originally developed to make sense of demographic differences in the risk of victimization; however, measures of lifestyles have had very limited success mediating the effects of demographic variables. Our research shows further that demographic differences appear to come from sources other than family and peer factors as well. Clearly, there is further room for improvement in our attempts to explain gender and racial differences in violent victimization.

The research on family and peer correlates of delinquency is too numerous to easily cite here, yet parallel research linking family and peer characteristics to general violent victimization has been comparatively scant. Our research indicates that these two contexts of adolescent life might be useful for understanding variation in risk of violent victimization. This importance suggests several new areas of inquiry for theory and research. The first new direction follows from the findings about family attachment. If the family determines vulnerability, then could family characteristics socialize children to be either victims or nonvictims in the long term? The recent extension of self-control theory to victimization suggests that this might be the case; however, because Gottfredson and Hirschi (1990) posited that differences in self-control stabilize in late childhood, the Add Health sample is not young enough to address this question about the long-term effects of family characteristics. Thus, the failure of the parental drinking measure to significantly influence risk might not say anything about the empirical status of self-control theory. Nevertheless, the possibility that vulnerability to crime is a time-stable personal trait clearly casts into question policies intended to reduce victimization by educating potential victims about what they can do to protect themselves from crime. This possibility also makes sense of the fact that crime prevention educational programs have been ineffective at making students safer from victimization (Finkelhor, Asdigian, & Dziuba-Leatherman, 1995). The family may therefore have broader relevance than the immediate-situation importance that we suggest here.

The second new direction relates to our results regarding the peer effect finding. Some of the major criminological theories have built themselves around explaining how peer associations cause crime (Akers, 1985). We explicitly adopted a situational interpretation for the peer effect on victimization; however, the extension of criminological theories that stress the socializing role of peer associations might offer new insights about how peer group membership matters when it comes to victimization. Although we do not necessarily endorse this approach, further extensions of criminological theory to victimization can add richness to an area that has long suffered from a lack of theoretical innovation and diversity. The vast literature on peers and delinquency, and the substantive debates about the meaning of the peer effects, may well apply to victimization. Even though our analyses did not exhaust the full range of risk factors for violent victimization, the results suggest avenues for further enriching theories of victimization.

References

Akers, R. L. (1985). Deviant behavior: A social learning approach. Belmont, CA: Wadsworth.

Akers, R. L. (1994). Criminological theories: Introduction and evaluation. Los Angeles: Roxbury.

Baron, S. W., Forde, D. R., & Kennedy, L. W. (2001). Rough justice: Street youth and violence. Journal of Interpersonal Violence, 16, 662–678.

Baron, S. W., Kennedy, L. W., & Forde, D. R. (2001). Male street youths' conflict: The role of background, subcultural, and situational factors. Justice Quarterly, 18, 759–790.

Bureau of Justice Statistics. (2003). Criminal victimization in the United States: 2001 statistical tables. Washington, DC: U.S. Department of Justice, Bureau of Justice Statistics.

Cohen, L. E., & Felson, M. (1979). Social change and crime rate trends: A routine activity approach. American Sociological Review, 52, 170–183.

Cohen, L. E., Kluegel, J. R., & Land, K. (1981). Social inequality and predatory criminal victimization: An exposition and test of a formal theory. American Sociological Review, 46, 505–524.

Coleman, J. S. (1980). Friendship and the peer group in adolescence. In J. Adelson (Ed.), Handbook of adolescent psychology (pp. 408–431). New York: John Wiley.

Corsaro, W., & Eder, D. (1990). Children's peer cultures. Annual Review of Sociology, 16, 197–220.

Decker, S. H., & Van Winkle, B. (1996). Life in the gang. Cambridge, UK: Cambridge University Press.

Erickson, M. L., & Jensen, G. F. (1977). Delinquency is still group behavior!: Toward revitalizing the group premise in the sociology of deviance. Journal of Criminal Law and Criminology, 68, 262–273.

Esbensen, F., Huizinga, D., & Menard, S. (1999). Family context and criminal victimization in adolescence. Youth & Society, 31, 168–198.

Farrington, D. P. (1987). Early precursors of frequent offending. In J. Q. Wilson & G. C. Loury (Eds.), From children to citizens, Volume III: Families, schools, and delinquency prevention (pp. 27–50). New York: Springer-Verlag.

Felson, M. (1986). Linking criminal choices, routine activities, informal control, and criminal outcomes. In D. B. Cornish & R. V. Clarke (Eds.), The reasoning criminal: Rational choice perspectives on offending (pp. 119–129). New York: Springer-Verlag.

Felson, M. (1998). Crime and everyday life. Thousand Oaks, CA: Pine Forge Press.

Finkelhor, D., Asdigian, N., & Dziuba-Leatherman, J. (1995). The effectiveness of victimization prevention instruction: An evaluation of children's responses to actual threats and assaults. Child Abuse and Neglect, 19, 141–153.

Glueck, S., & Glueck, E. (1950). Unraveling juvenile delinquency. New York: Commonwealth Fund.

Gottfredson, M. R., & Hirschi, T. (1990). A general theory of crime. Palo Alto, CA: Stanford University Press.

Haynie, D. L. (2001). Delinquent peers revisited: Does network structure matter? American Journal of Sociology, 106, 1013–1057.

Hindelang, M. J., Gottfredson, M. R., & Garofalo, J. (1978). Victims of personal crime. Cambridge, MA: Ballinger.

Hindelang, M. J., Hirschi, T., & Weis, J. G. (1981). Measuring delinquency. Beverly Hills, CA: Sage.

Hirschi, T. (1969). Cause of delinquency. Berkeley: University of California Press.

Hirschi, T. (1991). Family structure and crime. In B. J. Christensen (Ed.), When families fail . . . the social costs (pp. 43–65). Lanham, MD: University Press of America.

Horney, J., Osgood, D. W., & Marshall, I. H. (1995). Criminal careers in the short-term: Intra-individual variability in crime and its relation to local life circumstances. American Sociological Review, 60, 655–673.

Jensen, G. F., & Brownfield, D. (1986). Gender, lifestyles, and victimization: Beyond routine activity theory. Violence and Victims, 1, 85–99.

Lauritsen, J. L., Laub, J. H., & Sampson, R. J. (1992). Conventional and delinquent activities: Implications for the prevention of victimization among adolescents. Violence and Victims, 7, 91–108.

Lauritsen, J. L., Sampson, R. J., & Laub, J. H. (1991). Addressing the link between offending and victimization among adolescents. Criminology, 29, 265–291.

Loeber, R., & Stouthamer-Loeber, M. (1986). Family factors as correlates and predictors of juvenile conduct problems and delinquency. In M. Tonry & N. Morris (Eds.), Crime and Justice: A Review of Research (Vol. 7, pp. 29–149). Chicago: University of Chicago.

Miethe, T. D., & Meier, R. F. (1994). Crime and social context: Toward an integrated theory of offenders, victims, and situations. Albany: State University of New York Press.

Miethe, T. D., Stafford, M. C., & Long, J. S. (1987). Social differentiation in criminal victimization: A test of routine activities/lifestyles theories. American Sociological Review, 52, 184–194.

Miller, J., & Decker, S. H. (2001). Young women and gang violence: Gender, street offending and violent victimization in gangs. *Justice Quarterly, 18,* 115–104.

Piquero, A. R., & Hickman, M. (2003). Extending Tittle's control balance theory to account for victimization. *Criminal Justice and Behavior, 30,* 282–301.

Rebellon, C. J. (2002). Reconsidering the broken homes/delinquency relationship and exploring its mediating mechanisms. *Criminology, 40,* 103–135.

Rutter, M., & Giller, H. (1983). *Juvenile delinquency: Trends and perspectives.* New York: Guilford.

Sampson, R. J., & Laub, J. H. (1993). *Crime in the making: Pathways and turning points through life.* Cambridge, MA: Harvard University Press.

Sampson, R. J., & Wooldredge, J. (1987). Linking the micro- and macro- dimensions of lifestyle-routine activity and opportunity models of predatory victimization. *Journal of Quantitative Criminology, 3,* 371–393.

Sanders, W. (1994). *Gangbangs and drive-bys: Grounded culture and juvenile gang violence.* New York: Aldine de Gruyter.

Schreck, C. J. (1999). Criminal victimization and low self-control: An extension and test of a general theory of crime. *Justice Quarterly, 16,* 633–654.

Schreck, C. J., Wright, R. A., & Miller, J. M. (2002). A study of individual and situational antecedents of violent victimization. *Justice Quarterly, 19,* 159–180.

Singer, S. I. (1981). Homogeneous victim-offender populations: A review and some research implications. *Journal of Criminal Law and Criminology, 72,* 779–788.

Sparks, R. F. (1982). *Research on victims of crime.* Washington, DC: U.S. Government Printing Office.

Warr, M. (1996). Organization and instigation in delinquent groups. *Criminology, 31,* 17–40.

Warr, M. (2002). Companions in crime: The social aspects of criminal conduct. Cambridge, UK: Cambridge University Press.

Warr, M., & Stafford, M. C. (1991). The influence of delinquent peers: What they think or what they do? *Criminology, 29,* 851–866.

DISCUSSION QUESTIONS

1. What role do you think family and peers will have on victimization of young adults? What about older adults? Are these constructs relevant for understanding the victimization of persons who are not teens?

2. What are the policy implications of the findings that peers and family both matter in the occurrence of victimization?

3. How might the construct of low self-control impact the relationship between family, peers, and victimization? Is it possible that persons with low self-control may seek out delinquent peers? Given what you know about the development of self-control, how may the family be related to both self-control and victimization?

4. What other factors not identified in this article or in this section may be related to victimization? Why do you think the factors you identified would be important in understanding why some people are victimized?

---◈---

Introduction to Reading 3

We often think of gang members as being violent criminals. Terrance Taylor and his coauthors present an alternate view of gang members as being crime victims. They use data on eighth-graders in 11 locations (Kansas City, Missouri; Las Cruces, New Mexico; Milwaukee, Wisconsin; Omaha, Nebraska; Orlando, Florida; Philadelphia, Pennsylvania; Phoenix, Arizona; Pocatello, Idaho; Providence, Rhode Island; Torrance, California; and Will County, Illinois) to examine whether gang members are more likely to be victimized than non-gang members. Half the sample was female, and most lived in two-parent homes. The sample was ethnically and racially diverse (40% White, 27% African American, 19% Hispanic, 6% Asian, 2% Native American, and 6%

other). The authors found that gang members were more likely to be seriously victimized and that they experienced a greater frequency of victimization, even when other risk factors for victimization were taken into account. They also found that gang members have greater levels of risk factors and fewer protective factors, which may account for their increased risk for victimization.

Gang Membership as a Risk Factor for Adolescent Violent Victimization

Terrance J. Taylor, Dana Peterson, Finn-Aage Esbensen, and Adrienne Freng

Youth violent victimization has received substantial empirical inquiry (Acosta et al. 2001) and societal concern as a public health issue (National Institutes of Health [NIH] 2004). This has undoubtedly been sparked in part by the alarm about the "youth violence epidemic" during the late 1980s through the mid-1990s (Blumstein and Rosenfeld 1998, 1999; Cook and Laub 1998, 2002; Lynch 2002; Tolmas 1998), and may in large part explain why more than two-thirds of youth continue to report being concerned about violence as a serious problem facing our nation (Johnston, Bachman, and O'Malley 2003). Prior research has also found that some youth regularly make changes in their lifestyles to "manage everyday threats" associated with perceived risk of violent victimization (Irwin 2004). For example, 5.4 percent of youth included in the 2003 Youth Risk Behavior Surveillance System (YRBSS) did not attend school one or more days during the preceding month because of safety concerns on their way to, from, or at school (Centers for Disease Control and Prevention [CDC] 2004).

Although the youth violence epidemic was perceived as fairly rampant, extant evidence suggests that youth violence (Blumstein and Rosenfeld 1998, 1999; Cook and Laub 2002; Lynch 2002) and the percentage

of youth concerned about violence have all declined during the past decade (Johnston et al. 2003, Table 2.68). It is clear, however, that youth violence remains a salient issue. In fact, some scholars (Marcus 2005) contend that violence has become a part of everyday life for youth. Monitoring the Future 2003 data show that approximately 19 percent of high school seniors reported being threatened without a weapon or injury, 11 percent reported being threatened (but not injured) by a person with a weapon, 10 percent reported being injured by an unarmed person, and 4 percent reported being injured by someone with a weapon during the prior year (Johnston et al. 2003, Table 3.40). YRBSS data from the same period illustrate that approximately 33 percent of students had been in a physical fight, with 4 percent of students nationwide receiving injuries serious enough to require treatment by a medical professional (CDC 2004, Table 8). Additional data from the National Crime Victimization Survey (NCVS) show a nonfatal violent victimization (i.e., robbery, rape, assault) rate for youth aged 12 to 15 and 16 to 19 in 2003 of approximately 52 or 53 per 1,000, higher than any other ages. The highest rates were for simple assault (35 to 36 per 1,000) and *aggravated assault* (45 to 47 per 1,000), with much lower rates for *aggravated*

SOURCE: Taylor et al. (2007). Reprinted with permission of Sage Publications.

assault (9 and 12 per 1,000) and robbery (5 per 1,000; Bureau of Justice Statistics [BJS] 2003, Table 3.6).

One area that also has received considerable attention during the past two decades has been the role of gangs and gang members in youth offending and, more recently, victimization. Although no comprehensive statistics on the growth of youth gangs and gang members exists prior to the 1990s, surveys of law enforcement agencies suggest that the number of gangs and gang members increased during the 1980s into the mid-1990s (Curry, Ball, and Decker 1996; W. B. Miller 2001; Spergel and Curry 1990, 1993). Since 1996, it appears that there has been a steady decline in the prevalence of youth gangs (Egley 2002; Egley, Howell, and Major 2004), consistent with the general decline in youth violent offending and victimization. There is little agreement, however, between the meaning of these co-varying trends, with some (Blumstein 1995; Blumstein and Rosenfeld 1998, 1999) suggesting a causal linkage and others (Howell and Decker 1999; Maxson 1998; Maxson, Curry, and Howell 2002) stating otherwise. More agreement exists concerning the link between gang membership and violent offending: Findings from multiple sources have consistently demonstrated that youth gang members are more likely to offend and to offend at higher rates than their non-gang peers (Battin et al. 1998; Curry, Decker, and Egley 2002; Esbensen and Huizinga 1993; Esbensen and Winfree 1998; Huff 1998; Huizinga 1997; Loeber, Kalb, and Huizinga 2001; Thornberry et al. 2003). Fewer studies have explored the gang membership–victimization link, although a growing body of literature is being developed (J. Miller, 1998, 2001, 2002; J. Miller and Decker 2001; Peterson, Taylor, and Esbensen 2004; Savitz, Rosen, and Lalli 1980). For example, gang members continue to be at increased risk relative to non-gang members for homicide victimization (Howell 1999; Maxson 1999; Maxson et al. 2002).

The current study seeks to expand the extant literature on gang membership and violent victimization. Data were collected from youth attending public schools in 11 U.S. cities in 1995. Three specific questions are examined: (1) How do gang members and non-gang members differ in terms of violent victimization? (2) Assuming differences exist, what factors account for differences in the extent of violent victimization between gang members and non-gang members? and (3) Does gang membership remain a salient correlate of violent victimization, once other relevant factors are controlled? To examine these questions, we begin by describing the extant literature on risk and protective factors related to violent victimization with a particular emphasis on gang membership. We then discuss the main findings related to our research questions. We conclude with a discussion of the implications of the current study.

Risk and Protective Factors for Youth Violent Victimization

Although most commonly used in the public health arena (see Mercy and O'Carroll 1988), the risk and protective factor approach also has been highlighted as an important part of violence research within the field of criminology/criminal justice (see Moore et al. 1994). In light of this, several studies have identified a number of important risk and protective factors related to youth violent victimization, and different domains have been identified. Sampson and Lauritsen (1994), for example, suggest that risk and protective factors can be divided into three categories: individual, situational, and community. Others (Esbensen 2000; Hill et al. 1999; Howell 2003; Thornberry 1998) identify five categories: individual, family, peer, school, and community. Although no consensus exists, we focus here on five domains that we are able to examine with our data: individual, family, peer, school, and situational.

Individual risk and protective factors found to be associated with violent victimization include demographic characteristics (Hindelang, Gottfredson, and Garofalo, 1978), self-control (Piquero et al. 2005; Schreck, 1999; Schreck, Wright, and Miller 2002), and guilt and neutralizations (Agnew 1985; Ferraro 1983). In the family domain, parental monitoring of youth and youth's emotional attachment to parents protect against victimization (Esbensen and Huizinga 1991; Esbensen,

Huizinga, and Menard 1999; Schreck and Fisher 2004). Peer factors include concepts such as involvement with and commitment to prosocial and delinquent peers (Schreck and Fisher 2004; Schreck, Fisher, and Miller 2004). School risk factors include concepts such as school climate and youth's commitment to education (Gottfredson 2001; Gottfredson and Gottfredson 1985; Gottfredson et al. 2005; Welsh 2001; Welsh, Greene, and Jenkins 1999). Finally, the situational domain includes factors such as lifestyles and routine activities, including time spent away from the home, unstructured leisure time, and time spent with drugs and/or alcohol available (Cohen and Felson 1979; Cohen, Kluegel, and Land 1981; Hindelang et al. 1978; Osgood et al. 1996). Involvement in delinquent lifestyles has also been found to be a particularly salient situational risk factor for victimization (Esbensen and Huizinga 1991; Lauritsen, Sampson, and Laub 1991; Loeber et al. 2001; Rapp-Paglicci and Wodarski 2000; Shaffer and Ruback 2002; Zhang, Welte, and Wieczorek 2001). Extant research has generally found that, although conceptually distinct, risk factors for gang membership, offending, and victimization overlap (Esbensen 2000; Howell 2003; Huizinga and Jakob-Chien 1998; Loeber et al. 2001; Sampson and Lauritsen 1994). Given the growing body of work in the field, gang membership may be an additional risk factor for victimization deserving closer scrutiny.

Gang Membership and Violent Victimization

Gang members' increased involvement in activities representing delinquent lifestyles, relative to their non-gang peers, would suggest that their risk of violent victimization is also enhanced. Comparisons of gang and non-gang youth consistently and historically have produced significant differences in the prevalence and frequency of offending between these two groups. According to self-report surveys, gang youth account for approximately 70 percent of all self-reported violent offending in adolescent samples (Huizinga et al. 2003; Thornberry et al. 2003). In one study involving 15-year-old youth (Battin et al. 1998), gang members reported committing seven times as many violent acts during the past year as did youth without delinquent friends, and twice as many friends. Other studies (Curry et al. 2002) have found that simply being associated with a gang increases youth's involvement in delinquent behavior. Given the strong overlap between gang membership and offending and between offending and victimization, one may reasonably expect a connection between gang membership and victimization.

Gang membership itself may increase risk of individual violent victimization in a number of ways. Gang members may become the victim of predatory offending by others. Previous research has found, for example, that gang members are more likely than nonmembers to be involved in such delinquent acts as drug sales (Esbensen and Winfree 1998; Howell and Decker 1999; Howell and Gleason 1999; Huff 1998; Maxson 1995), which has been linked to increased likelihood of victimization (see Jacobs 2000). Additionally, gang members may be targets of retaliation from rival gangs. Sanders (1994) suggests that targeting members of rival gangs for shootings is an acceptable practice, although deliberately shooting at "innocents" is generally prohibited. These processes indicate that violence is a routine part of gang life (Decker and Van Winkle 1996:117). Stark evidence of this may be found in Decker and Van Winkle's St. Louis study. Of the 99 gang members interviewed in the early 1990s, 28 had died a violent death by mid-year 2003 (Decker, pers. comm., June 2, 2003).

Gang members may be at greatest risk of violence, however, from members of their own gang. Members may be required to participate in violent initiation rituals when entering or exiting a gang. Approximately two-thirds of the gang members in Decker and Van Winkle's (1996:69) ethnographic study of St. Louis gang members, for example, reported being "beat in" as part of their initiation process. Although the process varied from gang to gang, interviewees often recounted a process whereby prospective members were expected to fight against several current gang members who were arranged in a line or a circle. Gang members may also be subjected to harsh discipline from members of their own gangs for violating gang rules. Padilla's (1995:57) research on one gang organized around drug sales uncovered the use of violence by the gang to sanction members for violations of collective rules, referred to as "Vs." The process Padilla outlined is similar to one Decker and Van Winkle describe, under which violators are expected to walk through a line of other gang members who take turns beating on the transgressor.

Although an extensive body of literature on the topic has yet to be developed, published studies have begun quantitative examinations of the link between gang membership and victimization risk. An early study by Savitz et al. (1980) examined a sample of approximately 1,000 African American and White boys residing in Philadelphia. Their findings, contrary to the ethnographic studies and those examining homicide rates, showed no statistically significant differences between gang and non-gang youth in terms of fear of victimization or actual victimization experiences. These findings, however, may be tied to period or cohort effects, because the study was conducted long before the late 1980s' and early 1990s' elevations in the rates of youth violence and gang involvement. Consistent with ethnographic research, a recent study of adolescents in several cities conducted by Peterson and colleagues (2004) found that gang members experienced a greater likelihood of being victimized and more frequent victimization experiences than their non-gang peers. Additionally, victimization was found to peak in the year in which youth reported joining a gang, and youth who reported joining a gang for protection still experienced significantly more violent victimization than did non-gang members.

To this point, published studies examining the gang membership–violent victimization link have been descriptive or bivariate in nature. Rigorous quantitative examinations of these relationships, controlling for other known risk factors, is notably absent. Thus, we expand this line of inquiry by controlling for other known factors that increase risk of victimization. Importantly, we include involvement in delinquent activity to determine whether gang membership itself is related to victimization, or whether the link is contingent upon gang members' greater involvement in delinquency relative to other youth. Multiple regression analyses are also utilized to examine the unique impact of gang membership on youth violent victimization, not other factors.

Current Study

The current study uses information collected from a multi-site study of eighth-grade youth attending public schools (Esbensen 2002). During the spring of 1995, students in 11 locations (Kansas City, Missouri; Las Cruces, New Mexico; Milwaukee, Wisconsin; Omaha, Nebraska; Orlando, Florida; Philadelphia, Pennsylvania; Phoenix, Arizona; Pocatello, Idaho; Providence, Rhode Island; Torrance, California; and Will County, Illinois) completed group self-administered questionnaires requiring approximately 45 minutes to finish. All eighth-grade students in attendance on the day of survey administration and for whom parental consent was provided completed surveys. The final sample consisted of 5,935 eighth-grade public-school students, representing 42 schools and 315 classrooms.

Table 1 presents information about the study sample. Females (52 percent) comprise approximately half of the respondents, and most of the respondents live in two-parent homes (62 percent), that is, they indicated that a mother and father (including stepparents) were present in the home. The sample is ethnically diverse, with whites accounting for 40 percent of respondents; African Americans, 27 percent; Hispanics, 19 percent; Asians, 6 percent; Native Americans, 2 percent; and youth of Other racial/ethnic backgrounds, 6 percent. As would be expected with an eighth-grade sample, most are 14 years old (mean age = 13.82 years).

Measures

Self-Reported Victimization

Our dependent variable of interest is violent victimization. To measure violent victimization, respondents were asked to indicate whether any of the specified things had happened to them and, if so, how many times in the past 12 months. Three items were included to measure violent victimization: (1) been hit by someone trying to hurt you (i.e., simple assault), (2) been attacked by someone with a weapon or by someone trying to seriously hurt or kill you (i.e., aggravated assault), and (3) had someone use a weapon or force to get money or things from you (i.e., robbery).

Table 1 Sample Characteristics

Variable	Range	Percentage (n)	Mn (SD)
Gender			
Male		48 (2,830)	
Female		52 (3,054)	
Race			
White		40 (2,355)	
Black		27 (1,544)	
Hispanic		19 (1,098)	
Other		14 (835)	
Living arrangement			
Single parent		31 (1,833)	
Two parents		62 (3,628)	
Other		7 (417)	
Age (years)	13 to 15		13.82 (0.72)
Gang member		9 (522)	
Individual factors			
Impulsivity	1 to 5		2.85 (0.74)
Risk-seeking	1 to 5		3.06 (0.94)
Perceptions of guilt	1 to 3		2.31 (0.56)
Neutralization	1 to 5		3.11 (0.81)
Family factors			
Parental monitoring	1 to 5		3.72 (0.81)
Parental attachment	1 to 7		4.66 (1.26)
Peer factors			
Negative peer commitment	1 to 5		2.40 (.14)
Positive peer commitment	1 to 5		3.80 (1.12)
Prosocial peers	1 to 5		2.97 (0.80)
Delinquent peers	1 to 5		2.00 (0.86)
School factors			
School commitment	1 to 5		3.57 (0.77)
Situational factors			
Spent time with drugs/alcohol		31 (1,775)	
Spent time without adults		76 (4,398)	
Self-reported delinquency	0 to 204		15.53 (25.70)

Annual prevalence and Individual Victimization Rates (IVRs) were calculated from respondents' answers for each individual item. Annual prevalence is a dichotomous measure of whether a respondent reported being violently victimized one or more times during the prior year. The IVR is a continuous measure of the number of violent victimizations experienced by victims during the prior year. Individual items were truncated at 12 to compensate for outliers. In addition to the individual items, two composite measures—general and serious violent victimization—were also created. General violent victimization utilized all three individual items, whereas serious violent victimization excluded the hitting item.

Risk and Protective Factors

Nineteen independent variables representing gang membership, demographic characteristics, and key risk factors identified as correlates of victimization in prior research are included in the analyses. These independent variables include a number of measures of individual, familial, peer, school, and situational characteristics. A brief description of the variables is included below.

Gang membership. Gang membership is our primary variable of interest, and it consists of a single item measure of current gang membership. Youth who answered affirmatively to the question, "Are you now in a gang?" are classified as gang members, whereas youth who responded negatively are classified as non-gang members. We expect that gang membership is a key risk factor for violent victimization.

Individual Factors

Demographic characteristics. Youth were asked to report information about their demographic characteristics, including gender, race/ethnicity, age, and current living arrangement (whether they lived with their mother only, father only, mother and father, mother and another adult, father and another adult, other relatives, or some other living arrangement).

Self-control. Two subscales of the attitudinal self-control scale created by Arneklev et al. (1993) were incorporated into the survey. Impulsivity is a four-item scale measuring impulsive behavior, such as, "I often act on the spur of the moment without stopping to think" (alpha = .65). Risk-seeking is a four-item scale representing youth's willingness to engage risk-taking behavior, such as, "Sometimes I will take a risk just for the fun of it" (alpha = .82). Higher scores indicate greater levels of impulsivity and risk-seeking (i.e., lower levels of self-control), which we expect to be key risk factors.

Guilt and neutralizations. Guilt is measured using a 16-item scale (alpha = .94), with higher scores indicating higher levels of guilt for engaging in status offenses, minor offenses, serious offenses, and drug offenses. Youth were asked to report, for example, how guilty they would feel if they hit someone with the idea of hurting them. The neutralizations scale (alpha = .86) consists of 10 items tapping the respondent's belief that it is okay to engage in some deviant behaviors if extenuating factors are present, for instance, "It's okay to tell a small lie if it doesn't hurt anyone." Higher scores indicate a greater use of neutralizations for engaging in lying, stealing, and fighting behaviors. We expect that guilt is a key protective factor, whereas neutralizations are risk factors.

Family Factors

Parental monitoring. Parental monitoring is a scale consisting of four items measuring communication with parents (alpha = .73), including questions such as, "My parents know who I am with if I am not at home." Higher scores indicate greater levels of parental monitoring, a key protective factor.

Parental attachment. Attachment to parents is measured through two separate six-item semantic differential scales, one for mothers/mother figures (alpha = .84) and one for fathers/father figures (alpha = .88), including items tapping emotional attachment to parents. Higher scores indicate greater levels of attachment to parents, an important protective factor.

Peer Factors

Negative peer commitment. Commitment to negative peers is measured using a three-item scale (alpha = .84) including questions such as, "If your group of friends were getting you into trouble at home, how likely is it that you would still hang out with them?" Higher scores indicate greater levels of commitment to negative peers, a key risk factor.

Positive peer commitment. Commitment to positive peers is measured using a two-item scale (alpha = .77) including questions such as, "If your friends told you not to do something because it was against the law, how likely is it that you would listen to them?" Higher scores indicate greater commitment to positive peers, an important protective factor.

Prosocial peer involvement. Prosocial peer involvement is measured using an eight-item scale (alpha = .84), with higher scores indicating greater levels of involvement with peers who engage in prosocial behaviors such as religious activities. We expect that these are important protective factors.

Delinquent peer involvement. Delinquent peer involvement is measured using a 16-item scale (alpha = .94) tapping the proportion of the respondents' friends who engage in a variety of illegal activities. Higher scores indicate greater levels of involvement with peers who engage in delinquent activities, key risk factors.

School Factors

School commitment. Commitment to school was measured using a seven-item scale tapping the youth's orientation toward academic endeavors (alpha = .81), including questions such as, "I try hard in school." Higher scores indicate greater levels of commitment to school, an important protective factor.

Situational Risk Factors

Unsupervised leisure time. Respondents were asked to indicate whether they ever spent time with their current friends, not doing anything in particular, where no adults were present. Respondents who answered no received a score of 0, whereas those who answered yes received a score of 1.

Availability of alcohol and/or drugs. Respondents were asked to indicate whether they ever spent time with their current group of friends where drugs and alcohol were available. Youth who answered affirmatively to the question were scored 1, whereas youth who said no were scored 0.

Self-reported delinquency. Self-reported delinquency is measured using a 17-item summary index consisting of youth's frequency of self-reported involvement in various status, minor, property, and person offenses.

Analyses and Results

The Scope of Violent Victimization in the Sample

Table 2 presents the victimization prevalence and individual victimization rates for the total sample of youth and separately by gang status. Nearly half (48 percent)

Table 2 Prevalence and Individual Victimization Rates (IVR) by Gang Membership

	Total Sample (*n* = 5,935)		Non-Gang (*n* = 5,226)		Gang (*n* = 522)	
	Annual Prevalence (%)	Annual Individual Victimization Rates ($\bar{x}$)	Annual Prevalence (%)	Annual Individual Victimization Rates ($\bar{x}$)	Annual Prevalence (%)	Annual Individual Victimization Rates ($\bar{x}$)
Times been hit[a,b]	44	3.6	43	3.4	60	4.9
Times been robbed[a,b]	8	2.8	7	2.4	21	4.1
Times been attacked[a,b]	10	2.7	8	2.3	38	3.8
General violent victimization[a,b]	48	4.4	46	4.0	70	7.6
Serious violent victimization[a,b]	15	3.4	12	2.8	44	5.2

NOTE: *t*-test for independent samples comparisons of proportions or means

a. *p* < .05 for annual prevalence comparisons, non-gang and gang.

b. *p* < .05 for annual IVR comparisons, non-gang and gang.

of these youth had experienced one or more general violent victimizations during the past year, and approximately one-sixth of the youth (15 percent) had been the victim of serious violence. A closer examination indicated that simple assault (44 percent) was the most common form of violent victimization. Robbery and aggravated assault victimizations were less common. Approximately 1 in 10 youth (10 percent) had experienced one or more aggravated assault victimizations, and approximately 1 in 12 youth (8 percent) had been robbed during the past year. Victimized youth reported an average of 4.4 general violent victimization experiences and 3.4 serious violent victimization experiences during the past year. Simple assault victimizations were the most common (approximately 3.6 experiences during the past year), with fewer robberies and aggravated assaults (approximately 3 experiences during the past year).

We turn now to our first research question: How do gang and non-gang youth differ in terms of violent victimization? Table 2 presents information comparing gang members and their non-gang peers for each type of violent victimization. Results show that gang members were significantly more likely to be violently victimized during the past year, as well as to experience a significantly greater number of victimizations, than non-gang youth for each type of violent victimization. Substantively, the differences were quite large. For example, 70 percent of gang youth reported being the victim of general violence, compared with 46 percent of non-gang youth. Although most of the general violent victimization again consisted of assaults (60 percent of gang members, 43 percent of non-gang youth), differences in serious violent victimization were even more pronounced. Forty-four percent of gang youths reported being a victim of serious violence during the past year, with 38 percent of gang members reporting one or more aggravated assaults and 21 percent reporting one or more robberies during this time. Corresponding figures for non-gang members were 112 percent for any type of serious violent victimization, 8 percent for aggravated assaults, and 7 percent for robbery. Gang victims also experienced significantly more violent victimization incidents than did non-gang victims

during the prior year, as illustrated by the Individual Victim Rates. Thus, it appears that gang members were more likely than non-gang members to be violently victimized and, of those victimized, gang members experienced more annual violent victimizations than did their non-gang peers.

Gang Membership, Victimization, and Risk/Protection

We now turn to an examination of differences in levels of risk and protective factors between gang and non-gang youth. How, if at all, do these differences in victimization experiences represent a difference in exposure to the risk and protective factors of interest? Our comparisons include four groups: (1) non-gang non-victims, (2) non-gang victims, (3) gang non-victims, and (4) gang victims. Table 3 presents information on general violent victimization, whereas information about serious violent victimization is presented in Table 4.

Looking at the information presented in Tables 3 and 4, we offer the following general findings: (1) victims (regardless of gang involvement) report greater exposure to risk factors and lesser exposure to protective factors than do non-victims; (2) gang members (regardless of victimization experience) report greater exposure to risk factors and lesser exposure to protective factors than do non-gang members; and (3) gang members who are victims generally report the greatest exposure to risk factors and least exposure to protective factors, whereas youth who are neither gang members nor victims of violence generally report the least exposure to risk factors and greatest exposure to protective factors. We now examine these findings in more detail.

Two-way ANOVA analyses are used to examine the risk and protective factors, while taking into account gang membership and victimization status. The results of these analyses present a consistent picture. Controlling for victimization status, gang members exhibit significantly greater levels of each risk factor and lower levels of each protective factor than do non-gang members for general and serious violence.

Table 3   Mean Risk Factor Scores for General Violent Victimization Annual Prevalence—Controlling for Gang Membership

	Non-Gang (*n* = 5,226)			Gang (*n* = 522)		
	All Non-Gang (*n* = 5,226)	Non-Victims (*n* = 2,667)	Victims (*n* = 2,280)	All Gang (*n* = 522)	Non-Victims (*n* = 142)	Victims (*n* = 335)
Individual						
Impulsivity[a,b]	2.81	2.74	2.87	3.25	3.14	3.32
Risk-seeking[a,b]	2.99	2.81	3.16	3.70	3.53	3.80
Perceptions of guilt[a,b]	2.38	2.47	2.30	1.66	1.75	1.62
Neutralization[a,b]	3.02	2.88	3.16	3.90	3.72	4.02
Family						
Parental monitoring[a,b]	3.79	3.88	3.70	3.16	3.28	3.10
Parental attachment[a,b]	4.71	4.80	4.55	4.10	4.13	3.99
School						
School commitment[a,b]	3.64	3.74	3.55	2.95	3.07	2.87
Peer						
Positive peer commitment[a,b]	3.89	3.98	3.80	3.04	3.30	2.96
Negative peer commitment[a,b]	2.28	2.16	2.38	3.50	3.36	3.55
Pro-social peers[a,b]	3.04	3.12	2.96	2.36	2.47	2.32
Delinquent peers[a,b]	1.87	1.72	2.02	3.10	2.92	3.17
Situational						
Spend time without adults (%)[a,b]	75	70	79	88	86	90
Spend time with drugs/alcohol (%)[a,b]	26	19	33	76	67	81
Self-reported delinquency[a,b,c]	12.13	7.12	18.17	63.45	47.25	71.70

NOTE: Two-way ANOVA comparison of proportions or means

a. $p < .05$ main effect: gang membership

b. $p < .05$ main effect: victimization status

c. $p < .05$ interaction effect: gang membership by victimization status

Controlling for gang membership, victims exhibit significantly greater levels of each risk factor and lower levels of each protective factor than do non-victims for general and serious violence. Only in the case of self-reported delinquency are interactions between gang membership and victimization significant—for general and serious violent victimization, gang victims report the most extensive involvement in delinquency, followed by gang non-victims, non-gang victims, and non-gang non-victims. Substantively, these differences are quite large. On average, gang victims of general violence reported committing an average of nearly 72 offenses

Table 4 Mean Risk Factor Scores for Serious Violent Victimization Annual Prevalence—Controlling for Gang Membership

	Non-Gang (n = 5,226)		Gang (n = 522)	
	Non-Victims (n = 4,564)	Victims (n = 608)	Non-Victims (n = 274)	Victims (n = 219)
Individual				
Impulsivity[a,b]	2.77	3.04	3.17	3.37
Risk-seeking[a,b]	2.94	3.31	3.58	3.89
Perceptions of guilt[a,b]	2.42	2.17	1.76	1.56
Neutralization[a,b]	2.97	3.32	3.74	4.11
Family				
Parental monitoring[a,b]	3.83	3.50	3.33	2.99
Parental attachment[a,b]	4.71	4.38	4.17	3.93
School				
School commitment[a,b]	3.67	3.46	3.11	2.78
Peer				
Positive peer commitment[a,b]	3.93	3.62	3.29	2.77
Negative peer commitment[a,b]	2.23	2.54	3.35	3.66
Prosocial peers[a,b]	3.07	2.82	2.44	2.29
Delinquent peers[a,b]	1.82	2.25	2.86	3.37
Situational				
Spend time without adults (%)[a,b]	74	82	86	92
Spend time with drugs/alcohol (%)[a,b]	23	43	67	86
Self-reported delinquency[a,b,c]	10.14	27.98	46.23	86.79

NOTE: Two-way ANOVA comparison of proportions or means

a. $p < .05$ main effect: gang membership

b. $p < .05$ main effect: victimization status

c. $p < .05$ interaction effect: gang membership by victimization status

during the past year, compared with 47 offenses committed by gang non-victims, 18 offenses committed by non-gang victims, and 7 offenses committed by non-gang non-victims; corresponding offending rates for victims of serious violence were 87 offenses for gang victims, 46 offenses for gang non-victims, 28 offenses for non-gang victims, and 10 offenses for non-gang non-victims.

The Independent Relationship Between Gang Membership and Victimization Risk

The results thus far are consistent with what we would expect given extant literature, illustrating the increased violent victimization experienced by gang members relative to their non-gang peers as well as their greater levels of risk factors and lower levels of protective factors. They do not allow, however, a simultaneous examination of the effects of gang membership and other risk and protective factors on violent victimization. To address this issue, a series of regression analyses are conducted to examine the independent effect of gang membership on violent victimization prevalence and individual victimization rates, net other relevant factors. Separate logistic regression analyses are used to examine annual prevalence for general violent victimization and serious

violent victimization. Ordinary–least squares regression analyses are used to examine individual victimization rates of youth who had experienced one or more violent victimizations during the prior year.

The results of these analyses are somewhat surprising. Contrary to our expectations, gang membership is associated with a 26 percent decrease in the odds of having experienced one or more general violent victimizations during the past year, net other factors consistent with our expectations; however, gang membership is associated with a 50 percent increase in the odds of having experienced one or more serious victimizations during the prior year, net other factors. As clearly illustrated in the graphical displays of predicted probabilities of violent victimization by gang status net other factors (presented in Figure 1), the likelihood that youth experienced one or more general violent

Figure 1   Predicted Probabilities of Victimization Prevalence by Gang Membership, Net Other Factors

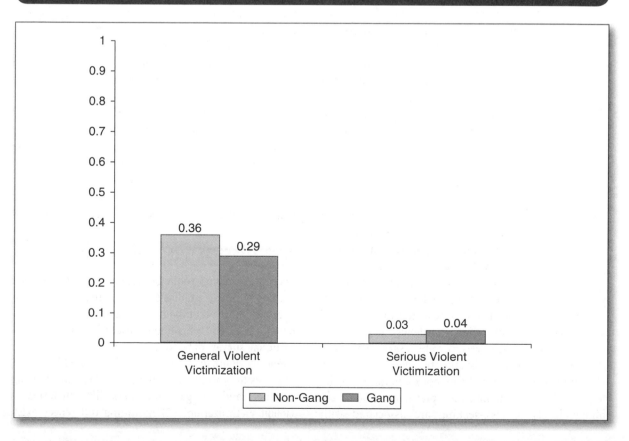

victimizations is greater than the likelihood that youth experienced one or more serious violent victimizations, regardless of whether youth belong to a gang. It should be noted, however, that the likelihood that youth (gang or non-gang) suffered one or more serious violent victimization experiences was small, and the magnitude of the differences between gang and non-gang youth for both types of victimization are modest.

A number of key demographic characteristics are also found to be significantly associated with the likelihood of having experienced one or more violent victimizations during the prior year, net other factors. Males are nearly twice as likely as females to have been violently victimized in terms of general violent victimization and serious violent victimization. Being Hispanic (relative to White) is associated with an 18 percent decrease in the odds of having experienced one or more general violent victimizations in the past year, but no significant race differences emerge for serious violent victimization. Compared with youth living with two parents, living in single-parent households is associated with a 25 percent increase in the odds of having been the victim of serious violence (although the groups are comparable in terms of general violent victimization), whereas living in "other" household configurations is associated with a 41 percent increase in the odds of having been the victim of general violent victimization and a 63 percent increase in the odds of having experienced one or more serious violent victimizations. These demographic differences also have relatively modest effects on violent victimization.

Several additional risk and protective factors are also found to be related to violent victimization, net other factors. The situational factors are the most consistent, with increased odds of violent victimization for youth who spent time without adults present, youth who spent time with drugs and/or alcohol present, and as youth reported greater involvement in delinquent activities. Guilt is also positively associated with general and serious violent victimization prevalence, net other factors. Other variables, however, exhibit less consistency in their relationships across the two violence indices. Higher levels of impulsivity are not significantly related to general violent victimization prevalence. Conversely, higher levels of risk-seeking are associated with

increased odds of general violent victimization prevalence but not serious violent victimization. Additionally, higher levels of parental attachment are associated with reduced odds of general violent victimization prevalence but not serious violent victimization, whereas the situation was reversed for levels of parental monitoring. Interestingly, once the effects of other relevant factors are controlled, only one of the peer variables is significant, and it is in an unanticipated direction: Greater commitment to delinquent peers is associated with reduced likelihood of general violent victimization.

An examination of individual victimization rates for victimized youth reveals slightly different patterns. Gang membership is found to have no statistically significant relationship with the average number of general or serious victimization incidents experienced by victims during the prior year, net other factors. Rates of general violent victimization are lower for African American and Hispanic youth (as compared to White youth) and higher for youth residing in "other" living arrangements (compared with youth living with two parents). No significant differences exist, however, for serious violent victimization. Increased levels of guilt and involvement in delinquency are also related to increased rates of general and serious violent victimization. Increased levels of parental attachment are associated with lower rates of general violent victimization, as is, interestingly, spending time where drugs and/or alcohol are present. It is important to note that few of the risk/protective factors are significantly associated with serious violent victimization rates. Rates of serious violent victimization are positively associated with levels of guilt, delinquent peers, and self-reported delinquency.

Discussion and Conclusions

Recent research on youth gang members has expanded the focus from an emphasis on their offending behavior to an examination of their victimization risks. The current study extends this line of research by using cross-sectional survey data from a sample of 5,935 youth attending eighth grade in public schools in 11 U.S. communities to examine three questions: (1) How do gang members and non-gang members differ in terms of violent victimization? (2) Assuming differences exist, what factors are related to differences in the extent of violent victimization between gang members and non-gang members? and (3) Does gang membership remain a salient correlate of violent victimization once other relevant factors, including delinquency involvement, are controlled? Gang members were found to be slightly more likely than non-gang members to experience one or more violent victimizations, especially serious violent victimizations, during the past year, and for those who were victimized, to experience more violent victimization episodes. Multivariate results illustrate that the likelihood that any youth (gang or non-gang) suffered one or more serious violent victimization experiences was small, however, and the magnitude of the differences between gang and non-gang youth were modest once the effects of other key risk factors were controlled.

The patterns regarding risk and protective factors for violent victimization among gang and non-gang youth were readily apparent. Gang youth exhibited significantly lower levels of protective factors and greater levels of risk factors than non-gang youth. This pattern was consistent for each of the risk and protective factors examined for general and serious violent victimization among the youth in our sample. This pattern was robust and differences were salient, holding even when victimization status was taken into account. Additionally, an important interaction between gang membership, involvement in self-reported delinquency, and victimization status was identified: Gang victims reported the most extensive involvement in delinquency, followed by gang non-victims, non-gang victims, and non-gang non-victims. Although perhaps not surprising given the link between offending and victimization identified in prior works, the large substantive differences between these groups suggest that delinquent offending is a particularly salient factor for violent victimization of youth.

Although bivariate results showed that gang members experienced more violent victimization (in terms of likelihoods and rates), gang membership itself was found to have discrepant effects on adolescent violent victimization, depending on which form of the dependent variable was examined, when other risk factors were controlled. Specifically, gang membership appears to suppress or protect against general violent victimization, increase risk of serious violent victimization, and exert no independent

effect on rates of victimization among victims of either index of violence, when important risk and protective factors are taken into account. The link between gang membership and violent victimization, then, appears to be quite complex, and many questions remain unanswered. Given the findings reported here, as well as a growing body of literature that attempts to disentangle offending and victimization, future research attempting to sort out these relationships is greatly needed.

Our findings suggest that gang membership is tied to other risk factors, particularly involvement in delinquent lifestyles, thereby accounting for gang members' increased levels of victimization relative to their non-gang peers. Despite that youth are hanging out with peers, unsupervised by adults and involved in delinquency, gang membership reduces the odds of general violent victimization; in the presence of these same lifestyle factors, however, gang membership increases odds of serious violent victimization. These findings, although seemingly incongruent, are consistent with other research indication, for example, that whereas gang members may experience enhanced levels of certain types of violent victimization (e.g., gang-related violent victimization), gang involvement may insulate members from other types of violent victimization (e.g., non-gang-related violent victimization; J. Miller 1998, 2001). One girl in Miller's (2001) study described: "'Sometimes you might be walkin' down the street with your flag out of your pocket . . . and . . . you might get shot'" (p. 161). But gang members in Miller's study also felt a level of protection from the gang: "In a group context, she felt more secure. 'When they sitting in park . . . it's people all around that just got people's backs, they just look out for each other'" (p. 159). Decker's (1996) threat hypothesis may be brought to bear as well, in that the threat of violence from rivals may quell lesser forms of violence and victimization among gang members; but when violence does erupt, it is likely to be more serious in nature.

Our results are consistent with notions that delinquent lifestyle factors mediate the direct relationship of gang membership on rate of violent victimization. To the extent that gang members, as our results indicate, are more likely than other youth to engage in unsupervised hanging out with peers, to hang out where drugs and/or alcohol are available, and to engage in a substantially greater amount of delinquent behavior, they experience a greater number of violent victimizations. Thus, the gang is a facilitating context, exposing youth to risky situations and increasing delinquency. Additional insight is gained from Irwin's (2004) findings that youth at moderate to high risk of victimization often turned to friends to manage everyday violence, whereas low-risk youth primarily utilized the approach of avoiding violent persons. These findings suggest that although gang membership itself may not directly impact the risk of youth violent victimization, gang members may turn to their peers—inside and outside of the gang—to handle disputes "informally." If such informal actions involve the use of violence, as several prior studies have shown, gang membership may exert an indirect effect on youth violent victimization.

Future research should continue to examine the context of gang members' violent victimization experiences within and outside of the gang setting. Furthermore, attempts to disentangle the relationships between gang involvement, involvement in delinquent lifestyles, and victimization should be undertaken. A return to group process research, as several (Klein 1995; Short and Hughes 2006) have argued, would provide insight into the mechanisms by which gang membership and situational factors result in violent victimization. We hope that qualitative inquiries on this topic continue while other studies begin to employ more rigorous quantitative analyses of risk factors related to gang membership and violent victimization.

References

Acosta, O. M., K. E. Albus, M. W. Reynolds, D. Spriggs, and M. D. Weist. 2001. "Assessing the Status of Research on Violence-Related Problems among Youth." *Journal of Clinical Child Psychology* 30:152–60.

Agnew, R. S. 1985. "Neutralizing the Impact of Crime." *Criminal Justice and Behavior* 12:221–39.

Battin, S. R., K. G. Hill, R. D. Abbott, R. F. Catalano, and J. D. Hawkins. 1998. The Contribution of Gang Membership to Delinquency beyond Delinquent Friends. *Criminology* 36:93–115.

Blumstein, A. 1995. "Youth Violence, Guns, and the Illicit-Drug Industry." *Journal of Criminal Law and Criminology* 86:10–36.

Blumstein, A., and R. Rosenfeld. 1998. "Explaining Recent Trends in U.S. Homicide Rates." *Journal of Criminal Law and Criminology* 88:1175–216.

———. 1999. "Trends in Rates of Violence in the U.S.A." *Studies in Crime and Crime Prevention* 8:139–68.

Bureau of Justice Statistics. 2003. "Criminal Victimization in the United States." In *The Sourcebook of Criminal Justice Statistics 2003,* edited by K. Maguire and A. L. Pastore. Washington, DC: Author. Retrieved May 17, 2006, from http//www.albany.edu/sourcebook/

Centers for Disease Control and Prevention. 2004. "Youth Risk Behavior Surveillance—United States, 2003." Surveillance Summaries, May 21, 2004. *Morbidity and Weekly Report,* 53 (No. SS-2).

Cohen, L. E., and M. Felson. 1979. "Social Change and Crime Rate Trends: A Routine Activity Approach." *American Sociological Review* 44:588–608.

Cohen, L. E., J. R. Kluegel, and K. C. Land. 1981. "Social Inequality and Predatory Criminal Victimization: An Exposition and Test of a Formal Theory." *American Sociological Review* 46:505–24.

Cook, P., and J. Laub. 1998. "The Unprecedented Epidemic in Youth Violence." pp. 27–64 in *Youth Violence, Crime and Justice: A Review of Research,* vol. 24, edited by M. Tonry and M. Moore. Chicago: University of Chicago Press.

———. 2002. "After the Epidemic: Recent Trends in Youth Violence in the United States." pp. 1–37 in *Crime and Justice: A Review of Research,* vol. 29, edited by M. Tonry and M. Moore. Chicago: University of Chicago Press.

Curry, G. D., R. A. Ball, and S. H. Decker. 1996. *Estimating the National Scope of Gang Crime from Law Enforcement Data.* Washington, DC: Office of Justice Programs, National Institute of Justice, Research in Brief (August).

Curry, G. D., S. H. Decker, and A. Egley, Jr. 2002. "Gang Involvement and Delinquency in a Middle School Population." *Justice Quarterly* 19:275–92.

Decker, S. H. 1996. "Collective and Normative Features of Gang Violence." *Justice Quarterly* 13:243–64

Decker, S. H., and B. Van Winkle. 1996. *Life in the Gang: Family, Friends, and Violence.* Cambridge, UK: Cambridge University Press.

Egley, A., Jr. 2002. "National Youth Gang Survey Trends from 1996–2000." *OJJDP Fact Sheet.* Washington, DC: U.S. Department of Justice, Office of Justice Programs, Office of Juvenile Justice and Delinquency Prevention.

Egley, A., J. C. Howell, and A. K. Major. 2004. "Recent Patterns of Gang Problems in the United States: Results from the 1996–2002 National Youth Gang Survey." pp. 90–108 in *American Youth Gangs at the Millennium,* edited by F.-A. Esbensen, S. G. Tibbetts, and L. Gaines. Long Grove, IL: Waveland Press.

Esbensen, F.-A. 2000. "Preventing Adolescent Gang Membership." *Juvenile Justice Bulletin.* Washington, DC: U.S. Department of Justice, Office of Justice Programs, Office of Juvenile Justice and Delinquency Prevention.

Esbensen, F.-A., and D. Huizinga. 1991. "Juvenile Victimization and Delinquency." *Youth and Society* 23:202–28.

———. 1993. "Gangs, Drugs, and Delinquency in a Survey of Urban Youth." *Criminology* 31:565–89.

Esbensen, F.-A., D. Huizinga, and S. Menard. 1999. "Family Context and Victimization." *Youth and Society* 31:168–98.

Esbensen, F.-A., and L. T. Winfree, Jr. 1998. "Race and Gender Differences between Gang and Non-Gang Youth: Results from a Multi-Site Survey." *Justice Quarterly* 15:505–26.

Ferraro, K. J. 1983. "Rationalizing Violence: Why Battered Women Stay." *Victimology: An International Journal* 8:203–12.

Gottfredson, D. C. 2001. *Schools and Delinquency*. New York: Cambridge University Press.

Gottfredson, G. D., and D. C. Gottfredson. 1985. *Victimization in Schools*. New York: Plenum Press.

Gottfredson, G. D., D. C. Gottfredson, A. A. Payne, and N. C. Gottfredson. 2005. "School Climate Predictors of School Disorder: Results from a National Study of Delinquency Prevention in Schools." *Journal of Research in Crime and Delinquency* 42:412–44.

Hill, K. G., J. C. Howell, J. D. Hawkins, and S. R. Battin-Pearson. 1999. "Childhood Risk Factors for Adolescent Gang Membership: Results from the Seattle Social Development Project." *Journal of Research in Crime and Delinquency* 36:300–22.

Hindelang, M., M. Gottfredson, and J. Garofalo. 1978. Victims of Personal Crime: An Empirical Foundation for a Theory of Personal Victimization. Cambridge, MA: Ballinger.

Howell, J. C. 1999. "Youth Gang Homicides: A Literature Review." *Crime and Delinquency* 45:208–41.

———. 2003. Preventing and Reducing Juvenile Delinquency: A Comprehensive Framework. Thousand Oaks, CA: Sage.

Howell, J. C., and S. H. Decker. 1999. "The Youth Gangs, Drugs, and Violence Connection." *OJJDP Juvenile Justice Bulletin*. Washington, DC: U.S. Department of Justice, Office of Justice Programs, Office of Juvenile Justice and Delinquency Prevention.

Howell, J. C., and D. K. Gleason. 1999. "Youth Gang Drug Trafficking." *OJJDP Juvenile Justice Bulletin*. Washington, DC: U.S. Department of Justice, Office of Justice Programs, Office of Juvenile Justice and Delinquency Prevention.

Huizinga, D. 1997. "Gangs and the Volume of Crime." Paper presented at the Annual Meeting of the Western Society of Criminology, Honolulu, Hawaii.

Huizinga, D., and C. Jakob-Chien. 1998. "The Contemporaneous Co-Occurrence of Serious and Violent Juvenile Offending and Other Problem Behaviors." pp. 47–67 in *Serious and Violent Juvenile Offenders: Risk Factors and Successful Intervention*, edited by R. Loeber and D. P. Farrington. Thousand Oaks, CA: Sage.

Huff, C. R. 1998. "Comparing the Criminal Behavior of Youth Gangs and At-Risk Youths." *NIJ Research in Brief*. Washington, DC: U.S. Department of Justice, National Institute of Justice.

Irwin, K. 2004. "The Violence of Adolescent Life: Experiencing and Managing Everyday Threats." *Youth and Society* 35:452–79.

Jacobs, B. A. 2000. *Robbing Drug Dealers: Violence beyond the Law*. New York: Aldine de Gruyter.

Johnston, L. D., J. G. Bachman, and P. M. O'Malley. 2003. "Monitoring the Future." In *The Sourcebook of Criminal Justice Statistics 2003*. Edited by K. Maguire and A. L. Pastore. Washington, DC: Author. Retrieved May 17, 2006, from http://www.albany.edu/sourcebook/

Klein, M. W. (1995). *The American Street Gang*. New York: Oxford University Press.

Lauritsen, J. L., Sampson, R. J., & Laub, J. H. 1991. "The Link between Offending and Victimization among Adolescents." *Criminology* 29:265–92.

Loeber, R., L. Kalb, and D. Huizinga. 2001. "Juvenile Delinquency and Serious Injury Victimization." *Juvenile Justice Bulletin*. Washington, DC: U.S. Department of Justice, Office of Justice Programs, Office of Juvenile Justice and Delinquency Prevention.

Lynch, J. P. 2002. "Trends in Juvenile Violent Offending: An Analysis of Victim Survey Data." *Juvenile Justice Bulletin*. Washington, DC: U.S. Department of Justice Office of Justice Programs. Office of Juvenile Justice and Delinquency Prevention.

Marcus, R. F. 2005. "Youth Violence in Everyday Life." *Journal of Interpersonal Violence* 20:442–47.

Maxson, C. L. 1995. "Street Gangs and Drug Sales in Two Suburban Cities." *NIJ Research in Brief*. Washington, DC: U.S. Department of Justice, Office of Justice Programs, National Institute of Justice.

———. 1998. "Gang Members on the Move." *Juvenile Justice Bulletin*. Washington, DC: Office of Juvenile Justice and Delinquency Prevention.

———. 1999. "Gang Homicide: A Review and Extension of the Literature." pp. 239–54 in *Homicide: A Sourcebook of Social Research*, edited by M. D. Smith and M. A. Zahn. Thousand Oaks, CA: Sage.

Maxson, C. L., G. D. Curry, and J. C. Howell. 2002. "Youth Gang Homicides in the 1990s." pp. 107–37 in *Responding to Gangs: Evaluation and Research*, edited by W. L. Reed and S. H. Decker. Washington, DC: U.S. Department of Justice, National Institute of Justice.

Mercy, J. A., and P. W. O'Carroll. 1988. "New Directions in Violence Prediction: The Public Health Arena." *Violence and Victims* 3:285–301.

Miller, J. 1998. "Gender and Victimization Risk among Young Women in Gangs." *Journal on Research in Crime and Delinquency* 35:429–53.

———. 2001. *One of the Guys: Girls, Gangs and Gender*. New York: Oxford University Press.

———. 2002. "Young Women in Street Gangs: Risk Factors, Delinquency and Victimization Risk." pp. 67–105 in *Responding to Gangs: Evaluation and Research*, edited by W. L. Reed and S. H. Decker. Washington DC: U.S. Department of Justice, National Institute of Justice.

Miller, J., and S. H. Decker. 2001. "Young Women and Gang Violence: Gender, Street Offending, and Violent Victimization in Gangs." *Justice Quarterly* 18:115–40.

Miller, W. B. 2001. *The Growth of Youth Gang Problems in the United States: 1970–98*. Washington, DC: U.S. Department of Justice, Office of Justice Programs, Office of Juvenile Justice and Delinquency Prevention.

Moore, M. H., D. Prothrow-Stith, B. Guyer, and H. Spivak. 1994. "Violence and Intentional Injuries: Criminal Justice and Public Health Perspective on an Urgent National Problem." pp. 167–216 in *Understanding and Preventing Violence*, vol. 4, *Consequences and Control*, edited by A. J. Reiss, Jr., and J. A. Roth. Washington, DC: National Research Council, National Academy Press.

National Institutes of Health. 2004. Preventing Violence and Related Health-Risking Social Behaviors in Adolescents: An NIH State-of-the-Science Conference." Retrieved May 17, 2006, from http://consensus.nih.gov/2004/2004Youth ViolencePreventionSOSO23html.htm

Osgood, D. W., J. K. Wilson, P. M. O'Malley, J. G. Bachman, and L. D. Johnston. 1996. "Routine Activities and Individual Delinquent Behavior." *American Sociological Review* 61:635–55.

Padilla, F. 1995. "The Working Gang." pp. 53–61 in *The Modern Gang Reader,* edited by M. W. Klein, C. L. Maxson, and J. Miller. Los Angeles: Roxbury.

Peterson, D., T. J. Taylor, and F.-A. Esbensen. 2004. "Gang Membership and Violent Victimization." *Justice Quarterly* 21:793–815.

Piquero, A. R., J. Macdonald, A. Dobrin, L. E. Daigle, and F. T. Cullen. 2005. "Self-Control, Violent Offending, and Homicide Victimization: Assessing the General Theory of Crime." *Journal of Quantitative Criminology* 21:55–71.

Rapp-Paglicci, L. A., and J. S. Wodarski. 2000. "Antecedent Behaviors of Male Youth Victimization: An Exploratory Study." *Deviant Behavior* 21:519–36.

Sampson, R. J., and J. L. Lauritsen. 1994. "Violent Victimization and Offending: Individual-, Situational-, and Community-Level Risk Factors." pp. 1–114 in *Understanding and Preventing Violence,* vol. 3, *Social Influences,* edited by A. J. Reiss, Jr., and J. A. Roth. Washington, DC: National Research Council, National Academy Press.

Sanders, W. B. 1994. *Gangbangs and Drive-Bys: Grounded Culture and Juvenile Gang Violence.* New York: Aldine de Gruyter.

Savitz., L., L. Rosen, and M. Lalli. 1980. "Delinquency and Gang Membership as Related to Victimization." *Victimology: An International Journal* 5:152–60.

Schreck, C. J. 1999. "Criminal Victimization and Low Self-Control: An Extension and Test of a General Theory of Crime." *Justice Quarterly* 16:633–54.

Schreck, C. J., and B. S. Fisher. 2004. "Specifying the Influence of Family and Peer Influences on Violent Victimization: Extending Routine Activities and Lifestyle Theories." *Journal of Interpersonal Violence* 19:1021–41.

Schreck, C. J., B. S. Fisher, and J. M. Miller, 2004. "The Social Context of Violent Victimization: A Study of the Delinquent Peer Effect." *Justice Quarterly* 21:23–47.

Schreck, C. J., R. A. Wright, and J. M. Miller. 2002. "A Study of Individual and Situational Antecedents of Violent Victimization." *Justice Quarterly* 19:159–80.

Shaffer, J. N., and R. B. Ruback. 2002. "Violent Victimization as a Risk Factor for Violent Offending among Juveniles." *Juvenile Justice Bulletin.* Washington, DC: U.S. Department of Justice, Office of Justice Programs, Office of Juvenile Justice and Delinquency Prevention.

Short, J. F., Jr., and L. A. Hughes. 2006. "Moving Gang Research Forward." pp. 225–38 in *Studying Youth Gangs,* edited by J. F. Short, Jr., and L. A. Hughes. Lanham, MD: AltaMira Press.

Spergel, I. A., and G. D. Curry. 1990. "Strategies and Perceived Agency Effectiveness in Dealing with the Youth Gang Problem." pp. 288–309 in *Gangs in America,* edited by C. R. Huff. Newbury Park, CA: Sage.

———. 1993. "The National Youth Gang Survey: A Research and Development Process." pp. 359–400 in *The Gang Intervention Handbook,* edited by A. Goldstein and C. R. Huff. Champaign, IL: Research Press.

Thornberry, T. P. 1998. "Membership in Youth Gangs and Involvement in Serious and Violent Offending." pp. 147–66 in *Serious and Violent Juvenile Offenders: Risk Factors for Successful Intervention,* edited by R. Loeber and D. P. Farrington. Thousand Oaks, CA: Sage.

Thornberry, T. P., M. D. Krohn, A. J. Lizotte, C. A. Smith, and K. Tobin. 2003. *Gangs and Delinquency in Developmental Perspective.* New York: Cambridge University Press.

Tolmas, H. C. 1998. "Violence among Youth: A Major Epidemic in America." *International Journal of Adolescent Medicine and Health* 10:243–59.

Welsh, W. N. 2001. "Effects of School and School Factors on Five Measures of School Disorder." *Justice Quarterly* 18:911–47.

Welsh, W. N., J. R. Greene, and P. H. Jenkins. 1999. "School Disorder: The Influence of Individual, Institutional, and Community Factors." *Criminology* 37:601–43.

Zhang, L., J. W. Welte, and W. F. Wieczorek. 2001. "Deviant Lifestyle and Crime Victimization." *Journal of Criminal Justice* 29:133–43.

DISCUSSION QUESTIONS

1. How does routine activities theory relate to the findings that gang members are more at risk of experiencing serious victimization than non-gang members?

2. What are the implications for the criminal justice system and for victims' services if gang members are also likely to be victims? Are gang members sympathetic victims? How would early victimologists likely have viewed gang members' victimization?

3. Taylor and colleagues found that gang members have higher levels of risk factors than do non-gang members. Do you think this is the case because they are gang members or that these risk factors may explain why they became involved in the gang to begin with?

4. How does this article help illuminate the complex relationship between victims and offenders and victimization and offending?

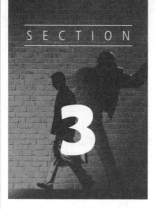

SECTION

3

Consequences of Victimization

et us revisit Polly, the young woman whose victimization was described in Section 2. When we left her, Polly was on her way back home after leaving a bar alone at night, and she was robbed and assaulted by two men. But Polly's story does not end there, and although the incident itself ended, Polly dealt with it for quite some time. Polly made it home safely; she entered her apartment, locked her door, and started to cry. She felt scared, alone, and her head was hurting. She told one of her roommates, who was home when she returned, what happened. Her roommate, Rachel, told her she should call the police and have someone look at her head. Polly was hesitant—after all, she did not know what to expect—but she really wanted to make sure that the men were caught, so she called the police and told the dispatcher what had occurred.

The police and emergency personnel arrived. She was taken to the hospital for her head injury and was released after receiving 10 stitches. Before she could go home, though, the police wanted to take her statement. They questioned her for more than an hour, asking minute details about what happened and about the offenders. They also asked her why she was walking home alone at night. The police officers left her with assurances that they would do everything they could to identify her attackers.

The days passed, and Polly had a hard time forgetting about the men and what had transpired. She was having a hard time getting out of bed. In fact, she missed several days of class. She found herself avoiding going out alone at night. She felt as though her life had taken an unexpected, unwanted, and frightening turn—one that she was worried would forever alter her life. Polly's concerns, like others', were most likely not unfounded.

Physical Injury

Clearly, when people suffer personal victimizations, they are at risk of **physical injury**. These injuries can include bruises, soreness, scratches, cuts, broken bones, contracted diseases, and stab or gunshot wounds. Some of these injuries may be temporary and short-lived, while others can be long-lasting or permanent.

According to data from the NCVS in 2006, 27% of assault victims sustained physical injuries. Those who experienced robbery were more likely to be injured; 35% of robbery victims suffered physical injury. A larger percentage of female victims were injured than male victims, although the differences were not large. For example, 29% of female assault victims compared with 25% of male assault victims reported being injured. There appears to be a difference in injury for racial groups as well. For both assault and robbery, injuries were present in a larger percentage of Black victims than White victims. The victim-offender relationship was also related to injury—incidents perpetrated by nonstrangers were more likely to result in injury than those perpetrated by strangers (46% for assault compared with 35% for robbery) (Bureau of Justice Statistics [BJS], 2006a).

The most serious physical injury is, of course, death. While the NCVS does not measure murder—remember, it asks people about their victimization experiences—the Uniform Crime Reports (UCR) can be used to find out the extent to which deaths are attributable to murder and nonnegligent manslaughter. In 2008, UCR figures showed that 14,180 murders were brought to the attention of the police. The majority of murder victims were male (78%), almost equal percentages were White and Black (49%), and 55% were murdered by an acquaintance. Almost three fourths of the homicides that involved a weapon were gun-related. Most of the homicides for which the circumstances were known resulted from an argument (FBI, 2009).

Mental Health Consequences and Costs

People differentially respond to trauma, including victimization. Some people may cope by internalizing their feelings and emotions, while others may experience externalizing responses. It is likely that the way people deal with victimization is tied to their biological make-up, their interactional style, their coping style and resources, and the context in which the incident occurs and in which they operate thereafter. Some of the responses can be quite serious and long-term while others may be more transitory.

Three affective responses that are common among crime victims are depression, reductions in self-esteem, and anxiety. The way in which **depression** manifests itself varies greatly across individuals. It can include symptoms such as sleep disturbances, changes in eating habits, feelings of guilt and worthlessness, and irritability. Generally, depressed persons will experience a decline in interest in activities they once enjoyed, a depressed mood, or both. For youth, depression is a common outcome for those who are victimized by peers, such as in bullying (Sweeting, Young, West, & Der, 2006). With the advent of technology and the widespread use of the Internet, recent research has explored online victimization and its effects. Online victimization is related to depressive responses in victims (Tynes & Giang, 2009).

Victimization may be powerful enough to alter the way in which a crime victim views him- or herself. Self-esteem and self-worth both have been found to be reduced in some crime victims, particularly female victims. In one study of youths in Virginia, Amie Grills and Thomas Ollendick (2002) found that, for girls, being victimized by peers was associated with a reduction in global **self-worth** and their self-worth was related to elevated levels of anxiety. There may also be a difference in crime's impact on self-appraisals based on the type of victimization experienced. For example, victims of childhood sexual abuse are likely to suffer long-term negative impacts to their **self-esteem** (Beitchman et al., 1992). Sexual victimization also has been linked to reductions in self-esteem (Turner et al., 2010).

Anxiety is another consequence linked to victimization. Persons who suffer from anxiety are likely to experience a range of emotional and physical symptoms. Much like depression, however, anxiety affects people differently. Most notably, anxiety is often experienced as irrational and excessive fear and worry, which may be coupled with feelings of tension and restlessness, vigilance, irritability, and difficulty concentrating. In addition, because anxiety is a product of the body's flight-or-fight response, it also has physical symptoms. These include a racing and pounding heart, sweating, stomach upset, headaches, difficulty sleeping and breathing, tremors, and muscle tension (Dryden-Edwards, 2007).

Although anxiety that crime victims experience may not escalate to a point where they are diagnosed with an anxiety disorder by a mental health clinician, victimization does appear to be linked to anxiety symptoms. For example, adolescents who experience victimization by their peers experience anxiety at higher levels than nonvictimized adolescents (Storch, 2003). The relationship between anxiety and victimization is likely complex in that victimization can lead to anxiety, but anxiety and distress are also precursors to victimization (Siegel, La Greca, & Harrison, 2009). Some victims do experience mental health consequences tied to anxiety that lead to mental health diagnoses.

Posttraumatic Stress Disorder

One of the recognized disorders associated with a patterned response to trauma, such as victimization, is **posttraumatic stress disorder (PTSD)**. Commonly associated with individuals returning from war and combat, PTSD is a psychiatric condition that recently has been recognized as a possible consequence of other traumatic events, such as criminal victimization. Currently classified by the American Psychiatric Association in the *DSM-IV-TR* as an anxiety disorder, PTSD is diagnosed based on several criteria outlined in detail in Table 3.1. A person must have

Table 3.1   *DSM-IV-R* Diagnostic Criteria for Posttraumatic Stress Disorder

1. **Stressor**: A person has been exposed to a traumatic event in which he or she has
   a. experienced, witnessed, or been confronted with an event(s) that involved actual or threatened death or serious injury, or threat to the physical integrity of oneself or others; and
   b. the response to the event included intense fear, helplessness, or horror.

2. **Intrusive recollection**: The trauma is reexperienced in at least one of the following ways:
   a. Recurrent and intrusive recollections of the event, such as images, thoughts, or perceptions
   b. Recurrent nightmares
   c. Acting or feeling as if the traumatic event were recurring, such as reliving the event, illusions, hallucinations, and flashbacks
   d. Intense psychological distress when exposed to cues that symbolize or resemble a component of the traumatic event
   e. Physiologic reactivity when exposed to cues that symbolize or resemble a component of the traumatic event

3. **Avoidance/numbing symptoms**: Regular avoidance of stimuli associated with the traumatic event and numbness of response. Three or more of the following symptoms must be present:
   a. Efforts to avoid thoughts, feelings, or conversations about the trauma
   b. Efforts to avoid activities, places, or people that cause the trauma to be remembered
   c. Inability to remember an important element of the trauma
   d. Significant reduced interest or participation in significant activities
   e. Feelings of detachment or estrangement from other people
   f. Lack of affect
   g. Lack of sense of future

(Continued)

Table 3.1 (Continued)

4. **Hyperarousal**: Persistent arousal symptomology. Must experience at least two of the following:
   a. Difficulty falling or staying asleep
   b. Irritability or emotional outbursts
   c. Problems concentrating
   d. Hypervigilance
   e. Exaggerated startle response

In order for PTSD to be diagnosed, the symptoms in Sections 2, 3, and 4 must be experienced for more than one month and must cause clinically significant distress or impairment in social, occupational, or other functional areas.

SOURCE: American Psychiatric Association (2000). Reprinted with permission from *The Diagnostic and Statistical Manual of Mental Disorders, Fourth Edition, Text Revision*. Copyright © 2000. American Psychiatric Association.

experienced or witnessed a traumatic event that involved actual or threatened death or serious injury to oneself or others, or threat to the physical integrity of oneself or others. The person must have experienced fear, helplessness, or horror in response to the event and then reexperienced the trauma over time via flashbacks, nightmares, images, and/or reliving the event. The person must avoid stimuli associated with the traumatic event and experience numbness of response, such as lack of affect and reduced interest in activities. Finally, PTSD is characterized by hyperarousal.

In order for PTSD to be diagnosed, symptoms must be experienced for more than 1 month and must cause clinically significant distress or impairment in social, occupational, or other functional areas (American Psychological Association, 2000). As you may imagine, PTSD can be debilitating and can impact a victim's ability to heal, move on, and thrive after being victimized. About 8% of Americans will experience PTSD, although women are more likely than men to experience this disorder (Kessler, Sonnega, Bromet, Hughes, & Nelson, 1995). The traumatic events most likely to lead to PTSD for men are military combat and witnessing a serious injury or violent death. Women, on the other hand, are most likely to be diagnosed with PTSD related to incidents of rape and sexual molestation (Kessler et al., 1995).

Although it is difficult to know how common PTSD is among crime victims, some studies suggest that PTSD is a real problem for this group. The estimate for PTSD in persons who have been victimized is around 25%. Lifetime incidence of PTSD for persons who have not experienced a victimization is 9% (Kilpatrick & Acierno, 2008). Depression also commonly co-occurs in victims who suffer PTSD (Kilpatrick & Acierno, 2003). Research has shown that victims of sexual assault, aggravated assault, and persons whose family members were homicide victims are more likely than other crime victims to develop PTSD (Kilpatrick & Tidwell, 1989). In support of this link, the occurrence of PTSD in rape victims has been estimated to be almost 1 in 3 (Kilpatrick, Edmunds, & Seymour, 1992).

Self-Blame and Learned Helplessness

Victims of crime may blame themselves for their victimization. One type of **self-blame** is **characterological self-blame**, which occurs when a person ascribes blame to a nonmodifiable source, such as one's character (Janoff-Bulman, 1979). In this way, characterological self-blame involves believing that victimization is deserved. Another type of self-blame is **behavioral self-blame,** which occurs when a person ascribes blame to a modifiable source—behavior (Janoff-Bulman, 1979). When a person turns to behavioral self-blame, a future victimization can be avoided as long as behavior is changed.

In addition to self-blame, others may experience learned helplessness following victimization. **Learned help-lessness** is a response to victimization in which victims believe that responding is futile and become passive and numb (Seligman, 1975). In this way, victims may not activate to protect themselves in the face of danger and, instead, put themselves at risk of subsequent victimization experiences.

⊠ Economic Costs

Not only are victimologists concerned with the impact that being a crime victim has on an individual in terms of health, but they are also concerned with the **economic costs** incurred by both the victim and the public. In this sense, victimization is a public health issue. Economic costs can result from property losses; monies associated with medical care; time lost from work, school, and housework; pain, suffering, and reduced quality of life; and legal costs. In 2008, the NCVS estimated the total economic loss from crimes at $17,397 billion. The NCVS also shows that the median dollar amount of loss attributed to crime was $125 (BJS, 2011). Although this number may appear to be low, it largely represents the fact that the typical property crime is a simple larceny/theft.

Direct Property Losses

Crime victims often experience tangible losses in terms of having their property damaged or taken. Generally, when determining **direct property losses,** the value of property that is damaged, taken, and not recovered, and insurance claims and administration costs are considered. According to the NCVS, in 2008, 94% of property crimes resulted in economic losses (BJS, 2011). In one of the most comprehensive reports on the costs of victimization—sponsored by the National Institute of Justice—Ted Miller, Mark Cohen, and Brian Wiersema (1996) estimated the property loss or damage experienced per crime victimization event. These estimates were used by Welsh and colleagues (2008) in their article, included in this section, on the costs of juvenile crime in urban areas. They found that arson victimizations resulted in an estimated $15,500 per episode. Motor vehicle theft cost about $3,300 per incident. Results from the NCVS show that personal crime victimizations typically did not result in as much direct property loss. For example, only 18% of personal crime victimizations resulted in economic loss. Rape and sexual assaults typically resulted in $100 of property loss or property damage. It is rare for a victim of a violent or property offense to recover any losses. Only about 29% of victims of personal crime and 16% of victims of property crime recover all or some property (BJS, 2011).

Medical Care

To be sure, many victims would gladly suffer property loss if it meant they would not experience any physical injury. After all, items can be replaced and damage repaired. Physical injury may lead to victims needing medical attention, which for some may be the first step in accumulating costs associated with their victimization. **Medical care costs** encompass such expenses as transporting victims to the hospital, doctor care, prescription drugs, allied health services, medical devices, coroner payments, insurance claims processing fees, and premature funeral expenses (Miller et al., 1996).

Results from the NCVS indicate that in 2008, 542,280 violent crime victims received some type of medical care (BJS, 2011). Of those victims who received medical care, slightly more than one third received care in the hospital emergency room or at an emergency clinic and 9% went to the hospital (BJS, 2011). Receiving medical care often results in victims incurring medical expenses. Almost 6% of victims of violence reported having medical expenses as a result of being victimized (BJS, 2011). About 63% of injured victims had health insurance or were eligible for public medical services (BJS, 2011).

Costs vary across types of victimization. For example, the annual cost of hospitalizations for victims of child abuse is estimated to be $6.2 billion (Prevent Child Abuse America, 2000). Medical treatment for battered women is estimated to cost $1.8 billion annually (Wisner, Gilmer, Saltman, & Zink, 1999). Per-criminal-victimization medical care costs also have been estimated. Assaults in which there were injuries cost $1,470 per incident. Drunk-driving victims who were injured incurred $6,400 in medical care costs (Miller et al., 1996).

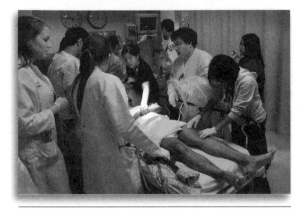

Gun violence is associated with substantial medical costs for victims. Although most crime victims do not require hospitalization, even if they are treated in the emergency room, a report on gun violence published by the Office for Victims of Crime showed that gunshot victims make up one third of those who require hospitalization (Bonderman, 2001). Persons who are shot and admitted to the hospital are likely to face numerous rehospitalizations and incur medical costs throughout their lifetimes. In 1994, the lifetime medical costs for all victims of firearm injuries totaled $1.7 billion. Spinal cord injuries are particularly expensive, with average expenses for first-year medical costs alone totaling more than $217,000. The average cost per victim of violence-related spinal cord injury is more than $600,000.

▲ **Photo 3.1** This victim of a gunshot wound received treatment in a trauma unit in Chicago, IL, on November 6, 2009.

Photo credit: © Getty Images

Mental Health Care Costs

When victims seek mental health care, this also adds to their total cost. It is estimated that between 10% and 20% of total **mental health care costs** in the United States are related to crime (Miller et al., 1996). Most of this cost is a result of crime victims seeking treatment to deal with the effects of their victimization. Between one quarter and one half of rape and child sexual abuse victims receive mental health care. As a result, sexual victimizations, of both adults and children, result in some of the largest mental health care costs for victims. The average mental health care cost per rape and sexual assault is $2,200, and the average for child abuse is $5,800. Victims of arson who are injured incur about $10,000 of mental care expenditures per victimization. Secondary victimization, which is discussed in detail in a later section, is also associated with mental health care costs. The average murder results in between 1.5 and 2.5 people receiving mental health counseling (Miller et al., 1996).

Losses in Productivity

Persons who are victimized may experience an inability to work at their place of employment, complete housework, or attend school. Not being able to do these things contributes to the total **lost productivity** that crime victims experience. In 2008, about 7% of persons in the NCVS who said they were violently victimized lost some time from work (BJS, 2011). About the same percentage of victims of property offenses lost time from work. Some victims are more prone to miss work than others. For example, almost one tenth of burglary victimizations cause victims to miss at least one day of work (BJS, 2011). Data from the NCVS show that 9% of robbery victimizations resulted in victims missing more than 10 days of work (BJS, 2011), while victims of intimate partner violence lost almost 8 million paid days of work annually (Centers for Disease Control and Prevention, 2003). Employers also bear some costs when their employees are victimized; victimized employees may be less productive, their employers may incur costs associated with hiring replacements, and employers may experience costs dealing with the emotional responses of their employees. Parents also may suffer costs when their children are victimized and they are unable to meet all their job responsibilities as a result of doing things such as taking the child to the doctor or staying home with the child (Miller et al., 1996).

Pain, Suffering, and Lost Quality of Life

The most difficult cost to quantify is the pain, suffering, and loss of quality of life that crime victims experience. When these elements are added to the costs associated with medical care, lost earnings, and programs associated with victim assistance, the cost to crime victims increases 4 times. In other words, this is the largest cost that crime victims sustain. For example, one study estimated the cost in out-of-pocket expenses to victims of rape to be slightly less than $5,100. The crime of rape, however, on average, costs $87,000 when its impact on quality of life is considered (Miller et al., 1996).

Another cost that crime victims may experience is a change in their routines and lifestyles. Many victims report that after being victimized, they changed their behavior. For example, victims of stalking may change their phone numbers, move, or change their normal routines. Others may stop going out alone or start carrying a weapon when they do so. Although these changes may reduce risk of being victimized again, for victims to bear the cost of crime seems somewhat unfair. Did Polly sustain any of these costs?

✂ System Costs

The victim is not the only entity impacted economically by crime. The United States in general spends an incredible amount of money on criminal justice. When including **system costs** for law enforcement, the courts, and corrections, the direct expenditures of the criminal justice system are more than $214 billion annually (BJS, 2006b). The criminal justice system employs more than 2.4 million persons, whose collective pay tops $9 billion. Obviously, crime is big business in the United States!

Insurance companies pay about $45 billion annually due to crime (U.S. News and World Report, 1996). The federal government also pays $8 billion annually for restorative and emergency services for crime victims (U.S. News and World Report, 1996). There are other costs society must absorb as a result of crime. For example, it costs Americans when individuals who are not insured or are on public assistance are victimized and receive medical care. The U.S. government covers about one fourth of health insurance payouts to crime victims. Gunshot victims alone cost taxpayers more than $4.5 billion dollars annually (U.S. News and World Report, 1996). These costs are not distributed equally across society. Some communities have been hit especially hard by violence—gun violence in particular. Some 96% of hospital expenses associated with gun violence at King/Drew Medical Center in Los Angeles are paid with public funds (Bonderman, 2001). To understand how expensive gun violence medical fees can be to the public, read the box item about James, who was shot and survived. We will discuss in Section 4 just how these costs are paid and who pays them.

THE STORY OF JAMES

James, 45, was shot in the knee on Sept. 9 as he sat in a car with another man, who died of his wounds at the scene. James' injuries, which also included a hole in the arm and fragments in the eye, were not near vital organs.

His knee looked bad when he came into Froedtert's trauma center, but that turned out to be just the beginning. The next day, he aspirated as a breathing tube was being inserted during surgery, and contents from his stomach got into his lungs.

"It is kind of like a chemical burn," trauma surgeon James Feeney said.

James had to stay in intensive care on a ventilator and be heavily sedated for almost two weeks, while his hospital charges ballooned.

For about a week, he was on drugs that essentially paralyzed most of his muscles. When he began to regain consciousness, he suffered another setback, called ICU psychosis. The maddening disorder is believed to be caused by a variety of factors in intensive care, including breathing tubes, lights, beeping noises, a lack of sleep and sensory deprivation or overload. It can make patients temporarily insane.

"His agitation was so severe every time we tried to take him off (the ventilator), he would get crazy and wild," Feeney said.

James eventually got out of intensive care and has improved dramatically. Doctors say they think they have saved his leg, although they don't know how functional it will be. They also don't know how much vision he lost. He is likely to need more surgeries on both the knee and the eye. James also will need extensive physical therapy.

"The truth is, a lot of these guys would have died 20 years ago before we had an organized system of trauma care," Feeney said.

After a call that started off as a man shot in the leg, he spent nearly six weeks in Froedtert. When he was discharged Oct. 20, the hospital charges—which will be billed to Medicare—topped $277,000.

Medicare caps reimbursement for shooting cases at $36,000, said Blaine O'Connell, Froedtert's chief financial officer. Medicaid and Milwaukee County's General Assistance Medical Program also pay only a fraction of the hospital's charges.

And for many uninsured patients, the hospital may collect even less, he said.

Ultimately, the losses on all those cases are factored into the rates the hospital must charge private insurers.

SOURCE: Diedrich and Fauber (2006). "Gunshot costs echo through economy. From hospitals to jails, price of violence adds up quickly," *Milwaukee Wisconsin Journal Sentinel Online* http://www.jsonline.com/news/milwaukee/29205944.html. Reprinted with permission.

Recurring Victimization

Another cost of victimization that is often not discussed or known is the real possibility that a person who is victimized once will be victimized again. In fact, persons who have been victimized are *more* likely to be victimized again than others who have not experienced *any* victimization. At first, this reality probably does not make sense. After all, if you were victimized, you may be likely to implement crime reduction strategies. For example, if you had your car broken into because you had valuables in plain view, would you keep items in your car again? You probably are shaking your head "no." So, why then are some people prone to being victimized not once, but again, and sometimes again and again and again? Before we can address that question, let us first find out the extent to which people are victimized more than once.

Extent of Recurring Victimization

To know the extent to which people experience more than one victimization, let us first identify what we mean. As seen in Table 3.2, **recurring victimization** occurs when a person or place is victimized more than once by any type of victimization. **Repeat victimization** occurs when a person or place is victimized more than once by the same type of victimization. **Revictimization** is commonly referred to when a person is victimized more than once by any type of victimization but across a relatively wide span of time—such as from childhood to adulthood. Revictimization has been most widely studied in terms of childhood sexual abuse and sexual assault in adulthood.

Now that we know what the terms mean, let's find out how often people and places are victimized more than once. Although most people and households in a given year are not victimized at all, some experience more than one victimization. The British Crime Survey, a victimization survey similar to the NCVS, revealed that 14% of burglary victims experienced two or more incidents during the same year (Nicholas, Povey, Walker, & Kershaw, 2005). Victims of personal crimes are also at heightened risk of experiencing more than one incident. Research shows that victims of intimate partner violence, rape, and assault are all at risk of experiencing a subsequent incident following their initial victimization.

These recurring victims also experience a disproportionate share of all victimization events. For example, 6% of the respondents in the British Crime Survey over 10 years experienced 68% of all the thefts that occurred. As noted in the article by Leah Daigle, Bonnie Fisher, and Francis Cullen (2008) in this section, 7% of college women had experienced two different sexual victimization incidents during the previous academic year, and these women experienced almost three fourths of all sexual victimizations.

Table 3.2 Terminology Related to Recurring Victimization

Type of Victimization	Type of Incidents Experienced	Length of Time Between Incidents
Recurring victimization	A victimization of any type followed by a victimization of any type Ex. A theft followed by an assault	Can be any time between incidents
Repeat victimization	A victimization followed by another victimization of the exact same type Ex. A theft followed by a theft	Generally, incidents occur relatively close to each other temporally in the same developmental period Ex. A college student is assaulted in May and assaulted in June of the same year
Revictimization	A victimization of any type followed by a victimization of any type Ex. A theft followed by an assault	Can be any time between incidents; generally refers to incidents that occur in different developmental time periods Ex. A person is abused as a child and then is raped as an adult

Characteristics of Recurring Victimization

Also interesting is that recurring victimization is likely to happen quickly. When examining the time between incidents, researchers have found that, often, little time transpires between incidents. For college women's sexual victimization, one study found that most subsequent incidents happened within the same month or 1 month after the initial incident (Daigle et al., 2008). What this means is that victims are at an increased risk of being victimized again in the time immediately following an initial incident. Over time, the risk of experiencing another victimization wanes.

What type of victimization are victims likely to experience if they experience more than one? Most likely, when a person is victimized a subsequent time, they will experience the same type of victimization previously experienced (Reiss, 1980). For example, a theft victim is likely to experience another theft if victimized a second time.

Theoretical Explanations of Recurring Victimization

We know, then, that recurring victimization is a reality many victims face, that it is likely to recur rather quickly if it does, and that the same type of victimization is likely to follow. But this picture of what recurring victimization "looks" like does not address why some people are victimized a single time and others find themselves victimized again.

There are two theoretical explanations that have been proffered to explain recurring victimization. The first is called **risk heterogeneity**. This explanation of recurring victimization focuses on qualities or characteristics of the victim. Those qualities or characteristics that initially place a victim at risk will keep that person at risk of experiencing a subsequent victimization if unchanged (Farrell, Phillips, & Pease, 1995). For example, remember Polly? Is there any quality or characteristic that placed her at risk for being accosted by the two men in the alley? You are probably thinking that her walking home at night may have been a risk factor for her. This was discussed in Section 2 about lifestyles and routine activities. Polly quite likely was victimized, at least in part, because she was seen by the two men as being a vulnerable target. In this way, walking home at night by herself placed her at risk. If Polly walks home at night by herself on other nights, she is again at risk of being victimized. In this way, Polly's walking home at night by herself placed her at risk of being victimized the first time, and it also places her at risk of being a victim in the future. What if she walked home because she could not afford a car? In other words, what if her social status or class placed her in a position that increased her vulnerability to crime victimization because she had to walk home at night rather than drive? This quality or characteristic would also fall into the explanation of risk heterogeneity. Also remember other factors, discussed in Section 2, that place individuals at risk of victimization more generally—living in disadvantaged neighborhoods and exposure to delinquent peers, for example. These factors, if left unchanged, will keep individuals at risk of subsequent victimization.

In contrast to the risk heterogeneity argument, the second theoretical explanation of recurring victimization is known as **state dependence.** According to state dependence, it is not the qualities or characteristics of a victim that are important for recurring victimization so much as what happens during and after the victimization (Farrell et al., 1995). How the victim and the offender act and react to the victimization event will predict risk of becoming a recurrent victim. In this way, the victim and offender are learning key information that will impact the likelihood of subsequent victimizations. For example, a victim of rape or other sexual victimization who resists or uses self-protective actions is less likely than those who do not to be victimized again (Fisher, Daigle, & Cullen, 2010). This reduction in risk is likely due to the victim learning that she has agency and control over her life. Protecting herself may even serve to empower her so that in the future she is able to identify and avoid risk. Likewise, the offender is likely learning that she is not an "easy" target and that victimizing her will not pay off in the future. In both scenarios, the victim is less likely to find herself the target of an offender. In Polly's case, it is difficult to know if she is likely to be victimized again based on a state-dependence explanation. Since she tried to resist and she called the police, she certainly is learning that she has some control over her life. If doing so empowers her, she likely will be less attractive as a target to offenders, and she may be less likely to find herself in risky situations—such as walking home at night alone. To be clear, neither of these explanations should be used to blame the victim or place responsibility for the victimization on the victim. The offender is responsible for his or her actions, and blame should rest there. These explanations are, however, tools to help understand why some people are targeted over and over again.

⊠ Vicarious Victimization

It is not only the victim and the system that are saddled with costs. The effects that victimization has on those close to the victim are also critical in understanding the total impact of crime. So far, we have discussed how a victim may need medical care, may seek mental health counseling, may lose time from work, and may have a less full life after being victimized. But what happens to those who love and care about these victims? Does witnessing a loved one go through victimization also exact a price?

The effects that victimization has on others are collectively known as **vicarious victimization**. Vicarious victimization has been most widely studied in regard to **homicide survivors**—people whose loved ones have been murdered—given the profound effect that homicide has on family members, even when compared with nonhomicide deaths. Homicide deaths are almost exclusively sudden and violent. Surviving family members often experience guilt about not being able to prevent the death. The involvement of the criminal justice system also adds an element to the response family members have, and there is often a feeling that others view the death as at least partly the victim's fault.

The studies on homicide survivors have largely found that they experience many of the same posttrauma symptoms that crime victims themselves experience. One study found that almost one quarter of homicide surviving family members developed posttraumatic stress disorder (PTSD) after the murder of their family member (as cited in Kilpatrick, Amick, & Resnick, 1990). The disorder and PTSD symptomology are often not transient, with homicide survivors exhibiting PTSD symptoms for up to 5 years following the murder (Redmond, 1989). Being a homicide survivor also may be related to greater PTSD symptoms than being a victim of a crime such as rape (Amick-McMullan, Kilpatrick, & Veronen, 1989). Also interesting, homicide survivors experience higher levels of PTSD than do family members who lose a loved one through means other than homicide, such as accidentally (Applebaum & Burns, 1991). PTSD is not the only psychological response that homicide survivors show. They also have higher levels of distress, depression, anxiety, and hostility than persons who have not experienced trauma (Thompson, Kaslow, Price, Williams, & Kingree, 1998).

In addition to psychological responses, homicide survivors may exhibit behavioral consequences. Parents whose children die via homicide are more likely to exhibit suicidal ideation than parents whose children commit suicide or die accidentally (Murphy, Tapper, Johnson, & Lohan, 2003). Other homicide survivors may exhibit lifestyle changes by avoiding places and activities—either because they are fearful or anxious, or because they no longer feel able to participate in activities that are reminiscent of times spent with their now-deceased loved one. Homicide survivors also evince feelings of vulnerability, loss of control, loss of meaning, and self-blame. As you can now be certain, criminal victimization has wide-reaching effects on the victim, the system, and others.

▲ **Photo 3.2**   A support group for family members of murdered people.

Photo credit: © iStockphoto.com/Alina Solovyova-Vincent

Another form of vicarious victimization occurs when a person is traumatized by the coverage violent acts receive through media or other outlets that provide information. This type of vicarious victimization is likely to occur when seven factors are present: (1) realistic threat of death to all members of the community; (2) extraordinary carnage; (3) strong community affiliation; (4) witnessing of event by community members; (5) symbolic significance of victims to community; (6) need for rescue workers; and (7) significant media attention (Young, 1989). Given these factors, traumatic events that do not directly affect a person or a person's loved ones may also cause harm such as PTSD. Events such as the terrorist attacks on September 11, 2001, are prime examples of traumas that can produce lasting, harmful consequences to people exposed to them. Other events, such as a serial killer operating in a community—as discussed by Michael Herkov and Monica Biernat (1997) in their article included in this section—may also be a form of vicarious victimization that can produce PTSD in community members.

✉ Reporting

All the consequences and outcomes we have discussed thus far are impacted by the victim **reporting** the offense to the police. Reporting may intensify some of these consequences, may moderate some of the impact, or may be somewhat unrelated to the victim's experiences after the incident occurs. Reporting is important for several reasons. One important factor about reporting to the police is that it is the first essential step in activating the formal criminal justice system. Without a report to the police, the victim is left to deal with the aftermath through other channels and the police will never begin an investigative process. Without this first critical step, it is extremely unlikely that an offender will ever be caught. When an offender "gets away" with crime, it can have important consequences. When this occurs, the offender is learning that he or she can continue to freely offend—perhaps even against the same person or household. Conversely, an arrest or real threat of arrest may deter potential offenders.

Victims may also be negatively impacted if they do not report. Many victims' services, as will be discussed in Section 4, are available only for victims who notify the police about their incident. For example, many district attorney's offices have victim advocates, whose job it is to help victims navigate the criminal justice system and assist them with other programs such as receiving victim compensation. The ability to utilize these services is typically conditioned on reporting, since the district attorney's office would not even know about a crime victim who did not first come forward.

With all these benefits to reporting, it is easy to forget that about half of all violent crime victims and just more than one third of property crime victims notify the police (BJS, 2006b). In 2006 slightly more than half of all robbery and aggravated assault victims and slightly less than one third of all rape and sexual assault victims reported their incidents to the police.

Reporting varies by crime type, but it also varies according to other characteristics (Hart & Rennison, 2003). Generally, violence against women and violence against older persons is more likely to come to the attention of the police than violence against men and younger persons. Victimizations that result in the victim suffering an injury are more likely to be reported than those that do not result in injury. When an offender is armed, perceived to be under the influence of alcohol and/or drugs, a stranger, and a non-gang member, the victim is more likely to call the police (Hart & Rennison, 2003).

Besides these incident characteristics, victims also give tangible reasons for not reporting their incidents to the police. Overall, the most common reasons given by victims of violence for why they do not report include that the victimization was a private or personal matter, that it was reported to another official, that the object was recovered/the offender was unsuccessful, or for fear of reprisal (BJS, 2006a). Table 3.3 shows the reasons victims give for not reporting to the police for different victimization types. But some victims do in fact bring their incidents to the attention of the police. Most commonly, victims of violence report their incidents to prevent future violence, to stop the offender, because it was a crime, and to protect others (BJS, 2006a). Table 3.4 shows the common reasons that victims do report for different victimization types.

For victims of property crime, the most common reasons given for not bringing the incident to the attention of the police are that the object was recovered/the offender was unsuccessful, feeling the police would not want to be bothered, or lack of proof. Property crime victims were motivated to report because they wanted to recover stolen property, because it was a crime, and to prevent further crimes against them by the offender (BJS, 2006a).

Table 3.3 Percentage of Reasons for Not Reporting Victimization to the Police

Reasons for Reporting	Type of Crime						
	Rape/ Sexual Assault	Robbery	Assault	Purse Snatching/ Pocket Picking	Household Burglary	Motor Vehicle Theft	Theft
Reported to another official	9.4	5.4	14.2	14.8	5.2	3.4	8.3
Private or personal matter	24.5	12.2	21.8	18.0	7.0	7.9	5.3
Object recovered; offender unsuccessful	10.0	11.2	18.5	17.5	21.4	16.3	28.7
Not important enough	2.0	1.9	7.1	3.7	6.2	0.0	4.4
Insurance would not cover	0.0	0.7	0.0	0.0	3.4	3.4	2.6
Not aware crime occurred until later	0.0	2.1	0.2	2.3	7.1	4.7	5.1
Unable to recover property; no ID number	0.0	0.0	0.2	7.4	4.8	1.0	6.5
Lack of proof	0.0	10.2	2.1	21.0	12.1	13.1	9.9
Police would not want to be bothered	4.8	11.4	5.8	3.3	10.9	3.1	10.5
Police inefficient, ineffective, or biased	0.0	17.8	3.1	0.0	6.9	15.3	3.8
Fear of reprisal	17.0	5.8	7.3	0.0	1.0	1.3	0.5
Too inconvenient or time-consuming	6.5	7.5	4.2	3.8	3.5	6.4	4.1
Other	25.9	13.8	15.6	8.3	10.3	24.2	10.4

SOURCE: BJS (2006a).

Table 3.4   Percentage of Reasons for Reporting Victimization to the Police

Reasons for Reporting	Type of Crime						
	Rape/ Sexual Assault	Robbery	Assault	Purse Snatching/ Pocket Picking	Household Burglary	Motor Vehicle Theft	Theft
Stop or prevent incident	15.4	8.6	28.5	10.3	11.3	5.3	9.3
Needed help due to injury	7.3	1.6	2.9	6.0	0.1	0.0	0.6
To recover property	2.4	15.2	0.3	30.9	18.0	35.5	23.6
To collect insurance	0.0	1.5	0.5	0.0	3.0	6.2	4.0
To prevent further crimes by offender against victim	16.8	20.0	21.9	10.7	12.3	5.4	8.0
To prevent crime by offender against anyone	2.5	13.2	8.2	3.3	7.9	6.8	6.5
To punish offender	13.8	5.9	6.6	0.0	5.5	5.0	4.5
To catch or find offender	11.6	9.3	4.2	9.9	8.6	6.7	6.6
To improve police surveillance	4.1	3.3	2.7	8.4	9.3	5.3	7.1
Duty to notify police	4.1	7.4	4.8	5.7	6.8	4.9	7.3
Because it was a crime	12.8	11.1	14.2	14.8	15.2	19.1	20.7
Other reason	4.1	2.1	3.9	0.0	1.5	0.9	2.1

SOURCE: BJS (2006a).

SUMMARY

- The potential consequences and costs to crime victims are plenty and occur over the short and long term. These costs include economic costs as well as costs to their functioning and health.
- A small proportion of crime victims experience physical injury, and most do not receive medical care. Victims of violence, particularly gun violence, are likely to need medical assistance. Female victims, Black victims, and those victimized by a nonstranger are more likely than other victims to experience an injury.
- Beyond physical injury, victims may need mental health care. Victims often experience mental health issues such as depression, anxiety, and posttraumatic stress disorder following their victimization. Victims of sexual assault, rape, and child abuse are the most likely to seek mental health care as a direct result of being victimized. Treatment for mental health issues is yet another cost that victims face.
- There are direct economic costs to victims as well. National Crime Victimization Survey data show that more than 90% of property crimes involve some economic loss to the victim. These economic costs include direct property losses in which a victim's property is stolen or damaged. They also include expenses related to medical care. Slightly less than

1 in 10 victims of violence incur medical expenses. Victims also lose money and productivity when they are unable to work, go to school, or complete housework. Almost 20% of victims of rape and sexual assault miss 10 or more days of work. Finally, victims may experience pain, suffering, and a reduced quality of life, all of which are difficult to quantify.

- Crime and victimization create costs to the system. The United States spends more than $214 billion annually on direct expenditures to operate the criminal justice system. Other elements of the economy are also hit by crime. Insurance companies make large payouts each year due to crime.
- One startling reality is that many victims will not suffer just one victimization but will find themselves victimized again in the future—becoming recurring victims. These victims may be particularly hard hit by the costs of victimization as they accumulate over time. Research on recurring victimization shows that it is likely to recur fairly quickly and that a person is prone to experiencing the same type of victimization as previously experienced. Two explanations have been developed to explain why someone is victimized more than once: state dependence and risk heterogeneity.
- It is not just the victim him- or herself who is pained by the event. Friends and family members may also experience costs when their loved ones are harmed. This is known as secondary victimization. Homicide survivors are more likely than others to experience posttraumatic stress disorder, distress, depression, and anxiety. They may find themselves unable or unwilling to participate in ordinary activities.
- Most criminal victimizations are not reported to the police, and crime reporting varies across crime type. Robbery and aggravated assault are the most common personal victimizations reported to the police. Females, older persons, and those injured are more likely than other victims to notify the police. Incident characteristics such as use of a weapon, the offender being under the influence of alcohol and/or drugs, and the offender being a non-gang member are related to reporting. Common reasons given for reporting are to stop the incident, to prevent the offender from offending again, and because it was a crime. Nonreporting is linked to the event being considered a personal/private matter, feeling the police would not want to be bothered, and being worried about reprisal.

DISCUSSION QUESTIONS

1. We will discuss in a later section who pays for the costs of victimization and how victims can be compensated. What do you think we should do for victims? Should their medical bills be paid? What about other costs? Who should be held accountable for paying those?

2. Why do people not report their victimizations to police? What barriers to reporting exist for crime victims? What are the implications of reporting or failure to report?

3. What costs did Polly experience as a result of her victimization? What long-term consequences do you think she may have to deal with?

4. Think about your own life and try to recall a time when you were victimized. Identify all the costs that came with your victimization. What short-term and long-term costs did you experience? Did you report the incident to the police?

KEY TERMS

physical injury	self-esteem	stressor
depression	anxiety	intrusive recollection
self-worth	posttraumatic stress disorder (PTSD)	avoidance/numbing symptoms

hyperarousal

self-blame

characterological self-blame

behavioral self-blame

learned helplessness

economic costs

direct property losses

medical care costs

mental health care costs

lost productivity

system costs

recurring victimization

repeat victimization

revictimization

risk heterogeneity

state dependence

vicarious victimization

homicide survivors

reporting

INTERNET RESOURCES

Cost of Crime (http://www.ncvc.org/ncvc/main.aspx?dbName=DocumentViewer&DocumentID=38710)
This website is part of the National Center for Victims of Crime and includes recent information on the costs of crime. It talks about victim compensation and the costs of property damage due to criminal activities. The website also includes links to several victim assistance programs and resources for the victims of crime.

"Addressing Predisposition Revictimization in Cases of Violence Against Women" (http://www.ojp.usdoj.gov/nij/topics/crime/violence-against-women/workshops/revictimization.htm)
This website includes summary information on a workshop hosted by the National Institute of Justice. This workshop was conducted to examine strategies, policies, and principles in place in 2005 and to focus research on victimization in the time period of predisposition (postarrest and prior to trial and/or sentencing).

Help for Crime Victims (http://www.ojp.usdoj.gov/ovc/help/hv.htm)
The Office for Victims of Crime has collected a list of websites that lend support and encouragement to homicide survivors and covictims. There is also a list of victim assistance and compensation programs in various areas.

National Center for PTSD (http://www.ptsd.va.gov/)
This website contains information on posttraumatic stress disorder (PTSD) in relation to the U.S. Department of Veterans Affairs. The center aims to help U.S. veterans and others through research, education, and training focused on trauma and PTSD. The website also has information for providers, researchers, and the general public on PTSD and its treatment.

Introduction to Reading 1

Using cost estimates from Miller et al.'s (1996) work, described in this section, Welsh and colleagues provided cost estimates for crimes committed by a cohort of boys who were about 7 years old at the beginning of the Pittsburgh Youth Study and 17 on average at the end of the study. The boys, who were drawn from public schools in metropolitan Pittsburgh, were asked about their own offending for seven types of crime: assault, rape/sexual assault, robbery, arson, larceny, burglary, and motor vehicle theft. Based on these self-reports, the authors were able to estimate the total cost of crimes for which these boys were responsible. The cost of crimes was estimated from Miller et al.'s work, but the cost estimates were adjusted to approximate what costs would be had they occurred in the county (Allegheny County) in which Pittsburgh is located and to reflect real or inflation-adjusted dollars (in year 2000). Even though they examine only a relatively small group of boys ($n = 503$), the authors found that these boys were collectively responsible for significant victim costs. They estimated that costs ranged from $89 million to $110 million. They also found that early onset offenders were responsible for greater costs than those boys whose onset into criminality occurred later.

Costs of Juvenile Crime in Urban Areas

A Longitudinal Perspective

Brandon C. Welsh, Rolf Loeber, Bradley R. Stevens,
Magda Stouthamer-Loeber, Mark A. Cohen, and David P. Farrington

The impact on society of juvenile crime, including damage to property, pain and suffering to victims of crime, and the involvement of the police and other agencies of the juvenile justice system can be converted into monetary terms. The damaged property will need to be repaired or replaced, and it is the victim who will often have to pay for this, as many crime victims do not have insurance. The pain and suffering that is inflicted on an individual from an assault or robbery can result in short-and long-term medical care, lost wages from not being able to work, as well as reduced quality of life from debilitating injuries, fear of repeat victimization, and counseling. Here

again it is the crime victim and also the victim's family, employer, and many services, such as Medicaid, welfare, and mental health that have to incur the costs associated with these services. Then there is the cost of the involvement of the police, the courts, and correction agencies. Although some of the costs that are incurred by the juvenile justice system go toward addressing the needs of victims, such as follow-up interviews by police and court-based victim assistance programs, the majority of the costs are directed at the processing of offenders, starting with the costs of police arrest to public defender, court appearances, serving a sentence, whether it be probation or incarceration, and

SOURCE: Welsh et al. (2008). Reprinted with permission.

aftercare programs on release into the community. There are also costs incurred by society in efforts to prevent juvenile crime, through various types of prevention programs.

Putting a price tag on crime can be viewed as a politically charged enterprise. On the political right, large dollar-cost estimates of the impact of crime are interpreted as justification for more punitive crime policies, whereas on the left such cost estimates are seen as yet another reason to invest in early intervention methods to ward off the future consequences of criminal activity. One only has to recall the debate surrounding the release of the National Institute of Justice's report, *Victim Costs and Consequences: A New Look,* by Miller, Cohen, and Wiersema (1996), which estimated that crime costs the nation $450 billion a year. (The cost of the criminal justice system was not included in this total.) In a *New York Times* article on the report, Butterfield (1996) interviewed a Republican congressman who said that it "demonstrates that the cost of building prisons and adding police are justified, in terms of the cost to our society." Although some Democrats shared this view, Butterfield (1996), quoting Alfred Blumstein and other criminologists, reported that they "expressed concern that the very high estimate made it easier to justify building expensive prisons and handing out longer sentences."

Since then, debate has not waned. Cook and Ludwig's (2000) book, *Gun Violence: The Real Costs,* was assailed in one review for inflating the costs of gun violence for the express purpose of drumming up political support for advocates of gun control measures (Kleck, 2001). Other reviews praised the book as an impartial, sophisticated study to quantify the societal impact of gun violence so as to allow for—as the book's authors intended—comparisons with other social problems (McDowell, 2001; Rosenfeld, 2001).

The present study may also be viewed as supporting some political agenda, namely, getting tough on juvenile crime, but it should not be so. At the end of this article, we explore some of the potential implications of our cost estimates. But these are grounded in research rather than partisan politics. Our approach is not meant to dismiss instrumental and emotional/symbolic dimensions of responses to crime and punishment (Freiberg, 2001; Garland, 1990) but rather to call for an economics-informed dialogue on the implications of the monetized social burden of crime. The main aim of this article is to assess the monetary cost to society of juvenile offending in urban settings. The focus is on one of the main types of costs of crime to society: costs associated with being a victim.

Costing Crime

Why put a price tag on the impact of crime? Miller, Cohen, and Rossman (1993) identified three main uses. First, it facilitates combining statistics on different crimes into a single, readily understood metric of dollars and cents. This allows for a comparison of the relative harm caused by different types of crime (Cohen, 1988, 2000). A second use is that this information can provide decision makers with guidance in allocating scarce resources, and a third use is that it can be used to carry out economic analysis of programs and policies to prevent or control crime. Cost-benefit and cost-effectiveness analyses are the two most widely used techniques of economic analysis. Each provides information that allows decisions to be made on whether to continue funding certain programs or to spend money on new programs. By quantifying the cost of a robbery, for example, one is able to say that if a crime prevention program prevented 10 robberies the saving is 10 times the cost of one robbery. In the case of a cost-benefit analysis, this saving of 10 times the cost of a robbery is compared to the cost of running the program, such as employee wages and benefits and rental of office space. If benefits are larger than costs, then the program has produced value for money. This may be a very powerful argument for continuing to fund a program (or eliminate a program if costs exceed its benefits).

Types of Costs

We have already mentioned some of the types of costs of crime, but not all of them. Economics typically distinguish between three main types of costs that are caused by crime:

1. Costs that offenders impose on victims and others.

2. Costs that society incurs to prevent or control crime.

3. Costs that offenders incur (Cohen, Miller, & Rossman, 1994).

This article focuses on the first type of crime costs—those that offenders impose on victims and other—that are largely made up of costs incurred by victims of crime. Costs to victims of crime can be classified into two main categories: tangible and intangible losses. Tangible or out-of-pocket victim costs include financial losses resulting from such things as damaged or stolen property, medical expenses (covered and not covered by insurance), and lost wages from not being able to work. Intangible victim costs include reduced quality of life, pain, suffering, and fear of being victimized again. Other parties that sometimes incur the costs of crime include the victim's family, witnesses to the crime, jury members, and society in general through, for example, increased insurance premiums.

Another component of victim costs is risk of death. This is more often applicable to crimes against the person, such as rape, robbery, and aggravated assault.

The monetary value of this is calculated by multiplying risk of death probabilities for the crimes in question by the "value of a statistical life."[1] Because the value of a statistical life includes both a tangible (wages) and intangible (quality of life) component, the risk of death also includes both tangible and intangible costs.

As illustrated in Table 1, for the violent crimes, intangible victim costs far outweigh tangible victim costs, whereas for property crimes this is reversed. In the case of robbery, for example, intangible costs are almost 2.5 times tangible costs ($6,700 vs. $2,700). For the property crime of burglary, tangible costs are $1,300 and intangible costs are $350. The reason for violent crimes having higher intangible victim costs than property crimes is that there is a greater likelihood of injury from violent crimes, and it is this form of harm that contributes to costly effects of pain, suffering, and reduced quality of life.[2] As noted above, some crimes also carry a risk of death. In the case of robbery, for example, adding the cost of risk of death to the tangible and intangible victim costs increases total victim costs of a typical robbery by two thirds, from $9,400 to $15,600.[3]

What Do We Know About the Costs of Juvenile Crime?

In recent years, there has been a great deal of research on the costs of crime. This research has covered a wide range of topics, including the cost of gun violence (Cook,

Table 1	Losses per Criminal Victimization			
Crime	**Tangible**	**Intangible**	**Risk of Death**	**Total**
Rape	$6,000	$96,000	$1,000	$103,000
Robbery	$2,700	$6,700	$6,200	$15,600
Aggravated assault	$1,800	$9,200	$29,700	$40,700
Burglary	$1,300	$350	NA	$1,650
Larceny	$440	NA	NA	$440
Motor vehicle theft	$4,100	$350	NA	$4,450

SOURCE: Adapted from Cohen (1998, p. 16, Table II).

NOTE: All costs are in 1997 dollars. NA = not applicable.

Lawrence, Ludwig, & Miller, 1999; Cook & Ludwig, 2000), the cost of violence against women (Laurence & Spalter-Roth, 1996), the cost to crime victims in general (Cohen, 1988; Cohen & Miller, 1998; Macmillan, 2000; Miller et al., 1996), the cost of and willingness of the public to pay for crime prevention programs (Cohen, Rust, Steen, & Tidd, 2004; Welsh & Farrington, 2000; Welsh, Farrington, & Sherman, 2001; Witte & Witt, 2001), the cost of crime to businesses (van Dijk & Terlouw, 1996), and the aggregate cost of crime to society (Anderson, 1999; see also Atkinson, Mourato, & Healey, 2003). The costs of juvenile crime have also been a topic that has been the subject of research by criminologists and economists, and what follows are estimates of juvenile crime costs from the leading studies.

Costs of a Criminal Career

Cohen (1998) reported that the typical criminal career over the juvenile and adult years costs society around $1.3 to $1.5 million (in 1997 dollars). Also estimated were the dollar costs of the associated problem behaviors of drug use and high school dropout, bringing the total societal cost of a high-risk youth to $1.7 to $2.3 million. Cohen drew on his previously published estimates of the cost of criminal offending to victims and the criminal justice system (Cohen, 1988, 1990; Cohen et al., 1994; Miller et al., 1993, 1996), as well as published reports on criminal careers, to calculate the monetized social burden of a life of crime.

Just focusing on the juvenile years, it was estimated that a typical juvenile criminal career imposes costs on society in the range of $80,000 to $325,000, or 6% to 22% of the total costs of a criminal career. Victim costs (tangible plus intangible) related to a juvenile criminal career were found to be three times greater than criminal justice costs ($60,000–$244,000 vs. $20,000–$82,000).

Delisi and Gatling (2003), using unit cost estimates developed by Cohen (1998), reported that the average career criminal costs society more than $1.1 million (in 2002 dollars). Cost estimates were based on the self-reported criminal history, from police contact to prison sentences, of 500 adult career criminals who were processed in a large urban jail located in the Western United States between 1995 and 2000. Unlike Cohen (1998), the authors did not report on the cost of a criminal career during juvenile years.

Some Important Criminal Career Parameters for the Calculation of the Costs of Crime. The cost of crime varies much with the frequency and severity of offending by individuals. Studies agree that early onset offenders, compared to late onset offenders, have a two-to-three-times higher risk of becoming tomorrow's chronic offenders (Loeber & Farrington, 2001a). Also it is generally accepted that offending levels by individuals are higher in urban compared to suburban or rural environments.

In addition, official records of delinquent offending based on arrest or court documents constitute an underestimate of actual offending as evident from self-reports (Loeber & Farrington, 1998). Thus, it can be expected that cost estimates resulting from delinquent offending need to be adjusted upward, once self-reported delinquency (SRD) is taken into account. Such estimates are especially needed to establish the costs of offending by chronic offenders and early compared to late onset offenders.

Aggregated Costs Of Juvenile Crime for States

Miller, Fisher, and Cohen (2001) examined the costs of juvenile violence in the Commonwealth of Pennsylvania in 1993.[4] The study was based on the violent offenses of homicide, rape, robbery, assault, and child physical and sexual abuse. The study used national victim cost data (Miller et al., 1996) adjusted for state wages and prices and state juvenile and adult justice system cost data. Violence by juveniles was estimated to cost $2.6 billion in victim costs and $46 million in perpetrator costs per year (in 1993 dollars). Juvenile perpetrator costs were made up of costs to the juvenile and adult justice systems, which included costs from probation, detention, juvenile treatment programs, and incarceration in adult prisons.

The study also reported on the costs of violence against juveniles that was committed by adults and other juveniles. Much higher victim costs were associated with violence committed by juveniles: $4.5 billion versus $2.6 billion. The main reason for this difference was because of a greater incidence of sexual abuse against juveniles committed by adults.

Other states, such as Florida and Washington, have also been the subject of studies that have estimated the costs of juvenile crime, examining the impact on both the justice system and crime victims (Florida Department of Juvenile Justice, 2000; Washington State Institute for Public Policy, 2002).

Aggregated Costs of Juvenile Crime for the Nation

The only national estimate of the costs of juvenile crime focuses on violent crime. Violent crime by juveniles was estimated to cost the United States $158 billion each year (Children's Safety Network Economics and Insurance Resource Center, 2000). This estimate includes some of the costs incurred by federal, state, and local government to assist victims of juvenile violence, such as medical treatment for injuries and services for victims. These out-of-pocket victim costs of juvenile violence came to $30 billion. But the majority of the costs of juvenile violence, the remaining $128 billion, were because of losses suffered by victims, such as lost wages, pain, suffering, and reduced quality of life. Missing from this $158 billion price tag are the costs of society's response to juvenile violence, which includes early prevention programs, services for juveniles, and the juvenile justice system. These costs are largely unknown.

Findings and Analysis

Total Number of Crimes

Between the ages of 7 and 17, boys in the youngest sample of the PYS [Pittsburgh Youth Study] self-reported around 12,500 of the seven types of serious crimes under study (see Table 2). More than two thirds (69.1%) of these crimes were assaults. Larceny was the second most common crime, accounting for one quarter (25.1%) of all self-reported crimes. Only one rape or sexual assault was reported.

Aggregate Costs

The costs of crime committed by 500 male juveniles between the ages of 7 and 17 years were estimated to range from a low of $89 million to a high of $110 million (in 2000 dollars). Like all cost figures presented here, this one is restricted to those costs associated with victims of crime, such as damaged property, lost wages from time off work, and pain and suffering. It is not known how much higher the aggregate cost would have been had we been able to estimate the cost of these 500 males to the juvenile justice system. In one study it was found that victim costs—the same types that we have used here—related to a juvenile criminal career were three times greater than juvenile justice costs (Cohen, 1998).

Costs of Violent Crime

Violent juvenile crime accounted for the largest share of the aggregate cost, between 92.7% and 94.1% or around $82 million to $103 million. As illustrated in Table 3, assault accounted for the overwhelming majority of the total cost associated with victims from property and violent crime (71.2% to 76.7%) and an even greater share of

Table 2 Self-Reported Crimes by Youngest Sample of the Pittsburgh Youth Study (PYS) Between Ages 7 and 17

Crime Type	Total Number
Arson	101
Burglary	276
Larceny	3,140
Motor vehicle theft	253
Assault	8,651
Rape/sexual assault	1
Robbery	92
Total	12,514

NOTE: Cases are weighted.

Table 3 Costs of Violent Crime by 500 Male Juveniles, Ages 7 to 17, in Urban Areas

Crime	Victim Costs	% of Total Victim Costs[a]
Assault	$63,369,000 to $84,110,000	71.2% to 76.7%
Homicide	$18,209,000	20.5% to 16.6%
Rape/sexual assault	$107,000	0.1% to 0.1%
Robbery	$779,000	0.9% to 0.7%
Total	$82,464,000 to $103,205,000	92.7% to 94.1%

NOTES: Risk of death has been taken out of the individual crime estimates because actual homicides are included. All costs are in 2000 dollars, and all costs, except homicide, are in present value (2.0% discount rate).

a. Of lower and upper bound total (violent plus property) victim costs

Table 4 Costs of Property Crime by 500 Male Juveniles, Ages 7 to 17, in Urban Areas

Crime	Victim Costs	% of Total Victim Costs[a]
Arson	$4,532,000	5.1% to 4.1%
Burglary	$454,000	0.5% to 0.4%
Larceny	$445,000	0.5% to 0.4%
Motor vehicle theft	$1,072,000	1.2% to 1.0%
Total	$6,503,000	7.3% to 5.9%

NOTE: Risk of death has been taken out of the individual crime estimates because actual homicides are included. All costs are in 2000 dollars, and all costs are in present value (2.0% discount rate).

a. Of lower and upper bound total (violent plus property) victim costs.

victim costs from violent crime (76.8% to 81.5%). At just over $18 million, homicide accounted for the next largest share of the total victim cost, with robbery, at a cost of $779,000, also accounting for less than 1%.

Costs of Property Crime

As shown in Table 4, victim costs from property crimes were estimated at $6.5 million or between 5.9% and 7.3% of the total cost to crime victims. Arson was estimated to be the costliest property crime at $4.5 million, accounting for two thirds (69.7%) of victim costs from all property crimes. Arson's share of the total (property and violent) victim cost was between 4.1% and 5.1%. Motor vehicle theft was the next costliest property crime committed by male juveniles in urban areas, followed by burglary and then larceny.

Tangible and Intangible Victim Costs

As noted above, there are different types of victim costs, and in the present study it was possible to assess tangible or direct and intangible or indirect losses. At $71 million to $84 million, intangible losses to crime victims, in the form of pain, suffering, and lost quality of life, made up more than three quarters (79.4% and 76.3%, respectively) of the aggregate cost (see Table 5). But not all crimes exact intangible losses to crime victims. In the case of larceny, only tangible losses could be counted because no estimates of intangible costs exist.

As shown in Table 5, intangible losses were higher than tangible losses for the four violent crimes, whereas the opposite was true for the four property crimes. For example, intangible losses made up 70.0% of victim costs from robbery and only 19.4% of victim costs from burglary. This is not at all surprising, because intangible losses are more likely to be incurred by victims of personal rather than property crimes.

Change in the Costs of Crime by Age

An assessment of the age-specific costs of crime by a cohort may provide important insights into the timing of introduction of early prevention programs or interventions in the teenage years. Figure 1 shows the mean average total victim costs (for all crimes) per juvenile from ages 7.5 to 17 years. It is necessary, however, to

Table 5 Tangible and Intangible Costs

Crime	Tangible Costs	Intangible Costs	Total
Assault	$7,478,000 to $15,140,000	$55,891,000 to $68,970,000	$63,369,000 to $84,110,000
Homicide	$6,373,000	$11,836,000	$18,209,000
Rape/sexual assault	$7,000	$100,000	$107,000
Robbery	$234,000	$545,000	$779,000
Arson	$2,465,000	$2,067,000	$4,532,000
Burglary	$366,000	$88,000	$454,000
Larceny	$445,000	$0	$445,000
Motor vehicle theft	$995,000	$77,000	$1,072,000
Total	$18,363,000 to $26,025,000	$70,604,000 to $83,683,000	$88,967,000 to $109,708,000

NOTE: Risk of death has been taken out of the individual crime estimates because actual homicides are included. All costs are in 2000 dollars, and all costs, except homicide, are in present value (2.0% discount rate).

treat the cohort as two distinct age groups: 7.5 to 10 (younger group)and 10.5 to 27 (older group). This was done because different crime types and numbers of crime types were monetized for the two groups in the SRA [Self-Reported Antisocial Behavior Scale] compared to the SRD (see above). It is important to point out that, for the younger group; this graphic displays the midpoint between the lower and upper bound average victim cost per juvenile. It is also important to note that, for the older group, subjects aged 10.5 and 11 were asked to report on crimes they committed in the last 6 months; whereas subjects aged 12 to 17 were asked about crimes they committed in the last year.

As shown in Figure 1, the average victim costs per juvenile were quite stable through age 10, ranging from a high of $9,000 (age 9) to a low of $7,000 (ages 9.5 and 10). This was not the case for the older group. At ages 10.5 and 11, costs totaled $199,000 and $15,000, respectively—for a total cost of $34,000 over the 1-year period. Subsequently, moving to the annual survey at age 12, costs were reduced to $21,000, increasing slightly through ages 14 and 15 ($24,000 each). However, costs decreased considerably at ages 16 ($12,000) and 17 ($15,000). As we noted earlier, one cannot compare total costs through age 11 to those

from ages 12 to 17 because of changes in the survey instrument. However, we could compare the two separate time periods to determine whether there are significant differences within each sample. This was done using panel regression models, where the dependent variable was total costs, and we controlled for each youth and each age. We find a statistically significant increase in costs for the first two time periods for the older group—ages 10.5 and 11. This is because of a significant increase in the number of assaults in the 10.5 and 11-year-old categories. We also find a statistically significant reduction in costs for ages 16 and 17. This drop in costs in the later years was caused by a substantial decrease in the number of assaults. Neither the increase in assaults at ages 10.5 and 11, nor the decrease in assaults at ages 16 and 17, appear to be the result of a few outliers. Instead, they appear to be overall trends in the data.

Costs of Early Versus Late Onset Offending

Past research has shown that early onset offenders have higher rates of serious offending than late onset offenders (Loeber & Farrington, 2001b). It stands to reason that these early starters should impose a greater

Figure 1 Costs of Crime Between Ages 4 and 17 Years

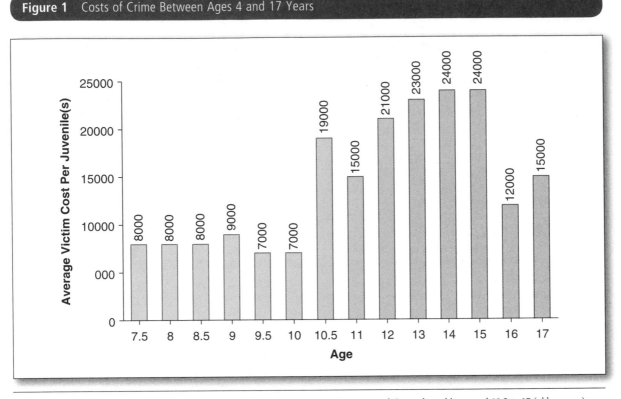

NOTE: Boys aged 7.5 to 10 (younger group) were administered the SRA questionnaire at 6-month intervals, and boys aged 10.5 to 17 (older group) were administered the SRD questionnaire at 12-month intervals (except at ages 10.5 and 11). All costs are in 2000 dollars.

financial burden on society over time. To investigate whether the costs of juvenile crime to victims are greater for early onset compared to late onset offenders, independent sample t-tests were used to assess between-group differences for average number of crimes per offender and average victim costs per offender. The early onset group includes subjects under age 13 (ages 10.5, 11, and 12) who committed at least one crime during this period. The late-onset group includes subjects aged 13 or older (ages 14–17) who committed at least one crime during this period only. Analyses were limited to these two age groups because they were both administered the SRD questionnaire.

It was found that subjects who began offending before age 13 had a significantly higher rate of serious offending over their juvenile years than those who began offending at age 13 or later (34.2 compared to 15.6 crimes per offender). As shown in Table 6, this translates into a significant difference in victim costs, with early starters causing, on average, $139,000 more than late starters.

Costs of Chronic Offending

Beginning with the Philadelphia Birth Cohort study (Wolfgang, Figlio, & Sellin, 1972), many prospective longitudinal studies of juvenile delinquency and later offending have found that a small number of offenders account for a substantial proportion if not a majority of offense (Loeber, Farrington, & Waschbush, 1998). In the present research this was no different, with 10.2% ($n = 34$) of the offending sample accounting for half (50.1%) of all self-reported offenses. Compared to the remaining offenders ($n = 298$), the chronics reported having committed, on average, nine times as many crimes (see Table 7).

Table 6 Costs of Early Versus Late Onset Offending

	Early Onset Offenders (*n* = 199)	Late Onset Offenders (*n* = 83)
Average number of crimes per offender	34.2	15.6
Average victim costs per offender	$224,000	$85,000

NOTE: All costs are in present value (2.0% discount rate) and in 2000 dollars. Both between-group differences are significant (*p* < .000).

Table 7 Costs of Chronic Offending

	Chronic Offenders (*n* = 34)	Other Offenders (*n* = 298)
Average number of crimes per offender	142.0	16.6
Average victim costs per offender	$793,000–$861,000	$101,000–$147,000

NOTE: All costs are in present value (2.0% discount rate) and in 2000 dollars. Both between-group differences are significant (*p* < .000).

Independent sample *t*-tests to assess the between-group difference for average victim costs per offender reveal that juvenile chronic offenders caused significantly higher average victim costs than the other group of offenders (see Table 7). At $793,000 to $861,000, the average victim costs per chronic offender were more than five to eight times higher than the average victim cost per nonchronic offender.

✕ Discussion and Conclusion

This study found that a typical cohort of 500 boys in an urban area, beginning in childhood through late adolescence, caused a substantial burden of harm to society in the form of victimization costs. Conservatively estimated, this harm ranged from a low of $89 million to a high of $110 million, with more than three quarters of this total resulting from crime victims' pain, suffering, and lost quality of life. Violent juvenile crime accounted for the largest share of the aggregate cost, with the bulk of these costs associated with assault, which is relatively less serious compared to homicide and robbery. It was also found that from an early age the cohort was responsible for substantial crime victim losses (expressed as average victim costs per juvenile), with these losses mounting in the teen years. Early onset offenders, compared to late onset offenders, had a significantly higher rate of serious offending, and this translated into a significant difference in crime victim costs. Also, chronic offenders—those accounting for half of all self-reported offenses—caused five-to-eight-times higher average victim costs than other offenders.

High Crime Costs Do Not Themselves Suggest a Policy Solution

We have identified the high costs associated with juvenile crime in our sample. However, the fact that juvenile offenders impose high costs on victims and society does not necessarily suggest a policy solution. Before settling on a policy recommendation, analysts need to assess both the costs of juvenile crime and the benefits of programs designed to reduce or mitigate its effects. Thus, without further information on the effectiveness of various policy alternatives, one should not use our findings to suggest that we must either "get tougher" on juvenile offenders or focus our attention exclusively on early prevention programs, or do both.

On the prevention side, there are already many direct research-based arguments that support the need for more spending on early prevention, such as results of meta-analyses and systematic reviews, cost-benefit analyses, and public opinion surveys (Cullen, Vose, Lero Jonson, & Unnever, 2007; Farrington & Welsh, 2003, 2007; Lösel & Beelmann, 2003; Welsh, 2003; Welsh & Farrington, 2000). On the punishment side, there is also some evidence that a movement toward harsher punishment in the United States reduced juvenile crime (Levitt, 1998).[5] Yet there is also growing evidence that alternatives to incarceration for some offenders might be more cost-beneficial than more punitive sanctions (see, e.g., Aos, Lieb, Mayfield, Miller, & Pennucci, 2004). Thus, one of the key policy issues to be considered in the face of scarce resources is the appropriate mix of prevention, punishment, and treatment.

This matter deserves an economic perspective. Miller et al. (2001), in discussing the policy implication arising from their findings on the costs of juvenile violence in Pennsylvania, noted that "Determining the appropriate balance among care of victims, perpetrators, and prevention of future victimizations requires exploring both unmet needs and public priorities in the face of scarce resources" (p. 7). By this the authors mean that any response to the high costs of crime needs to be driven by what society is (and is not) doing, how the public views the problem alongside other competing priorities, and what the government can afford relative to current expenditures in other areas. This is by no means an easy task, but it calls attention to the need for a systematic, knowledge-based approach in proposing a response to the high costs of juvenile crime.

In the absence of being able to report on original research that addresses each one of the factors raised by Miller et al. (2001), there is something to be said about the current state of the nation's response to juvenile crime that provides us with a starting point for thinking about implications for public policy arising from the findings reported here. In short, the high costs of juvenile crime need to be placed in context.

For many years, certainly during the follow-up of the cohort in this study and up to the present, the response to juvenile crime in this country has been increasingly punitive (McCord, Widom, & Crowell, 2001). This has involved juvenile courts delivering harsher dispositions, more juvenile offenders being transferred to adult court, a greater reliance on the use of confinement than treatment (Howell, 1997, 2003), and a growing number of juvenile offenders serving time in secure facilities. According to the Office of Juvenile Justice and Delinquency Prevention's Census of Juveniles in Residential Placement, the placement of juvenile offenders in juvenile correctional facilities grew by 43% between 1991 and 1999, from 76,000 to 109,000 (Sickmund, 2004). This overall increased punitiveness has also led many scholars to assert that the treatment and protection aims of the juvenile justice system have become more a matter of theory than of reality (Feld, 1998; Hagan & Foster, 2001).

At the same time, it cannot be said that this increased punitiveness has taken place amid an effort to strike a greater balance between prevention and repression (Vila, 1997; Welsh, 2005). In fact, just the opposite has occurred. Federal, state, and local resources for early prevention programs have been cut back, in many cases to help pay for the growing costs of these punitive responses (Butterfield, 2003b). Interestingly, at the same time, state and city governments faced with budget crises have been forced to reduce spending on some of the most costly punitive responses to adult and juvenile crime (Butterfield, 2003a, 2003c), some of which have been shown through cost-benefit analyses to be economically inefficient (Fass & Pi, 2002). In 2001, federal, state, and local government expenditures on the criminal justice system reached $167 billion, a 165% increase over 1982 expenditures (in inflation-adjusted dollars; Bauer & Owens, 2004).

Against the backdrop of these two trends—increasing punitiveness toward juvenile crime and declining resources for early prevention programs—some may view the high costs of juvenile crime in urban areas as a reason to allocate more public resources to prevention and intervention services. It may also be the case that one of the unmet needs in society's response to juvenile crime is to spend more on early prevention programs. There is evidence that the public has taken this view as well, because there appears to be growing demand for early prevention programs and little demand for increased use of incarceration (Cohen, Rust, & Steen, 2006).

What is required is a more balanced response. This balance is about improving the juvenile justice response to those that have already come in conflict with the law and expanding the role of early prevention and intervention service for children, teens, and families in greatest need.

The Need for Economic Evaluation Research

If there is indeed an interest in how best to reduce the high costs of juvenile crime, then it is also important to consider what are the most worthwhile or economically efficient measures that can be taken. This calls for economic evaluation research (e.g., cost-benefit analysis, cost-effectiveness analysis), which will aid in decisions on how best to use scarce public resources.

Although the number of cost-benefit analyses of early prevention and youth development programs is somewhat limited at present, one of the consistent findings to emerge from these studies is that these programs need only produce a modest level of crime reduction to fully pay back program costs and produce a dividend for government and crime victims (Aos, Phipps, Barnoski, & Lieb, 2001; Welsh & Farrington, 2000). This is particularly true for programs that are targeted at high-risk youth as opposed to the general population. This is largely explained by the large number of offenses committed (when a more realistic self-report estimate is used), the high unit cost of criminal offenses, and the relatively low cost per participant of running many of these programs. For example, in the case of the well-known Seattle Social Development Project (Hawkins, Catalano, Kosterman, Abbott, & Hill, 1999), which included modified classroom teaching practices, parent training, and child social skills training, Aos et al. (2001) estimated that it produced $4.25 in benefits for every dollar of cost ($18,524 in benefits divided by $4,355 in costs per program participant; in 2000 dollars), and for the program to break even with taxpayers (or government) and crime victims it needed to reduce crime by no more than 6%.[6]

Similar findings were obtained in Cohen's (1998) comparison of his cost estimates of a criminal career to Greenwood, Model, Rydell, and Chiesa's (1996) cost-effectiveness estimates of California's three strikes law and a number of alternative (nonpunitive) interventions to reduce crime. To be cost-beneficial, the following success rates were needed: 2% to 3% for home visit/day care, 3% to 5% for graduation incentives, and about 1 in 1,000 for delinquent supervision (Cohen, 1998, p. 30). Of course, the cost savings from preventing a criminal career are potentially far greater than from preventing a handful of offenses across a number of program participants. Substituting our cost estimates of the burden of harm caused to victims by chronic juvenile offenders (see Table 7) requires decreasing only marginally the success rates of these interventions (by about 1.8 times). This means that these nonpunitive interventions need only produce a modest level of crime reduction to pay back program costs and produce a dividend for society.

One of the first-known studies to investigate the cost-effectiveness of interventions for chronic or high-rate juvenile offenders was carried out by Rydell (1986). Two different strategies were compared: early developmental intervention (e.g., day care enrichment, parent training) and selective incapacitation (longer custodial sentences). On the basis of the finding that the predictability of high-rate offenders is between one third and one half, Rydell estimated that selective incapacitation could reduce crime by 5% to 7%. He concluded that

> To achieve that same reduction in crime, an early intervention program must reduce offense rates of treated offenders by 37% to 42%. Moreover, this analysis finds that the early intervention program can spend from $28,000 to $32,000 per person (total cost however long the treatment takes) and still cost no more than the selective incapacitation program. (Rydell, 1986, pp. 236, 238)

Another consistent finding to emerge from the literature on cost-benefit analyses of early prevention and youth development programs is that these programs provide important monetary benefits beyond reduced crime (Aos et al., 2004; Welsh & Farrington, 2000). These benefits can take the form of, for example, increased tax revenue from higher earnings, savings from reduced usage of social services, and savings from less health care utilization. In many cases, these nonprime benefits can account for a substantial portion of a prevention program's total benefits. In the case of the well-known Elmira (New York) nurse home visitation program (Olds et al., 1998), an independent cost-benefit analysis by Karoly et al. (1998) found that savings to the criminal justice system accounted for just 20% of total benefits, whereas reduction in welfare costs (57%), reduction in health care services (less than 1%), and tax revenue from increased employment (23%) made up the other benefits.

More cost-benefit and cost-effectiveness analyses need to be carried out to assess the independent and comparative value of early crime prevention, youth development, and juvenile justice programs. Research on the costs of juvenile crime should also be initiated on many fronts, including developing estimates of the costs that juvenile offending presents to the juvenile and criminal justice systems, testing our findings using other longitudinal surveys of the development of juvenile offending, and investigating the costs of female juvenile crime in urban areas and other settings.

The present study will not settle the debate on what the high costs of juvenile crime mean for public policy in this country. It does, however, offer to contribute to the knowledgebase on the monetary costs of juvenile crime, and through its focus on a real-life cohort covering the most crime-prone years offers a new look into how best we should be allocating scarce resources to achieve a safer, more sustainable society in the years to come.

Notes

1. This "should *not* [italics in original] be interpreted as the value of any one particular but instead is society's value of saving a 'statistical' life" (Cohen, 2001, p. 37).

2. Intangible costs are not actual out-of-pocket losses that people pay in dollars. Instead, they are the estimated monetary equivalent of pain, suffering, and lost quality of life.

3. The risk of death cost is based on the value of a statistical life of $3.4 million (in 1997 dollars; Cohen, 1998).

4. This was based on a larger report by the Joint State Government Commission (1995) that was prepared for the state's Task Force to Study the Issues Surrounding Violence as a Public Health Concern.

5. Levitt (1998) found that juvenile crime is responsive to harsher sanctions imposed by the criminal justice system.

6. This is based on Aos et al.'s (2001, p. 135) estimate that for the program to break even with taxpayers alone it needed to reduce crime by 27.6%. Program benefits to taxpayers ($3,898 per program participant), which were limited to savings to the criminal justice system, failed to cover the cost of running the program. This resulted in a benefit-to-cost ratio of 0.90. But when taxpayer benefits are added to crime victim benefits ($14,626 per program participant), for a benefit-to-cost ratio of 4.25, the percentage reduction in crime to break even for both parties is reduced by a factor of 4.72 times, from 27.6% to 5.85%.

References

Anderson, D. A. (1999). The aggregate burden of crime. *Journal of Law and Economics, 42*, 611–642.

Aos, S., Lieb, R., Mayfield, J., Miller, M., & Pennucci, A. (2004). *Benefits and costs of prevention and early intervention programs for youth.* Olympia: Washington State Institute for Public Policy.

Aos, S., Phipps, P., Barnoski, R., & Lieb, R. (2001). *The comparative costs and benefits of programs to reduce crime.* Version 4.0. Olympia: Washington State Institute for Public Policy.

Atkinson, G., Mourato, S., & Healey, A. (2003). The costs of violent crime. *World Economics, 4*, 79–94.

Bauer, L., & Owens, S. D. (2004). *Justice expenditure and employment in the United States, 2001.* Washington, DC: Bureau of Justice Statistics, U.S. Department of Justice.

Butterfield, F. (1996, April 22). Survey finds that crimes cost $450 billion a year. *The New York Times,* p. A8.

Butterfield, F. (2003a, June 7). As budgets shrink, cities see an impact on criminal justice. *The New York Times.* Available from http://www.nytimes.com

Butterfield, F. (2003b, March 9). Proposed White House budget cuts imperil a lifeline for troubled Oregon teenagers. *The New York Times,* p. A20.

Butterfield, F. (2003c, November 10). With cash tight, states reassess long jail terms. *The New York Times.* Available from http://www.nytimes.com

Children's Safety Network Economics and Insurance Resource Center. (2000). *State costs of violence perpetrated by youth.* (Updated on July 12, 2000). Retrieved August 30, 2002, from http://www.csneirc.org/pubs/tables/youth-viol.him

Cohen, M. A. (1988). Some new evidence on the seriousness of crime. *Criminology, 26*, 343–353.

Cohen, M. A. (1990). A note on the cost of crime to victims. *Urban Studies, 27*, 125–132.

Cohen, M. A. (1998). The monetary value of saving a high-risk youth. *Journal of Quantitative Criminology, 14*, 5–33.

Cohen, M. A. (2000). Measuring the costs and benefits of crime and justice. In G. LaFree (Ed.), *Measurement and analysis of crime and justice: Criminal justice 2000* (Vol. 4, pp. 263–315). Washington, DC: National Institute of Justice, U.S. Department of Justice.

Cohen, M. A. (2001). The crime victim's perspective in cost-benefit analysis: The importance of monetizing tangible and intangible crime costs. In B. C. Welsh, D. P. Farrington, & L. W. Sherman (Eds.), *Costs and benefits of preventing crime* (pp. 23–50). Boulder, CO: Westview.

Cohen, M. A., & Miller, T. R. (1998). The cost of mental health care for victims of crime. *Journal of Interpersonal Violence, 13,* 93–110.

Cohen, M. A., Miller, T. R., & Rossman, S. B. (1994). The costs and consequences of violent behavior in the United States. In A. J. Reiss, Jr., & J. A. Roth (Eds.), *Consequences and control: Understanding and preventing violence* (Vol. 4, pp. 67–166). Washington, DC: National Academy Press.

Cohen, M. A., Rust, R. A., & Steen, S. (2006). Prevention, crime control or cash? Public preferences toward criminal justice spending priorities. *Justice Quarterly, 23,* 317–335.

Cohen, M. A., Rust, R. A., Steen, S., & Tidd, S. T. (2004). Willingness-to-pay for crime control programs. *Criminology, 42,* 89–109.

Cook, P. J., Lawrence, B. A., Ludwig, J., & Miller, T. R. (1999). The medical costs of gunshot injuries in the United States. *Journal of the American Medical Association, 282,* 447–454.

Cook, P. J., & Ludwig, J. (2000). *Gun violence: The real costs.* New York: Oxford University Press.

Cullen, F. T., Vose, B. A., Lero Jonson, C. N., & Unnever, J. D. (2007). Public support for early intervention: Is child saving a "habit of the heart"? *Victims and Offenders, 2,* 109–124.

Delisi, M., & Gatling, J. M. (2003). Who pays for a life of crime? An empirical assessment of the assorted victimization costs posed by career criminals. *Criminal Justice Studies, 16,* 283–293.

Farrington, D. P., & Welsh, B. C. (2003). Family-based prevention of offending: A meta-analysis. *Australian and New Zealand Journal of Criminology, 36,* 127–151.

Farrington, D. P., & Welsh, B. C. (2007). Saving children from a life of crime: Early risk factors and effective interventions. New York: Oxford University Press.

Fass, S. M., & Pi, C. R. (2002). Getting tough on juvenile crime: An analysis of costs and benefits. *Journal of Research on Crime and Delinquency, 39,* 363–399.

Feld, B. C. (1998). Juvenile and criminal justice systems responses to youth violence. In M. Tonry & M. H. Moore (Eds.), *Youth violence. Crime and justice: A review of research* (Vol. 24, pp. 189–261). Chicago: University of Chicago Press.

Florida Department of Juvenile Justice. (2000). *The fiscal impact of reducing juvenile crime* (Management Rep No. 2000-12). Tallahassee: Bureau of Data and Research, Author.

Freiberg, A. (2001). Affective versus effective justice: Instrumentalism and emotionalism in criminal justice. *Punishment and Society, 3,* 265–278.

Garland, D. (1990). *Punishment and modern society: A study in social theory.* Chicago: University of Chicago Press.

Greenwood, P. W., Model, K. E., Rydell, C. P., & Chiesa, J. (1996). *Diverting children from a life of crime: Measuring costs and benefits.* Santa Monica, CA: RAND.

Hagan, J., & Foster, H. (2001). Youth violence and the end of adolescence. *American Sociological Review, 66,* 874–899.

Hawkins, J. D., Catalano, R. F., Kosterman, R., Abbott, R., & Hill, K. G. (1999). Preventing adolescent health-risk behaviors by strengthening protection during childhood. *Archives of Pediatrics and Adolescent Medicine, 153,* 226–234.

Howell, J. C. (1997). *Juvenile justice and youth violence.* Thousand Oaks, CA: Sage.

Howell, J. C. (2003). *Preventing and reducing juvenile delinquency: A comprehensive framework.* Thousand Oaks, CA: Sage.

Joint State Government Commission, General Assembly of the Commonwealth of Pennsylvania. (1995). *The cost of juvenile violence in Pennsylvania.* Staff report to the Task Force to Study the Issues Surrounding Violence as a Public Health Concern. Harrisburg, PA: Author.

Karoly, L. A., Greenwood, P. W., Everingham, S. S., Hoube, J., Kilburn, M. R., Rydell, C. P., et al. (1998). *Investing in our children: What we know and don't know about the costs and benefits of early childhood interventions.* Santa Monica, CA: RAND.

Kleck, G. (2001). Review of "Gun violence: The real costs." *Criminal Law Bulletin, 37,* 544–547.

Laurence, L., & Spalter-Roth, R. (1996). *Measuring the costs of domestic violence against women and the cost-effectiveness of interventions: An initial assessment and proposals for further research.* Final Report to the Rockefeller Foundation. Washington, DC: Institute for Women's Policy Research.

Levitt, S. D. (1998). Juvenile crime and punishment. *Journal of Political Economy, 106,* 1156–1185.

Loeber, R., & Farrington, D. P. (Eds.). (1998). *Serious and violent juvenile offenders: Risk factors and successful interventions.* Thousand Oaks, CA: Sage.

Loeber, R., & Farrington, D. P. (Eds.). (2001a). *Child delinquents: Development, intervention, and service needs.* Thousand Oaks, CA: Sage.

Loeber, R., & Farrington, D. P. (Eds.). (2001b). The significance of child delinquency. In R. Loeber & D. P. Farrington (Eds.), *Child delinquents: Development, intervention, and service needs* (pp. 1–22). Thousand Oaks, CA: Sage.

Loeber, R., Farrington, D. P., & Waschbush, D. A. (1998). Serious and violent juvenile offenders. In R. Loeber & D. P. Farrington (Eds.), *Serious and violent juvenile offenders: Risk factors and successful interventions* (pp. 13–29). Thousand Oaks, CA: Sage.

Lösel, F., & Beelmann, A. (2003). Effects of child skills training in preventing antisocial behavior: A systematic review of randomized evaluations. *Annals of the American Academy of Political and Social Science, 587,* 84–109.

Macmillan, R. (2000). Adolescent victimization and income deficits in adulthood. Rethinking the costs of criminal violence from a life-course perspective. *Criminology, 37,* 553–588.

McCord, J., Widom, C. S., & Crowell, N. A. (Eds.). (2001). *Juvenile crime, juvenile justice.* Panel on juvenile crime: Prevention, treatment and control. Washington, DC: National Academy Press.

McDowell, D. (2001). Review of "Gun violence: The real costs." *New England Journal of Medicine, 344,* 1484–1485.

Miller, T. R., Cohen, M. A., & Rossman, S. B. (1993). Victim costs of violent crime and resulting injuries. *Health Affairs, 12,* 187–197.

Miller, T. R., Cohen, M. A., & Wiersema, B. (1996). *Victim costs and consequences: A new look.* Washington, DC: National Institute of Justice, U.S. Department of Justice.

Miller, T. R., Fisher, D. A., & Cohen, M. A. (2001). Costs of juvenile violence: Policy implications. *Pediatrics, 107*(1). Retrieved March 17, 2002, from http://www.pediatrics.org/cgi/content/full/107/1/e3

Olds, D. L., Henderson, C. R., Cole, R., Eckenrode, J., Kitzman, H., Luckey, D., et al. (1998). Long-term effects of nurse home visitation in children's criminal and antisocial behavior. 15-year follow-up of a randomized controlled trail. *Journal of the American Medical Association, 280,* 1238–1244.

Rosenfeld, R. (2001). Review of "Gun violence. The real costs." *Journal of the American Medical Association, 286,* 605–607.

Rydell, C. P. (1986). The economics of early intervention versus later incarceration. In P. W. Greenwood (Ed.), *Intervention strategies for chronic juvenile offenders: Some new perspectives* (pp. 235–258). Westport, CT: Greenwood.

Sickmund, M. (2004). *Juveniles in corrections.* Washington, DC: Office of Juvenile Justice and Delinquency Prevention, U.S. Department of Justice.

van Dijk, J. J. M., & Terlouw, G. J. (1996). An international perspective of the business community as victims of fraud and crime. *Security Journal, 7,* 157–167.

Vila, B. (1997). Human nature and crime control: Improving the feasibility of nurturant strategies. *Politics and the Life Sciences, 16,* 3–21.

Washington State Institute for Public Policy. (2002). *The juvenile justice system in Washington State: Recommendations to improve cost-effectiveness.* Olympia, WA: Author.

Welsh, B. C. (2003). Economic costs and benefits of primary prevention of delinquency and later offending: A review of the research. In D. P. Farrington & J. W. Coid (Eds.), *Early prevention of adult antisocial behavior* (pp. 318–355). Cambridge, UK: Cambridge University Press.

Welsh, B. C. (2005). Public health and the prevention of juvenile criminal violence. *Youth Violence and Juvenile Justice, 3,* 23–40.

Welsh, B. C., & Farrington, D. P. (2000). Monetary costs and benefits of crime prevention programs. In M. Tonry (Ed.), *Crime and justice: A review of research* (Vol. 27, 305–361). Chicago: University of Chicago Press.

Welsh, B. C., Farrington, D. P., & Sherman, L. W. (Eds.). (2001). *Costs and benefits of preventing crime.* Boulder, CO: Westview.

Witte, A. D., & Witt, R. (2001). *What we spend and what we get: Public and private provision of crime prevention and criminal justice* (Working Paper No. 8204). Cambridge, MA: National Bureau of Economic Research.

Wolfgang, M. E., Figlio, R. M., & Sellin, T. (1972). *Delinquency in a birth cohort.* Chicago: University of Chicago Press.

DISCUSSION QUESTIONS

1. Given the high cost of crimes committed by juveniles, what should the response be by the juvenile justice system? Is there a cost benefit to investing in early intervention and prevention rather than increased punitiveness toward juveniles?

2. If you were the head of the department of juvenile justice in your state, how would you decide to direct monies in light of the costs of crime?

3. Why is it important to estimate the costs of crime for victims?

◈

Introduction to Reading 2

The recurrence of victimization of college women is the focus of Daigle, Fisher, and Cullen's (2008) paper. They use two data sources—the National College Women Sexual Victimization Study and the National College Women Violent Victimization Study—to provide estimates and descriptions of recurring and repeat victimization. Remember, repeat victimizations occur when a person, in this case a college woman, experiences more than one of the same type of victimization. The college women in both studies were asked in the spring of 1997 about victimization experiences that occurred since school began in the fall of 1996; thus, women answered questions about incidents that occurred, on average, over the course of 7 months.

The Violent and Sexual Victimization of College Women

Is Repeat Victimization a Problem?

Leah E. Daigle, Bonnie S. Fisher, and Francis T. Cullen

Although extensive research has documented that college women are at high risk of sexual assault, little is known about their *repeat* violent and sexual victimization or about how to prevent such incidents (see Gidycz, Coble, Lantham, & Layman, 1993; Gidycz, Hanson, & Layman, 1995). Studies have shown the impact of child and/or adolescent sexual abuse on revictimization in adulthood (see Breitenbecher, 2001), but these investigations have largely ignored repeat victimization within the single development period of early adulthood—a high-risk period for women. Furthermore, the likelihood of repeat rape and physical assault during adulthood by an intimate partner has been well documented (see Cattaneo & Goodman, 2005). This research, however, does not explicitly focus on different types of sexual (e.g., rape, coercion) and violent victimization (e.g., simple assault, robbery), potentially committed by different perpetrators, over a specific period in women's lives. Davis, Combs-Lane, and Jackson (2002) suggested as well that to understand the prevention of victimization, researchers should broaden their investigations to assess multiple incidents of sexual and physical assault that may occur within developmental periods.

Using two national-level samples of more than 8,000 female students, the current study makes four contributions to understanding college women's victimization. First, to address the existing gap in the extant literature, we assess the extent to which college women have experienced different types of repeat violent and sexual victimization during an academic year.

Second, we provide descriptive information on the *time course* for repeat sexual and violent victimization incidents. Studies of repeat property victimization have revealed that following an initial incident, subsequent victimization tends to recur quickly; there appears to be a delimited period of heightened risk for repeat victimization, which then decreases and eventually levels off (Farrell, 1995). We examine whether this time-course pattern generalizes to college women's repeat violent or sexual victimization incidents and also explore the implications of the findings for more effective campus prevention programs. Furthermore, researchers have explored whether victims are prone to experience victimizations of the same type; however, these analyses have not been conducted on sexual victimization incidents for college women. We examine a *crime-switch* matrix to depict the sequential pattern of sexual victimization incidents.

Third, we explore the preincident, situational, and postincident *characteristics of repeat incidents* and compare them to the characteristics of single incidents. Prior research has suggested factors of potential relevance. Thus, two preincident factors associated with victimization risk for women are knowing the perpetrator (Fisher, Cullen, & Daigle, 2005) and alcohol consumption (Abbey, Zawacki, Buck, Clinton, & McAuslan, 2004). Studies have also identified high-risk situations for sexual assault such as being in isolated locations (e.g., one's living quarters). The use of self-protective action is another situational factor that merits consideration; researchers have consistently reported that self-protective action can effectively thwart an attack (see Ullman, 1997).

SOURCE: Daigle et al. (2008). Reprinted with permission.

Fourth, research has revealed that women who are sexually assaulted generally do not report their victimization to the police or campus officials but do tend to tell other people, especially friends (Fisher, Daigle, Cullen, & Turner, 2003). Reporting, likely a key factor in preventing future victimization, has not been considered in the repeat victimization field. We also address this issue.

Results

The Extent of Repeat Victimization

Table 1 shows the proportion of college women who were victimized, how many times they were victimized, and the proportion of incidents that happened to these women.

The results show that a small proportion of women experienced a large percentage of all types of violent and sexual incidents during the academic year. Less than 1% of the women who experienced two or more violent incidents experienced 27.7% of all the violent incidents. More than 7.0% of the women experienced nearly three fourths (72.41%) of all the sexual victimization incidents. Noteworthy is that the most sexually

victimized, those 3.3% who experienced three or more sexual victimizations, experienced almost half (45.2%) of all sexual incidents.

Table 2 reports the rates of repeat victimization broken down by types of violent and sexual victimization. As shown, a much larger proportion of women experienced more than one sexual incident compared to those who experienced more than one violent incident. Close to half of the women, 47.3%, had been sexually victimized more than once by either the same type of sexual victimization or more than one type of sexual victimization since school began in the fall. A significantly smaller yet still substantial proportion of women, 14.4%, were violently victimized more than once by either the same type of violence or more than one type of violence during this time ($z = 8.653, p < .001$, two-tailed).

Repeat Violent Victimization. Also shown in Table 2 nearly 14% of the simple assault victims were repeat assault victims, and they experienced close to 28% of all the simple assaults. None of the robbery or aggravated assault victims were repeat victims.

Repeat Sexual Victimization. Three noteworthy results are evident in Table 2 regarding repeat sexual

---

Table 1 Proportion of College Women and Incidents by Number of Times Victimized

Number of Times Victimized	Type of Victimization					
	Violent			Sexual		
	College Women[a]	Incidents[b]		College Women[c]	Incidents[d]	
	%	%	*n*	%	%	*n*
0	95.1	0.0		84.5	0.0	
1	4.2	72.3	185	8.2	27.6	364
2	0.56	19.5	50	4.0	27.2	368
3 or more	0.14	8.2	21	3.3	45.2	596

a. *n* = 4,432
b. *n* = 256
c. *n* = 4,446
d. *n* = 1,318

Table 2   Rates of Violent and Sexual Repeat Victimization

Type of Victimization	Percentage of Victims	Number of Incidents	Percentage of Incidents	Percentage of Repeat Victims	Number of Repeat Incidents	Percentage of Repeat Incidents	Percentage of Incidents Experienced by Repeat Victims
Violent							
Robbery	0.38	17	17	6.6	0.0	0	0	0.0	0.0
Simple assault	3.5	153	183	71.5	13.7	21	30	16.4	27.9
Aggravated assault	1.3	56	56	21.9	0.0	0	0	0.0	0.0
Total	4.9	216	256	100.0	14.4	31	40	15.6	27.7
Sexual							
Rape	2.8	123	157	11.9	22.8	28	34	21.7	39.5
Sexual coercion	3.7	164	221	16.8	23.2	38	57	25.8	43.0
Unwanted sexual contact with force	5.0	221	296	22.5	22.6	50	75	25.3	42.2
Unwanted sexual contact without force	7.2	318	427	32.4	25.5	81	109	25.5	44.3
Threats	3.4	152	217	16.5	25.0	38	69	31.8	47.5
Total	15.5	691	1,318	100.0	47.3	327	627	47.6	72.4

victimization. First, repeat sexual victimization was common. From 22.8% (rape, sexual coercion, and unwanted sexual contact with force, respectively) to 25.5% (unwanted sexual contact without force) of the women were repeat victims. The percentage of repeat incidents was also striking. For example, nearly 22% of rapes were repeat rape incidents. Similarly, just more than 25% of sexual coercions and unwanted sexual contact with and without force were repeat incidents. Even more of the threats, 31.8% were repeat incidents.

Second, within *each* type of sexual victimization, a disproportionately small percentage of the victims experienced a large proportion of the incidents. For example, slightly less than one-fourth of the repeated rape victims experienced close to 40% of the rapes. Just more than 25% of the victims of unwanted sexual contact without force experienced 44.3% of these

incidents. A quarter of the repeat threat victims experienced 47.5% of the threats.

Third, women were more likely to repeatedly experience any type of sexual victimization compared to any type of violence. A significantly larger percentage of women experienced repeated rape (22.8%), sexual coercion (23.2%), unwanted sexual contact with (22.6%) or without force (25.5%), or threats (25.0%) compared to those who experienced repeated simple assault (13.7%; $p < .001$ for all pairs of two-tailed proportion test comparisons).

The Nature of Repeat Victimization

The Time Course of Repeat Victimization. Figure 1 presents the time course of repeated simple assault and each type of sexual victimization.[1] For each pair of repeated type of victimization, the difference in the number of months between the most recent incident and the next most recent incident was calculated. For example, if the most recent rape occurred in January and the rape before this one happened in November, then the number of months between the two incidents was 2 months. The number of months between paired repeat incidents that happened in the same month was 0.

As can be seen in Figure 1, there is an elevated risk of repeat violence or any type of repeat sexual victimization in a short time. In particular, this elevated risk is greatest within the same month. For example, 49% of the repeat rapes happened within the same month, as did 36% of the sexual coercions, 32% of the threats, 31% of physical assaults, and 28% of the unwanted sexual contacts with force.

The only exception to the increased risk of being victimized a second time within the same month was for unwanted sexual contact without force. The risk of unwanted sexual contact without force happening a second time was highest a month after the first incident. Of the repeat unwanted sexual contact without force incidents, 39% occurred in the following month, which is greater than the 25% that occurred within the same month.

As can also be seen in Figure 1, the risk of repeat sexual and violent victimization, with the exception of rape,[2] steadily declined over the passage of time when looking at the proportion of repeat incidents having had occurred 1, 2, 3, 4, 5, or 6 months apart. For example, 28% of the repeated unwanted sexual contacts with force incidents happened within the same month, compared to 23% that happened within 1 month; 20% within 2 months; 10% within 3 and 4 months, respectively; 7% within 5 months; and 2% within 6 months. This pattern suggests that the risk of a second violent or sexual victimization decreased after the passage of 1 month.

Figure 1   Time Course of Sexual and Violent Incidents

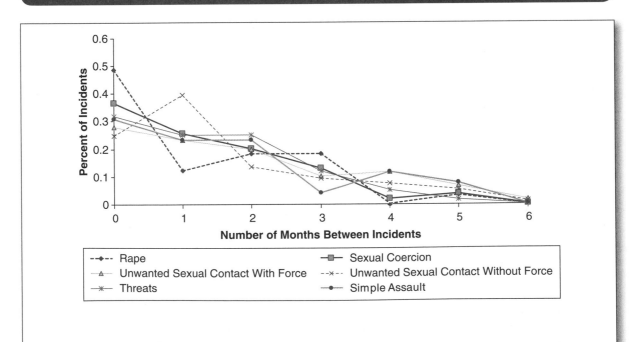

Victimization Crime-Switch Patterns. To examine the sequential pattern of type of victimization,[3] we constructed a crime-switch pattern matrix. For each victim of two or more incidents, we examined her sequentially paired incidents as to the type of victimization that composed the preceding—following incident pair (e.g., a rape followed by a rape, a sexual coercion followed by a rape).[4] As shown in Table 3, a matrix was constructed to reflect the total number of each type of victimization pairs that had occurred.

There is a significant relationship between the type of victimization that occurs in the preceding incident and the type of victimization that occurs in the following incident ($\chi^2 = 208.66$, $df = 16$, $p < .001$). Comparing the percentages in the diagonal (proneness) to those in the off-diagonal (crime switching) illustrates that regardless of type of sexual victimization, victims were most likely to have experienced the same type of victimization in consecutive incidents. More than half of the sexual contacts without force, 52%, were followed by another sexual contact without force. Almost 30% of all rapes were followed by a rape. There is also considerable proneness to sexual contact with and without force and threats among victims reporting these experiences.

Characteristics of Single and Repeat Incidents. The comparison of the characteristics of single and repeat violent and sexual incidents is presented in Table 4. From this table, it can be seen that few incident characteristics differ across single and repeat incidents. For simple assault, rape, and sexual coercion, none of the preincident, situational, or postincident characteristics were significantly different.

Some significant differences did emerge, however, for unwanted sexual contact with force, unwanted sexual contact without force, and threats. For unwanted sexual contact without force, single incidents were less likely to be committed by an intimate partner (3.7% vs. 14.5%) and more likely to be committed by someone known to the victim than repeat incidents (63.4% vs. 49.2%).

Overall, in a significantly larger proportion of the single incidents, women used self-protective action while the incident was going on compared to in the proportion of repeat incidents (84.8% vs. 73.5%; $Z = 3.795$, $p < .001$; results not presented in table). This result was also evident in the comparison of single incidents of rape, unwanted sexual contact with and without force incidents, and threats to repeat incidents. For example, in 86.1% of the single-incident rapes, the victim used protective action, compared to 74.2% of the repeat-rape incidents. Almost all (95.1%) of the single incidents of unwanted sexual contact with force involved the use of self-protective action by the victim, compared to 86.3% of the repeat incidents.

Repeat and single-rape incidents also differ in the types of self-protective action that were used. In single-rape incidents in which self-protective action was used, a larger proportion of women used two strategies—forceful physical and nonforceful verbal strategies, which have been shown to be effective in stopping the completion of rape—compared to the proportion who did in the first incident of a repeat-rape episode.[5] For example, in 71% of the single-incident rapes, women used forceful physical actions compared to only 14% of the repeat-rape incidents. Similarly, in 31% of the single incidents of unwanted sexual contact with force, women used nonforceful verbal actions compared to 26% of the repeat incidents (results not presented in table). Although the small number of cases limited meaningful statistical testing, the pattern indicates that women who experienced only a single rape or unwanted sexual contact with force incident did so because they used effective protective action to thwart the attack.

Discussion

Is Repeat Victimization a Problem?

Among victimized college women, repeat victimization is a common experience, striking from 14% to 26% of the women during an academic year. On any given campus, a relatively small proportion of women account for a disproportionate amount of the sexual and violent incidents. Furthermore, repeat victimization

Table 3 Victimization Crime-Switch Matrix: Sexual Victimization Pairs of Preceding and Following Incidents

Type of Sexual Victimization Reported as Following Incident	Type of Sexual Victimization Reported as Preceding Incident															
	Rape			Sexual Coercion			Sexual Contact With Force			Sexual Contact Without Force			Threats			
	%	n	Exp	%	n	Exp	%	n	Exp	%	n	Exp	%	n	Exp	Total
Rape	29.5	23	11	9.2	9	13	14.2	17	16	8.5	14	22	12.5	14	15	77
Coercion	20.5	16	14	50.0	48	17	4.2	5	21	13.9	23	29	6.3	7	19	99
Sexual contact with force	11.5	9	15	7.1	7	19	43.3	52	24	10.9	18	33	24.1	27	22	113
Sexual contact without force	19.2	15	24	25.5	25	30	21.7	26	37	52.1	86	21	21.4	24	34	176
Threat	19.2	15	15	9.2	9	19	16.7	20	23	14.6	24	31	35.7	40	21	108
Total	100.0	78		100.0	98		100.0	120		100.0	165		100.0	112		573

NOTE: Exp = expected

occurs quickly, with the risk of another victimization peaking in the time immediately following the initial victimization and then decreasing over time. The crime-switch analysis suggests that women are prone to experience repeat incidents of the same type of sexual victimization. Taken together, these findings are potentially disquieting. Repeat victimization does occur, is little understood, and is not systematically addressed by either sexual or violence prevention programs.

Why Does Repeat Victimization Occur?

At this exploratory stage, the cause of repeat victimization is unclear. However, three considerations merit attention. First, the incident-level risk factors for a single victimization are similar to the risk factors for repeat incidents. Our data indicate that the situational context in which victimizations occur—the victim-offender relationship, alcohol or drug consumption prior to the incident, location, and reporting—does not differ for single and repeat incidents. Those factors that increase the risk of victimization, if unchanged, will continue to be risk factors for subsequent

incidents. Accordingly, prevention efforts targeting the situational factors that are related to single victimizations may also be effective at preventing repeat episodes.

Second, one factor that did differ in the current study is that self-protective action was used more frequently in single incidents than in repeat incidents. In light of the existing research, this finding suggests that women who use self-protective action and who use effective actions may be less likely to experience a repeat victimization. It is possible that the ability to use self-protective action (whether physical or verbal) might reflect an underlying personal vulnerability that continues to make these women "attractive targets" for predatory men (Ullman, 1997). Failing to use self-protective actions, however, may also have a psychological impact on women that makes them feel less capable of preventing victimization in the future. By contrast, studies have cited beneficial psychological consequences for women who took self-defense training compared to those who did not. From this line of research, it is possible that the use of self-protective action is empowering, particularly because its use is related to an offense being attempted rather than

Table 4 Single-Incident and Repeat Violent and Sexual Incidents, Preincident, Situational, and Postincident Characteristics

	Type of Victimization																							
	Violent Victimization								Sexual Victimization															
	Simple Assault				Rape				Sexual Coercion				Unwanted Sexual Contact With Force				Unwanted Sexual Contact Without Force				Threats			
	Single[a]		Repeat[b]		Single		Repeat		Single		Repeat		Single		Repeat		Single		Repeat		Single		Repeat	
Incident Characteristic	%	n	%	n	%	n	%	n	%	n	%	n	%	n	%	n	%	n	%	n	%	n	%	n
Preincident characteristic																								
Victim-offender relationship																								
Intimate partner	28.3	30	29.8	14	8.3	3	21.3	13	20.3	14	32.3	31	3.7**	3	14.5	18	9.4	13	4.8	9	10.3	4	8.7	9
Someone known	47.2	50	46.4	22	77.8	28	70.5	43	66.6	46	55.2	53	63.4*	52	49.2	61	60.9	84	73.4	138	79.4	31	66.0	68
Stranger	24.5	26	23.4	11	13.9	5	8.2	5	13.0	9	12.5	12	32.9	27	36.3	45	29.7	41	21.8	41	10.3	4	25.2	26
Alcohol or drug consumption prior to incident by at least one party	48.2[c]	40	50.0	19	74.3	26	69.5	41	70.8	46	67.4	62	70.5	55	76.3	90	76.5	101	68.3	125	55.9	19	61.7	58
Situational characteristics																								
Living quarters	32.6	42	28.0	14	51.4	18	63.9	39	65.2	45	56.3	54	26.8	22	26.0	32	32.6	45	38.4	73	28.2	11	35.3	36
Used self-protective action	56.0	61	63.0	29	86.1*	31	74.2	46	63.2	43	60.4	58	95.1**	78	86.3	107	88.4**	122	75.3	143	87.2**	34	69.9	72
Postincident characteristics																								
Reported to authorities	29.0	31	23.4	11	8.3	3	4.8	3	0.0	0	0.0	0	1.2	1	1.6	2	0.7	1	1.6	3	7.7	3	5.8	6
Told someone	—[d]	—	—	—	61.1	22	69.4	43	65.2	45	69.8	67	81.5	66	75.8	94	75.9	104	72.0	136	62.1	28	62.1	64

a. Single incident

b. Repeat incident

c. Respondents were asked about whether or not the perpetrator had been drinking at the time of the incident

d. This information was not asked on the National College Women Violent Victimization survey.

*p < .10

**p < .05.

Two-tailed tests were performed for victim-offender relationship, living quarters. All of the other proportion tests were one-tailed tests.

completed (see Fisher et al., 2007; Ullman, 1997). Thus, a woman who actively tries to prevent her victimization from continuing may be more able or willing to protect herself after an initial victimization. This is not meant to imply "victim blaming" because the responsibility lies with the offender and no woman should have to fight off an attacker. Scientifically, however, there may be individual differences in the ability to use self-protection and in the effects of doing so— or not doing so. This phenomenon cannot be ignored but merits careful study.

Third, and related, the initial victimization might affect some women differently—in ways that are unmeasured by our data. Some victimized women may experience posttraumatic stress disorder (PTSD), depression, or self-blame that increases their vulnerability. Research on the effects of PTSD suggests that women who are sexually victimized may be less able to recognize risky situational cues and do not take protective action in such situations (Arata, 2000; Messman & Long, 2003). As a result, PTSD, depression, or self-blame may mediate the relationship between an initial sexual victimization and subsequent victimization (Ellis, Atkeson, & Calhoun, 1982). In this regard, future research is needed to collect individual-level and incident-level data longitudinally, including information on psychological factors that could make the risk of repeat victimization more likely to happen.

What Are the Policy Implications of Repeat Victimization?

The majority of efforts on college campuses to prevent victimization focus on either prevention of the initial incident or the response to an incident after it occurs (see Karjane, Fisher, & Cullen, 2005). Specifically, once an incident occurs, the focus is to have the person report the incident to campus police or authorities and to receive medical attention and/or psychological counseling. These are all valuable responses. A missing component to this response, however, may be the failure to take into account that violent and sexual victims are at an elected risk in the *near future* of a repeat

victimization. There needs to be *explicit* attention given to preventing a repeat incident among college women. Yet, little published empirical work has provided direct secondary prevention programs for this high-risk group.

Reducing victimization risk immediately following an initial victimization is challenging because those who are victimized do not generally report their victimization to the police or campus officials (see Fisher et al., 2003). Research shows, however, that most college women do disclose their victimization to a friend, including a roommate (Fisher et al., 2003). This finding suggests that a key to reaching victims and preventing a repeat victimization might be through students whose friends are victimized. Our results show that reporting to someone other than the police or campus officials does not differ from single and repeat incidents, meaning that single and repeat incidents are likely to be disclosed. Perhaps one of the reasons telling someone does not, at least in our data, facilitate a reduction in risk of subsequent victimization could be because the people who are told are not trained or educated to provide the kind of assistance and advise that would prove preventative.

Research has also shown that one of the key elements for reducing opportunity for victimization is capable guardianship (Tewksbury & Mustaine, 2003). In this regard, a friend of a victim, such as a roommate, could potentially serve as a "capable guardian." For college women, friends might furnish guardianship by not leaving the victim alone in risky social situations, by taking her to counseling or medical services to receive help with continuing vulnerabilities, and by encouraging her to learn about effective self-protective actions, such as enrolling in a self-defense course. This recommendation to foster social guardianship skills implies that colleges should conduct not only general victimization prevention seminars but also programs that specifically target the role that friends might play if someone they know is victimized. It will be important to raise consciousness that anyone who is victimized is at risk immediately thereafter for another victimization. Being a guardian is one possible way to assist a victim in reducing the chances of experiencing a repeat incident. Knowing whether or not this solution would be effective requires information beyond the scope of our cross-sectional research design.

Notes

1. Note that robbery and aggravated assault are not included in the time-course results because there were no repeat victims for either type of crime.

2. This exception needs to be qualified, as a small number of repeat rapes happened within 1 month of each other ($n = 4$), 2 months ($n = 6$), 3 months ($n = 6$), and 5 months ($n = 1$). None happened within 4 or 6 months of each other. Sixteen repeat rapes happened within the same month.

3. There were only five pairs of repeat violent victimization; hence, we did not include them in the crime-switch analysis.

4. A woman who experienced two incidents has one pair. For example, her rape in September is the preceding incident (incident #1) to her rape in October, which is the following incident (incident #2). These incidents constitute one pair. A woman who experienced three incidents has two pairs. For example, her rape in November is the preceding incident (incident #1) to her sexual coercion in December, which is the following incident (incident #2). These incidents comprise one pair. Her sexual coercion in December is the preceding incident (incident #2) to her unwanted sexual contact with force in April, which is the following incident (incident #3). These incidents comprise the second pair (Reiss, 1980).

5. Forceful physical actions included (a) attacked offender with a gun or knife; (b) attacked offender with other weapon; (c) used mace, pepper spray, or similar devices; and (d) used physical contact such as hitting, punching, or kicking against the offender (see Ullman, 1997). Nonforceful verbal action refers to the victim having used nonaggressive verbal responses with the offender. Nonforceful verbal action included three behaviors: (a) tried to reason or negotiate with the offender, (b) pleaded with or begged offender to stop, and (c) told the offender to stop. This measure is similar to the nonforceful verbal resistance measure utilized by Ullman (1997), who included the behaviors of pleading, talking, reasoning, begging, and crying.

References

Abbey, A., Zawacki, T., Buck, P. O., Clinton, A. M., & McAuslan, P. (2004). Sexual assault and alcohol consumption: What do we know about relationship and what types of research are still needed? *Aggression and Violent Behavior, 9,* 273–303.

Arata, C. M. (2000). From child victim to adult victim: A model for predicting sexual assault. *Child Maltreatment, 5,* 28–38.

Breitenbecher, K. H. (2001). Sexual revictimization among women: A review of the literature focusing on empirical investigations. *Aggression and Violent Behavior, 6,* 415–432.

Cattaneo, L. B., & Goodman, L. A. (2005). Risk factors for reabuse in intimate partner violence: A cross-disciplinary critical review. *Trauma, Violence, & Abuse, 6,* 141–175.

Davis, J. J., Combs-Lane, A. M., & Jackson, T. L. (2002). Risky behaviors associated with interpersonal victimization. *Journal of Interpersonal Violence, 17,* 611–629.

Ellis, E., Atkeson, B., & Calhoun, K. (1982). An examination of differences between multiple and single-incident victims of sexual assault. *Journal of Abnormal Psychology, 91,* 221–224.

Farrell, G (1995). Preventing repeat victimization. In M. Tonry & D. P. Farrington (Eds.), *Crime and justice: A review of research* (Vol. 19, pp. 469–534). Chicago: University of Chicago Press.

Fisher, B. S., Cullen, F. T., & Daigle, L. E. (2005). The discovery of acquaintance rape: The salience of methodological innovation and rigor. *Journal of Interpersonal Violence, 20,* 493–500.

Fisher, B. S., Daigle, L. E., Cullen, F. T., & Santana, S. (2007). Assessing the efficacy of the protective action sexual victimization completion nexus. *Violence and Victims, 22,* 18–42.

Fisher, B. S., Daigle, D. E., Cullen, F. T., & Turner, M. G. (2003). Reporting sexual victimization to the police and others: Results from a national-level study of college women. *Criminal Justice and Behavior, 30,* 6–38.

Gidycz, C. A., Coble, K., Latham, L., & Layman, M. J. (1993). Sexual assault experience in adulthood and prior victimization experiences: A prospective analysis. *Psychology of Women Quarterly, 17,* 151–168.

Gidycz, C. A., Hanson, K., & Layman, M. J. (1995). A prospective analysis of the relationships among sexual assault experiences. *Psychology of Women Quarterly, 19,* 5–29.

Karjane, H., Fisher, B. S., & Cullen, F. T. (2005). *Sexual assault on campus: What colleges and universities are doing about it.* Washington, DC: U.S. Department of Justice.

Messman, T. L., & Long, P. J. (2003). The role of childhood sexual abuse sequelae in the sexual revictimization of women: An empirical and theoretical reformulation. *Clinical Psychology Review, 23,* 537–571.

Reiss, A. J. (1980). Victim proneness in repeat victimization by type of crime. In S. E. Fienberg & A. J. Reiss (Eds.), *Indicators of crime and criminal justice: Quantitative studies* (pp. 41–53). Washington, DC: U.S. Department of Justice, Bureau of Justice Statistics.

Tewksbury, R., & Mustaine, E. E. (2003). College students' lifestyles and self-protective behaviors: Further considerations of the guardianship concept in routine activity theory. *Criminal Justice and Behavior, 30,* 302–327.

Ullman, S. E. (1997). Review and critique of empirical studies of rape avoidance. *Criminal Justice and Behavior, 24,* 177–204.

DISCUSSION QUESTIONS

1. Are you surprised to learn that repeat victimization is likely to recur quickly? What should colleges do to reduce the risk of repeat victimization, given that it happens so quickly?

2. What is a capable guardian? How do the findings relate to routine activities and lifestyles theory?

3. What other factors in addition to the ones used by Daigle, Fisher, and Cullen may place a college woman at risk of being repeatedly sexually victimized?

4. College women are very unlikely to be violently victimized (nonsexually victimized). Why are college women so much more likely to be sexually victimized than victimized in other ways?

5. Use the concepts of risk heterogeneity and state dependence to explain why some college women are repeat victims.

◈

Introduction to Reading 3

The effects of victimization are many, and people who are not directly victimized but are vicariously victimized may also experience negative effects. Vicarious victimization occurs when a person is traumatized by being exposed to violent acts through the media or in other ways. In their article, Herkov and Biernat (1997) examined the impact of vicarious victimization on 184 residents of Gainesville, Florida, where five college students were murdered by a serial killer, Ted Bundy. Respondents were studied 5 weeks ($n = 184$), 9 months ($n = 64$), and 18 months ($n = 30$) after the murders. They completed surveys regarding their experiences of posttraumatic stress disorder (PTSD) symptoms. The authors found that many experienced PTSD symptoms, and although they diminished over time, these symptoms still were present in some of the sample 18 months after the murders occurred.

Assessment of PTSD Symptoms in a Community Exposed to Serial Murder

Michael J. Herkov and Monica Biernat

Posttraumatic stress disorder (PTSD) represents a psychiatric syndrome in which an individual experiences a severe anxiety reaction to an overwhelming psychosocial stressor "outside the range of usual human experience" (American Psychiatric Association, 1987). Hallmarks of PTSD include reexperiencing of the traumatic event, avoidance of stimuli associated with the event, psychological numbing, and persistent symptoms of increased physiological arousal.

SOURCE: Herkov and Biernat (1997). Reprinted with permission.

Traditionally, the study of PTSD reactions has focused on individuals directly exposed to situations of extreme violence and carnage, such as military combat (Oei, Lim, & Hennessy, 1990) or rape (Burgess & Holmstrom, 1974; Calhoun & Atkeson, 1991). However, other traumatic events can lead to PTSD reactions. *DSM-III-R* includes in its description of qualifying traumatic events "destruction of one's home or community or seeing another person who has recently been or is being seriously injured or killed as the result of an accident or physical violence." Investigations of populations who have been exposed to natural disaster (Horowitz, Stinson, & Field, 1991) or accidents (Bromet, 1989) confirm that these stressors can also lead to PTSD reactions.

Recently, researchers have begun to examine whether people can experience psychological trauma through knowledge of or indirect exposure to violence. That is, can a person be traumatized through media or other information concerning violent acts such as murder, terrorism, or rioting in their community? Young (1989) describes this phenomenon as "vicarious victimization" and describes seven factors that can lead to increased likelihood of its occurrence, including a realistic threat of death to all members of the community, extraordinary carnage, strong community affiliation, witnessing of event by community members, symbolic significance of victims to the community (children, etc.), need for numerous rescue workers, and significant media attention.

Terr (1985) observed this phenomenon among family and friends of the children who had been taken hostage in the Chowchilla kidnapping case. Terr noted that while none of the family members of friends had been directly exposed to the incident, many appeared to have "caught" posttraumatic symptoms. These symptoms included frightening dreams and avoidance behaviors.

The purpose of the present study was to examine whether community residents in Gainesville experienced any posttraumatic symptoms following the serial murder of five college students. While once thought to be a rare event, recent Federal Bureau of Investigation statistics estimate that as many as 5,000 Americans a year may be murdered by serial killers (Holmes & DeBurger, 1985). The random and gruesome nature of the deaths, choice of students as victims, increased presence of rescue workers (i.e., police), and accompanying media coverage associated with the Gainesville murders meet virtually all of the conditions hypothesized by Young (1989) as setting the stage for community vicarious victimization. The presence of these symptoms in the community would provide additional empirical support for the concept of vicarious victimization.

Results and Discussion

Table 1 presents the frequency of PTSD symptoms reported by community residents across the three data collections. Symptoms are broken down into the categories in which they appear in *DSM-III-R*.

Table 1 Frequency of PTSD Symptoms at Initial Data Collection and 9- and 18-Month Follow-Ups

Symptom	Initial	9 Months	18 Months
Reexperience of the traumatic event			
Distressing thoughts	34%	22%	7%
Distressing dreams	4%	0%	0%
Feelings of danger	10%	13%	0%
Avoidance and emotional numbing			
Avoidance of activities	28%	22%	17%
Loss of interest in activities	7%	3%	0%
Feeling distant from others	9%	6%	3%
Restricted affect	3%	3%	0%
Sense of foreshortened future	35%	19%	17%
Symptoms of increased arousal			
Difficulty falling asleep	19%	19%	0%
Increased irritability	4%	5%	0%
Concentration difficulties	11%	8%	0%
Increased startle response	35%	34%	10%
Physiological reactivity	18%	25%	3%

As illustrated by Table 1, substantial numbers of subjects reported PTSD symptoms immediately following the murders. Symptoms most commonly reported were sense of foreshortened future (35%), increased startle response (35%), and distressing thoughts (34%), each representing a different symptom cluster.

Nine months following the murders, over one third of subjects were still reporting increased startle response and nearly one fifth reported distressing thoughts, avoidance of activities, difficulty falling asleep, and increased physiologic reactivity. By 18 months, however, virtually all symptoms had significantly decreased. In fact, only two symptoms (sense of foreshortened future and avoidance of activities) were reported by 10% or more of subjects. The most dramatic decreases were observed in the increased arousal and reexperiencing of the trauma categories. Thus, while symptoms appeared to resolve over time, they were by no means transient. This pattern of reduction of symptoms is similar to that observed in individuals who have been direct victims of violence (Calhoun & Atkeson, 1991). Thus, it would appear most subjects were able to experience "psychological healing," although the process took over 1 year.

These results are consistent with the meta-analysis study of post-disaster psychopathology conducted by Rubonis and Bickman (1991). In the meta-analysis of 32 studies, Rubonis and Bickman reported an overall psychopathology effect size of 174, with anxiety symptoms demonstrating the largest effect size (309). Similar to our results, these authors found that the psychopathology diminished over time.

Because of the severe attrition in subject participation across time, the possibility exists that the follow-up results reflect methodological artifacts rather than actual decrease of symptoms. That is, individuals most affected by the murders dropped out of the study. To control for this possibility, an analysis of the reported symptoms of the 30 subjects who responded to all three questionnaires was conducted.

As illustrated in Table 2, the pattern of observed symptoms of this subsample is remarkably similar to the results presented in Table 1, both in terms of initial symptoms and resolution of symptoms over time.

These data suggest that the observed decrease in subject-reported symptoms represents a real attenuation of the community's psychological response and is not solely due to attrition of the more distressed subjects.

Young (1989) hypothesized that vicarious trauma reactions are most likely to occur among individuals who closely identify with the actual victims. To test this hypothesis, psychological distress was analyzed by student status (all victims were college students) and gender (four of five victims were female). It was expected

Table 2 Endorsement of PTSD Symptoms by Subjects Responding to All Three Data Collections

Symptom	Initial	9 months	18 months
Reexperience of the traumatic event			
Distressing thoughts	27%	12%	4%
Distressing dreams	0%	0%	0%
Feelings of danger	4%	0%	0%
Avoidance and emotional numbing			
Avoidance of activities	31%	19%	16%
Loss of interest in activities	4%	0%	0%
Feeling distant from others	4%	4%	4%
Restricted affect	4%	4%	0%
Sense of foreshortened future	19%	15%	20%
Symptoms of increased arousal			
Difficulty falling asleep	8%	12%	0%
Increased irritability	4%	4%	0%
Concentration difficulties	4%	4%	0%
Increased startle response	19%	19%	7%
Physiological reactivity	12%	19%	4%

that students and women would experience the most severe psychological reactions. Comparison of students and non-students across all PTSD symptoms revealed significant differences on only two symptoms: avoidance of activities ($x^2(1) = 6.90$, $p < .009$) and sense of foreshortened future ($x^2(1) = 13.17$, $p < .004$). The finding that avoidance of activities was more often noted in students is not surprising in that this segment of the community is most likely to engage in high-risk activities (walking alone at night, going out to bars at night) that would tend to be avoided following a series of murders.

Gender, on the other hand, appeared to be closely related to psychological distress. As illustrated in Table 3, significantly more women reported PTSD symptoms following the murders than men. Specifically, women reported significantly more distressing thoughts, avoidance of activities, sense of foreshortened future, sleep difficulties, startle response, and physiological reactivity than men. Interestingly, the only symptom reported more frequently by men was increased irritability. While these findings may reflect the identification of Gainesville women with the student victims, another explanation may be that groups such as women that have been historical victims of violence may be more susceptible to traumatization through vicarious means.

While the above analyses provide information about the general presence of PTSD symptoms, they do not assess the magnitude of distress (i.e., the number of PTSD symptoms experienced by the individual subject). To evaluate the level of psychological distress reported by subjects, individual records were evaluated for presence of a PTSD "diagnosis" using *DSM-III-R* criteria. The term "diagnosis" is used only as a convenience to describe those subjects who reported PTSD symptoms from all *DSM-III-R* diagnostic clusters. The authors are clearly not asserting that these subjects experienced clinical PTSD and recognize that such a diagnosis would require a clinical interview, which was not part of the research. *DSM-III-R* decision rules were followed for all categories except C (avoidance/numbing). Because the questionnaire only sampled five of the seven category C criteria, the threshold for presence of this category was reduced from three to two criteria.

Table 3   Frequency of PTSD Symptoms by Gender at Initial Data Collection

Symptom	Males	Females	p
Reexperience of the traumatic event			
Distressing thoughts	23%	42%	.008
Distressing dreams	1%	7%	.09
Feelings of danger	5%	13%	.09
Avoidance and emotional numbing			
Avoidance of activities	7%	42%	.000
Loss of interest in activities	3%	9%	.14
Feeling distant from others	5%	11%	.333
Restricted affect	3%	4%	.67
Sense of foreshortened future	4%	55%	.000
Symptoms of increased arousal			
Difficulty falling asleep	11%	25%	.014
Increased irritability	7%	3%	.325
Concentration difficulties	5%	14%	.116
Increased startle response	12%	51%	.000
Physiological reactivity	1%	29%	.000

In the initial sample, 25 subjects (14%) endorsed sufficient criteria for a diagnosis of PTSD. While the lack of data on these subjects before the murders makes it difficult to judge the significance of this finding, epidemiological studies of PTSD in other communities indicate prevalence rates of between 1% and 3% (Davidson, Hughes, Blazer, & George, 1991; Helzer, Robins, & McEvoy, 1987). Thus, the magnitude of reported symptoms in the present sample is well above that expected in the community and suggests that the increase in PTSD prevalence reflects psychological sequela associated with the serial murders. However, it is important to note that the community studies of PTSD cited above utilized the Diagnostic Interview Schedule (Robins, Helzer, Croughland, Williams, & Spitzer, 1981), a structured interview with more stringent criteria than that used in the present study. Had such measures been used in the present study, the prevalence of PTSD may have been lower. Still, the finding that 14% of the present sample reported multiple symptoms from all PTSD symptoms clusters indicates that these individuals experienced significant psychological distress following the murders.

These PTSD subjects were further analyzed to determine how they differed from those individuals who did not meet diagnostic threshold or reported less magnitude of symptoms. Results indicate that the PTSD subjects were significantly younger ($F(157.24)$) = $2.35, p < .016$) and more likely to be female ($x^2(1) = 15.44$, $p < .000$) compared to the general sample. In fact, only one PTSD subject was a male.

These subjects also appeared to cope with their fears differently than the rest of the sample. For example, individuals who met diagnostic criteria for PTSD were much more likely to choose the coping behavior of purchasing a firearm than the general sample ($x^2(1) =$ $9.97, p < .001$) = $2.07, p > .10$). This is especially interesting in that virtually all of these subjects were women, a group not traditionally associated with firearm use. While this illustrates the intensity of fear experienced by these subjects, it is somewhat unsettling that the group with the most serious psychological reaction chose to protect themselves with the most deadly force.

Comparison between groups on other variables such as whether the murderer would kill again, was in police custody, or would eventually be caught were not related to diagnosis of PTSD. However, other variables such as home distance from the murder sites and status as a student, while not significant, showed relationship to the diagnosis. For example, 43% of the PTSD subjects lived within one mile of the murder sites while only 27% of the community respondents lived within that area. Similarly, a larger proportion of these subjects were students (41%) compared to the general sample (28%). Longitudinal study of these PTSD subjects indicated a substantial reduction in diagnoses. For example, in the 9-month follow-up sample, only four (6%) of the respondents still had symptoms sufficient to meet this diagnostic threshold. No subjects in the 18-month follow-up received a PTSD diagnosis. Thus, these subjects, similar to the overall sample, appeared to return to psychological equilibrium with time.

This study provides a model for understanding how communities respond to violence such as serial murder. While methodological concerns such as response rate, subject attrition, and absence of a control sample limit the generalizability of the findings, results do indicate that large numbers of Gainesville residents experienced significant psychological trauma following the student murders. Further, this distress was not transient, with many subjects requiring over a year to recover. These data clearly illustrate how a community can be traumatized by violence perpetrated on a few members. The nature of the killings, media response, and cohesion of a university town clearly met the criteria identified by Young (1989) as factors associated with vicarious victimization, thus making Gainesville particularly vulnerable to experiencing community vicarious victimization. In conclusion, this study suggests that the sense of community that binds people together also renders them vulnerable to the harm befalling only a few. It is apparently empirically true that communities are bound to each other emotionally and not simply geographically.

⊠ References

American Psychiatric Association. (1987). *Diagnostic and statistical manual of mental disorders* (3rd ed., revised). Washington, DC: Author.

Bromet, E. J. (1989). The nature and effects of technological failures. In R. Gist & B. Lubin (Eds.), *Psychosocial aspects of disaster* (pp. 61–85). New York: Wiley.

Burgess, A., & Holmstrom, L. L. (1974). Rape trauma syndrome. *American Journal of Psychiatry, 131*, 981–985.

Calhoun, K. S., & Atkeson, B. M. (1991). *Treatment of rape victims: Facilitating psychological adjustment.* New York: Pergamon.

Davidson, J. R., Hughes, D., Blazer, D. G., & George, L. K. (1991). Posttraumatic stress disorder in the community: An epidemiological study. *Psychological Medicine, 21*, 713–721.

Helzer, J. E., Robins, L. N., & McEvoy, M. A. (1987). Post-traumatic stress disorder in the general population: Findings of the Epidemiologic Catchment Area survey. *New England Journal of Medicine, 317*, 1630–1634.

Holmes, R. M., & DeBurger, J. E. (1985). Profiles in terror: The serial murderer. *Federal Probation, 49*, 29–34.

Horowitz, N. J., Stinson, C., & Field, N. (1991). Natural disaster and stress response syndromes. *Psychiatric Annals, 21*, 556–562.

Oei, T. P., Lim, B., & Hennessy, B. (1990). Psychological dysfunction in battle: Combat stress reactions and posttraumatic stress disorder. *Clinical Psychology Review,* 10(3), 355–388.

Robins, L. N., Helzer, J. E., Croughland, J. L., Williams, J. B. W., & Spitzer, R. L. (1981). *NIMH Diagnostic Interview Schedule Version III.* Public Health Service (PHS), publication ADM-T-42-3 (5-8-81). Rockville, MD: NIMH.

Rubonis, A. V., & Bickman, L. (1991). Psychological impartment in the wake of disaster: The disaster psychopathology relationship. *Psychological Bulletin, 109,* 384–399.

Terr, L. C. (1985). Psychic trauma in children and adolescents. In *Psychiatric Clinics of North America* (Vol. 8, No. 4, pp. 815–835). Philadelphia, PA: WB Saunders Company.

Young, M. A. (1989). Crime, violence, and terrorism. In R. Gist & B. Lubin (Eds.), *Psychology aspects of disaster* (pp. 61–85). New York: Wiley.

DISCUSSION QUESTIONS

1. What limitations are present in the study regarding their methodology? Do you think these results would apply in other scenarios and with other murders?

2. Do you think the effects of media coverage have diminished over time? Why or why not?

3. Does it surprise you that persons who share demographic characteristics of those who were murdered experienced the most symptoms? Why or why not?

4. What are the policy implications of the findings? In other words, what should be done for persons who live in communities disrupted by extreme acts of violence?

Victims' Rights and Remedies

L et's revisit Polly now that it has been a few days since she was victimized. Remember that Polly is a young undergraduate student who was accosted by two offenders as she was walking home. Her school bag was stolen, and she was assaulted. Unlike most victims, Polly called the police to report what had happened to her. She had to have 10 stitches at the hospital. Clearly a victim, she was still questioned by the police about why she was walking home alone at night. She very well may have felt victimized by this questioning—and we know that she had a hard time emotionally after being victimized. She found it hard to get out of bed, and she missed several classes—she even altered her schedule and stopped going out alone at night.

In Section 3, you considered the toll this victimization took on Polly—on her emotions and her lifestyle, and of course financially. As you know, Polly is not alone in suffering these costs. Many victims experience real costs and consequences. But how do victims deal with these outcomes? Are they left to recover on their own, or are services available to them? Whose responsibility is it to help crime victims? What happens when crime victims do not get the help they need and deserve? All these questions will be addressed in this section, and as you will see, a variety of rights and resources are available to crime victims today.

⚒ Victims' Rights

Once essentially ignored by the criminal justice system and the law, victims are now granted a range of rights. These rights have been given to victims through legislation and, in 32 states, through **victims' rights** amendments to state constitutions (National Center for Victims of Crime, 2009). The first such law that guaranteed victims' rights and protections was passed in Wisconsin in 1979; now, every state has at least some form of victims' rights legislation (Davis & Mulford, 2008). Despite each state having laws that afford victims' rights, they differ in whom the law applies to, when the rights begin, what rights victims have, and how the rights can be enforced. Common to all these state laws, however, is the goal of victims' rights—to enhance victim privacy, protection, and participation (Garvin, 2010).

Common Victims' Rights Given by State

Slightly less than half of U.S. states give *all* victims rights (Howley & Dorris, 2007). In all states, the right to compensation, notification of rights, notification of court appearances, and ability to submit victim impact statements before sentencing is granted to at least some victim classes (Deess, 1999). Other common rights given to victims in the majority of states are the right to restitution, to be treated with dignity and respect, to attend court and sentencing hearings, and to consult with court personnel before plea bargains are offered or defendants released from custody (Davis & Mulford, 2008). Other rights extended to victims are the right to protection and the right to a speedy trial. Importantly, some states explicitly protect victims' jobs while they exercise their right to participate in the criminal justice system. These protections may include having the prosecutor intervene with the employer on behalf of the victim or prohibiting employers from penalizing or firing a victim for taking time from work to participate (National Center for Victims of Crime, 2009). Some of these rights are discussed in more detail below, and others are discussed in separate parts of this section. To see an example of what rights a state grants, see the box on victims' rights in Virginia.

VICTIMS' RIGHTS IN VIRGINIA

The Victim Services Unit provides the following services to victims of crime:

- Advocacy on behalf of crime victims
- Notification of changes in inmate transfers, release date, name change, escape and capture
- Explanation of parole and probation supervision process
- Accompaniment to parole board appointments when requested by the victim
- Provide victims with ongoing support, crisis intervention, information, and referrals
- Training, education, and public awareness initiatives on behalf of victims of crime

Victims can register to be notified through Victim Information and Notification Everyday (VINE).

VINE is a toll-free, 24-hour, anonymous, computer-based telephone service that provides victims of crime two important features, information and notification. Victims may call VINE from any touch-tone telephone, any time, to check on an inmate's custody status. For inmate information, call 1-800-467-4943 and follow the prompts.

Victims may register with VINE for an automated notification call when an inmate is released, transferred, escapes, and to learn of an inmate's parole status if the inmate is parole eligible.

Victims of crime can address the Parole Board if they have any concerns regarding the release of an offender. Victims have the option of voicing their concerns through letters or through an in-person appointment with the Parole Board.

If victims would like a staff member of the Department of Corrections, Victim Services Unit to accompany them to the appointment, they can contact the Department of Corrections, Victim Services Unit.

SOURCE: Virginia Department of Corrections (2010). Reprinted with permission.

Notification

The right to **notification** allows victims to stay apprised of events in their cases. Notification is important for victims at various steps in the criminal justice process. In some jurisdictions, victims have the right to be notified when their offender is arrested and released from custody after arrest, such as on bail. Victims may also have the right to be notified about the time and place of court proceedings and any changes made to originally scheduled proceedings. Notification may also be given if the offender has a parole hearing and when the offender is released from custody at the end of a criminal sanction. Notification responsibilities may be placed on law enforcement, the prosecutor, and the correctional system. To make notification more systematic and reliable, some jurisdictions use automated notification systems to update victims (through letters or phone calls) about changes in their cases. These systems are often also set up so that a victim can call to receive updates. Victims of federal crimes can register to participate in the national automated victim notification system.

Participation and Consultation

One of the overarching goals of the victims' rights movement was to increase **participation and consultation** by victims in all stages of the criminal justice system. One way victims are encouraged to participate is by submitting or presenting a victim impact statement, which is discussed later in this section under "Remedies and Rights in Court." Other ways victims may participate is by consulting with judges and/or prosecutors before any plea bargains are offered or bail is set. Consultation may also occur before an offender is paroled or sentenced (Davis & Mulford, 2008).

Right to Protection

Victims may also need protection as they navigate the criminal justice process. Victims may be fearful of the offender and the offender's friends and family. Participation in the criminal justice system may, in fact, endanger victims. In response to this potential danger, many states include safety measures in their victims' rights, falling under the category of **right to protection**. For example, victims may be able to get no-contact or protective orders that prohibit the defendant from having any contact with the victim. Victims may also be provided with secure waiting facilities in court buildings. Victim privacy is also protected ever-increasingly in states; some disclose only minimal victim information in criminal justice records—such as law enforcement and court records (Davis & Mulford, 2008).

Right to a Speedy Trial

You have probably heard of offenders having a **right to a speedy trial**, but did you know that about half of all states also provide victims with this right? Although not as explicit as an offender's right, this right given to victims ensures that the judge considers the victim's interests when ruling on motions for continuance. In other words, in states that give victims this right, decisions about postponing a trial cannot be made without consideration of the victim. Some states also explicitly provide for accelerated dispositions in cases with disabled, elderly, or minor children victims (Davis & Mulford, 2008).

Issues With Victims' Rights

While victims' advocates have hailed the adoption of legislation and state-level amendments that give victims rights, the adoption of victims' rights has also come with problems. There has been some resistance to states and the federal government giving victims formal rights. Remember that criminal law is written in such a way as to make crimes harms against the state rather than the victim. Also think about how the U.S. Constitution provides widespread rights to those persons suspected of committing crimes. The U.S. Constitution does not currently include any language that provides victims with rights—but it does for persons suspected of committing crimes. Although this omission has been identified by some as deserving remedy, others argue that victims' rights do not have a place in our Constitution (Wallace, 1997). Concerns have also been expressed that providing victims with rights will create a burden on our already overburdened criminal justice system (Davis & Mulford, 2008).

Also problematic is what to do when victims' rights are not protected. What happens if a victim is not notified? Who is responsible? Does the victim have any recourse, legal or otherwise, when a right is violated? Many states do not have specific enforcement strategies in place in their victims' rights legislation, although states that have constitutional amendments generally have enforceable rights in the event that a state official violates a victim's constitutional rights. Victims may also seek a writ of mandamus, which is a court order that directs an agency to comply with a law (National Center for Victims of Crime, 2009). For other victims, although they are given rights on paper, there is little they can do if their rights are not protected. To remedy this, some states—such as California, in its passage of Marsy's Law—have passed legislation that is more comprehensive and includes language that gives victims the right to enforce their rights in court, called legal standing (National Victims' Constitutional Amendment Passage, n.d.). Some states have set up a designated agency to handle crime victims' complaints (National Center for Victims of Crime, 2009). Despite these developments, many state victims' bills of rights specifically note that when victims' rights are violated, the crime victim does not have the ability to sue civilly a government agency or official. Whatever the redress allowed to victims, you can probably see that for victims, not having their rights protected may feel like an additional victimization and one that they can do little about—at least not easily.

Federal Law

Thus far, we have discussed common rights that states grant to victims of crime, but the federal government has also recognized the importance of protecting the rights of crime victims. (See Table 4.1 for a timeline and brief description of key pieces of federal legislation related to victims' rights.) In 1982, the President's Task Force on Victims of Crime published a report that included 68 recommendations for how victims could receive recognition and get the rights and services they deserve. These recommendations led, in part, to the development of legislation that would grant victims their first federal rights. The first such piece of legislation passed was the **Federal Victim Witness Protection Act** (1982). This act mandated that the attorney general develop and implement guidelines that outlined for officials how to respond to victims and witnesses. Two years later, the **Victims of Crime Act** (1984) was passed to create the Office for Victims of Crime and to provide funds to assist state victim compensation programs. The funds are generated from fines and fees and from seized assets of offenders who break federal law. A critical step in victims' rights also occurred in 1990 with passage of the **Child Victims' Bill of Rights**, which extended victims' rights to child victims and witnesses. Child victims and witnesses were granted rights to have proceedings explained in language they can understand; to have a victims' advocate present at interviews, hearings, and trials; to have a secure waiting area at trials; to have personal information kept private unless otherwise specified by the child or guardian; to have an advocate to discuss with the court their ability to understand proceedings; to be given information about and referrals to agencies for assistance; and to allow other services to be provided by law enforcement. Also in 1990, the **Crime Control Act** and the **Victims' Rights and Restitution Act** were passed, creating a federal bill of rights for victims of federal crime and guaranteeing that victims have a right to restitution. Specifically, victims of federal crimes were given the right to

a. be reasonably protected from the accused;

b. reasonable, accurate, and timely notice of any public proceeding involving the crime or any release or escape of the accused and to not be excluded from such proceedings;

c. be reasonably heard at any public proceeding involving release, plea, or sentencing;

d. confer with the attorney for the government in the case;

e. full and timely restitution as provided by law;

f. proceedings free from unreasonable delay;

g. be treated with fairness and with respect for the victim's dignity and privacy.

The acts also provide that the court ensures that crime victims are afforded these rights.

The '90s also saw the adoption of the **Violent Crime Control and Law Enforcement Act** (1994), which included the implementation of the **Violence Against Women Act** (VAWA) that gave more than $1 billion to programs designed to reduce and respond to violence against women. It also increased funding for victim compensation programs and established a national sex offender registry (Gundy-Yoder, 2010). In 1996, the **Antiterrorism and Effective Death Penalty Act** was passed, making restitution mandatory in violent crime cases and further expanding compensation and assistance to victims of terrorism. Victims were given the right to provide victim impact statements during sentencing in capital and noncapital cases, and the right to attend the trials of their offenders was clarified via the **Victims' Rights Clarification Act** (1997).

Victims' rights were further expanded in the first part of the 21st century. The **Violence Against Women Act (2000)** was signed into law as part of the Victims of Trafficking and Violence Protection Act of 2000. It reauthorized some previous VAWA funding. This legislation also authorized funding for rape prevention and education, battered women's shelters, transitional housing for female victims of violence, and addressed violence against older women and those with disabilities. This act also expanded the federal stalking statute to include stalking over the Internet. In 2004, Congress passed the **Justice for All Act**, thus strengthening federal crime victims' rights and providing enforcement and remedies when there is not compliance. It also provided monies to test the backlog of rape kits.

Despite the provision and expansion of victims' rights at the federal level, there is still not a federal constitutional amendment. This lack of adoption may be somewhat surprising since the National Victims' Constitutional Amendment Network and Steering Committee was formed in 1987 and federal victims' rights constitutional amendments were introduced in both the House and the Senate in 1996. Additional victims' rights constitutional amendments were introduced in 1997, 1998, 1999, 2000, 2003, and 2004 (Maryland Crime Victims' Resource Center, 2007). To date, such an amendment has not been adopted.

▲ **Photo 4.1** U.S. President Bush shakes hands with Sen. Pat Leahy (D-VT) (2nd R), and Sen. Arlen Specter (R-PA) (R), during a bill-signing ceremony for the Violence Against Women and Department of Justice Reauthorization Act of 2005 at the White House on January 5, 2006.

⬚ Financial Remedy

In Section 3, you read about the substantial costs that victims face after being victimized. Some of these costs are financial. Victims may lose time from work, have hospital bills, seek and pay for mental health care, need a crime scene cleaned, or lose income from a loved one's death. To help assuage some of these costs, victims can apply for financial compensation from the state, receive restitution from the offender, or seek remedy civilly.

Table 4.1 Federal Legislation Pertaining to Victims' Rights

Legislation Timeline	Key Provisions
Federal Victim Witness Protection Act (1982)	• Provided for the punishment of anyone who tampers with a witness, victim, or informant • If victim provided address and telephone number, required notification for arrest of the accused, times of court appearances at which victim may appear, release or detention of accused, and opportunities for victim to address the sentencing court • Recommended federal officials consult with victims and witnesses regarding proposed dismissals and plea negotiations • Required that officials not disclose the names and addresses of victims and witnesses
Victims of Crime Act (1984)	• Established the Crime Victims Fund, which promoted state and local victim support and compensation programs • In 1998, amended to require state programs to include survivors of victims of drunk driving and domestic violence in eligibility for federal funds
Child Victims' Bill of Rights (1990)	Children who are victims or witnesses are provided these rights: • That proceedings be explained in language children can understand • A victims' advocate can be present at interviews, hearings, and trial • A secure waiting area at trial • Certain personal information kept private unless otherwise specified by the child or guardian • An advocate to discuss with the court their ability to understand proceedings • Information provided about agencies for assistance and referrals made to such agencies
Victims' Rights and Restitution Act (1990)	Provided victims with the right to • be reasonably protected from the accused; • reasonable, accurate, and timely notice of any public proceeding involving the crime or any release or escape of the accused and to not be excluded from such proceedings; • be reasonably heard at any public proceeding involving release, plea, or sentencing; • confer with the attorney for the government in the case; • be given full and timely restitution as provided by law; • have proceedings free from unreasonable delay; • be treated with fairness and with respect for the victim's dignity and privacy
Violent Crime Control and Law Enforcement Act (1994)	• Allocated $1.6 billion to fight violence against women • Included money for victims' services and advocates and for rape education and community prevention programs
Violence Against Women Act (1994)	• Provided $1 billion to programs designed to reduce and respond to violence against women • Increased funding for victim compensation programs and established a national sex offender registry
Antiterrorism and Effective Death Penalty Act (1996)	• Made restitution mandatory in violent crime • Expanded compensation and assistance to victims of terrorism
Victims' Rights Clarification Act (1997)	• Gave victims the right to provide victim impact statements during sentencing in capital and noncapital cases, and the right to attend the trial of their offender was clarified

(Continued)

Table 4.1 (Continued)

Legislation Timeline	Key Provisions
Violence Against Women Act (2000)	• Provided additional protections for immigrant victims of domestic violence • Authorized funding for rape prevention and education, battered women's shelters, transitional housing for female victims of violence, and addressed violence against older women and those with disabilities
Justice for All Act (2004)	• Provided additional federal protections of crime victims' rights • Provided funding to test the substantial backlog of DNA samples collected from crime scenes and convicted offenders

Victim Compensation

One way victims can receive financial compensation for their economic losses is through state-run **victim compensation** programs. First begun in 1965 in California, victim compensation programs now operate in every state. Money for compensation comes from a variety of sources. A large portion of funding comes from criminals themselves—fees and fines are collected from people who are charged with criminal offenses. These fees are attached to the normal court fees that offenders are expected to pay. In addition, the Victims of Crime Act of 1984 (VOCA) authorized funding for state compensation and assistance programs. Today, the VOCA Crime Victims Fund provides more than $700 million annually to states to assist victims and constitutes about one third of each program's funding (National Association of Crime Victim Compensation Boards, 2009). Not only did VOCA increase funding for state programs, it also required states to cover all U.S. citizens victimized within the state's borders, regardless of the victim's residency. It also required that states provide mental health counseling and that victims of domestic violence as well as drunk driving be covered.

Not all victims, however, are eligible for compensation from the Crime Victims Fund. Only victims of rape, assault, child sexual abuse, drunk driving, domestic violence, and homicide are eligible, since these crimes are known to create undue hardship for victims (Klein, 2010). In addition to the type of victimization, victims must meet other requirements to be eligible:

- Must report the victimization promptly to law enforcement; usually within 72 hours of the victimization unless "good cause" can be shown, such as being a child, incarcerated or otherwise incapacitated
- Must cooperate with law enforcement and prosecutors in the investigation and prosecution of the case
- Must submit application for compensation that includes evidence of expenses within a specified time, generally 1 year from the date of the crime
- Must show that costs have not been compensated by other sources such as insurance or other programs
- Must not have participated in criminal conduct or significant misconduct that caused or contributed to the victimization

Victims can be compensated for a wide variety of expenses, including medical care costs, mental health treatment costs, funeral costs, and lost wages. Some programs have expanded coverage to include crime scene clean-up, transportation costs to receive treatment, moving expenses, housekeeping costs, and child-care costs (Klein, 2010). Other expenses for which victims may be able to be compensated include the replacement or repair of eyeglasses or corrective lenses, dental care, prosthetic devices, and forensic sexual assault exams. Note that property damage and loss are not compensable expenses (Office for Victims of Crime, 2010) and only three states currently pay for pain and suffering (Klein, 2010). States have caps in place that limit the amount of money a crime victim may receive from the Crime Victims Fund, generally ranging from $10,000 to $25,000 per incident.

Although compensation clearly can provide a benefit for victims, there are some problems with current compensation programs. One problem is that only a small portion of victims who are eligible for compensation actually receive monies from these funds. The programs also do not seem to encourage participation in the criminal justice system. There is little evidence that persons who receive compensation are any more satisfied than others (Elias, 1984) or that they are more likely to participate in the criminal justice process (Klein, 2010).

Restitution

Unlike monies from crime victims' funds, **restitution** is money paid by the offender to the victim. Restitution is made by court order as part of a sentence—the judge orders the offender to pay the victim money to compensate for expenses. Much like compensation programs, expenses that may be recovered through restitution include medical and dental bills, counseling, transportation, and lost wages. Restitution can also be ordered to cover costs of stolen or damaged property, unlike in crime victim compensation programs. Restitution cannot be ordered to cover costs associated with pain and suffering; it is limited to tangible and documentable expenses.

Restitution has its benefits. It is based on the notion of restorative justice, which seeks to involve the community, the offender, and the victim in the criminal justice system. Paying restitution helps restore both the offender and the victim to their precrime status. Problematic, however, is that the offender must first be caught for restitution to be ordered. Often, crimes go unreported and offenders remain free from arrest. Even if an offender is arrested, it may be difficult for the court to determine an appropriate amount for restitution. How much money should be paid in restitution to a victim whose mother's engagement ring was stolen? The ring's worth to the victim may far outweigh the dollar amount a judge would require the offender to pay in restitution. In addition, many offenders lack sufficient funds to pay victims immediately, even when court ordered. As a result, restitution may not be met.

Civil Litigation

Although compensation and restitution programs may significantly aid victims in recouping crime victimization costs, not all economic costs may be covered. Recall, too, that neither program addresses pain and suffering costs (except for the three states that allow compensation for pain and suffering). To seek redress for these uncompensated costs, victims may pursue **civil litigation** against the offender. There are some key advantages afforded to a plaintiff (the person filing the lawsuit) in a civil suit. That person is a party to the lawsuit and is allowed to make key decisions regarding whether to accept a settlement—unlike in criminal court, where it is the state versus the defendant (National Crime Victim Bar Association, 2007). Persons can seek money for emotional as well as physical harm.

In addition, the burden of proof is different in the civil justice system. Liability must be proven by a fair preponderance of the evidence, not beyond a reasonable doubt, which is the standard of proof in the criminal justice system. If the court finds that the defendant is in fact liable, then the offender is held financially accountable for the harm caused to the defendant. Much like with restitution, however, the likelihood of the victim actually receiving the money awarded is tied to the offender being identified and the offender's ability to pay. Accordingly, it may be quite difficult for the victim to recover damages awarded. Also, the costs of entering into a civil lawsuit must be borne by the victim and can be quite expensive. The victim may have to hire an attorney, and civil lawsuits can sometimes drag on for years.

⊠ Remedies and Rights in Court

Rights are also afforded to crime victims in other phases of the criminal justice system. In the article included in this section, Meghan Stroshine Chandek and Christopher O. L. H. Porter (1998) discuss their findings from a study examining crime victims' satisfaction with the police. Although not discussed in detail in this section, police are often the first level of criminal justice with which crime victims interact. The response that victims receive from them may shape how they view the criminal justice system as a whole and may impact their future dealings (or not) with the system should they be victimized again. In addition to the police, the prosecutor and the courts also provide crime victims with rights. These rights are discussed below.

Victim Impact Statements

As previously discussed, the criminal trial involves two parties in an adversarial system that reflects crime as a harm against the state. As such, historically, victims seldom played more than the role of witness in the criminal trial. Not until the 1970s did victims receive rights that guaranteed them at least some voice in the criminal trial process. One of these rights was first adopted in 1976 in Fresno, California, and it gave the victim an opportunity to address the court through a **victim impact statement (VIS).** The VIS can be submitted by direct victims and by those who are indirectly impacted by crime, such as family members. The VIS is either submitted in writing or presented orally (victim allocution).

In the VIS, the harm caused is typically detailed, with psychological, economic, social, as well as physical effects included. Depending on the jurisdiction, the victim or others presenting a VIS may also provide a recommendation as to what the offender's sentence should be. Not only may the victim enter a VIS at sentencing, but most states allow for the victim to make a VIS at parole hearings as well. In some cases, the original VIS is included in the offender's file and will be considered during the parole process. In others, the victim is allowed to update the original VIS and include additional information that may be pertinent to the parole board. Less common, the victim may be allowed to make a VIS during bail hearings, pretrial release hearings, and plea bargaining hearings (National Center for Victims of Crime, 1999). Importantly, despite the victim's wishes, the VIS is used only as information and may impact the court's decision, but not always. As noted by the Minnesota Court of Appeals in *State v. Johnson* (2008), although the victim's wishes are important, they are not the only consideration or determinative in the prosecutor's decision to bring a case to trial.

There are many reasons to expect VISs to benefit victims. They give victims a right to be heard in court and allow their pain and experience to be acknowledged in the criminal justice process. As such, VISs may be therapeutic, especially if a victim's statement is referred to by the prosecutor or judge and if the victim's recommendation is in accordance with the sentence the offender receives. In addition to this potential therapeutic benefit, VISs may also provide valuable information to the court and criminal justice actors that allows them truly to understand the impacts criminal behavior has on victims. It may help the judge give a sentence that is more reflective of the true harm caused to the victim. Also it may prove beneficial to offenders to hear the impact of their crimes. Hearing the extent to which their actions hurt another person makes it more difficult for offenders to rationalize their behavior.

▲ **Photo 4.2** A victim delivers her victim impact statement in court during sentencing.

Photo Credit: © istockphoto.com/Rich Legg

SAMPLE VICTIM IMPACT STATEMENT GUIDE*

1. Please describe how this offense has affected you and your family.

2. What was the emotional impact of this crime on you and your family?

3. What was the financial impact of this crime on you and your family?

   (NOTE: Add "physical impact" for personal crimes.)

4. What concerns do you have, if any, about your safety and security?

5. What do you want to happen now?

6. Would you like an opportunity to participate in victim/offender programming (such as mediation/dialogue or victim impact panels) that can help hold the offender accountable for his/her actions? (NOTE: Only utilize this question if such programs are in place, and ensure that the victim has written resources that fully describe such programs.)

7. If community service is recommended as part of the disposition or sentence, do you have a favorite charity or cause you'd like to recommend as a placement?

8. Is there any other information you would like to share with the court regarding the offense, and how it affected you and your family?

   _____ Please check here if you would like to be notified about the status and outcome of this case.

*Allow as much space as is needed to complete the victim impact statement.

SOURCE: Justice Solutions (2002). Reprinted with permission.

Despite these proposed benefits, not all victims utilize the right to make a VIS. For example, recent data from Texas show that only 22% of VIS applications distributed to crime victims were returned to district attorney's offices (Yun, Johnson, & Kercher, 2005). The type of victimization for which VISs were submitted was most commonly sexual assault of a minor, followed by robbery (Yun et al., 2005).

Nonetheless, the reasons that victims in general do not make VISs are varied. They may not feel comfortable putting their feelings in writing or going to court and making a public statement; they may fear the offender and being retaliated against. Others may not be fully aware of their right to make a VIS or not know how to go about utilizing this right. Although it is certainly a victim's choice to make or not make a VIS, it may have an impact on the sentence the offender receives. Recent research shows that when VISs are made in capital cases, there is an increased likelihood that the offender will be sentenced to death (Blumenthal, 2009). Although a clear impact on noncapital offenses is not evident, research suggests that when VISs do impact sentencing, they do so in a punitive fashion (Erez & Globokar, 2010).

This may be good for the victim, but it does raise the issue of equal justice for offenders. Does an offender deserve a more severe penalty because a VIS is made? Conversely, do victims not deserve to have their offenders penalized as severely as others if they are not able or willing to make a VIS? This issue underlies some of the debate surrounding the use of VISs. The constitutionality of VISs has been questioned, particularly in capital cases. Current case law makes it constitutional for VISs to be made in capital cases. In *Payne v. Tennessee* (1991), the U.S. Supreme Court found that how the victim is impacted does not negatively impact the rights of the defendant—VISs are a way to inform the court about the harm caused. This decision allowed states to decide whether to allow VISs in capital cases.

The positive benefit for victims may be overstated in that making a VIS can be traumatizing for victims (Bandes, 1999). Victims may also be dissatisfied if their recommendations are not followed (Davis, Henley, & Smith, 1990; Erez, Roeger, & Morgan, 1994; Erez & Tontodonato, 1992). Furthermore, victims who make a VIS may not be likely to use and participate in additional criminal proceedings if they are victimized again, one of the key considerations in granting victims' rights (Erez & Globokar, 2010; Kennard, 1989).

Victim/Witness Assistance Programs

Victim/witness assistance programs (VWAPs) provide victims with assistance as they navigate the criminal justice system. These programs are designed to ensure that victims know their rights and have the resources necessary to exercise these rights. At its heart, however, is a goal to increase victim and witness participation in the criminal justice process, particularly as witnesses, with the notion that victims who have criminal justice personnel assisting them will be more likely to participate and to be satisfied with their experience.

These programs first began in the 1970s, with the first program established in St. Louis, Missouri, by Carol Vittert (Davies, 2010). Although not sponsored by the government, Vittert and her friends would visit victims and offer them support. Two years later, the first government victim assistance programs were developed in Milwaukee, Wisconsin, and Brooklyn, New York. Not long after, in 1982, the Task Force on Victims of Crime recommended that prosecutors better serve victims. Specifically, the task force noted that prosecutors should work more closely with crime victims and receive their input as their cases are processed. It also noted that victims need protection and that their contributions should be valued—prosecutors should honor scheduled case appearances and return personal property as soon as possible. To this end, VWAPs have been developed, most commonly administrated through prosecutors' offices but also sometimes run through law enforcement agencies. At the federal level, each U.S. attorney's office has a victim witness coordinator to help victims of federal crimes.

Today, these programs most commonly provide victims with background information regarding the court procedure and their basic rights as crime victims. Notification about court dates and changes to those dates is also given. They also provide victims with information regarding victim compensation and aid them in applying for compensation if eligible. A victim who wishes to make a VIS can also receive assistance from the VWAP in doing so. Another service offered by VWAPs is making sure the victims and witnesses have separate waiting areas in the courthouse for privacy. In some instances, VWAP personnel will attend court proceedings and the trial with the victim and their family.

Despite the efforts of VWAPs, research shows that some of the first of these programs did little to improve victim participation. The Vera Institute of Justice's Victim/Witness Assistance Project, which ran in the 1970s, provided victims with a wide range of services—day care for children while parents were in court, counseling for victims, assistance with victim compensation, notification of all court dates, and a program that allowed victims to stay at work rather than come to court if their testimony was not needed—to little "success" (Herman, 2004). An evaluation of the project showed that victims were no more likely to show up at court than those without access to these services. It was not until the Vera Institute developed a new program that provided victim advocates to go to court with victims that positive outcomes emerged. This program did, in fact, then have a positive influence on attendance in court (Herman, 2004). Few of the programs provide services identified in the research literature as most critical; instead, VWAPs are largely oriented toward ensuring that witnesses cooperate and participate in court proceedings rather than that crime victims receive needed services (Jerin, Moriarty, & Gibson, 1996). Victims' rights and the challenges that arise when trying to ensure that crime victims' rights are provided are discussed by Robert C. Davis and Carrie Mulford (2008) in their article in this section.

Family Justice Centers

Family Justice Centers have recently begun opening throughout the United States to better serve crime victims. Because crime victims often need a variety of services, family justice centers are designed to provide many services in "one stop." These centers often provide counseling, advocacy, legal services, health care, financial services, housing assistance, employment referrals, and other services (National Center on Domestic and Sexual Violence, 2011). The advantages of providing these services in one place are many—primarily, victims can receive a plethora of services without having to navigate the maze of health and social service agencies in their jurisdiction.

Restorative Justice

The traditional criminal justice system is adversarial, with the state on one side and the defense on the other attempting to determine if the offender did in fact commit a crime against the state. It is largely offender-centered—the offender's rights must be protected from investigation to conviction—and the victim traditionally has not been recognized as having a role beyond that of a witness, since crimes are considered harms against the state. Beginning in the 1970s, as discussed in Section 1, the victims' rights movement sought to garner a larger role for victims in the justice process and to ensure that victims are provided the services they deserved from the state and community agencies. Also during the 1970s, there was a movement in the criminal justice system to get "tough on crime." In doing so, more people were sentenced to prison and for longer and our correctional system moved away from a rehabilitation model to a justice model. No longer was the correctional system dedicated to "fixing" offenders—rather, its main focus became public safety by reducing crime. This reduction was thought to be achieved through the use of tough criminal sanctions rather than treatment for the offender. Although this experiment in incarceration is not over, another movement less focused on being punitive toward offenders within the criminal justice system also emerged during the 1970s—the **restorative justice** movement.

The restorative justice movement formally began in Canada in the 1970s, but some of its principles were in place long before. Our first "systems" of justice did not define crimes as harms against the state. As such, if a person was victimized, it was up to him or her or the family to seek reparation from the offender (Tobolowsky, 1999). It was essentially a victim-centered approach. As crimes were redefined as harms against the state (or the king), the system of justice that emerged was more offender-focused. Such a system was in place until the 1970s in the U.S., when people began to advocate for an increased role for the victim and for victims to receive rights similar to those of offenders. The restorative justice movement was an outgrowth of the attention given to the need for victims' rights and also the push-back from adoption of a crime-control model exclusively focused on punishment.

The restorative justice movement is based on the belief that the way to reduce crime is not by solely punishing the offender or by adhering to a strict adversarial system that pits the defendant against the state. Instead, all entities impacted by crime should come to the table and work together to deal with crime and criminals. In this way, the restorative justice movement sees crime as harm to the state, the community, and the victim (Johnstone, 2002). Accordingly, instead of offenders simply being tried, convicted, and sentenced without the victim and community playing more than a cursory role, the system should develop and adopt strategies to deal with crime that include all relevant parties. Instead of a judge or jury deciding what happens to the offender, the restorative justice movement allows for input from the offender, the victim, and community members harmed by the offense in making a determination of how to repair the harm caused by the offender. In this way, justice is not just handed down and does not just "happen;" it is a cooperative agreement. Simply stated, restorative justice is a process "whereby parties with a stake in a specific offence collectively resolve how to deal with the aftermath of the offence and its implications for the future" (Marshall, 1999, p. 5).

What types of programs meet this objective? Many of the programs in use today in the United States and throughout the world were adapted from or based on traditional practices of indigenous people, who, given their communal living situation, often have a stake in group members' ability to collaboratively resolve issues (Centre for Justice and Reconciliation, 2008). The most common types of programs are victim-offender mediation or reconciliation programs and restitution programs. Victim-offender mediation is discussed below, and restitution was discussed earlier in this section as a financial remedy for victims. Another program that is restorative in nature is face-to-face meetings between the victim and offender that do not involve formal mediation. **Family or community group conferencing** is also restorative. In this type of program, the victim, offender, family, friends, and supporters of both the victim and offender collectively address the aftermath of the crime, with the victim addressing how the crime impacted him or her, thus increasing the offender's awareness of the consequences of the crime (Centre for Justice and Reconciliation, 2008). Because supporters of both sides are present, it allows additional people with a stake in the process and outcome to give input. Victims and offenders report high levels of satisfaction with group conferencing (Centre for Justice and Reconciliation, 2008). Restorative justice is also practiced through **peacemaking** or **sentencing circles**. A circle consists of the victim, the offender, community members, victim and offender supporters, and sometimes members of the criminal justice community such as prosecutors, judges, defense attorneys, police, and court workers. The goals of the circle are to "build community around shared values" and to "promote healing of all affected parties, giving the offender the opportunity to make amends" and giving all parties a "voice and shared responsibility in finding constructive resolutions" (Centre For Justice and Reconciliation, 2008, p. 2). The circles are also designed to address the causes of criminal behavior. In sentencing circles, the parties work together to determine the outcome for the offender, while peacemaking circles are more focused on healing.

Victim-Offender Mediation Programs

Some victims may not wish to sit in the background and interact only on the periphery of the criminal justice system. Instead, they may wish to have face-to-face meetings with their offenders. As a way to allow such a dialogue between victims and offenders, **victim-offender mediation programs** have sprouted up throughout the United States, with more than 300 such programs in operation today (Umbreit & Greenwood, 2000). With the American Bar Association endorsing the use of victim-offender mediation and what appears to be widespread public support for these programs, victim-offender mediation is likely to become commonplace in U.S. courts (Umbreit & Greenwood, 2000). Victim-offender mediation is already widely used in other countries, with more than 700 programs operating in Europe (Umbreit & Greenwood, 2000).

Mediation in criminal justice cases most commonly occurs as a **diversion** from prosecution. This means that if an offender and victim agree to complete mediation and if the offender completes any requirements set forth in the mediation agreement, then the offender will not be formally prosecuted in the criminal justice system. In this way, offenders receive a clear benefit if they agree to and successfully complete mediation. Mediation can also take place as a condition of probation. For some offenders, if they formally admit guilt and are adjudicated, they may be placed on probation by the judge with the condition that they participate in mediation. In all instances, the decision to participate in victim-offender mediation programs is ultimately up to the victim (Umbreit & Greenwood, 2000). Most victims who are given the opportunity to participate in victim-offender mediation do so (Umbreit & Greenwood, 2000).

Victim-offender mediation programs are designed to provide victims—usually those of property crimes and minor assaults—a chance to meet with their offenders in a structured environment. The session is led by a third-party mediator whose job it is to facilitate a dialogue through which victims are able to directly address their offenders and tell them how the crime impacted their lives. The victim may also ask questions of the offender. To achieve the objectives of restorative justice, mediation programs in criminal justice use humanistic mediation, which is dialogue-driven rather than settlement-driven (Umbreit, 2000). The impartial mediator is there to provide unconditional positive concern and regard for both parties, with minimal interruption. As noted by Umbreit (2000), humanistic mediation emphasizes healing and peacemaking over problem solving and resolution. He notes,

> the telling and hearing of each other's stories about the conflict, the opportunity for maximum direct communication with each other, and the importance of honoring silence and the innate wisdom and strength of the participants are all central to humanistic mediation practice. (p. 4)

One tangible outcome often but not always stemming from victim-offender mediation is a restitution plan for the offender, which the victim plays a central role in developing. This agreement becomes enforceable in court, whereby an offender who does not meet the requirements can be held accountable.

What happens after an offender and victim meet? Do both offenders and victims benefit? What about the community? It is important to evaluate programs in terms of their effectiveness in meeting objectives, and victim-offender mediation programs have been assessed in this way. Collectively, this body of research shows many benefits to victim-offender mediation programs. Participation in victim-offender mediation has been shown to reduce fear and anxiety among crime victims (Umbreit, Coates, & Kalanj, 1994), including posttraumatic stress symptoms (Angel, 2005), and desire to seek revenge against or harm offenders (Sherman et al., 2005; Strang, 2002). In addition, both offenders and victims report high levels of satisfaction with the victim-offender mediation process (McCold & Wachtel, 1998; McGarrell, Olivares, Crawford, & Kroovland, 2000; see also Umbreit & Greenwood, 2000). Victims who meet with their offenders report higher levels of satisfaction than victims of similar crimes whose cases are formally processed in the criminal justice system (Umbreit, 1994a). In addition to satisfaction, research shows that offenders are more likely to complete restitution required through victim-offender mediation (Umbreit et al., 1994). More than 90% of restitution agreements from victim-offender mediation programs are completed within 1 year (Victim-Offender Reconciliation Program Information and Resource Center, 2006). Reduction in recidivism rates for offenders also has been found (Nugent & Paddock, 1995; Umbreit, 1994b). To learn more about victim-offender mediation programs, read the article by Patrick Gerkin (2009) in this section. In it, he discusses impediments for victim-offender mediation programs.

As you can see, our system has changed from victim-centered to entirely offender-focused and is now bringing the victim back into focus. Crime victims are afforded many rights in the criminal justice system. But, as you have seen, it is sometimes difficult for victims to exercise these rights, and they often have little recourse if their rights are not protected. These issues will certainly continue to be addressed as victims' voices are heard and their needs met.

SUMMARY

- Victims were first granted rights in the law in 1979. All states give the right to compensation, notification of rights, notification of court appearances, and ability to submit victim impact statements before sentencing.
- Other states may give the right to restitution, to be treated with dignity and respect, to attend court and sentencing hearings, and to consult with court personnel before plea bargains are offered or defendants released from custody. Other rights will also protect victims' employment status so they can testify against their offenders.
- There has been some resistance to states and the federal government giving victims formal rights. Although numerous federal acts have been passed with victims' rights in mind, there still is no victims' rights amendment in the U.S. Constitution.

- To help assuage some of the financial costs of a crime, victims can apply for financial compensation from the state, can receive restitution from the offender, or can seek remedy in civil court.
- A victim impact statement can be submitted by direct victims and by those who are indirectly impacted by crime, such as family members. In the victim impact statement, the harm that was caused is typically detailed, with psychological, economic, social, as well as physical effects included.
- Victim/witness assistance programs provide victims with guidance as they navigate the criminal justice system. These programs are designed to ensure that victims know their rights and have the resources necessary to exercise these rights. Another goal of these programs is to increase the likelihood that a witness or victim will interact with the criminal justice system.
- The restorative justice movement is based on the belief that the way to reduce crime is not solely by punishing the offender or by adhering to a strict adversarial system that pits the defendant against the state. Instead, all entities impacted by crime should come to the table and work together to deal with crime and criminals.
- To increase dialogue between offenders and victims, victim-offender mediation programs have emerged throughout the United States.

DISCUSSION QUESTIONS

1. Do you think it is the role of the criminal justice system to provide victims with rights? How else could we ensure that victims receive help?

2. What rights does the state in which you reside provide to crime victims? What rights do you think are most important?

3. Why would offenders be more likely to complete restitution in victim-offender mediation? Could it be used for other types of programs? Why or why not?

4. What types of services would Polly be eligible to receive? Explain.

KEY TERMS

victims' rights

notification

participation and consultation

right to protection

right to a speedy trial

Federal Victim Witness Protection Act (1982)

Victims of Crime Act (1984)

Child Victims' Bill of Rights (1990)

Crime Control Act (1990)

Victims' Rights and Restitution Act (1990)

Violent Crime Control and Law Enforcement Act (1994)

Violence Against Women Act (1994)

Antiterrorism and Effective Death Penalty Act (1996)

Victims' Rights Clarification Act (1997)

Violence Against Women Act (2000)

Justice for All Act (2004)

victim compensation

restitution

civil litigation

victim impact statement (VIS)

victim/witness assistance programs (VWAPs)

restorative justice

family or community group conferencing

peacemaking circle

sentencing circle

victim-offender mediation programs

diversion

INTERNET RESOURCES

Restorative Justice Online (http://www.restorativejustice.org/)

The restorative justice movement is concerned with repairing harm caused by crime. Restorative Justice Online provides information for criminal justice professionals, social service providers, students, teachers, and victims. It includes links to research as well as more general information. It also provides information for restorative justice around the world.

National Center for Victims of Crime Resource Library (http://www.ncvc.org/ncvc/main.aspx?dbID=DB_Links137)

The Center disseminates information online for crime victims and people working with crime victims or in the area of policy. In its resource library, you can find information on victim impact statements, statistics regarding the extent of various kinds of victimization, and information on how to assist lesbian, gay, bisexual, transgender, and queer victims, among other topics.

KlaasKids Foundation (http://www.klaaskids.org/vrights.htm)

States differ in the rights they provide to crime victims. This website provides links to information on each state's victims' rights.

National Association of Crime Victim Compensation Boards (http://www.nacvcb.org/links.html)

This website provides links to federal agencies and resources, national victim organizations, national and state criminal justice victim-related organizations, victim-related education links, state crime victim compensation boards, federal and state correctional agencies, victim service units, sex offender registries, and other resources. It is your go-to website for links related to crime victims.

Introduction to Reading 1

The authors used the theory of expectancy disconfirmation to understand victim satisfaction with the police. To do so, they conducted telephone surveys and collected official complainant records from a medium-sized Midwestern police department. The survey included 118 victims who experienced burglary or robbery and who reported their victimizations between May 15 and August 14, 1995. They found that crime victim expectations of the police role impacted satisfaction. In particular, they found support for expectancy disconfirmation theory—when victims felt that police did not meet their expectations, they were less satisfied than when their expectations were exceeded.

The Efficacy of Expectancy Disconfirmation in Explaining Crime Victim Satisfaction With the Police

Meghan Stroshine Chandek and Christopher O. L. H. Porter

The importance of studying crime victim satisfaction cannot be overstated, particularly in light of the community policing "revolution" of the 1980s and 1990s. This philosophy of policing encourages a police role that encompasses not only the more traditional crime-fighting mandate of the police, but also one that emphasizes service-oriented functions. Under a philosophy of community policing, it may be argued that the police are seen as having a "business orientation" (Ericson and Haggerty 1997). The police come to be viewed as consumer-oriented, providing a variety of services to their clientele—the public. As such, measures of police performance change to incorporate this dual emphasis on crime-fighting and service-oriented functions.

Under a crime-fighting mandate, measures of police performance are most often quantitative in nature. These quantitative indices of success may include the number of arrests made, the number of citations issued or clearance rates. When a police organization embraces a philosophy of community policing, however, other subjective and qualitative measures of performance must be considered to determine police effectiveness. Satisfaction with the police constitutes one such measure. As MacKenzie and Uchida (1994:286) state,

> It is evident that one important factor in effective policing is community acceptance of and interaction with the police. It is no longer considered appropriate to judge police effectiveness only by the number of arrests or convictions.

Put in the context of community policing, the study of satisfaction with the police takes on greater importance than it has had in the past.

SOURCE: Chandek and Porter (1998). Reprinted with permission.

The Philosophy of Community. Since the 1970s and early 1980s, a renewed interest in the plight of the crime victim has emerged (Karmen 1990). Criminologists and victimologists are particularly concerned that crime victims are not only being victimized by the perpetrator, but also by the very criminal justice system designed to assist them (Mann 1993). As a result, criminologists and victimologists have dedicated much of their efforts to studying the experiences of crime victims as their cases are processed by the criminal justice system. Through this research, many programs and initiatives have originated with the hope that crime victims not be further victimized, but rather treated with dignity and respect by the criminal justice system (Karmen 1990).

The theoretical basis of community policing also highlights the importance of studying crime victim satisfaction with the police. Under a philosophy of community policing, it is widely assumed that if the police can satisfy citizens (and thus crime victims), a variety of positive outcomes will result. A cornerstone of community policing is the belief that an improved relationship between the police and the public would ensue by forging a relationship that is cooperative rather than antagonistic—although recent research has not borne out this hypothesis (Frank et al. 1996; Reisig and Giacomazzi 1998).

The public may become a valuable resource by calling the police more frequently, providing the police with information and cooperating with police activities. If the police and the public work together, the police may become more effective and the public may become less fearful, more trusting and more willing to turn to the police when victimized. As Martin (1997:521) states,

> When victims perceive that their needs are met, police gain legitimacy, as does the criminal justice system. Victims are willing to use police as a resource.... The principles of community policing underscore the importance of police accountability to constituents which is reflected in victim satisfaction.

Literature Review

Crime Victim Satisfaction Research

Four primary studies have exclusively examined crime victim satisfaction with the police (Brandl and Horvath 1991; Poister and McDavid 1978; Shapland 1983). These studies have examined a variety of variables, such as crime victim demographic characteristics, type of crime, police response and case status variables, and the expectations of crime victims. This portion of the literature review is organized by the variable (or group of variables) under examination and summarizes both the consistency and inconsistency of the findings.

Crime Victim Demographic Characteristics. Studies that have examined the influence of age on satisfaction with the police have produced conflicting results. Poister and McDavid (1978) found that age was not related to reported levels of satisfaction with the police. Other studies, however, have shown that age significantly affects the levels of satisfaction with the police that are reported by crime victims (Brandl and Horvath 1991; Percy 1980). In Brandl and Horvath's (1991) study, older victim of serious property crimes (burglary or motor vehicle theft) were more likely to report higher levels of satisfaction with the police than were younger respondents. Percy (1980) also found older respondents more likely to report higher levels of satisfaction with the police.

Studies including the victim's race as an independent variable have likewise produced contradictory results. Poister and McDavid (1978) found race not to be significantly related to crime victim satisfaction with the police. Yet Percy's (1980) study found Whites to be significantly more likely to report higher levels of satisfaction than minority crime victims.

Of the studies exploring crime victim satisfaction with the police, only one found a statistically significant relationship between the gender of crime victims and their reported level of satisfaction with the police (Percy 1980). In this study, males were more likely to report higher levels of satisfaction with the police than were female crime victims.

Poister and McDavid (1978) explored the influence type of crime had on reported satisfaction levels of crime victims and found a significant relationship. In their study, higher levels of satisfaction were reported by persons who had fallen victim to more serious crimes. In addition, they found that "overall satisfaction varies with type of crime ... more than with [the] socioeconomic characteristics [of the victims]" (Poister and McDavid 1978:133).

Police Officer Investigative Activities/Effort. Two of the four studies discussed in this literature review examined the effect of certain police officer activities on crime victims' levels of satisfaction with the police (Brandl and Horvath 1991; Percy 1980). Brandl and Horvath (1991) found that when victims of property crimes perceived a greater degree of investigative effort, they were more likely to report being satisfied with the police.

Percy (1980) was also interested in the effect certain police officer activities—which could also be construed as indicators of police investigative effort—had on crime victims' satisfaction levels. This researcher found that some, but not all, of the actions he examined were significantly related to crime victims' reported level of satisfaction (making an arrest, comforting the victim and providing crime prevention information all significantly and positively influenced crime victims' satisfaction levels).

Police Officer Conduct. Brandl and Horvath (1991) measured police officer conduct using a "Professionalism Index." These authors found that police professionalism, as measured in their study, had the greatest impact on levels of crime victim satisfaction. As they stated,

> Thus, for all three crime types there was a strong and dependable relationship between the degree of professionalism perceived to have been exhibited by the investigating police officer and victim satisfaction; the more professional (courteous, understanding, concerned, and competent the officer was seen to be), the greater the likelihood of victim satisfaction. (Brandl and Horvath 1991:298)

Shapland (1983) also found police officer behavior crucial in determining levels of victim satisfaction. The major determinant of satisfaction for the crime victims in this study was attitude of the police officer(s). "Those police officers who appeared to be interested in what the victims said, took time to listen to them and seemed to take them seriously, promoted feelings of satisfaction in the victims" (Shapland 1983:235).

Expectations. Crime victim satisfaction studies that have looked at crime victims' expectations have done so in a very narrow and limited fashion. Percy (1980) examined the effects of response time on satisfaction levels in terms of the crime victims' expectations as they compared to the actual police response time. Results showed that when the response time of the police was faster than expected, the crime victim was more likely to be satisfied with the police. Brandl and Horvath's (1991) study produced concordant results finding that when police response time was faster than expected, the crime victim was more likely to report being satisfied with the police.

The Expectancy Disconfirmation Model

Attempts to understand the factors that play a role in consumer satisfaction with products have led consumer and marketing researchers to propose a model of consumer satisfaction: the Expectancy Disconfirmation model. Simply stated, this model postulates that consumer satisfaction is both a function of expectations regarding the attributes of a product and the actual attributes of the product (Cadotte, Woodruff and Jenkins 1987; Churchill and Surprenant 1982; Oliver 1977, 1980; Oliver and DeSarbo 1987). In other words, satisfaction is determined by how well a product measures up to consumers' expectations. The extent to which expectations are met or not met determines the extent to which expectations are disconfirmed—and thus the extent to which the consumer is satisfied (Oliver 1977).

According to Expectancy Disconfirmation theory, expectancy disconfirmation occurs when a consumer's expectations are not met. Three types of expectancy disconfirmation may be differentiated (Oliver 1980). Consumers experience *negative disconfirmation* when a product's attributes are less than expected. Consumers experience *zero disconfirmation* (or confirmation) when there is equality between a product's attributes and their expectations. Finally, consumers experience *positive disconfirmation* when a product's attributes exceed expectations. It is further proposed that a positive relationship exists between disconfirmation and satisfaction. In other words, the more a consumer's expectations are exceeded, the more highly satisfied he or she will be with the product.

The Expectance Disconfirmation model has received overwhelming support (Cadotte et al. 1987; Churchill and Surprenant 1982; Oliver 1977, 1980; Oliver and DeSarbo 1987). More recently, other variables have been added to provide a more comprehensive approach to examining consumer satisfaction (Oliver 1993; Spreng, Mackenzie and Olshavsky 1996). In addition, researchers have extended this model beyond product satisfaction to service satisfaction (Bitner 1990; Bolton and Drew 1991; Boulding et al. 1993). These efforts have produced findings similar to those for product satisfaction.

As potential "consumers" of police service, crime victims are believed to have expectations about such services. It is further believed that the police, through their performance, will confirm or disconfirm these expectations just as a product's attributes confirm or disconfirm consumers' expectations. It is hypothesized that the extent to which the police meet victims' expectations will, at least in part, determine victim satisfaction with the police. Expectancy Disconfirmation theory thus provides a promising framework for examining crime victim satisfaction with the police.

Hypotheses

Since crime victim satisfaction literature has produced contradictory findings as to the role that crime victim demographic characteristics and types of crime have on levels of satisfaction with the police, no specific hypotheses can be made with regard to the role these

variables may play in determining overall satisfaction with the police.

Crime victim research has highlighted the importance that police officer activities and conduct play in crime victims' satisfaction levels (Brandl and Horvath 1991; Shapland 1983). For the current study, two hypotheses can be made:

H1: The more investigatory activities the police perform, the more likely it is that crime victims will be satisfied.

H2: The more courteous the police are, the more satisfied crime victims will tend to be.

Both crime victim and consumer research have highlighted the role that the confirmations of expectation have in determining satisfaction levels (Brandl and Horvath 1991; Cadotte et al. 1987; Churchill and Surprenant 1982; Oliver 1980; Oliver and DeSarbo 1987; Percy 1980). Two specific hypotheses follow from using the Expectancy Disconfirmation model to understand the relationship between victim expectations and satisfaction:

H3: When crime victims' expectations are negatively disconfirmed, it is less likely they will be satisfied.

H4: When crime victims' expectations are positively disconfirmed, it is more likely they will be satisfied.

Measurements

Description, coding and values of the independent and dependent variables used in the analyses are shown in Table 1. Operationalizations of these variables are described below.

Police Officer Activities. Variables measuring police officer activities asked crime victims about their perceptions of officer investigative effort. During the telephone interview, respondents were asked whether the initial responding officer(s) did the following:

Table 1 Independent and Dependent Variables: Values and Descriptive Statistics

Variable	Values	N	%
Crime Victim Demographic			
Race	0 = Non-White	56	50.9
	1 = White	54	49.1
Gender	0 = Male	63	57.3
	1 = Female	47	42.7
Age	0 = 18–24 yrs.	26	23.6
	1 = 25–40 yrs.	45	40.9
	2 = 41–60 yrs.	28	25.5
	3 = 61+ yrs.	11	10.0
Type of Victim	0 = Robbery	64	58.2
	1 = Burglary	46	41.8
Police Response Variables			
Activity Scale	0 = 0/5 activates	1	0.9
	1 = 1/5 activates	7	6.4
	2 = 2/5 activates	22	2.0
	3 = 3/5 activates	34	30.9
	4 = 4/5 activates	26	23.6
	5 = 5/5 activates	20	18.2
Conduct Scale	0 = 0/5 behaviors	4	3.6
	1 = 1/5 behaviors	5	4.5
	2 = 2/5 behaviors	4	3.6
	3 = 3/5 behaviors	6	5.5
	4 = 4/5 behaviors	14	12.7
	5 = 5/5 behaviors	77	70.0
Crime Victim Expectation and Disconfirmation Variables			
Expectation Scale	0 = 0/5 activities expected	1	0.9
	1 = 1/5 activities expected	4	3.6
	2 = 2/5 activities expected	16	14.6
	3 = 3/5 activities expected	20	18.2
	4 = 4/5 activities expected	37	33.6
	5 = 5/5 activities expected	32	29.1
Disconfirmation Variable	0 = Negative disconfirmation	46	41.8
	1 = Confirmation	37	33.6
	2 = Positive disconfirmation	27	24.5
Dependent Variable			
Overall Satisfaction with the Police	0 = Dissatisfied	22	20.0
	1 = Satisfied	88	80.0

1. took notes;

2. made out a report;

3. attempted to find, locate or question additional witnesses;

4. searched for or collected evidence (e.g., fingerprints); or

5. provided information on available services or offered other advice.

Each question was asked separately; crime victims could answer each question with a response of *yes, no* or *uncertain*.

For the current analyses, the responses were summed to create a police Activity Scale that ranged in value from a low score of 0 to a high score of 5. Each value represents the number of affirmative responses crime victims gave. For example, a value of 1 on the police Activity Scale represents a case where the crime victim perceived the police officer to have done one of the five activities (e.g., takes notes, make out a report while on the scene). As evidenced in Table 1, 80.90

percent of surveyed victims answered that police officers performed some, but not all, of the activities.

Police Officer Conduct. In order to measure police officer conduct crime victims were asked whether the initial responding police officer(s)

1. was courteous or respectful,

2. was understanding,

3. was concerned,

4. took you seriously, and

5. listened to you.

Again, each question was asked separately, and subjects could reply *yes, no* or *uncertain.*

Responses were summed to create a police Conduct Scale ranging in value from 0 to 5. Zero represents a case where the police were not seen by the crime victim to have exhibited any of the above behaviors, and a value of 5 represents a case where the police were perceived to have exhibited all five behaviors. The descriptive statistics provided in Table 1 demonstrate that most crime victims (70%) perceived the police as exhibiting all the behaviors.

Crime Victim Expectations. Crime victims were asked a set of questions about their expectations that paralleled those regarding police activities, specifically whether they *expected* the police to

1. take notes;

2. make out a report;

3. attempt to find, locate or question additional witnesses;

4. search for or collect evidence (e.g., fingerprints); and

5. provide information on available services or offer advice.

Each question was asked independently, and respondents could answer *yes, no* or *uncertain.*

Reponses were summed to create an Expectation Scale that ranged in value from a low of 0 to a high score of 5. The fewer expectations the crime victim had with regard to police activities, the lower his or her score on the Expectation Scale. The frequencies for this variable show that, as a group, crime victims in this study tended to expect a lot from the police in terms of investigative activities: More than 62 percent of crime victims expected four or more activities from the police.

Expectancy Disconfirmation. To create this variable, the Expectation Scale was subtracted from the Activity Scale. This created values ranging from −5 to +5. In cases where the number of activities performed by the police were fewer than the number of activities expected by the victim, the result was negative disconfirmation. In other words, in these cases the police performed fewer activities than were expected by the crime victim. These instances are represented by negative values on the expectancy disconfirmation variable.

When the police performed exactly the same number of activities as the crime victim expected, crime victims experienced zero disconfirmation. These cases are represented by a zero value on the expectancy disconfirmation variable. In cases where the number of activities performed by the police exceeded the number of activities expected, the crime victim experienced positive disconfirmation. These cases are represented by a positive value on the expectancy disconfirmation variable. In these cases, the victim expected fewer activities than the police actually performed.

In this study (see Table 1), most crime victims (41.8%) experienced negative disconfirmation—they expected more investigatory activities than the police actually performed. Roughly 34 percent of victims experienced zero disconfirmation; their expectations and perceptions of police activity were the same. A little more than 24 percent of victims experienced positive disconfirmation, where their expectations were exceeded by the police.

Dependent (Y) Variable: Overall Satisfaction With the Police. The dependent variable in this study was operationalized as the crime victim's overall level of satisfaction with the police. This variable was measured by asking the respondent the following question: *Overall,* how satisfied were you with the way the police officer(s) handled the [entire] incident? The response options were *very satisfied, satisfied, uncertain, dissatisfied* or *very dissatisfied.* For analytic purposes, this variable was then recoded into a dichotomous variable consisting of the categories *satisfied* and *dissatisfied.* Neutral cases ($n = 8$) were dropped from the analyses. In this study, the great majority of crime victims (80%) were satisfied with the police.

Results

An analysis of variance (ANOVA) procedure was used to determine the difference between the mean levels of satisfaction across different values of the independent variables. The results of this analysis are contained in Table 2. In this study, the ANOVA procedure was used to determine whether satisfaction levels varied according to the different values of crime victim demographic variables, police response variables, and crime victim expectation and disconfirmation variables.

According to the results, analysis of crime victim demographic variables resulted in no significant findings. In other words, mean satisfaction levels of crime victims did not vary according to gender, race, age or type of victimization. There were, however, statistically significant differences in the means for the Activity Scale and the Conduct Scale.

The mean satisfaction levels of crime victims differed depending upon the level of investigative effort perceived to be demonstrated by the police. This was captured in the Activity Scale, where the police could be categorized as having performed between 0 and 5 of the following activities:

1. Taking notes

2. Making out a report

3. Attempting to find, locate or question additional witnesses

4. Searching for or collecting evidence

5. Providing information on available services or offering other advice

Table 2 portrays a distinct linear relationship between different values of the Activity Scale and satisfaction with the police. Crime victims were significantly more likely to be satisfied when the police were perceived as performing a greater number of activities.

Mean satisfaction levels of crime victims also differed depending upon how the crime victims perceived they were treated by the police. This was captured in the Conduct Scale, where police could be categorized as having exhibited between 0 and 5 of the following behaviors:

1. Acting courteous or respectful

2. Showing understanding

3. Being concerned

4. Taking the crime victim seriously

5. Listening to the crime victim

With the exception of the first category (in which all crime victims reporting that the responding officer exhibited none of the behaviors were satisfied), a general linear relationship was found between the Conduct Scale and satisfaction with the police. As the number of behaviors exhibited by the police increased, so too did the likelihood of crime victim satisfaction with the police.

It should be noted that the analysis of variance did not reveal a significant bivariate relationship between crime victim expectations (as measured in the Expectation Scale) and satisfaction with the police. While this may initially appear counterintuitive, the finding does not contradict the theoretical bases of this study. This study hypothesizes that it is expectations in *conjunction with* police officer activities that influence crime victims' satisfaction—not expectations alone.

This relationship is captured in the expectancy disconfirmation variable, which according to the ANOVA results was significantly related to crime

Table 2 Analysis of Variance (ANOVA) Results: Relationship Between Independent Variables and Satisfaction With the Police

Variable	Values	% Satisfied
Crime Victim Demographic Variables		
Gender	Male	76.1
	Female	85.1
Race	Non-White	75.0
	White	85.1
Age	18–24 yrs.	65.3
	25–40 yrs.	82.2
	41–60 yrs.	82.1
	61+ yrs.	100.0
Type of Victim	Robbery	78.1
	Burglary	82.6
Police Response Variables		
Activity Scale	0/5 activities	0.0*
	1/5 activities	40.0
	2/5 activities	50.0
	3/5 activities	50.0
	4/5 activities	64.2
	5/5 activities	93.5
Conduct Scale	0/5 behaviors	100.0*
	1/5 behaviors	14.2
	2/5 behaviors	63.6
	3/5 behaviors	82.3
	4/5 behaviors	92.3
	5/5 behaviors	100.0
Crime Victim Expectation and Disconfirmation Variables		
Expectation Scale	0/5 activities expected	100.0
	1/5 activities expected	100.0
	2/5 activities expected	100.0
	3/5 activities expected	90.0
	4/5 activities expected	72.9
	5/5 activities expected	68.7
Expectancy Disconfirmation Variable	Negative disconfirmation	56.5*
	Confirmation	95.5
	Positive disconfirmation	100.0

*$p(\mathrm{F}) < .05$

victims' satisfaction with the police. The impact of expectancy disconfirmation for satisfied crime victims was difficult to discern. Satisfied crime victims were distributed fairly evenly among the categories of the expectancy disconfirmation variable. The majority (39.8%) of satisfied crime victims fell in the confirmation group, where the police performed exactly the same number of activities as the crime victim had expected. Approximately one-third (30.7%) of crime victims whose expectations were exceeded by the police were satisfied. Surprisingly, almost 30 percent of crime victims who got less than they had expected (cases of negative disconfirmation) were still satisfied with the police. It would appear that for this group of crime victims, other variables played a larger role in determining their satisfaction with the police than did expectancy disconfirmation.

The utility of a theoretical framework lies in both its ability to predict and explain (Bacharach 1989). This study sought to determine whether using the Expectancy Disconfirmation model significantly increased our ability to predict crime victim satisfaction with the police. To make this determination, a multivariate regression analysis was necessary.

The dependent variable used in the current analyses comprised two categories—*satisfied* and *dissatisfied*. As such, the dependent variable was a dichotomous, categorical, ranked variable with no quantifiable difference between categories. Since the dependent variable was binary, linear regression models such as Ordinary Least Squares (OLS) were deemed inappropriate for use in the present endeavor. The data simply did not meet the assumptions required by linear regression models. According to Long (1997), when OLS is used with data with a binary dependent variable, possible consequences include: (1) heteroscedasticity, (2) residuals that are not distributed normally and (3) nonsensical probabilities. As such, it is much more appropriate to use a model designed for use with binary dependent variables. The Binary Probit regression is one such model.

Binary Probit analyses were conducted on two different models. Overall satisfaction with the police was regressed on crime victim demographic characteristics (race, gender, age and type of victim), the Conduct Scale and the expectancy disconfirmation variable. In this model, crime victim demographic characteristics were again found to be unrelated to satisfaction with the police. On the other hand, both the conduct of the police (as captured in the Conduct Scale) and the type of expectancy disconfirmation that crime victims experienced were significantly related to satisfaction levels. The more crime victims perceived the police to exhibit positive behaviors (e.g., being courteous and understanding), the more likely they were to be satisfied with the police. In addition, the more crime victims' expectations were exceeded by police activities, the more likely victims were to be satisfied.

Conclusions

Community policing has been characterized as having a "business orientation" (Ericson and Haggerty 1997). According to this analogy, members of the public are viewed as consumers of a variety of police services, and satisfaction with police service becomes an important means of measuring police effectiveness. One important group of police service consumers are crime victims.

The current study extended previous research in an attempt to reach a greater understanding of the factors influencing crime victim satisfaction with the

police. Using the Expectancy Disconfirmation model as a conceptual guide, this study was the first to specifically examine the central role that expectations play in determining satisfaction with the police. The statistical findings shed new light on satisfaction with the police and support each of the four hypotheses posited earlier in this paper. The first two hypotheses pertained to the perceived level of investigative effort and police conduct. According to the results, when police performed more activities at the scene (e.g., making a report, searching for evidence), crime victims were more likely to be satisfied, supporting Hypothesis 1. An increase in the number of positive behaviors exhibited by the police (e.g., being courteous, taking the victim seriously) led to the increased likelihood that crime victims would be satisfied, supporting Hypothesis 2.

Crime victim expectations about police officer investigatory activities and their subsequent type of disconfirmation appeared to exert significant influence on satisfaction levels. These data supported Hypothesis 3, which postulated that crime victims who experienced negative disconfirmation (or who received less than they expected) were less likely to be satisfied with the police. By the same token, crime victims whose expectations were positively disconfirmed (exceeded) tended to be more satisfied, providing support for Hypothesis 4.

The findings of this study clearly support previous research findings that police activities and conduct are important determinants of crime victim satisfaction (Brandl and Horvath 1991; Poister and McDavid 1978; Percy 1980; Shapland 1983). Furthermore, this study establishes that crime victim expectations and disconfirmation play an important role in determining reported levels of satisfaction with the police. It would appear that, as other researchers have discovered, using the Expectancy Disconfirmation model to explain satisfaction with service is efficacious (Bitner 1990; Bolton and Drew 1991; Boulding et al. 1993).

Implications

It is clear from the findings of this research that the conduct and activities of the police, as well as victim expectations of police investigatory activities, are related to crime victim satisfaction with the police. Police departments committed to practicing community policing may take several actions to improve satisfaction with the police.

This study, as well as previous ones (Brandl and Horvath 1991; Percy 1980; Poister and McDavid 1978; Shapland 1983), has demonstrated that police conduct is a crucial determinant of victim satisfaction. Police departments might implement training practices to encourage certain behaviors (being courteous or respectful). Generally speaking, police departments should develop training programs that help officers better interact with their consumers, particularly crime victims. These training programs could take the form of classes aimed at improving police officers' interpersonal skills or emphasizing the importance of interpersonal relations in policing. More specifically, these classes could educate police about the physical, economic and psychological hardships that victims experience and how they play an instrumental role in helping crime victims deal with such adversities.

Changing the means by which police departments evaluate (and by extension, reward) police officers might reinforce the training. The criteria by which officer performance is evaluated should be broadened to incorporate qualitative measures as well as traditional quantitative measure (such as arrests made or citations issued). Police organizations might actively seek citizen input regarding their dealings with police officers. Crime victims would be one logical source of such information. This change would certainly be in line with the goals of community policing (MacKenzie and Uchida 1994; Martin 1997). By changing evaluation systems in this way, police managers and administrators would send a strong message to the rank and file that their services, and the manner in which those services are delivered, is of vital importance. Though the rank and file may be strongly opposed to such "outsider scrutiny," the potential benefits may make it a risk worth taking.

Although this and other studies (Brandl and Horvath 1991; Percy 1980; Poister and McDavid 1978;

Shapland 1983) have found police investigative activities to influence levels of crime victim satisfaction, it is unrealistic to assume that the police should alter their manner of investigating crimes. The results of this study do indicate, however, that police departments might better satisfy crime victims by altering victim expectations of investigatory activities. The victims in this study tended to expect a great deal in terms of the investigative effort demonstrated by the police (62.7% expected the police to perform four or more investigatory activities). While research shows that certain police officer activities do not increase the likelihood of apprehending a suspect (Greenwood, Chaiken and Petersilia 1977), crime victims may well expect these activities to occur anyway and make satisfaction decisions based on whether the police perform them.

To alter crime victim expectations, the police who respond to a call might immediately inform victims what to expect in terms of their investigation. Furthermore, the police could provide victims with the reasoning behind their choices, which might also include a realistic assessment of their ability to solve the crime. By providing crime victims with clear instructions as to what to expect, it is possible that the police could create cases of confirmation or positive disconfirmation that may not have otherwise occurred.

⊠ **References**

Bacharach, S. B. 1989. "Organizational Theories: Some Criteria for Evaluation." *Academy of Management Review* 14(4): 496–515.

Bitner, M. J. 1990. "Evaluating Service Encounters: The Effects of Physical Surroundings and Employee Responses." *Journal of Marketing* 54:69–82.

Bolton, R. N. and J. H Drew. 1991. "A Multistage Model of Customers' Assessments of Service Quality and Value." *Journal of Consumer Research* 17:375–84.

Boulding, W., A. Kalra and V. A. Zeithaml. 1993. "A Dynamic Process Model of Service Quality: From Expectations to Behavioral Intentions." *Journal of Marketing Research* 30:7–27.

Brandl, S. G. and F. Horvath. 1991. "Crime Victim Evaluation of Police Investigative Performance." *Journal of Criminal Justice* 19:293–305.

Cadotte, E. R., R. B. Woodruff and R. L. Jenkins. 1987. "Expectations and Norms in Models of Consumer Satisfaction." *Journal of Marketing Research* 24:305–14.

Churchill, G. A. and C. Surprenant. 1982. "An Investigation into the Determinants of Customer Satisfaction." *Journal of Marketing Research* 19:491–504.

Ericson, R. V. and K. D. Haggerty. 1997. *Policing the Risk Society.* Toronto: University of Toronto Press.

Frank, J., S. Brandl, R. Worden and T. Bynum. 1996. "Citizen Involvement in the Coproduction of Police Outputs." *Journal of Crime and Justice* 19:1–30.

Greenwood, P. W., J. M. Chaiken and J. Petersilia. 1977. *The Criminal Investigation Process.* Lexington, MA: D.C. Heath.

Karmen, A. 1990. *Crime Victims: An Introduction to Victimology.* 2nd ed. Pacific Grove, CA: Brooks/Cole.

Long, J. S. 1997. *Regression Models for Categorical and Limited Dependent Variables.* Thousand Oaks, CA: Sage.

MacKenzie, D. L. and C. D. Uchida, eds. 1994. "Drug Policy Initiatives: The Next 25 Years." *Drugs and Crime.* Thousand Oaks, CA: Sage.

Mann, C. R. 1993. *Unequal Justice: A Question of Color.* Bloomington, IN: Indiana University Press.

Martin, M. E. 1997. "Policy Promise: Community Policing and Domestic Violence Victim Satisfaction." *Policing: An International Journal of Police Strategies and Management* 20(3): 519–31.

Oliver, R. L. 1977. "Effect of Expectation and Disconfirmation on Postexposure Evaluations: An Alternative Interpretation." *Journal of Applied Psychology* 62(4): 480–86.

———. 1980. "A Cognitive Model of the Antecedents and Consequences of Satisfaction Decisions." *Journal of Marketing Research* 17:460–69.

———. 1993. "Cognitive, Affective, and Attribute Bases of the Satisfaction Response." *Journal of Consumer Research* 20:418–30.

Oliver, R. L. and W. S. DeSarbo. 1987. "Response Determinants in Satisfaction Judgments." *Journal of Consumer Research* 14:495–507.

Percy, S. L. 1980. "Response Time and Citizen Evaluation of the Police." *Journal of Police Science and Administration* 8:785–86.

Poister, T. H. and J. C. McDavid. 1978. "Victim's Evaluations of Police Performance." *Journal of Criminal Justice* 6:133–49.

Reisig, M. D., and A. L. Giacomazzi. 1998. "Citizen Perceptions of Community Policing: Are Attitudes Toward Police Important?" *Policing: An International Journal of Police Strategies and Management* 21(3): 547–61.

Shapland, J. 1983. "Victim-Witness Services and the Needs of the Victim." *Victimology* 8:233–37.

Spreng, R. A., S. B. MacKenzie and R. W. Olshavsky. 1996. "A Reexamination of the Determinants of Consumer Satisfaction." *Journal of Marketing* 60(3): 15–32.

DISCUSSION QUESTIONS

1. In addition to expectancy disconfirmation theory, what else may impact how victims feel regarding how the police handle their cases? Do you think other factors may be more important?

2. What can police do to secure better victim satisfaction? Do you think police departments have placed much premium on achieving victim satisfaction?

3. The authors refer to crime victims as clients of the police. Given what you have read about crime victims, do you think this reference is accurate? Why or why not?

Introduction to Reading 2

In their article, Davis and Mulford (2008) provide an overview of the most common victims' rights provided to crime victims in the United States. The authors note the challenges that victims face as they try to assert these rights. They also discuss current developments in victims' rights and remedies, including compliance programs and victim law clinics, which are designed to increase agency compliance.

Victim Rights and New Remedies

Finally Getting Victims Their Due

Robert C. Davis and Carrie Mulford

During the last two decades, federal and state governments have dramatically expanded the rights of crime victims. Many forces have spurred this change, including activism by crime victims as well as national crime victimization surveys documenting surprisingly high rates of crime, yet low levels of crime reporting by victims (Tobolowsky, 1997; Young, 1999). In the early 1980s, President Reagan convened a Presidential Task Force on Victims of Crime to investigate crime trends and the treatment of victims by the criminal justice system. The Task Force's 1982 final report defined an agenda for restoring a balance between the rights of defendants and victims. It called for increased participation by victims throughout criminal justice proceedings and restitution in all cases in which victims suffer financial loss.

Even before the Task Force issued its report, however, Congress anticipated many of its recommendations in the 1982 Victim Witness Protection Act. This act authorized victim restitution and the use of victim impact statements at sentencing in federal cases. It also required the attorney general to issue guidelines for the development of further policies regarding victims and witnesses of crimes. Soon after, the 1984 Victims of Crime Act (VOCA) implemented more of the Task Force's recommendations on victim compensation. This second act by Congress redistributed monies levied from federal offenders to states, funding local aid to victims (Smith & Hillenbrand, 1999).

Recommendations by the Reagan Task Force regarding procedural rights for crime victims were at least as influential as those regarding restitution. First, Congress revised the Federal Rules of Criminal Procedure to require pre-sentence reports to include "any harm done to or loss suffered by any victim of the offense" along with "any other information that may aid the court in sentencing." Then, in the 1990 Victim Rights and Restitution Act, Congress gave crime victims in federal cases the right to notification of court proceedings and the right to attend them, the right to notice of changes in a defendant's detention status, the right to consult with prosecutors, and the right to protection against offender aggression. Under President Clinton, the 1994 omnibus Violent Crime Control and Law Enforcement Act gave victims in federal cases the right to speak at sentencing hearings, made restitution mandatory in sexual assault cases, and expanded funding for local victim services (Kelly & Erez, 1999; Kilpatrick, Beatty, & Howley, 1998).

In 2004, the Victim Rights Act, which provides for crime victims' rights in federal courts, was signed into law. As part of the 2004 act, victims of crime were given significantly expanded rights, including the right to be present and heard at public court proceedings involving release, plea, sentencing, or parole. The act also placed on the federal courts a duty to ensure that victims are afforded those rights. Previous federal law did not provide any means of enforcement and only recognized the right to be heard in federal district court for victims of violent crimes or sexual abuse.

The Rights of Victims Under State Laws

During the past 30 years, victims who previously had no rights to be notified or to participate in the criminal justice process have acquired statutory basic

SOURCE: Davis and Mulford (2008). Reprinted with permission.

rights and protections in every state, the first of which was passed in Wisconsin in 1979. Although every state has some form of victim rights legislation, the states differ in their definitions of who is eligible for rights, with some limiting rights only to victims of violent felonies or other subcategories of victims. About 40% of states extend rights to all classes of victims of crime (Howley & Dorris, 2007). States also differ in the types of rights provided. A study by researchers at the Vera Institute analyzed victim rights legislation from every state and then coded it using a standardized evaluation form (Deess, 1999). The study showed that rights to compensation, notification of court appearances, and submission of a victim impact statement before sentencing were provided in all states (at least for some victims). A majority of states also gave victims the right to restitution, to attend sentencing hearings, and the right to consult with officials before offers of pleas or release of defendants from custody. States vary widely in their eligibility requirements and organizational responsibility for the implementation of the rights (Goddu, 1993).

Notification

The right to notification is perhaps the most basic. Victims unaware of their rights and available services, or of the proceedings themselves, will be unable to exercise any rights they may have (Kilpatrick et al., 1998). Kilpatrick and his colleagues (1998) asked more than 1,300 victims to rank, in order of importance, 13 different rights. Three rights to notification ranked among the five most important, with the right to notification of suspect's arrest seen as "very important" by more than 97% of the victims interviewed—the highest rating overall. The researchers also categorized states as either strong-protection states or weak-protection states on the basis of the specificity, strength, and comprehensiveness of their victim rights to notification, participation, and restitution. They found that victims from strong-protection states were more likely to receive notification throughout criminal justice proceedings, including notice of arrests, trials, and parole hearings. Nonetheless, stronger legislation did not guarantee notification. Even in strong-protection states,

25% to 35% of victims did not receive required notification (Kilpatrick et al., 1998).

Participation and Consultation

The best-known form of participation by victims is the submission of a victim impact statement at the time of sentencing. Most states also authorize submission of a victim statement of opinion, a more subjective assessment by victims of the appropriate sentence. By 1997, 40 states had mandated that criminal justice officials consult victims before making decisions on bond, plea, sentencing, or parole (Kelly & Erez, 1999).

According to Kilpatrick et al.'s (1998) study, victim impact statements are the most frequent form of victim participation, submitted by more than 90% of people informed of their right to do so. In strong-protection states, the survey found that participation frequently went beyond these statements, with victims significantly more likely than those within weak-protection states to have input during bond hearings, provide testimony in court, and submit victim impact statements at parole hearings (Kilpatrick et al., 1998).

One fear voiced by opponents of expanded victim rights has been that more participation by victims would lead to harsher sentencing by judges. Injecting personal statements into the sentencing decision would reduce uniformity in sentencing, introduce a greater degree of arbitrariness, and result in harsher treatment of convicted offenders across the board (Abramovsky, 1986; Talbert, 1988). There is little research on this question, and the few studies that have examined it have produced inconsistent results (Erez & Tontodonato, 1990; Davis & Smith, 1994). This inconsistency may be the product of weak commitment to the use of these statements and other input from victims. For example, the study by Davis and Smith (1994) found that although prosecutors and judges endorsed victim impact statements in theory, many resisted integrating them into their established routines.

Compensation and Restitution

Crime victims can incur medical costs associated with physical or emotional trauma, repair, and replacement costs associated with property crime, and opportunity costs of time they lose, measured in lost income. In theory, victims can recover these costs either from offenders required to pay restitution or through public compensation. Nearly all states authorize corrections officials to require restitution from offenders as a condition of parole. In addition, officials in a majority of states have the authority to order offenders to pay restitution as part of a suspended sentence or work release (Smith & Hillenbrand, 1999).

Likewise, public victim compensation programs in the large majority of states are funded by fees or charges paid by the offender. Since 1984, the federal VOCA legislation has encouraged states to institute victim compensation programs, and in 1988, VOCA was amended to ensure that victims of domestic violence and drunk driving were not excluded from compensation. States whose programs meet VOCA requirements can draw on federal subsidies that cover up to 40% of their payments to victims. All states limit eligibility to victims who report crime to the police and help prosecute offenders (Smith & Hillenbrand, 1999).

In practice, a minority of victims appear to receive compensation or restitution. A 1991 study found that less than a third of victims of violent crime were encouraged by a criminal justice official to file for compensation (McCormack, 1991). Kilpatrick et al.'s (1998) survey found that fewer than 20% of people eligible for restitution received it. Moreover, contrary to expectation, judges in states with stronger victim protections were significantly less likely to order victim restitution for economic losses than were judges from states with weaker victim rights protections.

Right to Protection

Criminal justice officials are increasingly concerned with the safety of victims and witnesses, especially in cases involving intimate partner violence or gang crimes. A majority of states provide victims with some right to protection including information about measures to take in the event of intimidation by the defendant, no contact orders, and separate and secure waiting facilities in court buildings. An increasing number of states are protecting victims from possible intimidation by limiting disclosure of victim information in law enforcement or court records or by not requiring that they provide their address or place of employment in testimony given in open court (Howley & Dorris, 2007).

Right to a Speedy Trial

Approximately half of states provide victims the right to a speedy disposition or trial. Generally, such legislation requires the court to consider the interests of victims in ruling on motions for continuance (Howley & Dorris, 2007). In some states, the law also provides for accelerated dispositions in cases involving victims who are elderly, disabled, or minor children.

The Challenge of Implementing Victim Rights Legislation

One of the most frequent conclusions from empirical research in victim rights is that despite the scope of federal and state legislation, criminal justice systems do not honor these rights. Consequently, some states began to develop state-level victim services offices that serve as both oversight agencies monitoring compliance and centers for referrals and linkages to victim services organizations. In the early 1990s policymakers in Wisconsin created the Wisconsin Victim Resource Center, a body that functions to enforce the new victims' rights laws (Office for Victims of Crime, 1998). Still, as Kilpatrick and his colleagues observed after their 1998 survey, even within states with strong victim rights legislation, many victims are not notified about key hearings and proceedings, many are not given the opportunity to be heard, and few receive restitution. Although victims in these states generally fared better than those in states with weak victim rights legislation, as many as one third of victims in strong-protection states were not afforded the opportunity to exercise certain rights.

Kilpatrick et al.'s (1998) findings have been echoed in other recent studies. A study of Texas law enforcement officers revealed that only 25% fulfilled their statutory obligation to notify victims of the state compensation program (Fritsch, Caeti, Tobolowsky, & Taylor, 2004). An audit conducted in Florida found that between one quarter and one half of victims did not receive information on victim rights from first responders and that agencies were not consistently documenting compliance with victim notification requirements (Auditor General, 2001). Similarly, a survey of Oregon crime victims found that 30% to 60% were denied rights to be notified of court dates, parole hearing dates, and restitution (Regional Research Institute for Human Services, 2002).

Officials in criminal justice agencies responsible for victims argue that state legislatures often do not provide funding to implement victim rights statutes. Criminal justice officials surveyed by the American Bar Association in 1989 were quite happy with their state's victim rights legislation, believing that it increased victims' satisfaction with officials and the criminal justice system, increased victims' willingness to cooperate, increased information for making case decisions, and improved their job satisfaction. But a major source of dissatisfaction was with the lack of resources provided to implement the legislation (Smith & Hillenbrand, 1999). These complaints were echoed in the Kilpatrick et al. (1998) study; in which, state and local officials indicated that inadequate funding, training, and enforcement of rights still present problems. According to this study, only 39% of local officials from strong-protection states and 27% of those from weak-protection states felt that funding for victim rights implementation was sufficient. A more recent study by the Vera Institute found that a majority of prosecutors believed that victim rights laws had imposed significant costs on their offices and other criminal justice agencies, requiring them to hire new staff and spend more money to contact victims by mail and by phone (Davis, Henderson, & Rabbitt, 2002).

Few states provided some form of remedy in their original victim rights legislation for victims whose rights were not honored by criminal justice officials. In fact, at one point, more than 15 states banned legal challenges to case resolutions or other redress for denial of victim rights by including clauses stating that the violation of a right did not create a civil cause of action against any government agency or official and that the failure to provide a right to a victim could not be used as a ground for appeal (Tobolowsky, 1997).

Those concerned with the rights of victims have come to realize that it is not enough to grant rights to victims: These rights must be backed up with ways to ensure that the agencies and officials responsible for informing victims of rights actually provide that information to victims. Beloof (2005) argues that there are three obstacles to full ability of victims to exercise their rights: government discretion to deny rights, lack of a way to enforce rights, and appellate court discretion to deny review. He further argues that victims will achieve real rights when they get legal standing to defend their rights, when appellate courts can void court decisions made in violation of victim rights, and when review of violations is a matter of right. These remedies have been the focus of new legal and programmatic efforts on behalf of victims.

Constitutional Amendments

In 1982, California was the first state to pass a constitutional amendment enumerating crime victims' rights that included the right to public safety. The Reagan Task Force helped to spur the adoption of constitutional amendments by other states with its recommendation for a federal constitutional amendment to ensure victim rights. As a step toward building support for a federal constitutional amendment, members of the Task Force helped launch a 1986 National Victims' Constitutional Amendment Network of crime victim advocacy groups pursuing state-level victim rights and state constitutional amendments (Young, 1999). Today, 32 states have adopted constitutional amendments regarding victim rights (Howley & Dorris, 2007).

Constitutional amendments provide greater assurance that victim rights will actually be observed. Because amending a state constitution is more difficult than passing a statute, this approach gives victim rights a greater degree of permanence. Also constitutional amendments take precedence over conflicting statutory provisions; courts have generally honored rights contained in constitutional amendments as indicative of the public will (Howley & Dorris, 2007). Finally, constitutional protection for victim rights makes enforcement potentially more likely. Court decisions in some states have held that victim rights amendments give victims' legal standing to pursue their rights.

However, Beloof (2005) argues that constitutional amendments alone are inadequate because most state constitutional rights of victims are silent about available review and remedies. Moreover, government discretion typically curtails victim standing to enforce constitutional rights. Contrary to the intent of constitutional amendments, Beloof cites cases in some states in which state court judges have declared victim constitutional rights to be advisory rather than mandatory.

Compliance Programs

Victim rights compliance programs are responsible for educating criminal justice agencies and often the public about victims' rights. Crime victim compliance programs currently exist in 13 states. In most of the states where compliance programs exist, the programs were established as part of the enabling legislation that accompanied the state constitutional amendments. However, there is a great deal of variety in the autonomy and authority among the programs. Most of the compliance programs are located in either the governor's office or in the state's Department of Justice. At least a couple of states (e.g., Alaska and Connecticut) created independent "watchdog" agencies to oversee the enforcement of victims' rights. Most states rely on legislative appropriations and or VOCA funds to support their victim rights compliance programs.

Many programs conduct trainings, either on their own or in conjunction with other mandatory training programs. The programs also field calls from victims and the general public and provide information about victim rights and referrals to other victim services, including compensation. All the crime victim compliance programs receive complaints from victims when an individual feels that his or her rights have been violated. Most of the programs have a formal complaint process, whereby the victim fields a complaint and the compliance officer (also known as compliance coordinators or ombudsman) communicates the complaint to the individual or agency against whom the complaint was directed. Usually, the complaint is investigated by the compliance officer or a compliance board, and the officer or board determines if the compliance programs differ in how violations are handled. In most states, the compliance officer communicates the violation to the agency against whom the complaint was made and attempts to educate the agency about victims' rights. Some states file annual reports to the legislature or other commissions or boards in which the offending agencies are named, whereas other are not permitted to name specific agencies (e.g., South Carolina). Mandated training (e.g., Maryland) and corrective action plans (e.g., Colorado) are available to some compliance programs. Alaska is the only state that has the authority to set fines or apply for criminal misdemeanor charges for serious infraction. Connecticut has used press conferences and public release of infractions as mechanisms to gain compliance.

Several victims' rights compliance programs provide an arbitration role and do not consider themselves to be either victim advocates or agency advocates, whereas others consider themselves to be victim advocates (Edwards, Myrus, & Felix, 2006). In only a couple of states (e.g., Connecticut and Alaska) are the compliance programs given the authority to appear in court on behalf of victims.

Victim Rights Clinics

In an effort to promote the enforcement of victims' rights, as well as awareness and education in the area of crime victim rights, the National Crime Victim Law Institute (NCVLI) was established in 2000. In 2002, to help address the enforcement of victims' rights through direct pro bono representation of victims in the court process, the Office for Victims of Crime within the U.S. Department of Justice entered into a cooperative agreement with NCVLI to establish legal clinics in several jurisdictions. The clinic demonstration project was created to advocate for the expansion of the enforcement of victims' rights in the criminal justice system and to educate legal professionals about victim rights law. Then in 2006, Congress appropriated monies to support NCVLI in its efforts to enforce crime victims' rights in federal jurisdictions through federal clinics.

The first state clinic was funded in Arizona, followed by clinics in California, Maryland, New Mexico, and South Carolina the following year. In 2005, four additional clinics were added, including state clinics in Idaho, New Jersey, and Utah and a federal clinic in Arizona. Two additional federal clinics, one in South Carolina and one in Maryland, were added in 2007. NCVLI continues to expand and provide funding to the existing clinics. NCVLI serves as an umbrella organization and, in addition to funding the clinics, provides them with training and technical assistance, including programmatic and financial monitoring, and legal support and research.

The eight state clinics all have the same basic mission and goals but operate through a variety of models. Two of the clinics, California and Idaho, are operated as law school clinics, with supervision of law students by the clinic director. Several clinics were developed within existing victim service organizations and employ staff attorneys to represent victims (e.g., Maryland, South Carolina, and Arizona). Two other clinics developed within nonprofit organizations that serve targeted victim populations (e.g., Utah's clinic was initially located within the Salt Lake Rape Recovery Center and New Mexico's clinic is located within the DWI Resource Center). New Jersey is the only clinic that was developed independently, specifically to enforce the rights of crime victims. The various models offer different strengths and weaknesses (Small, Roman, Owens, & Shollenberger, 2006). For example, the clinics that are operated through law schools have better access to law students, so are better able to fulfill the goal of educating future lawyers in victim rights law. However, it is more difficult for these clinics to access client populations in need of representation in victim rights law cases. On the other hand, a state clinic that is located in a reputable victim service organization, such as Maryland that has well-established relationships with victims and other victim serving organizations, has ready access to victims but less interaction with law students.

One of the goals of the clinics is to represent victims in cases that have the best chance of making a significant impact to future victims. Some of the clinics have embraced this goal more fully than others (Small et al., 2006). Some of the clinic directors find it difficult to turn away any victims in need of representation, whereas other clinics would like to be more selective, but do not have a large enough case load to justify turning cases away. NCLI provides training and assistance to the clinics in selecting the most significant cases. The issues involved in the selection of significant cases are jurisdiction specific, but include such things as rights to be present, be heard, receive notice, and rights to privacy.

The NCVLI organization also hosts an annual conference on crime victim rights law, a cluster meeting of the clinic directors, and a membership organization, the National Alliance of Victims' Rights Attorneys (NAVRA). As indicated on NCVLI's Web site, NAVRA is an "alliance of attorneys committed to the protection, enforcement, and advancement of crime victims' rights nationwide." NAVRA membership allows attorneys to use the list server, receive conference call training on crime victim rights issues, and receive case updates. A federally funded evaluation of NCVLI and the state clinics began in January 2008.

Conclusion

The process of securing rights for crime victims has been a long and uneven one. Federal and state government efforts to pass legislation providing rights to crime victims helped to give victims greater involvement in criminal cases. However, legislation was not enough to guarantee victims' rights in the process. It soon became clear that those charged with assisting victims to exercise their rights were not always complying with statutes and that there was little recourse when they did not.

To try to ensure that victims received rights in practice, new approaches were adopted. Constitutional amendments were efforts to make victim rights fundamental and more enforceable. Compliance programs and victim rights clinics are more recent initiatives designed to improve compliance with victim rights statutes. These programs attempt to ensure that more victims are able to exercise their rights through training of criminal justice officials, representation of victims in individual cases, and filing complaints against agencies or individuals who deny victims their rights. A current evaluation will examine the extent to which the clinics are making the exercise of victim rights more universally accepted.

References

Abramovsky, A. (1986). Crime victim rights. *New York Law Journal, 3*, 1, 3.

Auditor General. (2001). *The provision of victim services pursuant to section 960.001, Florida statutes: Operational audit* (Report No. 02-044). Tallahassee, FL: Office of the Florida Auditor General.

Beloof, D. E. (2005). The third wave of crime victims' rights: Standing, remedy, and review. *Brigham Young University Law Review, 2005*, 256–370.

Davis, R. C., Henderson, N. J., & Rabbitt, C. (2002). *Effects of state victim rights legislation on local criminal justice systems.* New York: Vera Institute of Justice.

Davis, R. C., & Smith, B. E. (1994). The effects of victim impact statements on sentencing: A test in an urban setting. *Justice Quarterly, 11*, 453–469.

Deess, P. (1999). *Victims' rights: Notification, consultation, participation, services, compensation, and remedies in the criminal justice process.* New York: Vera Institute of Justice.

Edwards, J., Myrus, E., & Felix, T. (2006). *Evaluability assessment: State compliance projects.* Unpublished report completed for the National Institute of Justice, Department of Justice.

Erez, E., & Tontodonato, P. (1990). The effects of victim participation on sentencing outcomes, *Criminology, 28*, 452–474.

Fritsch, E. J., Caeti, T. J., Tobolowsky, P. M., & Taylor, R. W. (2004). Police referrals of crime victims to compensation sources: An empirical analysis of attitudinal and structural impediments. *Police Quarterly, 7*, 372–393.

Goddu, C. R. (1993). Victim rights or a fair trial wronged? *Buffalo Law Review, 41*, 245–272.

Howley, S., & Dorris, C. (2007). Legal rights for crime victims in the criminal justice system. In R. C. Davis, A. J. Lurigio, & S. Herman (Eds.), *Victims of crime* (3rd ed., pp. 299–314). Thousand Oaks, CA: Sage.

Kelly, D. P., & Erez, E. (1999). Victim participation in the criminal justice system. In R. C. Davis, A. Lurigio, & W. Skogan (Eds.), *Victims of crime* (2nd ed., pp. 231–244). Thousand Oaks, CA: Sage.

Kilpatrick, D. G., Beatty, D., & Howley, S. S. (1998). *The rights of crime victims: Does legal protection make a difference?* Washington, DC: U.S. Department of Justice, National Institute of Justice.

McCormack, R. (1991). Compensating victims of violent crime. *Justice Quarterly, 8*, 329–346.

Office for Victims of Crime. (1998). *Victims' rights compliance efforts: Experience in three states.* Washington, DC: Author.

Presidential Task Force on Victims of Crime. (1982). *Final report.* Washington, DC: Government Printing Office.

Regional Research Institute for Human Services. (2002). *Oregon crime victims' needs assessment: Final report.* Portland: Oregon Department of Justice.

Small, K., Roman, C., Owens, C., & Shollenberger, T. (2006). *Evaluability assessment of the State and Federal Clinics and System Demonstration Project: Final report.* Unpublished report for the National Institute of Justice, Department of Justice.

Smith, B. E., & Hillenbrand, S. W. (1999). Making victims whole again: Restitution, victim-offender reconciliation programs, and compensation. In R. C. Davis, A. Lurigio, & W. Skogan (Eds.), *Victims of crime* (2nd ed., pp. 245–256). Thousand Oaks, CA: Sage.

Talbert, P. A. (1988). Relevance of victim impact statements to the criminal sentencing decision. *UCLA Law Review, 36*, 199–232.

Tobolowsky, P. M. (1997). Constitutionalizing crime victim rights. *Criminal Law Bulletin, 33*, 395–423.

Young, M. A. (1999). Victim rights and services: A modern saga. In R. C. Davis, A. Lurigio, & W. Skogan (Eds.), *Victims of crime* (2nd ed., pp. 194–210). Thousand Oaks, CA: Sage.

DISCUSSION QUESTIONS

1. Why are the victim rights granted in most states the ones that are most common? What rights are not commonly provided but should be?

2. Why do you think so few victims receive compensation?

3. What barriers are in place to prevent agency personnel from meeting the requirements of victim rights legislation or victim rights amendments?

4. What else can be done to ensure that agencies provide victims' rights that are guaranteed to them?

Introduction to Reading 3

One of the programs derived from the restorative justice movement is victim-offender mediation. At its heart, restorative justice requires that offender, community, and victim needs be accounted for in programs. Gerkin (2009) investigates whether victim-offender mediation programs are truly restorative, given the roles each party plays. He addresses the barriers to successful programs through observing victim-offender mediations and examining agreements reached. His is a particularly useful piece in that it outlines the process of victim-offender mediation, includes questions typically asked by mediators, and contains documents presented to participants.

Participation in Victim-Offender Mediation

Lessons Learned From Observations

Patrick M. Gerkin

Restorative justice, in its many forms, has emerged as one of several competing philosophies to the approach of crime and justice in numerous countries throughout the world (Van Ness & Strong, 2006). An often cited definition provided by Tony Marshall (1996) states that restoratives justice is "a process whereby all the parties with a stake in a particular offence come together to resolve collectively how to deal with the aftermath of the offence and its implications for the future" (p. 37). There are many programs that claim to be part of the restorative justice movement, and as many if not more names to denote these different varieties. *Restorative justice* is used as an umbrella term to describe any number of programs that view crime and the response to crime through a restorative lens. Victim-offender mediation programs (VOMP), victim-offender reconciliation programs (VORP), family group conferencing, community reparative boards, sentencing circles, and sentencing panels are just a few of the names now used to denote restorative programs. In addition to these, there are a number of multiform programs that might include some combination of aspects from the programs previously listed. Declan Roche (2003) states, "Although this range illustrates confusion about the meaning and application of restorative justice, there remain four fundamental ideals: personalism, reparation, reintegration, and participation" (p. 60). According to Roche, the programs that attempt to integrate all four basic ideals represent the driving force of restorative justice.

One form of restorative justice that has seen continued growth is victim-offender mediation. In the United States, cases may be referred to victim-offender mediation programs from a variety of sources, including judges, law enforcement officers, probation officers, victim advocates, prosecutors, and defense attorneys.

The goals of victim-offender mediation as reported by Bazemore and Umbreit (2001) include

> Supporting the healing process of victims by providing a safe, controlled setting for them to meet and speak with offenders on a strictly voluntary basis.

SOURCE: Gerkin (2009). Reprinted with permission of Sage Publications on behalf of Georgia State University Research Foundation.

Allowing offenders to learn about the impact of their crimes on the victims and take direct responsibility for their behavior.

Providing an opportunity for the victim and offender to develop a mutually acceptable plan that addresses the harm caused by the crime. (pp. 2–3)

This list of goals put forth by Bazemore and Umbreit is far from exhaustive. The outcomes and processes compiled in this list do not speak to the restorative nature of the intended mediation outcomes such as meeting needs, empowering victims and offenders, recognition, and reintegration. These are the outcomes that make justice restorative.

The goal of this research is to extend the knowledge about victim-offender mediation as a restorative process. The findings are derived from the amalgamation of data collected through observations of the victim-offender mediation process and analysis of agreements produced within.

Dennis Sullivan and Larry Tifft (2001) liken restorative justice to needs-based justice. In needs-based justice, "we seek to create and apply restorative values and meet needs in a harm situation" (p. 101). Meeting the needs of the parties involved is how the situation is made right. This includes meeting the needs of not only the victims but also the offenders and the community. Thus, restorative justice represents a shift from a rights-based or deserts-based justice system to a needs-based justice system. Sullivan and Tifft state:

When we examine what is required to embrace a restorative approach to justice, we see a political economy in which the needs of all are met, but met as they are defined by each person. Such an approach towards justice puts a great premium on the participation of everyone, and on the expression of the voice of each. In other words, the well-being of everyone involved in a given social situation is taken into account: that is, everyone involved is listened to, interacted with, or responded to on the basis of her or his present needs. (pp. 112–113)

Justice begins with identifying the needs of the persons involved. This concept can be difficult to grasp because it lies outside of the dominant retributive paradigm. In a needs-based system the thoughts and feelings of all people are vital. The psychological and emotional needs of victims and offenders are going to vary from person to person, which is part of the reason why participation in the restorative process is so significant. The only way to uncover victim and offender needs is to provide them with opportunities to communicate exactly what those needs are.

Another goal of restorative justice is to empower the participants. This is accomplished by involving the participants in the process of achieving justice. Harris (2003) states:

Empowerment is achieved in part through active participation in the creation of the outcome produced by the restorative justice response to harm. Everyone needs to feel that they are in control of their own lives. Only then can they give to others, participate in intimate relationships, make contributions to community life, engage in cooperative activities, and exercise leadership. These capacities are to be valued and nurtured in everyone. Learning to accept responsibility for ourselves and our actions only when we have opportunities for choice and occasions to find and use our power. (p. 134)

The literature contained herein demonstrates why participation is so vital to restorative practices and needs to be examined as a topic of research in the evaluation of restorative justice. The four fundamental ideals of restorative practices—personalism, reparation, reintegration, and participation—identified by Roche (2003) implicitly suggest that the individuals who are all too often only subjects of the justice process need to participate in the restorative vision of justice. Furthermore, the goals of empowerment, recognition, and meeting needs cannot be met without the active participation of the individuals involved in the social practice of restorative justice.

Critiques of Restorative Practices

Recently, several critiques have emerged with specific focus on the power dynamics evident in restorative practices. One such work (Pavlich, 2005) examines the designations of victim and offender, suggesting that participants are encouraged by the mediators and in some ways by the mediation process to play particular roles in restorative justice. Pavlich is concerned with the ways in which one's response to the events that bring them to mediation is governed by the process.

> Victims do not exist *sui generis,* in and of themselves; that is, they do not exist in any absolute abstract sense, but are produced through rituals, rules and techniques of power embedded in such social practices as restorative justice techniques. One is not, in essence, a victim; more contentiously, one becomes a victim by participating in contexts designed to create particular forms of the victim identity. (p. 52)

According to Pavlich, these governmentalities create roles for both victims and offenders that dictate not only what is expected of them as participants but also what is not acceptable. Consequently, Pavlich suggests that participants are limited in terms of the types of participation allowed in mediation.

Arrigo, Milovanovic, and Schehr (2005) claim that master signifiers in the restorative process, such as reconciliation, healing, restitution, community, and responsibility, force victims to explain their experiences within this master discourse. "For victims and offenders, VOM discursive practices only offer the opportunity to locate experiences of pain, hurt, confusion, regret, retribution, and the like, within a master discourse" (p. 105). Caging the participants within this master discourse means the participants are robbed of the opportunity to fully articulate their experiences with the harm produced. They go on to state, "Lost in this more scripted process is the opportunity for more genuine self-disclosure, more authentic healing;

occasions that would otherwise facilitate the subject to speak his or her own 'true' words" (p. 106).

Another recent criticism leveled at restorative justice is that it appears to exist both in opposition to and within the criminal justice system. Restorative practices rely on the police, the courts, and even criminal law to set the restorative process into motion. They also make use of many of the concepts used by criminal justice practitioners. The terms *victim* and *offender* are used to describe the two parties who meet for mediation. These terms are familiar to the individuals who occupy these positions and their usage cements restorative practices within the confines of the traditional criminal justice system.

> We are suggesting that to conceive and speak of others in terms of identity-fixing and identity-separating categories such as offender and victim is itself a source of harm because these designations are personally deconstructive and non-integrative. By using them, we force upon the person harmed and the person responsible of the harm a fixed, false identity. (Sullivan & Tifft, 2001, p. 80)

Defining the situation in this way creates power relationships that must be acknowledged and that shape the behavior of the parties involved. Perhaps restorative justice is more coercive than conventional justice. "A far worse imbalance will emerge with the offender finding himself or herself not only lined up in defense against the state but also against the victims and perhaps some new entity or presence put there to represent the 'community'" (Harris, 2004, p. 34).

Finally, Howard Zehr (1990) states, "In the aftermath of crime, victims' needs form the starting point for restorative justice. But one must not neglect offender and community needs" (p. 200). The process of achieving justice begins with the needs of the parties involved, including victims, offenders, and the community. However, Sullivan and Tifft (2001) have noted that victim and offender needs exist on two separate levels. We pay close attention to the victim's psychological and emotional needs, and yet we often do not recognize the offender's psychological and emotional needs. Instead we focus on needs such as employment, housing, and education. There is little doubt that these needs are significant; however, as Sullivan and Tifft (2001) note,

> by focusing on this level of needs alone we do not show the same level of concern for them as those who have been harmed. This is true even when the former might also be suffering from isolation and disorientation, and require the same psychological care and emotional support that those they harmed require. (p. 83)

By addressing offender needs in this fashion, the retributive justice system often neglects the offender's other needs and as a result does little to address the issues that may cause one to engage in the harm-producing behavior in the first place. Often we find offenders are victims themselves in many ways. They are victims of violence, aggression, and neglect and may lack the emotional support and care networks that support their own psychological and emotional needs.

Harris (2004) acknowledges this as one significant challenge to restorative practices. She states:

> Equality refers to the basic, yet radical, idea that all persons have equal value as persons. Once we develop a true comprehension of the basic sameness that flows from equality, we find it impossible to justify doing to others what we do not want done to ourselves. A commitment to equality thus carries with it a commitment to mutual care for the growth and welfare of all. (p. 132)

To deny that offenders also have needs would be to deny offenders the opportunity to heal and to have their harms repaired. As such it would cease to be a true needs-based justice. This research examines impediments to victim and offender participation in the social practice of restorative justice. One of the key elements in the restorative justice process is meaningful participation. The ability of restorative practices to achieve the desired outcomes depends in part on the ability of the parties involved to act as participants in the restorative process. Roadblocks to participation represent roadblocks to the practice of restorative justice and consequently to restorative outcomes. Yet this aspect of restorative justice remains a largely unexamined research topic. This research extends the body of knowledge about victim and offender participation in restorative justice. This research fills another void as it joins only a handful of studies reporting results from research based on observations of restorative processes (see Karp, 2001).

Specific attention is paid to the extent that the mediators and/or the mediation process itself encourage the participants to play a particular role in the processing of their case through a victim-offender mediation program. Through an examination of the agreements produced within the mediations observed, this research also examines the power that the participants have to determine the outcomes of the mediation process.

Methods

The unit of analysis for this research includes mediations processed at a Balanced and Restorative Justice (BARJ) Center. The BARJ center opened in 2000 and today they operate a victim-offender mediation program for the delivery of restorative justice to local communities. With an average caseload of more than 400 cases per year, 409 in 2003 and 405 in 2004, this center is an exceptional site for the evaluation of victim-offender mediation as a form of restorative justice.

This BARJ center serves four counties, although a vast majority of the cases come from the county where the BARJ program is located. This county has a population of approximately 175,000. The U.S. Bureau of the Census data from the year 2000 indicate that the population was 81% White, 14% African American, 2% multiracial, and the remaining population was composed of less than 1% American Indian or Alaska native, Asian Indian, Chinese, Filipino, and Korean, or some other race. The population in the year 2000 was 49% male and 50% female. The median age in years was 36, with those 18 and older constituting 72% of the population.

The BARJ center handles mediations for both juvenile and adult cases, although a majority of the cases, more than 90%, involve juvenile offenders. All mediations occur at the BARJ center located in the heart of the downtown area adjacent to the county courthouse.

The program receives referrals from two sources. The first source of referrals is the court system. In court-referred cases, the mediation is used as a form of diversion. The second source of referrals is city police officers. These types of referrals are on the rise, accounting for approximately 25% of all cases processed by the BARJ center. The process is voluntary for both victim and offender. On referral to mediation, the BARJ center director establishes contact with the determined victim by telephone to inquire about participation in mediation. If the determined victim agrees to participate, the BARJ center director establishes contact with the offender to inquire about participation. If both parties agree to participate, the BARJ center director determines the appropriate program for the participants.

Referrals to the BARJ center are assigned to either a victim-offender mediation or a family group conference. Assignment into one of these two programs is made at the discretion of the BARJ center director, who screens the cases using the police reports, comments from the juvenile court or the arresting officer, and discussion with the victim and offender over the phone. According to the director, typical considerations for determining assignment include the seriousness of the harm, the restitution amount, the number of victims and offenders, and the perceived level of preparation required for the participants. The most significant variables of consideration are the seriousness of the harm and the need for participant preparation.

In some cases the BARJ center director determines that one or both participants need to be prepared before their case can be processed. These cases are referred to the family group conference program and involve more contact time between the BARJ center staff and the individual participants. Victims can be emotionally distraught, hostile, or simply have many questions about the restorative process. The extra preparation time allows the BARJ staff members to answer questions and to explain the mediation process and rules in more detail on a day prior to the scheduled conference. Offenders may also be curious about the process. Other offenders receive extra preparation time when the staff wants to ensure the individual is willing to take responsibility and that he or she will not be disrespectful to any of the individuals involved or to the restorative process. The need for preparation with either victim or offender is evaluated over the phone by the BARJ center director.

In family group conferences, a member of BARJ center staff, not the assigned mediator, meets with the offender and/or victim individually on a day prior to their scheduled mediation. This meeting allows for one of the BARJ center case managers to spend time preparing the participation(s) for the mediation. On the day of the mediation, the victim and offender start the mediation together, as they have already been briefed about the mediation process and are prepared to participate.

Cases in which the BARJ center director determines that the participants require less preparation are scheduled for the victim-offender mediation program. In the case of victim-offender mediations, the victim and offender will attend their premediation session with the assigned mediator, on the same day as their mediation. Offenders are asked to arrive ½ hr earlier than the victims for their premediation meeting. The premediation meetings are held individually and then the participants are brought together to start the mediation.

The BARJ center uses a standard introduction for the premediation sessions for both the victim-offender mediations and the family group conference. An outline of the process can be found in Appendix A. In addition to covering the information contained in the introductory outline, all participants are afforded the opportunity to ask questions about the process, restorative justice, their case, or any concerns they have about their scheduled mediation.

At the mediation itself, the participants are seated with their supporters on either side of a table with the mediator seated between the parties, at the head of the table. Procedurally, there is no difference between the victim-offender mediations and the family group conferences once the mediation begins. In late 2006, the BARJ center combined these programs and now offers a single program that they refer to as victim-offender mediation.

The BARJ center receives referrals for a wide array of criminal behavior; however, more than 60% of the cases processed are property crimes. The second leading cause of referral is for assaults, including young children and siblings. The BARJ center does not handle retail fraud cases, domestic violence cases between partners, or child abuse cases. They do process cases in which a child is abusive toward his or her parent(s) or sibling(s). Felony cases are extremely rare but have been referred to the center for mediation.

The BARJ center currently has 73 active mediators. All 73 have completed a 40-hr mediator training module following the BADGER model of mediation, and approximately half of them have additional victim-offender mediation training (see Appendix B). BADGER is an acronym that suggests an outline for the mediation process.

The 73 mediators who volunteer at the BARJ center come from six counties in this Midwestern state, although a large majority of them reside in the BARJ center's home county. There are 35 male and 38 female mediators, with a median age of approximately 55. The ages range from 32 to 80. However, the youngest and oldest are both extremes. The vast majority of the mediators are White; less than 10% of the mediators are of a racial or ethnic minority, mostly African Americans.

All mediations were observed by one researcher, who compiled detailed notes. All premediation sessions were also observed for each case processed as victim-offender mediation but not for those processed as a family group conference, as they occurred on different days than the conference itself. The observations were completed between May and July 2005.

Participant consent was obtained by the researchers on the date of the scheduled mediation. For victim-offender mediations, consent was gained prior to the premediation meeting between the mediator and the participants. In the case of family group conferences, consent was obtained prior to the beginning of the mediation itself. In the case of juvenile participants, both the juvenile and their adult guardian were invited to participate and asked to provide consent. All participants were given the opportunity to raise questions before signing the consent document. The consent document was created and approved as part of a proposal submitted to the Human Subjects Institutional Review Board at Western Michigan University. To alleviate some apprehension, the participants were told the researcher was there to observe the process and its outcomes, not the individual participants involved. Each participant invited to participate in the study gave consent for the researcher to observe their mediation.

The researcher was not seated at the same table as the participants during the mediation and did not participate in any of the mediations in any way. A total of 14 mediations were observed in which 17 agreements were produced and collected. In addition to the observations, post mediation survey data from 119 victims and 130 offenders were collected. The findings reported herein are derived from the observations and agreements produced by the respective mediations.

I must express some caution about the conclusions of this research because of the study's limitations. In particular, the small number of cases observed at this BARJ center makes it rather difficult to draw generalizable conclusions about this program, let alone about restorative practices as a whole. However, to the extent that all of these mediators were provided the same training and that each mediator follows the same procedural guidelines, one can assume these mediations would be representative of the mediation process at this BARJ center.

Despite these limitations, the results have much to offer. As Presser (2006) suggests about her observations of mediation, "It provides much-needed qualitative data on what goes on during victim-offender mediation, and thus offers a snapshot of restorative justice practice in situ" (p. 317). We must continue to evaluate restorative practices beyond the level of participant satisfaction and the ability to create agreements. Just because a program claims to be restorative, we cannot simply regard it as so and assume the outcomes will be restorative in nature. Restorative practices are a work in progress. Evaluations such as these can help shape the future of restorative justice. They can inform practitioners about what works and about the obstacles that stand in the way of achieving a justice that satisfies and restores people, repairs relationships, reintegrates participants, and meets needs. This research offers a firsthand examination of the interactions that take place in victim-offender mediations.

Findings

The findings of this research have been divided into three separate sections. The first section, titled The Mediations, includes background information about the participants, the process, and the mediations themselves. The second section, titled Participation, delineates the levels of participation observed in the mediations for victims and offenders and offers some explanation for the levels observed. This is followed by the section titled Effects of Power Imbalance on Level of Participation, which discusses the ability of the participants involved to influence the stipulations of the agreements produced by their respective mediations. The final section, titled Barriers to Participation, addresses one of the obstacles to participation for both victims and offenders and explains how traditional notions of justice can account for a lack of participation.

The Mediations

Twenty offenders and 16 victims participated in the 14 mediations observed. Eighteen of the 20 offenders were juveniles and 14 of the 16 victims were adults. Three of the cases observed had multiple offenders and two of the cases had multiple victims. Table 1 provides the demographic information of the mediation participants in the mediations observed.

Sixteen of the 20 offenders were accompanied by at least one supporter for their mediation. Twelve of the offenders had one supporter present. In 10 of those 12 cases the supporter present was the offender's mother, whereas the other two included a brother and a father. All of these offenders were juveniles. Four of the offenders had two members of their social network present. In each of these cases, the members present were the offender's mother and grandmother. All of these offenders were juveniles. The remaining four offenders had no members of their respective social networks present. This includes two mediations involving family members as both victim and offender and two others cases in which the offenders had no supporters present.

Victims were far less likely to have supporters present. Of the 17 victims involved in the mediations, only two had supporters present. In both of these cases the victim was a juvenile and the supporter present was the victim's mother. In three other cases there were two victims present, thus creating an opportunity for the victims to support one another, but no other supporters were present.

The crimes for which the individuals came together for mediation ranged from property crimes to violent personal crimes, including one status offense also. There were seven cases of breaking and entering, five cases of arson, three cases of assault, two cases of larceny and malicious destruction of property, and one case each of mail fraud, trespass, receiving stolen

Table 1 Demographics of Mediation Participants

	Female	Male	White	Black	Interracial	Adult	Juvenile
Victim	9	7	11	3	2	14	2
Offender	2	18	9	10	1	2	18

NOTE: For further information, including a case-by-case breakdown of demographic information regarding both victim and offender, see Appendix C.

property, and truancy. The five cases of arson and five of the seven cases of breaking and entering were the result of one case involving multiple offenders. The two cases of malicious destruction were also part of one case involving multiple offenders. One other offender was charged with two offenses, trespass and assault.

Five of the 14 mediations observed were considered family group conferences by the BARJ center and the remaining nine were considered victim-offender mediations. There were no differences in terms of the number of participants or the mediation process used in the family group conferences and the victim-offender mediation programs. For this reason, each of the encounters observed are referred to as mediations.

Participation

Following the lead of Karp, Sweet, Krishenbaum, and Bazemore (2004), I have categorized victim and offender participation into three groups: high, medium, and low. My classifications into one of these three categories were based on my observations of the mediations, with specific attention to the participants' contributions. The specific characteristics of participation identified and used to determine one's level of participation are outlined in Appendix C.

Victim Participation. There were a total of 16 victims in the 14 mediations observed. Eight of the victims were observed as having a high level of participation. Three of the victims were placed in the medium participation category and the remaining five victims were categorized as having a low level of participation. A further examination of these results revealed several interesting patterns of behavior.

Of the eight victims observed to have participated at a high level, seven of them had a preexisting relationship with their identified offender. A preexisting relationship was identified when the victim and offender knew one another and had some form of social relation prior to the mediation. In two of the mediations, the victim and offender were family members. Other examples include juveniles who were patrons of the business they harmed and were on a first-name basis

with the owners or managers; a juvenile in mediation with an adult administrator from the school the boy attended; and a juvenile who was part of the same circle of friends with the girl who harmed her. A preexisting relationship between the victim and offender appears to be one source of strong victim participation.

A preexisting relationship was also an indicator of victim lecturing in the mediations observed. Victim lecturing was identified when the victims talked down to the offenders and addressed the offender as a superior or authority figure. This included reprimands and disapproval of what victims had identified as bad behaviors and warnings about consequences for further bad behavior. The victims were found to have lectured the offender in four different mediations. All four of these mediations involved adult victims and juvenile offenders, and three of the four involved a preexisting relationship. The school administrator lectured his former student as though he was in his own office, talking down to the juvenile and issuing numerous warnings. Another victim of a breaking and entering lectured for several minutes at his juvenile offenders about the value of hard work, and one of the parents involved with her child spent a significant portion of the mediation lecturing her child about responsibility. The victim's sex was not a predictor of lecturing, as two of those who lectured were male and two were female.

Victim lecturing is a powerful indicator of the power differentials within the relationship between the participants in the mediation. Victim lecturing sends a message to the offender that he or she is occupying a subservient status in the mediation process.

Of the four victims who demonstrated a low level of participation, one was a juvenile. Two of the other participants identified as having a low level of participation were involved in cases with multiple victims. In fact, there were only two cases that involved multiple victims and each of these cases had one victim who showed a low level of participation. In each of those cases, the other victim involved showed a higher level of participation, one high level and the other a medium level, respectively.

Juveniles were identified as victims in only two of the mediations. In one of those mediations, the juvenile demonstrated a high level of participation. This mediation involved juvenile girls as both victim and offender. The other case involved two juvenile males and both parties demonstrated a low level of participating. In this mediation, the juvenile victim's mother participated in the mediation extensively, even contributing to the stipulations of the agreement produced within the mediation.

There was only one mediation involving adults in the roles of victim and offender. This mediation involved two adult victims and one adult offender. The offender was a male and the identified victims were male and female. In this mediation, neither victim showed a high level of participation. The male victim demonstrated a medium level of participation whereas the female was observed to have participated at a low level.

Despite the relatively modest participation levels demonstrated by victims in the mediations observed, one stage of the mediation was dominated exclusively by victims. In the mediation's agreement-writing stage, victims were provided more opportunities to participate and consequently to identify needs and to have those needs addressed within the agreements produced. One of the tactics used to accomplish this end was selective facilitation. Selective facilitation is a tactic used by mediators to steer the mediation in one direction or another. It was used in virtually all of the mediations observed, to maneuver toward some issues and away from others. The use of selective facilitation can be accomplished in numerous ways. For example, the mediation can be moved into the agreement-writing stage when the mediator is satisfied with the interaction that has occurred by asking a series of questions. These questions are also indicative of what the mediators seek from the mediation participants. Selective facilitation was used in the mediations observed to elicit specific contributions from the victims, particularly in the late stages of the mediation where the participants are asked to identify their respective needs and to contribute to the creation of an agreement that will help them meet those needs. Typical questions posed by the mediators to the victims included

1. What can we do to make this right?

2. What would you like to see done in this situation?

3. What needs to be done to repair this situation?

4. What needs have been created by this harm? and

5. What would you like to see done here?

Although these questions are not standard, they are examples of the questions posed to the victims during the agreement-writing stage. This is not a criticism of the mediations observed. It demonstrates a commitment to identifying and meeting the victims' needs. Without this line of questioning it would be difficult to identify victim needs and to create an agreement that addresses them. These questions direct victim responses to issues that are important in the mediation's agreement-writing stage and demonstrate to the victims their ownership in the agreement created in response to harm they experienced.

Offender Participation

Seven offenders from four mediations were placed in the high participation category. In each of these cases, the offenders spoke at length about their involvement in the case. Each of these offenders answered questions and contributed significantly to the substance of the mediation, providing detail and even initiating conversations. In each of these mediations, there was very little parental involvement in the mediation process.

In five other mediations, the six offenders were placed in the medium participation category. These offenders contributed, but largely responded to questions, and rarely initiated conversations. When these offenders did respond to questions, they often provided very little detail. Significant parental involvement was noted in one of these mediations. The parent answered questions about the minor involved and offered information about the child, the offense, and the believed causes of his or her actions.

Finally, seven offenders from five mediations were placed in the low participation category. These offenders were virtually nonparticipatory. These offenders often responded to questions with one-or-two-word answers, if at all. They spent most of the time staring at the table or floor, looking out the window, and/or doodling on the scraps of paper provided by the BARJ staff.

Despite the high level of participation observed in cases of seven offenders, their participation was not consistently demonstrated throughout the various mediation stages. The high level of participation was common in the early stage where offenders spoke about their involvement in the harm and provided answers to the questions posed by their victims, but disappeared in the later mediation stages. Specifically, offender participation dissipated with the start of the agreement-writing stage.

Victim lecturing also affected the offender's level of participation. In three of the four cases in which the victim lectured the offender, the offender's participation was low. In all three cases, juvenile offenders had a visible response to the victim's lecture. Each of these offenders responded by lowing their head and falling completely silent, only responding to questions posed. In each of these cases there was a preexisting relationship with clear power differentials among the participants. In each case, the dominant party in the preexisting relationship delivered the lecture.

In one other case, the offender was observed to have a medium level of participation. In this case, the victim did not lecture the offender throughout the case but only during the agreement-writing stage. Despite the victim lecture, there was no preexisting relationship. However, this mediation, like the others involving victim lectures, involved an adult victim and a juvenile offender.

None of the offenders identified or expressed any needs in their own words or from their own perspective. Others, including parents and victims, spoke about what they believed the offenders needed, most often citing a lesson to be learned from the situation. In these cases, after mediation the agreement-writing stage was dominated almost without exception by the victims. As previously suggested, the questions posed to the participants are crucial. They indicate what the participants' responses and contributions should be in the various mediation stages. The line of questioning used

to address the offender in the agreement-writing stage was very different from what was used with the victims. Questions posed to the offenders included

1. Can you do this?

2. Does this sound fair to you?

3. Do you think you can do this?

These questions hardly amount to participation and certainly do not allow offenders to identify needs of their own; they simply ask the offenders to acquiesce to the victims' needs. Consequently, offender needs go unacknowledged and unaddressed, and offender participation in this mediation stage is rather limited.

In just two instances, the offender offered his or her own plan for how to repay his or her victim, to make the situation right. In one of the mediations, a boy wanted to work for a local marina to pay the restitution his victim was seeking. The offender had been apprehended by the police at the marina after stealing items from another business. In this instance, no damage was done at the marina and nobody was present as a marina representative to field such a request, so the offender's idea was dismissed. In another instance, two offenders requested to work off their restitution for the victim at his business. The business owner denied their request noting that such an arrangement would violate state labor laws. He further noted that it would be too dangerous. Eventually the participants left the BARJ center with the case unresolved and no agreement completed. The parties could not agree on the amount of restitution to be paid.

The point is not that the decisions to refuse these suggestions were wrong, for they appear to be quite logical. The point is that the ideas were rejected without exploring the more broad implications of the offer to make the victims whole again in hopes of repairing the situation. Each of the juveniles was trying to express what each believed to be the right thing. They hoped to work to repay the individual they harmed, and the ideas they suggested were not given full consideration.

The mediators' actions were directly responsible for the levels of participation exhibited by the offenders.

The mediators' questions were used by the offenders as an indication of the level of involvement they should have in the mediation process. The offenders responded to the questions with varying levels of detail, but because they were never asked to provide input to the agreements, they were unable to do so.

Effects of Power Imbalance on Level of Participation

Perhaps nowhere was the victim's power in the mediation process more evident than in the agreement-writing stage. A pattern that emerged within the agreements produced was that the victims often created stipulations in the agreement that far exceeded the scope of the harm they experienced. They often acted as the victim and judge. I am not suggesting they imposed guilt, but many took advantage of the opportunity to impose a sentence. Many of the victims created stipulations within the agreements that appear to go beyond making the situation right or meeting their needs.

One case involved a juvenile who had stolen some money from her mother's purse. Contained in the agreement was a laundry list of items, including the following:

1. (Offender's name) agrees that her friends will not call after 9:00 p.m.

2. (Offender's name) will respect the curfew hour established by her mother.

3. (Offender's name) agrees she will perform chores in a timely manner when requested to do so by her mother.

4. (Offender's name) agrees there will be no visitors in the home unless (Mother's name) is present.

These stipulations were in addition to finding a job, paying restitution, and a host of other items. The scope of this agreement goes beyond the harm the parties came together to discuss and it demonstrates the power victims hold in mediation, particularly in the agreement-writing stage. The purpose of the agreement is to repair the harm inflicted by the offender's action, to make the victim whole again. In this case, the mother used the agreement as a means to address a number of issues with her daughter's behavior, issues that have little to do with the harm created by her daughter's actions.

In another agreement produced by the mediation, the stipulations included the following:

1. (Offender's name) will fill out five job applications until he gets a job.

2. At the end of the week, (Offender's name) will send (Victim's name) copies of these applications (by way of parole officer).

3. Once a full-time job is obtained, (Victim's name) would like to see (Offender's name) maintain that job for at least a six-month period without any absence or tardiness.

If the offender was able to accomplish these tasks, he would not have to pay restitution to the victim. If he failed to do so, the offender was expected to pay the restitution in full.

Stipulations in various other agreements included to maintain a certain grade point average, enroll in the school band, perform various chores whenever asked by one's parents, read 20 books, flush the toilet, provide food and water to the dog daily, and keep one's room clean. Overall, I found the agreements went beyond the scope of the harm created in 7 of the 17 agreements produced.

The point here is not whether these stipulations are good or bad for the offenders. In fact, most of these stipulations appear to be suggested and written into the agreement with the offenders' best interest in mind. These stipulations may be in line with the current juvenile justice philosophy of promoting prosaic activities to help rehabilitate the juvenile offender and to insulate them from further trouble, but they also demonstrate the victim's power in the mediation process. In many ways this cements the practice of mediation alongside other forms of state-sponsored justice—a justice system in which the individuals involved do not participate.

Many of these stipulations go well beyond the harm created by the offenders' actions. They do not stem from any need identified by the victims but rather from the victims' personal feelings about what they believe the offender needs. What these data demonstrate is that victims hold too much power in the agreement-writing stage. They can effectively impose their will on the offender by individually creating the agreement and including stipulations that extend beyond the scope of the harm created by the offenders' actions.

There was only one case where the participants failed to reach an agreement. In this case it was the parents of the two juvenile offenders involved who disputed the victim's request for financial compensation. During the agreement-writing stage, the two parents actually went so far as to tell their children to be quiet as they negotiated the amount of restitution with the victim. The two boys sat silently as their parents debated with the victim to reach a dollar amount acceptable to each party. The victim and the offenders' parents could not agree on an amount of restitution and the individuals eventually left without an agreement in place. This was the only case where the victim's requests were not agreed to by the individuals responsible for the harm. However, it was not the offenders themselves who rejected the victim's request; it was their parents, acting on their behalf.

Barriers to Participation

In his book titled *Changing Lenses* (1990), Howard Zehr argues that for one to envision the use of restorative justice, one must first be able to examine crime through a restorative lens. This requires that one "change lenses" that allows one to see crime and the potential responses to crime in a new way. The problem, however, is that many people are not even aware that such a lens exists, thus making it virtually impossible for those people to view crime and responses like restorative justice in this fashion.

I noted throughout my observations that both juvenile and adult participants were relatively unclear about the purpose and goals of mediation. The individuals were skeptical of participation and juvenile offenders often expressed an interest in the notes that mediators made throughout the mediation. They appeared to

perceive the mediators as an authority figure, similar to a judge, able to make decisions and hand out judgments. One participant, when asked what the worst possible outcome of the mediation would be, stated, "To go to juvie [juvenile detention]." Her answer is very telling about her knowledge of the restorative vision of justice. Her bigger fear was to be sent to juvenile detention, a common result of a juvenile's interaction with the retributive criminal justice system. This suggests she assumed that the mediators of her case had the authority to make such a determination, which they did not. It is also an indication of the girl's beliefs about restorative justice and its home within the criminal justice field. The girl sat unprepared to be part of creating the justice herself—waiting for justice to be done to her and for someone to send her to juvenile detention.

Situating restorative justice within the criminal justice field gives victims and offenders an expectation about the process and outcomes. The restorative vision of justice shares very little with traditional forms of criminal justice, yet the participants' only knowledge of restorative justice is that of criminal justice. Because they have knowledge about criminal justice, they often believe this knowledge to be accurate about restorative justice as well. In some ways they are right. After all, the participants are appearing for mediation because of state intervention in their lives, for a crime as defined by the state. It does not matter how restorative justice practitioners choose to define their actions. Offenders are still referred to mediation because they have violated a rule or law, and therefore the state has intervened in their lives. Would these people be in mediation if not for the order of the state or a referral from the police? How can restorative justice stand in opposition to a system that it is a part of and in some ways strengthens?

Of the 20 offenders involved in the 14 mediations, only 7 were considered to have provided a high level of participation and these offenders demonstrated a high level of participation in only a portion of the mediation. Six others were placed in the medium category and the remaining seven offenders were considered to have participated at a low level. What causes such minimal participation in a process that is designed for and encourages offender participation? I argue one cause is they are not aware of the principles of restorative justice and that they are supposed to be actively involved in creating the mediation's outcome.

Victims are also unprepared and unaware of what the restorative process entails. They often consider the mediator to be the administrator of justice. The victims I observed commonly questioned the mediators or looked to them for guidance about decisions regarding the agreement. In two of the three cases involving community service, the victims asked the mediators to help them provide a number of hours to be completed.

Both victims and offenders lack a restorative vision of justice. We cannot assume that one premediation meeting is going to be enough to overcome years of experience with retributive forms of justice. For many, the retributive form has been internalized and it may take more than one day or one meeting to provide victims and offenders with a restorative vision that is so essential to their participation in restorative justice. When participants lack an understanding of the principles of restorative justice, they become subjects of the process rather than participants. This adds yet another layer to the power dynamics within victim-offender mediation. Both victims and offenders appear to perceive the mediator as an authority figure: offenders waiting for the mediator to hand out justice and victims looking to the mediator for guidance in producing the mediation's outcome. However, restorative justice is about ownership of the problem and the solution resting with the individuals involved. Yet without this knowledge, both victims and offenders perceive the mediator as an authority figure and the power obstacle to individual participation. The participants appear overpowered by the process. This can be attributed to their experiences and knowledge about traditional criminal justice processes within the United States' retributive criminal justice system and their belief that restorative justice is a part of it.

Conclusion

Meaningful participation is central to restorative processes like victim-offender mediation. Low levels of participation make it difficult for victim-offender mediation to achieve the fundamental goals of empowering, recognizing, repairing the harm, meeting needs, and reintegrating the participants. I attribute the low level of participation exhibited by the participants to two underlying causes. The first cause is a problem in the implementation of victim-offender mediation as a restorative process. The second and closely related cause is the power dynamics evident in the mediations observed that are the result of the aforementioned flaw in the mediation process implementation. One other form of domination that appears to affect both victims and offenders equally is their expectations about restorative justice.

My observations exemplify part of the criticism lodged against restorative justice by Pavlich (2005). Pavlich suggests restorative justice communicates to offenders that they are responsible for their harms and limits their involvement to an account of their responsibility in the harm produced. They can participate all they want in answering questions for their victims, explaining their involvement in the harms, and even offering an explanation for their actions. In a sense we ask the individual to acknowledge their responsibility and then sit idly by as the victim tells what they believe will make the situation right. This puts the offender in a difficult situation. They are not asked to contribute to the agreement, and having accepted responsibility for the harm they are given few viable options but to agree.

The agreement-writing stage of the mediation is perhaps the most important stage for participation on both sides. It is here that the individuals involved in the mediation come to own the response to the harm produced and ensure their needs are considered in the response. Maintaining a high level of participation throughout the mediation process is essential for the mediation to achieve the intended outcomes. An offender's agreement to the stipulations put forth in the agreement hardly amounts to full participation in the creation of the agreement.

Sustaining a high level of meaningful participation among the individuals involved requires a process that encourages and elicits participation throughout the process. This explains the higher level of participation among the victims in the agreement-writing stage as there is a premium placed on their participation. A high level of participation is elicited from the victims by the mediators through the questions posed, yet the same concern for the level of participation among offenders is not apparent.

Not only is victim participation elicited by the mediators but their authority also goes unchallenged. The victims have broad discretion to create restrictions on the behaviors of the offenders, who appear to be powerless in their ability to influence the agreement produced by their mediation. Moreover, the restrictions placed on the offenders go well beyond the harm experienced by the victims and appear to be a rather simple diagnosis of offender needs from the victim's point of view.

Offenders should not feel as though they are present only to accept responsibility and feel obliged to acquiesce to the victim's desires. The consequences of this are devastating for the practice of restorative justice. Without participation, offenders are less likely to feel empowered and to identify their needs, and consequently are less likely to have their needs addressed. This means restorative justice cannot be conceived of as needs-based justice. When it fails to identify and meet the offenders' needs, even when succeeding to do so for victims, the process is not needs based. In a needs-based response, the needs of one are not placed before another.

The power dynamics found in victim-offender mediation are further complicated by preexisting relationships in which there is a clear subordinate. The dynamics of the preexisting relationship carry over into the interaction between the participants in the mediation. When the participants enter the mediation with a preexisting relationship and the previous relationship is defined by a clear power imbalance, the power dynamics do not disappear. Instead, the parties enter the mediation in those same positions, and the individual with less power becomes less able to fully participate and influence the mediation's outcome.

Finally, both victims and offenders perceive restorative justice as a form of criminal justice. Because participants do not possess a restorative lens to look through, they are often misguided by their assumptions about the restorative process. Their views are confirmed or strengthened in some way by the mediation process itself. The designation of the individuals involved as victim and offender is familiar to the participants and helps establish their views of restorative justice as criminal justice.

Furthermore, restorative justice processes are only initiated in the wake of some behavior identified by the state as crime. Restorative justice then does not challenge the state's authority to define crime but strengthens it. "Restorative justice thus conceptually and practically subordinates itself to the very criminal justice system it claims to escape" (Pavlich, 2005, p. 35).

Much has already been communicated when they retain their retributive notions of criminal justice that have been internalized by years of living within a society that chooses to deal with crime in this fashion. They expect to be a spectator as someone, usually a judge, makes decisions about their fate.

Thus, what we have are participants who are largely unprepared in way of participation in a process that necessitates their participation for success. The situation is akin to placing someone into a foreign culture where common practices stand very much outside their own cultural norms (of which they know very little about) and asking them to participate. As in the situation described, full and knowledgeable participation is unlikely. It would take weeks, if not months, for the individual to learn about the culture, to dissocialize from their own culture, and to be resocialized before meaningful participation could occur. Yet there is an expectation within the practice of restorative justice for people to be prepared to examine the situation through a restorative lens when they simply do not possess one. Perhaps this expectation is unreasonable.

Implications

The implications of this research for restorative justice practitioners are many. This research suggests that all participants need to be prepared to participate in restorative processes. Participants must first come to see crime and the response to it through a restorative lens. If one is not able to view the situation through a restorative lens, he or she will be unable to view the restorative outcomes that are desired. Preparation involves helping individuals develop a restorative lens, making restorative outcomes a reasonable solution in the participant's eyes.

A second policy implication deals with the administration of restorative practices. The restorative approach to justice entails a political economy in which the needs of all individuals are met, but met as they are defined by each person (Sullivan & Tifft, 2001). Insofar as the victim's needs become the sole focus of restorative practices, without concern for the offender's needs, a needs-based justice is not achieved. When this happens, the restorative process makes possible—and even encourages—victim domination. In the practice of mediation, this translates into victims being asked to contribute more and to victims' desires becoming the sole focus of the agreements produced. This was evident in the mediations observed, particularly in the agreement-writing stage. Consequently, restorative outcomes like empowerment and meeting needs are less likely for offenders. A process or strategy that encourages offender participation, at least to the extent that it encourages victim participation throughout the mediation, is necessary.

One final implication for practitioners would be to acknowledge the power dynamics inherent in social practices such as restorative justice. Specifically, this research has discovered that preexisting power differentials among individuals tend to be reconvened in the restorative justice setting. When the power differentials manifest themselves in a reduced level of participation for the overpowered, it reduces the potential for restorative outcomes.

Finally, this research demonstrates a need for further inquiries about levels of participation in restorative practices and the need for preparation before participation. Ideally, this research will involve both observation and interviews with participants in restorative practices. Restorative practices are social events. They are very amenable to observation as a method of inquiry. Observations allow for an examination of these social events in their natural environment, and interviews provide an opportunity for those who have participated to use their own voice to articulate their experience with the restorative process.

✎ Appendix A. The Restorative Justice Center

Victim-Offender Mediation and Family Group Conferencing: Introduction

Welcome. . . . Thank you for participating. . . .

(Check name and address)

_____ and I are unpaid volunteers trained through the SCAO.

> We are nonjudgmental.
>
> We do not tell you what to do.
>
> We assist you in coming to an agreement.

Explain CONFIDENTIALITY (Sign forms)

Our purpose here today is sixfold:

1. Examine what happened. (Victim first)

2. Help the offender understand the harm done to:

   a. the victim

   b. the victim's family

   c. the community

   d. the offender's family

   e. the offender

3. Help the victim understand the offender's motives.

4. To the extent possible, identify what needs to be done to repair the harm.

5. To the extent possible, arrange compensation for the victim and the community.

6. To the extent possible, reconnect the offender to the families and the community.

RULES:

1. No interruptions

2. Civility

3. Destruction of notes to ensure confidentiality

QUESTIONS?

DO YOU ACCEPT THESE RULES?

⊠ **Appendix B**

B—BEGIN THE MEDIATION DISCUSSION

    Case intake

    Room preparation

    Who participates

    Opening statement

A—ACCUMULATE THE INFORMATION

    Assumption

    Bias awareness

    Listening/questioning/note-taking skills

D—DEVELOP THE AGENDA

    Identify the issues

    Frame in neutral language

    Order for discussion

G—GENERATE MOVEMENT

    Process the issues

    Persuasive techniques

E—ESCAPE TO CAUCUS (if necessary)

    Purpose

    Order

    Closing

R—RESOLVE THE CONFLICT

    Writing the agreement

    No agreement

    Closing the mediation

⊠ **Appendix C. Summary of Mediation Participants**

	Race	Sex	Status	Preexisting Relationship
Mediation No. 1				
Victim(s)	White	Male	Adult	Owner
Offender(s)	White	Female	Adult	Owner
	White	Male	Juvenile	Consumer
Mediation No. 2				
Victim(s)	White	Female	Adult	No
Offender(s)	White	Male	Juvenile	No
Mediation No. 3				
Victim(s)	White	Male	Adult	No
Offender(s)	White	Male	Juvenile	No

(Continued)

(Continued)

	Race	Sex	Status	Preexisting Relationship
Mediation No. 4				
Victim(s)	Interracial	Male	Juvenile	No
Offender(s)	Black	Male	Juvenile	No
Mediation No. 5				
Victim(s)	Interracial	Female	Juvenile	Same friends
Offender(s)	Black	Female	Juvenile	Same friends
Mediation No. 6				
Victim(s)	White	Male	Adult	No
Offender(s)	White	Male	Juvenile	No
Mediation No. 7				
Victim(s)	White	Female	Adult	No
Offender(s)	Black	Male	Juvenile	No
Mediation No. 8				
Victim(s)	White	Male	Adult	No
Offender(s)	White	Female	Adult	No
	White	Male	Juvenile	No
Mediation No. 9				
Victim(s)	White	Female	Adult	Manager
Offender(s)	Black	Male	Juvenile	Consumer
Mediation No. 10				
Victim(s)	Black	Male	Adult	Administrator
Offender(s)	Black	Male	Juvenile	Student
Mediation No. 11				
Victim(s)	Black	Female	Adult	Mother[a]
Offender(s)	Interracial	Male	Juvenile	Family friend
Mediation No. 12				
Victim(s)	White	Female	Adult	Mother
Offender(s)	White	Male	Juvenile	Son

	Race	Sex	Status	Preexisting Relationship
Mediation No. 13				
Victim(s)	Black	Female	Adult	Mother[a]
Offender(s)	Black	Female	Juvenile	Daughter
Mediation No. 14				
Victim(s)	White	Male	Adult	Neighbor/owner
Offender(s)	White	Male	Juvenile	Neighbor
	White	Male	Juvenile	Neighbor

a. Indicates lecturing present in the mediation and is placed by the individual who lectured.

There were adult victims and juvenile offenders in 11 of the 14 mediated cases. Furthermore, the victim and offenders had a previous relationship in that the victim held a superior position over the offender in four of the cases. Males were the offenders in all but two cases, although there was much greater variation among the victims. The victim was a female in nine of the mediated cases, and males were the victims in seven.

⊠ Appendix D

High Participation

Those individuals identified as having a high level of participation demonstrated conscious participation in the mediation by making and maintaining eye contact with the other participants. They demonstrated focus to the events and questions they were being asked by contributing not only often, but at length in the mediation process. Their contributions included both detail and substance in the mediation process. They responded to questions with direct answers that provided relevant information about the subjects of the questions. Above and beyond their level of responsiveness to questions posed, these individuals took an active role in determining the subject and direction of the mediation by initiating conversations.

Medium Participation

Those participants identified as providing medium participation were less attentive in the mediation and in the substance provided. They made eye contact sporadically throughout the mediation, but never consistently. These participants responded with limited detail and were less likely to initiate dialogue if at all. They responded to questions, although they offered little detail and often failed to address the subject of the question posed. These individuals acted almost exclusively in a responsive fashion, speaking only when asked to do so.

Low Participation

Those offenders characterized as having low participation were almost nonparticipatory. Despite their presence in the room, they showed no interest and offered very little in terms of substance to the mediation. These offenders responded only to questions and failed at times to even do this. Their eyes remained fixed on the table for most of the mediation, they doodled on the pads of paper provided, looked out the window, and generally showed a lack of interest in the mediation. When these offenders contributed it was often with one-or-two-word answers, and they offered very little detail and substance to the mediation.

⊠ References

Arrigo, B. A., Milovanovic, D., & Schehr, R. C. (2005). *The French connection in criminology: Rediscovering crime, law, and social change.* Albany: SUNY Press.

Bazemore, G., & Umbreit, M. (2001). *A comparison of four restorative conferencing models.* Washington, DC: U.S. Dept. of Justice and Delinquency Prevention.

Harris, M. K. (2004). An expansive, transformative view of restorative justice. *Contemporary Justice Review, 7*(1), 117–141.

Harris, N. (2003). Evaluating the practice of restorative justice: The case of family group conferencing. In L. Walgrave (Ed.), *Repositioning restorative justice* (pp. 121–135). Cullompton, Devon, UK: Willan.

Karp, D. (2001). Harm and repair: Observing restorative justice in Vermont. *Justice Quarterly, 18*(4), 727–757.

Karp, D., Sweet, M., Krishenbaum, A., & Bazemore, G. (2004). Reluctant participants in restorative justice? Youthful offenders and their parents. *Contemporary Justice Review, 7*(2), 199–216.

Marshall, T. F. (1996). The evolution of restorative justice in Britain. *European Journal on Criminal Policy and Research, 4*(4), 31–43.

Pavlich, G. (2005). *Governing paradoxes of restorative justice.* Portland, OR: Cavendish.

Presser, L. (2006). The micro-politics of victim offender mediation. *Sociological Inquiry, 76*(3), 316–342.

Roche, D. (2003). *Accountability in restorative justice.* New York: Oxford University Press.

Sullivan, D., & Tifft, L. (2001). *Restorative justice: Healing the foundations of our everyday lives.* Monsey, NY: Willow Tree.

Van Ness, D., & Strong, K. (2006). *Restoring justice* (3rd ed). Cincinnati, OH: Anderson.

Zehr, H. (1990). *Changing lenses.* Scottsdale, PA: Herald.

DISCUSSION QUESTIONS

1. Based on Gerkin's findings, do you think that victim-offender mediation is truly restorative in nature? Why or why not?

2. How can the impediments to successful programs be addressed to better improve victim-offender mediation?

3. After reading this article, do you think that victim-offender mediation may be more useful for certain victims and for certain crimes? Which ones?

4. Do you see any issues surrounding offender rights in the victim-offender mediation programs that were observed? What about issues with victim rights? Explain.

Sexual Victimization

On March 5, 2010, a woman reported to the police,

> We went downtown to Velvet Elvis, where we first saw our friends. We then saw Ben [Roethlisberger] and his friends/bodyguards. We went to take pictures (group) with him and then we left him alone. We went to The Brick where they happened to be, and we continued to have casual conversation, he even made crude, sexual remarks. They ended up leaving, and we went to Capital, where they also were. Ben asked us to go to his "VIP" area (back of Capital) we all went with him. He said there were shots for us, numerous shots were on the bar, and he told us to take them. His bodyguard came and took my arm and said come with me, he escorted me into a side door/hallway, and sat me on a stool. He left and Ben came back with his penis out of his pants. I told him it wasn't OK, no, we don't need to do this and I proceeded to get up and try to leave. I went to the first door I saw, which happened to be a bathroom. He followed me into the bathroom and shut the door behind him. I still said no, this is not OK, and he then had sex with me. He said it was OK. He then left without saying anything. I went out of the hallway/door to the side where I saw my friends. We left Capital and went to the first police car we saw.

The complaint of rape against Pittsburgh Steelers quarterback Ben Roethlisberger was investigated, but not less than 2 weeks later, a lawyer for the woman asked the prosecutor to drop the investigation into the events of March 5. He said,

> What is obvious in looking forward is that a criminal trial would be a very intrusive personal experience for a complainant in this situation, given the extraordinary media attention that would be inevitable. The media coverage to date, and the efforts of the media to access our client, have been unnerving, to say the least. (*The Smoking Gun*, 2010)

Subsequently, Ben Roethlisberger was not criminally charged in this case. But what really happened that night? Was the woman raped? Did she willingly engage in sex? This case is a common example of the inherent problems with knowing, defining, and proving when a rape or other type of sexual victimization occurs. The ways in which law enforcement and the media reacted to the victim are also telling of the challenges that victims face when coming forward. This chapter covers these issues—it defines what sexual victimization is, describes the extent and effects of sexual victimization, and details how the criminal justice system deals with victims.

What Is Sexual Victimization?

A unique form of victimization that has been widely studied because of its pervasiveness and negative effects is sexual victimization. Generally speaking, **sexual victimization** encompasses any type of victimization involving sexual behavior perpetrated against an individual. These behaviors can range from forced penetration to surreptitiously videotaping an individual undressing. Sexual victimization can take a toll on victims, with effects ranging from physical injury to psychological trauma to risk of additional sexual victimizations.

Rape

Originally defined in common law as unlawful carnal knowledge (i.e., vaginal penetration) of a woman who is not the perpetrator's wife by force and against her will, more contemporary definitions of rape have been developed. **Rape** can now be perpetrated by and against both females and males, and includes other types of penetration such as oral, digital, and anal. Most states' laws no longer exclude husbands.

Each state has its own legal definition of what behavior constitutes rape, but most states share some commonalities. First, rape occurs when there is nonconsensual contact between the penis and the vulva or anus, or when there is penetration of the vulva or anus. Rape also occurs when there is contact between the mouth and penis, vulva, or anus, or penetration of another person's genital or anal opening with a finger, hand, or object. Second, in order to be considered rape, force or threat of force must be used. Third, contact or penetration must occur without the consent of the victim or when the victim is unable to give consent—such as when asleep or under the influence of drugs or alcohol.

There are different types of rape, including forcible rape, drug or alcohol facilitated rape, incapacitated rape, and statutory rape—all of which involve unique circumstances. A **forcible rape** is one in which the offender uses or threatens to use force to achieve penetration (Fisher, Daigle, & Cullen, 2010). Another type of rape is **drug or alcohol facilitated rape**, which occurs when a person is raped while under the influence of drugs or alcohol after being deliberately given alcohol, a drug, or other intoxicant without his or her knowledge or consent (Fisher et al., 2010). A third type of rape is **incapacitated rape**. This type of rape occurs when a victim cannot consent because of self-induced consumption of alcohol, a drug, or other intoxicant. Incapacitated rape also occurs when a person is unable to consent due to being unconscious or asleep (Fisher et al., 2010).

Another type of rape, statutory rape, is somewhat unique in that it does not involve force. Rather, **statutory rape** occurs when a person who is under the proscribed age of consent engages in sex. Because the person is legally unable to give consent to have sex, such activity is illegal. The minimum age that most states set is between 14 and 18 years old.

Some states also consider the difference in age between the offender and the victim in determining whether it is statutory rape (Daigle & Fisher, 2010). In this way, some statutory rape laws are age-graded; that is, offenders have to be a certain number of years older than victims to activate the laws. For example, in Alabama, statutory rape occurs if the offender is at least 16 years old, the victim is less than 16 but more than 12, and the offender is at least 2 years older than the victim ("Sexual Assault Laws of Alabama," n.d.). Statutory rape laws that allow offenders who are close in age—such as two students enrolled in high school—to be convicted of sex offenses have recently come under fire in some states as not targeting sexual predators. For example, in the following box, the case of Genarlow Wilson is discussed. Georgia has since revamped its laws so that a case like Wilson's will not be handled in the same way. These new laws are commonly referred to as "Romeo and Juliet" laws.

Genarlow Wilson was a 17-year-old star athlete and honors student in 2003 when he attended a New Year's Eve Party in Douglasville, Georgia, and had consensual sex with a 15-year-old girl. He was found guilty of felony child molestation and sentenced to 10 years in prison. This case sparked public outrage, and as a result, Georgia lawmakers changed the law, which now makes consensual sex between teens a misdemeanor. However, the change was not retroactive, so Wilson remained behind bars.

A judge overturned Wilson's case and he was going to be set free, but Georgia Attorney General Thurbert Baker announced he would appeal that decision, which effectively kept Wilson in jail. Baker said he filed the appeal to resolve "clearly erroneous legal issues," saying that the judge did not have the authority "to reduce or modify the judgment of the trial court."

On July 20, the Georgia Supreme Court heard the appeal. However, it wasn't until October 27 that the court ruled in a 4–3 decision that the new law "represents a seismic shift in the legislature's view of the gravity of oral sex between two willing teenage participants." The justices talked about a "sea change" in attitudes on sex, and they concluded the 10-year sentence was "grossly out of proportion to the severity of the crime." This led to Genarlow Wilson's release after 2 years in prison for child molestation.

SOURCE: Excerpt from www.americaiswatching.org/genarlowwilson.html

Sexual Victimization Other Than Rape

Although rape is a sexual offense that garners widespread media attention, there are other types of sexual victimization. Perhaps surprisingly, these other forms are much more common than rape. They include sexual coercion, unwanted sexual contact, and noncontact sexual abuse. Each of these has unique elements, but all have been shown to cause negative outcomes for victims. See Table 5.1 for a description of the major types of sexual victimization discussed in this chapter.

Sexual Coercion

Similar to rape, **sexual coercion** involves intercourse or penetration with the offender's penis, mouth, tongue, digit, or object. The key difference between rape and sexual coercion has to do with the means the offender uses. Instead of using force or threats of force, the offender coerces the victim into having sexual intercourse. Coercion entails emotional or psychological tactics, such as promising reward, threat of nonphysical punishment, or pressuring/pestering for sex (Fisher & Cullen, 2000). For example, if the offender threatens to end the relationship unless the victim engages in sex, the offender is sexually coercing the victim. If a professor threatens to lower a student's grade if the student does not engage in sex, the student is being sexually coerced. If a person uses continued "sweet talk" and sexual advances, sufficiently pressuring the victim into compliance, the victim is being sexually coerced. In each of these scenarios, if the offender was unsuccessful, the behavior would be classified as an attempted sexual coercion.

Unwanted Sexual Contact

Not all sexual victimizations involve force or penetration. **Unwanted sexual contact** occurs when a person is touched in an erogenous zone, but it does not involve attempted or completed penetration. It may or may not involve force. What kinds of actions may be classified as unwanted sexual contact? Unwanted contact such as touching, groping, rubbing, petting, licking, or sucking of the breasts, buttocks, lips, or genitals constitutes

Table 5.1 Definitions of Different Types of Sexual Victimization

Type of Sexual Victimization	Definition
Forcible rape	The offender uses or threatens to use force to achieve penetration.
Drug or alcohol facilitated rape	A person is raped while under the influence of drugs or alcohol after being deliberately given alcohol, a drug, or other intoxicant without his or her knowledge or consent.
Incapacitated rape	A victim cannot consent because of self-induced consumption of alcohol, a drug, or other intoxicant. Incapacitated rape also occurs when a person is unable to consent due to being unconscious or asleep.
Statutory rape	A person who is under the proscribed age of consent engages in sex. Because the person is legally unable to give consent to have sex, such activity is illegal.
Sexual coercion	The offender coerces the victim into having sexual intercourse. Coercion entails emotional or psychological tactics, such as promising reward, threat of nonphysical punishment, or pressuring/pestering for sex.
Unwanted sexual contact	A person is touched in an erogenous zone, but it does not involve attempted or completed penetration. It may or may not involve force.
Coerced sexual contact	The offender uses psychological or emotional coercion to touch, grope, rub, pet, lick, or suck the breasts, lips, or genitals of the victim.
Unwanted sexual contact with force	The offender uses force or threatens to use force to touch the victim in an erogenous zone.
Visual abuse	A perpetrator uses visual means. These may include showing the victim pornographic materials, sex organs, or taking photographs or video of the victim while she or he is nude or having sex, without the victim's consent.
Verbal abuse	A perpetrator says or makes sounds that are intentionally condescending, sexual, or abusive.

unwanted sexual contact (Fisher & Cullen, 2000). This contact can be above or under clothing. If the offender uses psychological or emotional coercion, then the unwanted sexual contact is **coerced sexual contact**. If the offender uses force or threatens to use force, the behavior is classified as **unwanted sexual contact with force**.

Noncontact Sexual Abuse

Not all actions that people may consider victimizing involve touching or penetration. Other forms of sexual victimization, categorized as **noncontact sexual abuse**, are visual or verbal. **Visual abuse** occurs when a perpetrator uses unwanted visual means. The perpetrator may send pornographic images or videos via text messaging or e-mail or may post images to social networking sites or the Internet (Fisher & Cullen, 2000). **Verbal abuse** occurs when a perpetrator says something or makes sounds that are intentionally condescending, sexual, or abusive (Fisher & Cullen, 2000). For example, a perpetrator may make sexist remarks, may make catcalls or whistles in response to a victim's appearance, or may make noises with sexual overtones. A person asking inappropriate questions about another person's sex or romantic life is also considered verbal abuse.

⊠ Measurement and Extent of Sexual Victimization

Uniform Crime Reports

As previously discussed, one of the most widely used sources of data on crime and victimization is the Uniform Crime Reports (UCR), which show the amount of crime known to law enforcement in a given year. To be included in the UCR, a victim must report his or her victimization or the police must somehow become aware that a crime has occurred. As we will discuss below, rape and other forms of sexual victimization are often not reported to the police. In addition, the UCR uses a fairly restrictive definition of rape. Rape in the UCR is limited to forcible rape, which is defined as the carnal knowledge of a female forcibly and against her will (Federal Bureau of Investigation, 2010). Attempts and assaults to commit rape by force or threat of force are also included. Note that only females can be the victims of forcible rape and that sexual assaults other than rape (e.g., forced touching) are not included. Just less than 89,000 rapes were reported to law enforcement in 2009. The rate of rape in 2009 was 28.7 offenses per 100,000 inhabitants (Federal Bureau of Investigation, 2010).

National Crime Victimization Survey

Because many persons do not report being raped or sexually victimized to the police, a more accurate portrayal of the extent of rape and other sexual victimizations can be found from the National Criminal Victimization Survey (NCVS). This survey of households is used to find out about individuals' victimization experiences, including rape and sexual assault. The NCVS is different from the UCR in regard to rape and sexual victimization in several ways. First, the NCVS includes estimates of the extent of sexual assault as well as rape. Second, the NCVS is a self-report survey; its estimates include incidents that have been reported to the police and also those that were not. Third, both male and female victims of rape and sexual assault are included. According to the NCVS, in 2008 there were 140,620 completed and attempted rapes and 107,600 sexual assaults. This equates to a rape/sexual assault rate of 1.0 per 1,000 persons 12 years old and older (Rand, 2009).

National Violence Against Women Survey

Another source of information on the extent of rape is the National Violence Against Women Survey (NVAWS) (Tjaden & Thoennes, 2000). Conducted between November 1995 and May 1996, 8,000 men and 8,000 women were surveyed via telephone about abuse they experienced in the previous 12 months and during their lifetimes. The NVAWS included questions about rape, and both men and women were asked about their rape experiences. For the survey, rape was defined as "an event that occurred without the victim's consent, that involved the use or threat of force to penetrate the victim's vagina or anus by penis, tongue, fingers, or objects, or the victim's mouth by penis" (Tjaden & Thoennes, 1998, p. 13). Results from the NVAWS revealed that 18% of women had been raped (completed or attempted) during their lifetime and 0.3% had experienced a rape (completed or attempted) during the previous 12 months. Men were less likely to report being raped—3% of men reported a rape (completed or attempted) during their lifetime and 0.1% had been raped (completed or attempted) in the previous 12 months (Tjaden & Thoennes, 2000).

Sexual Experiences Survey

Other research has examined college females and their experiences of sexual victimization and rape. The first national-level study of college women's sexual victimization was conducted by Mary Koss and her colleagues (Koss, Gidycz, & Wisniewski, 1987). To measure the extent to which females were sexually victimized, Koss developed the **Sexual Experiences Survey (SES)**, a 10-item survey designed to measure rape, sexual coercion, and sexual contact.

The survey items used to measure these sexual victimizations were behaviorally specific. Behaviorally specific questions are those that include descriptive language and examples of the behaviors that constitute rape or sexual victimization. The survey did not rely on the victim to know whether she had been raped; instead, she was asked whether certain behaviors occurred. Koss found that more than half the women in the study had experienced some form of sexual victimization since the age of 14. More than 15% of the women had been raped, almost 15% had experienced a sexual contact, and almost 12% had been sexually coerced. She also found that 16.6% had been raped (attempted or completed) during the previous 12 months, showing that college women faced a real risk of being raped. The SES underwent a redesign in 2007 (Koss et al., 2007).

National College Women Sexual Victimization Study

Conducted in the spring of 1997 by Bonnie Fisher, Francis Cullen, and Michael Turner (1998), the **National College Women Sexual Victimization Study (NCWSV)** is a nationally representative study of college women. In this study, 4,446 college females were asked about their experiences of rape and sexual victimization since school began in fall 1996. The types of sexual victimization asked about included completed and attempted rape, attempted and completed sexual coercion, attempted and completed unwanted sexual contact with and without force, threats, and stalking. Two important methodological aspects of the NCWSV should be discussed. First, the study used a two-step measurement strategy, similar to the one employed by the NCVS. In the first stage, individuals were asked a set of behaviorally specific screening questions designed to capture the experience of rape and sexual victimization. For example, women were asked, "Since school began in fall 1996, has anyone *made* you have sexual intercourse by using *force or threatening to harm* you or someone close to you? Just so there is no mistake, by intercourse, I mean putting a penis in your vagina" (Fisher, Cullen, & Turner, 2000, p. 6). If a woman affirmatively responded to any of the screening questions, she then continued to the second stage to fill out an incident report. Incident reports were filled out for each incident of rape or sexual victimization that she indicated had occurred. The incident report, similar to the NCVS, included detailed information about the incident, such as the relationship of the offender to the victim, whether alcohol or drugs were used by the victim or offender, the location of the incident, and victim responses such as reporting and using self-protective action, among other things.

In total, 15.5% of the women had experienced at least one sexual victimization during the academic year. Rape was the least likely to occur, although 2.5% experienced either a completed or an attempted rape. Unwanted sexual contact was most common—10.9% reported experiencing at least one incident. College women also reported experiencing noncontact sexual abuse. More than half the women in the study reported having general sexist remarks made in front of them, and slightly more than 1 in 5 reported receiving obscene telephone calls or messages. Just more than 6% were exposed to pornographic pictures or materials when they did not consent to see them.

National Study of Drug or Alcohol Facilitated, Incapacitated, and Forcible Rape

Dean Kilpatrick, Heidi Resnick, Kenneth Ruggiero, Lauren Conoscenti, and Jenna McCauley (2007) recently conducted a national-level study of three types of rape: the **National Study of Drug or Alcohol Facilitated, Incapacitated, and Forcible Rape**. Forcible rapes involve unwanted oral, anal, or vaginal penetration. Drug or alcohol facilitated rapes also involve unwanted oral, anal, or vaginal penetration, but the victim must also have stated that she thought the offender purposefully gave her alcohol or drugs without her permission or attempted to get her drunk. This type of rape also could occur if the victim was too drunk to control her behavior or if she was unconscious (Kilpatrick et al., 2007). Incapacitated rapes also involve unwanted oral, anal, or vaginal penetration, but the victim must have voluntarily used drugs or alcohol or have been passed out or too drunk or high to control her behavior.

The study included almost 2,000 adult women between the ages of 18 and 34 and about 1,000 adult women aged 35 and older. Asking about incidents that occurred in the past year and during their lifetime, the study found that 18% of women had been raped during their lifetime. Most of the rapes, 14.5%, were forcible rapes, and 5% were drug or alcohol facilitated or incapacitated rapes. Twelve-month estimates were 0.52% for forcible rape and 0.42% for drug or alcohol facilitated or incapacitated rape.

Risk Factors for and Characteristics of Sexual Victimization

Although anyone can be sexually victimized, some people are at greater risk than others. As you probably surmise, females are more likely than males to be sexually victimized, at all stages of life. Males are relatively unlikely to be sexually victimized, but when they are, they are likely to be under the age of 12 (Tjaden & Thoennes, 2006). Females face the greatest risk of being sexually victimized during their late teens and early 20s—often spanning the years women are enrolled in college. For both genders, risk of being sexually victimized wanes over time. Besides gender and age, socioeconomic status and location of residence are also related to risk of sexual victimization. Persons of lower socioeconomic status, persons who are unemployed, and persons who live in large urban areas face a greater risk of being sexually victimized than others (Rennison, 1999). Black persons have higher rates of sexual victimization than others (Rennison, 1999).

Beyond these demographic characteristics, other factors differentially place persons at risk of being sexually victimized. As discussed in Section 3, one of the prevailing theories of victimization is lifestyles/routine activities theory. According to this theory, victimization is ripe to occur when motivated offenders, lack of capable guardianship, and suitable targets coalesce in time and space. This theory applies to sexual victimization as well as other predatory victimizations. Remember that motivated offenders are thought to be omnipresent—but a person has to be in proximity to these motivated offenders to be victimized. For sexual victimization, this can be achieved through dating (being alone with a potential offender), going to parties (particularly those with a high concentration of males, such as fraternity parties), or frequenting bars. Lack of capable guardianship also places people at risk for sexual victimization. Capable guardianship can come in the form of social guardianship and physical guardianship. Social guardianship is created when persons are present who can protect a person from being sexually victimized. Having a roommate or going out at night with others can provide social guardianship. Possessing mace, pepper spray, or an alarm system can provide physical guardianship. When people lack social guardianship and physical guardianship, they are more likely to be victimized. Finally, victims who are deemed suitable by offenders are more likely than others to be sexually victimized. What factors do you think would make a person "suitable"? Factors such as being female, visibly intoxicated, and alone may increase risk of sexual victimization.

A risky lifestyle also has been linked to sexual victimization risk. What constitutes a risky lifestyle is up for debate, but participating in particular activities certainly increases risk for sexual victimization. For example, alcohol use has been intimately linked to sexual victimization. According to data from the NVAWS, almost 20% of women who were raped as adults had used alcohol and/or drugs at the time of the victimization. Offenders are also likely to be under the influence of alcohol or drugs when perpetrating sexual victimizations (Tjaden & Thoennes, 2006). For college women, alcohol also plays an important role. College females who have a greater propensity for substance use are more likely to be sexually victimized than others (Fisher et al., 2010). As noted in the reading by Antonia Abbey (2002) in this section, alcohol also plays a role in sexual victimization perpetration and outcomes, especially for college students.

Characteristics of Sexual Victimization

In addition to knowing what factors place an individual at risk of being sexually victimized, it is also instructive to know what the "typical" sexual victimization and rape looks like. This is not to discount the seriousness of such victimizations but to understand the commonalities that many of them share.

Offenders

There is not one template for rape and sexual victimization offenders—they can be anyone. Data from the UCR indicate that those arrested for committing forcible rape are most commonly White males between the ages of 18 and 30 years old (Federal Bureau of Investigation, 2009). Females compose less than 2% of all those arrested for forcible rape.

Even more studied is the victim-offender relationship. Researchers have wanted to know whether people are more likely to be raped or sexually victimized by strangers or by people who are known to them. In doing so, researchers found that rape and sexual victimization was relatively unlikely to be perpetrated by a stranger. Indeed, they found that rape rarely begins with an offender jumping out of the dark in a secluded place in a blitz-style attack, identified as "real rape" (Estrich, 1988). Instead, it often occurs in the context of a date or other social situation by someone with whom the victim is acquainted or on a date.

Injury

Although being raped or sexually victimized can be traumatizing, fortunately, most victims do not suffer serious physical injury. Slightly more than one third of rape victims in the NVAWS reported that they experienced some type of physical assault in addition to the rape, such as being slapped or hit (Tjaden & Thoennes, 2006). Even when such injuries are incurred, only about one third of those victims who report injury seek medical treatment for their injuries (Rand, 2008).

Weapon Use

Offenders are also relatively unlikely to have or use a weapon during the perpetration of a rape or sexual assault. About one third of rapes and sexual assaults involve any weapon. When a weapon is used, it is most likely to be a firearm (Rand, 2009).

⊠ Responses to Sexual Victimization

Acknowledgment

After experiencing a rape, a victim may go through various emotions—anger, fear, sadness, just to name a few. As a victim copes with the rape, one of the things he or she may do is think about the event itself and attempt to define it. A victim may feel immediately that he or she was raped (referred to as acknowledgement); on the other hand, a victim may see the incident as a horrible misunderstanding. It may not make sense that a person who is raped may not define it as such, but research shows that, in fact, many victims do not label their experience as rape. In fact, less than half the women (47.4% of completed rape victims) who participated in the NCWSV discussed earlier who met a legal definition of rape labeled the incident a rape (Fisher, Cullen, & Turner, 1998).

Labeling an incident as rape may be important for several reasons. A victim may not get help from family, friends, or professionals if she or he does not think what happened was rape. The police are also unlikely to be notified of the incident if the victim does not think what happened was a crime. So why, then, do some victims not label or define their rape as such? It may be that victims are unsure how to label the incident (Littleton, Axson, Breitkopf, & Berenson, 2006), see it as a miscommunication (Layman, Gidycz, & Lynn, 1996), or think it was a crime other than rape (Layman et al., 1996). Women may also be less likely to label their experiences as rape when the perpetrator is someone they know, such as their date or boyfriend (Koss, 1985). In Koss's (1988) study, only 10.6% of college women who were raped believed that they were not victimized. In other words, almost all rape victims felt that what had happened to them was "wrong" or victimizing, even if they did not label it as rape.

Reporting

Acknowledgment of rape is likely closely tied to reporting rape or sexual victimization to the police. As previously noted in the discussion about the extent of rape and sexual victimization, rape is one of the most underreported crimes. In fact, results from the NCVS suggest that less than half of persons who are raped and sexually assaulted report their experiences to the police (Rennison, 2002). Rapes and sexual assaults of college women are also unlikely to come to the attention of the police. In the NCWSV, less than 5% of the rapes were reported to the police (Fisher et al., 1998). In Kilpatrick et al.'s (2007) study, it was found that 10% of college women who experienced a drug or alcohol facilitated or incapacitated rape reported their incident to police. Slightly more (18%) victims of forcible rape told the police about what happened.

Why would a person who has experienced such a serious crime not report the incident to the police? One reason may be tied to acknowledgment: Victims who are not sure the incident was a crime are unlikely to report (Fisher, Daigle, Cullen, & Turner, 2003). Similarly, victims may be unsure that harm was intended on the part of the offender. Another reason victims do not report may be that they want to keep the event private (Rennison, 2002). Other victims have noted fear of the offender seeking reprisal (Rennison, 2002). Suspiciousness of the police and fear that they may be biased also drive victims' decisions not to report to the police (Rennison, 2002).

Resistance/Self-Protective Action

Not all victim responses happen after the incident has been completed. Instead, many victims of rape and other sexual attacks report that they tried to do something during the course of the incident either to stop it from occurring or being completed or to protect themselves. When this occurs, it is said that the victim used a **resistance strategy** or some type of **self-protective action**. Self-protective actions generally are classified into one of four types: forceful physical, nonforceful physical, forceful verbal, or nonforceful verbal (Ullman, 2007). **Forceful physical strategies** include actions such as shoving, punching, or biting the offender. **Nonforceful physical strategies** are passive actions, such as fleeing or pulling away. Self-protective actions may also be verbal. **Forceful verbal strategies**, such as yelling, are active and are used either to scare the offender or to attract the attention of others. Pleading with, talking to, and begging the offender are examples of nonaggressive, **nonforceful verbal strategies** (Ullman, 2007).

You may wonder whether it is wise to fight an attacker who is trying to rape or sexually victimize you. Although it is not effective in every situation, research shows that most women do use some type of self-protective action. In about two thirds of all rape and sexual assault victimizations recorded in the NCVS in 2007, the victim used some type of self-protective action (BJS, 2010). Rape and sexual assault victims were most likely to say that they resisted their offender, scared or warned their offender, or ran away or hid from their offender. Generally, nonforceful verbal strategies have been shown to be ineffective in reducing the chances of a rape being completed (Clay-Warner, 2002; Fisher, Daigle, Cullen, & Santana, 2007). The type of self-protective action used is important. Research shows that if a victim is trying to stop the incident from occurring (i.e., keep it from being completed) and prevent injury, then the level of self-protective action should match the offender's efforts (Fisher et al., 2007). This concept, known as the **parity hypothesis**, states that a victim's use of self-protective action should be on par with the offender's attack. In other words, if the offender is using physical force, the victim's most effective defense will be forceful physical self-protective action.

◪ Consequences of Sexual Victimization

One reason that rape and sexual victimization have received so much attention in research and in the media is because of their often pernicious effects. Victims frequently experience serious consequences, some temporary and others long lasting. How a victim responds to being sexually victimized varies according to numerous factors, such as age and maturity of the victim, social support for the victim, the relationship of the offender to the victim, how the victim defines the incident, whether the victim reports to the police and how the system responds, whether the victim discloses to others and how they respond, the severity of the victimization, the level of injury, and the overall view of the community regarding sexual victimization.

Physical, Emotional, and Psychological Effects

As previously noted, most people who are sexually victimized do not suffer serious physical injury; however, possible physical effects include pain, bruises, cuts, scratches, genital/anal tears, nausea, vomiting, and headaches (National Center for Victims of Crime, 2008). Victims may experience these effects immediately following the event or as long-term consequences.

In addition to physical effects, victims may experience emotional and psychological effects related to their sexual victimization. Some victims experience depression, which may lead to suicidal ideation (Stepakoff, 1998). They may experience anger, irritability, feelings of guilt, and helplessness. Posttraumatic stress disorder also has been linked to sexual victimization—victims may have nightmares, flashbacks, exaggerated startle responses, and difficulty concentrating. Victims also may experience reductions in self-esteem or become more negatively self-focused (McMullin, Wirth, & White, 2007).

Behavioral and Relationship Effects

Sexual victimization may also impact a person's behavior. Sexual victimization has been linked to delinquency and criminal behavior (Widom, 1989), compulsive behavior, as well as substance use and abuse (Knauer, 2002). Some victims change their lifestyles by becoming more isolated, spending more time alone. Victims may also exhibit self-mutilating behavior or attempt suicide (Minnesota Department of Health, 1998).

Persons who are sexually victimized may also find that they have a difficult time navigating their personal relationships and have problems in their sexual functioning. Victims may find it difficult to enter into or maintain intimate relationships and to engage in parenting and other nurturing behaviors. Finally, victims may also evince changes in their sexuality—they may avoid sex, have difficulty becoming aroused, experience intrusive thoughts, and have difficulty reaching orgasm (Maltz, 2001). Conversely, some victims may increase their levels of sexual activity.

Costs

A detailed account of the costs that victims of crime may incur was presented in Section 3, so all the costs associated with sexual victimization will not be reviewed here. It should be noted, however, that sexual victimization carries with it many costs. As a consequence of the problems—physical, emotional, psychological, behavioral, and relationship—that victims experience, many seek assistance from mental health professionals. About one third of the female victims and one fourth of the male victims who reported rape in the NVAWS sought mental health counseling (Tjaden & Thoennes, 2000). About 12% of victims of rape and sexual assault in the NCVS missed time from work as a result of their victimization (Maston, 2010). Data from the NCVS (Klaus & Maston, 2008) show that 18% of rape and sexual assault victims missed more than 10 days of work following the incident. Between one quarter and one half of all rape and child sexual abuse victims receive mental health care. As a result, sexual victimizations of both adults and children result in some of the largest mental health care costs for victims. The average mental health care loss per rape or sexual assault is $2,200, and the average for child abuse is $5,800. All told, the estimate of total economic loss a rape victim experiences is $87,000 on average when impact on quality of life is considered (Miller, Cohen, & Wiersema, 1996).

Recurring Sexual Victimization

As noted in Section 3, persons who are victimized once are at increased risk of experiencing subsequent victimizations. This relationship holds true for victims of sexual victimization. Research shows that individuals who experience childhood sexual abuse are at risk of being sexually victimized in adolescence and adulthood (Breitenbecher, 2001). Subsequent victimization may also occur relatively quickly. Research on college women shows that they are at risk of being repeatedly sexually victimized even during the course of a single academic year. In fact, data from the NCWSV show that more than 7% of the college women surveyed experienced more than one sexual victimization incident over the approximate 7-month recall period (Daigle, Fisher, & Cullen, 2008). These women experienced more than 72% of all sexual victimizations reported by their sample. What is especially alarming is that when a subsequent sexual victimization did occur, it was likely to occur within the same month or in the month immediately following the initial incident.

⊠ Legal and Criminal Justice Responses to Sexual Victimization

Legal Aspects of Sexual Victimization

Up until the mid-1960s and 1970s, traditional legal definitions of rape limited the offense to forceful unlawful carnal knowledge of a woman without her consent. What, then, did such definitions exclude? These definitions limited rape to incidents perpetrated against women and involving vaginal penetration. Under common law, husbands could not be charged with raping their wives. Calls for rape law reform resulted in changes to these laws: Now, both males and females can be victims and perpetrators of rape; other forms of penetration, such as digital and anal penetration, are included; and the marital exemption has been removed in all 50 states. This movement to change laws also applied to other criminal sexual victimizations: Such victimizations can be perpetrated by both genders, and married persons can be held liable for offending against their partners.

Before rape law reform, victims were often required to produce corroborating evidence for the incident to be prosecuted in court. Because rape is often a crime that occurs in private without additional witnesses, it can be difficult for victims to prove that they did not consent to have sex. For this reason, victims were often asked to provide corroborating evidence—such as injury, presence of a weapon, presence of semen, timely reporting to the police, and proof that the victim provided at least some degree of physical resistance. Remember, however, that a relatively small portion of victims are seriously physically injured, few incidents involve a weapon, and the most common perpetrator of such an offense is an acquaintance or someone known to the victim. These types of rape are the most difficult to prove, and the victim is likely to have difficulty proving, beyond her word, that a rape took place.

Corroboration is no longer a requirement to further reduce the stress the victim may feel in the criminal justice process, rape shield laws also have been enacted. **Rape shield laws** prohibit the defense from using a victim's previous sexual conduct in court. In some circumstances, a victim's past may be used, but this is determined in a closed hearing, not in front of a jury. Generally, such information can be used only if doing so is necessary to establish the facts of the case. Another recent change involves prohibiting the use of polygraph examinations of victims. Some police departments and prosecutors have required that rape and sexual assault victims submit to a polygraph examination before they will investigate their claims or initiate prosecution. A few states have passed legislation to prohibit law enforcement officers and prosecutors from requiring these tests.

Violence Against Women Act (1994)

As part of the Omnibus Crime Bill of 1994, Congress passed the **Violence Against Women Act** (VAWA), which was designed to provide women protection from systematic violence, including sexual violence. The act provided $1.6 billion in funding for education, research, treatment of victims, and improvement of state criminal justice system response to female victims of violence. It also provided funding to improve victim services and to create more shelters for female victims of domestic violence, among other services. Through VAWA, the collection of crime statistics on violence against women was stressed, as well as the protection of college women and immigrant women and children. Interstate domestic violence and sexual assault crimes were identified as prosecutable federal offenses, and guarantees of interstate enforcement of protection orders were included.

The VAWA legislation was renewed in 2000. This renewal identified dating violence and stalking as additional crimes against women deserving of legal protections, created a legal assistance program for victims of domestic violence and sexual assault, promoted supervised visitation programs for families affected by violence, and granted additional protections to immigrants experiencing intimate partner violence, sexual violence, and stalking. The VAWA legislation was further reauthorized in 2005, with several key additions. It continued to focus on serving underserved populations, such as immigrant women and women with disabilities, and created cultural and language-specific services in communities to this end. It broadened services to include children and teenagers instead of just adults, created the first federal funding stream for rape crisis centers, protected victims from being evicted based on their victimization status, and emphasized prevention of violence. Related to policing and prosecution, it included a provision that polygraph examinations not be required of sexual assault victims as a condition of charging or prosecution. Although elements of the VAWA legislation have been met with some resistance from the courts, its impact on the way in which the criminal justice system responds to violence against women is evident.

HIV and STD Testing

A common fear among victims of rape is that they may contract HIV or other sexually transmitted diseases from their attackers. This fear is particularly elevated when a person is raped by a stranger (Resnick, Monnier, & Seals, 2002). To assuage this fear, most states have implemented policies that allow for or require convicted sex offenders to submit to HIV testing. Some states, such as Wisconsin and Georgia, allow for testing if the victim requests it and if probable cause for the victim's exposure is established (RAINN, 2009; Ritsche, 2006). See Table 5.2 for a detailed description of Georgia's policy on HIV testing for criminal offenders. Other states allow for pretrial testing. The Centers for Disease Control and Prevention has estimated the likelihood of contracting HIV from an HIV-positive person. When consensual vaginal sexual intercourse occurs, there is between a 0.1% and 0.2% chance of contracting HIV. Estimates are higher for consensual rectal sexual intercourse—between 0.5 and 3.0% (Centers for Disease Control and Prevention, 2006). Risk of exposure is likely higher for nonconsensual activity; however, those people who are most likely to be infected with HIV (men who have sex with men, intravenous drug users) are not the most likely rape offenders.

Sex Offender Registration and Notification

In 1996, the U.S. Congress passed Megan's Law, which requires that sex offenders convicted in federal court register a current address with criminal justice agencies. Following this enactment at the federal level, all states now have laws that require at least some types of sex offenders to register with state agencies in order for these agencies to keep track of where sex offenders live (Chon, 2010). With this registration also come restrictions on where sex offenders can live. Registered sex offenders often cannot live within a certain distance—such as 1,000 feet—of a school or other place where children congregate (Chon, 2010). Notification also allows for potential employers, community residents, organizations, and people who work with potential victims to be notified that a person living in the community is a sex offender. Notification does not always have to be overt; it can be achieved by making registries readily available and accessible, such as through the Internet or through a law enforcement agency (Chon, 2010). Notification can also occur through the distribution of fliers, door-to-door visits, and letters. Information such as the offender's name, address, description, and the charged offense is typically provided (Chon, 2010).

Table 5.2 Georgia's Law Regarding HIV Testing for Sex Offenders

Crimes and offenders	Individuals arrested or convicted of rape, sodomy, aggravated sodomy, child molestation, incest, or statutory rape, or other sexual offense that involves significant exposure
Is testing required or available?	Testing is available (but not mandatory) upon arrest of offender of enumerated offenses at the request of the victim upon showing of probable cause that person committed the crime and that significant exposure occurred.
	Testing is required upon a verdict or plea of guilty or nolo contendere to any AIDS-transmitting crime.
When does testing occur?	Upon court order after arrest or within 45 days following a guilty verdict or plea
What is the process?	In the case of an arrest, the victim, or the parent or legal guardian of a minor or incompetent victim, makes a request to the prosecuting agency to request that the alleged offender voluntarily submit to a test.
	If the person arrested declines, the court, upon a showing of probable cause, may (but is not required) to order the test to be performed.
To whom is the information disclosed?	To the victim, or to any parent or guardian of any such victim who is a minor, by the Department of Community Health
What other services for victims are available in connection with the testing?	None specified
Other	The cost of the test shall be borne by the victim or by the person arrested, in the discretion of the court.
Source and/or applicable references	Ga. Code Ann. §§ 17-10-15, 24-9-47, 31-22-9.1

SOURCE: RAINN (2011). Reprinted by permission of RAINN, Rape, Abuse and Incest National Network.

Police Response

The way in which police respond to victims of sexual assault and rape is critical to how the victims interact—or not—with the criminal justice system. As previously noted, victims of rape and sexual assault are unlikely to report their victimizations to the police. One common reason victims do not report is a lack of trust in the police or a belief that the police will not take the incident seriously. Unfortunately, events shown in the media suggest that this fear is not completely unfounded. For example, as reported by Barton and Vevea (2010), Gregory Below was arrested and charged in 2010 with 32 counts relating to sexual assault, stalking, kidnapping, and assault involving seven different women. One of the women reported to the police that Below met her at a club and directed her to take him to find crack cocaine and then to a vacant apartment. She told the police that, while in the apartment, he beat her with a closet rod and sexually assaulted her. She said he threatened to kill her and throw her body in the river if she did not comply with his orders. One of the officers ran a background check on her that showed a recent drug charge, and that officer admitted that he did not believe her claims, despite her visible injuries. It would be 2 years and several more victims later before Below was finally arrested. Those victims may have been spared had the police responded differently to the earlier victim's claims.

The Below case speaks to a larger issue with which police and victims must deal—suspicion of victims making **false allegations**. For a report to be considered false, sufficient evidence must establish that a sexual assault did not happen. An allegation is not considered false if the investigation fails to prove that a sexual assault occurred or if there is a lack of evidence to prove it occurred. Such cases would be deemed **baseless**— not false (Lonsway, Archambault, & Berkowitz, 2007). Unfortunately, statistics of false allegations often erroneously use baseless cases as evidence. In the Uniform Crime Reporting program, police departments label baseless and false allegations using one term—**unfounded**. To know what percentage of rape and sexual assault complaints are actually false, researchers need to determine if a claim is merely baseless or actually false. In using this standard, a study of complaints of sexual assault in Portland, Oregon, showed that 1.6% were false. This percentage was less than that for stolen vehicles, 2.6% (Raphael, 2008). Estimates provided by the San Diego, California, police show that 4% of their reports for sexual assault were false, while a study of British police showed that only 2% of reported sexual assault cases were false (Lonsway et al., 2007). It seems, then, that false reports of rape and sexual assault are not typical.

Quite likely a larger problem than false allegations is the lack of action police may take in cases or their not responding to victims in a sensitive manner. To deal with victim concerns and encourage positive police response, many police departments have implemented new policies and practices. Many police departments now have special investigatory **sex crime units.** These units staff officers who are specially trained in how to respond to victims of sex crimes, including training in crisis counseling and in investigatory techniques germane to such cases. Police departments sometimes have **victim/witness assistance programs** that provide guidance to victims during the investigation and criminal justice process. People working in these programs may go with the victim to the hospital for an exam, accompany her or him to court, provide transportation to court, assist with filing compensation claims, and assist her or him in receiving counseling. Another important function they may serve is providing notification when the offender is released from police custody. The way in which police respond to victims is particularly important given that research has shown that when victims are treated with empathy and receive support, they are more likely to cooperate with criminal justice personnel, to remember more details about the incident, and to receive psychological benefit (Meyers, 2002).

Medical-Legal Response

Victims of rape and sexual assault may find themselves at the hospital either to seek medical treatment for their injuries or to receive a forensic examination. This process can be quite daunting. In the past, rape victims would often find that they were not given priority in emergency rooms and were forced to wait for long periods of time in crowded waiting rooms, unable to eat, drink, or urinate until after they were examined. Staff at the hospital often were not trained in how to conduct forensic evidence collection and were sometimes insensitive regarding the special needs of sexual assault victims (Littel, 2001). To combat these problems, the **sexual assault nurse examiner (SANE)** program was developed in the mid- to late 1970s. SANEs are forensic, registered nurses. They perform forensic examinations of sexual assault victims that include collecting information about the crime, performing a physical examination to evaluate and inspect the victim's body, collecting and preserving all evidence, collecting urine and blood samples, providing the victim with prophylactic medications for the prevention of sexually transmitted diseases, and providing the victim with referrals (Buschur, 2010). Evidence is collected through swabbing, debris collection, and photo documentation (Buschur, 2010). For example, at least two photos of external injuries should be taken and swabs should be collected if body fluids are present on the victim (Buschur, 2010). The release of evidence to the police is done at the consent of the victim in cases that already have been reported to the police or if the victim agrees to report (with the exception of cases in which medical personnel must mandatorily report, such as child abuse cases). SANEs may also conduct evidentiary exams of suspects in sexual assault and rape cases. Rebecca Campbell, Debra Patterson, and Lauren Lichty (2005), in their article included in this section, further discuss the role of SANEs in sexual assaults. They pay particular attention to research examining the effectiveness of SANEs.

In addition to SANEs, many communities have **sexual assault response teams (SARTs)** that work to coordinate responses to sexual assault and rape victims. Started in the early to mid-1970s, these teams often consist of individuals who work together to ensure that victims of sexual assault receive assistance in navigating the medical and criminal justice systems. Most SARTs include individuals from the prosecutor's office, local law enforcement agency, advocacy groups, and forensic examiners as core members (Howton, 2010). When a victim seeks medical treatment or reports his or her incident to a law enforcement agency, the SART in that jurisdiction (if one exists) is notified and activated. Individuals then work to ensure that the victim receives appropriate medical care, including a forensic medical exam, and is treated with dignity and respect by individuals in the criminal justice system. They further assist the victim in receiving additional services such as counseling.

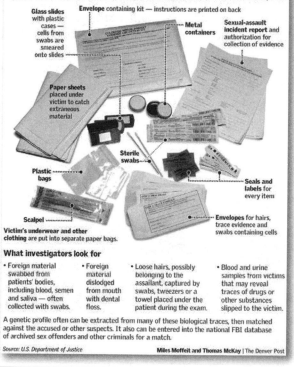

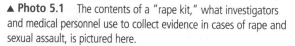

▲ **Photo 5.1** The contents of a "rape kit," what investigators and medical personnel use to collect evidence in cases of rape and sexual assault, is pictured here.

Photo credit: Miles Moffeit and Thomas McKay, "The 'black box' of sex-crime investigations," July 24, 2007, *The Denver Post*. Reprinted with permission.

Prosecuting Rape and Sexual Assault

The prosecutor is the key actor in charging and trying any criminal case. The prosecutor has discretion in deciding what cases to charge and in which cases to offer plea bargains to the defendant. Historically, prosecutors tend to move forward with prosecuting those cases that are easiest to prove legally—those that have a clear victim and strong evidence to prove the defendant's guilt. It may not be surprising, then, to learn that prosecutors have been reluctant to try rape and sexual assault cases since they often involve a lack of evidence and are "he said, she said"-type cases. Although the decision to charge or not is most commonly impacted by legal factors, characteristics of the victim also play a role. This may be the case even more so in rape and sexual assault cases, in which eyewitnesses and physical evidence are uncommon. As such, prosecutors may evaluate how a jury or judge will perceive the victim in terms of his or her background, character, and behavior.

In a recent study on charging decisions of prosecutors on sexual assaults, Spohn and Holleran (2004) found that in cases in which the perpetrator was a nonstranger, prosecutors were less likely to charge when the victim had engaged in risk-taking behavior at the time of the incident. Similarly, prosecutors were unlikely to charge if the victim's character or reputation was questioned and if the victim and offender had been or were intimate partners. For incidents involving strangers, on the other hand, victim characteristics and behaviors did not predict charging decisions. Rather, stranger-perpetrated sexual assaults were charged based on legal factors, such as the presence of physical evidence and whether the perpetrator used a gun or knife. As noted by Judy Shepherd (2002) in her article included in this section, prosecutors are not the only ones who make tough decisions in rape cases. She discusses her own experience of being a juror for a rape trial in Alaska. She notes that jurors' beliefs in rape myths impact the way in which they evaluate information presented at trial and make decisions regarding guilt. Rape myths are stereotypes or false beliefs that people hold about rape offenders, victims, and rape in general that serve to justify male sexual aggression against females (Lonsway & Fitzgerald, 1994). An example of a rape myth is that women lie about consensual sex afterward and call it rape.

 ## Prevention and Intervention

▲ **Photo 5.2** A self-defense class is pictured here. Some prevention programs incorporate self-defense, building on the knowledge that using resistance has been shown to thwart the completion of sexual victimization.

Photo credit: © Washington Post/Getty Images

Most programs designed to reduce the occurrence of rape and sexual victimization target college students. One reason for this is an amendment to the Clery Act (discussed in detail in Section 9) that requires colleges and universities to develop sexual assault prevention policies. Another reason is the attention that has been given to sexual victimization among college students and the ability to target this at-risk population easily.

Some of the most effective programs include teaching women how to assess situations as risky, how to acknowledge situations as potentially leading to rape, and how to act with resistance. These programs often include self-defense training. Self-defense training has been shown to reduce the likelihood that college women will be raped. It also increases their use of self-protective behaviors, use of assertive sexual communication, and belief in their ability to resist offenders (for a review, see Daigle, Fisher, & Stewart, 2009). Increasing college women's ability to recognize risk also appears to be effective in reducing rape (Marx, Calhoun, Wilson, & Meyerson, 2001). Bystander programs are another type of prevention program that look promising. These programs focus on training men and women to be agents of change. People are taught to intervene when they hear sexist comments or see high-risk behavior and instructed on how to react after a rape occurs (Banyard, Plante, & Moynihan, 2007). With the development of prevention programs, reducing the occurrence of sexual victimization may be possible, but targeting the reasons why offenders engage in this type of behavior also will be necessary. Until the individual and societal causes can be addressed, it is imperative that the extent, characteristics, and causes of sexual victimization be understood so that victims can be responded to in helpful ways and potential victims can minimize their risk of being victimized.

SUMMARY

- Sexual victimization is any type of victimization that involves sexual behavior perpetrated against an individual. There are many types of sexual victimization, such as forcible rape, drug or alcohol facilitated rape, incapacitated rape, statutory rape, sexual coercion, unwanted sexual contact, coerced sexual contact, unwanted sexual contact with force, visual abuse, and verbal abuse.
- Rape is measured using both Uniform Crime Reports (UCR) and the National Criminal Victimization Survey (NCVS). The UCR is dependent on people reporting rape as a crime, and the definition of rape according to the UCR is carnal knowledge of a woman against her will. This means that male victims of sexual assault and rape are not documented in the UCR. The UCR also does not measure other types of sexual victimization, such as unwanted sexual contact or sexual coercion. About 89,000 rapes were reported to law enforcement in 2009.
- The NCVS looks at both male and female victims and both rape and sexual assault. The NCVS does not rely on victims to report the crime. According to the NCVS, in 2008 there were 140,620 completed and attempted rapes and 107,600 sexual assaults.
- Another source of information on the extent of rape is the National Violence Against Women Survey (NVAWS). It was conducted via telephone, with 8,000 males and 8,000 females interviewed.
- The Sexual Experiences Survey is a 10-item survey designed to measure rape, sexual coercion, and sexual contact.
- The National College Women Sexual Victimization Study is a nationally representative study of college women. About 16% of college women reported experiencing some type of sexual victimization during the academic year.
- The National Study of Drug or Alcohol Facilitated, Incapacitated, and Forcible Rape is a national-level study of those three types of rape.
- Certain risk factors may place a person at higher risk of victimization. Females are more likely to be sexually victimized than males. Age can also determine when a person is at the highest risk of victimization. A person's socioeconomic status and where he or she lives are also related to risk of sexual victimization. Persons who are Black have higher rates of sexual victimization than others.
- Lifestyles/routine activities theory relates to sexual victimization. According to this theory, victimization is ripe to occur when motivated offenders, lack of capable guardianship, and suitable targets coalesce in time and space. For sexual victimization, this can be achieved through dating (being alone with a potential offender), going to parties (particularly those where many males are present, such as fraternity parties), or frequenting bars.
- White, young males are the most common perpetrators of sexual victimization according to the UCR and NCVS.
- Sexual victimization is traumatic, but most victims do not suffer physical injury and few of those who do suffer injuries seek medical help. Rapes that involve a stranger are more likely to result in injury. Offenders are also unlikely to use or have a weapon during the perpetration of a rape or sexual assault.
- It is hard sometimes for a victim to feel and acknowledge that he or she is the victim of rape. However, it is an important first step to label the incident as rape so the victim can receive necessary help from family, friends, and professionals.
- Rape is one of the most underreported crimes. Rapes and sexual assaults of college women are the most unlikely to come to the attention of the police. There are several reasons that reporting of rape is so low, including fear of reprisal or belief that the police will be biased. Victims also cite wanting to keep the matter private as a reason not to report.

- A resistance strategy or self-protective action is something victims of rape or other sexual attacks do during the course of the incident either to stop it from occurring or being completed or to protect themselves. There are four types of self-protective action: forceful physical strategies, nonforceful physical strategies, forceful verbal strategies, and nonforceful verbal strategies.
- There are many physical, emotional, psychological, behavioral, and relationship effects caused by rape and sexual victimization. There are also significant financial costs associated with rape.
- Repeat sexual victimization occurs when people who have been victimized once experience a subsequent victimization.
- Several legal reforms have been enacted to help rape victims. Rape shield laws, the Violence Against Women Act (1994), sex offender registration, and required HIV testing of offenders are all laws that help victims of sexual crimes navigate the criminal justice system. Certain units such as sexual assault nurse examiners and sexual assault response teams work to assist victims as they deal with the medical and legal system.

DISCUSSION QUESTIONS

1. What are some issues with requiring HIV testing of offenders? Do you agree with this reform?

2. Investigate your state's laws for rape and sexual assault. How are rape and sexual assault defined? Who can be a victim? Who can be an offender? What are the proscribed punishments for committing these acts?

3. Why is measurement so important in determining accurate estimates of the extent of rape and other types of sexual victimization?

4. With the widespread use of the Internet and technology, how might the nature of sexual victimization change?

5. Given what you know about reporting, use of self-protective actions, and recurring sexual victimization, how can we prevent sexual victimization?

KEY TERMS

sexual victimization

rape

forcible rape

drug or alcohol facilitated rape

incapacitated rape

statutory rape

sexual coercion

unwanted sexual contact

coerced sexual contact

unwanted sexual contact with force

noncontact sexual abuse

visual abuse

verbal abuse

Sexual Experiences Survey (SES)

National College Women Sexual Victimization Study (NCWSV)

National Study of Drug or Alcohol Facilitated, Incapacitated, and Forcible Rape

resistance strategy

self-protective action

forceful physical strategies

nonforceful physical strategies

forceful verbal strategies

nonforceful verbal strategies

parity hypothesis

rape shield laws

Violence Against Women Act (1994)

false allegations

baseless allegations

unfounded

sex crime units

victim/witness assistance programs

sexual assault nurse examiner (SANE)

sexual assault response team (SART)

INTERNET RESOURCES

University of Pittsburgh at Johnstown: Sexual Victimization (http://www.upj.pitt.edu/2273/)

This website is part of the University of Pittsburgh at Johnstown's counseling center and provides information about the types of rape, the definition of rape, and how consent is defined. It also discusses the steps a person should take if he or she is raped and some of the side effects of rape, including posttraumatic stress disorder. The website also links to other articles by the college's center: two on date rape drugs, including the use of alcohol as a drug that leads to incapacitated rapes, and another on sexual assault prevention.

The Sexual Victimization of College Women (http://www.ncjrs.gov/pdffiles1/nij/182369.pdf)

One of the most at-risk groups for sexual victimization is college women, and this report—put together by Bonnie Fisher, Francis Cullen, and Michael Turner discusses findings from the NCWSV. They also discuss stalking, since that is a crime seen frequently on college campuses. This is a comprehensive report on the sexual victimization of college women and provides a wide range of information on the topic.

Male Survivor: Overcoming Sexual Victimization of Boys & Men (http://www.malesurvivor.org/default.html)

While sexual victimization is widely studied as a problem women face, this website examines and provides resources and support for men who were sexually victimized as children, adolescents, or adults. It examines common myths of sexual victimization, such as the belief that males cannot be victims and that sexual crimes committed against males are always perpetrated by homosexual males. The website also includes survivor stories and publications from researchers examining the topic of male sexual victims.

RAINN: Rape, Abuse, and Incest National Network (http://www.rainn.org/)

RAINN is the nation's largest anti-sexual assault organization. Its website provides information about local counseling centers and how to help a loved one who may be the victim of sexual abuse. The website also lists statistics, reporting, and tips on how to reduce the risk of becoming a victim of sexual assault or rape. You can also learn about volunteering for RAINN, donating money, and becoming a student activist.

The Date Safe Project (http://www.datesafeproject.org/)

This website provides information for parents and students on how to be safe while dating. There is also information about curricula and classroom exercises on dating, hooking up, and parties. There are also sexual assault survivor stories. The Date Safe Project provides parents, educators, educational institutions, students, military installations, community organizations, state agencies, and federal government with resources, educational materials, and programming addressing consent, healthy intimacy, sexual education, sexual assault awareness, bystander intervention, and support for sexual assault survivors.

National Sexual Violence Resource Center (http://www.nsvrc.org/)

This is a comprehensive collection and distribution center for information, statistics, and resources related to sexual violence. It serves as a resource for coalitions, rape crisis centers, allied organizations, and others working to eliminate sexual assault. The Center does not provide direct services to sexual assault victims but, rather, supports those who do.

Introduction to Reading 1

To address the specific medical needs that victims of sexual assault face and to assist in evidence collection, many jurisdictions now have sexual assault nurse examiners (SANEs). Campbell et al. (2005) provide an overview of the historical development of SANE programs and a thorough discussion of the effectiveness of SANE programs. They specifically tailor their discussion to the goals that SANE programs are trying to attain, such as creating settings that can address both physical and medical needs of victims, providing comprehensive and consistent medical services, improving the quality of forensic evidence collection, increasing prosecution rates, and creating community change. Gaps in the literature and ways in which SANE programs can better achieve positive outcomes are also addressed.

The Effectiveness of Sexual Assault Nurse Examiner (SANE) Programs

A Review of Psychological, Medical, Legal, and Community Outcomes

Rebecca Campbell, Debra Patterson, and Lauren F. Lichty

In sexual assault nurse examiner (SANE) programs, specially trained forensic nurses provide 24-hour-a-day, first-response medical care and crisis intervention to rape survivors in either hospitals or clinic settings. This article reviews the empirical literature regarding the effectiveness of SANE programs in five domains: (a) promoting the psychological recovery of survivors, (b) providing comprehensive and consistent postrape medical care (e.g., emergency contraception, sexually transmitted disease [STD] prophylaxis), (c) documenting the forensic evidence of the crime completely and accurately, (d) improving the prosecution of sexual assault cases by providing better forensics and expert testimony, and (e) creating community change by bringing multiple service providers together

to provide comprehensive care to rape survivors. Preliminary evidence suggests that SANE programs are effective in all domains, but such conclusions are tentative because most published studies have not included adequate methodology controls to rigorously test the effectiveness of SANE programs. Implications for practice and future research are discussed.

Rape survivors encounter significant difficulties seeking help from their communities after an assault.[1] Fewer than half of rape victims treated in hospital emergency departments (EDs) receives basic services, such as information about the risk of pregnancy, emergency contraception to prevent pregnancy, and information on the risk of STDs/HIV. Furthermore, most rape cases are not prosecuted by the criminal

SOURCE: Campbell et al. (2005). Reprinted with permission.

justice system. During the past three decades, victim advocates have developed local, state, and national reforms to try to address these problems (see Martin, 2005, for a review). For example, many hospitals will now allow advocates to be present with survivors during their ED treatment to provide emotional support and advocate on their behalf for needed medical services. Rape crisis centers have been instrumental in creating new policies to standardize forensic evidence collection. Often referred to as "rape kits," these protocols were developed so that all survivors who seek postassault medical care can have the evidence of the crime thoroughly documented. In addition, many states have dramatically reformed their criminal sexual assault laws, dropping antiquated requirements that made prosecution nearly impossible. These reforms have undoubtedly had a profound impact on the lives of countless rape victims, and yet most survivors still do not receive adequate medical care and most rape cases are not prosecuted.

Also, within the past 30 years, another reform effort emerged—this one led by the nursing profession with support and collaboration from rape crisis centers. Concerned about the quality of care that survivors were receiving in hospital EDs, nurses across the country became trained in forensic evidence collection so that they, rather than doctors, could provide postassault medical care. Consistent with the basic tenets of nursing practice, these alternative programs sought to provide health care while also attending to the emotional needs of rape survivors. Sexual assault nurse examiner (SANE) programs were created in communities throughout the United States whereby specially trained forensic nurses would provide 24-hour-a-day, first-response medical care and crisis intervention to rape survivors in either hospitals or clinic settings.[2] With increased attention to collecting forensic evidence with state-of-the-art techniques, many SANEs hoped that the prosecution of rape would increase as well. The first SANE programs emerged in the 1970s, and they expanded rapidly throughout the 1990s. Now numbering nearly 450 programs nationwide (International Association of Forensic Nurses [IAFN], 2005), SANEs offer survivors and their communities an alternative

model of care, one that emphasizes comprehensive, multisystem service delivery. Although there are still far more communities without SANE programs and an unknown number of survivors who are still struggling for medical care and legal justice, it is important to examine whether SANE programs have made a positive difference in rape survivors' postassault help-seeking experiences.

The purpose of this article is to review the literature regarding the effectiveness of SANE programs as a reform effort. The literature on SANE programs is largely descriptive, with numerous articles detailing how SANE programs have been created, what kinds of problems they have encountered, and how they have resolved those issue (e.g., Ahrens et al., 2000; Aiken & Speck, 1995; Antognoli-Toland, 1985; Arndt, 1988; Cornell, 1998; Fulginiti et al., 1996; Hatmaker, Pinholster, & Saye, 2002; Ledray, 1992, 1995, 1996; Lenehan, 1991; O'Brien, 1996; Rossman & Dunnuck, 1999). Similarly, there is a substantial body of work on the technical aspects of forensic evidence collection and the administration of SANE programs (e.g., Hohenhaus, 1998; Ledray, 1997a, 1997b, 2000; Ledray & Barry, 1998; Ledray & Netzel, 1997; O'Brien, 1998; Sievers & Stinson, 2002). Other authors have already written syntheses regarding these aspects of SANE programs (Hutson, 2002; Lang, 1999; Ledray, 1999; Littel, 2001). Therefore, the goal of this article is to advance the literature by focusing on the growing empirical literature regarding the effectiveness of SANE programs in multiple domains. What do we know about the success of these programs?

To set the stage for examining the effectiveness of SANE programs, we will begin by briefly reviewing the research on rape survivors' experiences with hospital EDs to uncover what is problematic about this "old" approach to postassault care and what SANE programs sought to change. Then, we will provide an overview of SANE programs, examining how their current structure, function, and operations attempt to provide a more comprehensive and survivor-centered model of care. With this background, we will then review the empirical literature on the effectiveness of SANE programs in five domains.[3] First, SANE programs strive to create settings that address survivors' emotional needs as well as their health concerns. As such, this article will review the evidence on how SANE programs may help survivors' psychological recovery from the rape. Second, we will examine whether SANE programs provide more consistent and comprehensive medical services than what survivors receive in traditional hospital ED care. Third, another founding goal for many SANE programs was to improve the quality of forensic evidence collection, so the empirical literature on nurses as forensic evidence collection specialists will be reviewed. Fourth, by documenting the physical evidence of sexual assault so carefully, it is possible that SANE programs may increase prosecution rates in their communities, and the few studies that have explicitly tested this hypothesis will be examined. Finally, the creation and maintenance of SANE programs often requires a coordinated community effort between multiple social systems and agencies. As such, the evidence regarding how SANE programs function as catalysts for community change will be examined. This article will conclude by exploring the implications of these findings for practitioners and policymakers as well as outlining recommendations for future research on the effectiveness of SANE programs.

Why Are SANE Programs Needed?

When rape survivors seek help after an assault, they are most likely to be directed to the medical system—specifically, hospital EDs (Resnick et al., 2000). Although most victims are not physically injured (Ledray, 1996), survivors are sent to the hospital anyway, primarily for forensic evidence collection (Martin, 2005). The survivor's body is a crime scene and because of the invasive nature of sexual assault, a medical professional, rather than a crime scene technician, is needed to collect the evidence. The "rape exam" or "rape kit" usually involves plucking head and pubic hairs; collecting loose hairs by combing the head and pubis; swabbing the vagina, rectum, and/or mouth to collect semen, blood, or saliva; and obtaining fingernail clippings and scrapings in the event the victim scratched the assailant. Blood samples

may also be collected for DNA, toxicology, and ethanol testing. Throughout this process, medical professionals must take extreme care so as not to taint or destroy the evidence (Ledray, 1999).

Martin (2005) noted that many ED physicians are reluctant to do these exams largely because they do not feel that this is a medical procedure that requires their expertise. Instead, they believe that they should be treating other patients with emergent health threats. This reluctance on the part of ED physicians to do rape forensic evidence collection manifests as long wait times for survivors as most spend 4 to 10 hours in the ED before they are examined (Littel, 2001). During this wait, victims are not allowed to eat, drink, or urinate so as not to destroy physical evidence of the assault (Littel, 2001; Taylor, 2002). ED physicians also do not like doing evidence collection because if subpoenaed to testify in court, they would be challenged on their qualifications, training, experience, and ability to conduct the exam (Ledray, 1999; Littel, 2001). Indeed, most ED personnel lack training specifically in forensic evidence collection, and as a result, many rape kits collected by ED doctors are done incorrectly and/or incompletely. Even ED physicians with forensic training usually do not perform forensic exams frequently enough to maintain their proficiency (Littel, 2001).

Forensic evidence collection is often the focus of hospital ED care, but rape survivors have other medical needs such as injury detection and treatment, information about the risk of pregnancy and emergency contraception to prevent pregnancy, and information on the risk of STDs and prophylaxis. However, numerous studies have found that fewer than half of rape victims treated in hospital EDs receive these basic services. For example, most rape survivors receive a medical exam and forensic evidence collection kit (70% to 81%; Campbell, 2005; Campbell, Wasco, Ahrens, Self, & Barnes, 2001). Yet only 40% of the survivors in the National Victim Survey (National Victim Center, 1992) and 49% of the women in Campbell et al.'s (2001) sample of urban rape survivors received information about the risk of pregnancy. With respect to emergency contraception to prevent pregnancy, accounts from victims indicate that 28% to 38% of women receive this service (Campbell, 2005; Campbell et al., 2001), but analyses of hospital records have found lower rates of 20% to 28% (Amey & Bishai, 2002; Rovi & Shimoni, 2002; Uttley & Petraitis, 2000; see also Smugar, Spina, & Merz, 2000). Approximately one third of rape survivors receive information about the risk of STDs/HIV from the assault, and between 34% and 57% obtain medication to treat STDs (Amey & Bishai, 2002; Campbell, 2005; Campbell et al., 2001; National Victim Center, 1992; Rovi & Shimoni, 2002).

In addition to gaps in service delivery, it appears that rape survivors are often treated insensitively by hospital ED staff. These negative experiences with social system personnel have been termed "the second rape" (Madigan & Gamble, 1991), "the second assault" (Martin & Powell, 1994), or "secondary victimization" (Campbell & Raja, 1999; Williams, 1984). For example, it is not uncommon for hospital ED staff to question victims about their prior sexual histories, what they were wearing at the time of the assault, what they did to "cause" the assault, why they were with the assailant in the first place (if they knew the rapist), and why they trusted the assailant (if they knew the rapist). Medical professionals may view these questions as necessary and appropriate, but rape survivors report that they are very upset and distressed by such questioning. Campbell (2005) found that as a result of their contact with ED doctors and nurses, most rape survivors stated that they felt bad about themselves (81%), guilty (74%), depressed (88%), nervous/anxious (91%), violated (94%), distrustful of others (74%), and reluctant to seek further help (80%) (see also Campbell & Raja, 2005, for replicated rates). Similarly, Campbell et al. (1999) found that victims of non-stranger rape who received minimal medical services but encountered high secondary victimization had significantly elevated levels of posttraumatic stress symptomatology. These rape survivors were doing worse than the victims who did not seek medical services at all. These findings suggest that when victims place their trust in the medical system for help after a rape, they risk the possibility of additional distress.

The History and Current Operations of SANE Programs

To address these health care gaps for rape victims, the nursing profession created Sexual Assault Nurse Examiner (SANE) programs. These alternative service programs were designed to circumvent many of the problems of traditional hospital ED care by having specially trained nurses, rather than doctors, provide 24-hour-a-day, first-response care to sexual assault victims in either hospital or non-hospital settings. Nurses were also interested in learning the intricacies of forensic evidence collection and expert witness court testimony (Ledray & Arndt, 1994; Littel, 2001). The first SANE programs emerged in the 1970s in Memphis, Tennessee (1976), Minneapolis, Minnesota (1977), and Amarillo, Texas (1978) (Ledray & Arndt, 1994). In 1992, the first international meeting of SANEs was held with representatives from programs across the United States and Canada, and the International Association of Forensic Nurses (IAFN) was formed (Littel, 2001). Forensic nursing was identified as a specialty by the American Nurses Association (ANA) in 1995. Rapid development of SANE programs occurred in the mid-1990s as knowledge about SANE programs spread (Littel, 2001).

Currently there are nearly 450 SANE programs throughout the United States and its territories (IAFN, 2005).[4] Most SANE programs (75% to 90%) are hospital-based (e.g., housed within EDs or clinic settings), but some are located in community settings (10% to 25%; e.g., rape crisis centers or medical office buildings; Campbell et al., 2005; Ledray, 1997a). Nearly all programs serve adolescents and adults, and approximately half serve pediatric victims as well (IAFN, 2005). SANE programs are staffed by clinicians (usually registered nurses or nurse practitioners) who have typically completed 40 hours of classroom training, which includes instruction in evidence collection techniques and the use of specialized equipment, chain-of-evidence requirements, expert testimony, injury detection and treatment, pregnancy and STDs screening and treatment, rape trauma syndrome, and crisis intervention. An additional 40 to 96 hours of clinical training is also needed (e.g., performing pelvic exams on non-rape survivors, observing SANEs complete exams, courtroom observation), and specialized continuing education is often required by local programs (IAFN, 2005; Ledray, 1997b, 1999).

This extensive training formed the foundation for an alternative model of postassault care. In outlining a national protocol for forensic and medical evaluation of sexual assault victims, Young, Bracken, Goddard, and Matheson (1992) stated, "The broad goals of the national model protocol are to minimize the physical and psychological trauma to the victim and maximize the probability of collection and preserving physical evidence for potential use in the legal system" (p. 878). To address survivors' psychological needs, SANEs strive to preserve victims' dignity, ensure that they are not retraumatized by the exam, and assist them in regaining control by letting them make decisions throughout the evidence-collection process. Many SANE programs work with their local rape crisis centers so that rape victim advocates can also be present for the exam to provide emotional support (Hatmaker et al., 2002; Lang, 1999; Littel, 2001; Rossman & Dunnuck, 1999; Seneski, 1992; Smith, Homseth, Macgregor, & Letourneau, 1998; Taylor, 2002). To attend to survivors' physical health needs, most SANE programs offer emergency contraception for sexual assault victims who are at risk of becoming pregnant and prophylactic antibiotics to treat STDs that may have been contracted in the assault (Lang, 1999; Taylor, 2002). Although the risk of contracting HIV from a rape is typically low (Ledray, 1999), it is a primary concern for most victims. As such, most SANE programs provide victims with information about degree or risk and testing options (Lang, 1999; Ledray, 1999).

With respect to the forensic evidence collection itself, most SANE programs use specialized forensic equipment that is not often used in traditional hospital ED care, such as a colposcope, which allows for the detection of micro, lacerations, bruises, and other injuries in sexual assault victims (Voelker,

1996). A camera can be attached to the colposcope to photo document genital injuries (Lang, 1999). Toludine blue is also used by some SANEs in the detection of genital trauma by enhancing the visualization of microlacerations (Ledray, 1999). The forensic evidence collected by the SANEs is typically sent to the state crime lab for analysis, and the results are forwarded to the prosecutor's office. If a case is prosecuted, the SANE may provide factual or expert witness testimony at the trial (Ledray, 1998; Ledray & Barry, 1998). In factual witness testimony, a SANE provides information as to what exactly occurred in her or his interactions with the victim (e.g., what evidence was collected, what injuries were sustained, etc.). If a SANE is reviewed by the court and deemed to be an expert, she or he can testify not only about evidence collected and the facts of the case but also about opinions and conclusions that can be drawn from evidence.

Yet when a SANE provides testimony, either factual or expert, "[she] is not an advocate, she is a witness" (Ledray, 1998, p. 287). This raises a potential role conflict for SANEs as they need to care for the psychological well-being of their patients which may involve advocacy, and yet, from a legal perspective, they need to be unbiased. Although it is possible to provide empathic care and emotional support without "biasing" the forensic or legal components of the case, in the event a case does go to trial, it is preferable that the SANE not be viewed as a victim advocate (Ledray, 1998, 1999; Ledray & Barry, 1998; Littel, 2001). Consequently, as SANEs have become more involved in court testimony, many programs have revised their policies and procedures to ensure adequate attention to survivors' emotional needs without compromising SANEs' credibility as witnesses. For instance, Smith et al. (1998) resolved this role conflict by involving rape crisis center advocates to provide emotional support while the nurses complete the medical-legal examination and maintain the chain of evidence. This "division of labor" was deemed necessary so that the SANEs could be effective, unbiased expert witnesses in court. In addition, rape victim advocates can offer survivors confidentiality, whereas SANEs may have to testify about their communications with victims (Littel, 2001).

SANEs provide extensive psychological, medical, and legal services for rape survivors, but truly comprehensive care involves the efforts of many service providers, including law enforcement personnel, crime lab staff, prosecutors, and rape crisis center staff. As such, many SANE programs today operate as part of multidisciplinary response teams (e.g., Sexual Assault Response Teams [SARTs]) or coordinated community response initiatives (Hutson, 2002; Littel, 2001). Historically, many SANE programs were created by the sole or primary initiative of individual nurses seeking to make change in their communities, but now it is becoming more common that multidisciplinary coordinating committees work together to create SANE programs (Campbell et al., 2005; Hutson, 2002). These steering groups are often charged with creating initial policies and procedures and ensuring cooperation, rather than competition, between agencies (Hutson, 2002). Recognizing the importance of collaboration, three states currently require all SANE programs who apply for state funding to have a multidisciplinary team to oversee the implementation of their SANE program (Littel, 2001). Many SANE programs continue to work closely with the members of the multidisciplinary committee to review cases and verify that survivors received comprehensive care (Littel, 2001).

⊠ The Effectiveness of SANE Programs

Because the work of SANE programs is multifaceted, addressing psychological, medical, and legal concerns, defining and measuring "success" or "effectiveness" is complex. For example, some SANE programs have made it a goal to improve prosecution of sexual assault cases in their communities, whereas other have noted that the rape prosecution is influenced by many factors, only one of which is the presence and quality of forensic evidence. Consequently, some programs have not defined success by prosecution rates. Therefore, the evaluation of individual SANE programs must reflect the specific goals and missions of that program, but it may be useful to consider multiple indices of success when evaluating the collective work of SANE programs as a reform effort. In this section of the article, the empirical literature on SANE programs will be examined to evaluate the success of SANE programs in five domains: (a) promoting the psychological recovery of survivors, (b) providing comprehensive and consistent medical care, (c) documenting the forensic evidence of the crime completely and accurately, (d) improving the prosecution of sexual assault cases by providing better forensics and expert testimony, and (e) creating community change by bringing multiple service providers together to provide comprehensive care to rape survivors (see Table 1).

Psychological Effectiveness

Although the forensic aspect of SANEs' work typically receives the most attention by the legal and medical communities, it is the commitment to victims' psychological well-being that defines how SANEs work with their patients throughout all aspects of care. Putting this point in perspective, Ledray, Faugno, and Speck (2001) noted that a SANE is a compassionate and supportive RN who is also a skilled forensic technician. Although emotional care is a primary goal of SANE programs, there have been few studies that have systematically evaluated the psychological impact of SANE programs. In a study of the Memphis SANE program, Solola, Scott, Severs, and Howell (1983) found that 50% of victims in their study were able to return to their usual vocation within 1 month, and in 3 to 6 months, 85% felt secure alone in public areas. After 12 months, more than 90% of the survivors were entirely free of their initial assault-related anxieties and emotional discomposure. Unfortunately, this publication did not provide sufficient details regarding the methodology of this study to assess whether the recovery gains were attributable to the SANE program or to "normal" recovery processes. Other research suggests that, at the very least, rape survivors perceive SANEs as

Table 1 Research Findings on the Effectiveness of Sexual Assault Nurse Examiner (SANE) Programs

Studies	Major Findings
Psychological effectiveness	
Solola, Scott, Severs, & Howell (1983)	More than 90% of survivors treated in the Memphis SANE program were not experiencing assault-related anxiety.
Malloy (1991)	Victims treated in the Minneapolis SANE program identified the nurses listening to them as one thing that helped them the most during their crisis period.
Ericksen et al. (2002)	Victims treated at a Canadian SANE program felt respected, safe, in control, believed and supported, cared for by people with expertise, informed, and cared for beyond the hospital because they received the option for follow-up care.
Medical/health care effectiveness	
Crandall & Helitzer (2003) Ciancone, Wilson, Collette, & Gerson (2000)	STD prophylaxis and emergency contraception were more routinely provided in the SANE program as compared to the traditional hospital ED care.
Derhammer, Lucente, Reed, & Young (2000)	After a SANE program was implemented, victims were significantly more likely to be given a complete physical exam than before the SANE program was created.
Forensic effectiveness	
Ledray & Simmelink (1997) Sievers, Murphy, & Miller (2003)	SANE-collected kits were more thorough and had fewer errors than the non-SANE kits.
Legal effectiveness	
Crandall & Helitzer (2003)	Police filed more charges of sexual assault post-SANE as compared to pre-SANE. The conviction rate for charged SANE cases was also significantly higher, resulting in longer average sentences.
Community change effectiveness	
Crandall & Helitzer (2003)	The working relationships and communication between medical and legal professionals improved substantially after the implementation of a SANE program.

helpful and supportive. In an evaluation of the Minneapolis SANE program, Malloy (1991) surveyed 70 patients in crisis and found that 85% of the survivors identified the nurses listening to them as one thing that helped them the most during their crisis period.

In the most in-depth study on this topic, Ericksen et al. (2002) conducted semistructured qualitative interviews with eight survivors who were treated in a Canadian "specialized sexual assault service," which included both specially trained physicians and SANEs. The primary goal of this study was to understand what it meant to survivors to receive this kind of care. Using latent content analysis, the authors identified nine major themes in the participants narratives: (a) They felt they were met and they were treated with dignity and respect,

(b) they felt the presence of the nursing staff—they provided information about what to expect and listened to the survivors, (c) they felt safe—the caregivers were women and were sensitive in their care, (d) they appreciated how they were physically touched—the nurses held their hands during the exam, (e) they felt in control—they were given options and were not pushed toward certain choices, (f) they felt reassured—they felt believed and supported by the staff, (g) they felt they were cared for by people with expertise—their care providers knew what they were doing, (h) they felt informed—they were given information and the staff were careful not to overwhelm them with too much information, and (i) they felt cared for beyond the hospital—they received follow-up care or the option for follow-up care. These descriptive data provide insight into how and why SANE programs may be psychologically beneficial to rape survivors, but as the authors of this study also noted, there is a need for larger-scale studies on the short-term and long-term psychological impact of SANE programs on survivors' recoveries.

Medical/Health Care Effectiveness

Many rape survivors treated in hospital EDs do not receive needed medical services, which was another problem that SANE programs sought to address. As with the literature on psychological outcomes, there are few published reports documenting rates of medical service delivery in SANE programs, but available data suggest that victims treated in SANE programs receive consistent and broad-based medical care. In a national survey of SANE program staff, Ciancone, Wilson, Collette, and Gerson (2000) found that 97% of programs reported that they offer pregnancy testing, 97% provide emergency contraception, and 90% give STD prophylaxis. The SANE Program staff indicated that services such as conducting STD cultures, HIV testing, toxicology, and ethanol screening are not routinely performed but are selectively offered to survivors.[5] These rates of service delivery are substantially

higher than what has been found in studies of traditional ED care (e.g., Amey & Bishai, 2002; Campbell, 2005; Campbell et al., 2001; Rovi & Shimoni, 2002). However, Ciancone et al.'s data were collected from SANE program staff about what they say they provide in their programs rather than from individual survivors regarding the actual services they received. Coming closer to a direct assessment of service delivery, Derhammer, Lucente, Reed, and Young (2000) examined chart records before and after a SANE program was implemented in a hospital ED and discovered that in only 11% of the pre-SANE cases were survivors given a complete physical exam (both external exam and internal vaginal exam). This percentage jumped significantly to 95% after the SANE program was implemented. Unfortunately, this study did not document rate changes for other medical services, such as emergency contraception and STD prophylaxis, pre-SANE to post-SANE.

In the most comprehensive and methodologically rigorous study to date on medical service delivery in SANE programs, Crandall and Helitzer (2003) compared the services received for sexual assault cases seen at the University of New Mexico's Health Sciences Center for the 2 years prior to the inception of a SANE program (1994 to 1996) ($n = 242$) and 4 years afterward (1996 to 1999) ($n = 715$). Statistically significant changes in medical services delivery rates were found from pre-SANE to post-SANE. For example, the rate of pre-SANE pregnancy testing in this hospital was 79% and increased to 88% post-SANE. Providing emergency contraception was also more common after the SANE program was created (66% to 87%). STD prophylaxis was also more routinely provided in the SANE program as compared to the traditional hospital ED care (89% to 97%). Given the quasi-experimental design of this study, these increases are likely attributable to the implementation of the SANE program, but it is worth noting that the pre-SANE rates of service provision found at this hospital were already substantially higher than what has been found in prior studies of medical service delivery. For instance, service delivery rates for emergency contraception in hospital EDs are typically 20% to 38%, and at the University of New Mexico's Health Sciences Center, they were 66% before the SANE program even started. Even though this hospital may have already been providing reasonably comprehensive care to rape survivors, their rates of service delivery still significantly increased post-SANE. However, it is not clear whether a SANE program could make such headway in hospitals with lower starting rates of service delivery.

Forensic Effectiveness

SANE programs emerged not only because traditional ED care did not pay adequate attention to survivors' emotional and medical health needs but also because the forensic evidence collection itself needed to be improved. ED physicians receive either no training or only minimal training in forensics, which has raised concern among victim advocates that the evidence of sexual assault is not being adequately documented (Ledray, 1999; Littel, 2001). SANEs sought to address this issue through extensive training and practice in forensic techniques. However, since taking on this new role, SANEs throughout the country have been challenged by both the medical and legal communities as to whether they were qualified and skilled enough to perform this task (DiNitto, Martin, Norton, & Maxwell, 1986; Littel, 2001). The clinical case study literature suggests that SANEs are not only competent in forensic evidence collection, but they are actually better at it because of their extensive training and experience. For example, Cornell (1998) noted that "With the [SANE] program, physicians are removed from the role of witness. Now evidence is collected more consistently and adequately" (p. 46). Similarly, Littel (2001) noted that SANE programs have "greatly improved the quality and consistency of collected evidence" (p. 7). Yet clinical case reports, though remarkably consistent in their conclusions, do not provide definitive evidence of the effectiveness of SANEs in forensic evidence collection. Empirical studies that directly compare the evidence collected by SANEs and physicians on objective criteria would better inform the debate about whether nurses are competent forensic examiners.

To date, there have been only two such comparative studies conducted in the United States. First, Ledray and Simmelink (1997) reported the finding from an audit study of rape kits sent to the Minnesota Bureau of Criminal Apprehension. Twenty-seven kits conducted by SANEs were compared to 73 kits collected by physicians or non-SANEs with respect to completeness of specimens collected, documentation, and maintenance of chain of custody. Overall, the SANE-collected kits were more thorough and had fewer errors than the non-SANE kits. For example, with respect to completeness of evidence, 96% of the SANE kits versus 85% of non-SANE kits collected the swabs to match the recorded orifice of penetration, 92% of the SANE kits versus 15% of non-SANE kits contained an extra tube of blood for alcohol and/or drug analysis, and in 100% of the SANE kits versus 81% of non-SANE kits the blood stain card was properly prepared. In addition, the chain of evidence was broken in some non-SANE kits but was always maintained in SANE kits. Although these descriptive data suggest that the SANEs' evidence collection was more thorough and accurate, inferential statistics were not reported so it was not known whether these differences were statistically significant.

A larger-scale study by Sievers, Murphy, and Miller (2003) explicitly tested differences between SANE and non-SANE kits and also found support for better evidence collection by SANEs. Specifically, this study compared 279 kits collected by SANEs and 236 by doctors/non-SANEs on 10 quality-control criteria and found that in 9 of 10 categories, the SANE-collected kits were significantly better. The kits collected by SANEs were significantly more likely than kits collected by physicians to include the proper sealing and labeling of specimen envelopes, the correct number of swabs and other evidence (pubic hairs and head hairs), the correct kind of blood tubes, a vaginal motility slide, and a completed crime lab form. The Sievers et al. study provides the strongest evidence to date that SANEs collect forensic evidence correctly and, in fact, do so better than physicians. However, it is important to note that training and experience, not job title or professional degree, are the likely reasons behind these findings. Further under-scoring the link between experience and evidence quality, DiNitto et al. (1986) reported that prosecutors in Florida were "satisfied with evidence collected by nurse examiners, crediting the training of the nurse examiners . . . prosecutors tended to be more pleased with the quality of a physician's evidence *when the examiner had conducted many exams and thus had perfected the techniques*" (p. 539, emphasis added). Because SANEs have made it a professional priority to obtain extensive forensic training and practice, it is not surprising that both case study and empirical data suggest that they are better forensic examiners than physicians and nurses who have not completed such training.

Legal Effectiveness

SANEs provide law enforcement personnel and pros-ecutors with detailed forensic evidence documenting crimes of sexual assault, which raises the question: Do SANE programs have an impact on prosecution rates in their communities? As with the literature on the quality of forensic exams, case studies suggest that SANE programs increase prosecution (Aiken & Speck, 1995; Cornell, 1998; Hutson, 2002; Littel, 2001; Seneski, 1992). For example, there are reports that SANE programs specifically increase the rate of plea bargains because when confronted with detailed forensic evidence collected by the SANEs, assailants will decide to plead guilty (often to a lesser charge) rather than face trial (Aiken & Speck, 1995; Ledray, 1992; Littel, 2001; Seneski, 1992). Other reports indi-cate that when cases do go to trial, the expert witness testimony provided by SANEs is instrumental in obtaining convictions (O'Brien, 1996; Smith, 1996, as cited in Ledray, 1999).

Yet there have been few studies that have empiri-cally tested the hypothesis that SANE programs increase prosecution. Studies that report the prosecu-tion rates for SANE programs rarely include a com-parison group (e.g., rates before and after the SANE program was implemented or comparisons to another community without a SANE program). However there

is already an extensive literature on "typical" rates of prosecution in communities without SANE programs. For example, arrest rates in rape cases have been found to vary between 25% (Frazier & Haney, 1996) to 49% (Spohn & Horney, 1992).[6] Prosecution rates are quite variable, with published findings ranging from 14% (LaFree, 1980) to 35% (Spohn & Horney, 1992) to 56% (Spohn, Beichner, & Davis-Frenzel, 2001). Drawing from these published reports, it can be infor-mative (though not conclusive) to compare arrest and prosecution rates in communities with SANE pro-grams to these figures from communities without SANE programs.

For example, Solola et al. (1983) examined the legal outcomes for 621 victims who were treated in the Memphis SANE program in 1980. Police reports were filed in 573 of these cases (92%), and 124 resulted in an arrest and successful prosecution (22% of reported cases). However, 135 cases were still pend-ing at the time this study was conducted, and if the rates of arrest and prosecution are examined only in closed cases, the prosecution rate was 28%. In either analysis, the prosecution rates of 22% or 28% are still higher than what has been found for non-SANE cases (typically 14% to 18%; some as high as 35% to 56% on average). Similarly, in her case study of the Santa Cruz County SANE program, Arndt (1988) noted that 42% of sexual assaults involving victims 14 years and older resulted in arrests of the perpetrators and 58% of child molestation cases resulted in arrest, which again is higher than what is found for cases that do not involve SANE programs (typically 25% to 44%; some as high as 49% on average). Ledray (1992) reported that of 417 rape cases in Minneapolis in 1990, 193 were presented by police to the county attorney (46%). Of those 193 cases, 60 were not charged by the prosecutor (31%), 65 defendants pleaded guilty (34%), 14 went to trial (7%) (6 found guilty, 8 found not guilty), and the outcomes in the remaining 54 cases were not reported.

As noted previously, a stronger methodological design would include a direct comparison of legal out-comes for SANE cases versus non-SANE cases, and to date, there has been only one such study. Crandall and

Helitzer's (2003) comparison of legal outcomes in a New Mexico jurisdiction before and after the implementation of a SANE program found that significantly more victims treated in the SANE program reported to the police than did before the SANE program was launched in this community (72% versus 50%) and significantly more survivors had evidence collection kits taken (88% versus 30%). Police filed more charges of sexual assault post-SANE as compared to pre-SANE (7.0 charges/perpetrator versus 5.4). The conviction rate for charged SANE cases was also significantly higher (69% versus 57%), resulting in longer average sentences (5.1 versus 1.2 years). These data provide the strongest evidence yet that SANE programs can have a beneficial impact on the prosecution of sexual assault cases. However, as was noted previously, this New Mexico community may be somewhat atypical in its pre-SANE responses to sexual assault survivors. The pre-SANE conviction rates were substantially higher than published reports and post-SANE numbers were higher still, which raises the question whether such effects are possible in communities with lower starting conviction rates.

Community Change Effectiveness

The effectiveness of SANE programs in multiple domains—psychological, medical, forensic, and legal—suggests that something profoundly different happens when survivors are treated in these alternative programs. SANE programs' successes may be attributable not only to the work of the individual nurses but also to the kind of community-level change that comes about in forming and sustaining a SANE program (Ahrens et al., 2000). As discussed previously, rape survivors need help from multiple service providers, and SANE programs provide a structure for comprehensive, integrated care. Some programs may deliberately identify community change as a founding goal and purpose, but others may find that such change happens along the way as part of the process of implementing a SANE program. Indeed, case reports from local SANE programs suggest that these programs increase interagency collaboration and cooperation,

which improves care for survivors (Hatmaker et al., 2002; Selig, 2000; Smith et al., 1998).

In the only empirical study of the effectiveness of SANE programs in creating community change, Crandall and Helitzer (2003) interviewed 28 key informants from health care, victim services, law enforcement, and prosecution who had been involved in the care of sexual assault survivors both before and after a SANE program was implemented in their community. The informants stated that before the SANE program, community services were disjointed and fractionalized, but afterward care for survivors was centralized because there was a point of convergence where multiple service providers could come together to help victims. Informants also noted that the SANE program increased the efficiency of law enforcement officers by reducing the amount of time they spent waiting at the medical facility. As a result, officers could spend more time investigating the case. Moreover, the informants believed that police officers were better able to establish positive rapport with survivors, which increased the quality of victim witness statements.

In addition to improving the services provided to survivors, the informants indicated that since the SANE program was implemented, working relationships and communication between medical and legal professionals had improved substantially. For instance, prior to SANE, law enforcement had difficulty communicating with health care providers because their working relationship lacked consistency. The SANE program created standardized response protocols and hosted regular interagency meetings to review cases and engage in ongoing quality improvement. One important benefit of this direct communication was that officers were able to identify more quickly and accurately trends in similar assaults and perpetrator types, which was instrumental in discovering a pattern rapist in their community.

Whereas the collaborative relationship between the medical and legal communities greatly benefited from the emergence of a SANE program in this community, the results were not so clear-cut for the relationship between the SANE program and the local rape crisis center. The advocates interviewed in

Crandall and Helitzer's (2003) study had "mixed emotions" about their work with the SANE program. Advocates believed that the SANEs felt that they could do the advocates' job and that services of the advocates were duplicative and unnecessary. Ironically, the advocates felt that before the creation of the SANE program, hospital ED personnel had valued their role, but now the SANEs sometimes acted as though the advocates were in their way. It is interesting to note that the health care informants had a different perspective. The medical providers stated that before the SANE program, they felt that advocates were in their way while they were trying to treat victims, but post-SANE they perceived the advocates as helpful and supportive to victims. It is possible that the process of creating the SANE program highlighted the need for multiple service providers to work together to provide care for survivors. Hospital personnel may not have fully appreciated the need for specific attention to survivors' emotional well-being, but with the emergence of the SANE program, this issue was highlighted. However, in this community it appears that the emergence of the SANE program called into question whether rape victim advocates or SANEs should have the primary responsibility for the emotional care of rape survivors. As noted previously, because SANEs may testify in court, their communication with victims is not confidential, but advocates can provide confidential services. This suggests that there is a need for advocates, and they can work together with SANEs to provide an emotionally supportive setting for care. The issue of confidentiality was not examined in Crandall and Helitzer's study, but it is an important factor to consider in future work on the relationship between SANE programs and rape crisis centers.

◈ The Future of SANE Programs: Implications for Practice, Policy, and Research

The current literature on SANE programs consists primarily of case study reports, with few empirical studies that have tested the effectiveness of SANE programs in multiple domains. Yet from the information that is available, it appears that SANE programs promote the psychological recovery of rape survivors, provide comprehensive medical care, obtain forensic evidence correctly and accurately, and facilitate the prosecution of rape cases. Through this work, SANE programs can be instrumental in creating interagency collaborative relationships that improve the overall community response to rape. However, such conclusions are tentative because most published studies have not included adequate methodological controls or comparisons to rigorously test the effectiveness of SANE programs. Nevertheless, these preliminary findings can be helpful to SANE practitioners and policymakers because they indicate that this approach to treating sexual assault survivors has merit and should continue for further evaluation and analysis.

Specifically, this information may help practitioners with two primary issues: launching new programs and developing "benchmarks" of effectiveness for established programs. First, knowing that SANE programs have the potential to be effective in multiple domains may be instrumental in starting new SANE programs. It is often difficult to obtain broad-based community support for new initiatives because it is not yet known if the effort will be successful. Although the full impact of creating a SANE program in an individual community cannot be known prior to implementation, the literature suggests that many SANE programs have been able to address the psychological, medical, and legal needs of rape survivors. Although policy analysts have noted that empirical research is not always convincing in policy decisions (Weiss, 1983), the fact that there is independent evidence demonstrating promising effects is probably more persuasive than individual's beliefs that such an effort is worthwhile. Why should a local community launch a new SANE program? Because there is ample evidence that the "old" model of traditional hospital ED care is not only incomplete but also potentially revictimizing, and there is emerging evidence that the "new" model created by SANE programs addresses major gaps in service delivery for sexual assault survivors. However, communities should also consider cost-benefit issues as some hospitals serve very few sexual assault victims, and as such, a designated SANE may not be cost effective for that community.

Second, the literature on SANE programs can also serve as a reference for expected or desired outcomes in established programs. Once a program is launched and the challenges of implementation have been resolved, many community stakeholders will want to know whether the program is effective. As this review illustrated, effectiveness can be defined in multiple ways, and the research findings suggest many possible positive outcomes or benchmarks. If a program discovers, for example, that prosecution rates are not improving in their community, it is helpful to know that the literature suggests other SANE programs have been instrumental in increasing prosecution. This creates an opportunity for professional dialogue to identify what worked in one program and consider how those elements could be successfully transplanted to another program.

Although the current literature on SANE programs can provide practitioners and policymakers with useful information, this review suggests that there is a pressing need for more methodologically rigorous research on the effectiveness of SANE programs. Because most studies are small in scale and have not been replicated, it is important that neither researchers nor practitioners overstate what SANE programs can accomplish. From a methodological perspective, future research on the effectiveness of SANE programs needs to attend to three primary issues. First, larger-scale studies are needed whereby the experiences of more survivors in more programs are analyzed. The case study literature on SANE programs contains multiple studies from a small number of programs (e.g., Minneapolis, Memphis). To evaluate the effectiveness of SANE programs as a reform effort, it is necessary to review data from many more programs. Second, most studies in the literature do not include comparison groups, which must be addressed in future research. Comparisons are needed over time (e.g., before and after SANE programs are implemented) as well as between comparable communities with and without SANE programs. Moreover, SANE programs are remarkably diverse (e.g., hospital versus community based), and although there are common elements that define them, unique elements of individual programs need to be compared. Third, longitudinal evaluations are needed that follow survivors through the process of receiving care in a SANE program and then link those experiences to short-term and long-term outcomes.

From a substantive perspective, future research is needed on the underlying processes that contribute to the effectiveness of SANE programs. If future studies can replicate the positive findings in the current literature, it is important to explore the mechanisms leading to those effects. With respect to psychological recovery, it is not yet known how SANE programs contribute to survivors' emotional well-being. Is it that SANE programs do not "re-rape" victims, causing secondary victimization, and hence survivors have less distress? Is it that SANE programs provide coordinated care and referrals to counseling services for survivors? Furthermore, what is the unique positive contribution of the SANE vis-à-vis the rape victim advocate who is also present and attending to the survivors' well-being? These issues of process are equally important when examining prosecution outcomes. For example, if prosecution rates are higher in communities with SANE programs, why is that? Is it because the quality of the evidence is stronger (as is suggested by Sievers et al., 2003) or because the expert testimony of the SANE is compelling (as is suggested by Ledray & Barry, 1998), or because SANE programs provide survivors with emotional support and resources that are needed to withstand the lengthy process of prosecution (as is suggested by Seneski, 1992)? Or did the SANE program create collaborative networks that finally enabled disparate social systems to work together toward a common outcome? Understanding the mechanisms by which SANE programs are having positive psychological and legal effects is an important next step for the field.

Finally, future research would benefit from stronger collaborations between sexual assault researchers and SANE program practitioners. The work of SANE programs is remarkably complex, and research and evaluation projects would benefit tremendously from the diversity of perspectives that come from collaborative partnerships. In studying the effectiveness of SANE programs, a research protocol must be sensitive to the safety and confidentiality needs of survivors (see Sullivan & Cain, 2004). Moreover, it is particularly important, for both ethical and methodological reasons, that evaluations of SANE programs do not interfere with the actual provision of services. Researchers and SANE program staff need to work together to address these practical issues. Indeed, Mouradian, Mechanic, and Williams (2001) went further to recommend that researchers and practitioners should work together on all aspects of a research project, from design to dissemination. Collaborative research can be very time-consuming but ultimately can produce methodologically rigorous research that answers important policy question (Riger, 1999). Not so coincidentally, that is exactly what is needed in future research on the effectiveness of SANE programs.

Implications for Practice, Policy, and Research

- The literature on SANE programs may help practitioners advocate for the development of new SANE programs because there is preliminary "proof" that this alternative model of care reflects a substantial improvement over traditional hospital ED services. The literature on SANE programs provides practitioners with benchmarks for effectiveness and desired outcomes.
- There is a pressing need for more methodologically rigorous research on the effectiveness of SANE programs to make more concrete statements regarding the functioning and impact of SANE programs.
- Future studies need to include comparisons between communities with and without SANE programs and between SANE programs with different structures, functions, and operations.

- Longitudinal evaluations are needed that follow survivors through the process of receiving care in a SANE program to assess both short- and long-term outcomes.
- Future research is needed to understand the underlying processes and mechanisms that contribute to SANE programs having positive psychological and legal effects.
- Collaboration between sexual assault researchers and SANE program practitioners would improve research and evaluation projects by increasing the awareness of and sensitivity to the complexities of SANE programs.

Notes

1. Throughout this review, the terms "victim" and "survivor" will be used interchangeably. Some researchers and advocates have called for using the term "survivor" rather than "victim" to emphasize the strength required to recover from rape; others recommend the term "victim" to refer to those who have been recently assaulted and the term "survivor" to refer to those further along in recovery. In this article, the terms are used interchangeably to reflect both the violent nature of this crime (hence "victim") and the long-term work of recovering from such violence (hence "survivor"). In addition, this review focuses on female survivors of sexual assault. Although epidemiological data suggest that both females and males are raped, females are at substantially higher risk for assault. As a result, most research to date has focused on female rape survivors, so it is not known if the research findings summarized in this article apply to populations of male victims.

2. Some programs use the term "forensic nurse examiner" (FNE) rather than "sexual assault nurse examiners" (SANE). Because most programs use the term "SANE," we will use this terminology throughout this review.

3. The studies included in this review were limited to those that focused exclusively on SANE programs (as opposed to other coordinated care initiatives, such as Sexual Assault Response Teams [SARTs]) and were specific to work of SANEs (as opposed to counselors, advocates, or other staff who worked in or with the SANE program).

4. Although there are nearly 450 SANE programs nationwide, which is a substantial number, it is important to note that there are still far more hospitals that perform rape exams without SANEs.

5. These services are not routinely provided because they raise complicated legal issues for survivors that must be examined

and resolved on a case-by-case basis. For example, if a survivor was tested for HIV at the time of the exam and the results were positive, it is possible that she may be required to notify the rapist of her HIV-positive status and may be a risk for being sued by the rapist. Many SANE programs have decided to discuss the implication of these kinds of services with survivors and allow victims to decide if and how they wish to proceed.

6. Spohn and Horney (1992) examined rape arrest and prosecution rates in six jurisdictions and found widely varying figures. The rate of 49% reflects the average arrest rate across the six cities.

References

Ahrens, C. E., Campbell, R., Wasco, S. M., Aponte, G., Grubstein, L., & Davidson, W. S. (2000). Sexual assault nurse examiner (SANE) programs: An alternative approach to medical service delivery for rape victims. *Journal of Interpersonal Violence, 15,* 921–943.

Aiken, M. M., & Speck, P. M. (1995). Sexual assault and multiple trauma: A sexual assault nurse examiner (SANE) challenge. *Journal of Emergency Nursing, 2,* 466–468.

Amey, A. L., & Bishai, D. (2002). Measuring the quality of medical care for women who experience sexual assault with data from the National Hospital Ambulatory Medical Care Survey. *Annals of Emergency Medicine, 39,* 631–638.

Antognoli-Toland, P. (1985). Comprehensive program for examination of sexual assault victims by nurse: A hospital-based project in Texas. *Journal of Emergency Nursing, 11,* 132–135.

Arndt, S. (1988). Nurses help Santa Cruz sexual assault survivors. *California Nurse, 84*(5), 4–5.

Campbell, R. (2005). What really happened? A validation study of rape survivors' help-seeking experiences with the legal and medical system. *Violence & Victims, 20,* 55–68.

Campbell, R., & Raja, S. (1999). The secondary victimization of rape victims: Insights from mental health professionals who treat survivors of violence. *Violence & Victims, 14,* 261–275.

Campbell, R., & Raja, S. (2005). The sexual assault and secondary victimization of female veterans: Help-seeking experience with military and civilian social systems. *Psychology of Women Quarterly, 29,* 97–106.

Campbell, R., Self, T., Barnes, H. E., Ahrens, C. E., Wasco, S. M., & Zaragoza-Diesfeld, Y. (1999). Community services for rape survivors: Enhancing psychological well-being or increasing trauma? *Journal of Consulting and Clinical Psychology, 67,* 847–858.

Campbell, R., Townsend, S. M., Long, S. M., Kinnison, K. E., Pulley, E. M., Adames, S. B., et al. (2005). Organizational characteristics of sexual assault nurse examiner (SANE) programs: Results from the national survey. *Journal of Forensic Nursing, 1,* 57–64.

Campbell, R., Wasco, S. M., Ahrens, C. E., Self, T., & Barnes, H. E. (2001). Preventing the "second rape": Rape survivors' experiences with community service providers. *Journal of Interpersonal Violence, 16,* 1239–1259.

Ciancone, A., Wilson, C., Collette, R., & Gerson, L. W. (2000). Sexual assault nurse examiner programs in the United States. *Annals of Emergency Medicine, 35,* 353–357.

Cornell, D. (1998). Helping victims of rape: A program called SANE. *New Jersey Medicine, 2,* 45–46.

Crandall, C., & Helitzer, D. (2003). *Impact evaluation of a sexual assault nurse examiner (SANE) program* (NIJ Document No. 203276). Washington, DC: National Institute of Justice.

Derhammer, F., Lucente, V., Reed, J., & Young, M. (2000). Using a SANE interdisciplinary approach to care for sexual assault victims. *Journal on Quality Improvement, 26,* 488–495.

DiNitto, D., Martin, P. Y., Norton, D. B., & Maxwell, M. S. (1986). After rape: Who should examine rape survivors? *American Journal of Nursing, 86*(5), 538–540.

Ericksen, J., Dudley, C., McIntosh, G., Ritch, L., Shumay, S., & Simpson, M. (2002). Clients' experiences with a specialized sexual assault service. *Journal of Emergency Nursing, 28,* 86–90.

Frazier, P. A., & Haney, B. (1996). Sexual assault cases in the legal system: Police, prosecutor, and victim perspectives. *Law and Human Behavior, 20,* 607–628.

Fulginiti, T. P., Seibert, E., Firth, V., Quartertone, J., Harner, L., Webb, J., et al. (1996). A SANE experience in a community hospital: Doylestown's first six months. *Journal of Emergency Nursing, 22,* 422–425.

Hatmaker, D., Pinholster, L., & Saye, J. (2002). A community-based approach to sexual assault. *Public Health Nursing, 19,* 124–127.

Hohenhaus, S. (1998). Sexual assault: Clinical issues, SANE legislation and lessons learned. *Journal of Emergency Nursing, 24,* 463–464.

Hutson, L. A. (2002). Development of sexual assault nurse examiner programs. *Emergency Nursing, 37,* 79–88.

International Association of Forensic Nurses (IAFN). (2005). *Database of the International Association of Forensic Nurses.* Retrieved on April 8, 2005, from http://www.forensicnurse.org

LaFree, G. (1980). Variables affecting guilty pleas and convictions in rape cases: Toward a social theory of rape processing. *Social Forces, 58,* 833–850.

Lang, K. (1999). *Sexual assault nurse examiner resource guide for Michigan communities.* Okemos: Michigan Coalition Against Domestic and Sexual Violence.

Ledray, L. E. (1992). The sexual assault nurse clinician: A fifteen-year experience in Minneapolis. *Journal of Emergency Nursing, 18,* 217–222.

Ledray, L. E. (1995). Sexual assault evidentiary exam and treatment protocol. *Journal of Emergency Nursing, 21,* 355–359.

Ledray, L. E. (1996). The sexual assault resource service: A new model of care. *Minnesota Medicine, 79,* 43–45.

Ledray, L. E. (1997a). SANE program locations: Pros and cons. *Journal of Emergency Nursing, 23,* 182–186.

Ledray, L. E. (1997b). SANE programs staff: Selection, training and salaries. *Journal of Emergency Nursing, 23,* 491–495.

Ledray, L. E. (1998). Sexual assault: Clinical issues, SANE expert and factual testimony. *Journal of Emergency Nursing, 24,* 284–287.

Ledray, L. E. (1999). *Sexual assault nurse examiner (SANE) development and operations guide.* Washington, DC: Office for Victims of Crime, U.S. Department of Justice.

Ledray, L. E. (2000). Is the SANE role within the scope of nursing practice? On "pelvics," "colposcopy," and "dispensing of medications." *Journal of Emergency Nursing, 26,* 79–81.

Ledray, L. E., & Arndt, S. (1994). Examining the sexual assault victim: A new model for nursing care. *Journal of Psychosocial Nursing, 32,* 7–12.

Ledray, L. E., & Barry, L. (1998). SANE expert and factual testimony. *Journal of Emergency Nursing, 24,* 3.

Ledray, L. E., Faugno, D., & Speck, P. (2001). SANE: Advocate, forensic technician, nurse? *Journal of Emergency Nursing, 27,* 91–93.

Ledray, L. E., & Netzel, L. (1997). DNA evidence collection. *Journal of Emergency Nursing, 23,* 156–158.

Ledray, L. E., & Simmelink, K. (1997). Efficacy of SANE evidence collection: A Minnesota study. *Journal of Emergency Nursing, 23,* 75–77.

Lenehan, G. (1991). Sexual assault nurse examiners: A SANE way to care for rape victims. *Journal of Emergency Nursing, 17,* 1–2.

Littel, K. (2001). Sexual assault nurse examiner programs: Improving the community response to sexual assault victims. *Office for Victims of Crime Bulletin, 4,* 1–19.

Madigan, L., & Gamble, N. (1991). *The second rape: Society's continued betrayal of the victim.* New York: Macmillan.

Malloy, M. (1991). *Relationship of nurse identified therapeutic techniques to client satisfaction reports in a crisis program.* Unpublished master's thesis, University of Minnesota, Minneapolis.

Martin, P. Y. (2005). *Rape work: Victims, gender, and emotions in organization and community context.* New York: Routledge.

Martin, P. Y., & Powell, R. M. (1994). Accounting for the second assault: Legal organizations framing of rape victims. *Law and Social Inquiry, 19,* 853–890.

Mouradian, V. E., Mechanic, M. B., & Williams, L. M. (2001). *Recommendations for establishing and maintaining successful researcher-practitioner collaborations.* Wellesley, MA: Wellesley College, National Violence Against Women Prevention Research Center.

National Victim Center. (1992). *Rape in America: A report to the nation.* Arlington, VA: Author.

O'Brien, C. (1996). Sexual assault nurse examiner program coordinator. *Emergency Nurses Association, 22,* 532–533.

O'Brien, C. (1998). Light staining microscope: Clinical experience in a sexual assault nurse examiner (SANE) program. *Journal of Emergency Nursing, 24,* 95–97.

Resnick, H. S., Holmes, M. M., Kilpatrick, D. G., Clump, G., Acheron, R., Best, C. L., et al. (2000). Predictors of post-rape medical care in a national sample of women. *American Journal of Preventive Medicine, 19,* 214–219.

Riger, S. (1999). Working together: Challenges in collaborative research on violence against women. *Violence Against Women, 5,* 1099–1117.

Rossman, L., & Dunnuck, C. (1999). A community sexual assault program based in urban YWCA: The Grand Rapids experience. *Journal of Emergency Nursing, 25,* 424–427.

Rovi, S., & Shimoni, N. (2002). Prophylaxis provided to sexual assault victims seen at U.S. emergency departments. *Journal of American Medical Women's Association, 57,* 204–207.

Selig, C. (2000). Sexual assault nurse examiner and sexual assault response team (SANE/SART) programs. *Nursing Clinics of North America, 35,* 311–319.

Seneski, P. (1992). Multi-disciplinary program helps sexual assault victims. *The American College of Physician Executives,* pp. 417–418.

Sievers, V., Murphy, S., & Miller, J. (2003). Sexual assault evidence collection more accurate when completed by sexual assault nurse examiners: Colorado's experience. *Journal of Emergency Nursing, 29,* 511–514.

Sievers, V., & Stinson, S. (2002). Excellence in forensic practice: A clinical ladder more for recruiting and retaining sexual assault nurse examiners. *Journal of Emergency Nursing, 28,* 172–174.

Smith, K., Homseth, J., Macgregor, M., & Letourneau, M. (1998). Sexual assault response team: Overcoming obstacles to program development. *Journal of Emergency Nursing, 24,* 365–367.

Smugar, S. S., Spina, B. J., & Merz, J. F. (2000). Informed consent for emergency contraception: Variability in hospital care of rape victims. *American Journal of Public Health, 90,* 1372–1376.

Solola, A., Scott, C., Severs, H., & Howell, J. (1983). Rape: Management in non-institutional settings. *Obstetrics & Gynecology, 61,* 373–378.

Spohn, C., Beichner, D., & Davis-Frenzel, E. (2001). Prosecutorial justifications for sexual assault case rejection: Guarding the "gateway to justice." *Social Problems, 48*(2), 206–235.

Spohn, C., & Horney, J. (1992). *Rape law reform: A grassroots revolution and its impact.* New York: Plenum.

Sullivan, C. M., & Cain, D. (2004). Ethical and safety considerations when obtaining information from or about battered women for research purposes. *Journal of Interpersonal Violence, 19,* 603–618.

Taylor, W. K. (2002). Collecting evidence for sexual assault: The role of the sexual assault nurse examiner (SANE). *International Journal of Gynecology and Obstetrics, 78,* S19–S94.

Uttley, L., & Petraitis, C. (2000). Rape survivors and emergency contraception: Substandard care at hospital emergency rooms. *Prochoice Ideas,* pp. 3–4.

Voelker, R. (1996). Experts hope team approach will improve the quality of rape exams. *Journal of the American Medical Association, 275,* 973–974.

Weiss, C. H. (1983). Ideology, interest, and information: The basis of policy decisions. In D. Callahan & B. Jennings (Eds.), *Ethics, the social sciences, and policy analysis* (pp. 213–245). New York: Plenum.

Williams, J. E. (1984). Secondary victimization: Confronting public attitudes about rape. *Victimology: An International Journal, 9,* 66–81.

Young, W., Bracken, A., Goddard, M., & Matheson, S. (1992). Sexual assault: Review of a national model protocol for forensic and medical evaluation. *Obstetrics & Gynecology, 80,* 878–883.

DISCUSSION QUESTIONS

1. Prior to SANE programs, how were sexually assaulted women handled by hospitals' emergency departments, by nurses, and by doctors? Why were SANE programs developed?

2. How helpful have SANE programs been in the community? What have SANE programs created and how could they possibly change the effects of completed rape?

3. Do you think SANE programs are effective for women who have been the victims of rape? In what ways have they been shown to be effective?

4. How have SANE programs aided the criminal justice system?

5. If SANE programs continue to expand, what else could be suggested to improve them?

Introduction to Reading 2

Rarely has someone had access, as did Shepherd, to a courtroom in a rape trial that resulted in a retrial. She was chosen to serve as a juror on a rape and burglary trial. The case was eventually retried, and she attended almost all of the second trial, including jury selection. She used her evaluation and behind-the-scenes experience as a juror as a way to make sense of how rape myths—commonly held beliefs about what rape is and what rape victims "look like" that diminish the responsibility of the offender—shape juror deliberations and case outcomes. This article helps illuminate the reasons why rape cases brought to trial may not result in convictions and why similar evidence may be evaluated differently based on changes in the composition of the jury.

Reflections on a Rape Trial

The Role of Rape Myths and Jury Selection in the Outcome of a Trial

Judy Shepherd

This article reviews arguments and jury deliberations from a rape trial that took place in spring 1999 and was retried 7 months later. It presents the circumstances of the case, the evidence and arguments of the prosecution and defense, discussions among jurors during the first trial, and the outcome of each trial. It also raises questions about the treatment of sexual assault victims in the courts, the

SOURCE: Shepherd (2002). Reprinted with permission.

effect of jury selection on the outcomes of trials, and the persistence of myths regarding women and sexual assault in American society.

Sexual assault continues to be the most underreported violent crime in the United States. According to a report by the Bureau of Justice Statistics (Rennison, 1998), only 31% of all rapes and sexual assaults were reported to law enforcement officials in 1998 compared to 62% of all robberies, 57.6% of all aggravated assaults, and 40.3% of all simple assaults. Even with such underreporting, 330,000 women aged 12 and older were the victims of rape, attempted rape, or sexual assault in the United States in 1998, a 7.1% increase from 1997 (Rennison, 1998).

The common rationale for such underreporting of this serious crime is the treatment that victims receive from societal institutions, especially the legal system. The difficulty of bringing a rape case to trial and of obtaining a conviction for this crime has been well documented. For example, in 1984, Russel found that "less than 1% of rapes and attempted rapes result in convictions in the U.S." (as cited in Ward, 1995, p. 196). Furthermore, a 3-year investigation of state rape prosecutions by the Committee on the Judiciary, U.S. Senate (1993) revealed:

Ninety-eight percent of rape victims will never see their attacker apprehended, convicted and incarcerated;

Over half (54 percent) of all rape prosecutions result in either a dismissal or an acquittal;

A rape prosecution is more than twice as likely as a murder prosecution to be dismissed and 30 percent more likely to be dismissed than a robbery prosecution;

Approximately 1 in 10 rapes reported to the police results in time served in prison; 1 in 100 rapes (including those that go unreported) is sentenced to more than 1 year in prison;

Almost one-quarter of convicted rapists are not sentenced to prison, but instead, are released on probation;

Nearly one-quarter of convicted rapists receives a sentence to a local jail—for only 11 months (according to national estimates);

Adding together the convicted rapists sentenced to probation and those sentenced to local jails, almost half of all convicted rapists are sentenced to less than 1 year behind bars. (p. 1)

This article presents an in-depth case study of a rape trial that occurred in Alaska in the spring and fall of 1999, with particular attention to the jury selection process and the reliance on rape myth arguments throughout the deliberations. It also points to areas for further research and advocacy regarding attitudes toward rape and the treatment of rape victims in this society.

✒ Review of the Literature

The acceptance of the myths about rape, which are commonly help beliefs that shift the blame for a sexual assault from the assailant to the victim and serve to minimize the prevalence and seriousness of rape (Stout & McPhail, 1998), has been the focus of many studies. Common myths include the beliefs that "victims are lying, victims are malicious, sex was consensual, and rape is not damaging. . . . The underlying assumptions about rape suggest that women are essentially responsible for male sexual behavior" (Ward, 1995, p. 25). Ward (1995), who studied attitudes toward rape on college campuses, found in 1980 that only 36% of those surveyed disagreed with the statement that rape is provoked by women's appearance and behavior, and 60% maintained that women who go out alone put themselves in a position to be raped. In a 1991 attitude survey by Halcomb and others (as cited in Ward, 1995),

24% of the respondents agreed with the statement, "women frequently cry rape falsely" and 22% agreed that rape is often provoked by the victim, 22% agreed a woman could prevent a rape if she really wanted to, 32% agreed that some women ask to be raped and may enjoy it, and 29% agreed that if a woman says no to having sex, she means maybe or even yes. (p. 45)

Several studies have demonstrated that gender is correlated with the acceptance of rape myths. According to Ward's (1995) review of the literature on rape attitudes, "Studies show men are more accepting of rape myths than women (Margolin et al., 1989), more tolerant of rape (Hall et al., 1986), and have less empathy towards victims (Bradley et al., 1991)" (p. 45). Ward also cited Giacopassi and Dull's 1986 study that found that men were more likely to agree that normal men do not commit rape and that women were more likely to disagree with the statement that "women who ask men out are probably looking for sex, that women say no but mean yes, and that date rape should not be considered as serious as stranger rape" (p. 46). After reviewing studies on attitudes toward rape, Ward concluded, "The sensitive issue of coercive sex between people who know each other, the most common form of sexual violence, appears to be trivialized more frequently by men" (p. 46).

It is also important to note that

the danger of false rape complaints has been vastly overrated. The police find the number of false rape charges to be comparable to the level of false charges brought in other types of crimes. There are rare occasions when individuals falsely accuse others of crimes, but evidence suggests that the episodes are no more frequent in rape cases than in other serious cases. (Hans & Vidmar, 1986, p. 206)

And as Stout and McPhail (1998) noted,

Although false charges of rape are often widely publicized, FBI statistics (as cited in Lonsway & Fitzgerald, 1994) suggest that only 2% of rape charges are false; this rate is lower than or comparable to the rate for other felonies. (p. 261)

Educational level has also been correlated with the acceptance of rape myths, as noted in Ward's (1995) review of studies of rape. Burt (as cited in Ward, 1995), who sampled approximately 600 adults in Minnesota, found that "education exerted a direct effect on the rejection of stereotyped, prejudicial views of rape. Better educated respondents were less willing to endorse such statements as, 'in the majority of rapes, the victim is promiscuous or has a bad reputation'" (p. 47). Other studies on educational level found similar results. Jeffords and Dull (as cited in Ward, 1995) found that supporters of marital rape legislation in Texas were more likely to be female, single, young, and well educated, and Williams (as cited in Ward, 1995), in a survey of 1,000 San Antonio residents, found education to be the most powerful predictor of attitudes toward rape.

A review of the literature on jurors' attitudes, based on mock juries or posttrial interviews, demonstrated that jurors are influenced by the prior relationship of the victim and assailant as well as the victim's character. In reviewing Kalving and Zeisel's studies on jury trials, Epstein and Langenbahn (1994) noted "not only that juries are prejudiced against the prosecution in rape case, but also that they were extremely lenient with defendants if there was any suggestion of 'contributory behavior' on the part of the victim" (p. 66). One contributing behavior that clearly affects perceptions of rape is the consumption of alcohol. According to a study by Richardson and Campbell (as cited in Ward, 1995), "People are more likely to see intoxication as contributing to the woman's responsibility in sexual assault" (p. 76). A study by LaFree (as cited in Hans & Vidmar, 1986), which included posttrial interviews with 331 jurors who heard cases of forcible sexual assault, found that none of the measures of evidence, including eyewitnesses, the number of prosecution witnesses and exhibits, the use of a weapon, or injury to the victim, affected jurors' beliefs about the defendant's guilt or innocence prior to deliberations. However, jurors were affected by the characteristics of the victim and defendant. When the victim held a blue-collar job, when she reportedly had sexual intercourse outside marriage, or when she drank or used drugs, jurors were more likely to believe the defendant was innocent. Jurors who had conservative attitudes about sex roles were especially likely to believe the defendant was not guilty of rape when they learned that the victim used drugs or alcohol. Thus, in cases where the victim's word was a primary issue, jurors were influenced more by the character of the victim than by hard evidence, even corroborative evidence.

Another factor that has been found to contribute to the outcomes of rape trials is whether physical force was used. Deitz (as cited in Ward, 1995) found in jury simulation studies that guilty verdicts are less likely to be rendered in rape cases when there is no evidence that the victim resisted, and Wyler (as cited in Ward, 1995) noted that "women who resist attempted rape are perceived as less responsible and less to blame for their assault than those who do not resist" (p. 77). Also, Williams (as cited in Ward, 1995) found that "when the victim is acquainted with the rapist, the latter is less likely to be charged or convicted" (p. 110).

In light of these studies, Hans and Vidmar (1986), who extensively studied the jury system, noted:

> The result of these studies on jury decisions in rape cases, taken together, are troubling in some respects. Widespread adherence to rape stereotypes and myths make it difficult not only for victims who fail to match the pristine picture of the ideal victim, but also for [the defendant] whose courtroom appearance and lifestyle make him seem like a rapist. (p. 214)

All the studies on jurors' attitudes just reviewed were either with mock juries made up of university students, in which no challenges and dismissals were involved, or posttrial interviews with jurors. The case study reported in this article is unique in that I served as a juror and thus had the opportunity to participate in and note (immediately after the deliberations) the jurors' arguments and the dynamic of the jury, which were not recorded or open to the public.

⊠ Method

In spring 1999, I was chosen to serve as a juror on a rape and burglary (forcible entry) trial. Because I teach in both the Social Work Department and the Women's Studies Program at the University of Alaska, the lack of challenges

to my serving as a juror was a surprise. My service as a juror gave me a unique opportunity to learn firsthand about the court system, to become knowledgeable about court proceedings in a rape trial, to become aware of the treatment of jurors and the dynamics of juries, and to be a participant in a jury's deliberations. This trial lasted 6 days, with jury deliberations covering 2 days.

At the end of each day of jury deliberations, I went directly to my office and recorded as precisely as I could information on arguments and proceedings of the trial and discussions that took place during the deliberations. I recorded only arguments and comments presented during the deliberations but no information about specific jurors, and I did not link comments made during the deliberations to any particular juror.

When this case was retried 7 months later, I attended almost every day of the 9-day trial, including the jury selection proceedings. Doing so afforded me the opportunity to ascertain how the makeup of juries is affected by peremptory challenges and to check the accuracy of details in my notes from the first trial regarding the presentations of the defense's and prosecution's cases as well as to record any differences in evidence presented during the two trials. All this information gave me the opportunity to check the validity of my impressions as a participant observer during the first rape trial and to gain a fuller picture of the case, the court proceedings, and the outcome of the trial. Because I could not take down verbatim quotes during the first trial, I used statements made by the attorneys for the defense and prosecution during the jury selection proceedings and opening and closing statements at the second trial to present exact quotations. The arguments presented by the defense and prosecution were consistent in the two trials. The only significant difference between the cases in the two trials was the amount of expert testimony and evidence presented on DNA in the second trial.

⊠ The Case

Description

The alleged rape and burglary (forced entry) that was the focus of this trial took place in fall 1998 in a primarily Athabascan Indian village in Alaska. The village is not on the road systems and has a population of 150 to 200 people. It is a wet village, meaning that alcohol can be purchased and consumed within the village boundaries. In this remote village, the only law enforcement presence is one village public safety officer (VPSO), whose job is to keep order in the community. The VPSO does not carry a gun and does not make arrests or investigate felony crimes. In the case of an allegation of a serious crime, such as a rape, the VPSO would take the victim's statement and then call state troopers, who would fly into the village to investigate the crime. In this village, routine health care is provided by a health aide, a local resident who is trained in basic first-aid techniques. The health aides in the villages are instructed in procedures to follow in cases of alleged rape and are given rape kits to use during their examinations of victims. The kits include swabs for collecting evidence and procedures to follow so that evidence is not contaminated.

The incident that was the focus of this trial took place on a weekend of celebrations in the village that included a softball tournament and a wedding and brought many out-of-town visitors to the village. The alleged crime was the rape of a 66-year-old Alaska Native woman from the village where the incident occurred. The alleged victim had lived her entire life in the village, had never received any formal schooling, was the mother of 12 children and a grandmother, and recently had back surgery and walked with a slight limp.

The alleged assailant was a 55-year-old Alaska Native man from a neighboring village who had known the alleged victim since childhood and who occasionally hunted and fished with her husband and brother. He stated that he was in the village where the attack took place to visit his brother who lived in the village and to partake in the celebrations.

The Prosecution's Case

According to the alleged victim, she had been visiting the homes of friends and relatives on the evening

before the assault and had consumed some alcoholic beverages along with her friends. In the evening, she returned to her home alone (her husband was out of town fishing) and locked the door to her house and went to bed. At around 5:00 a.m., someone knocked on her door. Thinking it was her brother who had planned to come over for coffee, she opened the door. According to the alleged victim, the alleged assailant pushed her into the house and into the bedroom, pulled off her pants, raped her, and then left her house. The alleged victim stated that she felt dirty and showered and burned the clothes she had been wearing along with the trash. When her grandson came over to do laundry later in the day, he found her lying on the couch looking depressed. He asked her what was wrong, and she told him that she had been raped and asked him to get the VPSO.

The VPSO took the alleged victim's statement in which she identified the alleged assailant and then drove her to the village health clinic, where she was given a pelvic examination. A swab from her vaginal area was taken and subsequently sent to the crime lab in Anchorage as possible evidence. The alleged victim was later sent by plane to the hospital in the nearest urban center for an examination with a culpascope, a machine that takes pictures of the inside of the vagina to see if internal bruising, which may be consistent with forced sexual intercourse, is present. The alleged victim underwent a second culpascope examination 9 days after the first examination. The second examination, a standard procedure in the case of sexual assault, is used to determine whether any bruising that was present in the first examination is also present 9 days later. If the bruising is not present in the second examination, it is assumed that a trauma, such as a sexual assault, caused the bruising, which has subsequently healed. If the bruising or anomaly in the vaginal area is still present in the second examination, it is assumed that this is a normal condition for the woman examined and was not the result of trauma to the vaginal area.

In the courtroom, the alleged victim identified the alleged assailant as the man who had entered her home and raped her. This was the same man she identified to the VPSO, the village health aide, and the hospital nurse.

The evidence presented by the prosecuting attorney included a chart showing the match between the accused assailant's DNA and the semen that was on the swab taken during the initial examination of the alleged victim. The DNA analysis was done by the crime lab in Anchorage using a six-marker test. The alleged assailant accused another man, who he said had sexual intercourse with the alleged victim, but DNA profile precluded this possibility. The prosecuting attorney explained that an Athabascan database establishing the statistical probability of another DNA match in the Athabascan population had not been established; however, research on neighboring Alaska Native populations showed that the likelihood of a similar DNA profile using the six-marker test would be in the range of 3,000 to 1.

The prosecuting attorney also showed full-color photographs and a television-screen image of the alleged victim's vaginal area taken from the culpascope examination, which showed severe internal bruising. The nurse who examined the alleged victim testified that the bruising evident in the pictures was consistent with a sexual assault. The bruising in the vaginal area was not evident 9 days later, demonstrating that such bruises were not normal for this woman.

The Defense's Case

The accused assailant maintained that he "never touched that woman," and the defense attorney claimed this was a case of mistaken identity and an inadequate targeted investigation by the VPSO and state troopers. The defense attorney discredited the alleged victim's identification of the alleged assailant, stating that she had been drinking and thus would have difficulty identifying anyone. The questions that the defense attorney asked the alleged victim included, "Weren't you drunk? Weren't you obnoxious? Did you drink this much or this much? Is 'My back hurts' all you said to the assailant?"

The defense attorney also discounted the utility of DNA evidence, noting that it gave information only on a DNA match, but there was always a possibility that there were other matches. He also focused on the lack of established DNA probability ratios for Athabascan Indians and challenged the statistical background of the state's DNA expert and her credibility as an expert witness. He further argued that the culpascope examination provided no useful information because there was a strong possibility that a 66-year-old woman would not lubricate during sexual intercourse, and thus the bruising apparent in the culpascope pictures could have been the result of vigorous consensual sex. He also questioned the credibility of the nurse who explained the culpascope pictures because of the length of training she had received on the culpascope.

Witnesses who were called by the defense included a woman (who appeared to be intoxicated on the stand) who stated that the alleged assailant had slept on her living room floor on the night of the attack and the VPSO's wife, who testified that she saw the alleged assailant knock on the alleged victim's door the morning of the attack. The defense asked her what the man was wearing to determine if it was the same man the alleged victim identified. The VPSO's wife stated emphatically that the man she saw knocking on the alleged victim's door was the same man she saw the next day at the softball field and was the alleged assailant who was present in the courtroom, only he was wearing a different jacket on the morning she saw him at the alleged victim's house.

In his concluding remarks, the defense attorney maintained there were too many unanswered questions in this case. He stated:

We're not here to say [alleged victim] didn't have sex with someone. What she did and who she did it with is her business. Maybe she doesn't want to reveal that. We're saying this man didn't do it. He had no reason to hurt that lady. He didn't break in to physically assault her or hurt her. This wasn't like breaking in to jimmy a door. No one forced their way into this house. Her husband was away. She partied. One way she partied was she got drunk. She got pretty good and drunk. She was so drunk she said it happened on Friday morning but didn't report it till 15 hours later. She may have had sex with somebody when she was passed out, and she may think it was [defendant], but she is wrong.

According to Epstein and Langenbahn (1994), defense attorneys use the following three basic strategies in rape cases: consent, identification, and denying that the crime occurred. In the consent defense, the attorney acknowledges that the defendant engaged in sexual relations with the complainant but argues that the complainant consented. In the identification defense, the attorney neither denies nor acknowledges that rape occurred but claims that the accused was not the attacker. In the third defense, the attorney argues either that the alleged acts do not constitute rape or that no such acts occurred.

In this case, the defense used the identification strategy by claiming that this was a case of mistaken identity. He attempted to establish that the alleged assailant had on different clothes than the man who had been seen by the VPSO and his wife knocking on the alleged victim's door. He noted that DNA testing is not an accurate test and that there was a likelihood of a similar DNA profile. He called a witness who stated that the alleged assailant was asleep on her floor along with several others the morning of the attack, and he claimed that the state trooper had too quickly arrested the alleged assailant without looking for other possible suspects. The defense attorney also noted that the alleged victim was drunk and that the bruising evidenced in her vaginal area could be the result of "vigorous sex," not necessarily sexual assault. Thus, in accordance with the literature on public perceptions of good rapes versus bad rapes, the defense attorney attempted to present this case as a dubious or bad rape, an acquaintance rape in the alleged victim's home where there was no sign of a physical struggle and where the alleged victim had consumed alcohol.

Outcome of the First Trial

The jury deliberated on this case for approximately 12 hours over the course of 2 days. The outcome was a deadlocked jury, meaning that no consensus was reached. Deadlocked juries occur in about 1 in 20 cases (Hans & Vidmar, 1986). With a deadlocked or hung jury, the alleged assailant would go free unless the prosecution thought that there was a strong enough case to go forward with a retrial and the alleged victim agreed to undergo a second trial.

Jury Selection

To understand this trial's outcome, one must first consider the jury selection process and resultant makeup of the jury. The jury selection process for the first trial lasted a day and a half. In this process, the names of 14 jurors (12 jurors and 2 alternates) were chosen at random out of a pool of approximately 40 people. Each of the 14 potential jurors gave information on his or her place of residence, occupation, spouse's occupation, number of children and ages, birthplace, interests, involvement in prior lawsuits, previous experience as a juror, and whether he or she knew anyone associated with the trial. The potential jurors were each interviewed by the prosecution and defense attorneys.

Potential jurors can be dismissed in two ways. They can be released for cause, meaning that because of prior knowledge of the case, a relationship with someone associated with the trial, or previous experiences that may prejudice them, they could be deemed unable to be objective and thus would be dismissed. They can also be dismissed from a case through peremptory challenges. In criminal cases in Alaska, each lawyer is allowed 10 peremptory challenges (and an additional challenge for each alternate on a case) in which potential jurors can be dismissed from the case without stating a cause. In this case, the defense attorney first asked questions of all potential jurors as a group. The following examples of the questions he asked illustrate the criteria that the defense used to select jurors who were favorable to his case and his attempt to build his case during the jury selection process: Do you feel when police investigate crimes they have an obligation to be thorough and investigate both sides? How many know enough about fingerprint evidence to know it might be useful in an investigation? Raise your hand if you feel fibers and hair are useful to an investigation. Raise your hand if you have ever had mistaken identify happen to you. Do any of you personally know of anyone who when they are really drunk has made a claim that is fantastic or unbelievable? Do you feel police investigators have a duty to produce evidence they know exists? Raise your hand if you know what the letters DNA stand for. Have any of you had special courses in the fields of biology? Any particular courses in DNA? Any particular training in statistics? Is there anybody that cannot accept the proposition that the accused does not have the burden of proving anything? Have you ever had to rely on lab tests and later found out the lab test was wrong? Anybody here ever heard the phrase "There are lies, damn lies, and statistics"?

The prosecuting attorney's questions focused on whether anyone had been on a jury and if so, whether the jury had reached a verdict. He also asked the potential jurors about their views on drinking.

In the first trial, those who were dismissed by the defense attorney included a woman who had written a master's thesis on DNA, an individual related to a police officer, a lawyer and relative of a lawyer, and a middle-aged Alaska Native woman. The prosecuting attorney dismissed anyone who had a prior negative experience with the courts; the prosecution's other reasons for dismissals were not clear to me. The jury that remained was made up of 8 men and 4 women. All the jurors were non-Native and Caucasian and currently resided in the urban center where the trial was held; 2 of the jurors (both female) had college degrees.

Because I was a potential juror, I did not have the opportunity to take notes on all who were selected and dismissed during this trial. However, during the retrial of this case, I kept notes on all the potential jurors and compared the initial and final juror seatings. From this analysis, I found that in the retrial, the defense dismissed significantly more women (6) than men (3) and that of the 8 individuals who were dismissed,

7 were in occupations that required a college degree. Thus, in keeping with the literature on the believability of rape myths (that level of education and gender are the best predictors of acceptance of rape myths), the final jury seated after the defense and prosecution challenges would be expected to be more likely to believe rape myths than the initial jurors who were randomly selected.

Jury Deliberations

In the first deadlocked jury, 5 jurors voted for a guilty verdict (3 women and 2 men), and 7 voted for acquittal (1 woman and 6 men). However, during most of the jury deliberations, 2 female jurors held out for a guilty verdict while others argued either for an acquittal or were undecided. Throughout the deliberations, 7 of the 8 male jurors sat at one end of the table, and all 4 female jurors and 1 male juror sat together at the other end. At the final vote, the 3 female and 1 male jurors who sat together voted guilty, and 6 of the 7 male jurors who sat together voted for acquittal.

The jurors who voted for acquittal agreed with the defense attorney's arguments. Many thought that the alleged victim was not credible because she had consumed alcoholic beverages and suspected that she was lying to cover up consensual sex. Most of the jurors agreed with the defense attorney that both the DNA evidence and the pictures taken from the culpascope examination should not be considered in this case because DNA tests show only a probable match and the severe bruising evident in the alleged victim's vaginal area could have been the result of rigorous consensual sex. Also, many jurors believed that the state did a sloppy job of investigation and that a targeted investigation had occurred. The sentiment among some jurors was that the VPSO's wife started spreading the word around the village that the man she saw knocking at the alleged victim's door that morning committed the rape because "she wanted to be a big cheese" and was "the perfect police officer's wife." Some jurors believed that she told her husband her feelings, which he then told the state trooper, and that the trooper immediately arrested the alleged assailant upon entering the village.

Examples of statements made during jury deliberations in the first trial are presented next, organized in relation to some of the commonly held rape myths presented in Stout and McPhail (1998). The jurors' comments demonstrate arguments that were used in and affected the outcome of the trial. It is important to remember that in this case, the alleged victim identified the alleged assailant consistently, and the alleged assailant maintained that he never touched the woman. Also, no one other than the alleged assailant made any claims that the alleged victim had slept with anyone else, and the man that the alleged assailant claimed had sex with the alleged victim had a DNA profile that excluded him as a sexual partner.

1. Women routinely lie about rape for their purposes: "She had sex with someone else and said it was him to cover it up." "She claimed rape so her husband wouldn't get mad." "It wasn't [the defendant] but someone with close DNA."

2. Only bad women are raped: "She was drunk." "How could she recognize who it was?"

3. You can't rape an unwilling woman: "When asked what she said to him, she said 'My back is hurting.' Why didn't she just say no?" "She didn't fight him off."

4. Women who are raped must have provoked the rape by leading men on or dressing provocatively: "She had consensual sex with him and wanted to cover it up so her husband wouldn't get mad." "She encouraged him at [name's] house and later he came over and it went too far." "'Don't, stop' can mean two different things."

5. Most rape is committed by African American men against European women: This myth was not evident in this trial, but racism was apparent as can be seen in such comments as, "They were all soused and lying." "They were all soused; it just depends which drunk you want to believe." "Want to know my personal experience with Natives and sex? They all cover up for one another." "I lived in a village; I know how they party."

6. Most women secretly desire rape and enjoy it: "He was on top of her, and then she started feeling guilty and worried her husband would find out."

7. It can be called rape only if the assailant is a stranger who has a weapon and causes great physical injury: "She had no bruises."

8. Our society abhors rape and gives rapists long and harsh sentences: "We could ruin a guy's life." "If there is a reasonable doubt, we are required to give a verdict of not guilty." "I think he's guilty, but I don't feel comfortable passing a guilty verdict and knowing he's going to prison."

Discussion of the Outcome of the First Trial

The outcome of this trial was a shock to me because I found the alleged victim to be believable (she was a 66-year-old grandmother who consistently identified the alleged assailant, who was reported to be extremely distraught by all who came in contact with her after the assault, and who broke down in tears on the stand when discussing the sexual assault). I also thought that the state had provided sound scientific data that a sexual assault had occurred and that the alleged assailant was linked in several ways to the crime. During the trial, I thought that without scientific tests, the prosecution would have had great difficulty getting a conviction in this case but that with DNA evidence linking the alleged assailant to the crime and with pictures taken during the culpascope examination showing severe bruising of the victim's vaginal area, a conviction would be the outcome. The fact that both the DNA evidence and the results of the culpascope examination were disregarded was surprising. In regard to the pictures showing serious vaginal bruising being disregarded because of the alleged victim's age and lack of lubrication, I asked the other jurors, "Why would a woman who just had recent back surgery and who bruised so severely have consensual sex?" Their response was that she was too drunk to care or feel any pain. Thus, this jury's verdict was consistent with LaFree's (as cited in Hans & Vidmar, 1986) findings that jurors may disregard even corroborative evidence if they believe that the alleged victim's character is questionable.

The jurors' fascination with a targeted investigation and the idea of mistaken identity was also surprising. Throughout the jury deliberations, I thought that sexism was evident because many jurors discredited both the crime lab expert ("Who does she think she is strolling in here with a suit and briefcase?") and the female nurse who did the culpascope examination ("Why did the state bring a nurse? A doctor would have had instant credibility."). Similarly, many jurors thought that the VPSO's wife, who stated she saw the alleged assailant knock on the alleged victim's door, contributed to a targeted investigation although neither attorney implied or even mentioned this possibility. Most of the jurors did not consider the alleged victim to be believable, believing that she was lying to cover up other sexual escapades or consensual sex with the alleged assailant. Most of the jurors thought that the state did not prove its case because fingerprints were not taken, clothing and bedclothes were not tested for semen, and other suspects were not considered, although a DNA specimen was taken during the vaginal examination and the alleged victim consistently identified the alleged assailant.

In conclusion, one could say that in this sexual assault case, most jurors thought there was reasonable doubt that the alleged victim had been sexually assaulted. Rather, they believed that the alleged victim either had consensual sex with the alleged assailant or consensual sex with someone else but was not raped and did not suffer harm. When statements made during the jury deliberation were considered in regard to common rape myths, it became apparent that almost every myth was validated by some jurors and used as an argument for acquittal. Many male jurors could identify on some level with the alleged assailant, as was evidenced by comments such as these: "Mistaken identify happened to me once"; "'Don't, stop' can mean two different things, and it's hard to know which"; and

"Would you want to ruin a man's life?" The lack of gravity about this sexual assault trial was apparent in such jurors' comments as the following: "Why don't they have Playboy magazines here to read?" in reference to reading materials supplied in the jurors' quarters. Other comments that trivialized the case included "They were all soused; it just depends which drunk you want to believe" and "They all cover up for one another." At the end of the deliberations, when the final vote had been taken, a male juror stated, "Seven to five, we still kicked ass."

The outcome of this trial raises some serious questions regarding our judicial system in general and sexual assault trials in particular. The first concern is with the jury selection process. If this jury were indeed a representative sample of the community and a true jury by peers, the outcome would be disturbing in terms of prevalent attitudes toward women and sexual assault. As I mentioned earlier, throughout the jury deliberations, it was apparent that the majority of jurors strongly held many rape myths. Unfortunately, it is obvious that this jury was neither a randomly selected cross-section of the community nor a jury of peers. Potential jurors were excluded if they knew anything about DNA or were familiar with the law or law enforcement officers, more women than men were excused, and the only Native woman who was selected as a potential juror was excused. The result was a jury consisting of twice as many men as women, with only two jurors in occupations requiring college degrees and no Alaska Natives or residents of rural villages.

In 1999, Supreme Court Justice Sandra Day O'Connor called for a review of lawyers' rights to exclude possible jurors without giving a reason or for cause because they heard about the case from the media. She said that these practices give the impression of "unrepresentative juries." O'Connor warned that

the use of unlimited "cause" challenges to prospective jurors, coupled with extensive media coverage of some cases, leaves some courts to search out the most ignorant and poorly informed citizens to serve as jurors in high-profile cases, because only those citizens are likely to have avoided forming any opinion. ("O'Connor Urges Examination," 1999, p. A-8)

Furthermore, in this case, both the alleged assailant and the alleged victim were from small rural villages, but the jurors were all non-Natives living in an urban area. Such a jury allows for stereotypes and suppositions that would probably not enter into the deliberations of a true jury of one's peers. Blatantly racist comments, including suppositions about Natives' alcohol consumption and sexual practices, were made, as were comments about small villages and the way people gossip and stick together. It is important to note that felony trials in interior Alaska are routinely scheduled in the urban center, although the defense can request that a trial be moved to a regional center closer to the village. In a regional center, however, it would probably be difficult to select jurors who had no prior knowledge of the case or anyone involved in it.

Another issue of concern in this trial was the treatment of the alleged victim, who was asked grilling questions about her alcohol consumption. In addition, although the defense attorney said in his concluding statement that he would not go into the sex life of a 66-year-old woman, he implied that the jurors should consider it (which they clearly did), asking such questions in the retrial as, "Can you tell this jury that absolutely you did not have sex with anyone there?" Full-color pictures of the alleged victim's genital area taken during the culpascope examination were passed around to the jurors and displayed on two television screens with the caption, "Genital Area of [alleged victim]." If a 66-year-old grandmother is treated this way and suspected of lying to cover up sexual escapades, one wonders what would be included in the court proceedings and jury deliberations of a date rape trial of a young woman.

⊠ The Retrial

Seven months after the original trial, a retrial was held, conducted by same judge with the same prosecuting and defense attorneys. The jury was different in its gender makeup (7 men and 7 women), and one of the jurors was married to an Alaska Native woman. At the retrial, I took detailed notes on all the potential jurors who were called and questioned by the prosecuting and defense attorneys to ascertain how peremptory challenges changed the makeup of the jury.

Of the initial randomly selected pool of 14 jurors in the sectional background related to current occupation, there were 2 undergraduate college students, 1 doctoral student, 1 accountant, and 3 school teachers. No Alaska Natives were included in this initial pool. Both the defense and prosecution dismissed 9 jurors each, which meant that 32 potential jurors were reviewed for this case.

The nine potential jurors who were dismissed by the defense in the second trial were six Caucasian women and three Caucasian men, eight of whom were either college students or in careers that required college degrees. Of the nine jurors who were dismissed by the defense, eight were in occupations that require a college degree: three college students, one high school math teacher, two accountants, one social worker, and one engineer. Thus, there was a high level of educational attainment in that seven potential jurors were seeking or had completed postsecondary degrees. The defense also dismissed an Alaska Native woman. The nine who were dismissed by the prosecution included six men and three women. Occupational status did not seem to matter in the prosecution's dismissals as much as attitudes toward drinking (two persons were dismissed who believed that drinking was wrong) and prior experience with the courts either for driving while intoxicated, child custody, or past service as a juror on a criminal trial. After peremptory challenges, the final jurors included 7 men and 7 women. Two of the men were school teachers, but no other jurors were in occupations in which an educational degree beyond the secondary level was required. Thus, peremptory challenges in this case changed the juror pool in terms of its gender makeup and educational level as determined by current occupational status. As one female observer during the jury selection process stated, "They sure don't want any smart women on that jury, do they?"

Additional evidence presented by the state in the second trial included a database for the probability of a DNA match in the Athabascan population, a more sophisticated DNA analysis done by a Seattle laboratory with results presented by its director (a man with a Ph.D.), a local respected (male) physician's corroboration of the nurse's culpascope conclusions, and a young girl who said the assailant made lewd comments to her on the morning of the alleged rape. The defense again used the mistaken identity argument and attempted to discredit the alleged victim because she was drunk and had not fought off her assailant. The prosecution meticulously presented the DNA evidence showing the probability of another matching DNA profile in the Athabascan population to be in the range of 1 to 2.5 million.

After fewer than 3 hours, the jury in the second trial found the alleged assailant guilty of both first-degree rape and first-degree burglary. Jurors' comments to the judge on returning to the jurors' room after the verdict had been given indicated that the DNA evidence convinced them because this was argued as a case of mistaken identity. However, in both trials, some jurors questioned why the defense did not use the argument that this was a case of consensual sex. In both trials, some jurors stated that there would not have been a case if the defense had argued consensual sex (i.e., the alleged victim's testimony and evidence of bruising from the culpascope examination would not have mattered).

⚔ Conclusion

Participation as a juror in this 1999 sexual assault trial was a disconcerting and eye-opening experience both in terms of the jury selection process and the sexist, racist remarks that were evident in the jury's deliberations, which are not open to the public or recorded. Because this was a review of only one trial in one location, it is possible that the deliberations and outcome of the trial can be attributed merely to the poor job of jury selection and case presentation by the prosecuting attorney or to the uniqueness of the region where the trial took place. This would be a comforting thought and might be the case. On the other hand, in light of the previously mentioned findings that (a) almost all rape victims

never see their attackers caught, tried, and imprisoned; (b) about 25% of convicted rapists never go to prison; and (c) another 25% receive sentences in local jails, where the average sentence is 11 months, the outcome of this trial does not appear to be an aberration. Rather, it seems consistent with the outcomes of other sexual assault trials, and thus an examination of jury selection and deliberations in this trial can perhaps contribute to an understanding of why the rates for reporting of and convocation for rape are so low in the United States.

Involvement as a juror in the first trial led me to conclude that there are several areas that people who are concerned about violence against women must focus on. First, the court system needs to be monitored in regard to the treatment of rape victims and the representativeness of jurors. Gender, educational background, and racial and class representation are important considerations for a true trial by peers. Ten peremptory juror challenges coupled with challenges for cause can dramatically alter the composition of juries and affect the outcomes of trials. As Ward (1995) noted, "legal analysts frequently argue that on many occasions the evidence presented at a rape case does not reliably predict a verdict as trial outcome is based more on jurors' attitudes about rape" (p. 111).

Second, more research is necessary in relation to factors that affect the outcomes of sexual assault trials and the sentencing of assailants, and this research should be widely publicized. Third, rape victims still need to know clearly what they will face in court in terms of the continued prevalence of rape myths, peremptory challenges, and the state's need to prove the case beyond a "reasonable doubt."

Finally, and of utmost importance, there is a need for more education about sexual respect and sexual assault in the American educational system and workplace. Rape myths are still persistent in our society in spite of the efforts of women's groups and feminist researchers. As Stout and McPhail (1998) stated, "Changes in laws have made it somewhat easier for rapists to be prosecuted and for rape victims to be protected, yet if the jury still believes in rape acceptance myths, all is lost" (p. 283). Rape myths serve "to blame women for the rape and shift the blame from the perpetrators to the victim and allow men to justify their sexual aggression. Accepting rape myths also serves to minimize the seriousness and prevalence of rape" (Stout & McPhail, 1998, p. 206). Educational programs in schools, workplaces, and universities must strive to reach a broad audience, which includes those who are the most likely to hold rape myths. From a more societal perspective,

> rape is not an isolated symptom to be plucked out of society. It is an act that is often supported, condoned, tolerated, encouraged, and regulated by a patriarchal society that gives man a sense of entitlement and privilege. The conditions in society that allow rape to flourish must be confronted. (Stout & McPhail, 1998, p. 284)

This case demonstrated that DNA evidence, culpascope pictures of bruising consistent with sexual assault, and the victim's identification of the assailant can all be readily disregarded by jurors who believe common rape myths that blame the victim and minimize the seriousness of the crime. As I noted previously, members of both juries stated that in this case, a defense argument of consensual sex would have been readily believed. Only through careful monitoring of legal procedures that include the selection of a jury of one's peers and through widespread education efforts regarding sexual assault can we expect to see a change in both the rate of reporting and prosecution for rape.

References

Committee on the Judiciary, U.S. Senate. (1993). *The response to rape: Detours on the road to equal justice.* Washington, DC: Government Printing Office.

Epstein, J., & Langenbahn, S. (1994). *The criminal justice and community response to rape.* Washington, DC: Department of Justice, National Institute of Justice.

Hans, C., & Vidmar, N. (1986). *Judging the jury.* New York: Plenum.

O'Connor urges examination of jury challenges. (1999, May 16). *Fairbanks Daily News-Miner,* p. A-8.

Rennison, M. (1998). *Criminal victimization 1998, changes 1997–98 with trends 1993–98: Bureau of Labor Statistics, National Victimization Survey.* Retrieved October 9, 2001, from http://www.ojp.usdoj.gov/bjs/pub/cv98/pdf

Stout, K., & McPhail, B. (1998). *Confronting sexism and violence against women: A challenge for social work.* New York: Longman.

Ward, C. (1995). *Attitudes toward rape: Feminist and social psychological perspectives.* Thousand Oaks, CA: Sage.

DISCUSSION QUESTIONS

1. In addition to impacting juror deliberations, how else might rape myths affect a person's reaction to finding out that someone he or she knows has been raped? How might this impact victim disclosure?

2. Were you surprised by the outcome of the first trial? Why do you think the jury came to the conclusion it did? What do the different outcomes tell you about the criminal justice system in general?

3. How does jury selection impact case outcomes? Does the desire to have a fair and just trial for the defendant sometimes negatively impact the victim? How so?

Introduction to Reading 3

In her article, Antonia Abbey (2002) reviews the literature on the relationship between alcohol use and sexual assault among college students. In doing so, she discusses the cognitive and situational mechanisms through which alcohol impacts males and their perpetration of sexual assault. The reasons why alcohol use by females is also linked to sexual assault victimization are also discussed. Importantly, she notes how this relationship can inform policy to reduce sexual assault among college students.

Alcohol-Related Sexual Assault

A Common Problem Among College Students

Antonia Abbey

Alcohol-related sexual assault is a common occurrence on college campuses. A college student who participated in one of our studies explained how she agreed to go back to her date's home after a party: "We played quarter bounce (a drinking game). I got sick drunk; I was slumped over the toilet vomiting. He grabbed me and dragged me into his room and raped me. I had been a virgin and felt it was all my fault for going back to his house when no one else was home." A male college student who forced sex on a female friend wrote that, "Alcohol loosened us up and the situation occurred by accident. If no alcohol was consumed, I would never have crossed that line."

This article reviews the literature on college students' sexual assault experiences. First, information is provided about the prevalence of sexual assault and alcohol-involved sexual assault among college students. Then theories about how alcohol contributes to sexual assault are described. After making suggestions for future research, the article concludes with a discussion of prevention and policy issues.

Incidence and Prevalence of Sexual Assault Among College Students

The term *sexual assault* is used by researchers to describe the full range of forced sexual acts including forced touching or kissing; verbally coerced intercourse; and physically forced vaginal, oral and anal penetration. The term *rape* is typically reserved for sexual behaviors that involve some type of penetration due to force or threat of force; a lack of consent; or inability to give consent due to age, intoxication or mental status (Bureau of Justice Statistics, 1995; Koss, 1992). Less than 5% of adolescent and adult sexual assault victims are male, and when men are sexually assaulted, the perpetrator is usually male. Thus, most research focuses on female victims and male perpetrators.

Rates of Sexual Assault Reported by College Women

The most methodologically rigorous study of sexual assault prevalence was completed by Koss et al. (1987), who surveyed 6,159 students from 32 colleges selected to represent the higher education enrollment in the United States. They used 10 behaviorally specific questions to assess women's experiences with forced sexual contact, verbally coerced sexual intercourse, attempted rape and rape since the age of 14. In this survey, 54% of the women had experienced some form of sexual assault. Fifteen percent of the women had experienced an act that met the legal definition of completed rape; an additional 12% had experienced attempted rape. Of these women, 17% had experienced rape or attempted rape in the previous year. Only 5% of the rape victims reported the incident to the police; 42% told no one about the assault.

Similar prevalence rates have been found in studies conducted at colleges throughout the United States (Abbey et al., 1996a; Copenhaver and Grauerholz, 1991; Mills and Granoff, 1992; Muehlenhard and Linton, 1987). Most of these studies have been cross-sectional. In the prospective study that followed students for the longest period of time, Humphrey and White (2000) surveyed women from one university beginning in the fall of their first year and ending in the spring of their fourth year. Annual prevalence rates were alarmingly high, although they declined slightly each year. In their first year of

SOURCE: Abbey (2002). Reprinted with permission from *Journal of Studies on Alcohol, Supplement, 14*, pp. 118–128, 2002 (presently *Journal of Studies on Alcohol and Drugs*). Copyright by Alcohol Research Documentation, Inc., Rutgers Center of Alcohol Studies, Piscataway, NJ 08854.

college, 31% of the women experienced some type of sexual assault; 6.4% experienced completed rape. In their fourth year of college, 24% of the women experienced a sexual assault; 3.9% experienced completed rape. Greene and Navarro (1998) reported that none of the college women in their prospective survey reported their sexual assault to any college official. Women who reported their sexual assaults to authorities often labeled their treatment by the system as "a second rape." Awareness of the derogatory manner in which many victims are treated deters others from reporting.

A few studies have focused on prevalence rates among minority students. Rates of sexual assault experienced by black, Hispanic, Asian and white college women appear to be relatively comparable (Abbey et al., 1996a; Koss et al., 1987; Mills and Granoff, 1992).

Rates of Sexual Assault Reported by College Men

College men acknowledge committing sexual assault, although at lower rates than these acts are reported by women. In Koss et al.'s (1987) national study, 25% of the college men surveyed reported committing some form of sexual assault since the age of 14; 7.7% reported committing an act that met the standard legal definition of rape since the age of 14. Similar results have been found by other researchers (Abbey et al., 1998; Kanin, 1985; Muehlenhard and Linton, 1987; Rapaport and Burkhart, 1984). About two thirds of college men who acknowledge committing sexual assault report being multiple offenders (Abbey et al., 1998). Koss and her colleagues (Koss, 1988; Koss et al., 1987) suggested that college men report rates lower than college women do because many men view the woman's nonconsent as vague, ambiguous or insincere and convince themselves that their forcefulness was normal seduction not rape.

Prevalence of Alcohol-Related Sexual Assault

On average, at least 50% of college students' sexual assaults are associated with alcohol use (Abbey et al., 1996a, 1998; Copenhaver and Grauerholz, 1991; Harrington and Leitenberg, 1994; Presley et al.,

1997). Koss (1988) reported that 74% of the perpetrators and 55% of the victims of rape in her nationally representative sample of college students had been drinking alcohol. Most studies do not include sufficiently detailed questions to determine if the quantity of alcohol consumed is an important factor. An exception is a study by Muehlenhard and Linton (1987), which compared the characteristics of dates that did and did not involve sexual assault. Sexually assaultive dates were not more likely than nonassaultive dates to involve drinking; however, heavy drinking was more common on sexually assaultive dates.

Typically, if either the victim or the perpetrator is drinking alcohol, then both are. For example, in Abbey et al. (1998), 47% of the sexual assaults reported by college men involved alcohol consumption. In 81% of the alcohol-related sexual assaults, both the victim and the perpetrator had consumed alcohol. Similarly, in Harrington and Leitenberg (1994), 55% of the sexual assaults reported by college women involved alcohol consumption. In 97% of the alcohol-related sexual assaults both the victim and the perpetrator had consumed alcohol. The fact that college sexual assaults occur in social situations in which men and women are typically drinking together makes it difficult to examine hypotheses about the unique effects of perpetrators' or victims' intoxication.

In general, alcohol consumption is more common among whites than blacks (Caetano et al., 1998). Thus, not surprisingly, alcohol-related sexual assaults appear to be more common among white college students than among black college students (Abbey et al., 1996a; Harrington and Leitenberg, 1994). Rates of alcohol-related sexual assault have not been examined in other ethnic groups.

Approximately 90% of the sexual assaults reported by college women are perpetrated by someone the victim knew; about half occur on a date (Abbey et al., 1996a; Koss, 1988). Only about 5% involve gang rapes. The most common locations are the woman's or man's home (this includes dormitory rooms, apartments, fraternities, sororities and parents' homes) in the context of a date or party. Alcohol-involved sexual assaults more often occur among college students who know each other only casually and who spent time together at a party or bar (Abbey et al., 1996a; Ullman et al., 1999).

⊠ Explanations for the Relationship Between Alcohol Consumption and Sexual Assault

The fact that alcohol consumption and sexual assault frequently co-occur does not demonstrate that alcohol causes sexual assault. The causal direction could be the opposite; men may consciously or unconsciously drink alcohol prior to committing sexual assault to have an excuse for their behavior. Alternatively, other variables may simultaneously cause both alcohol consumption and sexual assault. For example, personality traits, such as impulsivity, or peer group norms may lead some men both to drink heavily and to commit sexual assault.

It is likely that each of these causal pathways explains some alcohol-involved sexual assaults. A complex behavior such as sexual assault has multiple determinants both across different perpetrators and for any one

perpetrator. Abbey (1991) proposed seven different explanations for the relationship between alcohol and sexual assault. An expanded version of this model is described below and is summarized in Figure 1 (for a more thorough review, see Abbey et al., 1996b). This model focuses on the most common type of sexual assault that occurs between men and women who know each other and are engaged in social interaction prior to the assault, the prototypic college sexual assault situation. As can be seen in the figure, a combination of preexisting beliefs and situational factors contribute to acquaintance sexual assault. Alcohol has independent and synergistic effects. Some general information about causes of acquaintance rape are described below because alcohol often exacerbates dynamics that can arise without alcohol.

Two general caveats are needed before the literature supporting each element of the model is reviewed. First, there are personality characteristics (e.g., impulsivity, low empathy) and past experiences (e.g., childhood sexual abuse, delinquency) that have been consistently linked to sexual assault perpetration. This literature has been extensively reviewed elsewhere (Seto and Barbaree, 1997; White and Koss, 1993). Consequently, this article focuses on attitudinal and

Figure 1   Conceptual Model of Alcohol-Related Acquaintance Sexual Assault

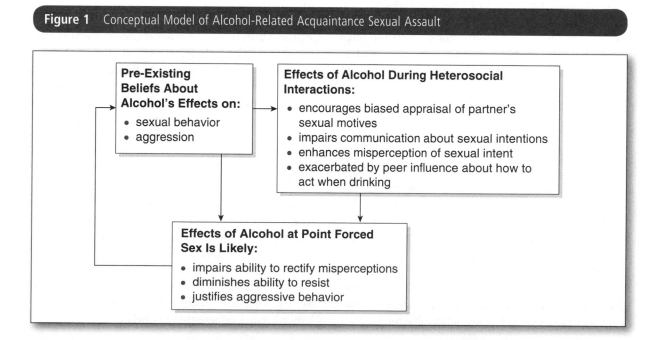

situational factors that interact with alcohol consumption to increase the likelihood of sexual assault occurring among college students. These factors are more likely to be amenable to change, and suggestions for prevention and policy initiatives are made at the end of this article.

A second important caveat concerns the relationship between explanations and causal responsibility. As the quotes at the beginning of this article indicate, perpetrators often use alcohol to excuse sexual assault perpetration, whereas victims often feel guilty because they were drinking. However men are legally and morally responsible for acts of sexual assault they commit, regardless of whether or not they were intoxicated or felt that the woman had led them on previously. The fact that women's alcohol consumption may increase their likelihood of experiencing sexual assault does not make them responsible for the man's behavior, although such information may empower women when used in prevention programs.

Traditional Gender Role Beliefs About Dating and Sexuality

American gender role norms about dating and sexual behavior encourage men to be forceful and dominant and to think that "no" means "convince me." Men are expected to always be interested in sex, whereas women learn that they should not appear too interested in engaging in sexual activities or they will be labeled "fast" or "promiscuous." Women are expected to set the limits on sexual activities and are often held responsible when men overstep them (Clark et al., 1999; Werner and LaRussa, 1985). Men often interpret a woman's sexual refusal as a sign that they should try harder or a little later rather than that they should give up. Although such beliefs may sound outdated, surveys of college students consistently find that men are expected to initiate sexual relations and that women are expected to set the limits on how much sexual activity occurs (Clark et al., 1999; Wilsnack et al., 1997).

Both men and women agree that there are circumstances that make forced sex acceptable. For example, McAuslan et al. (1998) asked college students to indicate the extent to which it was acceptable for a man to

verbally pressure or force a date to have sexual intercourse. More than half the men thought verbal pressure was acceptable if she kissed him, if they had dated a long time or if he felt she had led him on. More than 20% thought verbal pressure was acceptable if either of them was drinking alcohol or if they met at a bar. Force was viewed as less acceptable than verbal pressure, although 17% of men accepted force as a strategy under some circumstances. Overall, fewer women than men perceived pressure or force as acceptable, although the rank ordering of circumstances was comparable for both genders. Malamuth (1989) asked college men how likely it was that they would rape a woman if they were certain that there would be no negative consequences. On average, one third of college men indicated that they would be at least somewhat likely to rape a woman if they could be certain they would not be caught. The data from these two lines of research are disturbing because they demonstrate how commonly held beliefs set the stage for date rape and why it is so seldom perceived as a crime. As is described in more detail below, these beliefs are more likely to be acted on when men have been drinking alcohol.

Men's Expectations About Alcohol's Effects

Men anticipate feeling more powerful, sexual and aggressive after drinking alcohol (Brown et al., 1980; George and Norris, 1991; Presley et al., 1997; see the first box in Figure 1). These expectancies can have a power of their own, independent of the pharmacological effects of alcohol. Expectancies tend to become self-fulfilling (Snyder and Stukas, 1999). Thus, if a man feels powerful and sexual after drinking alcohol, then he is more likely to interpret his female companion's friendly behavior as being a sign of sexual interest, and he is more likely to feel comfortable using force to obtain sex. In one study, college men who had perpetrated sexual assault when intoxicated expected alcohol to increase male and female sexuality more than did the college men who perpetrated sexual assault when sober (Abbey et al., 1996b). Although these cross-sectional results do not demonstrate causality, they suggest that beliefs about alcohol's effects may have encouraged these students' behavior.

Several studies have demonstrated that college men who thought they were drinking alcohol were more sexually aroused by depictions of forcible rape than college men who did not think they had consumed alcohol (George and Marlatt, 1986; George and Norris, 1991). Actual alcohol consumption did not affect these men's sexual arousal. George and Marlatt argued that the belief that one has consumed alcohol provides justification for engaging in socially inappropriate sexual behavior. If a man can say to himself, "I did that only because I was too drunk to know what I was doing," then he does not have to label himself as deviant.

Stereotypes About Drinking Women

Many college men perceive women who drink in bars as being sexually promiscuous and, therefore, appropriate targets for sexual aggression (Kanin, 1985; Martin and Hummer, 1989). For example, a college man who reported sexually assaulting a woman in one of our studies justified his behavior by writing, "She was the sleazy type . . . the typical bar slut."

In vignette studies, women who drink alcohol are frequently perceived as being more sexually available and sexually promiscuous than women who do not drink alcohol. For example, George et al. (1995) asked college students to read a vignette about a couple on a date. A woman who drank several beers was perceived as being more promiscuous, easier to seduce and more willing to have sex than a woman who drank cola. College students believe that dates are more likely to include sexual intercourse when both participants drink alcohol (Corcoran and Thomas, 1991).

Alcohol as a Sexual Signal

The studies reviewed above involve clearly consensual sexual situations. Other authors have asked college students to evaluate vignettes that depict forced sex between dating partners. Even when force is clearly used, the mere presence of alcohol leads many students to assume the woman wanted sex. For example, Norris and Cubbins (1992) found that nondrinking college women and men were most likely to view a depiction of acquaintance rape as consensual when both members of the couple had been drinking alcohol. Norris and Kerr (1993) found that nondrinking college men who read a forced sex vignette indicated that they were more likely to behave like the man in the story when the man had been drinking alcohol than when he was sober. Finally, Bernat et al. (1998) asked college men to listen to a depiction of a date rape and evaluate at what point the man was clearly forcing sex. Men who had previously committed sexual assault and who thought the couple had been drinking alcohol required the highest degree of female resistance and male force to decide the man should stop. In combination, these studies suggest that when forced sex occurs after a couple has been drinking together, men, and sometimes women, are much less likely to recognize that the woman does not want to have sex. The results of these studies are not due to pharmacological effects of alcohol because sober individuals made these judgments. Instead, these studies suggest how strongly men equate drinking with a woman and having sex with her.

Men's Misperceptions of Women's Sexual Intent

Men frequently perceive women's friendly behavior as a sign of sexual interest, even when it is not intended that way. In a series of studies with college women and men, Abbey and her colleagues (Abbey, 1982; Abbey et al., 2000) have demonstrated that men perceive women as behaving more sexually and as being more interested in having sex with their male partner than the women actually are. Male observers make judgments similar to those made by male actors (Abbey, 1982), indicating that these are general gender differences in perceptions of women's behavior. Cues used to convey sexual interest are often indirect and ambiguous; thus it is easy to mistake friendliness for flirtation. For example, when an opposite sex acquaintance is very attentive, this might be a sign of sexual attraction. Alternatively, it might be a sign of politeness or merely an active interest in the topic of conversation.

Men usually feel responsible for making the first move because of gender role expectations about who initiates dating and sexual relations. Due to the embarrassment associated with rejection, these initial moves are usually subtle. For example, the man may stand close or ask the woman to slow dance or suggest they go to his apartment to talk. If he perceives an encouraging response (she does not back away or she agrees to dance or she goes to his apartment), then he will make another move (e.g., rub her back, tell her his roommates are not home). Both men and women are used to this indirect form of indicating sexual interest and usually manage to make their intentions clear and save face if their companion is not interested (Abbey, 1987). However, because the cues are vague, miscommunication can occur. Also, college men expect to have intercourse much earlier in a relationship than women do (Roche and Ramsbey, 1993); hence men are likely to initiate sexual advances before women expect them.

The man's alcohol consumption enhances the likelihood that misperception will occur and will escalate to the point that he forces sex (see second box in Figure 1). Alcohol consumption disrupts higher-order cognitive processes such as abstraction, conceptualization, planning and problem solving, making it difficult to evaluate complex stimuli (Hindmarch et al., 1991; Peterson et al., 1990). When intoxicated, people have a narrower perceptual field; they are less able to attend to multiple cues and instead tend to focus on the most salient cues (Chermack and Giancola, 1997). Steele and Josephs (1990) labeled this phenomenon "alcohol myopia." Thus, if an intoxicated man is sexually attracted to his female companion, it is easy for him to interpret any friendly cue as a sign of her desire to have sex with him and to ignore or discount any cue that suggests she is not.

Muehlenhard and Linton (1987) compared the characteristics of college students' dates that did and did not involve sexual assault. Men believed that dates on whom they had forced sex had "led them on" to a greater extent than did dates on whom they had not forced sex. Similarly, women who had experienced forced sex on a date were more likely than those who had not to believe that the man felt "led on," although women reported that this had not been their intention. In a more focused examination of the relationships between misperception, alcohol consumption and sexual assault, Abbey et al. (1998) found that the more frequently college men had misperceived a woman's sexual intentions and the more frequently they were

drinking alcohol when they misperceived a woman's intentions, the more frequently they had committed sexual assault.

Alcohol's Effects on Men's Willingness to Behave Aggressively

If a man feels that he has been led on or teased by his date he may feel justified forcing sex when sober (McAuslan et al., 1998). However, research consistently indicates that alcohol increases the likelihood that individuals will behave aggressively, especially if they feel as if they have been threatened or harmed (see third box in Figure 1). Experimental studies demonstrate that intoxicated men retaliate strongly if they feel threatened or provoked (Taylor and Chermack, 1993). Furthermore, once they begin behaving aggressively, it is difficult to make intoxicated men stop unless nonviolent cues are extremely salient.

In the case of sexual assault, a man may feel his aggressiveness is justified if he believes his partner encouraged his sexual interest and that once led on a man has a right to sex. Intoxication limits one's ability to consider the long-term negative consequences of behavior because it limits one's focus to short-term immediate cues. Thus an intoxicated man is likely to focus on his sexual arousal and sense of entitlement rather than the potential pain and suffering of his victim or the possibility that he will be punished. An alcohol-induced sense of disinhibition and reduction in anxiety and self-appraisal makes it easier for men to use physical force to obtain sex (Ito et al., 1996).

Alcohol's Effects on Women's Ability to Assess and React to Risk

A woman who is drinking alcohol experiences the same types of cognitive deficits as a man does. Thus, if a woman feels that this is a platonic relationship or that she has made it clear that she is not interested in sexual intercourse at this point in time, alcohol will make her less likely to process potentially contradictory cues and realize that her partner is misperceiving her. For example, imagine a man and a woman who have been dating several weeks. After seeing a movie together, the man may suggest going back to his apartment for a drink. His underlying message is "let's go back there to have sex" but he does not say that directly. The woman may respond, "Well, I guess I could come back for one drink, but I really can't stay long." Her underlying message is "I'd like to get to know you better but I'm not spending the night." However, she is also being indirect. Cognitive deficit theories (Steele and Josephs, 1990; Taylor and Chermack, 1993) suggest that when drinking it is very easy to focus only on the part of the message that one wants to hear. In this example, the man may hear only the confirming part of the message, "I'll come to your apartment," and ignore the disconfirming part of the message, "I won't stay long." In contrast, the woman focuses on the message she wants to hear, "I want to spend more time with you," rather than the message the man is trying to send, "I want to be alone with you so we can have sex."

In their study of college sexual assault victims, Harrington and Leitenberg (1994) examined whether alcohol consumption was related to consensual sexual activity prior to the assault. Overall, 74% of the women had engaged in kissing or another form of sexual contact prior to the forced sex. Victims who were intoxicated were more likely to have engaged in consensual sexual activities with the man than were sober victims. This finding supports the argument described above. Intoxicated women are less likely to realize that by kissing the man they are encouraging him to expect sexual intercourse. A woman in one of our studies wrote, "Alcohol put me in the mood for petting, kissing, holding and hugging, and he may have interpreted that as going further with sexual activity."

In addition, if and when a woman realizes that she has been misperceived, she must decide how to respond. Norms of female politeness and indirectness regarding sexual communication are so well internalized that some women find it difficult to confront a man directly, especially if they like him and hope to continue the relationship (Lewin, 1985). Unfortunately, if the woman is not direct and forceful about her lack of interest in sex, her companion is likely to perceive her behavior as flirtation or coyness, rather than as a refusal. Even a direct "no" is often interpreted as "try later" (Byers and Wilson, 1985); thus repeated, direct refusals are often needed for a woman to make her intentions clear to a

persistent man. The longer a man continues to believe that consensual sex will occur, the more likely it is that he will feel justified forcing sex because he feels that he has been led on (McAuslan et al., 1998; Muehlenhard and Linton, 1987).

Testa and Livingston (1999) interviewed sexual assault victims, half of whom were college students. Women who were drinking at the time of the sexual assault reported that their intoxication made them take risks that they normally would avoid. For example, they felt comfortable taking a ride home from a party with a man they did not know well or letting an intoxicated man into their apartment. These women indicated that alcohol made them feel comfortable in situations that they usually would have perceived as dangerous. Norris et al. (1996) observed that when interacting with men on dates or at parties women must often "walk a cognitive tightrope" (p. 137). Women want men to like them and have been socialized to wear revealing clothes, act friendly and assume responsibility for maintaining positive social relationships by laughing at men's jokes, complimenting them and appearing interested in what they have to say. However, women also realize that sexual assault is common and that they must be on the alert to be assured that they can trust the man with whom they are interacting. Thus women's affiliation and safety motives are in conflict. On a date or with friends at a party or bar, women (and men) typically assume they can trust their companions. Being intoxicated allows women to let down their guard and focus on their desire to have fun and be liked rather than on their personal safety. Thus alcohol myopia may lead women to take risks they would not normally take.

Alcohol's Effects on Women's Ability to Resist Effectively

Alcohol's effects on motor skills may limit a woman's ability to resist sexual assault effectively. There is some evidence that attempted as opposed to completed rapes are more common among sober than intoxicated victims, suggesting that sober victims are more able to find a way to escape or resist effectively (Abbey et al., 1996b). For example, a woman in one of our studies wrote, "I was very drunk and could not drive or get away from him even though we were in my car." Harrington and Leitenberg (1994) found that acquaintance rape victims who reported being at least somewhat drunk were less likely to use physical resistance strategies than were victims who were not drunk.

Many men who have committed sexual assault realize that it is harder for women to resist sexual advances when intoxicated; thus they try to get their female companion drunk as a way of obtaining sex (Kanin, 1985; Mosher and Anderson, 1986). Three-quarters of the college date rapists interviewed by Kanin indicated that they purposely got a date intoxicated to have sexual intercourse with her. Playing drinking games has been related to sexual victimization (Johnson et al., 1998). Women drink more than usual when playing drinking games and men may use these games to get women drunk with the hope of making it easier to have sex with them.

Alcohol's Effects on Perceptions of Responsibility

Alcohol consumption is sometimes used as a justification for men's socially inappropriate behaviors (Berglas, 1987). Of the college date rapists interviewed by Kanin (1984), 62% felt that they had committed rape because of their alcohol consumption. These men believed that their intoxicated condition caused them to initially misperceive their partner's degree of sexual interest and later allowed them to feel comfortable using force when the women's lack of consent finally became clear to them. These date rapists did not see themselves as "real" criminals because real criminals used weapons to assault strangers. Figure 1 (first box) includes a feedback loop between feeling that alcohol justifies aggressive behavior and preexisting beliefs about alcohol's effects. Once a man has used intoxication to justify forced sex, he is more likely to believe that alcohol causes this type of behavior and to use this as an excuse in the future.

In contrast, women tend to feel more responsible for sexual assault if they had been drinking alcohol (Norris, 1994). Women are often criticized for losing control of the situation, not communicating clearly, not resisting adequately and failing in their gatekeeper role.

In one of our surveys, a woman replied to a question about if the assault was avoidable, "Yes, if I had not been intoxicated . . . I would have been more in control of myself and the situation."

Other people also tend to blame intoxicated women for sexual assault. For example, Richardson and Campbell (1982) asked male and female college students to read a story about a college woman raped by a guest while cleaning up after a party. Both male and female students perceived the victim as more responsible when she was intoxicated. The woman was also perceived as less likable and moral when she was drunk; however, alcohol consumption did not affect these judgments about the male. A more recent study (Hammock and Richardson, 1997) replicated the findings regarding the victim's alcohol consumption. Victims of sexual assault were held more responsible by male and female college students when they were intoxicated. These findings may help explain why less than half of college student sexual assault victims tell anyone about what happened (Koss et al., 1987). They may anticipate being blamed rather than supported.

Several other studies have found that judgments about sexual assault vignettes depend on whether both the victim and perpetrator were drinking or only the victim was drinking. For example, Stormo et al. (1997) found that when both the man and the woman were equally intoxicated, drinking women were held more responsible for sexual assault; in contrast, drinking men were held less responsible. However, a sober man was judged to be more responsible when he assaulted an intoxicated woman, perhaps because he was seen as taking advantage of her. It is noteworthy that observers sometimes derogate men for taking advantage of an intoxicated woman, although many sexual assault perpetrators seem to experience no remorse about using this strategy to obtain sex (Kanin, 1985; Mosher and Anderson, 1986).

Peer Environments That Encourage Heavy Drinking and Sexual Assault

For some drinkers, alcohol provides a justification for engaging in behaviors that are usually considered inappropriate. This excuse-giving function is only effective if one's peer group shares the same beliefs. The peer group norms in some college social environments, including many sororities and fraternities, accept getting drunk as a justification for engaging in behaviors that would usually be embarrassing. The peer norms for most fraternity parties are to drink heavily, to act in an uninhibited manner and to engage in casual sex (Martin and Hummer, 1989; Norris et al., 1996). Although researchers have focused on Greek organizations, heavy episodic drinking and forced sex are not condoned by all fraternities or all members of fraternities. Other types of formal (e.g., athletic groups) and informal college peer networks can encourage drunken excess and inappropriate behavior.

Martin and Hummer (1989) argued that many fraternities create a social environment in which sexual coercion is normalized because women are perceived as commodities available to meet men's sexual needs. Alcohol is used to encourage reluctant women to have sex. One fraternity man stated that at parties, "We provide them [Little Sisters] with 'hunch punch' and things get wild. We get them drunk and most of the guys end up with one" (p. 465). With no remorse or guilt, this fraternity man described his plans to get one particular woman drunk by serving her punch without letting her know it was spiked for the challenge of having sex with a "prim and proper sorority girl" (p. 465).

Research has also been conducted with sorority women to determine the types of social pressure that they experience. Norris et al. (1996) found that most sorority women know that the emphasis at many fraternity parties is on heavy drinking and casual sex. In focus groups, they articulated warning signs such as getting too drunk or receiving attention from specific men who have a reputation for forcing sex. However, most of these women believed that they were "too smart to be raped" (p. 132). Thus these sorority women recognized that being drunk makes women easy targets, yet they thought they were better than other women at staying alert when drunk. These sorority women also seemed unwilling to report sexual assault when it occurred. They thought that the Greek system received too much negative press; thus they felt responsible to be positive about it.

Summary of Research Regarding Alcohol's Role in College Sexual Assaults

Alcohol increases the likelihood of sexual assault occurring among acquaintances during social interactions through several interrelated pathways. These pathways include beliefs about alcohol, deficits in higher-order cognitive processing and motor impairments induced by alcohol and peer group norms that encourage heavy drinking and forced sex. There is a synergistic relationship between men's personality traits (e.g., low empathy, high impulsivity), attitudes (e.g., believe forced sex is sometimes acceptable, believe women are coy about their sexual intentions and enjoy forced sex) and alcohol's effects. If a man believes forced sex is acceptable and women cannot be trusted, he may be comfortable raping when sober. Alcohol makes it even easier for men to feel comfortable forcing sex because alcohol myopia helps them focus solely on their desire to have sex rather than on the woman's signs of refusal and pain. Although data have been presented to support each of these arguments, causality cannot be firmly established because each study had methodological limitations. In combination, however, these studies demonstrate the many ways in which alcohol consumption can contribute to sexual assault.

Directions for Future Research

Given how many sexual assaults occur in high school and how many high school students report heavy episodic drinking, long-term longitudinal studies are needed that follow youth from early adolescence into adulthood. Prospective research allows potential causes, such as stereotypes about drinking women, alcohol expectancies and usual alcohol consumption, to be measured prior to the experience of college sexual assault.

More precise measurement is needed of the amount of alcohol consumed in sexual assault situations. Because most researchers assess only whether or not any alcohol was consumed, it is impossible to evaluate whether perpetrators or victims were intoxicated at the time of the assault. The effects of one glass of wine with dinner are likely to be very different from the effects of 10 beers consumed within a 2-hour period at a party. Another difficult measurement issue concerns how to enhance the accuracy of drunken recall. If a woman was so drunk she was unconscious when she was raped, it may be impossible for her to fully and accurately describe what occurred. Methodological studies are needed that focus on how best to ask questions to enhance accurate recall of events that occurred under various levels of intoxication.

In-depth qualitative studies are necessary to better understand the precise role of alcohol in sexual assault. These studies need to include students from different cultural and ethnic backgrounds. Research with minority students, students at commuter schools and gay students is needed. A few authors have focused on Greek organizations and athletes; however, students with other interests and lifestyles also need to be represented in qualitative research.

Alcohol administration studies are required because only when participants are randomly assigned to drink an alcoholic or nonalcoholic beverage can one be certain that differences in their behavior are due to alcohol rather than other factors such as prior drinking history. Because sexual assault cannot be an outcome in laboratory studies, appropriate proxies must be used. Some researchers have exposed participants to pornography as a proxy for sexual assault (George and Marlatt, 1986; Hall and Hirschman, 1994). Other researchers have asked participants to evaluate written or audio depictions of sexual assault when intoxicated or sober (Bernat et al., 1998; Norris and Cubbins, 1992). Whenever participants read stories about sexual assault, there is a concern that they may not respond in the same way that they would to an event in their own lives. Research that helps explain how other people react to sexual assault victims is important in its own right because victims are so often blamed by others.

Many of the studies that have informed theory about alcohol's role in sexual assault have examined general aggressive and sexual behavior. Additional research in these areas can be used to develop prevention and treatment programs. For example, research can investigate the circumstances under which men are most willing to aggress against a female confederate (Taylor and Chermack, 1993) or delineate the types of cues that intoxicated men are most likely to misperceive (Abbey et al., 2000).

Prevention and Policy Implications

There are many potential prevention and policy implications that stem from this review. The suggestions provided here are derived from the literature; however, they have not been evaluated. It is crucial that colleges develop evaluation plans so that they can determine the effectiveness of the programs they utilize.

One simple, but important, policy implication that derives from this review is that the individuals on campus who are responsible for programs on the prevention of alcohol misuse must work in conjunction with those individuals responsible for programs on the prevention of sexual assault. Most acquaintance rape prevention programs discuss alcohol as a risk factor, but many do not emphasize it (Bohmer and Parrot, 1993). In a similar manner, programs that describe responsible drinking do not typically emphasize sexual assault as a consequence of heavy drinking. Programs on prevention of alcohol misuse can provide students with the precise definition of sexual assault in their state and information about the prevalence of alcohol-related sexual assault among college students. These programs can also explain that alcohol is not legally considered a mitigating factor for sexual assault and that having sex with someone too intoxicated to give consent is legally rape.

Most research currently being conducted to explain alcohol's effects on behavior focus on the role of alcohol-induced cognitive deficits in producing a variety of risky, socially disapproved of behaviors. According to alcohol myopia theories (Steele and Josephs, 1990; Taylor and Chermack, 1993), alcohol causes people to focus on the most salient cues in the situation and ignore or minimize peripheral cues. In the domain of sexual assault, the assumption is that the man's immediate sexual arousal and anger are much more salient than the potential risk of being accused of sexual assault.

This argument suggests that increasing the salience, explicitness and centrality of inhibitory information should be an effective prevention strategy. If the costs of sexual assault are obvious, undesirable and immediate, then intoxication-driven sexual assaults are less likely to occur because the potential perpetrator cannot forget about the likely, undesirable consequences. This suggests that colleges need strong, consistent, well-publicized policies that no one can ignore. Men need to know that "no means no" and that forced sex is a crime that the university will not tolerate. Students needs to know how to report sexual assault to university authorities, how cases will be evaluated and what the sanctions are for the perpetrators and organizations that facilitated the assault. The campaign to reduce driving while intoxicated has used a similar approach by making the legal and social consequences of driving while intoxicated more salient and serious, and it has been successful in reducing the incidence of this crime (Voas et al., 1998).

The second predominant theory regarding how alcohol exerts its effects concerns the role of people's beliefs about alcohol. If students believe that alcohol makes them do wild and crazy things that they would not do otherwise, then they are much more likely to act out when drinking. The policy implications of this research are twofold. First, educational efforts are needed to change student's alcohol expectancies and to emphasize negative consequences such as making bad decisions, feeling embarrassed the next day and doing poorly in school. Second, these programs have to make it clear that intoxication does not excuse illegal or immoral behavior, so claiming "I did it because I was drunk" will not reduce the consequences. General interventions designed to challenge college students' expectancies about alcohol's effects have been effective in reducing alcohol consumption (Darkes and Goldman, 1993), suggesting that those specifically targeted at expectancies regarding sex and aggression may also be beneficial.

Many college women realize that getting drunk at a fraternity party puts women at risk of being sexually assaulted (Norris et al., 1996). However, a sense of personal invulnerability leads women to believe that they are too smart for it to happen to them. These college women are not unique; many psychological studies have demonstrated that young people feel personally invulnerable to the consequences of a wide variety of risky behaviors (Weinstein and Klein, 1996). Prevention programs that strip away some of this sense of personal invulnerability are necessary so that women will take more precautions. Optimism is in many ways psychologically adaptive; thus programs must avoid scare tactics that make women feel helpless and unable to trust any man. Although the rates of sexual assault are very high, the probability of any one date or party involving sexual assault is low. Thus women must be able to enjoy themselves most of the time, but remain alert for men that are trying hard to get them to drink alcohol, take drugs or accompany them to an isolated location.

Women sometimes seem to feel that it is easier to give in than to fight a sexually coercive man. Lewin (1985) quoted a college woman who wrote, "I feel that I had to go through with the complete sex act because of a feeling of pressure.... I felt perhaps I would let him down and as a result he would like me less ... in fact never spoke to me after the experience.... I should have been as selfish as he was" (p. 184). Some authors have suggested that a passive response is most likely if the man is a current or past boyfriend who feels that he is entitled to sex (Testa and Livingston, 1999). The myth that it is impossible for a sexually aroused man to control himself still seems to be believed by many male and female college students. These findings about some women's reluctance to be forceful with sexually persistent men have prevention and policy implications. Educational programs for women need to encourage them that they have the right to refuse sex at any time, with anyone, regardless of their relationship or previous degree of sexual interaction. In addition, women need to know that being verbally and physically assertive are often effective resistance strategies and that when they are drunk they will have a harder time effectively resisting. Educational programs for men need to teach them to take subtle signs of disinterest seriously. If a woman says "no, I don't want to do that now," that comment should be enough to stop their sexual advances; a woman should not have to scream or kick to get her point across. Many female and male college students engage in sexual activities they later regret, because they are uncomfortable being straightforward in sexual communications. Programs that help students learn to talk about sex with potential sex partners are needed. Because alcohol makes it easy to ignore subtle signals, men need to be particularly careful when they are drinking to communicate their sexual desires clearly and to obtain active consent from a woman before engaging in sex.

Prevention programs should begin in middle school, as dating relationships begin to develop. College students are still open to new ideas; thus sexual assault prevention messages need to be provided to male and female college students early and frequently. New students can be provided with information at orientation about the many consequences of heavy drinking, including sexual assault. Programs need to be interesting and to use a variety of modalities including videos, theater groups, role-playing and coed discussion groups. Peer leaders are crucial to demonstrate that other students share these concerns. Special efforts need to be made with Greek organizations, sports teams and other large social groups to enlist their support in prevention efforts. Students are motivated by their peers' beliefs. Demonstrating that not all members of Greek organizations or athletes approve of heavy drinking or forced sex can empower more students to show their disapproval. Conducting needs assessment surveys and focus groups with students on campus can provide information that helps tailor prevention programs to the specific needs of students at that institution. Faculty, staff and administrators need to be well informed so that they can support program efforts. Women who report being sexually assaulted after drinking heavily at a party need to know that they will be treated with respect and concern by campus personnel, or they will continue to keep this crime a secret.

References

Abbey, A. Sex differences in attributions for friendly behavior: Do males misperceive females' friendliness? J. Pers. Social Psychol. 42: 530–838, 1982.

Abbey, A. Misperceptions of friendly behavior as sexual interest: A survey of naturally occurring incidents. Psychol. Women Q. 11: 173–194, 1987.

Abbey, A. Acquaintance rape and alcohol consumption on college campuses: How are they linked? J. Amer. Coll. Hlth 39 (4): 165–169, 1991.

Abbey, A., McAuslan, P. and Ross, L. T. Sexual assault perpetration by college men: The role of alcohol, misperception of sexual intent, and sexual beliefs and experiences. J. Soc. Clin. Psychol. 17: 167–195, 1998.

Abbey, A., Ross, L. T., McDuffie, D. and McAuslan, P. Alcohol and dating risk factors for sexual assault among college women. Psychol. Women Q. 20: 147–169, 1996a.

Abbey, A., Ross, L. T., McDuffie, D. and McAuslan, P. Alcohol, misperception, and sexual assault: How and why are they linked? In: Buss, D. M. and Malamuth, N. M. (Eds.) Sex, Power, Conflict: Evolutionary and Feminist Perspectives, New York: Oxford Univ. Press, 1996b, pp. 138–161.

Abbey, A., Zawacki, T. and McAuslan, P. Alcohol's effects on sexual perception. J. Stud. Alcohol 61: 688–697, 2000.

Berglas, S. Self-handicapping model. In: Blane, H. T. and Leonard, K. E. (Eds.) Psychological Theories of Drinking and Alcoholism, New York: Guilford Press, 1987, pp. 305–345.

Bernat, J. A., Calhoun, K. S. and Stolp, S. Sexually aggressive men's response to a date rape analogue: Alcohol as a disinhibiting cue. J. Sex Res. 35: 341–348, 1998.

Bohmer, C. and Parrot, A. Sexual Assault on Campus: The Problem and the Solution, Lexington, MA: Lexington Books, 1993.

Brown, S. A., Goldman, M. S., Inn, A. and Anderson, L. R. Expectations of reinforcement from alcohol: Their domain and relation to drinking patterns. J. Cons. Clin. Psychol. 48: 419–426, 1980.

Bureau of Justice Statistics. Violence against Women: Estimates from the Redesigned Survey, NCJ 154348, Washington: Department of Justice, 1995.

Byers, E. S. and Wilson, P. Accuracy of women's expectations regarding men's responses to refusals of sexual advances in dating situations. Int. J. Women's Stud. 4: 376–387, 1985.

Caetano, R., Clark, C. L. and Tam, T. Alcohol Hlth Res. World 22: 233–241, 1998.

Chermack, S. T. and Giancola, P. R. The relation between alcohol and aggression: An integrated biopsychosocial conceptualization. Clin. Psychol. Rev. 17: 621–649, 1997.

Clark, C. L., Shaver, P. R. and Abrahams, M. F. Strategic behaviors in romantic relationship initiation. Pers. Social Psychol. Bull. 25: 707–720, 1999.

Copenhaver, S. and Grauerholz, E. Sexual victimization among sorority women: Exploring the links between sexual violence and institutional practices. Sex Roles 24 (1–2): 31–41, 1991.

Corcoran, K. J. and Thomas, L. R. The influence of observed alcohol consumption on perceptions of initiation of sexual activity in a college dating situation. J. Appl. Social Psychol. 21: 500–507, 1991.

Darkes, J. and Goldman, M. S. Expectancy challenge and drinking reduction: Experimental evidence for meditational process. J. Cons. Clin. Psychol. 61: 344–353, 1993.

George, W. H., Cue, K. L., Lopez, P. A., Crowe, L. C. and Norris, J. Self-reported alcohol expectancies and post-drinking sexual inferences about women. J. Appl. Social Psychol. 25: 164–186, 1995.

George, W. H. and Marlatt, G. A. The effects of alcohol and anger on interest in violence, erotica, and deviance. J. Abnorm. Psychol. 95: 150–158, 1986.

George, W. H. and Norris, J. Alcohol disinhibition, sexual arousal, and deviant sexual behavior. Alcohol Hlth Res. World 15: 133–138, 1991.

Greene, D. M. and Navarro, R. L. Situation-specific assertiveness in the epidemiology of sexual victimization among university women. Psychol. Women Q. 22: 589–604, 1998.

Hall, G. C. and Hirschman, R. The relationship between men's sexual aggression inside and outside the laboratory. J. Cons. Clin. Psychol. 62: 375–380, 1994.

Hammock, G. S. and Richardson, D. R. Perceptions of rape: The influence of closeness of relationship, intoxication and sex of participant. Viol. Vict. 12: 237–246, 1997.

Harrington, N. T. and Leitenberg, H. Relationship between alcohol consumption and victim behaviors immediately preceding sexual aggression by an acquaintance. Viol. Vict. 9: 315–324, 1994.

Hindmarch, I., Kerr, J. S. and Sherwood, N. The effects of alcohol and other drugs on psychomotor performance and cognitive function. Alcohol Alcsm 26: 71–79, 1991.

Humphrey, J. A. and White, J. W. Women's vulnerability to sexual assault from adolescence to young adulthood. J. Adolesc. Hlth 27: 419–424, 2000.

Ito, T. A., Miller, N. and Pollock, V. E. Alcohol and aggression: A meta-analysis on the moderating effects of inhibitory cues, triggering events, and self-focused attention. Psychol. Bull. 120: 60–82, 1996.

Johnson, T. J., Wendel, J. and Hamilton, S. Social anxiety, alcohol expectancies and drinking-game participation. Addict. Behav. 23: 65–79, 1998.

Kanin, E. J. Date rape: Unofficial criminals and victims. Victimology 9: 95–108, 1984.

Kanin, E. J. Date rapists: Differential sexual socialization and relative deprivation, Arch. Sexual Behav. 14: 219–231, 1985.

Koss, M. P. Hidden rape: Sexual aggression and victimization in a national sample of students in higher education. In: Burgess, A. W. (Ed.) Rape and Sexual Assault II, New York: Garland, 1988, pp. 3–25.

Koss, M. P. The underdetection of rape: Methodological choices influence incidence estimates. J. Social Issues 48: 61–75, 1992.

Koss, M. P., Gidycz, C. A. and Wisniewski, N. The scope of rape: Incidence and prevalence of sexual aggression and victimization in a national sample of higher education students. J. Cons. Clin. Psychol. 55: 162–170, 1987.

Lewin, M. Unwanted intercourse: The difficulty of saying no. Psychol. Women Q. 9: 184–192, 1985.

McAuslan, P., Abbey, A. and Zawacki, T. Acceptance of pressure and threats to obtain sex and sexual assault. Paper presented at the SPSSI Convention, Ann Arbor, Michigan, 1998.

Malamuth, N. M. The attraction to sexual aggression scale: Part two. J. Sex Res. 26: 324–354, 1989.

Martin, P. Y. and Hummer, R. A. Fraternities and rape on campus. Gender Soc. 3: 457–473, 1989.

Mills, C. S. and Granoff, B. J. Date and acquaintance rape among a sample of college students. Soc. Work 37: 504–509, 1992.

Mosher, D. L. and Anderson, R. D. Macho personality, sexual aggression, and reactions to guided imagery of realistic rape. J. Res. Pers. 20: 77–94, 1986.

Muehlenhard, C. L. and Linton, M. A. Date rape and sexual aggression in dating situations: Incidence and risk factors. J. Counsel. Psychol. 34: 186–196, 1987.

Norris, J. Alcohol and female sexuality: A look at expectancies and risks. Alcohol Hlth Res. World 18: 197–201, 1994.

Norris, J. and Cubbins, L. A. Dating, drinking, and rape: Effects of victim's and assailant's alcohol consumption on judgments of their behavior and traits. Psychol. Women Q. 16: 179–191, 1992.

Norris, J. and Kerr, K. L. Alcohol and violent pornography: Responses to permissive and nonpermissive cues. J. Stud. Alcohol, Supplement No. 11, pp. 118–127, 1993.

Norris, J., Nurius, P. S. and Dimeff, L. A. Through her eyes: Factors affecting women's perception of and resistance to acquaintance sexual aggression threat. Psychol. Women Q. 20: 123–145, 1996.

Peterson, J. B., Rothfleisch, J., Zelazo, P. D. and Pihl, R. O. Acute alcohol intoxication and cognitive functioning. J. Stud. Alcohol 51: 114–122, 1990.

Presley, C. A., Meiman, P. W., Cashin, J. R. and Leichliter, J. S. Alcohol and Drugs on American College Campuses: Issues of Violence and Harassment, Carbondale, IL: Core Institute, Southern Illinois University at Carbondale, 1997.

Rapaport, K. and Burkhart, B. R. Personality and attitudinal characteristics of sexually coercive college males, J. Abnorm. Psychol. 93: 216–221, 1984.

Richardson, D. and Campbell, J. L. Alcohol and rape: The effect of alcohol on attributions of blame for rape. Pers. Social Psychol. Bull. 8: 468–476, 1982.

Roche, J. P and Ramsbey, T. W. Premarital sexuality: A five-year follow-up study of attitudes and behaviors by dating stage. Adolescence 28: 67–80, 1993.

Seto, M. C. and Barbaree, H. E. Sexual aggression as antisocial behavior: A developmental model. In: Stoff, D. M., Breiling, J. and Maser, J. D. (Eds.) Handbook of Antisocial Behavior, New York: John Wiley & Sons, 1997, pp. 524–533.

Snyder, M. and Stukas, A. A., Jr. Interpersonal processes: The interplay of cognitive, motivational, and behavioral activities in social interaction. Annual Rev. Psychol. 50: 273–303, 1999.

Steele, C. M. and Josephs, R. A. Alcohol myopia: Its prized and dangerous effects. Amer. Psychol. 45: 921–933, 1990.

Stormo, K. J., Lang, A. R. and Stritzke, W. G. K. Attributions about acquaintance rape: The role of alcohol and individual differences. J. Appl. Social Psychol. 27: 279–305, 1997.

Taylor, S. P. and Chermack, S. T. Alcohol, drugs, and human physical aggression. J. Stud. Alcohol, Supplement No. 11, pp. 78–88, 1993.

230 SECTION 5 SEXUAL VICTIMIZATION

Testa, M. and Livingston, J. A. Qualitative analysis of women's experiences of sexual aggression: Focus on the role of alcohol. Psychol. Women Q. 23: 573–589, 1999.

Ullman, S. E., Karabatsos, G. and Koss, M. P. Alcohol and sexual assault in a national sample of college women. J. Interperson. Viol. 14: 603–625, 1999.

Voas, R. B., Wells, J., Lestina, D., Williams, A. and Greene, M. Drinking and driving in the United States: The 1996 national roadside survey. Accid. Anal. Prev. 30: 267–275, 1998.

Weinstein, N. D. and Klein, W. M. Unrealistic optimism: Present and future. J. Social Clin. Psychol. 15: 1–8, 1996.

Werner, P. D. and LaRussa, G. W. Persistence and change in sex role stereotypes. Sex Roles 12 (9–10): 1089–1100, 1985.

White, J. W. and Koss, M. P. Adolescent sexual aggression within heterosexual relationships: Prevalence, characteristics and cause. In: Barbaree, H. E., Marshall, W. L. and Hudson, S. M. (Eds.) The Juvenile Sex Offender, New York: Guilford Press, 1993, pp. 182–202.

Wilsnack, S. C., Plaud, J. J., Wilsnack, R. W. and Klassen, A. D. Sexuality, gender and alcohol use. In: Wilsnack, R. W. and Wilsnack, S. C. (Eds.) Gender and Alcohol: Individual and Social Perspectives, New Brunswick, NJ: Rutgers Center of Alcohol Studies, 1997, pp. 250–288.

DISCUSSION QUESTIONS

1. What can college campuses do to curb alcohol-involved sexual assault?

2. Can you think of other reasons why alcohol is so common in sexual assaults of college women?

3. What are the cognitive effects that alcohol has on people? How are these effects linked to sexual assault? What are the behavioral or situational impacts of alcohol? How are they linked to sexual assault?

4. Do you think that alcohol will have the same relationship to sexual assault among non-college students? Why or why not?

Intimate Partner Violence

or most of us, what goes on behind closed doors in our intimate relationships remains private. Except for the details we choose to share with others, the give-and-take, the ups and downs, and the good and bad times are largely shared by only us and our partners. For others, these details are sometimes made public and, in some circumstances, are widely broadcast for the world to know. Unfortunately, it is usually those times involving fighting that become public fodder. For Mel Gibson, an Oscar-winning actor and director, his "rants" at then live-in girlfriend Oksana Grigorieva were exposed. Several phone conversations between him and Oksana were taped and then posted to the Internet, where anyone with an Internet connection could listen to what was said by both parties. Below is a transcript from one of their conversations. It was edited lightly to censor profanity, and some lines were omitted.

M: I'm sick of your bull****! Has any relationship ever worked with you? Nooo!

O: Listen to me. You don't love me, because somebody who loves does not behave this way.

M: Shut the f*** up. I know . . . *(cross-talking)* because I know absolutely that you do not love me and you treat me with no consideration.

O: One second. One second. Can I please speak? . . .

O: You just enjoy insulting me, that's all.

M: F*** you, I so f***** do, because you hurt me so bad.

O: I didn't do . . . I don't . . .

M: You insult me with every look, *(garbled)* every f***** heartbeat, you selfish harpy.

O: I did not do anything, and I apologized for nothing . . .

O: I wanted to peace. I wanted to have peace.

M: Keep peace.

O: Because you are unbalanced!

O: You need medication! . . .

M: I need a woman! . . . I need a f***** woman. *(panting)* I don't need medication. You need a f***** bat in the side of the head. All right? How 'bout that? You need a f***** doctor. You need a f***** brain transplant. You need a f*****, you need a f***** soul. I need medication. I need someone who treats me like a man, like a human being. With kindness, who understands what gratitude is, because I f***** bend over backwards with my balls in a knot to do it all for her and she gives me sh**, like a f***** sour look or says I'm mean. Mean? What the f** is that? This is mean! Get it? You get it now? What mean is? Get it? *(panting)* You f***** don't care about me. I'm having a hard time, and you f***** yank the rug, you b****, you f***** selfish b****. *(panting)* Don't you dare hang up on me.

O: I can't listen to this anymore.

M: You hang up, I'm coming over there.

O: I'll call the police.

M: What?

O: I'll call the police.

M: You f***** c***. I'm coming to my house. You're in my house, honey.

O: Yes, but you, honey, don't call me honey. You just . . .

M: *(screaming)* You're in my house! So I'll call the police and tell them there's someone in my house. How 'bout that?

Now that you have read this transcript, think about how you would characterize their conversation to others. Is this just a typical, everyday fight that any couple might have? Is it abusive? Are both parties contributing equally? What kind of effect might this kind of conversation have on Oksana? On Mel? On their young child? This section will discuss the variations of intimate partner violence, its causes and effects, and how the criminal justice system and other agencies respond to victims.

Defining Intimate Partner Violence and Abuse

In order to understand intimate partner violence (IPV), we must first define what it entails. First, we must identify what an intimate partner is. An **intimate partner** can be a husband or wife, an ex-husband or ex-wife, a boyfriend or girlfriend, or a dating partner. **Violence**—the intentional physical harm of another person—that occurs between these people includes overt **physical violence** such as hitting, slapping, kicking, punching, and choking. Throwing objects at another person is also physical violence. In short, any intentional physical harm that results in pain is physical violence. When it comes to intimate partners, however, physical violence may not encompass all the harm done. For example, yelling at and verbally degrading a partner may also be seen as violent—or at least as **emotional abuse** (Payne & Gainey, 2009). The transcript from Mel Gibson's phone conversation is, quite likely, indicative of emotional abuse. This emotional form of violence also includes threats of harm, restraint of normal activities or freedom, and denial of access to resources (National Research Council, 1996). Violence within intimate relationships can also be sexual in nature. **Sexual violence** includes unwanted sexual contact, sexual coercion, and rape. As discussed in Section 5, sexual violence often involves someone known to the victim, including current or former intimate partners. Given these varied forms of IPV and abuse, examples of this type of victimization are plenty. Is it abusive when a man does not provide his wife any mode of transportation and insists that she stay at home during the day, calling throughout the day to ensure that she is home and alone? Many would consider this isolating and controlling behavior abusive, although not violent. What about when a couple is fighting and the woman shoves her boyfriend? Even though the shoving may not have created serious injury or even really hurt her boyfriend, the woman did, in fact, use physical violence.

These are not the only ways in which IPV has been defined and described. Michael Johnson, for instance, put forth that there are two major types of IPV. The first type, called **intimate terrorism**, is rooted in the need for power and control, of which the abuse is but one element. Intimate terrorism involves severe, persistent, and frequent abuse that tends to escalate over time. This type of IPV is likely to result in serious injury and to be seen by criminal justice professionals and social service agencies. It is the type of IPV that has been viewed as most problematic and deserving of money and research attention to reduce its occurrence and pernicious effects. The second type of IPV is called **situational couple violence,** or common couple violence. Instead of resulting from a desire for power and control, situational couple violence occurs when conflict gets out of hand and results in violence. It could start with a "run-of-the-mill" disagreement and then turn violent. This type of IPV tends not to result in serious injury or to be a part of a larger pattern of persistent and frequent abuse; it also is unlikely to come to the attention of criminal justice and social service agencies. Two additional types of IPV also have been identified: violent resistance and mutual violent control. Violent resistance occurs when a person is violent but not controlling; instead, the person's partner is the violent and controlling one in the relationship. In mutual violent control, both people are violent and controlling.

These descriptions and definitions show how IPV is currently viewed. Historically, however, IPV was not viewed in these terms. Originally, IPV was defined only as physical violence perpetrated by husbands against their wives, but this definition has evolved over the past 40 years to include emotional and sexual violence. We also now recognize that IPV is not exclusive to married couples and that both men and women can be perpetrators and victims. But let's discuss how we got to this point. There is a bit of controversy regarding just how tolerant or intolerant society has been toward men's violence against their wives. Some have argued that such violence was essentially tolerated, if not condoned, given that males were considered dominant (Dobash & Dobash, 1979). The man was the head of the family and was able to use power and control, even violence, to control his wife and children. Presumably, minor forms of violence were permitted as long as they were used by men to maintain their positions of dominance. It is not apparent, however, just how pervasive IPV was during this time. As you may imagine, there were no national-level studies conducted in which people reported victimization and perpetration. Instead, we can look to laws and their usage during this time as a guide. Importantly, there have been laws in place since the 1600s that specifically prohibited violence against wives in America. By the 1870s, most states had adopted such laws. Punishment of wife abusers was generally informal, with vigilante groups often taking the matter into their own hands. In addition, punishments such as public shaming were often used. For example, a man who assaulted his wife in Portland during the early 20th century could face flogging (as cited in Felson, 2002).

Despite these laws, it does appear that some trial and appellate courts tolerated minor forms of violence by husbands against their wives (Pleck, 1987), but most courts did not tolerate physical violence in any form (Felson, 2002). When courts failed to convict or uphold convictions of men who had abused their wives, it was mainly done not with the view that violence against women by their husbands was acceptable but with the view that the courts should not intervene based on the principle of privacy (Felson, 2002). In the same vein, females were rarely arrested or brought to court for abusing their husbands. As a whole, then, courts have routinely rejected the notion that men have the right to physically assault their wives.

As with many other forms of victimization, IPV really took center stage in policy and research in the 1970s. The women's rights movement was central in focusing attention on women as victims of IPV. During this time, feminists argued that IPV was a reflection of the subjugation of women and that the male-dominated criminal justice system did little to protect women. In response to these claims, an outgrowth of the women's rights movement was to open domestic violence shelters for battered women and to provide women with the assistance they needed to get out of abusive relationships. Since then, efforts to identify, describe, prevent, and respond to IPV have expanded dramatically. Specific responses by the criminal justice system and other social service agencies are discussed later in this section.

Measurement and Extent

Now that you know what IPV is, you are probably wondering how common it is for partners, who supposedly care about each other, to engage in violence against each other. As you might imagine, it can be difficult to know exactly how frequently such behavior occurs given that it often occurs in private. In addition, people may be reluctant to call the police when the perpetrator is someone close to them; thus, official data sources may underestimate the extent to which violence between intimate partners occurs. Much like for other types of victimization, one of the most widely used research methodologies is to employ surveys that ask people to self-report their victimizations and perpetrations. The findings from such studies are addressed in this section.

National Crime Victimization Survey

Remember from our discussion in Section 2 that the **National Crime Victimization Survey (NCVS)** is a survey of U.S. households in which individuals are asked about their victimization experiences during the previous 6 months. Individuals who report experiencing a victimization event complete an incident report for each event. Within this detailed incident report, individuals are asked to identify their relationship with the perpetrator. Violent incidents perpetrated by spouses or ex-spouses, boyfriends or girlfriends, and former boyfriends or girlfriends are considered in the NCVS to be IPV. Data from the NCVS indicate that the rate of IPV declined from 1993 to 2008 (Catalano, Smith, Snyder, & Rand, 2009). Nonetheless, in 2008, females experienced 552,000 nonfatal IPV victimizations, making the IPV rate for females 4.3 victimizations per 1,000 females age 12 or older. Males were less likely to report IPV victimizations. Males experienced 101,000 nonfatal IPV victimizations. This equates to a rate of 0.8 victimizations per 1,000 males age 12 or older (Catalano et al., 2009). For both males and females, the most common type of IPV reported in the NCVS was simple assault, which composes more than half of all IPV victimizations (Catalano et al., 2009).

Conflict Tactics Scale

Developed in the 1970s, the **Conflict Tactics Scale (CTS)** is designed to measure levels and use of various conflict tactics. Created by Murray Straus and Richard Gelles, along with other colleagues, this scale was constructed to examine conflict in intimate relationships. The CTS since has been revised into the revised Conflict Tactics Scale (CTS-2) (Straus, Hamby, Boney-McCoy, & Sugarman, 1996). Conflict was seen by Straus and Gelles as being inevitable in close interpersonal interactions, but it is not conflict itself that matters but the ways in which couples resolve it. The ways in which couples resolve conflict are known as conflict tactics, and the CTS-2 examines the use of conflict tactics in three domains: physical assault, psychological aggression, and negotiation. It also includes items to measure injury and sexual coercion of and by an intimate partner.

The CTS-2 comprises 78 questions that measure these domains. Respondents are asked about the frequency of occurrence of each item during the past year and are given eight response options, ranging from *never* to *more than*

20 times in the past year. Items are presented in pairs so that individuals are asked about victimization and perpetration of the same item together. An example of a question from the CTS-2 is shown in the following box. Respondents are asked about how often their partners push or shove them. Although this item is designed to measure physical assault, other items ask about the other domains. Since its development, the CTS (and now the CTS-2) has become the most widely used survey instrument to measure the occurrence of IPV.

EXAMPLE OF CTS-2 QUESTIONS

Please circle how many times you did each of these things in the past year, and how many times your partner did them in the past year. If you or your partner did not do one of these things in the past year, but it happened before that, circle "7."

How often did this happen?

1 = Once in the past year

2 = Twice in the past year

3 = 3–5 times in the past year

4 = 6–10 times in the past year

5 = 11–20 times in the past year

6 = More than 20 times in the past year

7 = Not in the past year, but it did happen before

0 = This has never happened

My partner pushed or shoved me.

1   2   3   4   5   6   7   0

My partner punched or hit me with something that could hurt.

1   2   3   4   5   6   7   0

SOURCE: Straus et al. (1996).

Straus and Gelles used the CTS in the National Family Violence Surveys in 1978 and 1985, both national samples. They found that in about 1 in 8 couples, the husband perpetrated at least one violent act in the previous 12 months. Females were also found to perpetrate violence—with females perpetrating IPV in 12% of couples in the study. This finding that males and females were essentially equally likely to use violence in their relationships is one of the major contributions of Straus and Gelles' work. This finding goes against conventional wisdom that males are violent and females are passive in their relationships. You are probably not surprised, then, to learn that the CTS has been criticized. It does not include spousal homicide, which does not reflect sexual symmetry (Payne & Gainey, 2009). In addition, critics point out that checklists such as the CTS do not consider the motives, meanings, and consequences of violence. Females may be using violence in relationships in response to initial violent acts by males. Moreover, even when women hit or strike their partners, they are less likely to cause serious injury compared with a man hitting or striking a female. Even without considering the context, the CTS can be used to uncover the extent to which partners use various conflict tactics, even if the underlying processes that create the conflict and response are not captured (Straus, 2007). Other concerns surround the underreporting of male violence and the overreporting of female violence by survey participants, which could contribute to the gender symmetry found (Dragiewicz, 2010), although a meta-analysis of studies using the CTS found that males are more likely to under-report (Archer, 1999). Despite these concerns, the CTS has made major contributions to the field.

National Violence Against Women Survey

Recall from Section 5 that the **National Violence Against Women Survey (NVAWS)** was a telephone survey of 8,000 women and 8,000 men aged 18 and older (Tjaden & Thoennes, 2000). In the study, participants were asked about violence they had experienced. Specifically, they were asked about psychological and emotional abuse perpetrated by current or former spouses and cohabiting partners, and rape, physical assault, and stalking they had experienced during their lifetime. The survey was conducted, in part, to focus on violence against women and on the victim-offender relationship. Of those surveyed, 22% of the women and 7% of the men said they had been physically assaulted by their current or former intimate partner during their lifetime. Respondents were also asked about IPV occurring in the 12 months prior to the survey. More women (1.3%) than men (0.09%) experienced IPV during that time period. Other gender differences emerged in the NVAWS: Women were more likely than men to report being injured and to report their victimizations to the police. Moreover, women reported experiencing a greater number of physical assaults by the same partner (recurring victimization) than did men. Women who were victimized averaged 6.9 physical assaults by the same partner compared with an average of 4.4 assaults for male victims.

⊠   Who Is Victimized?

IPV is experienced by people in all walks of life; however, certain people are at greater risk than others. For example, age is a factor in IPV risk. Younger adults are more likely to experience IPV than are older adults, although IPV perpetrated against older women has received particular attention recently. Race is also a characteristic that has been studied. Findings from the NCVS indicate that Black females are at a greater risk of experiencing IPV than are White females. The data show that the rate of IPV against Black females is 2.5 times higher than that against women of other races (Rand & Rennison, 2004). When examining race and age in the NCVS, this difference appears to be driven by the difference in rates across race for women between the ages of 20 and 24, where the variation between rates for Black and White women is most marked. It should be noted that the NVAWS data also revealed higher rates of victimization among Black women but that this difference was eliminated when other variables, such as sociodemographic and relationship variables, were included (Tjaden & Thoennes, 2000). In other words, factors such as low income, unemployment, and marital status account for the relationship between IPV and race. Finally, much research shows that females are more likely to experience IPV than are males, although this finding has been a point of debate and deserves special attention.

Gender and Intimate Partner Violence

Who is more likely to perpetrate and who is more likely to be victimized by IPV—males or females? A close reading of the previous sections on the extent of IPV indicates mixed findings. The NCVS, police records, emergency room data, and the NVAWS all show that females are more likely than males to be victims of IPV and that males are the dominant perpetrators. Results from studies using the CTS suggest that this gender gap in IPV perpetration and victimization is not so evident. In a meta-analysis examining gender differences in IPV, women were found to be slightly more likely than men to engage in IPV and to do so more frequently (Archer, 2000). So, which is correct? Let's examine the findings a little more closely.

Criminal victimization survey data, police records, and emergency room data all measure more severe forms of IPV, while the CTS includes measures of slapping and pushing, which are relatively unlikely to cause injury. Serious, injurious IPV, which males are more likely to engage in than females, is rare and unlikely to be revealed through surveys. Reliance on surveys, particularly those using the CTS to measure IPV, may underestimate serious physical violence. In addition, men are less likely to seek medical attention and are less likely to report their victimization to police. Men are also less likely to define their victimization as criminal—which hinders reporting. Insofar as studies depend on males defining their experiences as criminal or on reporting to the police or seeking medical assistance, the extent of IPV victimization for males is likely to be underestimated.

So where else are gender differences evident? First, women are more likely than men to be violently victimized, when they are, by someone they know and by intimate partners. Males are more likely to experience violence at the hands of strangers and nonintimates (Rand, 2009). Keep in mind that males are more likely to be violently victimized—but males are generally "safe" from violence in the home or with their dating partners. Females, on the other hand, are "safe" in general—females are simply less likely to experience any type of violent victimization. When they do, however, it is more likely to be IPV than it is for males. Second, it seems that severe IPV is more a male enterprise. Males are more likely to cause injury in IPV incidents. This may be because men are generally bigger and stronger than women and can more easily physically harm their partners when engaging in violence. Third, males also may use tactics in IPV incidents that are more likely to result in violence. Women are likely to throw objects at their partners, push, and shove—all behaviors that may, indeed, cause injury. Keep in mind, however, that these behaviors are less likely to cause serious damage than punching and using weapons. Fourth, males are more likely to engage in intimate terrorism—the kind of violence that is rooted in power and control and that often results in serious physical injury. Women are more likely to be participators in situational or common couple violence. Indeed, gender differences are greatest for physical aggression and less so for verbal aggression (Bettencourt & Miller, 1996). These distinctions between male and female IPV as well as a discussion of intimate terrorism and situational couple violence and two other types of IPV are explored by Michael Johnson (2006) in his article included in this section.

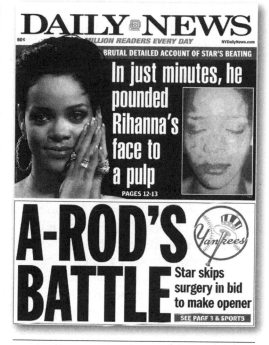

▲ **Photo 6.1** Anyone can be a victim of intimate partner violence—but most victims do not have their pictures plastered on the news. This picture shows the injuries that singer Rihanna suffered at the hands of her then-boyfriend singer Chris Brown. How do you feel about victims' having their injuries displayed and written about in this manner?

Photo credit: © NY Daily News via Getty Images

Special Case: Same-Sex Intimate Partner Violence

So far, we have discussed IPV within heterosexual relationships, but people in homosexual relationships are also at risk of experiencing IPV. The kinds of behavior that are classified as IPV are largely the same for homosexual and heterosexual people. One act, however, is unique to homosexual persons: "outing an individual." If a person is not out to family, friends, coworkers, or in the community, an abuser may threaten to reveal this secret.

Although there is now an awareness of same-sex IPV, it is difficult to measure and hard to know exactly to what extent it occurs. Most of the research relies on small convenience samples rather than large-scale representative samples. Renzetti and Miley (1996) estimate that between 22% and 46% of lesbians have been in an abusive relationship, although this could mean that lesbians experienced IPV in a heterosexual relationship. Data from the NVAWS show that about half of lesbians experienced physical violence in their current intimate relationships (Neeves, 2008). Homosexual men are also at risk of being the victims of IPV, and some estimate that homosexual men are more likely than homosexual women and heterosexual men to experience IPV (Tjaden, Thoennes, & Allison, 1999) and that homosexual men experience IPV at rates similar to those of heterosexual women (Baum & Moore, 2002).

Even with the awareness of same-sex IPV, victims may still be treated differently than victims of opposite-sex IPV. There is a real fear among homosexual victims of not being believed or being demeaned or insulted by law enforcement (Jaquier, 2010). For these reasons, victims of same-sex IPV may be more reluctant to report than others. In addition, resources that are now widely available to victims, especially female victims, of IPV are not necessarily available for same-sex IPV victims. In some states, laws are written such that the partners must be opposite-sex, married, or civilly united; thus, same-sex couples who live together or are dating may find they are not protected under the same domestic violence laws (Allen, 2007). Same-sex IPV victims may be able to secure an order of protection, but doing so requires that a police report is filed—something victims may be particularly unwilling to do—and some states do not grant lesbians and gay men this remedy (Allen, 2007). Gay male victims of IPV may also have difficulty seeking shelter, since most domestic violence shelters admit only females. Finally, most domestic violence programs do not offer programming specifically designed to address same-sex IPV (Allen, 2007).

Special Case: Stalking

In recent years, a special type of victimization often tied to discussions of IPV is stalking. Although definitions of stalking vary across jurisdictions, a working definition for this term is a course of conduct that is unwanted and harassing that would cause a reasonable person to be fearful (National Center for Victims of Crime, n.d.). Not recognized as a crime until the 1980s in the United States, stalking is now recognized in all 50 states and by the federal government as a criminal act. State stalking laws differ, but generally, a victim must show that the behavior constitutes a course of conduct; in other words, the offending behavior must have occurred more than once. Also, the behavior must be unwanted and harassing and must cause some level of emotional distress without a legitimate purpose. Finally, the offender must be a credible threat to the victim, the victim's family, or the victim's friends (Mustaine, 2010). To this end, stalkers may show up at the victim's home or work; follow or track the victim; send unwanted letters, cards, or e-mails; damage the victim's property or home; monitor Internet, computer, and phone use; post information on the Internet or spread rumors; or threaten to hurt the victim, the victim's family, or the victim's pet (National Center for Victims of Crime, n.d.).

Estimates of stalking indicate that 1 in 12 women and 1 in 45 men will be stalked sometime during their lifetime. Each year, 3.4 million adults are stalked, with more than three fourths of these victims being stalked by someone they know. Young adults aged 18 and 19 are the most common victims of stalking, with individuals between 20 and 24 also experiencing high rates (Baum, Catalano, Rand, & Rose, 2009). Persons who are divorced or separated are at the highest risk for being stalked, as are persons who are poor (Baum et al., 2009). Almost 30% of stalkers are current or former intimate partners (Baum et al., 2009). Stalking can be long-term and involve frequent contact—almost half of all stalking victims experience at least one unwanted contact per week, and 11%

of stalking victims have been stalked for 5 years or more, although victims most commonly report being stalked for 6 months or less (Baum et al., 2009).

The types of behaviors that stalkers engage in vary. Almost two thirds of victims received unwanted phone calls or messages. About one third of victims had rumors spread about them, received unwanted letters or e-mails, were followed or spied on, had the offender show up at various places, or had the offender wait for them (Baum et al., 2009). About one fourth of stalking victims reported being stalked through some type of technology, such as e-mail or instant messaging. Global positioning systems (GPS) can be used to monitor victims (10% of victims report), and video and digital cameras or listening devices are also used in a small percentage of stalking cases (8% of victims report) (Baum et al., 2009).

Given the unique circumstances of stalking, it often carries significant consequences for victims. Many victims fear the uncertainty that comes with being stalked, not knowing what will happen from one moment to the next. They also report fearing that the stalking will not stop (Baum et al., 2009). In addition to fear, stalking victims experience higher levels of anxiety, insomnia, social dysfunction, and severe depression than do members of the general population (Blaauw, Winkel, Arensman, Sheridan, & Freeve, 2002). As with other victims, stalking may result in loss of time at work and changes in behavior. For example, about 1 in 8 victims of stalking miss time from work, and 1 in 7 change their residence as a result of the stalking (Baum et al., 2009). About 21% of stalking victims reported changing their day-to-day activities, 13% changed their route to school or work, and 2.3% altered their appearance (Baum et al., 2009). Victims also took action to protect themselves. More than 17% changed their phone number, while 8% installed caller ID or call block (Baum et al., 2009). More than 13% changed their locks or installed a security system (Baum et al., 2009). Stalking also can come with economic costs. About 3 in 10 stalking victims report expenses related to dealing with their stalking, with 13% spending more than $1,000 (Baum et al., 2009).

⊠ Risk Factors and Theories for Intimate Partner Violence

You may be wondering at this point why people are victimized by their intimate partners. Keep in mind that there is not a singular reason why IPV occurs, but researchers have identified many factors that produce IPV. The most common factors are discussed below.

Stress

While the family and the home are seen as places of sanctuary, the family also can cause stress. Given the close physical interaction that families share, it is probably not hard to imagine scenarios that involve stress—money is tight, children are fussy, schedules are hectic. Although stress does not always result in IPV, it certainly can. How couples deal with stress and conflict can dictate the couples' likelihood of experiencing IPV.

Cohabitation

Although IPV is commonly thought of as occurring within marital relationships, it most commonly occurs within **cohabitation** relationships. In fact, the rate of IPV is highest among couples living together, and these couples experience more severe forms of violence (Wallace, 2007). There are several reasons why cohabiting couples experience more incidents of and more severe IPV. One is that individuals in these relationships may be less committed and perceive less stability—two qualities that may increase the amount of conflict in the relationship (Buzawa, 2007). Another explanation may be that these couples feel more social isolation and abuse may be more likely to occur and continue without the support of family and friends (Stets & Straus, 1990).

Power and Patriarchy

Power can be defined as a person's ability to impose his or her will on another. For couples, this likely involves everyday life decisions. Both men and women usually have power in relationships—for example, a man may make more money and a woman may make more decisions about the household. It is when power is abused that it may result in IPV. Power can be used to isolate and scare and take advantage, which is abusive. **Patriarchy** is defined as a form of social organization in which the man is dominant and is allowed to control women and children. Patriarchy may allow IPV by men and even encourage it by giving men a "green light" to exercise their dominance. The research on power and patriarchy and how it relates to IPV is mixed (Coleman & Strauss, 1986). There is some evidence that patriarchal norms and structural inequality are positively related to IPV rates (Yllo & Straus, 1990). Men who hold patriarchal values have been shown in other research to be more violent against women (Sugarman & Frankel, 1996). Some research shows that when couples share power and are egalitarian, there is less IPV. Conversely, other research shows that men who believe in traditional sex roles are less likely to be abusive toward their intimate partners (Bookwala, Frieze, Smith, & Ryan, 1992; Rosenbaum, 1986). In fact, Felson (2002) contends that patriarchy has little applicability to the dyadic relationships between intimate partners.

Social Learning

According to the traditions of **social learning** theorists, criminal behavior is learned behavior. People can learn by observing others engaging in crime and by having their own criminal behavior reinforced. Social learning theory has helped understand why IPV occurs. People who grow up in homes where IPV occurs are more likely than others to be both perpetrators and victims of IPV (Foshee, Bauman, & Linder, 1999; Riggs, Caulfield, & Street, 2000). It is not just witnessing IPV that can have negative effects on later behavior; individuals who are abused or neglected as children have been shown to be more likely to engage in IPV later in life (Heyman & Smith, 2002; Widom, 1989) and to be victimized (Riggs et al., 2000). Beyond exposure to and experiencing violence, social learning theory also addresses how definitions, values, and norms favorable to IPV play a role in IPV. Individuals who have positive expectations about the use of violence are more likely to perpetrate IPV (Foshee et al., 1999).

Risky Lifestyle

According to lifestyles theory, those individuals who engage in **risky lifestyles** expose themselves to people and situations that are likely to increase their victimization risk (Hindelang, Gottfredon, & Garofalo, 1978). Although this theory has been used to explain risk of victimization, it has been less widely adopted by researchers studying IPV. Two ways in which risky lifestyles have been linked to IPV are through associating with known criminals and using alcohol and drugs.

Associating With Known Criminals

According to lifestyles theory, the more time people spend in the company of offenders, the more likely they are to become victims themselves. Although females are generally less likely to be victimized than males, when they spend time with criminal others, their risk of victimization increases—similar to how it increases for males. As explained by Kristin Carbone-Lopez and Candace Kruttschnitt (2010) in their article in this section, females who have criminal mates are more likely to be the victims of IPV than are females with prosocial intimate partners.

Alcohol and Drugs

In Section 2, we discussed the role that alcohol plays in victimization. It should come as no surprise, then, to learn that alcohol also has been studied in connection with IPV. A wealth of research shows that IPV offenders use illegal drugs and/or drink excessively more than others (Scott, Schafer, & Greenfield, 1999). Findings from the NVAWS show that binge drinkers (people who have five or more drinks in a setting) are 3 to 5 times more likely to engage in IPV against a female than those who abstain from drinking (Tjaden & Thoennes, 2000). Some researchers argue that the relationship between alcohol use and IPV perpetration is mediated by attitudes that support the use of IPV (Buzawa, 2007).

Alcohol use by the victim is also important to examine. Alcohol use is related to IPV victimization in several ways. First, women who have histories of IPV have higher rates of substance abuse than women without such histories (Coker, Smith, Bethea, King, & McKeown, 2000). Second, many victims report having used alcohol or drugs prior to being victimized (Greenfeld & Henneberg, 2000). Using drugs or alcohol may make them more vulnerable or may affect them cognitively and behaviorally such that they initiate or respond to conflict differently than they would when not under the influence. Third, alcohol and drug use may be likely after an IPV victimization. In this way, alcohol or drug use may be a consequence of IPV rather than a cause.

▨ Consequences of Intimate Partner Violence

Negative Health Outcomes

One of the most obvious physical effects of IPV is, of course, **injury**. IPV is one of the most common reasons why women present in the emergency room (ER). More than one third of women who seek care in the ER for violence-related injuries do so because they were victimized by a current or former spouse, a boyfriend, or a girlfriend (Rand, 1997). Often, these injuries that bring women to ERs are quite serious; in one study of women who sought care in a metropolitan ER, almost 3 in 10 required hospital admission and more than 1 in 10 required major medical treatment (Berios & Grady, 1991). In addition to injury, female victims of IPV are more likely to experience frequent headaches and to suffer from gastrointestinal problems (Family Violence Prevention Fund, 2010). IPV may also result in back pain, gynecological disorders, pregnancy difficulties, sexually transmitted diseases, central nervous system disorders, and heart or circulatory conditions (Centers for Disease Control and Prevention, 2010).

Death

Although it may seem surprising, in 2007, intimate partners committed 14% of all homicides (Catalano et al., 2009). The risk of being murdered by an intimate partner is greater for females than for males—70% of victims killed by an intimate partner are female. A sizeable portion of fatalities of women are caused by IPV (Catalano, et al., 2009). In recent years, about one third of female murder victims were killed by an intimate partner (National Coalition Against Domestic Violence, 2007). In contrast, of those men who were murdered, only 5% were killed by their current or former spouse or boyfriend (Catalano et al., 2009). Fatal outcomes of IPV are also found for abused women who are pregnant. IPV is the leading cause of homicide and injury-related deaths among pregnant women (Frye, 2001). Also of note is that Black females are more likely than White females to be murdered by a spouse, boyfriend, or girlfriend (Catalano et al., 2009).

Psychological/Emotional Outcomes

IPV also commonly carries significant psychological and emotional consequences. Remember that IPV can come in the form of emotional or psychological abuse—by definition, it creates emotional and psychological harm. Being physically abused can also cause such harm. Victims of IPV are more likely to experience depression, anxiety, sleep disorders, and posttraumatic stress disorder (Centers for Disease Control and Prevention, 2010). Some of the psychological consequences are quite severe—battered women have higher rates of suicide than do nonbattered women (Stark, 1984). Attempting suicide and considering suicide are also more common among women experiencing IPV than among other women (Coker et al., 2002). Suicide ideation and attempt was about 6 to 9 times more likely for adolescent girls who had been sexually or physically hurt by a dating partner than for girls who did not report such harm (Silverman, Raj, Mucci, & Hathaway, 2001).

Revictimization

One of the consequences of being a victim of IPV is that additional IPV events commonly occur, called **revictimization**. As noted in the discussion of the NVAWS, the typical victim of IPV experiences more than one victimization. Women in the study reported experiencing, on average, 6.9 physical assaults by the same partner. Males also experienced revictimization—averaging 4.4 assaults by the same partner. Research using official records of rearrest, victim interviews, offender interviews, court records, probation records, and shelter records have supported this finding that many victims are abused again and many offenders reoffend (see Bennett Cattaneo & Goodman, 2005). For example, in a study that examined victims of arrested batterers, 28% experienced physical abuse, threats, or unwanted contact during the 3-month follow-up period (Bennett Cattaneo & Goodman, 2003). In a study of men convicted of domestic violence, 16% were rearrested or had a complaint made against them by the same victim (Taylor, Davis, & Maxwell, 2001).

This pattern of revictimization is not a new revelation. The **cycle of violence** was first described by Lenore Walker in 1979. Through interviews with battered women, she identified a common pattern of abuse that involves different phases. In the first phase, the **tension building phase**, the abuser and victim interact in a close relationship that likely involves positive and charming behavior on the part of the abuser. This period of tranquility does not last as day-to-day pressures and more serious events generate tension. These periods of tension and stress precede what may at first be minor violence. The woman is likely to attempt to appease the abuser, "tiptoeing" around and trying not to incite the abuser's violence. This tension, though, continues until the abuser explodes into a fit of rage. When this occurs, the second phase begins. The **acute battering phase** involves the abuser engaging in major and often serious physically assaultive behavior. This phase is followed by the **honeymoon phase.** The abuser is calm and loving and most probably begging his partner for forgiveness. He likely is making promises not to engage in violence again. If the relationship continues, Walker proposes that the cycle of violence is likely to start over. Remember, though, not all abuse is repetitive and most abuse is not the explosive type of violence that Walker's cycle of violence concerns.

⊠ Why Women Do Not Leave Abusive Relationships

The cycle of violence is but one reason why women do not leave abusive relationships. The ups and downs of these relationships, which often include some positive "high" points, can be confusing emotionally. Remember that often the person who is abused sees positive traits in the abuser—probably loved and maybe still loves him. Even when women do make a decision to leave, many return to their abusers. In fact, women leave an average of six times before permanently severing their relationships (Okun, 1986). But many women do, in fact, eventually leave.

But why is leaving so difficult? As mentioned, people in relationships often feel a love for and commitment to the other person that they define as important. This commitment may be even more important if the two are married, and marriage and relationships may be defined as more important than personal safety, especially for women

(Barnett & LaViolette, 1993). They may also share children; thus, leaving may mean splitting the family and taking children away from the other parent.

Some women may not be financially capable of leaving their abusers. Women may be economically dependent on their abusers and feel that they are unable to leave for financial reasons. They may not be employed and may lack the resources necessary to establish independence. In addition, abused women may be isolated from their friends and family and lack a support system that could help them as they try to leave. Abusers often purposefully isolate their partners from social networks, thus making it difficult for them to leave (National Center for Victims of Crime, 2008). Finally, women may be embarrassed, ashamed, and scared. Feeling scared may be warranted. Abused women often have been living a life characterized by a pattern of abuse. Even if they want to leave, research shows that women are more likely to be killed after leaving their husbands than while living with them (Wilson & Daly, 1993).

⊠ Criminal Justice System Responses to Intimate Partner Violence

As noted, the major developments in recognizing and responding to IPV first occurred in the 1970s. Between 1975 and 1980, 44 states passed laws concerning IPV, mainly focusing on the prevention of and protection from IPV and providing victims needed resources (Escobar, 2010). Today, every state has some type of IPV law. These laws and how they impact criminal justice and social service agency responses are discussed in the following subsections.

The Police Response

Traditionally, police were reluctant to become involved in IPV cases. They often did not make arrests in these types of cases, instead allowing privacy to trump police involvement. The reasons for this lack of formal response are varied. Police may not want to intervene in IPV cases because of their perception that such a case will not result in a conviction (Buzawa & Buzawa, 1993). They also may view what happens behind closed doors and within families as beyond the purview of the criminal justice system. Police also may not want to respond to or make arrests in these cases since they are thought to be quite dangerous for the police (Kanno & Newhill, 2009), although others have disputed this claim (Hirschel, Dean, & Lumb, 1994; Pagelow, 1997). Also, victims may request that officers not make an arrest or may tell officers that they will not follow through with charges; hence, police have been reluctant to arrest (Wallace, 2007).

The ability of police to make arrests, even when they are willing to do so, may also be limited. Historically, police have been unable to make an arrest for a misdemeanor without a warrant. A **misdemeanor** is a crime that usually is less serious than a felony and carries a maximum penalty of a year in jail. Police could make an arrest for a misdemeanor only if they had a warrant or if they had witnessed the event. As you might imagine, police often do not arrive on the scene of a crime until after the violence occurs and, thus, would not be able to make an arrest without first getting an arrest warrant. This inability of police to make arrests changed during the 1980s, when research was conducted on the utility of arrests in IPV cases.

In 1984, Lawrence Sherman and Richard Berk conducted the **Minneapolis Domestic Violence Experiment** to examine the deterrent effect of arrest on domestic violence perpetrators. To make it a true experiment, the researchers set up the study so that the police randomly responded to calls for domestic violence in one of three ways: The police could make an arrest that would result in one night of incarceration, separate the parties for a period of 8 hours, or advise the couple using the officer's judgment. For eligible calls, the officer would take a color-coded card indicating which of the three response options he or she was to use for the call. The study ran from March 17, 1982, to August 1, 1982, and covered 314 eligible domestic violence calls. For each call, biweekly follow-up interviews were conducted

▲ **Photo 6.2**   Minneapolis, MN, was the location of the famous Minneapolis Domestic Violence Experiment that effectively changed the way that domestic violence cases were policed throughout the country.

Photo credit: ©Thinkstock/Comstock

with the victims for 6 months, although only 62% of the victims participated in the initial follow-up interview and slightly less than half participated in all 12 interviews. The researchers also collected police reports during the 6-month follow-up to check for additional domestic violence incidents.

What did the researchers find? From both the victim interviews and police records, the researchers concluded that those who were arrested were less likely than others to commit additional domestic violence offenses. Only 10% of those who were arrested committed a subsequent offense, according to police records, as compared with 24% of those who were separated and 19% of those who were advised. Interview data also indicated that arrest was most effective, with only 19% of those arrested committing additional domestic violence. One third of those who were separated and 37% of those advised by officers committed additional offenses, according to victim interviews.

Based on these findings and lawsuits against the police for negligence and unequal treatment of female victims, advocates for mandatory arrest in domestic violence cases were given ample evidence for their position. Not long after, many states began revamping their domestic violence laws to allow police to make arrests in misdemeanor domestic violence cases without a warrant. Other states enacted mandatory arrest policies. **Mandatory arrest policies** are those that proscribe arrest by police officers when there is probable cause that a crime was committed and enough evidence exists for an arrest. Note that mandatory arrest policies may create a situation in which a victim does not want the offender to be arrested and the police make an arrest regardless. By 1992, seven states had adopted mandatory arrest policies (Dobash & Dobash, 1992). Today, more than 20 states have mandatory arrest policies (Iyengar, 2007).

It is interesting that mandatory arrest policies began to be adopted so quickly following the Minneapolis Domestic Violence Experiment because replication studies conducted in five jurisdictions—Charlotte, Dade County, Colorado Springs, Milwaukee, and Omaha—provided different results. In only one location, Dade County, did police records show that arrest had a deterrent effect. In Omaha, Charlotte, and Milwaukee, arrest deterred offenders initially but caused an escalation of domestic violence over time. One of the most important findings from the replication studies was that arrest differentially impacted offenders. Specifically, it was found that arrest of employed offenders did produce a deterrent effect, likely because these offenders had a lot to lose. Conversely, unemployed offenders were more likely to engage in violence (Maxwell, Garner, & Fagan, 2001). From the replication studies, it was concluded that mandatory arrest policies may not be a "one-size-fits-all" solution. Instead, they may be beneficial for some offenders and actually detrimental for others in terms of recidivism.

As states look to adopt policies for their law enforcement officers' responses to IPV, they have other options in addition to requiring arrest. Some states have **pro-arrest** or **presumptive arrest policies**. These policies require arrest but limit this requirement to specific situations in which certain criteria are met. That is, there is a presumption of arrest. These policies are similar to mandatory arrest policies in that the decision to arrest may be independent from the desires of the victim. Consent on the victim's part is not required. When deciding not to make an arrest, an officer may be required to provide a written justification for doing so (Payne & Gainey, 2009).

Less strict policies on policing IPV are **permissive arrest policies**. These policies do not mandate or presume that an arrest will be made by law enforcement when warranted. Rather, they allow police to use their discretion to best determine how they should respond in IPV situations.

In addition to these policies used by police departments, another outgrowth of the movement to handle IPV incidents differently has been for officers to make dual arrests. In a dual arrest, both the offender and the victim are arrested. This practice of arresting both the victim and the offender is tied to the belief that many offenders are also victims and that IPV is part of mutual, common-couple violence. When officers arrive on the scene and there is evidence that both parties engaged in violence, they can make a dual arrest. Many states have guidelines as to how and when dual arrests can be made so that victims are not erroneously arrested. Some states have policies that require the officer to arrest only the aggressor; others require that police officers provide written justification for making a dual arrest (Hirschel, Buzawa, Pattavina, Faggiani, & Reuland, 2007). Although dual arrest does not always occur, a large-scale study of dual arrest using National Incident-Based Reporting System data found that it does appear to be more common in homosexual intimate partner incidents and when the offender is female (Hirschel, 2008). As you might imagine, arresting a victim may create a situation in which the victim is afraid to call for assistance if needed in the future for fear of being arrested again. Moreover, victims may not want to participate in the prosecution of the offender if they were also arrested (Bui, 2001).

Court Responses

Law enforcement officers are not the only ones who have an important job in responding to and dealing with incidents of IPV. After an incident has been reported to and investigated by the police, the prosecutor is then given the police file to determine whether formal charges will be brought against the offender. Early criticisms of the prosecutor's job as it relates to IPV centered on the belief and research showing that prosecutors rarely brought formal charges against IPV offenders, even after arrest (Chalk & King, 1998; Hartman & Belknap, 2003). Some have concluded that there is a "widespread underprosecution of domestic violence cases" (Sherman, 1992, p. 244). The current rate of prosecution for these types of cases is in dispute, but a recent review of more than 135 studies in more than 170 jurisdictions found that on average about one third of reported offenses and more than 60% of arrests resulted in the filing of charges by the prosecutor. In addition, about one third of the arrests and more than half the prosecutions resulted in the defendant being found guilty (Garner & Maxwell, 2009). The authors noted that there was a large amount of variation across jurisdictions in terms of prosecution and conviction rates, which implies that making blanket statements regarding prosecution and conviction of all IPV cases is troublesome. Additional research is also clearly needed in this area to help resolve the conflicting viewpoints about prosecution and conviction rates.

What factors do prosecutors consider when deciding whether to prosecute a person for IPV? Research shows that prosecutors are most likely to move forward with prosecution when the victim has been visibly physically injured (Jordan, 2004). Other research has shown that prosecutors consider whether the victim or the defendant had been drinking or under the influence of drugs at the time of the incident (Jordan, 2004; Rauma, 1984; Schmidt & Steury, 1989) and the history between the victim and the offender (Schmidt & Steury, 1989). Finally, victim willingness to participate in the process has been shown to be related to prosecutorial decision making in IPV cases (Dawson & Dinovitzer, 2001).

Let's discuss victim participation and how it impacts prosecution. As it may seem, one reason that prosecution and subsequent conviction may be difficult to achieve is lack of participation on the part of the victim. The victim may not want to testify against the offender. He or she may still be involved with the offender. They may share children. The victim may be fearful of what will happen if he or she testifies in court. The victim may have calculated that he or she cannot "afford" to move forward in the criminal justice process. Whatever the reason, some jurisdictions have **no-drop prosecution** policies—the victim is not able to drop the charges against the offender and the prosecutor's discretion in deciding to charge is curtailed. After adoption of such policies, rates of dismissal have dropped (Davis, Smith, & Davies, 2001). Despite this seemingly positive result of no-drop prosecution policies, there is concern that they may be harmful to victims in that the offender may retaliate against the victim and forced participation may be victimizing. Also concerning is that victims may be reluctant to call the police if they are aware of the no-drop policy. In some states, victims may not have to testify against their abusers if they are legally married to them. The laws that provide this exception are known as **spousal or marital privilege laws**. In states that have these laws, the number of times the option not to testify against the abuser can be used varies—in Maryland, it can be used only once (Bune, 2007).

More recently, the court system has responded to IPV by adopting special courts that handle domestic violence cases. Designed to increase coordination among criminal justice and social service agencies, these courts hold criminals accountable and consider the needs of victims (Gover, MacDonald, & Alpert, 2003). These courts often have a therapeutic focus, which allows for the offender to get help identifying the causes of his or her abusive behavior and controlling this behavior in the future (Gover et al., 2003). Although evaluations of domestic violence courts are not widespread, evidence suggests that participating in the court and its associated programs generally reduced recidivism against the same victim (Goldkamp, Weiland, Collins, & White, 1996) and rearrests (Gover et al., 2003).

Legal and Community Responses

The criminal justice system is not the only entity that addresses the needs of victims of IPV. Legal and community services also have been developed to assist victims.

Protective Orders

One way victims can seek protection from their abusers is by obtaining a **protective order** or a restraining order. Both types of order are designed so that victims will be protected from their offenders. Victims of IPV may get an order in hopes of ensuring their protection. These orders may include stay-away orders, no-contact orders, or peaceful contact orders. Some states differentiate between protective and restraining orders, with the criminal court issuing protective orders and the civil court issuing restraining orders. This is not the arena to attempt to differentiate how every state operates in this area, but commonalities will be discussed.

Once an offender is arrested, the judge in a criminal court may issue a no-contact order at the offender's arraignment or first court appearance. Some states require the issuance of a no-contact order in IPV cases. The order stays in effect for a specified amount of time (typically 2 weeks) or until it is amended by the court, sometimes at the request of the victim and agreement of the prosecutor (Hartman & Alligood, 2010). If a victim would like to acquire a restraining order against the offender, the victim is generally required to go before the court and ask for a temporary order in an "emergency hearing." The judge determines whether an order should be issued. If the victim would like to extend the order, then a second, more comprehensive hearing is conducted (Hartman & Alligood, 2010). Even though a defendant (the offender) may be unaware of a hearing for a temporary order, the offender is notified if the victim requests a more permanent order. Depending on the state, if the order is granted, it stays in place for the length of time set by statute. For example, in the California Superior Court, a civil restraining order in IPV situations can stay in place for up to 5 years.

Similar to the disparate procedures and orders states have in place, enforcement of protective orders and restraining orders varies from state to state. Despite this, states are mandated under the 1994 Violence Against Women Act to receive full faith and credit from courts for orders granted in other states. Most states have also passed their own full faith and credit laws that require enforcement of other states' protective orders as though they were enacted in that state. So what do states do when a person violates a protective order? Most states have criminal sanctions that can be implemented when an offender violates an order. A person who violates an order may be charged with a felony, a misdemeanor, or contempt of court (Office of Justice Programs, 2002). Some states have laws that make a violation a new offense, and some require that anyone found guilty of violating a protective order serve time in confinement (Office of Justice Programs, 2002). In other states, a violator may face bail forfeiture or revocation of bail, pretrial release, or probation (Office of Justice Programs, 2002).

Although protective orders are available in all states for the protection of IPV victims, most eligible victims do not secure protective orders. Research from the NVAWS shows that 17.1% of female victims of assault perpetrated by intimate partners received protective orders and 36.6% of female victims of intimate partner-perpetrated stalking received protective orders (Tjaden & Thoennes, 2000). Women who get orders of protection may be different from those who do not—they are more likely to be employed full-time, to have been injured by the abuse, to have experienced sexual coercion by the offender, and to be severely depressed, with more serious symptoms of mental health problems (Wolf, Holt, Kernic, & Rivara, 2000). The reading by Karla Fischer and Mary Rose (1995) included in this section further identifies the reasons why women decide to get protective orders.

Also important to consider is the effectiveness of protective orders. A body of research shows that protective orders reduce recidivism. Even so, these studies show that orders are violated between 20% and 40% of the time (Jordan, 2004). Victims also report a variety of feelings regarding protective orders, with some feeling satisfied and having a sense of security after receiving a protective order (Keilitz, Hannaford, & Efkeman, 1997). Others report being frustrated with the timely and confusing process they had to go through in order to secure a protective order (Ptacek, 1999). These feelings may be tied to the fact that victims often have to be the ones to notify the police if a protective order is violated—meaning that decision making and enforcement is largely up to victims. This, quite understandably, may be a role that few victims relish.

Domestic Violence Shelters

Another resource available to victims of IPV is domestic violence shelters. First opened for women in 1974 in Minnesota, domestic violence shelters offer a place of refuge for victims. They provide a range of services, including short-term room and board, emergency clothing and transportation, counseling, assistance with the legal process, 24-hour crisis lines, programs for children, and help with seeking employment. Most domestic violence shelters allow residents to stay more than 30 days in residence, and about one third allow stays of 60 days or more (Lyon & Lane, 2009). It appears that domestic violence shelters are indeed meeting the needs of victims. In a recent study, it was found that almost three fourths of domestic violence victims reported that the assistance they received from shelters was very helpful (Lyon, Lane, & Menard, 2008).

The services shelters offer may be lacking when it comes to some victims. Lesbian women are more likely to believe that shelters are for heterosexual women and to have negative experiences with shelters than are other women (Lyon et al., 2008). Men also are unlikely to seek assistance from domestic violence shelters as compared with women; shelters were traditionally designed to service only female victims, although recent efforts to assist male victims have been made. The first domestic violence shelter exclusively for men opened in Minnesota in 1993. This shelter housed 50 men within the first 6 months of opening (Cose, 1994). Minority women may also be reluctant to utilize domestic violence shelters. They may perceive that their particular needs are not met, especially if the staff is largely White (Lyon et al., 2008). More generally, victims may experience problems with other residents, and the lack of privacy may be difficult (Lyon & Lane, 2009).

Health Care

Protective orders and domestic violence shelters are common resources for victims of IPV, but health care professionals are also in a unique position to help. Recall that IPV is one of the major reasons that women seek medical assistance from ERs and is a key reason why women become the victims of homicide. Because of the injuries caused by IPV and the fact that health care professionals see women on a regular basis, they are in a position to help women who are victimized, even if the violence is not the reason the woman is seeking medical treatment. As such, many health care organizations recommend that patients be screened for IPV (Waalen, Goodwin, Spitz, Petersen, & Saltzman, 2000). Screening can allow for victims to be identified and then subsequently referred to services. Screening typically involves the medical professional asking questions either from a screening instrument or other instrument designed to assess whether a patient is a victim of IPV. Despite this innovative approach, screening rates remain low. One study found that only 13% of victims of acute IPV were asked by a doctor or nurse about violence, even though they had presented at the emergency department (Krasnoff & Moscati, 2002).

Can you think of ways in which improved screening by health care professionals may be achieved? This holistic approach to IPV intervention appears to be a critical link, so improving health care screening is needed, particularly for women. It may be a way for victimizations that often remain hidden to be revealed in a safe, controlled environment. It may be the step victims need to get assistance or referrals to resources that can help them stop the cycle of violence.

SUMMARY

- There are two major types of intimate partner violence (IPV)—intimate terrorism and situational or common couple violence. Within these types, IPV can be physical, emotional/psychological, or sexual.
- Since IPV often occurs in private, it is somewhat difficult to measure. That being the case, it is hard to know the true extent to which people are victimized by intimate partners.
- Three commonly used resources on IPV statistics are the National Crime Victimization Survey (NCVS), the Conflict Tactics Scale (CTS), and the National Violence Against Women Survey (NVAWS). Depending on the resource, estimates of IPV vary. Both the NCVS and NVAW indicate that more females than males are victimized. Research using the CTS shows that gender symmetry may exist.
- Young people, Black women, and those with lower socioeconomic status are the most likely to be victimized.
- IPV can also occur in same-sex relationships. By some estimates, homosexual men experience higher rates of IPV victimization than do homosexual women or heterosexual men.
- Stalking is a special type of victimization. Stalking is a course of conduct that is unwanted and harassing and that would cause a reasonable person to be fearful. The three most common types of stalking are unwanted phone calls or messages, unwanted letters or e-mails, or stalking using new technology such as e-mail or instant messaging. The main fear with stalking is the uncertainty of what will come next.
- Female victims of IPV tend to experience more severe and more frequent violence than do male victims.
- Stress, cohabitation, power and patriarchy, social learning, risky lifestyles, and alcohol use and/or drug use are all risk factors for IPV victimization.
- In addition to injury and death, victims of IPV may also experience psychological and emotional consequences. Posttraumatic stress disorder, depression, suicide ideation, suicide attempts, suicide completion, anxiety, and sleep disorders all have been linked to IPV victimization.

- Revictimization can also occur. Court reports, victim surveys, and official records all demonstrate that victims of IPV are at risk of experiencing additional IPV incidents.
- Police historically have been reluctant to deal with IPV cases. The Minneapolis Domestic Violence Experiment changed how police respond to IPV cases. Results from the study showed that arresting offenders reduced recidivism, although replication studies did not find this clear result. Instead, the replication studies showed that arrest impacted offenders differentially, with those who were employed being deterred but others not.
- Courts have responded to the problem of IPV by instituting no-drop prosecution policies and adopting domestic violence courts.
- Processes of obtaining a protective order vary from state to state, but in all states, protective orders against offenders can be obtained by victims of IPV.
- Domestic violence shelters also serve victims of IPV by providing safe and secure living arrangements. These shelters also link victims with other social services to help them transition out of abusive relationships.
- Health care professionals also may be able to help victims of IPV. Screening instruments have been developed to help health care officials identify and refer patients who have been victimized.

DISCUSSION QUESTIONS

1. On what side of the gender symmetry debate do you fall? When females are offenders, do you think it is largely in response to male aggression? When female victims do not apply for protective orders, do you think it encourages recurring IPV? Why or why not?

2. What measures should the criminal justice system, social service agencies, and the health care community take to reduce IPV victimization?

3. Given the fact that females traditionally have been considered the "true" victims of IPV, how do you think this affects the criminal justice system's response when males report crime, especially in instances where there are few or no visible physical injuries?

4. What are the advantages and disadvantages of mandatory arrest policies for IPV?

KEY TERMS

intimate partner violence

physical violence

emotional abuse

sexual violence

intimate terrorism

situational couple violence

National Crime Victimization Survey (NCVS)

Conflict Tactics Scale (CTS)

National Violence Against Women Survey (NVAWS)

cohabitation

power

patriarchy

social learning

risky lifestyles

injury

revictimization

cycle of violence

tension building phase

acute battering phase

honeymoon phase

misdemeanor

Minneapolis Domestic Violence Experiment

mandatory arrest policies

pro-arrest policies

presumptive arrest policies

permissive arrest policies

no-drop prosecution

spousal or marital privilege laws

protective order

screening

INTERNET RESOURCES

Office on Violence Against Women (http://www.ovw.usdoj.gov/index.html)

Housed in the U.S. Department of Justice, the Office on Violence Against Women administers both financial and technical assistance to help communities develop programs, policies, and practices with the goal of ending domestic violence, dating violence, sexual assault, and stalking. On its website, you can find information on these types of violence against women, information about grant opportunities and funded projects, and help for persons who have been victimized. It also provides up-to-date information about federal legislation and policies directed at violence against women.

"Teen Dating Violence" (http://www.cdc.gov/violenceprevention/intimatepartnerviolence/teen_dating_violence.html)

Recently, research on intimate partner violence has uncovered that teens are at risk of experiencing violence within their dating relationships. For information on teen dating violence, including what it is, facts about teen dating violence, why it occurs, and its consequences, visit the Centers for Disease Control webpage on teen dating violence. Also included are links to various resources for more information on and assistance with teen dating violence.

WomensLaw.org (http://www.womenslaw.org/)

This website is a resource for persons who are experiencing or who have experienced intimate partner violence. It includes links and information about state and federal laws addressing intimate partner violence as well as how victims can better navigate the criminal justice system. People can find information on how to apply for protective orders, how to prepare for court, and how to find assistance safely in their communities.

Menweb (http://www.menweb.org/battered/)

Men can also be the victims of intimate partner violence. One Internet resource for men is Menweb, which provides information for men who have experienced intimate partner violence. It also includes links to news items and research that highlight intimate partner violence perpetrated by women against men. Men's stories of victimization are also presented.

Introduction to Reading 1

Before Johnson's (2006) article, most attention given to intimate partner violence focused on severe battering of women. Johnson proposes that there are essentially four different types of intimate partner violence—intimate terrorism, violent resistance, situational couple violence, and mutual violent control. In this provocative piece, he discusses gender symmetry and asymmetry in the perpetration of intimate partner violence and notes how the different types of intimate partner violence are based on the dyadic control context of the violence—who is controlling and who is violent. He finds that intimate terrorism is perpetrated largely by males within the context of a desire of control and that violent resistance is largely the domain of females. He further notes how sampling differences contribute to the extent and nature of estimates of intimate partner violence found in the literature.

Conflict and Control

Gender Symmetry and Asymmetry in Domestic Violence

Michael P. Johnson

The central argument of this article is that there are four major types of intimate partner violence and the failure to distinguish among them has left us with a domestic violence literature that, to a large extent, may be uninterpretable. The types of domestic violence (situational couple violence, intimate terrorism, violent resistance, and mutual violent control) are defined conceptually in terms of the control motives of the violent member(s) of the couple, motives that are identified operationally by patterns of controlling behavior that indicate an attempt to exercise general control over one's partner. With respect to implications for the question of gender symmetry, these types of domestic violence differ dramatically. In heterosexual relationships, intimate terrorism is perpetrated almost exclusively by men, whereas violent resistance is found almost exclusively among women. The other two types are gender symmetric. With respect to

the general importance of distinguishing among types of violence, I believe that they have different causes, different patterns of development, different consequences, and that they require different forms of intervention.

Resolving the Gender Symmetry Debate

The long-standing argument in the family literature regarding the gender symmetry of intimate partner violence takes the form of a disagreement about the nature of heterosexual intimate partner violence, as if heterosexual partner violence were a single phenomenon. One side of the debate, generally referred to as the feminist perspective (Kurz, 1989), presents compelling empirical evidence that heterosexual intimate partner violence is largely a problem of men assaulting female

SOURCE: Johnson (2006). Reprinted with permission.

partners (Dobash, Dobash, Wilson, & Daly, 1992). The other side, generally taken in the family violence perspective, presents equally compelling empirical evidence that women are at least as violent as men in such relationships (Straus, 1999). How can they both be right?

In 1995, I published an article that argued that the answer to this question is that: (a) partner violence is not a unitary phenomenon, (b) the two groups of researchers generally use different sampling strategies, (c) the different sampling strategies tap different types of partner violence, and (d) these types differ in their relationship to gender (Johnson, 1995). I argued further that the types probably also differ with respect to their causes, the nature of the violence itself, the development of the violence during the course of a relationship, its consequences, and the type of intervention required. If these arguments are correct, then it follows that we cannot draw any conclusions about the nature of partner violence from studies that do not distinguish among types of partner violence (Johnson & Ferraro, 2000). Nevertheless, studies continue to be published regularly that treat partner violence as a unitary phenomenon, many of them claiming to provide further evidence on the gender symmetry issue. For example, Archer's (2000) influential meta-analysis of the evidence regarding gender symmetry, in spite of citing my 1995 article, essentially ignored the proposed distinctions among types of violence and concluded that women are slightly more violent than men in heterosexual partnerships.

Here is the basic argument of the 1995 article. With regard to sampling, I argued that general differences in sampling strategies were the major source of the ostensible inconsistencies between the feminist and family violence data. In general, the studies that demonstrated the predominance of male violence used agency data (courts, police agencies, hospitals, and shelters), whereas the studies that showed gender symmetry involved so-called representative samples. I argued that both of these sampling strategies are heavily biased: the former through its use of biased sampling frames (agencies), the latter through refusals. Although the biases of agency sampling frames have generally been taken to be obvious (Straus, 1990b),

representative sample surveys have mistakenly been assumed to be unbiased. The final samples of so-called random sample surveys are, of course, not random, due to refusals. I estimated, for example, that the refusal rate in the National Family Violence Surveys was approximately 40% rather than the 18% usually reported (Johnson, 1995). Could there be two qualitatively different forms of partner violence, one gender symmetric and overselected in general surveys, the other committed primarily by men and overselected in agency samples?

To address this question, I identified a number of agency-based studies that used the Conflict Tactics Scales (CTS; Johnson, 1995) to assess the nature of partner violence in agency samples and compared them with general surveys that used the same instrument.[1] My conclusion from this comparison was that the two sampling strategies identified partner violence that differed not only in gender symmetry but also in frequency of per-couple incidents, escalation, severity of injuries, and mutuality. Agency samples identified partner violence that was more frequent, more likely to escalate, more severe, less likely to be mutual, and perpetrated almost entirely by men. This gender-asymmetric pattern resonated for me with feminist analyses of partner violence as one tactic in a general pattern of controlling behaviors used by some men to exercise general control over "their" women (Dobash & Dobash, 1979; Pence & Paymar, 1993; Stark & Flitcraft, 1996).

The asymmetry of such control contrasted dramatically, it seemed to me, with the family violence perspective's predominantly symmetric image of partner violence as a matter of conflict. I hypothesized that there were two qualitatively different forms and/or patterns of intimate partner violence—one that was part of a general strategy of power and control (intimate terrorism), the other involving violence that was not part of a general pattern of control, probably a product of the escalation of couple conflict into violence (situational couple violence).[2] Furthermore, I argued that, on one hand, couples involved in situational couple violence would be unlikely to become agency clients because such situationally provoked violence would not in most cases call for police intervention, emergency room visits, protection from abuse orders, or divorce. On the other hand, couples involved in intimate terrorism would be unlikely to agree to participate in general surveys because victims fear reprisals from the batterer, and batterers fear exposing themselves to intervention by the police or other agencies. Although these arguments seemed reasonable enough, my 1995 literature review provided no direct evidence of their validity because none of the studies reviewed made such distinctions among types of violence.

◪ Distinguishing Among Types of Heterosexual Intimate Partner Violence

First, I want to discuss one approach to assessing differences in the extent to which an individual's violence toward his or her partner is embedded in a general context of control, then I will move on to the implications of taking into account the behavior of the partner. The basic idea is to look at a variety of nonviolent, controlling behaviors to identify individuals who behave in a manner that indicates a general motive to control. Note that (a) this moves the focus from the nature of any one encounter between the partners to a search for patterns of behavior in the relationship as a whole, and (b) it most certainly is not based in the nature of the violent acts themselves. Some critics have argued that I am simply making the old distinction between more serious and less serious violence. I disagree; the distinction lies in the degree of control present. I assume that there is considerable variability in the nature of the violent acts involved in controlling and noncontrolling violence, a variability that would lead to considerable overlap between them in terms of the "seriousness" of the violence. Although I do have some hypotheses about average differences among types of violence in terms of the nature of the violent acts involved, the types themselves are defined by the degree of control, not by characteristics of the violence. Thus, if we want to make these distinctions, surveys need to ask questions not just about violence but

also about the use of a variety of other control tactics in the relationship.

Before I proceed to illustrate this general measurement strategy with data from Frieze's 1970s Pittsburgh study, however, I need to move the 1995 discussion from the individual to the dyadic level. This initial distinction between *intimate terrorism* and *situational couple violence* was focused too narrowly on the behavior of one violent partner. If one considers the behavior of both people in the relationship, one can identify four basic types of individual violence. First, an individual can be violent and noncontrolling and in a relationship with a partner who is either nonviolent or who is also violent and noncontrolling. This is what I called *situational couple violence*. Second, one can be violent and noncontrolling but in a relationship with a violent and controlling partner. Given that the behavior of the partner suggests an attempt to exert general control, I labeled this type of violence *violent resistance*. Third, one can be violent and controlling and in a relationship with a partner who is either nonviolent or violent and noncontrolling. This is the pattern I have called *intimate terrorism*. Finally, a violent and controlling individual may be paired up with another violent and controlling partner. I have labeled this *mutual violent control*. To make these distinctions, one would have to ask questions regarding a variety of control tactics in addition to violence, ask them with regard to both partners, and do so in a data set that was likely to include representatives of each of the four types.

I was able to find one such data set, Irene Frieze's data from interviews with 274 married and formerly-married women living in southwestern Pennsylvania in the late 1970s (Frieze, 1983; Frieze & Browne, 1989; Frieze & McHugh, 1992). Her mixed sampling design seemed likely to represent the major types of violence—it included women selected from shelters and courts (an agency sample) and a matched sample of women who lived in the same neighborhoods (a general survey sample).

It might be useful to know some of the general characteristics of the sample, keeping in mind that these will be determined to a large extent by the biases of the initial court and shelter populations because the "general" sample was matched by neighborhood and is, therefore, likely to be similar in general demographics. The sample was predominantly White (86% White, 14% Black) and working class (56% had a high school education or less; only 21% had finished college). About one third of the women worked full-time and about one half were full-time homemakers. Median age at the time of the interview was 32 years; however, ages ranged from 18 to 83 years. Median age at marriage was 21 years but ranged from 15 to 59 years. As this is clearly not a representative sample, it would not be useful to make any general statements about the prevalence of violence; of course, all of the women in the court and shelter samples had experienced violence from their partners, as had 34% of the general sample. Only one third of the women in the court and shelter samples were still with their husbands, compared with about three fourths of those in the general sample.

From the lengthy interview protocols, I identified a number of items tapping control tactics that did not involve violence toward one's partner. Seven measures were created to tap control tactics analogous to those identified by Pence and Paymar (1993): threats, economic control, use of privilege and punishment, using children, isolation, emotional abuse, and sexual control.

Threats

Each measure of threats (one for husbands, the other for wives) is the mean of two items with 5-point response formats ranging from *no, never* (1) to *often* (5). The first item is: "Has your husband (Have you) ever gotten angry and *threatened* [emphasis in survey instrument] to use physical force with you (him)?" The second item was: "Is he (Are you) ever violent in other ways (such as throwing objects)?" For wives' reports of their husband's behavior, the mean of this variable is 2.72 (between "once" and "two or three times"), the standard deviation 1.51, and the range from 1.00 to 5.00. Cronbach's alpha for the two-item scale is .74. For wives' reports of their own behavior, the mean is 1.99 ("once"), the standard deviation 1.05, the range is from 1.00 to 5.00, and alpha is .46.

Economic Control

Economic control is the average of two dichotomized items. The first asks, "Who decides how the family money will be spent in terms of major expenses?" It was dichotomized with a high score indication that either "husband (wife) makes entire decision" or "husband (wife) has deciding vote." The second item asked for an open ended response to, "How much money do you (does your husband) have to spend during an average week without accounting to anyone?" The dichotomization cut point was chosen to make this second item more an indicator of control than of disposable income: A response of US$10 or less indicated high control, one of more than $10 indicated low control. For husbands' economic control, the two-item scale has a mean of 1.36, a standard deviation of .39, ranges from 1.00 to 2.00, and has an alpha of .46. For wives, the mean is 1.20, the standard deviation .27, the range from 1.00 to 2.00, and alpha is .12.

Use of Privilege and Punishment

This scale is the mean of six items, each of which indicates that the target person uses one of the following tactics to get his or her spouse to do what he or she wants.[3] The six items were: (a) "suggests that you should do something because he knows best or because he feels he is an expert at a particular thing," (b) "stops having sex with you," (c) "threatens to leave you," (d) "emotionally withdraws," (e) "suggest[s] that you should do something because other people do," and (f) "restricts your freedom." The response format for all items addresses frequency, ranging from *never* (1) to *rarely* (3) to *always* (5). For husbands, the scale has a mean of 2.03 ("rarely"), a standard deviation of .81, ranges from 1.00 to 4.83, and has an alpha of .76. For wives, the mean is 1.92 ("rarely"), the standard deviation .62, the range is from 1.00 to 4.19, and alpha is .65.

Using Children

There are three items in this data set that get at a spouse's use of the children to get his or her way with his or her partner. Two of them involve responses to the question, "When your husband is angry with you, how does he show it?" The two relevant response options were "Directs his anger to the children or pets" and "Uses physical violence with the children." The third item is, "Does he ever try to get what he wants by doing any of the following to you? How often?" One of the actions listed is, "uses physical force against the kids to get what he wants from you," with the five response options ranging from *never* to *always*. This item was dichotomized between *never* and *rarely,* and the three items were averages. For wives' reports of their husbands' behavior, the mean was 1.19, the standard deviation .30, the range from 1.00 to 2.00, and alpha equal to .68. For wives' reports of their own behavior, the mean was 1.12, the standard deviation .21, the range from 1.00 to 2.00, and alpha equal to .41.

Isolation

The measure of isolation is the mean of two items with 5-point response formats ranging from *never* to *always.* The items are: "Does your husband know where you are when you are not together?" and "Are there places you might like to go but don't because you feel your husband wouldn't want you to—How often does this happen?" For wives' reports of their husbands' behavior, the mean of this measure is 3.32 (between *sometimes* and *usually*), the standard deviation is .77, the observed range is from 1.00 to 5.00, and alpha is equal to .09. For wives' reports of their own behavior, the mean is 2.64 (between *rarely* and *sometimes*), the standard deviation is .84, the range from 1.00 to 5.00, and alpha equals .06.

Emotional Abuse

The three-item emotional abuse scale includes one item that gets at active abuse (sex is sometimes unpleasant because "He compares you unfavorably to other women"), and two "passive abuse" items that indicate that he never or rarely praises, and never or rarely is "nice to you in other ways (smiling, concerned with how you are feeling, calling you affectionate names, etc.)." All three items are dichotomies. For husbands, the mean is 1.25, the standard deviation is .33, the scale ranges from 1.00 to 2.00, and alpha is .57.

For wives, the mean is 1.08, the standard deviation is .21, the range is from 1.00 to 2.00, and alpha is .48.

Sexual Control

There are two items in the sexual control scale, tapping whether sex is ever unpleasant because "he forces me to have sex when I don't want to," or "he makes you do things you don't want to do." Both items are dichotomies. For husbands, the mean is 1.22, the standard deviation .36, the range is 1.00 to 2.00, and alpha is .70. For wives, the mean is 1.02, the standard deviation .01, the range is 1.00 to 2.00, and alpha is .35.

⧗ Types of Intimate Partner Violence

Each of these measures was standardized and entered into a cluster analysis. The clustering algorithm was Ward's method, an agglomerative approach that selects each new case to add to a cluster based on its effect on the overall homogeneity of the cluster, and which therefore tends to produce tightly defined clusters, rather than strings (Aldenderfer & Blashfield, 1984, pp. 43–45). Euclidean distance was the measure of dissimilarity. The index of dissimilarity for one-cluster through 15-cluster solutions was examined.

The results indicated a two-cluster solution as optimal, and the pattern was quite simple, with one cluster exhibiting a high average on all seven of the control tactics, the other being relatively low on all seven. The results of the cluster analysis were then used to distinguish controlling from noncontrolling violence at the individual level, that is, for each husband and wife. Finally, data regarding both partners were used to distinguish among types of individual violence, as defined above.

Table 1 presents data only on violent individuals, that is, those who had been violent in their relationship at least once (67% of the husbands and 54% of the wives). The question tapping violence was: "Has he (Have you) ever actually slapped or pushed you (him) or used other physical force with you (him)?" The table places individual violence within its dyadic control context, distinguishing among four types of individual violent behavior.[4] The first row, "intimate terrorism," refers to relationships in which only one of the spouses is violent and controlling. The other spouse is either nonviolent or has used violence but is not controlling. In this data set, that violent and controlling spouse is the husband in 94 of the 97 cases. The second row refers to cases in which the focal spouse is violent but not controlling, and his or her partner is violent and controlling. I call it *violent resistance,* and it is almost entirely a woman's type of violence in this sample of heterosexual relationships. Of course, that is because in these marriages almost all of the intimate terrorism is perpetrated by men, and in some cases the wives do respond with violence, although rarely are they also controlling (see mutual violent control below). In the third row, we have "situational couple violence," individual noncontrolling violence in a dyadic context in which neither of the spouses is violent and controlling. It is close to gender symmetric, at least by the crude criterion of prevalence. (Data on the frequency and consequences of men's and women's situational couple violence, not shown, indicate that by other criteria men are more violent than women even within situational couple violence.) The last row, "mutual violent control," refers to controlling violence in a relationship in which both spouses are violent and controlling. There were only five such couples (10 individuals) in this data set, and among heterosexual couples such violence is by definition gender symmetric.

Now, let's take a quick look at the characteristics of the violent acts involved in men's intimate terrorism and situational couple violence, to see if they fall in line with what I found in my 1995 literature review. The violence in men's intimate terrorism was quite frequent; the median number of incidents in these marriages is 18.[5] Violence was much less frequent in men's situational couple violence with a median of three violent events. Intimate terrorism was reported to escalate in 76% of the cases, situational couple violence in only 28%.[6] In terms of injuries, the violence of intimate terrorism was sometimes severe in 76% of the cases, compared with 28% of the cases of situational couple violence.[7] In men's intimate terrorism, wives rarely respond with violence; the median of the ratio of the number of times the wife had been violent to the number of times the husband had been violent is .17. In men's situational couple violence, the median ratio was .40. These data do not leave much doubt that intimate terrorism and situational couple violence are not the same phenomenon.[8]

Finally, let me nail down my sampling argument, that general survey samples tap primarily situational couple violence, whereas agency samples give access primarily to intimate terrorism. Again, looking only at male violence, Table 2 shows that the general sample

Table 1   Individual Violent Behavior by Gender (Violent Individuals Only, as Reported by Wives, Percentages)

	Husbands	Wives	n
Intimate terrorism	97	3	97
Violent resistance	4	96	77
Situational couple violence	56	44	146
Mutual violent control	50	50	10

Table 2   Who Is Finding Whom? (Violent Husbands Only, as Reported by Wives, Percentages)

	Survey Sample (n = 37)	Court Sample (n = 34)	Shelter Sample (n = 43)
Mutual violent control	0	3	0
Intimate terrorism	11	68	79
Violent resistance	0	0	2
Situational couple violence	89	29	19

NOTE: For simplicity, Table 2 presents data only on husbands. For wives, the general survey sample is dominated by situational couple violence, and the court and shelter samples are dominated by violent resistance.

typical of family violence research includes almost nothing but situational couple violence, with only 11% of the violence being intimate terrorism. In stark contrast, the court and shelter samples that are typical of feminist research get at violence that is predominantly intimate terrorism.

What kind of mistakes can this generate? The article on the "battered husband syndrome" that initiated the gender symmetry debate (Steinmetz, 1977) is an excellent example of the danger of thinking that a general sample provides information about what we conventionally mean by "domestic violence," that is, intimate terrorism. Steinmetz told anecdotes about true husband battering, involving violent women who are intimate terrorists in the sense described above. However, she then cited general survey data, which we now know represent primarily situational couple violence, to make her case that such battering is as frequent a problem as is wife battering. The data she presented, in fact, have nothing to do with husband battering. Serious as husband battering may be in each particular case, as a general phenomenon it is dramatically less frequent than wife battering.

Similarly, one can err by assuming that the patterns observed in agency samples describe all partner violence. It is common for shelter workers to argue in educational programs that violence always escalates, that if he hit you once he'll do it again, and that it will get worse. Such a pattern is, as shown above, much more true of intimate terrorism than of situational couple violence, and of course intimate terrorism is what we generally see in our shelter work. However, most couples who experience violence, including those in our audiences, are involved in situational couple violence.[9] For those audience members, we are providing an inaccurate picture of the likely course of their relationships.

Dramatic as these sampling biases are, Table 2 also shows that neither type of male violence is found exclusively in one type of sample, implying that it is possible to study the differences between these two types of violence in a variety of research settings. First, the 11% of men's violence that is intimate terrorism in the general sample indicates that with large enough samples, it may be possible to study situational couple violence and intimate terrorism with survey data. To do that, of course, we need to include questions that will allow us to distinguish one from the other. Second, because women do bring cases of situational couple violence to the courts (29%) and to shelters (19%), researchers in those contexts would be able to study the effects of various intervention strategies on the two types of violence. Again, however, it won't happen unless we gather information that will allow us to make these distinctions.

⊠ Implications for Measurement

The evidence I have just presented regarding the dramatic differences between intimate terrorism and situational couple violence (in terms of gender, per-couple frequency of incidents, escalation, severity of injury, and reciprocity) should serve as a warning that until further notice we have to assume that the answers to all of our important questions about domestic violence may be different for the different forms of violence. How do we collect and analyze data to allow us to make the necessary distinctions?

The general strategy for making such distinctions involves two steps. First, and most critically, we must gather information on a variety of control tactics for both members of the couple. Not all surveys do this. It was fortunate that Irene Frieze had the foresight to collect such data in the 1970s. I have been using a number of other data sets to try to operationalize these distinctions and have had to dance around major shortcomings. The recent National Violence Against Women Survey (Johnson & Leone, 2000; Tjaden & Thoennes, 2000) asks a number of control tactics questions about the attacker but gathers no information about the behavior of the respondent. Susan Lloyd's (Leone, Johnson, Cohan, & Lloyd, 2001; Lloyd & Taluc, 1999) data from a poor neighborhood in Chicago have the same shortcoming. The National Family Violence surveys (Straus & Gelles, 1990) ask about the behavior of both partners but do not assess a variety of control tactics. We need to insist that every study of partner violence asks questions about the control tactics and the violence used by both partners.

However, what questions to ask? Whatever the drawbacks of the Conflict Tactics Scales (CTS) may be (Dobash et al., 1992), they have made the considerable contribution of having provided a standard approach to assessing violence. As yet, there appears to be no standard approach to the assessment of control, although a number of scholars have developed scales that are reasonable candidates, including Marshall (1996) on psychological abuse, Stets (1993) on control, Tolman (1989) on psychological maltreatment, Pence and Paymar (1993) on power and control, Straus and his colleagues' (Straus, Hamby, Boney-McCoy, & Sugarman, 1996; Straus, Hamby, Boney-McCoy, & Sugarman, 1999; Straus & Mouradian, 1999) revised Conflict Tactics Scales and Personal and Relationships Profile, and most recently, Cook and Goodman (this issue [2006]) in their Coercion and Conflict in Abuse Experience Scale. It would be very helpful if some consensus could be developed on a standard set of control measures.

When we have the data, we face the task of using it to create a typology of controlling and noncontrolling violence. In my early work, I approached this problem through the use of cluster analysis; one could similarly use latent class analysis for dichotomous items. Cluster analysis and class analysis identify clusters of individuals who have similar profiles of responses to a set of items, and they provide statistical criteria for choosing the optimal number of groups (clusters or classes) for describing a particular data set. The problem is that they do not provide clear-cut criteria for distinguishing the members of one group from those in others, and there is therefore no straightforward way of comparing the results of cluster analyses from different samples. Thus, in more recent work, I have simply created a control scale and dichotomized it to identify high- and low-control individuals. I have found that in the three secondary data sets with which I have been working there are fairly strong relationships between the cluster solutions and a simple scale score created out of the same control items. This approach has the advantage of providing clear criteria for placement into the four

types of violence so that comparisons can be made across studies. In a series of analyses of available data sets, my colleagues and I have shown that the violence types are related as expected to characteristics of the actual violence (frequency, escalation, severity, and mutuality) and to various physical and psychological effects on the victim (posttraumatic stress symptoms, general healthy, injury, interference with work) and on the relationship (Johnson, 1999; Johnson, Conklin, & Menon, 2002; Johnson & Leone, 2005; Leone, Johnson, & Cohan, 2003; Leone, Johnson, Cohan, & Lloyd, 2004).

Whatever specific approach we choose, I believe it is critical that we reassess conclusions that have been drawn from a literature that has failed to make these critical distinctions. We have to be wary of findings that treat situational couple violence, intimate terrorism, violent resistance, and mutual violent control as a single phenomenon. Situational couple violence, intimate terrorism, violent resistance, and mutual violent control simply cannot have the same causes, developmental trajectory, consequences, or prognosis for effective intervention. If we want to understand partner violence, to intervene effectively in individual cases, or to make useful policy recommendations, we must make these distinctions in our research.

Postscript

Since the original version of this article was presented in 2000 (Johnson, 2000) at a National Institute of Justice workshop, considerable progress has been made in research on different types of intimate partner violence. Janel Leone and I, using the National Violence Against Women data, have shown that victims of intimate terrorism are attacked more frequently and experience violence that is less likely to stop. They are more likely to be injured, to exhibit more of the symptoms of posttraumatic stress syndrome, to use painkillers (perhaps also tranquilizers), and to miss work.

They have left their husbands more often and, when they do leave, are more likely to acquire their own residence (Johnson & Leone, 2005). Using data from a study of a poor neighborhood in Chicago, my colleagues and I found that victims of intimate terrorism reported poorer general health, a greater likelihood of visiting a doctor, more psychological distress, and a greater likelihood of receiving government assistance (Leone et al., 2004). Leone's dissertation (Leone et al., 2003), using data from the Chicago Woman's Health Risk Study, demonstrated that victims of intimate terrorism were more likely than victims of situational couple violence to seek formal help (e.g., police, medical, counseling), and that this association was mediated by other factors, including violence severity, injuries, and perceived social support. They were not more likely to seek help from family or friends. Alison Cares and I (Johnson & Cares, 2004), using Frieze's Pittsburgh data, have shown that childhood experiences of family violence are strongly related to adult perpetration of intimate terrorism, but not to situational couple violence.

Amy Holtzworth-Munroe and her colleagues (Holtzworth-Munroe, 2000, 2002; Holtzworth-Munroe, Meehan, Herron, Rehman, & Stuart, 2003; Holtzworth-Munroe, Rehman, & Herron, 2000) have continued their line of research that uses cluster analysis to distinguish among types of perpetrators. They generally found three clusters, one corresponding to situational couple violence (which they label *family only*), and two others that appear to correspond to two types of intimate terrorists (*borderline/dysphoric* and *generally violent/antisocial*). The clusters differ, among other things, in terms of the personality of the perpetrators, the pattern of violence, the perpetrators' attitudes toward women, and the general level of control exercised in their relationships (with the high control scores of borderline/dysphoric and generally violent/antisocial perpetrations supporting my assumption that these men are involved in intimate terrorism, whereas the family-only perpetrators are more likely involved in situational couple violence).

Graham-Kevan and Archer (2003a, 2003b) collected data on violence and control from British samples of women residing at Woman's Aid refuges and their partners, male and female students, men attending male treatment programs for domestic violence and their partners, and male prisoners and their partners. Their cluster analyses confirm the distinction between intimate terrorism and situational couple

violence and show support for my finding regarding differences in the frequency and severity of violence.

In spite of this progress, the gender symmetry debate continues, ignoring these demonstrations that there are different types of intimate partner violence and that data that do not distinguish among them are problematic at best. Papers continue to be published using general survey data to demonstrate that men and women are equally violent, and implying through their generalizations about *violence* or *domestic violence* or *intimate partner violence* that men and women are equally likely to be involved in intimate terrorism (e.g., Capaldi & Owen, 2001; Moffitt, Robins, & Caspi, 2001; Straus & Ramirez, 2002). Even recent reviews of the literature that debunk the myth of gender symmetry continue to focus on measurement issues as the major source of contradictory findings (Kimmel, 2002; Saunders, 2002). Frustrating as this is to me, all I can do is assume that we are in the midst of a gradual recognition that domestic violence is not a unitary phenomenon, and that we are close to a tipping point at which we will see a dramatic decline in the number of published studies that simply compare violent with nonviolent relationships without making distinctions among types of violence. The evidence is accumulating that I was right when I argued in the early 1990s that "If we want to understand partner violence, to intervene effectively in individual cases, or to make useful police recommendations, we must make these distinctions in our research" (Johnson, 1993, n.p.).

◿ Notes

1. Much of the gender debate in the literature has centered on the inadequacies of the Conflict Tactics Scales (CTS) as a means to measure intimate partner violence (Dobash, Dobash, Wilson, & Daly, 1992; Straus, 1990a). Comparing studies that all used the CTS eliminated this potential source of bias.

2. The terminology I have used has changed somewhat over the years, although the definitions have remained the same. The 1995 article referred to *patriarchal terrorism* and *common couple violence*. I soon abandoned the former term because it begged the question of men's and women's relative involvement in this form of controlling violence. It also implied that all such intimate terrorism was somehow rooted in patriarchal structures, traditions, or attitudes.

I still believe that most intimate terrorism is perpetrated by men in heterosexual relationships and that there are women intimate terrorists in heterosexual and same-sex relationships (for descriptions of intimate terrorism in lesbian relationships, see Renzetti, 1992). Furthermore, it is not clear that all intimate terrorism, even men's, is rooted in patriarchal ideas or structures. I later abandoned *common couple violence* in favor of *situational couple violence* because the former terminology implied to some readers that I felt that such violence was acceptable. I also prefer the new terminology because it more clearly identifies the roots of this violence in the situated escalation of conflict.

3. At this point, I will stop reporting alternative forms of the question, unless it seems necessary for clarity.

4. There has been some confusion regarding the role of frequency or severity of violence in the construction of this typology. Because this is a typology of violent behavior, all members of the four types have been violent toward their partner. However, the distinctions among the types are based entirely on control context, not frequency or severity of violence.

5. The interview question was, "Can you estimate how many times, in total, he was violent with you?"

6. The question was, "Did he become more violent over time?"

7. Severity of violence was assessed in a section of the interview dealing with "the time your husband was the most violent with you (him)." The question was, "How badly were you hurt?" It was an open-ended question with probes, coded into the following categories: (a) force, not hurt; (b) no physical injury; (c) simply injury; (d) severe, no trauma; (e) severe, some trauma; and (f) permanent injury.

8. Once again, I want to be clear that there is no tautology involved here. The types are defined in terms of control context, not the frequency or nature of the violence. The hypothesized differences in characteristics of the violence are derived from theory. It is assumed that attempts by husbands to exert general control over their wives will be met by considerable resistance in the United States, a cultural context in which marriage is seen by most women as an egalitarian partnership. Thus, the intimate terrorist will in some cases turn to violence repeatedly and escalate its severity to gain control. Furthermore, given the size and strength differences between men and women, a woman faced with a partner who is determined to gain the upper hand by whatever means will not in the long run be likely to continue to resist physically but will turn to other tactics to gain control of her life (Campbell, Rose, Kub, & Nedd, 1998).

9. If my arguments regarding the biases of various types of sampling strategies are correct, it is almost impossible to develop precise estimates of the incidence of the various types of violence. I come to the conclusion that most partner violence is situational couple violence in the following way, based on figures in my 1995 article. First, accepting my evidence that almost all of the partner violence in general surveys is situational couple violence, we can use the figures from the National Family Violence Surveys to estimate the incidence of situational couple violence. Second, extrapolating from agency data in two states that keep excellent shelter statistics, we can develop an estimate of the incidence of intimate terrorism. Those figures, which may be found in the 1995 article, suggest that there is probably 3 times as much situational couple violence as intimate terrorism, which would mean that 75% of women experiencing violence from their male partners are experiencing situational couple violence.

References

Aldenderfer, M. S., & Blashfield, R. K. (1984). *Cluster analysis.* Newbury Park, CA: Sage.

Archer, J. (2000). Sex differences in aggression between heterosexual partners: A meta-analytic review. *Psychological Bulletin, 126,* 651–680.

Campbell, J. C., Rose, L., Kub, J., & Nedd, D. (1998). Voices of strength and resistance: A contextual and longitudinal analysis of women's responses to battering. *Journal of Interpersonal Violence, 13,* 743–762.

Capaldi, D. M., & Owen, L. D. (2001). Physical aggression in a community sample of at-risk young couple: Gender comparisons for high frequency, injury, and fear. *Journal of Family Psychology, 15,* 425–440.

Cook, S. L., & Goodman, L. A. (2006). Beyond frequency and severity: Development and validation of the Brief Coercion and Conflict Scales. *Violence Against Women, 12,* 1050–1072.

Dobash, R. E., & Dobash, R. P. (1979). *Violence against wives: A case against patriarchy.* New York: Free Press.

Dobash, R. P., Dobash, R. E., Wilson, M., & Daly, M. (1992). The myth of sexual symmetry in marital violence. *Social Problems, 39,* 71–91.

Frieze, I. H. (1983). Investigating the causes and consequences of marital rape. *Signs, 8,* 532–533.

Frieze, I. H., & Browne, A. (1989). Violence in marriage. In L. Ohline & M. Tonry (Eds.), *Family violence* (pp. 163–218). Chicago: University of Chicago Press.

Frieze, I. H., & McHugh, M. C. (1992). Power and influence strategies in violent and nonviolent marriages. *Psychology of Women Quarterly, 16,* 449–465.

Graham-Kevan, N., & Archer, J. (2003a). Intimate terrorism and common couple violence: A test of Johnson's predictions in four British samples. *Journal of Interpersonal Violence, 18,* 1247–1270.

Graham-Kevan, N., & Archer, J. (2003b). Physical aggression and control in heterosexual relationships: The effect of sampling. *Violence and Victims, 18,* 181–196.

Holtzworth-Munroe, A. (2000). A typology of men who are violent toward their female partners: Making sense of the heterogeneity in husband violence. *Current Directions in Psychological Science, 9*(4), 140–143.

Holtzworth-Munroe, A. (2002). Standards for batterer treatment programs: How can research inform our decisions? *Journal of Aggression, Maltreatment and Trauma, 5,* 165–180.

Holtzworth-Munroe, A., Meehan, J. C., Herron, K., Rehman, U., & Stuart, G. L. (2003). Do subtypes of martially violent men continue to differ over time? *Journal of Consulting and Clinical Psychology, 71,* 728–740.

Holtzworth-Munroe, A., Rehman, U., & Herron, K. (2000). General and spouse-specific anger and hostility in subtypes of martially violent men and nonviolent men. *Behavior Therapy, 31,* 603–630.

Johnson, M. P. (1993, November). *Violence against women in the American family: Are there two forms?* Paper presented at the Pre-Conference Theory Construction and Research Methodology Workshop, annual meeting of the National Council on Family Relations, Baltimore, MD.

Johnson, M. P. (1995). Patriarchal terrorism and common couple violence: Two forms of violence against women. *Journal of Marriage and the Family, 57,* 283–294.

Johnson, M. P. (1999, November). *Two types of violence against women in the American family: Identifying patriarchal terrorism and common couple violence.* Paper presented at the annual meeting of the National Council on Family Relations, Irvine, CA.

Johnson, M. P. (2000, November). *Conflict and control: Symmetry and asymmetry in domestic violence.* Paper presented at the National Institute of Justice Gender Symmetry Workshop, Arlington, VA.

Johnson, M. P., & Cares, A. (2004, November). *Effects and noneffects of childhood experiences of family violence on adult partner violence.* Paper presented at the annual meeting of the National Council on Family Relations, Orlando, FL.

Johnson, M. P., Conklin, V., & Menon, N. (2002, November). *The effect of different types of domestic violence on women: Intimate terrorism vs. situational couple violence.* Paper presented at the annual meeting of the National Council on Family Relations, Houston, TX.

Johnson, M. P., & Ferraro, K. J. (2000). Research on domestic violence in the 1990s: Making distinctions. *Journal of Marriage and the Family, 62,* 948–963.

Johnson, M. P., & Leone, J. M. (2000, July). *The differential effects of patriarchal terrorism and common couple violence: Findings from the National Violence Against Women Survey.* Paper presented at the Tenth International Conference on Personal Relationships, Brisbane, Australia.

Johnson, M. P., & Leone, J. M. (2005). The differential effects of intimate terrorism and situational couple violence: Findings from the National Violence Against Women Survey. *Journal of Family Issues, 26*(3), 322–349.

Kimmel, M. S. (2002). "Gender symmetry" in domestic violence: A substantive and methodological research review. *Violence Against Women, 8,* 1332–1363.

Kurz, D. (1989). Social perspectives on wife abuse: Current debates and future directions. *Gender & Society, 3,* 489–505.

Leone, J. M., Johnson, M. P., & Cohan, C. L. (2003, November). *Help-seeking among women in violent relationships: Factors associated with formal and informal help utilization.* Paper presented at the annual meeting of the National Council on Family Relations, Vancouver, Canada.

Leone, J. M., Johnson, M. P., Cohan, C. M., & Lloyd, S. (2001, June). *Consequences of different types of domestic violence of low-income, ethnic women: A control-based typology of male-partner violence.* Paper presented at the International Network on Personal Relationships, Prescott, AZ.

Leone, J. M., Johnson, M. P., Cohan, C. M., & Lloyd, S. (2004). Consequences of male partner violence for low-income, ethnic women. *Journal of Marriage and Family, 66,* 471–489.

Lloyd, S., & Taluc, N. (1999). The effects of male violence on female employment. *Violence Against Women, 5,* 370–392.

Marshall, L. L. (1996). Psychological abuse of women: Six distinct clusters. *Journal of Family Violence, 11*(4), 379–409.

Moffitt, T. E., Robins, R. W., & Caspi, A. (2001). A couples analysis of partner abuse with implications for abuse-prevention policy. *Criminology and Public Policy, 1,* 5–26.

Pence, E., & Paymar, M. (1993). *Education groups for men who batter: The Duluth model.* New York: Springer.

Renzetti, C. M. (1992). *Violent betrayal: Partner abuse in lesbian relationships.* Thousand Oaks, CA: Sage.

Saunders, D. G. (2002). Are physical assaults by wives and girlfriends a major social problem? A review of the literature. *Violence Against Women, 8,* 1424–1448.

Stark, E., & Flitcraft, A. (1996). *Women at risk: Domestic violence and women's health.* Thousand Oaks, CA: Sage.

Steinmetz, S. K. (1977). The battered husband syndrome. *Victimology, 2,* 499–509.

Stets, J. E. (1993). Control in dating relationships. *Journal of Marriage and Family, 55*(3), 673–685.

Straus, M. A. (1990a). The Conflict Tactics Scales and its critics: An evaluation and new data on validity and reliability. In M. A. Straus & R. J. Gelles (Eds.), *Physical violence in American families: Risk factors and adaptations to violence in 8,145 families* (pp. 49–73). New Brunswick, NJ: Transaction Publishers.

Straus, M. A. (1990b). Injury and frequency of assault and the "representative sample fallacy" in measuring wife beating and child abuse. In M. A. Straus & R. J. Gelles (Eds.), *Physical violence in American families: Risk factors and adaptations to violence in 8,145 families* (pp. 75–91). New Brunswick, NJ: Transaction Publishers.

Straus, M. A. (1999). The controversy over domestic violence by women: A methodological, theoretical, and sociology of science analysis. In X. B. Arriaga & S. Oskamp (Eds.), *Violence in intimate relationships* (pp. 17–44). Thousand Oaks, CA: Sage.

Straus, M. A., & Gelles, R. J. (Eds.). (1990). *Physical violence in American families: Risk factors and adaptations to violence in 8,145 families.* New Brunswick, NJ: Transaction Publishers.

Straus, M. A., Hamby, S. L., Boney-McCoy, S., & Sugarman, D. B. (1996). The revised Conflict Tactics Scales (CTS2): Development and preliminary psychometric data. *Journal of Family Issues, 17*(3), 283–316.

Straus, M. A., Hamby, S. L., Boney-McCoy, S., & Sugarman, D. B. (1999). *The personal and relationships profile (PRP)*. Durham: University of New Hampshire, Family Research Laboratory.

Straus, M. A., & Mouradian, V. E. (1999, November). *Preliminary psychometric data for the Personal and Relationship Profile (PRP): A multi-scale tool for clinical screening and research on partner violence*. Paper presented at the annual meeting of the American Society of Criminology, Toronto, Canada.

Straus, M. A., & Ramirez, I. L. (2002, July). *Gender symmetry in prevalence, severity, and chronicity of physical aggression against dating partners by university students in Mexico and the USA*. Paper presented at the annual meeting of the International Society for Research on Aggression, Montreal, Canada.

Tjaden, P., & Thoennes, N. (2000). Prevalence and consequences of male-to-female intimate partner violence as measured by the National Violence Against Women Survey. *Violence Against Women, 6*, 142–161.

Tolman, R. M. (1989). The development of a measure of psychological maltreatment of women by their male partners. *Violence and Victims, 4*(3), 159–177.

DISCUSSION QUESTIONS

1. What evidence is there that surveys may represent a different picture of intimate partner violence than data collected from agencies?

2. As society develops and women gain more equality, do you think that intimate terrorism rates should decline? Why or why not?

3. What are the policy implications for intervention and prevention if there are four types of intimate partner violence?

4. What can you conclude about gender symmetry and asymmetry in intimate partner violence perpetration and victimization after reading this article?

◈

Introduction to Reading 2

One key line of inquiry in the intimate partner violence literature is understanding the causes of intimate partner violence. Using lifestyles theory and assortative mating, Carbone-Lopez and Kruttschnitt (2010) investigate why incarcerated women have experienced intimate partner violence. They propose that women who are themselves criminal may choose mates who are similarly criminal, thus putting them at risk of being victimized by that mate. Further, lifestyles theory would suggest that having a lifestyle that is risky will put women in contact with criminal others, hence increasing their chances of being the victims of intimate partner violence. To study these propositions, the authors use data from the Women's Experiences of Violence study, which includes a sample of 206 women who were incarcerated in the Hennepin County Adult Detention Facility in Minnesota. Women were interviewed and asked about the previous 36 months before their incarceration, using a life events calendar. They were asked about their and their partners' substance use and their partners' involvement in crime along with their experiences of intimate partner violence.

Risky Relationships?

Assortative Mating and Women's Experiences of Intimate Partner Violence

Kristin Carbone-Lopez and Candace Kruttschnitt

Violence against women has come to be viewed as a serious social and public health problem in the United States and most Western industrialized nations. Overwhelmingly, though, research has focused on a particular ideal type of battered or victimized woman (Loseke, 1992). Women whose behaviors are not consistent with this portrait of a "battered woman" (e.g., women who use alcohol or drugs or who are involved in criminal activity) have been given less scholarly attention. Nevertheless, practitioners and policymakers have long argued that women offenders have substantial histories of violent victimization (see, e.g., Wellish & Falkin, 1994), and their claims now have empirical support. Research generally indicates that the risk of victimization among incarcerated women is much greater than that for women within the general population; a national survey of jails and prisons in the United States indicates that between 40% and 57% of female inmates were physically or sexually abused prior to their current sentence and, for the majority of these women, their abusers were intimate partners (Wolf Harlow, 1999; also see Chesney-Lind, 2002; Dugan & Castro, 2006; Snell, 1994).[1] Other research seeking to document women's pathways into and out of crime also highlights their histories of victimization, particularly within intimate relationships (e.g., Leverentz, 2006; Richie, 1996).

The attention given to the concentrated risk of intimate partner violence (IPV) among female offenders has not, however, produced an adequate understanding of the phenomenon. We think this omission is particularly noteworthy given the instrumental role criminological research has played in documenting and explaining the links between offending and victimization (Jensen & Brownfield, 1986; Sampson & Lauritsen, 1990), even in the case of female offenders. For example, we know that women who are regularly involved with street drugs, prostitution, and gangs are at elevated risk for victimization because of the individuals they associate with and the criminogenic situational contexts in which they are involved (e.g., Baskin & Sommers, 1998; Maher, 1997; Miller, 2001). Although this offender victimization link has been fleshed out within the context of routine activities and lifestyle theories, no theories of IPV have attempted to integrate women's criminality into an understanding of their risk of victimization (Kruttschnitt, McLaughin, & Petrie, 2004). By and large, explanations of general victimization include a focus on characteristics and behavior of the victims, whereas theories of violence against women, and IPV in particular, focus predominantly on the offenders; or as Meier and Meithe (1993) suggest, such theories "concentrate only on accounting for the pool of motivated offenders" while neglecting the "victim side of the equation" (p. 495).

One of the reasons for this focus may be an overwhelming reluctance on the part of researchers who study violence against women to "blame the victim."[2] Indeed, following Amir's (1971) application of the concept of victim precipitation to rape (and the subsequent controversy it inspired), studies of victims' behaviors have been confined primarily to street violence or violence involving more distal victims and offenders. Yet to advance our understanding of criminal victimization, we must be willing to look at all sides of the equation, including the characteristics

SOURCE: Carbone-Lopez and Kruttschnitt (2010). Reprinted with permission.

and activities of victims of IPV that may increase their risk of victimization.

To this end, the following research uses data from a sample of incarcerated women to examine the relationship between their criminal activities and their experiences with violent victimization by intimate partners. In other words, we seek to explain the high rates of victimization, specifically by intimate partners, among incarcerated women. We use lifestyle and assortative mating perspectives to examine the role that women's risky behaviors (e.g., alcohol and drug use) and their involvement with criminal men play in their experiences of IPV. First, however, we briefly review existing research on the relationship between women's offending and their victimization experiences. As we demonstrate, these inquiries have focused primarily on the role of early victimization on juvenile and adult offending among women.

Blurred Boundaries

Scholars, seeking to place women's involvement in criminal offending within the context of their life histories, draw attention to the blurred boundaries between offending and victimization. Explaining girls' involvement in delinquency, Chesney-Lind and Shelden (2004), for example, highlighted the role of victimization, and particularly sexual abuse, as an important and uniquely gendered background factor (Giordano, Deines, & Cernkovich, 2006). Similarly, other have found that women engage in drug use or criminal activity as a way of coping with the violence they experience at the hands of their partners or as a survival strategy (see, e.g., Gilfus, 1992; Moe, 2004). In an examination of the intersection between race, victimization, and offending, for example, Richie (1996) provided support for a theory of "gender entrapment" in which women's paths to committing crimes are greatly influenced by their victimization experiences, both within the family of origin and within intimate relationships.

Less attention has been paid to the issue of how criminality itself increases the risk of victimization within the context of an intimate relationship. Particular lifestyles may be intricately linked with partner violence

through the routine activities and behaviors that are associated with criminal activity or using illicit drugs. Such lifestyles, which include bartering over drugs and visiting locations in which drugs are sold and consumed, are dangerous for women (El-Bassel, Gilbert, Schilling, & Wada, 2000). In this regard, Armstrong and Griffin's (2007) recent analysis of the effects of short-term changes in life circumstances on female offenders' risk of victimization is particularly instructive. They found women's risk of victimization, by any offender, increased in those months in which they were in an unstable living situation or used drugs or engaged in criminal activity. However, because they focused on overall risk of victimization, without accounting for the victim-offender relationship, their results are unable to directly link women's risk of intimate violence with lifestyle indicators. Although other research suggests that particular lifestyles increase women's chances of establishing intimate relationships with male offenders (see, e.g., Covington, 1985; Leverentz, 2006; O'Brien, 2001; Sterk, 1999), to date there has been relatively little attention paid to such relationships and, more important, the way in which they may contribute to women's risk of victimization. To more fully understand this relationship, we draw on theories of victimology and relationship formation to suggest (a) that women offenders will have a greater likelihood than nonoffenders of pairing with criminogenic males and consequently (b) the likelihood of IPV is higher among women offenders who pair with offenders relative to those who pair with nonoffenders.

Theoretical Framework

Lifestyle Explanations for Victimization

Arguably, the theoretical perspective that has played the greatest role in the study of criminal victimization is that of lifestyle or routine activities theory. Lifestyle theories of victimization were initially used by Hindelang, Gottfredson, and Garofalo (1978) to attempt to explain young Black males' elevated risks of victimization. They argued that particular patterns of daily (or routine) activities and lifestyle choices foster contact between potential offenders and victims and, consequently, increase the risk of victimization. To the degree that persons share sociodemographic characteristics with potential offenders, they are more likely to interact with them, subsequently enhancing their risk of violence. Also known as the "principle of homogamy" (Sampson & Lauritsen, 1990), the overall premise is that associating with criminals may be dangerous.

According to lifestyle theory, because of gendered role expectations wherein women traditionally stay closer to home, undertake domestic caretaking activities, and have fewer criminal associates, their risks of victimization should be much lower than those of males (Meier & Miethe, 1993). Although there is ample evidence that women do generally have lower rates of victimization than men, when they engage in deviant or criminal activities, their risks of victimization are elevated just as male offenders' risks are.[3] As previously noted, research indicates that drug-using women (Armstrong & Griffin, 2007; Sterk, 1999), prostitutes (Maher, 1997), and female gang members (Miller, 2001; Peterson, Taylor, & Esbensen, 2003) are particularly likely to experience victimization relative to nonoffenders, simply because their lifestyles place them in frequent contact with criminogenic individuals.

Assortative Mating

One explanation for the disproportionate violence experienced by incarcerated and criminal women can be developed using the theory of assortative mating. This theory suggests that individuals enter into romantic relationships (or mate) with others who share their characteristics and preferences (Vanyukov, Neale, Moss, & Tarter, 1996). Research on this phenomenon, also known as "homophily" (McPherson, Smith-Lovin, & Cook, 2001), provides substantial evidence of homogeneity with respect to various sociodemographic, behavioral, and intrapersonal characteristics among married and cohabiting couples in the United States. In other words, individuals tend to become romantically involved with others who are like them and who engage in similar activities. For example, couples tend to be matched on characteristics such as race (e.g., Kalmijn, 1994;

Spanier & Glick, 1980), education (e.g., Blackwell, 1998; Mare, 1991), and age (e.g., Jepsen & Jepsen, 2002).

There is also some evidence that assortative mating may not be confined to personal or demographic characteristics. As Simons, Stewart, Gordon, Conger, and Elder (2002) suggest, deviant persons are likely to meet, date, and subsequently fall in love with individuals who are also involved in delinquent or criminal behavior. Indeed, several studies highlight matching based on substance abuse (e.g., Vanyukov et al., 1996; Yamaguchi & Kandel, 1993); levels of sensation seeking, which is associated with antisocial behavior (Caspi & Herbener, 1990; Moffitt, Caspi, Rutter, & Silva, 2001); and psychopathological tendencies (Merikangas, 1982).

Although individuals may be more likely to become involved in romantic relationships with others who are like them, the mechanisms by which this occurs may be less about personal choices and more about environment and the availability of potential mates. Thus, there are three possible explanations for these assortative mating patterns. First, as Engfer, Walper, and Rutter (1994) suggest, individuals may be more likely to meet and become romantically involved with one another because they share similar backgrounds. Second, some individuals actively select similar partners who exhibit their same values or who enjoy the same activities and thus will not disapprove of their lifestyle. A final way in which partners may become matched on deviant lifestyles is that conventional individuals may avoid social contact with individuals involved in crime and delinquency. As such, individuals' fields of potential partners are constrained and dominated by persons who look and act much like themselves. Sterk's (1999) research on women who use crack in Atlanta, Georgia, provides evidence of this type of matching; she found that as women's drug habits increased, they felt increasingly uncomfortable around individuals who did not share their values. As a result, their "pool of available partners" tended to become limited to men who were also drug users.[4]

Following from this, women who are themselves involved in criminal or illegal activities may be more likely to be in intimate relationships with men involved in similar (i.e., criminal) activities.[5] Yet how would such matching increase women's risks of experiencing IPV? Prospective studies find that criminally involved partners increase the likelihood of "stormy and contentious" relationships (Laub, Nagin, & Sampson, 1998; Sampson & Laub, 1993), and evidence suggests that such stormy relations may be linked to the violent proclivities of male offenders. For example, they are more likely than nonoffenders to be involved in partner abuse (Fagan & Browne, 1994; Holtzworth-Monroe & Stuart, 1994; Hotaling, Straus, & Lincoln, 1989; Moffitt, Krueger, Caspi, & Fagan, 2000), and perpetrators of violence against women commonly have histories of violence and conduct problems outside of their intimate relationships (Capaldi & Clark, 1998; Farrington, 1994; Giordano, Millhollin, Cernkovich, Pugh, & Rudolph, 1999; Kruttschnitt et al., 2004). Accordingly, women's criminality may increase their odds of partnering with male offenders (Giordano, Cernkovich, & Rudolph, 2002; Leverentz, 2006), which in turn may increase their risk of IPV.

Research Questions

Drawing on lifestyle explanations of victimization and the theory of assortative mating, the following research examines IPV victimization among a sample of incarcerated women. Examining violence experienced by an incarcerated population is crucial, not only because of previous studies documenting high rates of victimization among female offenders but also because national surveys (e.g., the National Crime Victimization Survey) exclude individuals who are institutionalized. Moreover, individuals involved in illegal activities are desirable targets of violence because they are not likely to call police; as such, the violence they experience rarely appears in official crime statistics. Our sample, therefore, includes women who are not likely to be included in crime and victimization statistics, although they are at very high risk for violence.

In the following analyses, we focus on the victimization experiences of women offenders and examine the role of women's involvement in risky activities, including their relationships with criminal or offending men, in their experiences of partner violence. We have

two primary objectives: (a) to describe incarcerated women's involvement in intimate relationships with partners who are also engaged in criminal behavior; and (b) to assess the effects of women offenders' intimate relationships with criminal partners on their risk of IPV, net of other risk factors, including their own criminal behavior.

Method

Sample and Procedures

The research described here is part of a larger, multisite study of women offenders, called Women's Experiences of Violence. This study examines the personal, situational, and community-level factors that are associated with women's experiences of violence, both as offenders and as victims.[6] The racially diverse sample of 206 women was drawn from the female population incarcerated in the Hennepin County Adult Detention Facility in Minnesota, a short-term jail (postsentencing) that houses males and females in separated buildings.

Because of high turnover rates and short jail stays, which precluded true random sampling, women serving straight sentences (i.e., not weekend or "shock" sentences) were selected from rosters of the total jail population based on nearest approaching release dates. Interviews were conducted in private rooms, away from correctional staff and other inmates. Trained interviewers outlined the research objectives of the study and assured women that they would be guaranteed confidentiality. After reviewing the research procedures, women willing to participate signed a consent form. For their participation, women were given a small incentive.

Results

Table 1 presents descriptive information on the individual characteristics of the 162 women offenders who reported intimate relationships within the 36-month calendar period. Not surprisingly, they have much in common with other populations of incarcerated women (see Kruttschnitt & Gartner, 2003; Slocum, Simpson, & Smith, 2005). For example, women of color are substantially overrepresented in comparison to their representation in the general population in Minnesota and the women generally report low educational attainment (36% had less than high school education while another 33% had only a high school diploma or GED). The convicting offenses (not shown) of the women involved in intimate relationships tended to be drug-related or alcohol-related crimes (30%), property crimes (27%), and prostitution (19%).

When examining the characteristics of the relationships that these women were involved in, we first note that a large proportion of the relationships (43%) involved violence.[7] The relationships had lasted 54 months on average, and the majority (68%) of them involved cohabitation at some point. Not surprisingly, these women's relationships were marked by extensive substance use, both their own and that of their partners. In slightly more than one quarter of these relationships, women were using alcohol either every day or nearly every day, and in 40% of relationships, they reported frequent illicit drug use by their partners; one third of women's intimate partners used alcohol frequently and nearly one third used illicit drugs every day or nearly every day. In terms of the extent to which women in the sample were intimately involved with partners who also engaged in criminal behavior, 26% of the relationships were with partners engaged in property offending and more than one third (37%) involved partners who manufactured or distributed illicit drugs. Thus, partnering with a male offender was not an unusual occurrence among these women. We see that partner substance use has both substantive and significant effects on the likelihood of IPV. First, it is not surprising that age emerges as a significant covariate of IPV. Relationships involving younger women are more likely to be violent; as women age, the odds that they will encounter IPV within a particular relationship decrease by 4% each year. Second, we also see that

Table 1 Sample Characteristics in Women's Experience of Violence in Minnesota

Variable	%	M	SD
Respondent characteristics[a]			
Age		34.6	8.6
Race			
White	40		
Black	33		
American Indian	19		
Mixed or other race	9		
Education			
Up to 9th grade	11		
10th through 11th grade	25		
High school/GED	33		
Some college	25		
College degree	6		
Childhood sexual assault victimization	20		

Variable	%	M	SD
Relationship and partner characteristics[b]			
Cohabitation	68		
Relationship length (in months)		53.9	69.5
Risky activities during relationship[b]			
Respondent substance use			
Frequent alcohol use	26		
Frequent illicit drug use[c]	40		
Partner substance use			
Frequent alcohol use	33		
Frequent illicit drug use[c]	31		
Partner criminal involvement			
Property crime	26		
Drug market	37		
Intimate partner violence victimization[b]	43		

a. $n = 162$ respondents

b. $n = 219$ relationships

c. Illicit drugs include crack cocaine, powder cocaine, heroin, speed, meth, or other amphetamines.

partner's substance use mediates the effect of women's own use on their risk of IPV; it is the partner's frequent use of alcohol and drugs—not the women's—that is associated with violence within a relationship. Relationships in which women's partners frequently used alcohol have a 350% increase, and those in which partners used illicit drugs every day or nearly every day have a 186% increase, in the odds that they will involve violence. This suggests that the lifestyles and routine activities of women's partners play a larger role than do women's own risky activities in their likelihood of encountering violence within a particular relationship. In other words, who women were intimately involved with was much more important than their own behavior (including their use of alcohol and illicit drugs) in increasing their likelihood of IPV.[8]

We find that, net of these risk factors, women's relationships with criminogenic partners were also at significantly greater risk of IPV as compared to those relationships with nonoffenders. Partner involvement in various property crimes, yet not in drug dealing, increased the odds of IPV within the relationship by 144%. In addition, partner's frequent use of alcohol and drugs had an independent effect on the risk of violence within a relationship; partner's alcohol use increased the odds of violence by 335% and drug use increased the odds of violence by 111%. In sum, although many of the well-known risk factors of victimization are associated with the experiences of IPV among the women in our sample, there appears to be evidence for our initial claim that romantic involvement with criminal and substance-using partners may put women at great risk of victimization.

Discussion and Implications

In this work, we examined the role of women's risky activities and their involvement in risky relationships (i.e., with criminal partners) on their likelihood of experiencing IPV. In particular, we attempted to explain the relatively high rate of violence among incarcerated female populations. Building on both lifestyle and assortative mating theories, we argued that women offenders involved in intimate relationships with criminogenic partners are at greater risk of violence than their counterparts who partner with nonoffenders. We now summarize our main findings, in some cases drawing on the rich narrative data collected regarding specific incidents of violence that women reported occurring within their relationships.[9]

Overall, our findings are consistent with our hypothesis and previous research that highlights the elevated risk of IPV within this marginalized population (i.e., Dugan & Castro, 2006). These findings extend previous research with incarcerated women that generally focuses only on lifetime experiences of IPV by demonstrating that, even within a relatively short time frame (the previous 36 months), more than one third of women offenders' intimate relationships involved violence and 12 women experienced violence within multiple intimate relationships during that time frame.

Second, and in line with previous research, we find that early experiences of victimization—in our case, we focused specifically on childhood sexual abuse—heighten women's vulnerability to future victimization as well. The fact that the women in our sample were all involved in criminal offending suggests that crime may mediate the pathway from early to later victimization experiences. As others have argued (i.e., Chesney-Lind, 1997; Chesney-Lind & Shelden, 2004), childhood victimization may lead women to become involved in criminal activities and subsequently increase their risk of victimization in adulthood. On the other hand, there is also evidence that early experiences of violence are also associated with decreased efficacy in selecting and maintaining appropriate relationships. As a result, women with early victimization experiences may be at greater risk of becoming involved in relationships that "re-enact" the childhood abuse (Griffin et al., 2005). Indeed, a number of the women in our sample described what they saw as the connection between their childhood assaults and their adult relationships. For example, one woman noted that her stepfather's sexual abuse of her eventually led to her involvement at the age of 15 with a man who was in his 30s, who became the father of her children and was violent throughout their relationship.

Third, our data also indicate considerable matching between partners in terms of involvement in criminal activities as well as use of alcohol and drugs (also see Leverentz, 2006). Many of the women in our incarcerated sample were involved in intimate relationships with men who were also involved in crime and used drugs or alcohol. This matching had serious implications on the women's risk of violence. Although not all of the women who described violent incidents linked them to their partners' or their own involvement in crime, a considerable number did so. Women described violent incidents that occurred because of their hesitation or unwillingness to become involved in their partners' criminal activity. For example,

> The reason it happened was he was sentenced to the work release and I was supposed to stay at his house while he was gone and sell drugs.... He had bagged and cut up huge amounts of drugs so that all I had to do was sell them. Then the next day I left. When he ran into me [after his release], he acted like everything was just fine and he wasn't mad at me. Once we got to his house basically he just started beating the shit out of me. Talking about how much money he lost that time, how we could have been such a good team ...

Even when women went along with the activity, they were not always immune from violence. For example, one woman described an incident that actually occurred in the course of co-offending with her partner and that stemmed from a dispute over how to spend their ill-gotten gains:

> We were in the car. We went on a lick [committing crimes] and made some money that day. I was trying to get us a car. He wanted to use the money for drugs. He left off somewhere. Me and my friend went on a lick. He saw me in the car with her and came up to curse me out. I was trying to close the window. He reached in and punched me in the mouth. I picked up a bottle and swung at him and missed him....

Finally, another woman, who had been involved with two violent partners during the calendar period, attributed much of the violence within each relationship to the drugs—not just as the reason why her partners would commit the violence but also as the reason why she would put up with and stay with them. She said, "Bottom line: If I wasn't such a junkie, I could avoid them [violent incidents]. The way to avoid them was to not get high." Thus, our findings are consistent with prior research that shows that criminal networks and substance abuse elevate risks of victimization. We find that partner criminality is significantly related to the likelihood of violence within a relationship, even controlling for a woman's early experiences of sexual victimization, her own level of substance abuse, and her own use of violence.

Considerable attention has been paid recently to the potential benefits of romantic relationships and marriages in producing desistance, most notably for male offenders (Horney et al., 1995; Sampson & Laub, 1993; Warr, 1998). Although less is known about how such relationships affect women, there are indications that the impediments to recidivism among women are both individual and social (Eaton, 1993) and that the effect of an intimate relationship on desistance hinges on the prosocial or antisocial values of the partner (Giordano et al., 2002). At the very least, our findings suggest that marriage, or marriage-like relationships, in some cases may be a risk factor, both for future criminal involvement and for future victimization, particularly for women. It seems important, then, to recall the qualifier that Laub and colleagues (1998) added with respect to the role of marriage in producing desistance; that it must be a "good" marriage.

Unfortunately, female offenders, like those who comprise our sample, may be particularly constrained in their choice of the kinds of partners who are likely to produce good marriages. As Leverentz (2006) noted, women offenders often emerge from and return to neighborhoods of concentrated criminality and high rates of incarcerated men. This creates a shortage of heterosexual men that can (or will) function as intimate partners; this "availability factor" seems to be most critical within the Black community (Potter, 2008). If these communities provide few eligible noncriminally involved males for nonoffending females, the selection pool will be even more dismal for women re-entering the community after a period of incarceration.

Accordingly, we feel compelled to offer two caveats pertaining to the relevance of assortative mating theory for understanding women offenders' high rates of IPV. First, to pair, two individuals must meet. Although the theory of assortative mating as traditionally used generally assumes equal access to a large pool of potential mates, we know that in selected neighborhoods hit hard by the hyper-incarceration trend of the last few decades, this is not true. Numerous studies have demonstrated that high incarceration rates decrease the number of eligible males and their likelihood of marriage when they return from prison (Darity & Myers, 1994; Lynch & Sabol, 2004; Western, 2006). Thus, the choices that women offenders have for partners will be conditioned by their environment or the neighborhoods to which they return. As suggested, given the disproportionate effect that such increased incarceration rates have had on women from our sample, particularly from Black communities, they may face additional obstacles in finding partners when returning to these communities after jail.

Following from this, our second point is that existing research on assortative mating may overemphasize the role of choice and selection in pairing. For women offenders, finding a high-quality or "ideal" partner may be difficult, if not impossible, to realize and, as such, matching into a somewhat stable relationship likely takes precedence over matching into an ideal relationship (see Giordano et al., 2002). In other words, they may not be able to attract their preferred match, ending up with someone who is outside their acceptability range but who resembles them demographically and in terms of their personality and behavior.

Unfortunately, we do not have information on how or why women in our sample became romantically involved with their partners. As Engfer and colleagues (1994) suggested, it is likely that some of these offenders sought out similarly minded romantic partners, either because of the added incentives that such individuals brought to the relationships (e.g., drugs, cash,

opportunities for offending) or because prior experience showed them that relationships with nonoffenders are likely to fail because of their vastly different social sample; however, it is equally likely that they had few opportunities to meet and become romantically involved with more conventional, or noncriminally involved, partners. And because so many had experienced childhood abuse, their ability to perceive danger in relationships may have been compromised (Chu, 1992).

Despite these caveats, we think assortative mating can provide a framework for understanding women offenders' experiences of IPV, but our analyses are only the first step in understanding these experiences. We hope that future research will extend our analysis in several ways. For example, in subsequent analyses, it would be important to obtain independent verification of women's partners' criminality. Although we have no reason to believe that women would lie to us about this, it may well be that some women had incomplete or inaccurate information about the extent of their partners' criminality. As such, it is likely that our data underestimate the extent to which these men were actually involved in crime; this would mean that an even larger proportion of female offenders' relationships are potentially with criminal partners. Furthermore, our sample was composed only of active offenders. Our goal, however, was not to provide a general theory of IPV but rather to begin to explain what previous research has consistently demonstrated: that rates of IPV are much higher among women in prison and jail than women in general (Chesney-Lind, 2002; Dugan & Castro, 2006; Snell, 1994; Wolf Harlow, 1999). Yet future research should also consider how partner criminality and involvement in risky activities contribute to IPV among samples of nonoffenders and how social conditions shape the partner choices of women of different races. In Richie's (1996) groundbreaking work with battered Black women, she found that although the women in her sample had a strong sense of racial and ethnic identity, it left them vulnerable to their partners' violence. In many cases, the women excused the actions of their partners because of the "harsh realities of African American life in this country" (p. 62). Furthermore, they were unlikely to seek help from police or other public services because they were viewed as the "opposition" (also see Potter, 2008). We suspect that, to some degree, these concerns and reluctance to involve others may apply to other women of color as well; however, this is an empirical question that remains to be answered.

Finally, although these analyses offer support for the proposition that women's risky activities, and especially those of their partners, influence their risk of intimate violence, they do not address how these activities operate to increase the potential for such violence. For example, violence may stem from one or both partners' use of drugs or involvement in the drug economy. Detailed accounts of violent incidents from the women in our study suggest that, in many cases, disputes that began when the drugs ran out turned violent. One woman described such a situation: "We were drinking and high and trying to go find more drugs. We were arguing about something and I went to get out of the car and ... he ran me over with the car." In a separate incident, another woman indicated that her partner also became violent after the drugs ran out:

> We were getting high. We ran out. I wanted to go home and he didn't. I refused and started walking home; he took my coat and said he wouldn't give it back until I went out on the street. So he hit me a couple times to give me some momentum. . . .

These narratives suggest that further analyses should investigate the situational context of partner violence when both parties are involved in criminal activity and substance use.

Notwithstanding limitations, we feel that our research has important policy implications, particularly in light of the attention being given to the collateral consequences of crime and incarceration. Although policy initiatives cannot tell women whether and with whom they can become romantically involved, treatment providers and probation officers frequently warn women against fraternizing with known criminals. Because offending and incarceration are not spread evenly across communities, this creates unrealistic conditions

for most women offenders who will likely be returning to residential areas with high concentrations of offenders or former offenders (Rose & Clear, 1998). Perhaps, then, the more sanguine approach would be to focus on individual offenders' histories of both substance abuse and domestic violence and warn women about their risks of experiencing IPV from known offenders. Knowledge of these risks could provide another catalyst or "hook for change," part of the process that will be necessary if they are to craft new lives (Giordano et al., 2002), even in severely disorganized and disadvantaged communities. The effects of policies designed to curb criminality, then, need to be based on the current realities of family life and communities in which families are residing. Without such a reality check, we are doomed to observing and documenting problems such as the substantial histories of IPV among female offenders, rather than understanding and correcting them.

Notes

1. In contrast, estimates of women's lifetime experiences of intimate partner violence (IPV) from a noninstitutional, general population sample are reported to be 22.1% (Tjaden & Thoennes, 2000).

2. There are, however, important exceptions to this. For example, Schwartz and Pitts (1995) apply routine activities theory to the study of women's experiences of sexual assault by acquaintances on college campuses. Focusing on the third premise of routine activities theory, suitable targets, they find that more frequent alcohol use and friendships with "motivated offenders" (those males who get women drunk to have sex with them) increase college women's risk of sexual assault.

3. There has been some suggestion that involvement in criminal activities may be the most victimogenic of all lifestyle characteristics (Jensen & Brownfield, 1986). Considerable evidence suggests that involvement in serious criminal offending, as well as more minor acts of deviance, increases one's risk of personal victimization (Jensen & Brownfield, 1986; Lauritsen, Sampson, & Laub, 1991; Sampson & Lauritsen, 1990).

4. Research on women's desistance from crime identifies the social isolation that drug users experience and the associated problems as well. Women trying to kick their drug habits are generally encouraged to sever ties with former or current partners—as they are highly likely to be involved in drugs and criminal activity—specifically to facilitate their own recovery (Giordano, Cernkovich, & Rudolph, 2002; Leverentz, 2006).

5. An interesting focus of recent criminological research is the way in which assortative mating influences stability in offending across the life course. Simons, Stewart, Gordon, Conger, and Elder (2002), for example, found that adolescent involvement in delinquent behavior increased the likelihood of a young woman's involvement with an antisocial romantic partner, which was also strongly associated with her later involvement in crime and delinquency. Our research attempts to go beyond a social influence argument (i.e., that criminally involved partners contribute to and encourage criminal behavior in one another) and instead examines the effect that these pairings have on the nature of the relationship itself and the likelihood of intimate violence.

6. Data were also collected in Toronto, Canada (Rosemary Gartner, principal investigator) and Baltimore, Maryland (Sally Simpson, principal investigator). The three cities involved in this study vary in a number of important ways (e.g., size, ethnic composition, crime rates, and availability of handguns) that likely have implications for understanding both the situational and community contexts of women's experiences of violence.

7. We also calculated the number of women in our sample who were victims of IPV. We found that the majority of the women (51%) had experienced IPV within a relationship in the previous 36 months; also 12 women reported IPV in multiple relationships within that time frame.

8. We are grateful to one reviewer who suggested that we examine interaction terms between a woman's substance use during a relationship and that of her partner. We created an interaction term representing relationships in which both partners frequently used alcohol and a second interaction that indicates relationships in which both partners frequently used illegal drugs. Our findings suggest that relationships in which partners were matched on drug use were more likely to be violent. Whether this pattern reflects differences in the pharmacological effects of these substances or the difficulties in obtaining them remains an interesting area for further research.

9. The narratives were collected as part of the study's focus on specific victimization events that may have occurred in women's lives within the prior 36 months. Women were asked to provide a description of each violent incident, and the interviewers attempted to capture women's responses verbatim.

✂ References

Amir, M. (1971). *Patterns in forcible rape.* Chicago: University of Chicago Press.

Armstrong, G. S., & Griffin, M. L. (2007). The effect of local life circumstances on victimization of drug-involved women. *Justice Quarterly, 24,* 80–105.

Baskin, D. R., & Sommers, I. B. (1998). *Casualties of community disorder: Women's careers in violent crime.* Boulder, CO: Westview.

Blackwell, D. L. (1998). Marital homogamy in the United States: The influence of individual and paternal education. *Social Science Research, 27,* 159–188.

Capaldi, D. M., & Clark, S. (1998). Prospective family predictors of aggression toward female partners for at-risk young men. *Developmental Psychology, 34,* 1175–1188.

Caspi, A., & Herbener, E. (1990). Continuity and change: Assortative mating and the consistency of personality in adulthood. *Journal of Personality and Social Psychology, 58,* 250–258.

Chesney-Lind, M. (2002). Imprisoning women: The unintended victims of mass imprisonment. In M. Maure & M. Chesney-Lind (Eds.), *Invisible punishment: The collateral consequences of mass imprisonment* (pp. 79–94). New York: Free Press.

Chesney-Lind, M. (1997). *The female offender: Girls, women, and crime.* Thousand Oaks, CA: Sage.

Chesney-Lind, M., & Shelden, R. G. (2004). *Girls, delinquency, and juvenile justice* (3rd ed.). Belmont, CA: Wadsworth.

Chu, J. A. (1992). The revictimization of adult women with histories of child abuse. *Journal of Psychotherapy Practice and Research, 1,* 259–269.

Covington, J. (1985). Gender differences in criminality among heroin users. *Journal of Research in Crime and Delinquency, 22,* 329–354.

Darity, W. A., & Myers, S. L., Jr. (1994). *The Black underclass: Critical essays on race and unwantedness.* New York: Garland.

Dugan, L., & Castro, J. (2006). Comparing predictors of violent victimization for NCVS women with those for incarcerated women. In K. Heimer & C. Kruttschnitt (Eds.), *Gender and crime: Patterns in victimization and offending* (pp. 171–194). New York: New York University Press.

Eaton, M. (1993). *Women after prison.* Buckingham, UK: Open University Press.

El-Bassel, N., Gilbert, L., Schilling, R., & Wada, T. (2000). Drug abuse and partner violence among women in methadone treatment. *Journal of Family Violence, 15*(3), 209–228.

Engfer, A., Walper, S., & Rutter, M. (1994). Individual characteristics as a force in development. In M. Rutter & D. Hay (Eds.), *Development through life: A handbook for clinicians* (pp. 79–111). Oxford, UK: Blackwell Scientific.

Fagan, J., & Browne, A. (1994). Violence between spouses and intimates: Physical aggression between women and men in intimate relationships. In A. R. Reiss, Jr. & J. A. Roth (Eds.), *Understanding and preventing violence* (Vol. 3, pp. 115–292). Washington, DC: National Academy Press.

Farrington, D. (1994). Childhood, adolescent, and adult features of violent males. In L. R. Huesman (Ed.), *Aggressive behavior: Current perspectives* (pp. 215–240). New York: Plenum.

Gilfus, M. E. (1992). From victims to survivors to offenders: Women's routes of entry and immersion into street crime. *Women and Criminal Justice, 4,* 63–89.

Giordano, P. C., Cernkovich, S. A., & Rudolph, J. L. (2002). Gender, crime, and desistance: Toward a theory of cognitive transformation. *American Journal of Sociology, 107,* 990–1064.

Giordano, P. C., Deines, J. A., & Cernkovich, S. A. (2006). In and out of crime: A life course perspective on girls' delinquency. In K. Heimer & C. Kruttschnitt (Eds.), *Gender and crime: Patterns in victimization and offending* (pp. 17–40). New York: New York University Press.

Giordano, P. C., Millhollin, T. J., Cernkovich, S. A., Pugh, M. D., & Rudolph, J. L. (1999). Delinquency, identity, and women's involvement in relationship violence. *Criminology, 37,* 17–37.

Griffin, S., Ragin, D. F., Morrison, S. M., Sage, R. E., Madry, L., & Primm, B. J. (2005). Reasons for returning to abusive relationships: Effects of prior victimization. *Journal of Family Violence, 20,* 341–348.

Hindelang, M., Gottfredson, M. R., & Garofalo, J. (1978). *Victims of personal crime: An empirical foundation for a theory of personal victimization.* Cambridge, MA: Ballinger.

Holtzworth-Monroe, A., & Stuart, G. L. (1994). Typologies of male batterers: Three subtypes and the differences among them. *Psychological Bulletin, 116,* 476–497.

Horney, J. D., Osgood, W. D., & Marshall, I. H. (1995). Criminal careers in the short term: Intra-individual variability in crime and its relations to local life circumstances. *American Sociological Review, 60,* 655–673.

Hotaling, G. T., Straus, M. A., & Lincoln, A. J. (1989). Intrafamily violence, and crime and violence outside the family. In L. Ohlin & M. Tonry (Eds.), *Family violence* (pp. 315–375). Chicago: University of Chicago Press.

Jensen, G. F., & Brownfield, D. (1986). Gender, lifestyles and victimization: Beyond routine activity. *Violence and Victims, 1,* 85–99.

Jepsen, L. K., & Jepson, C. A. (2002). An empirical analysis of the matching patterns of same-sex and opposite-sex couples. *Demography, 39,* 435–453.

Kalmijn, M. (1994). Assortative mating by cultural and economic occupational status. *American Journal of Sociology, 100,* 422–452.

Kruttschnitt, C., & Gartner, R. (2003). Women's imprisonment. In M. Tonry (Ed.), *Crime and justice: A review of research* (Vol. 30, pp. 1–81). Chicago: University of Chicago Press.

Kruttschnitt, C., McLaughlin, B. L., & Petrie, C. V. (2004). *Advancing the federal research agenda on violence against women.* Washington, DC: National Academy of Sciences Press, steering committee for the workshop on issues in research on violence against women.

Laub, J. H., Nagin, D. S., & Sampson, R. J. (1998). Good marriage and trajectories of change in criminal offending. *American Sociological Review, 63,* 225–238.

Lauritsen, J. L., Sampson, R. J., & Laub, J. H. (1991). The link between offending and victimization among adolescents. *Criminology, 29,* 265–292.

Leverentz, A. M. (2006). The love of a good man? Romantic relationships as source of support or hindrance for female ex-offenders. *Journal of Research in Crime and Delinquency, 43,* 459–488.

Loseke, D. R. (1992). *The battered woman and shelters: The social construction of wife abuse.* Albany: State University of New York Press.

Lynch, J. P., & Sabol, W. J. (2004). Effects of incarceration on informal social control in communities. In M. Pattillo, D. Weiman, & B. Western (Eds.), *Imprisoning America: The social effects of mass incarceration* (pp. 135–164). New York: Russell Sage.

Maher, L. (1997). *Sexed work: Gender, race and resistance in a Brooklyn drug market.* Oxford, UK: Oxford University Press.

Mare, R. D. (1991). Five decades of educational assortative mating. *American Sociological Review, 56,* 15–32.

McPherson, M., Smith-Lovin, L., & Cook, J. M. (2001). Birds of a feather: Homophily in social networks. *Annual Review of Sociology, 27,* 415–444.

Meier, R. F., & Miethe, T. D. (1993). Understanding theories of criminal victimization. *Crime and Justice: A Review of Research, 17,* 459–499.

Merikangas, K. (1982). Assortative mating for psychiatric disorders and psychological traits. *Archives of General Psychiatry, 39,* 1173–1180.

Miller, J. (2001). *One of the guys: Girls, gangs and gender.* Oxford, UK: Oxford University Press.

Moe, A. M. (2004). Blurring the boundaries: Women's criminality in the context of abuse. *Women's Studies Quarterly, 32,* 116–138.

Moffitt, T. E., Caspi, A., Rutter, M., & Silva, P. A. (2001). *Sex differences in antisocial behavior: Conduct disorder, delinquency, and violence in the Dunedin longitudinal study.* Cambridge, UK: Cambridge University Press.

Moffitt, T. K., Krueger, R. F., Caspi, A., & Fagan, J. (2000). Partner abuse: How are they the same? How are they different? *Criminology, 38,* 199–232.

O'Brien, P. (2001). *Making it in the "free world": Women in transition from prison.* Albany: State University of New York Press.

Peterson, D., Taylor, T. J., & Esbensen, F. (2003, March). *Gang girls, gang boys, and the victimization dimension.* Paper presented at the annual meeting of the Academy of Criminal Justice Sciences, Boston.

Potter, H. (2008). *Battle cries: Black women and intimate partner abuse.* New York: New York University Press.

Rajah, V. (2007). Resistance as edgework in violent intimate relationships of drug-involved women. *British Journal of Criminology, 47,* 196–213.

Richie, B. E. (1996). *Compelled to crime: The gender entrapment of battered Black women.* New York: Routledge.

Rose, D. R., & Clear, T. R. (1998). Incarceration, social capital and crime: Implications for social disorganization theory. *Criminology, 36,* 441–479.

Sampson, R. J., & Laub, J. H. (1993). *Crime in the making: Pathways and turning points through life.* Cambridge, MA: Harvard University Press.

Sampson, R. J., & Lauritsen, J. L. (1990). Deviant lifestyles, proximity to crime, and the offender-victim link in personal violence. *Journal of Research in Crime and Delinquency, 27,* 110–139.

Schwartz, M. D., & Pitts, V. L. (1995). Exploring a feminist routine activities approach to explaining sexual assault. *Justice Quarterly, 12,* 9–31.

Simons, R. L., Stewart, E., Gordon, L. C., Conger, R., & Elder, G. H., Jr. (2002). A test of life-course explanations for stability and change in antisocial behavior from adolescence to young adulthood. *Criminology, 40,* 401–434.

Slocum, L. A., Simpson, S. S., & Smith, D. A. (2005). Strained lives and crime: Examining intra-individual variation in strain and offending in a sample of incarcerated women. *Criminology, 43,* 1067–1110.

Snell. T. L. (1994). *Women in prison: Survey of state prison inmates, 1991* (Bureau of Justice Statistics Special Report NCJ 145321). Washington, DC: Government Printing Office.

Spanier, G. B., & Glick, P. C. (1980). Mate selection differentials between whites and blacks in the United States. *Social Forces, 58,* 707–725.

Sterk, C. E. (1999). *Fast lives: Women who use crack cocaine.* Philadelphia: Temple University Press.

Tjaden, P., & Thoennes, N. (2000). *Full report of the prevalence, incidence, and consequences of violence against women.* Washington, DC: U.S. Department of Justice.

Vanyukov, M. M., Neale, M. C., Moss, H. B., & Tarter, R. E. (1996). Mating assortment and the liability to substance abuse. *Drug and Alcohol Dependence, 42,* 1–10.

Warr, M. (1998). Life course transitions and desistance from crime. *Criminology, 36,* 183–216.

Wellish, J., & Falkin, G. P. (1994). *San Antonio programs.* Washington, DC: U.S. Department of Justice, National Institute of Justice.

Western, B. (2006). *Punishment and inequality in America.* New York: Russell Sage.

Wolf Harlow, C. (1999). *Prior abuse reported by inmates and probationers* (Bureau of Justice Statistics Special Report NCJ 172879). Washington, DC: U.S. Department of Justice.

Yamaguchi, K., & Kandel, D. (1993). Marital homophily on illicit drug use among young adults: Assortative mating or marital influence? *Social Forces, 72,* 505–528.

DISCUSSION QUESTIONS

1. Is the theory of assortative mating supported in this article? How can the findings regarding assortative mating be used by policymakers and practitioners?

2. Beyond substance use, what other lifestyle factors may impact why a woman in this sample experienced intimate partner violence?

3. Why would a partner's risky activities better explain a woman's experience of intimate partner violence than her own risky behavior? What is an alternate explanation for this finding?

4. How does a woman's involvement in criminal activity impact her decision to report her victimization to the police? Given that the women in this sample had been arrested and were incarcerated, what could the criminal justice system do to help?

◈

Introduction to Reading 3

One question people commonly ask is why women stay in abusive relationships. Another question one may ask is why women choose to use, or not use, legal resources such as orders of protection. Research has examined who is statistically more likely to invoke the formal criminal justice process, but Fischer and Rose (1995) add to this literature. In their article, they discuss the reasons battered women finally do seek out orders of protection and the barriers these women perceive in doing so. The article is particularly useful in that it incorporates statistics with real-life stories from persons who have experienced intimate partner violence. It highlights that the decision to seek an order of protection is a process rather than an isolated choice.

When "Enough Is Enough"

Battered Women's Decision Making Around Court Orders of Protection

Karla Fischer and Mary Rose

Different reasons compel battered women to seek court orders of protection. Battered women decide to invoke the legal system by determining that they have had "enough." At this point women must often confront significant barriers to obtaining court orders, most of which involve symbolic and tangible fears. A number of motivations, however, seem to counteract these fears. In addition, women may experience psychological benefits from orders by gaining some measure of control in their lives. A greater understanding of the factors involved in such decision making may assist legal authorities who deal with battered women.

> Throughout my whole marriage, people would say to me, "That's enough, you've just had enough." But when is enough enough? I never knew . . . and when has he had plenty of chances? . . . Everybody else was going, "I wouldn't let it go that far, that's enough, come on, that's enough." And I would sit and cry and go "Okay, but I, I just don't know, because I've got a family." It's just like no one knows, unless you're there in the situation . . . where was the drawing point. . . . It's different for me because he raped me, and that was it for me. I meant that, it's different for me because he raped me, and that was it he just actually did something so bad that I was out.

As researchers in domestic violence (e.g., Berger, Fischer, Campbell, and Rose 1994) and law (e.g., Mahoney 1991) struggle to draw some conceptual circle around abusive behavior, women in chronically abusive relationships engage in a parallel psychological process of identifying or rejecting the label of "abused" (cf. Browne 1991). The above sample narrative of a battered woman reflects an internal hesitation: Is this situation bad enough to justify leaving? This hesitation can often be the starting point in women's decisions to consider outside help for the escalating violence in their lives, and it may also signal an acknowledgment of the negative impact of violence on their own as well as their children's physical and emotional well-being (Kirkwood 1993).

The search for the invisible "enough" line compels battered women to confront the serious obstacles that stand in the way of a successful escape from an abusive relationship. These barriers to leaving, well-documented and described by other researchers, include the economic and practical difficulties of leaving, the social and psychological ties between the abuser and the victim, as well as fear of retaliation by the abuser (e.g., Browne 1987; Ferraro and Johnson 1983; Fischer 1992; Gelles and Straus 1988; Kirkwood 1993; Martin 1976; Pagelow 1984; Sullivan and Davidson 1991).

Women may leave battering relationships in a number of ways, one of which may necessitate the assistance of the legal system through a court order of protection. Although rooted in the factors connected to leaving the abuser, invoking the legal system involves an additional conscious decision-making process. In this article, we explore the different reasons that battered women choose to involve the law to enforce their decisions to leave their abusers. Illustrated primarily

SOURCE: Fischer and Rose (1995). Reprinted with permission.

through narrative analysis, the themes that underlie battered women's rationales for seeking court protection teach us about the meaning and impact of the law in battered women's lives. We end by discussing the psychological benefits of obtaining an order of protection, which provides additional insight on the role of law in interventions in domestic violence.

⊠ Battered Women's Initial Reasons for Obtaining Court Orders

Our work with battered women who seek court protection orders began through a study of women's experiences with court orders, conducted by the first author in a medium-sized Midwestern county court. Obtaining a court order of protection involves a two-step process. First, there is an "emergency," or ex parte, hearing; here, the victim is usually granted a short-term restraining order that does not require that the abuser be notified of the legal process. Next, the woman must return to court to obtain a more permanent order; this occurs after the legal papers, including notice of the second hearing, have been serviced on the abuser. One piece of this research surveyed a large sample of battered women ($n = 287$) who were waiting at the courthouse after their temporary orders had been approved and their paperwork was being processed. In a brief questionnaire, they were asked to report their reasons for obtaining the order and their expectations for what that order would do for them.[1] Their collective answers to this questionnaire suggest a complex portrait of victims: They are choosing to leave their abusers after fairly extensive periods of abuse, which they view as escalating. Although the victim doubts that the order itself will evoke change in the abuser, instead expecting his pattern of abuse to continue, they have faith that the legal system will protect them from further violence.

Nearly all women acknowledged that they had had "enough," stating that one reason for obtaining the order was that they were "tired of the abuse" (92%). Most of these women equated that with making a change in their lives (87%). For the majority of battered women,

the order of protection was a last resort, after other sources of help seeking had failed (75%). Smaller numbers targeted specifically the abuser's failure to seek help for either his alcohol or drug problem (45%) or for his violent behavior (53%). Many women also pointed to the role that others had in encouraging their decisions to obtain the order, citing family and friends (74%), the police (50%), or other public agencies (30%).

The women's perceptions of the place of the violence in their relationships seemed to be crucial to their decision to invoke court protection. Walker (1979) argued that most battering relationships occur in a "cycle," where the abuse occurs after a period of tension building and is followed by a "honeymoon," where the couple reconciles and the abuser is apologetic for what he did. Gradually cycles become more frequent and the abuse within them more severe. The women in our sample appeared to have enhanced sensitivity to changes in the abuse, including some perspective on its cyclical nature. The women had been in fairly long-term relationships (average of 7 years), where the abuse had been occurring for an extended period of time (4 years, on average). In addition, most women (66%) had at least one child by their abuser. The majority of women acknowledged that the abuse was becoming both more severe (60%) and more frequent (59%). About half the women (43%) believed that the physical violence was becoming worse, whereas many more (85%) targeted the emotional abuse as the source of the escalation. Many women also were concerned because they believed that abuse was beginning to affect their children (50%) or that the batterer was assaulting the children (15%). Therefore, in spite of strong social and psychological connection to their long-term relationships, these women were able to recognize the escalating and expanding abuse in their lives. In choosing to invoke the law's protection as they left their assailants, they broke the cycle of violence that (at least in part) kept them trapped in these relationships.

The implicit message behind an order of protection, from those women who sought it, seems to be "I can leave you, and you can't hurt me for it." They believe the legal system will stand behind them and reinforce that message. At the time they made the decision to obtain the order, 98% said that they would stay out of this relationship permanently. Most (85%) were confident of their ability to provide financially for themselves and their children without his assistance. They felt their decisions to obtain the orders were good ones (91%), and expressed feeling more in control of their lives (98%) and the relationship (89%). Their overall positive outlook can be best understood in the context of the women's faith in the criminal justice system to protect them, as nearly all (95%) reported feeling confident that the police would respond rapidly to violations of the order. Thus what seems to have changed for these women is not the threat of future violence in their lives, but the belief that some outside intervention would be available to them.

Barriers to Obtaining Orders of Protection

The reasons that these women offer at the moment they actually obtain their order are only one part necessary to understand their decision-making process. The second piece of our research has been to examine this process more closely. In a second study, a subsample ($n = 83$) of the original population that completed the initial questionnaire was interviewed shortly after their hearing regarding their permanent orders.[2] The themes that emerged from these narratives speak to the simultaneous presence of a wariness of the legal system, with a desperate need for life-change.

Because women anticipated that they might need to enforce their orders, foremost on many women's minds was the perceived difficulty of calling the police. This act involved an emotional toll and corresponding guilt attached to the image of their abuser's arrest. Several women put this rather simply:

It's hard to tell the police to take your husband away.

I had a hard time even calling the police on him because I was so emotionally attached.

I thought, I can't put someone in jail. I just can't...women [should] not have so much control of putting the men in jail.

On the other hand, imagining calling the police gave some women a sense of ultimate power and control. This was occasionally reflected in an unrealistic assessment of police response to violations of orders:

...but, you know, just at any moment [e.g., if she runs into him], if I just pick up the phone and call the police and that's just [an] automatic warrant for his arrest. It's not my warrant, it's the judge and it's the attorney's warrant.

For one woman, calling the police at some possible future time invoked fear that escalated to the point where she had to drop her order:

I spent money on it. I spent the time. I really needed it, but, you know, when you get something like that ... you have to work with the law ... you have to be the one to dial the telephone. See, and I'm a chicken shit. And I couldn't dial those 3 numbers [911]. I didn't want to cause a scene....

Fear of calling the police if the order was violated was often connected to perceived negative consequences to obtaining court protection. Some women specifically mentioned, for example, that they had to overcome their fear of leaving the relationship permanently because they knew that obtaining the order would signal the end of their relationship:

I guess part of it that was real intimidating is because I knew that when I did this that there would be a divorce. I knew that I was sealing my divorce papers as I was doing this.
    ...I mean, I kept telling myself I can do this on my own, I can raise the kids, take care of the house, and be all right as a person and manage without him. And then when it

was really coming down to it, I guess I was scared. I mean, I know that I can do it. But it was like facing reality . . . [and] the prospect of being alone.

For other women, the fears that were associated with court protection were less symbolic and more tangible. It is not uncommon for perpetrators to threaten their victims if abuse is revealed (Browne 1987). In this sample, some women were afraid of retaliation from their abusers, either because of prior threats around police involvement or a belief that the order itself would escalate the abuser's violence. Women occasionally reported that their abusers saw themselves as "above the law"; for example, that "no order was going to do anything for him." As illustrated by the following narrative, these women often take pride in swallowing their retaliation fright and in choosing to invoke court protection:

He had threatened to smash my face in if I ever called the police. . . . You know, for me to really go out and do it is real significant to me. To go out and really take the initiative to really do something for myself.

The last set of fears that appear to be barriers for battered women obtaining court protection pertain to their fear of the actual court experience. This fear is at least partly located in stereotypes that only people who have engaged in criminal behavior belong in court; in addition, women experience feelings of shame about why they are appearing before a judge. Women's common characterizations of their actual court experiences were "scary" and "intimidating":

Custody was never an issue but I was so scared. I thought, "God, the judge is going to think I'm an awful person, too. . . . Please don't let him take my child from me."

But just the process, the legal process is really frightening. And if somebody isn't frightened by that whole process, they must be really callous or used to it or something.

I was scared to death. . . . I was embarrassed. I was so deeply hurt to think that my life had come to this.

In discussing their reaction to their court experience, women occasionally remarked that this fear of the court process can be so overwhelming as to cause traumatic dissociative reactions. As Herman (1992) observed: "If one set out by design to devise a system for provoking intrusive post-traumatic symptoms, one could not do better than a court of law" (p. 72). For example one woman in our study relayed:

I could stand up there and start crying, you know, and I really didn't know if I was going to be able to handle it. . . . When I walked in there and I sat down, that was me physically in there, but that was not me . . . I was not there that day. . . . I was scared to death.

For many women, fear of encounters with the legal system was the result of previous negative experiences with the law, sometimes involving police responses to prior orders of protection (which a minority of women in our sample—17%—had obtained). The common thread among these women's narratives was that the law, either in response to the battering itself or to violations of orders, failed to care about the violation of their rights and rarely imposed consequences on the abusers for their actions:

. . . they don't care. . . . They treat women like property anyway. The cops' attitudes are basically like that because I called the police on him several times. . . . He knew the law's attitude, too. He knew that.

He was taken to jail. I went to the hospital, had the X-rays and he was home before I was. . . . It made me feel, as far as he was concerned, that the [previous order] was not protecting me. Because every time I called, I mean they turned around and let him right back out. . . . I don't care even that one day that I let him in to use the phone, he was still in violation the minute he threw my head against the wall. And what the cops put me through, it made me feel like, unless I was seriously almost dead, I wouldn't call them again.

In sum, one of the major barriers to obtaining orders of protection that battered women report is fear. Some of these fears are tangible, borne out of threats of retaliation from the abuser or previous negative encounters with the legal system. However, what emerged in many women's narratives was also a more symbolic sense of dread that surrounds the imposition of a public solution to what they consider to be a private problem. In other words, some women view their use of the law as representative of a failure to negotiate the relationship on their own. This view of courts as an arena for those who cannot "handle" a private, relational issue has been documented among other populations who use courts to settle disputes (cf. Merry 1990). Coupled with highly specific and reasonable fears, admitting to a perceived failure on the part of women in this study makes engaging the legal system an extremely difficult step to take.

Although part of their decisions, battered women's reports of the barriers we have discussed are obviously not complete obstacles to obtaining court orders. The decision-making process seems to be capped by a final motivating force or event, one that requires them to move beyond these obstacles and seek court protection in the face of what might be overwhelming fears for other battered women who do not make this choice.

The Underlying Process of Decisions: The Final Motivators

Our earlier characterization of the initial reasons battered women reported as significant in their decision to invoke legal protection indicates that these women have been in fairly long-term relationships in which the abuse has been occurring for an extended period of time. They recognize that the abuse is escalating and beginning to have a negative impact on their children,

and they feel like the court is their last resort to being able to successfully end the violence in their lives. Sometimes the frustration invoking what they see as a last resort turns into resentment for needing to have a judge enforce what they believe to be an inherent right:

> I mean it seems to me that it's kind of silly that a person even has to have a piece of paper, you know, you start thinking about life, liberty and the pursuit of happiness, and it's like, you should be entitled to that anyway. . . . I mean it seems to me that legally anybody who comes to your house and you say, "Go away," if they don't go away, they should be arrested.

Consistent with this bitterness that they are not being given that which they are naturally entitled, some women emphasize the need to have the law act, in one form or another, as a "loudspeaker." The law was deliberately chosen because it was the only form of communication to which the abuser would listen, guaranteeing that the message would be heard. As one woman said, who saw one alternative to this particular legal method:

> People like that, people like [my abuser], you, you can't handle outside the legal system. . . . There's no rules, and if he doesn't have rules that he has to follow then that opens me up to a lot of things for him to do.

The rules that women wished the loudspeaker of the law to communicate varied, but most centered on those illustrated by the following narratives. Women wanted their abusers to learn that society is intolerant of abuse, occasionally hoping that these messages would lead to change:

> I want him to find out. . . . That it's not just me. Society takes it seriously. And you don't put up with it.

> . . . with the law being behind me and now he knows that the law is involved in our situation maybe that will help change his personality. . . . I think [obtaining the order] brought it to his attention.

It's . . . not only that I do not like it but that it's also not right. It's not legal. It's not only that it's not morally fair. . . . I have felt great comfort in that. To know that it's just not simply my choosing that it's not okay, but it really, really, isn't okay.

Presumably these women had to admit to themselves that they could not stop the abuse. They chose the law as an attempt to enhance the resources they needed, most pressing of which was to communicate the basic message that he is not allowed to abuse her. As important as it is for the loudspeaker of the law to send the message that abuse is not acceptable, it also seems that a critical factor for these women is that the message is in fact heard by the abuser.

Similar to the loudspeaker notion, some women expressed the motivating factor for seeking their order of protection as the need to have a public record of abuse. Part of this was an underlying desire to break the silence of abuse by "making someone else aware of my situation." These women believed that the legal papers on file at the courthouse, which detailed the abuse they had experienced, would speak for them and their efforts to stop the abuse if something were to ever happen to them.[3] A second narrative elaborates on the importance for those close to her to have the documentation that an order of protection allows:

I knew that in order to keep my own sanity and feel better and feel safe, that I needed to do something legally. That even though I did take a legal course of action, at least if it didn't stop, at least there is enough people around me that if something seriously, seriously happened to me because of him, that he was not going to get away with anything. And the only way to do that is to go a legal route and have documentation. . . . There are laws to protect all of us. . . . I just knew I was in serious trouble with this guy and in order to feel safe I took the legal route.

Other women felt that the order served a more specific documentary purpose: proof to the police or other actors in the criminal justice system that they were serious in their conviction to end the abuse or the relationship. As one woman said, "[an order of protection] says, 'I'm not a willing victim.'" Their hope was that having an order of protection would bring about more effective police response:

That if anything happens, that if you call the police they'll be there right away because they know something else has happened before.

I was scared that it would happen all over again and that if it did that the police wouldn't respond. That was actually the biggest thing.

I just wanted to make sure that I was, had a fairly safe environment for myself and the kids . . . to where, if, if he were to try and to, you know, do some damage . . . at least I knew the police would come out. And somebody would believe me. Like, you know, I didn't fall down on the ice and do this to [my broken] arm.

These remarks indicate that having public documentation of abuse through an order of protection may lead to a concomitant heightening of women's expectations that the system will in fact protect them. When violations of orders are not recognized and punished, some women express outrage. For instance, one woman attempting to enforce her order of protection was angered by the unanticipated, uncooperative response she found in the state attorney's office. Fearing for her safety, she informed the state attorney that she had given his name to her family and stated: "I want my children to sue you when they don't have their mom."

Finally, many women mentioned in some way that what drove them to the law was the need to have some measure of control in their lives. This is supported by the quantitative findings (reported above), where women reported an enhanced sense of control over both the relationship and other aspects of their lives after they obtained the order. For some, this control took the form of a communication to the abuser that they had truly had enough and were transforming their personal territory into a safety zone:

> Well he [the abuser] didn't think I'd do it, which I didn't because I, you know, just felt too ashamed about it. I didn't want anyone to know.... So I had to draw the line and show what I would do.

> I knew at the beginning that I needed to do something to get control again ... to feel like I had a say or I wasn't going to put up with you doing this to me. I'm not going to put up with you coming to my house or calling my house or ... being violent with me anymore or abusive or threatening or any of those things....

> Control ... when it comes to your physical safety, to me, that's where you, like, draw the line. I mean, although maybe it's not more important [than emotional abuse], but somehow to me that was just a real simple decision to make....

> [The order] was something to hide behind ... it was my strength. I had control. I was calling the shots. Not that I was calling the shots, but it made, there, it was something in the middle, like a ref, like a coach or something, ... "Watch where you're stepping."

For these women, the core motivation to invoke the law seems to be reclamation of what the abuse has systematically stripped from them: their control over their activities, their bodies, and their lives.

In addition to the need for control, we have identified other primary motivators to obtaining an order as the need for the law to act as a "loudspeaker" and the need to have a public record of abuse. These motivators often supersede any number of barriers women face when deciding whether to seek court orders of protection. The barriers that were the most salient to the women in our study were the emotional costs of being forced to call the police to have the abuser arrested, a fear of retaliation for seeking legal protection, fear of ending the relationship permanently, and perceptions

of the scary and intimidating environment of the courthouse. Some women also needed to trust that the legal system would work this time, despite prior negative experiences.

Our vision of the process of obtaining a court order for battered women starts with acknowledging that the abuse they have experienced is enough—enough to justify leaving and enough to require the assistance of the law. The process then involves moving past a number of specific fears about the public nature of court protection and recognizing the presence of compelling motivators to seek court protection. The end result is that women feel a need to have the legal system both approve and reinforce their decisions to leave their abusers. They see the legal system as a force larger than themselves and as having tangible power over their abusers, power they themselves have had stripped from them as a result of the abuse they suffered. Women who go through the difficult decision-making process to end abuse and seek legal protection often discover a number of psychological benefits beyond the practical assistance of a court order of protection.

⊠ The Fallout From Decision Making: The Psychological Benefits of an Order of Protection

How can acquiring a piece of paper that prohibits the abuser from further derogating his victim with violence be psychologically meaningful to the battered women who actually seek these court orders? The narratives of the women answering this question point, in different and indirect ways, to the symbolic value of that piece of paper. The order of protection becomes, for many women, a symbol of her own internalized strength; it represents the time she stood up to her abuser and told him, through the judge who signed her order, that she refused to "take it" anymore. As nearly all battered women are silenced about the abuse and the impact it has on their lives (Fischer, Vidmar, and Ellis 1993), the legal system becomes an enabling factor that allows them to find their own voices again.

To illustrate this aspect of the symbolic meaning of court orders of protection, consider the following narratives:

> Once I got the order I thought, it's time to start all over. When you go to court to get the order and you walk out with it in your hand, you feel like you have a little bit of power over your life again. Most abusers are bigger and stronger than you are, and the order gives you a little bit more of an edge . . . it makes you feel as if you ate a can of spinach, like POPEYE.
>
> What I needed for myself was not to feel like a victim anymore. After so long of just taking it and taking it I needed to be able to show myself as much as show him that I was tired of being a victim. To me I still have the order of protection, because that feeling, of fighting back and speaking out, will never leave me. I still carry the order with me and, in my mind, it's still valid. [She dropped her order of protection the day after she obtained it.]
>
> I mean it's like you're more confident. . . . It makes me feel like I've started, I mean, it's like I'm my old self again, before I got married. . . . I mean, you just, it's like your whole life is in this little fist right here. And it's like you open it up and . . . it's just like you're spreading your wings all over again.
>
> It was significant because I put myself ahead of him. . . . I let him run my life . . . as much as I resented it, I allowed it . . . and this is the first thing I did that I outwardly said, I'm more important than you are.

The orders may serve as a symbol for feeling better about themselves, as a turning point for change, or as a vision of a better life in the future. Woven through the text of these particular symbols were images of strength: That piece of paper becomes a psychological as well as legal victory, reflecting a determined woman rather than a weak, passive victim.

The legal system also more directly invokes the process of the reclamation of "voice." Voice, in this context, is intertwined with the function of the order of protection identified as a "loudspeaker." When battered women reported that the loudspeaker had worked, the end result was the benefit of feeling that the message had actually been received. Women repeatedly expressed that they felt as if they had constantly told him that they do not like the abuse, they are not going to stand for it, they told him to leave them alone, and that, in fact, he cannot abuse them. Whereas their voices were silenced or simply not heard, the force behind the legal paper communicated for them, in language to which the abuser was forced to respond:

> It showed him just how serious I was about this: I'm not all talk and no show. It wasn't a control issue for me—I felt like I needed to make him understand how scared I and the kids were—He would not hear what I was saying, so I needed the law to say it for me.
>
> You always have the idea about going back into the relationship and making it work. But how can you make it work when it hasn't been fixed? I want him to realize that he has not been fixed. I just want him to know that he has no right to do these things to me.
>
> I think the order made him take note . . . he's real strict about following the law so if I sent him a legal paper then he's like, she's serious, I'm just going to leave her alone because I don't want to get into trouble.

Why might such a piece of paper, these judge-validated court orders, have such a powerful symbolic impact? The answer to this question returns us to the issue of the nature of battering relationships. When one pays attention to the relationship context of battering, a common thread involves a systematic pattern of domination and control by the abuser (Fischer et al. 1993). He typically accomplishes this by creating an ever-changing structure of rules that the victim must follow or be punished for violating. Over time he may need to do less and less to concretize the rules, as she may become so adept at survival that she self-censors her own and possibly the children's behavior in frantic attempts to avoid displeasing him. The abuser gets to define who she is, what she believes in, what her individual needs are. She is nothing unless he tells her that she is, and she has nothing unless he gives it to her. Her life and herself are completely socially constructed by him.

Intervention by the legal system interrupts this pattern of control and domination because it intervenes at the level of the relationship rather than the individual. An order of protection does not simply dictate that the abuser's behavior must change by stopping the abuse; it also structures other dimensions of the relationship, such as how and when the couple may contact each other, how and when visitation is to take place, and how property is to be temporarily divided. The battered woman, perhaps for the first time in a very long while, has the opportunity to have her vision for the structure of the relationship validated through the signature of a judge. Thus the order of protection comes to symbolize that all the relationship's rules can be broken: The batterer's hold over his victim through violence, the victim's silent compliance, and the batterer's power to define his victim no longer exist.

Implications for the Legal System

Throughout this article, we have described women's decisions to seek court orders of protection as a process that involves choosing to say "Enough!" along with a mix of fear, dread, and a strong desire for control and change. Although obtaining orders of protection may result in positive psychological outcomes, these outcomes result from intervening in a complicated relational context that the abuser has previously dominated. In short, both our quantitative and qualitative data suggest that women's decisions to pursue legal means are neither easy nor lightly undertaken. Instead, battered women encounter a difficult period of indecision and uncertainty about how best to draw lines and resolve conflicts with someone with whom they are still intimately connected. Women may emerge from this process with high hopes and great needs—making

them extremely vulnerable—and this result has a number of implications for the legal system.

First, given that women have high expectations for protection, poor treatment by legal representatives can easily discourage women's attempts to gain control of their lives. In addition, Browne (1991) pointed out the difficult consequences of being labeled a "victim" in our society, and how fears regarding other people's responses often decrease victim's abilities to disclose their problems. Thus judgmental or blaming attitudes from authorities are likely to undermine women's motivations to move away from their abusive relationships. Battered women's self-blame and concerns about "how I must look" were common themes in our narratives, and it is of primary importance that legal personnel not exacerbate women's own (often exaggerated) sense of responsibility and shame for their situations.

Of course, vulnerability to the reactions of legal representatives also means that positive encounters are especially salient to victims. In fact, our narratives contained vivid and enthusiastic descriptions of police officers who were helpful or even instrumental in some women's lives. In an objective sense, these officers were simply doing their jobs. However, as we have described, many women perceive the legal system as a force far more powerful than themselves. Therefore, coming from such an authority, words of support for the woman and genuine condemnation for the abuser's actions may have significant impact. For instance, one woman described how a supportive police officer had given her a piece of paper with information about orders of protection months before she decided to seek one. She held onto this paper, underlining relevant sections, and came to view it as reassuring while she struggled to decide how to end the abuse in her life.

Finally, it is important for legal actors to understand that the needs of battered women may not always be consonant with legal mandates and frameworks at any given point in time. To the great frustration of judges and police, women will often drop their orders of protection (approximately 50% nationwide; see Kinports and Fischer 1993). Women described the humiliating difficulties they encountered as a result of dropping their orders:

...he [the judge] didn't want to serve [the order] because the first two, the first one I dropped and then the second one he went through counseling and I dropped it again. So the judge says, "Well we're not gonna serve this third one because you've dropped these others. You didn't do anything."

... [The judge] put a big guilt trip on me that if I had kept the last one he had gave me, I wouldn't have got hurt this time, and on and on. And he really just tore me up when I went back.

Nevertheless, in describing their reasons for dropping orders, most women simply say that the order had given them what they needed. If one assumes that what is needed in these relationships is not only cessation of violence, but also that the women have some measure of control, then it makes sense that an order might fill this need, without resulting in a permanent court-ordered separation between the woman and the abuser. Instead, the "loudspeaker" may have communicated the necessary message and, in essence, the woman feels safer. Understanding the larger context of decisions about orders of protection may reduce some of legal personnel's frustrations. Central in the context of court orders is that seeking them, just like leaving the abusive relationship, is not a single event, but a process that occurs over time. One woman in our sample stated these thoughts succinctly.

You know, I don't really know how, though, to make it better or make the courts and the police understand that ... even if we take [the abuser] back, it still doesn't give them the right to abuse us. And that, they can lecture us all they want, we're still going to take them back until we've had our fill of it. So if they'd just be there for us, and tend to the abuser and make him feel like a fool.

In short, what seems most important for authorities who interact with battered women is to understand that their primary function may be as communicators both to women and for women. Battered women need reassurance and support that they should not have to tolerate violence and emotional abuse in their lives. In addition, authorities will often need to send this message to the abuser, both directly and through deliberate and consistent response to women's requests for legal intervention. Rather than condemn women's decisions to drop court protection or return to the abuser as a step backward or a willingness to be abused, police and court officers can conceptualize this event as a necessary step in the process of leaving. Many women who dropped their orders felt that it was important to give him "one more chance" to behave, without the threat of jail. When the abuse begins again, they take what they have learned from vacating their court protection and try again. Indeed, battered women who had prior orders of protection are far less likely than first-timers to drop their orders (Fischer 1992). Finally, the legal system must be responsive to and patient with the complex process involved in defining "enough." Perhaps the greatest potential role for criminal justice personnel is to kick-start the intrinsic "enough" analysis by communicating to battered women each and every time they intervene that they are willing to do everything in their power to ensure that the abuse they have already endured is more than enough.

Notes

1. These women sought court orders of protection from a Midwestern, medium-sized urban county court between November 1990 and December 1991 with the assistance of the local battered women's shelter. Although women could obtain restraining orders without the shelter's assistance, more than two thirds of all orders filed in this county were clients (not residents) of the shelter. Two thirds of the research participants were White, slightly less than half were married, and three fourths lived in the urban area of the county. Participants had a mean age of 30.5 years, and most had at least one child (87%) that was fathered by the abuser (63%). Whereas there was great variability in socioeconomic status, the sample was predominantly poor and working class (30% received public assistance and those who worked earned a mean monthly salary of $685). Participants' abuse histories suggested a fairly extensive pattern of physical, emotional, and sexual abuse, with 50% reporting injuries resulting from the abuse and 35% reporting

being abused four times per week or more. These sample descriptives are fairly consistent with those reported in prior studies of orders of protection. Fagan and colleagues (e.g., Grau, Fagan, and Wexler 1984) concluded that women who sought orders of protection were younger, and typically in shorter, less violent marriages than other battered women. Horton, Simonidis, and Simonidis (1987) studied two samples of battered women who sought restraining orders, and reported that most victims were young (mean age of 30 years), married, receiving public assistance (50%), and had an average relationship length of 5 years. Chaudhuri and Daly (1992), in a study of a small sample of battered women seeking court protection, reported that the participants had an extremely short relationship length (average 18 months) and were young, married, minority, and employed (80%). Our sample, therefore, reflects a somewhat more diverse population of battered women who seek restraining orders, particularly in terms of socioeconomic status and relationship length. Although abuse-history information reported from the other studies was sketchy, this sample may also represent a more severely abused group of battered women. The questionnaire that participants completed at the courthouse was developed after extensive pilot testing and each section demonstrated adequate internal consistency in the sample.

2. Participants for this study were recruited from a subsample of women in the sample. Eighty-two percent of the sample available were successfully recruited for the study, which consisted of a 3-hour (on average) semistructured interview conducted by the first author and two trained research assistants. The interviews were audio-taped and later transcribed.

3. It could be argued that a recent example of this appeared in the highly publicized O. J. Simpson trial. Evidence of Mr. Simpson's abuse of Nicole Brown Simpson came from apologetic letters he had written her, which she saved in a safe-deposit box.

References

Berger, Allison, Karla Fischer, Rebecca Campbell, and Mary Rose. 1994. "Defining Domestic Violence: How Battered Women Draw the Line."

Browne, Angela. 1987. *When Battered Women Kill.* New York: Free Press.
———.1991. "The Victim's Experience: Pathways to Disclosure." *Psychotherapy* 28:150–6.

Chaudhuri, Molly and Kathleen Daly. 1992. "Do Restraining Orders Help? Battered Women's Experience with Male Violence and Legal Process." pp. 227–52 in *Domestic Violence: The Changing Criminal Justice Response,* edited by E. S. Buzawa and C. G. Buzawa. Newbury Park, CA: Sage.

Ferraro, K. J. and J. Johnson. 1983. "How Women Experience Battering: The Process of Victimization." *Social Problems* 30:325–39.

Fischer, Karla. 1992. *The Psychological Impact and Meaning of Court Orders of Protection for Battered Women.* Ph.D. dissertation: University of Illinois, Urbana-Champaign.

Fischer, Karla, Neil Vidmar, and Rene Ellis. 1993. "The Culture of Battering and the Role of Mediation in Domestic Violence Cases." *Southern Methodist University Law Review* 46:2117–73.

Gelles, Richard J. and Murray A. Straus. 1988. *Intimate Violence.* New York: Simon & Schuster.

Grau, Janice, Jeffrey Fagan, and Sandra Wexler. 1984. "Restraining Orders for Battered Women: Issues of Access and Efficacy." *Women and Politics* 4:12–28.

Herman, Judith. 1992. *Trauma and Recovery.* New York: Basic Books.

Horton, A., K. Simonidis, and L. Simonidis. 1987. "Legal Remedies for Spousal Abuse: Victim Characteristics, Expectations, and Satisfaction." *Journal of Family Violence* 2:265–79.

Kinports, Kit and Karla Fischer. 1993. "Orders of Protection in Domestic Violence Cases: An Empirical Assessment of the Impact of the Reform Statutes." *Texas Journal of Women and the Law* 2:163–276.

Kirkwood, Catherine. 1993. *Leaving Abusive Partners.* London: Sage.

Mahoney, Martha. 1991. "Legal Images of Battered Women: Redefining the Issue of Separation." *Michigan Law Review* 90:1–93.

Martin, Del. 1976. *Battered Wives.* San Francisco: Volcano.

Merry, Sally. 1990. *Getting Justice and Getting Even.* Chicago: University of Chicago Press.

Pagelow, Mildred D. 1984. *Family Violence.* New York: Praeger.

Sullivan, Cris and William S. Davidson. 1991." "The Provision of Advocacy Services to Women Leaving Abusive Partners: An Examination of Short-Term Effects." *American Journal of Community Psychology* 19:953–60.

Walker, Lenore E. 1979. *The Battered Woman.* New York: Harper & Row.

DISCUSSION QUESTIONS

1. What factors are related to a woman's initial decision to seek an order of protection?

2. What barriers hinder women in seeking orders of protection?

3. Given the reasons why women do and do not seek orders of protection, how can the criminal justice system and social services better assist women?

Victimization at the Beginning and End of Life

Child and Elder Abuse

Rosienell Adams, a 27-year-old woman, was recently convicted of aggravated child abuse against her 22-month-old son. The son had scars, burns, and other injuries, but one burn was particularly gruesome: a third-degree burn on his hand in the exact shape of an iron. In court, a woman with the Children's Hospital Intervention and Prevention Services Clinic testified that if the burn had been accidental, the scar would show evidence of a "glancing" burn rather than the outline of the iron, which is consistent with the iron being held on the child's hand. After the burn occurred, the mother did not seek medical care for her son, even though the child's father drove her to the hospital. She simply pretended to get medical attention (*The Gadsden Times*, 2011).

This case is an example of child abuse. Had it occurred to a person over the age of 60, it would qualify as a case of elder abuse. This section is devoted to these special victims—those who are very young and those who are older. These two types of victimizations are discussed in the same section because they are defined by the victim's position in the life course. Accordingly, the ways in which policy and actors respond to the victim and victimization are tied to the stage in the life course. Also, the causes and consequences of child abuse and elder abuse have a direct tie to the victim's stage in the life course.

Child Maltreatment

If the case of physical child abuse perpetrated by Rosienell Adams had occurred prior to the late 19th century, it might have gone without intervention. In fact, child maltreatment was not much of a concern to the public or the justice system until an 1875 case involving a 9-year-old girl, Mary Ellen. She was abused and neglected by her caregivers, but a concerned neighbor intervened to get her help. This neighbor, Mrs. Wheeler, quickly found that no policies or laws were in place to address issues of child maltreatment. As such, she did not receive the assistance she had hoped for. She turned to Henry Bergh, the founder and president of the Society for the Prevention of Cruelty to Animals (SPCA). He intervened on her behalf in court and was able to get Mary Ellen removed from the custody of her abusers and placed in a children's home. After this case, Bergh formed the New York Society for the Prevention of Cruelty to Children (Payne & Gainey, 2009; Pfohl, 1977). This development, coupled with the House of Refuge movement—which called for institutions to provide care for maltreated children—and the development of the first Juvenile Court in 1899 in Cook County, Illinois, served to create services that would address, at least in part, maltreated children.

It would take another 50 years, during the 1940s and 1950s, for child maltreatment to again receive meaningful attention. This time, the attention came from the medical community. Pediatric radiologists began to notice in X-rays broken bones that they believed were attributable to child abuse. These X-rays provided empirical evidence to support the occurrence of child abuse. In the 1960s, the *Journal of the American Medical Association* published a set of presentations on battered child syndrome, further contributing to the recognition of child abuse as a real issue and a medical condition (Payne & Gainey, 2009). During the 1970s, these developments spurred widespread concern about child maltreatment that led to the passage of the first federal legislation addressing child abuse.

What Is Child Maltreatment?

Child maltreatment can take two major forms: abuse and neglect. Abuse occurs when a person causes harm to a child (a person under the age of majority in a state, usually 18 years). It consists of actions done to a child rather than what a person fails to do for a child. Abuse can be physical, but it can also be emotional or sexual. What distinguishes these three types of abuse? Physical abuse may at first appear to be self-explanatory, but it merits some discussion to distinguish the differences between physical abuse and punishment. **Physical abuse** involves injury or physical harm of a child. Examples of physical abuse include hitting, punching, burning, slapping, and cutting. It may result from a person's intention to hurt a child, but it also may result from a misuse of discipline, such as a person using a belt to punish a child (Saisan, Smith, & Segal, 2011). When physical discipline is used not to teach a child right from wrong but, instead, out of anger and to make a child live in fear, it crosses the line from discipline into abuse (Saisan et al., 2011). Table 7.1 displays examples of physical abuse.

Emotional abuse may not carry physical marks, but it also can be quite harmful for children. When you think of **emotional abuse,** what first comes to mind likely is name calling, yelling, threatening, or bullying. These are overt and aggressive acts of emotional abuse. Emotional abuse can take less aggressive forms. For example, belittling, shaming, and humiliating a child are forms of emotional abuse. In addition, ignoring, rejecting, or limiting physical contact with a child constitutes emotional abuse (Saisan et al., 2011). As you may imagine, these behaviors can cause psychological harm for a child and may be detrimental to the child's social development. For a detailed description of emotional abuse, see Table 7.1.

The third type of abuse is sexual abuse of a child. **Child sexual abuse** can be active and involve physical touching but does not have to involve touching. Exposing children to sexual content or situations can be sexually abusive, as can involving children in pornography or prostitution. See Table 7.1 for a description of what child sexual abuse includes. The Child Abuse Prevention and Treatment Act defines child sexual abuse as

> the employment, use, persuasion, inducement, enticement, or coercion of any child to engage in, or assist any other person to engage in, any sexually explicit conduct or simulation of such conduct for the purpose of producing a visual depiction of such conduct; or the rape, and in cases of caretaker or inter-familial relationships, statutory rape, molestation, prostitution, or other form of sexual exploitation of children, or incest with children. (Sec. 111.42 U.S.C. 5106g)

Table 7.1	Types of Child Maltreatment
Physical	Deliberate attempt to harm a child. It can be distinguished from discipline by one question: Is the action intended to teach the child right from wrong or to create a life of fear for the child? Examples: • Hitting • Burning • Slapping (Saisan et al., 2011)
Emotional	Damaging to the child's mental health and/or social development; often leaves psychological marks Examples: • Belittling, shaming, and humiliating a child • Name calling and making negative comparisons to others • Telling a child he or she is "no good," "worthless," "bad," or "a mistake" • Frequent yelling, threatening, or bullying • Ignoring or rejecting a child as punishment—using the silent treatment • Withholding hugs, kisses, and other forms of physical contact from the child • Exposing the child to violence or the abuse of others (Saisan et al., 2011)
Neglect	Parents or caregivers do not provide the child's basic needs, such as adequate food, clothing, hygiene, and supervision. Examples: • Physical: failure to provide necessary food or shelter, or lack of appropriate supervision • Medical: failure to provide necessary medical or mental health treatment • Educational: failure to educate a child or attend to special education needs • Emotional: inattention to a child's emotional needs, failure to provide psychological care, or permitting the child to use alcohol or other drugs (Child Welfare Information Gateway, 2008b)
Sexual	Rape, sexual assault, molestation, prostitution, or sexual exploitation of a child; also includes incest with children. Involving children in sexually explicit conduct or simulation for the purpose of producing a visual depiction of conduct is also child sexual abuse (The Child Abuse Prevention and Treatment Act).

SOURCE: Saison, J., Smith, M., & Segal, J. (2010). "Child abuse and neglect: Recognizing and preventing child abuse." Reprinted with permission from Helpguide.org ©2001–2010. All rights reserved. For Helpguide's Series on Abuse, visit www.Helpguide.org.

Another way children can be maltreated is not abuse in that it is not active and does not result from the direct, intentional, and active harming of a child. **Neglect** results when a child's basic needs are not met. Children are, by definition, reliant on their caregivers for almost everything—food, shelter, transportation, love, and care. Legally, caregivers are required to meet these basic needs and are bound by law to make sure that children are getting a certain level of care. When this does not happen, children are said to be neglected. As such, neglect can be physical, medical, educational, or emotional, depending on the required need not being met. For example, if food is not being provided to a child, that is physical neglect. If a child routinely misses school because of a parent, it is educational neglect. See Table 7.1 for a description of the types of neglect.

Measurement and Extent of Child Maltreatment

It is difficult to know the true extent of child maltreatment in the United States. Child maltreatment is hard to detect for several reasons. Children may be too young to verbalize what is happening to them, even if they are capable of understanding that they are being mal-treated. This is one of the reasons why child abuse goes undetected—children are not likely to tell anyone that they are being harmed. Another is that parents, caregivers, and other family members are often the perpetrators. These people are unlikely to tell the police or other authorities that abuse or neglect is occurring if they are the perpetrators. Others who may be concerned about the child may not know that maltreatment is occurring, so they may not report either.

Nonetheless, what we do know about the extent of child maltreatment mainly comes from official data sources. Because, as you will read later in this section, all states require that certain people report suspected child maltreatment to authorities, official statistics reflect these reports. The **National Child Abuse and Neglect Data System (NCANDS)** pro-vides an analysis of annual data on child abuse and neglect reports made to state child protective service agencies, including those in the District of Columbia and Puerto Rico (U.S. Department of Health and Human Services [DHHS], 2010). Once a referral is made to the child protective service agency, the agency determines whether the case will be screened in to determine if maltreatment has occurred or if a child is at risk (DHHS, 2010). The case can also be screened out if the agency determines that no maltreatment has occurred or that the child is not at risk (DHHS, 2010).

▲ **Photo 7.1**   Peggy McMartin Buckey and her son Ray Buckey in court during their trial for sexually molesting children at the McMartin pre-school that they operated. After one child alleged sexual abuse, police sent a letter to the parents of the other children who attended the pre-school. Eventually, the Buckeys were charged, along with others working at the pre-school, with abuse of over 40 children. Peggy was acquitted at trial on all charges, and the jury was deadlocked on several counts against Ray, but the case was later closed. The interviewing techniques used on the children came under severe scrutiny.

Photo credit: ©Bettmann/Corbis

Another resource for data on maltreated children is the **National Incidence Study (NIS)**. Most recently, data from the fourth NIS (NIS-4) were released. This study includes information on cases investigated by Child Protective Services but also cases that were identified by professionals in the community. To receive information on these cases, 10,791 professionals (called sentinels) submitted data forms on children with whom they had contact who were victims of abuse and/or neglect during the study period (3-month reference period in either 2005 or 2006) (Sedlak et al., 2010).

According to the NCANDS, in 2009 there were about 3.3 million referrals in the United States. These referrals dealt with the alleged maltreatment of about 6 million children (DHHS, 2010). Most of these referrals, 61.9%, were screened in for determination by child protective service agencies (DHHS, 2010). Of the referrals that were screened in, about one fourth resulted in Child Protective Services determining that the child was a victim of maltreatment, with 22% of these resulting in a substantiated disposition (DHHS, 2010). Results from the NIS-4 study indicate that more than 1.25 million children during 2005–2006 experienced maltreatment that resulted in demonstrable harm (Sedlak et al., 2010). When using a less restrictive definition, the NIS-4 study estimated that 1 in 25 children expe-rienced maltreatment (even if they were not yet harmed by maltreatment but the sentinel thought the maltreatment endangered the child or if a child protective services investigation indicated maltreatment) (Sedlak et al., 2010). Neglect accounted for the majority of cases reported in the NCANDS—more than three fourths suffered from neglect. Next most common was physical abuse (17.8%), followed by sexual abuse (9.5%) and psychological maltreatment (7.6%) (DHHS, 2010).

As with other official data sources, these estimates of child maltreatment are probably low since child abuse and neglect are likely to go undetected and unreported to Child Protective Services and/or law enforcement. One way to tap into this "dark figure of crime" is through surveys. As noted before, children may not be able to identify whether they have been victimized. To address this problem, researchers in one study asked questions directly of children aged 10 years and older but asked the caregiver who was "most familiar with the child's daily routine and experiences" questions about victimization experiences of children younger than 10 (Finkelhor, Turner, Ormrod, & Hamby, 2009, p. 2). Results from this study showed that 10.2% of the children had been maltreated by a significant adult in their life during the previous year (Finkelhor et al., 2009).

The most serious consequence of child maltreatment is death. Fortunately, most cases of child maltreatment do not result in fatalities, but some children lose their lives far too early at the hands of an abuser. In 2009, an estimated 1,770 children died as a result of child maltreatment (DHHS, 2010). Unlike maltreatment reported to NCANDS, generally, about one third of child fatalities were attributed exclusively to neglect, while one third were caused by multiple forms of maltreatment (DHHS, 2010).

Who Are Victims of Child Maltreatment?

Based on reports to child protective service agencies, the "typical" victim of child maltreatment is quite young. In fact, victims from birth to 1 year old had the highest rate of victimization (DHHS, 2010). More than 80% of children who died from child maltreatment were under the age of 4 (DHHS, 2010). Victims were almost equally likely to be males as females (48.2% compared to 51.1%). As you may imagine, however, female children are more likely than male children to be the victims of child sexual abuse (Sedlak et al., 2010). Male children, on the other hand, are at greater risk of dying from child maltreatment than are female children (DHHS, 2010). White children made up the largest percentage of reports (44.0%), but African American (22.3%) and Hispanic (20.7%) children were disproportionately represented in reports, given their composition in the population (DHHS, 2010). Similarly, the NIS-4 study revealed rates of maltreatment for Black children to be higher than those for both White and Hispanic children (Sedlak et al., 2010).

Who Perpetrates Child Maltreatment?

To understand the causes of child maltreatment, it is particularly instructive to identify the perpetrator of the abuse. Data from the NCANDS show that biological parents are the most likely perpetrators of child maltreatment reported to Child Protective Services (DHHS, 2010). More than 80% of child maltreatment incidents in 2009 were perpetrated by parents. Mothers are more likely to be perpetrators of child maltreatment than are fathers, except in cases of child sexual abuse (DHHS, 2010). This difference may be attributable to the fact that mothers are likely to be responsible for the majority of caregiving responsibilities. Of the incidents not perpetrated by parents, other relatives make up the largest group, followed by unmarried partners of parents (DHHS, 2010). Most perpetrators are White (48.5% in 2009) and between the ages of 20 and 39 (66.4% in 2009) (DHHS, 2010).

It may be difficult to imagine that a parent could take the life of a child, but of the child maltreatment incidents that resulted in death, more than three fourths were perpetrated by one or both parents (DHHS, 2010). Of these, 27.3% were perpetrated by a mother acting alone, 14.8% by a father acting alone, and 22.5% by both a mother and father (DHHS, 2010). Slightly less than 5% of child fatalities were caused by a parent's partner (DHHS, 2010).

Risk Factors for Child Maltreatment

For other types of victimization, we have examined risk factors tied to the victim that place him or her at more risk of being victimized as compared with others. The very nature of being a child may place children at risk. That is, children's vulnerability in terms of size and not being able to protect themselves may make them targets. In addition, children who are "fussy" and who have difficult temperaments are more prone to abuse than other children (Rycus & Hughes, 1998). Examining the family structure and caregivers for factors tied to an increased risk for child abuse is also instructive, since traditional explanations of victimization such as routine activities and lifestyles theory do not apply.

Familial Risk Factors

Several factors about a child's family are tied to increased risk of maltreatment. One key factor consistently found in the lives of children who are maltreated is that they come from families living in poverty. This is not to say that all maltreated children live in poverty, but poverty and child maltreatment are correlated. Poverty is often tied to unemployment and a lack of general resources needed to meet the needs of children, often creating stress that leads to violence and neglect (Bond & Webb, 2010). In the NIS-4 study, children with parents who were unemployed and who lived in low socioeconomic status households had higher rates of maltreatment than other children (Sedlak et al., 2010).

In addition to poverty, the structure of the family impacts risk of child maltreatment. Children who live with two biological parents are at lowest risk of being maltreated, while children who live with a single parent who have a cohabiting partner are at greatest risk (Sedlak et al., 2010). Those children who live with a single parent with a cohabiting partner have a rate of abuse more than 10 times that of other children (Sedlak et al., 2010). In particular, boyfriends of mothers are responsible for a disproportionate share of child maltreatment. One study found that boyfriends account for half of all child abuse committed by nonparents (Margolin, 1992). In addition, family size is connected to maltreatment. Children who live in households with four or more children are at greater risk of maltreatment than children who live in households with two children (Sedlak et al., 2010). The role of fathers in particular also has been explored. See the reading by Neil B. Guterman and Yookyong Lee (2005) in this section for an overview of the role of fathers in child abuse.

Individual Risk Factors

Factors specific to the individual also relate to risk of child abuse. According to social learning theory, abusive behavior is learned behavior. In this way, people abuse or neglect their children because they experienced or witnessed such behavior when they were younger. Research shows that young teen mothers are particularly vulnerable to abusing and neglecting their children (Afifi & Brownridge, 2008). This relationship may be due to the fact that young mothers do not have the resources to meet the needs of their children. Also, they may not have the maturity to deal with the stresses of full-time parenthood. Both factors may lead to abuse and neglect.

Particularly in regard to neglect, parents may also simply lack the knowledge of what children need. In this way, they may not be equipped to provide the nurturing, food, and medical care required (Cantwell, 1999). In addition, parents may lack judgment for making good parenting choices, such as deciding when a child is old enough to be left alone (Cantwell, 1999). Under these circumstances, parents' behavior may meet the definitions of neglect.

Other individual risk factors include substance use by parents. Like criminal behavior and aggression, generally, child abuse and neglect are often linked to substance abuse. Between one third to two thirds of all child maltreatment cases involve substance use (DHHS, 1999). Substance abuse is tied to violent behavior but is also linked to an inability or unwillingness to provide physical or emotional support to children (Wilcox & Dew, 2008). Parents with addictions often fail to place their children before their addiction and may leave their children alone and unsupervised or in the care of untrustworthy people (Bond & Webb, 2010). Substance abuse often co-occurs with other problems such as mental illness, unemployment, stress, and impaired family functioning, all of which may increase risk for child maltreatment (Child Welfare Information Gateway, 2009). Parental mental illness may also place children at risk. Depression, particularly, has been linked to child maltreatment (Conron, Beardslee, Koenen, Buka, & Gortmaker, 2009). Depression may lead to poor parenting in that parents suffering from depression may be emotionally unavailable to their children (Weinberg & Tronick, 1998).

Consequences of Child Maltreatment

It is often said that children are resilient—that they bounce back from adversity to overcome even the worst of situations. This is often the case. For some children, being abused or neglected will not seemingly leave lasting scars. But for others, the effects of maltreatment are deep-rooted and long-lasting, impacting multiple areas of their lives.

Physical, Cognitive, and Developmental Effects

We have already discussed that abuse and neglect may cause death, but there are other physical effects that could be present. One outcome that has recently garnered attention is **shaken baby syndrome**, which causes brain hemorrhages, skull fractures, and retinal hemorrhages (U.S. National Library of Medicine, 2011). It is estimated that between 1,200 and 1,400 children in the United States alone are injured or killed each year by shaking (U.S. National Library of Medicine, 2011).

Shaken baby syndrome is a specific consequence of a specific action—shaking a baby—but other outcomes of maltreatment can impact development for children. Depending on the severity of abuse and neglect, it may cause impairments to brain development, which can effect cognitive, language, and academic abilities (Watts-English, Fortson, Gibler, Hooper, & De Bellis, 2006). Children who are neglected may not have their basic developmental needs met and may suffer from developmental delays. Consider that parents who are neglectful are less likely to provide an enriched environment that would allow children to flourish. Moreover, neglected children are less likely to be getting the necessary food and nutrients that a brain needs to develop normally. Accordingly, neglected children are more likely to experience developmental delays (Child Welfare Information Gateway, 2008a).

Psychological Effects

It is common for victims of all types of child maltreatment to experience guilt, shame, fear, and anger in response to their victimization. Children may blame themselves for the abuse or neglect or, if their family dissolves, for the breakup of their family unit. At least in some children, abuse and neglect has been associated with panic disorder, anxiety, depression, posttraumatic stress disorder, attention-deficit hyperactivity disorder, reactive attachment disorder, anger, and dissociative disorders (Child Welfare Information Gateway, 2008a).

Although all types of maltreatment may cause psychological harm, they are associated more commonly with different types of outcomes. For example, physical abuse may lead to internalizing emotions, which may lead to anxiety and depression (Bond & Webb, 2010). Neglect is commonly associated with children being unable to trust and being socially withdrawn (Bond & Webb, 2010). The psychological effects of sexual abuse often evince themselves in interpersonal relationships later in life. Sexually abused children often have difficulty developing and maintaining healthy sexual relationships (Browne & Finkelhor, 1986). In addition, child sexual abuse victims are more likely to experience anxiety, depression, anger, and problems with substance abuse than non-abused children (Browne & Finkelhor, 1986). Child sexual abuse is also linked to suicidal ideation (Martin, Bergen, Richardson, Roeger, & Allison, 2004).

Effect on Criminality and Other Behaviors

As discussed in Section 2, research has uncovered a victim-offender overlap in that people who are victims are often criminal and vice versa. In this way, one of the risk factors for criminality is being a crime victim. This relationship holds true for being a victim of child abuse as well. Research consistently has shown that persons who suffer child abuse are more likely to be involved in delinquent and criminal behavior than others. Being abused and neglected increases the risk of arrest for violent and general juvenile offenses by 1.9 times (Widom, 2000). This effect carries over to adulthood—abused and neglected children are 1.6 times more likely to be arrested as adults (Widom, 2000). Being sexually abused as a child increases the chances that a person will offend later in life as well. Sex offenders are more likely to have a history of sexual abuse than non-sex offenders (Jesperson, Lalumiere, & Seto, 2009). To read about the role child abuse plays in later intimate partner violence perpetration and victimization, see the reading by Anu Manchikanti Gómez (2010) included in this section.

The effect of maltreatment on crime and delinquency may be gendered in that, in retrospective studies of offenders, the relationship between child maltreatment and offending appears to be more relevant for females than males. In fact, female inmates reported more childhood maltreatment than did male inmates (McClellan, Farabee, & Crouch, 1997).

Criminal and delinquent behaviors are just one type of maladaptive behavior that is associated with child maltreatment. Many other behavioral consequences have been linked to child maltreatment. There appears to be a clear link between abuse and neglect in childhood and later use of alcohol and illegal drugs. It is not surprising, then, to learn that a large proportion of people in drug treatment programs report that they were abused as children (Kang, Magura, Laudet, & Whitney, 1999; Rounds-Bryant, Kristiansen, Fairbank, & Hubbard, 1998). Generally, abused and neglected children are more likely to have problems in school (Kelley, Thornberry, & Smith, 1997). In addition, these children are more likely to experience teen pregnancy (Kelley et al., 1997). Children who are sexually abused are more likely to engage in risky sexual behavior and to contract a sexually transmitted disease (Johnson, Rew, & Sternglanz, 2006). Beyond consequences tied to sexual health, child sexual abuse victims are more likely to have eating disorders (Smolak & Murnen, 2002), to engage in self-mutilating behaviors (Fliege, Lee, Grimm, & Klapp, 2009), and to experience revictimization (Classen, Pales, & Aggarwa, 2005).

Effect on Adult Poverty

The link between child maltreatment and later delinquency and criminal behavior has been explored, but other outcomes have also been investigated. Generally, this research has highlighted that child maltreatment can have long-lasting negative effects. Being maltreated as a child has been shown to increase a person's chances of living in poverty and being unemployed. In one study, adults who had experienced maltreatment in childhood had two times the risk of being unemployed (Zielinski, 2009). Physical abuse in childhood was linked to a 60% increase in the risk of living in poverty in adulthood (Zielinski, 2009).

Responses to Child Maltreatment

Legislation

As previously noted, the landmark federal legislation defining child maltreatment, the **Child Abuse Prevention and Treatment Act** (Pub. L. 93-247), was passed in 1974. This act provided definitions of child abuse and neglect and, as noted below, required states to pass mandatory reporting laws for suspected cases of child maltreatment in order to receive certain federal funding. It also established a national clearinghouse for information on child abuse, promoted research on child abuse and neglect, and provided grant and other monies to support research and programs designed to address child maltreatment.

The mandatory reporting requirement was especially important given the difficulty in detecting child abuse. Currently, every state in the United States has some form of **mandatory reporting law** for suspected child abuse cases. These laws require that certain individuals report to authorities (e.g., law enforcement, a child protective agency, Department of Health and Human Services) if they suspect that a child is a victim of maltreatment. Mandatory reporters typically include individuals who work with children, such as teachers, day-care workers, law enforcement, mental health care providers, social workers, and school personnel. Some states require film developers to report. States differ regarding the standard of knowledge that triggers the duty to mandatorily report. In some states, it is a "reasonable suspicion"; in others, it is a "reasonable cause to believe"; and in others, a person is required to "know or suspect" (Child Welfare Information Gateway, 2010b). When a person does not report suspected child maltreatment when required to do so, he or she could be held criminally liable. Most typically, this is a misdemeanor that carries a fine as punishment (Child Welfare Information Gateway, 2010b). Civil liability, however, can attach for non-reporting. To further encourage reporting, the Child Abuse Prevention and Treatment Act requires states to have legislation that provides immunity from prosecution, both criminally and civilly, for individuals who report suspected child abuse in "good faith." See the following box for Kentucky's mandatory reporting law for suspected cases of child abuse. In 2009, most reports were made by education personnel (16.5%) and legal and law enforcement personnel (16.4%) (DHHS, 2010). The next most common reporters were anonymous reporters (8.9%) and parents (6.8%) (DHHS, 2010).

KENTUCKY MANDATORY REPORTING REQUIREMENTS REGARDING CHILDREN

Who must report?	Any person must report abuse or neglect of a child.
Standard of knowledge	Any person who knows or has reason to believe that a child is being neglected or abused must report.
Definition of applicable victim	An "abused or neglected child" is any person who is under the age of 18, and whose health or welfare is harmed or is threatened with harm when his parent, guardian, or other person exercising custodial control or supervision of the child • inflicts physical or emotional injury or allows such to be inflicted; • creates or allows to be created a risk of physical or emotional injury; • engages in conduct that makes the parent unable to care for the needs of the child (e.g., alcohol or drug abuse); • continuously or repeatedly fails or refuses to provide essential parental care and protection for the child, considering the age of the child; • commits or allows to be committed sexual abuse, exploitation, or prostitution of the child; • creates or allows to be created a risk of sexual abuse, exploitation, or prostitution of the child; • abandons or exploits the child;

	• fails to provide the child with adequate care, supervision, food, clothing, shelter, and education or medical care necessary to the child's well-being (exemption is made for parents who elect not to provide medical treatment for religious reasons); or • fails to make sufficient progress on a court-approved case plan that would allow the child to be returned to the custody of the parent such that the child remains in foster care for 15 of the most recent 22 months.
Who to report to?	• Local law enforcement: Department of Kentucky State Police • Department of Social Services of the Cabinet for Human Resources • State or County Attorney

SOURCE: RAINN (2009). Reprinted by permission of RAINN, Rape, Abuse and Incest National Network.

Another approach to reducing child abuse through legislation has been through **safe haven laws**. These laws allow mothers or custodial parents who are in crisis to relinquish their babies anonymously to designated places so that their babies are protected and can receive medical care until a permanent home is found. The mother or custodial parent is protected from criminal prosecution if he or she follows the requirements of the safe haven laws. As of May 2010, 49 states and Puerto Rico had adopted safe haven laws (Child Welfare Information Gateway, 2010a). Most states allow infants (up to 72 hours old in some states, up to 1 month in others, and other ages stipulated in other states) to be dropped off at hospitals, health care facilities, law enforcement agencies, emergency service providers, or fire stations (Child Welfare Information Gateway, 2010a). As you may imagine, the goal of these laws is to reduce infanticide and abandonment of babies and to ensure that parents who feel they are incapable of caring for their infants know they can, without penalty to themselves, leave their children in the safe hands of others.

Criminal Justice System Response

Despite the difficulty in detecting child abuse, when the criminal justice system is made aware of suspected cases of child abuse, the cases are often treated seriously. Recent research analyzing 21 studies on prosecutions of child abuse cases revealed that they are treated similarly to other violent crimes (Cross, Walsh, Simone, & Jones, 2003). For cases that were carried forward with charges, conviction rates averaged 94% (Cross et al., 2003).

One reason why conviction rates may be high is because of the protections afforded victims in the courtroom. Recognizing the special care that maltreated children may need from the criminal justice system, courtroom assistance is provided to them in the form of a **guardian ad litem (GAL)**. A GAL appears on behalf of the child in court to represent the child's best interests. In child abuse and neglect cases involving allegations against a parent, GALs are often required since, by definition, the child's interests conflict with those of the parent. Many times the GAL will be a court-appointed special advocate, a volunteer with the National Court Appointed Special Advocates Association (2010). In addition to having court-appointed advocates, children may also be allowed to testify through videotape, closed-circuit television, or a two-way mirror rather than testify in the courtroom with the suspect present (Bennett, 2003). In addition, the judge may choose to clear the courtroom of all observers or of the suspect during the child's testimony (National Center for Prosecution of Child Abuse, 2010). Anatomically correct dolls may also be used to facilitate testimony so that the child can use the doll to demonstrate abusive behavior (National District Attorneys Association, 2008). In addition, the courtroom's furniture may be reorganized to make it more comfortable for testifying (Chon, 2010).

Elder Maltreatment

What Is Elder Maltreatment?

Maltreatment of the elderly is discussed in the same section as maltreatment of children because victimization of these two groups is similar in many ways. Both children and the elderly can be abused because of their status—they rely on others for their care. Because of this special status, persons who are charged with their care bear a special responsibility. When they do not meet this responsibility, they may be committing abuse or neglect.

Generally, maltreatment of a person over the age of 60 is considered elder maltreatment. Similar to child maltreatment, elder maltreatment comes in many forms. **Physical elder abuse** is nonaccidental harm against an elderly person that causes pain, injury, or impairment (National Center on Elder Abuse, 2011). Hitting, beating, pushing, shoving, slapping, kicking, punching, and burning are examples of physical abuse. In addition, it is elder physical abuse to force-feed, inappropriately administer drugs, inappropriately use physical restraints, and use physical punishment against elderly persons (National Center on Elder Abuse, 2011). It is also elder maltreatment to emotionally abuse an elderly person. **Emotional or psychological elder abuse** occurs when emotional pain or distress is caused by intimidation, humiliation, ridicule, blaming, verbal assaults, insults, and harassment (National Center on Elder Abuse, 2011). Less direct forms of emotional or psychological elder abuse include treating an elderly person like an infant; isolating an elderly person from family, friends, or activities; and refusing to engage or speak with the elderly person

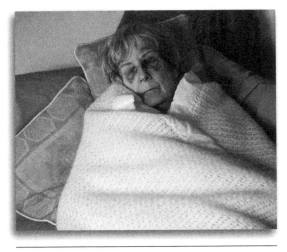

▲ **Photo 7.2** An 83-year-old woman who was physically abused by her home health assistant rests at home after being released from the hospital.

Photo credit: ©istockphoto/Mary Hope

(National Center on Elder Abuse, 2011). Older adults can also be sexually abused. Unlike children, who cannot give consent to engage in sexual conduct, adults can as long as they are cognitively able to do so. **Sexual elder abuse** occurs when sexual contact occurs with an elderly person without their consent. This contact can be physical touching (e.g., penetration, kissing) but can also be nonphysical, such as showing an elderly person pornographic images against his or her will or coercing a woman to undress against her will (National Center on Elder Abuse, 2011).

Neglect of elderly persons is also a form of maltreatment. When an elderly person is under someone's care and that person fails to fulfill a caretaking obligation, it is **neglect** (Center on Elder Abuse, 2011). One of these obligations may be fiduciary in that persons may be responsible for paying for an elderly person's care. In addition, in-home service providers may be neglectful if they fail to provide necessary care (National Center on Elder Abuse, 2011). Neglect at its most extreme comes in the form of abandonment. **Abandonment** occurs when an elderly person is deserted by a person who has assumed responsibility to provide his or her care or by a person who has physical custody of the elderly person (National Center on Elder Abuse, 2011).

Two additional types of maltreatment are specific to elderly persons. One is **financial exploitation**, which involves the illegal or improper use of an elderly person's property, assets, or funds (National Center on Elder Abuse, 2011). This misuse can be carried out by anyone, not just a caretaker. Examples of financial exploitation include cashing an elderly person's check without permission, stealing money, coercing an elderly person into signing a document such as a will, and forging an elderly person's signature (National Center on Elder Abuse, 2011). Elderly persons are also at risk of having a conservatorship, guardianship, or power of attorney misused. A power of attorney is a legal document that grants a person the ability to act on another's behalf (Stiegel, 2008). An example of **power of attorney abuse** is provided in following box. Seniors should also be wary of health care fraud and abuse. This type of elder maltreatment is perpetrated by nurses, doctors, health care workers, and professional care workers. It includes such behaviors as not providing health care but charging as though it were provided, double-billing or overcharging for medical services, overmedicating, undermedicating, recommending fraudulent remedies, and Medicaid fraud (Robinson, de Benedictis, & Segal, 2011).

ABUSE OF DURABLE POWER OF ATTORNEY: CASE EXAMPLE

Helen was an 85-year-old woman who was in poor health. Her daughter, Susan, was designated as her durable power of attorney, but Susan took advantage of this role. She sold Helen's home and placed the money from the sale into bank accounts in Helen's name, but a year later, she withdrew all the money. Instead of using the money for Helen's care, Susan used it for her own business ventures and personal expenses. Once Helen realized that the money was gone, she contacted the civil courts, but she could not afford a civil attorney and did not find much help with the free legal services program for elders. She also received little help from the Adult Protective Service Agency. Helen lost all hope. Six weeks later, Helen died.

SOURCE: Stiegel (2008). Copyright ©American Bar Association, 2008.

Although not a form of maltreatment committed by others, you should also be aware that older adults may self-neglect. **Self-neglect** occurs when elderly persons refuse or fail to provide the care they need for themselves or to take appropriate safety precautions. Examples of self-neglect are not eating or drinking, refusing to take necessary medication, and lack of personal hygiene (National Center on Elder Abuse, 2011).

Measurement and Extent of Elder Maltreatment

It is difficult to know to what extent elderly people are maltreated in the United States each year. One reason that a true picture of elder maltreatment is hard to paint is because many elderly persons may not report their victimization to law enforcement. This lack of reporting can be due to several reasons (Hodge, 1999). Elderly persons may not want to report if the perpetrator is a family member or loved one. They also may be dependent on the care of the perpetrator and be fearful that if they report, the person will retaliate against them. A real worry may also be that they will not have anyone to care for them and will be forced to live in an institutionalized setting. In addition, they may be unaware of the abuse, especially in cases of financial or medical abuse. It is also possible that elderly persons feel shame, humiliation, and embarrassment regarding their maltreatment that impedes their seeking assistance. Finally, they may be poor witnesses in that they may not recall specific details, may be confused about what happened, and may forget about their victimization (Hodge, 1999).

Reports From Adult Protective Services

Because reporting to the police is limited, to understand the extent of elder maltreatment, we must look to resources outside of police reports, although official reports to other agencies are often used. For example, similar to what is done for child abuse, if elder maltreatment is suspected, some people are required to report this suspicion to Adult Protective Services in their jurisdiction. Others may use Adult Protective Services as a reporting resource even if not required to report (e.g., a concerned neighbor). The **National Elder Abuse Incidence Study** was conducted in 1996 using reports made to adult protective service agencies in a sample of 20 counties in 15 states and reports from sentinels—trained individuals who have frequent contact with the elderly (Tatara, Kuzmeskus, Duckhorn, & Bivens, 1998). It measured the extent to which domestic elder maltreatment (maltreatment of elderly persons in noninstitutionalized settings) occurred against individuals aged 60 or older during 1996.

From this study, it was estimated that 449,924 elderly persons experienced abuse and/or neglect (Tatara et al., 1998). Few of these victims (16%) had their cases reported to and substantiated by Adult Protective Services (Tatara et al., 1998). In 1996, there were 236,479 reports of suspected maltreatment made to Adult Protective Services (actual reports, not including reports from sentinels). Almost half the cases were substantiated after Adult Protective Services investigated (Tatara et al., 1998). Most of these cases involved maltreatment by another (62%), as opposed to self-neglect (Tatara et al., 1998).

Also using information from Adult Protective Services, Teaster and colleagues (2006) produced estimates of elder maltreatment for the 2003 fiscal year. These estimates were produced based on survey responses of Adult Protective Services contacts in each state, who responded about elder maltreatment cases in their records. In 2003, 253,426 reports of maltreatment against persons aged 60 and older were made to Adult Protective Services. Of these reports, 76% were investigated and 35% were substantiated (Teaster et al., 2006).

Estimates Derived From Surveys

Elder maltreatment estimates also have been derived through survey research. In the first large-scale random sample survey of elder maltreatment, Pillemer and Finkelhor (1988) found that the prevalence rate of elder maltreatment was 32 persons per 1,000. The National Crime Victimization Survey is not the best resource for elder maltreatment victimization statistics. Although it does include elderly persons in its sample, it asks questions only of persons who are not institutionalized, reports only estimates broken down for persons in certain age groups, and measures only criminal victimization. To examine elder victimization, the latest survey provides estimates for persons aged 65 and older. Using this data, the violent victimization rate in 2007 for persons aged 65 and older was 3.3 per 1,000, and the property crime victimization rate was 75.0 per 1,000 (Bureau of Justice Statistics, 2010). Persons in this age category had the lowest victimization rates.

Other efforts have been made to produce estimates of the extent of elder maltreatment in the United States using surveys. One national-level study was conducted using the **National Social Life, Health, and Aging Project**. This sample included 3,005 individuals aged 57 to 85 who lived in the community. They were asked if they had experienced verbal, financial, or physical maltreatment during the previous year. The results of the study do not provide estimates of maltreatment perpetrated by non-family members, but 9% of the sample reported verbal maltreatment, 3.5% reported financial maltreatment, and 0.02% reported physical maltreatment by family members (Laumann, Leitsch, & Waite, 2008).

Special Case: Elder Maltreatment in Institutions

Slightly more than 5% of the population aged 65 and older live in nursing homes or assisted living facilities (Orel, 2010). Although these facilities are charged with caring for people who cannot find this care in their own homes, sometimes people are abused or neglected there. According to Adult Protective Services records, the majority of incidents of elder maltreatment occur in domestic settings. Only 6% of substantiated reports of elder maltreatment occurred in long-term care settings, and 2% occurred in other locations, which included assisted living facilities (Teaster et al., 2006). Another source of data regarding elder maltreatment in long-term care facilities is the **Long-Term Care Ombudsman Program**. Every state is required to have an ombudsman program under the federal Older Americans Act. The Long-Term Care Ombudsman Program receives complaints about elder maltreatment in long-term care facilities. In 2008, more than 271,000 complaints about long-term care were made to the program (in 2008, the Ombudsman Program investigated more than 271,000 complaints made by 182,506 individuals and provided information on long-term care to another 327,000 people). Of these complaints, 12,916 were for abuse, gross neglect, or exploitation (Administration on Aging, 2008). To see how victimization of the elderly compares to victimizations of other persons, see the reading by Ronet Bachman and Michelle L. Meloy (2008) included in this section.

Special Case: Intimate Partner Violence of Older Women

As women get older, if they remain in abusive relationships, it makes sense that these relationships will remain abusive without intervention. It is also possible that women enter into new relationships that are abusive. Although research shows that older women are less likely to experience intimate partner violence than their younger counterparts, older women do still face some risk. One study reported that almost 2% of women over the age of 55 in the sample reported physical abuse in their intimate relationships (Zink, Fisher, Regan, & Pabst, 2005). Similar percentages reported experiencing sexual abuse (Zink et al., 2005). Most commonly, the person who perpetrated the abuse was the woman's spouse (Zink et al., 2005). This research suggests that older women may be at risk of elder maltreatment not only at the hands of family members but also by their intimate partners.

Who Are Victims of Elder Maltreatment?

Adult Protective Services reports and survey data show that females make up a greater proportion of victims of elder maltreatment than do males (Bureau of Justice Statistics, 2010; Tatara et al., 1998; Teaster et al., 2006). This disproportionate representation of females in victimization statistics is unlike most types of victimization. Why do you think this is the case? There may be several reasons. One reason is that people may be more watchful over older women and may be more likely to report suspected cases of maltreatment. Another reason may be due to the longer life expectancy of women—they make up a larger proportion of the elderly population than do males (Zink et al., 2005). Another reason may be tied to the victim-offender relationship. As you will learn in the following section, perpetrators of elder maltreatment are often spouses, and elderly females are at risk of experiencing domestic violence at the hands of their partners. Some of the maltreatment presented in the statistics may be domestic violence.

As mentioned, elder maltreatment includes maltreatment of individuals over the age of 60; however, not all persons in this age group share similar risks of experiencing maltreatment. From Adult Protective Services reports, it seems that older persons are at greatest risk. In fact, slightly less than half (43%) of substantiated cases of maltreatment were against individuals who were 80 years old or older (Teaster et al., 2006). Along with gender and age, another demographic characteristic to consider is race. In both surveys and in official Adult Protective Services data reports, White elderly persons are shown to be the majority of the victims (Bureau of Justice Statistics, 2010; Laumann et al., 2008; Teaster et al., 2006).

Characteristics of Elder Maltreatment Victimization

Most cases of investigated and substantiated elder maltreatment are of self-neglect (Teaster et al., 2006). Next most common are caregiver neglect cases, followed by financial exploitation cases (Teaster et al., 2006). Results from the

National Incidence Study were slightly different. Although self-neglect cases were most common, physical abuse cases were the next most commonly substantiated, followed by abandonment, emotional/psychological abuse, financial/material abuse, and neglect (Tatara et al., 1998). These differences are likely due to temporal differences in study time period and in study design. From the National Social Life, Health, and Aging Project, it was found that verbal maltreatment committed by family members was more common than both financial maltreatment or physical maltreatment perpetrated by family members; physical maltreatment was the least commonly experienced type of elder maltreatment out of the three types examined (Laumann et al., 2008).

Who perpetrates elder abuse? Adult Protective Services data show that females make up the majority of suspected perpetrators, although these statistics were based on reports from only 11 states (Teaster et al., 2006). Most often, an adult child was the perpetrator, constituting one third of all alleged perpetrators in substantiated elder maltreatment cases (Teaster et al., 2006). Other family members were the alleged perpetrators in 22% of the cases, and 11% of the alleged perpetrators were spouses or intimate partners (Teaster et al., 2006). More than three fourths of the alleged perpetrators were under the age of 60; about one fourth were between the ages of 40 and 49 (Teaster et al., 2006).

Risk Factors for Elder Maltreatment

Although it may be hard to imagine how anyone could harm or neglect an elderly person, certain risk factors place a person at risk of being a perpetrator or a victim of elder maltreatment.

Perpetrator Risk Factors

If the family has suffered from domestic violence, elder maltreatment may be an outgrowth of this domestic violence pattern. As persons age and roles change, the parents who were once the abusers may become the abused by children who hold resentment. In this way, the children may be continuing a cycle of violence (Fedus, 2010). Even if abuse was not present in the family's history, adult children who have taken on a caregiving role may find it to be quite stressful. This stress may overwhelm them to the point that they become abusive or neglectful (Bonnie & Wallace, 2003). This explanation is rooted in **dependency theory**—as dependency of the elderly person increases, stress will increase, as will the risk of abuse and neglect (Bonnie & Wallace, 2003). Children who take on this care-taker role may think it entitles them to access their parents' financial resources, even if they do not legally have access to these resources. In this way, financial exploitation may occur.

Risk factors for institutional abuse are also tied to the perpetrator. Research on abuse in nursing homes shows that nursing aids compose the largest group of abusers (Payne & Cikovic, 1995). Nursing aids typically work long shifts, often for little pay. They usually have to care for many patients on their shift—elderly persons who require high levels of supervision and who have high needs—even though they may have received minimal training. The demands placed on them are, indeed, great. Without the necessary training and skills, abuse and neglect are likely unfortunate consequences.

Routine Activities Theory

Older persons typically do not have risky lifestyles. They may, however, be seen as vulnerable or suitable targets. They may be unable to physically protect themselves from victimization and, thus, may be easily harmed. Likewise, some older people may be suitable targets for scam artists, given their waning cognitive ability or lack of techno-logical sophistication. In this way, people may take advantage of them. These risk factors have been linked in research to elder maltreatment. Elders who are unable to care for themselves are more likely to be abused than others (Tatara et al., 1998). Similarly, about 60% of substantiated elder abuse victims experienced some degree of confusion (Tatara et al., 1998). Persons in nursing homes are also likely to experience cognitive deficits, and about half of them are confined to a bed or wheelchair (Orel, 2010). These characteristics make these elderly persons more vulnerable to abuse but also increase the burden of caring for them. That is, the stress of caring for individuals who are unable to care for themselves or who have cognitive impairments is likely high; thus, caretakers who are not equipped to care for these people are more likely than others to become frustrated and respond in abusive and aggressive ways.

Older persons, especially if they live alone, also may lack capable guardianship. Consider that family members make up a large portion of perpetrators of elder maltreatment—even if elderly persons do have guardianship, it may not be capable if the people who are supposed to be providing guardianship are the ones mistreating them. In this way, routine activities theory may be applicable to elder maltreatment. When motivated offenders, suitable targets, and lack of capable guardianship coalesce in time and space, elder maltreatment is likely to occur.

Responses to Elder Maltreatment

In recognition of elder maltreatment as a serious form of victimization that warrants special consideration in the law and the criminal justice system, special laws and criminal justice programs have been developed.

Legislation

The **Older Americans Act of 1965** (amended in 2006, Pub. L. 109-365) provided protection of rights for older Americans, among other things. One of these protections involved the creation of the State Long-Term Care Ombudsman Program discussed earlier. In addition, in order to receive certain funds, state agencies must develop and enhance programs to address elder maltreatment. It also provides grants for states to create elder justice systems. More recently, the **Elder Justice Act** (Pub. L. 111-148) was signed into law in 2010. This law provides $400 million over 4 years for Adult Protective Services, monies for grants to detect and prevent elder abuse, to establish and support forensic centers relating to elder maltreatment, to support the Long-Term Care Ombudsman Program, and to enhance long-term care staffing.

States also have legislation to protect elderly persons. Most abuse perpetrated against elderly persons is illegal in all states because existing statutes make these behaviors illegal regardless of the victim's age. Emotional/psychological abuse and neglect is also illegal, depending on the behavior of the perpetrator, the outcome for the victim, and the nature of the relationship between the perpetrator and the victim. For example, most states impose a legal duty on adult children to care for and protect their elderly parents (Stiegel, Klem, & Turner, 2007). Once a person has taken on the role of caregiver, that person has a duty to provide care or seek assistance in providing care. Failing to do so can lead to that person being held criminally liable (Stiegel et al., 2007).

Along with criminalizing abuse and neglect, all states have Adult Protective Services agencies to which people can report suspected cases of elder maltreatment. Forty-eight states require at least some persons to report suspected elder maltreatment, while the other two (Colorado and South Dakota, as of December 2009) encourage but do not mandate reporting (RAINN, 2011b). States vary as to who is a mandatory reporter, but people who have regular contact with older adults, such as physicians and health care workers, are mainly mandatory reporters. Reports can be made to Adult Protective Services or to other agencies such as law enforcement. See the following box for Alabama's mandatory reporting law for elder maltreatment.

ALABAMA'S MANDATORY REPORTING REQUIREMENTS REGARDING ELDERS/DISABLED

Who must report?	• Physicians • Other practitioners of the healing arts • Caregivers
Standard of knowledge	Reasonable cause to believe that any protected person has been subjected to physical abuse, neglect, exploitation, sexual abuse, or emotional abuse
Definition of applicable victim	A "protected person" is anyone over 18 years of age • who has an intellectual disability or developmental disability (including but not limited to senility); or • who is mentally or physically incapable of adequately caring for himself or herself and his or her interests without serious consequences to himself or herself or others.
Who to report to?	• The county department of human resources or the chief of police of the city and county, or • the sheriff of the county if the observation is made in an unincorporated territory Reports of a nursing home employee who abuses, neglects or misappropriates the property of a nursing home resident shall be made to the Department of Public Health.

SOURCE: RAINN (2011a). Reprinted by permission of RAINN, Rape, Abuse and Incest National Network.

Criminal Justice System Response

The criminal justice system has also responded to the problem of elder maltreatment. The criminal justice community has joined forces to develop multidisciplinary and multiagency teams to investigate and prosecute elder maltreatment. For example, the Northeast Healthcare Law Enforcement Association (NHLEA) consists of law enforcement managers from federal, regional, and state elder, patient abuse, and health care fraud law enforcement units. The NHLEA enforces, investigates, and prosecutes maltreatment against elderly persons and has developed a criminal offender elder abuse database to track offenders convicted of elder and patient abuse (Hodge, 1999). Other jurisdictions have Financial Abuse Specialist Teams designed to investigate financial exploitation of elderly persons. Specialized prosecution units also have been instituted. San Diego has an Elder Abuse Prosecution Unit that specializes in the prosecution of elder maltreatment and provides training to law enforcement in how to investigate elder abuse and training to banks in how to identify financial exploitation (Hodge, 1999).

As you can see, the causes and consequences of both child maltreatment and elder maltreatment are situated in the life course, but child abuse has garnered more widespread attention than has elder abuse. Nonetheless, with the passage of the Elder Justice Act and the aging of the population, coupled with longer life spans, elder maltreatment may become a policy and programmatic hot bed.

SUMMARY

- The two major forms of child maltreatment are abuse and neglect.
- The difference between discipline and physical abuse of a child is the intent to harm versus the intent to teach the child right from wrong.
- Because children often cannot verbalize how they are being abused and their parents or caregivers often are the abusers, it is difficult to know the extent of child maltreatment in the United States.
- Data sources such as the National Child Abuse and Neglect Data System and National Incidence Study (NIS) provide most of the information known about child maltreatment.
- Results from the NIS-4 indicate that more than 1.25 million children experienced maltreatment during 2005–2006 that resulted in demonstrable harm.
- Death is the most serious outcome of child maltreatment, and one third of child fatalities in 2009 were attributed to neglect.
- Victims of child maltreatment are almost equally likely to be male and female, with children under age 4 making up 80% of fatalities in 2009.
- White children made up the largest percentage of reports of child maltreatment, but African American and Hispanic children were disproportionately represented in reports, given their composition in the population.
- Biological parents of the child are most likely to be the abusers, with mothers more likely than fathers to be the perpetrator.
- Risk factors for child maltreatment include child's temperament and vulnerability. Family risk factors include living in poverty, stress, single-headed households, and living with a single parent with a cohabiting partner. Parenting risk factors include witnessing or experiencing violence as a child or adolescent, substance use/abuse, depression, and mental illness.
- There are many consequences of child maltreatment. Children who are maltreated are at heightened risk of experiencing cognitive and developmental disorders and suffering from psychological disorders. They also may experience problems in school and have problems in their sexual relationships. Later in life, maltreated children are at risk of living in poverty and being unemployed.
- Children who are victims of maltreatment are likely to be involved in criminal behavior, delinquency, and to use alcohol and drugs.
- The Child Abuse Prevention and Treatment Act made it mandatory for states to report suspected cases of child maltreatment in order to receive certain federal funding.
- Mandatory reporting laws require certain people to report suspected cases of child abuse.
- Safe haven laws protect mothers and custodial parents from criminal prosecution so they can give their babies up if they feel they cannot raise them.
- The criminal justice system recognizes that children need special attention and certain court-appointed advocates when they testify in court.
- In addition to abuse and neglect, financial exploitation is a form of maltreatment specific to the elderly. Elderly persons are also at risk of having conservatorship, guardianship, or power of attorney misused.
- Elders are also capable of self-neglect when they refuse or fail to provide care for themselves, such as not eating or drinking, refusing to take necessary medication, and lack of personal hygiene.
- The Long-Term Care Ombudsman Program is mandatory in every state, allowing for elder maltreatment complaints to be reported.

- In 1996, Adult Protective Services received 236,479 reports of suspected maltreatment.
- In 2006, only 6% of substantiated reports of elder maltreatment occurred in long-term care settings, and 2% occurred in other locations, which included assisted living facilities.
- Females make up a greater proportion of victims of elder maltreatment than do males.
- Females are also the most likely perpetrators of elder maltreatment. Adult children are most likely to be the perpetrators of elder abuse.
- Most cases of investigated and substantiated elder maltreatment are cases of self-neglect. Next most common are caregiver neglect cases, followed by financial exploitation cases.
- As dependency of the elderly person increases, stress will increase, as will the risk for abuse and neglect.
- Research on abuse in nursing homes shows that nursing aids compose the largest group of abusers.
- Elderly persons may be unable to physically protect themselves from victimization and, thus, may be easily harmed. Some older people may be suitable targets for scam artists, given their waning cognitive ability or lack of technological sophistication.
- Almost all states have mandatory reporting laws for suspected cases of elder abuse.
- The Elder Justice Act, signed into law in 2010, provides $400 million to Adult Protective Services for detection and prevention of elder abuse and/or maltreatment.
- The criminal justice system has the Northeast Healthcare Law Enforcement Association, which consists of officers and officials who enforce, investigate, and prosecute maltreatment offenders.
- Overall, child and elder maltreatment share some similarities, such as the types of abuse and risk factors, and show some differences in responses by the criminal justice system and typical perpetrators.

DISCUSSION QUESTIONS

1. What are some issues with generating accurate estimates of the extent of child maltreatment and elder maltreatment?

2. Why do you think parents are the most likely to be perpetrators of their children's maltreatment? Why are adult children the most likely perpetrators of elder maltreatment?

3. What is the victim-offender overlap as it pertains to child maltreatment?

4. Why do you think child maltreatment has been linked with adult poverty?

5. Why do you think females are more likely to be victims of elder maltreatment than are males? Why are male children more likely to be killed through child maltreatment than are female children?

KEY TERMS

physical abuse

emotional abuse

child sexual abuse

neglect

National Child Abuse and Neglect Data System (NCANDS)

National Incidence Study (NIS)

shaken baby syndrome

Child Abuse Prevention and Treatment Act

mandatory reporting law

safe haven laws

guardian ad litem (GAL)

physical elder abuse

emotional or psychological elder abuse

sexual elder abuse

abandonment

financial exploitation

power of attorney abuse

self-neglect

National Elder Abuse Incidence Study

National Social Life, Health, and Aging Project

Long-Term Care Ombudsman Program

dependency theory

Older Americans Act of 1965

Elder Justice Act

INTERNET RESOURCES

Child Welfare Information Gateway (http://www.childwelfare.gov)

This website provides a great deal of information about child maltreatment. It includes data, statistics, and laws on child maltreatment, along with information on how child maltreatment can be prevented. It also has links to resources in Spanish.

National MCH Center for Child Death Review (http://www.childdeathreview.org/)

This is a resource center for state and local death review teams that investigate child deaths. Links for each state's child mortality data are provided so that you can see the most common causes of death, including those attributable to child maltreatment.

National Center on Elder Abuse (http://www.ncea.aoa.gov)

This website provides background information about elder maltreatment, including definitions and statistics. Nursing home abuse, a type of elder maltreatment, is discussed. Links to information about how to report elder maltreatment and about Adult Protective Services are also provided.

National Committee for the Prevention of Elder Abuse (http://www.preventelderabuse.org/)

This website provides information about the different types of elder maltreatment, including sexual violence, domestic violence, and financial exploitation. Links to publications and current news items related to elder maltreatment are also included.

Introduction to Reading 1

Official statistics show that mothers are suspected in the majority of all child maltreatment cases. Despite this statistic, researchers have identified that fathers often play an integral role in the etiology of child abuse and neglect. To address this issue, Guterman and Lee (2005) discuss the important role that fathers play in cases of physical child abuse and neglect. They pay particular attention to the role of fathers given the fact that fathers are overrepresented in cases of severe physical child abuse. In doing so, they consider how the presence or absence of the father contributes to maltreatment, how the role of the father impacts the family economically and how this impacts maltreatment, how the age of the father impacts maltreatment, how substance use by the father impacts maltreatment, how the father's childhood impacts maltreatment, and how father/mother interaction impacts maltreatment.

The Role of Fathers in Risk for Physical Child Abuse and Neglect

Possible Pathways and Unanswered Questions

Neil B. Guterman and Yookyong Lee

Fathers and Maltreatment: Overrepresented and Underconsidered

Consideration of the role that fathers play in the risk for future physical child abuse and neglect is long overdue. A growing body of evidence has pointed out that fathers, as well as father figures, are highly overrepresented as perpetrators of physical child abuse, particularly in its most severe forms (e.g., Brewster et al., 1998; Krugman, 1985; Margolin, 1992). For example, Sinal et al.'s (2000) review of inflicted closed head injury (shaken baby syndrome) cases in North Carolina reported that 44% were perpetrated by fathers and 20% were perpetrated by mothers. Similarly, a review of child-maltreatment-related fatalities in the state of

Missouri reported that while 21% of identified perpetrators were biological mothers, 23% were biological fathers, and 44% were unrelated males in the household (Stiffman, Schnitzer, Adam, Kruse, & Ewigman, 2002). Given that fathers provide, on the whole, substantially less direct child care than mothers (Margolin, 1992; Yeung, Sandberg, Davis-Kean, & Hofferth, 2001), these proportions of fathers and possible father surrogates as perpetrators of severe child abuse appear as rather startling.

Despite the overrepresentation of fathers as perpetrators in severe physical child maltreatment, concern for the role of fathers and fathering in the etiology of physical child abuse and neglect has, until very recently, remained largely in the background in child maltreatment research. Lately, however, interest in the role of

SOURCE: Guterman and Lee (2005). Reprinted with permission.

fathers regarding physical child abuse and neglect has grown in conjunction with increased acknowledgement of the major sociological shifts taking place in relation to father's roles in American families more generally. Many have noted the dramatic increase in mothers' participation in the American labor force and the changing nature of both gender relations and child care provision in the United States (Cabrera, Tamis-LeMonda, Bradley, Hofferth, & Lamb, 2000; Marsiglio, Amato, & Day, 2000).

Given these shifts, the role and even the presence of fathers in the rearing of their children have occupied a growing degree of the public's recent attention. Interest in developmental research and public discourse in the 1980s and 1990s initially focused on the connection between fathers' presence (or, more accurately, their absence) and child development outcomes. However, developmental researchers have more recently unpacked the father presence-versus-absence dichotomy, refuting preconceived notions that fathers, particularly those viewed as high risk, are uninvolved in parenting and examining more elaborately the role of fathering and fathers' involvement in family life and child well-being (e.g., Danziger & Radin, 1990; Field, 1998; Hossain, Field, Pickens, Malphurs, & Del Valle, 1997; Phares & Compas, 1992; Vandell, Hyde, Plant, & Essex, 1997; Vogel, Boller, Fardell, Shannon, & Tamis-LeMonda, 2003).

With respect to physical child abuse and neglect risk, prior research has implicated the influence of fathers rather indirectly. For example, a host of prior studies has observed single parenthood (i.e., single motherhood) as a risk factor for physical child abuse and neglect (e.g., Dubowitz, Hampton, Bithoney, & Newberger, 1987; Gelles, 1989; Schloesser, Pierpont, & Poertner, 1992) but has rarely assessed the nature of father's absence or nonresident fathers' involvement as shaping risk or protective elements for future physical child abuse or neglect (Dubowitz, Black, Kerr, Srarr, & Harrington, 2000). Similarly, one of the most clearly established empirical patterns found in the literature is the association between low socioeconomic status and risk for child abuse and neglect, especially for physical neglect (e.g., National Research Council, 1993). To date, such research has emphasized the economic and psychosocial stressors that accompany a family's low socioeconomic status rather than the specific economic role played by fathers, as differentiated from that played by mothers, in heightening or reducing physical child abuse and neglect risk.

Prior research examining social network relationships and physical child abuse and neglect has also rather indirectly suggested an important role for fathers by documenting mothers' problematic relationships with their significant others as a correlate of child maltreatment (e.g., Coohey, 1995; Corse, Schmid, & Trickett, 1990; Kirkham, Schinke, Schilling, & Meltzer, 1986; Straus & Kantor, 1987). Despite such links, there remains little precise understanding of the specific ways that mothers' relationships with fathers shape the family system to potentially heighten physical child abuse and neglect risk or, conversely, to potentially lower such risk in a protective fashion.

In the absence of a well-elaborated empirical base clarifying the role of fathers in physical child abuse and neglect risk, many theoretical lenses have been drawn on to help explain the role of fathers as relevant to physical child abuse and neglect. Each theoretical model places certain elements of fathering in the foreground while placing others in the background and none accounts for the complete array of fathering factors that might explain their roles in a comprehensive way.

To illustrate, sociobiological theory, for example, emphasizes adaptive behaviors that increase the likelihood of passing on one's genes to future generations and highlights that parents invest and divest effort in child rearing and children based on their genetic closeness (df. Malkin & Lamb, 1994; Radhakrishna, Bou-Saada, Hunter, Catellier, & Kotch, 2001). Sociobiological theory has been applied as a lens from which to understand higher rates of physical abuse and neglect by stepfathers or mothers' boyfriends; however, it is limited in helping to provide an understanding as to why known biological fathers are also overrepresented as perpetrators of physical abuse and neglect, or why biological mothers maltreat their children who carry reproductive value.

Other theories such as feminist and economic theories also help in understanding various roles that fathers may play within the family. Feminist theory has placed emphasis on father's power as related to their gender and the potential abuse of their power in the family context. It has particularly been applied to shed light on the high co-occurrence observed between domestic violence and child maltreatment (Margolin, 1992). Economic theories have been principally helpful in emphasizing the economic role fathers play in family life, providing a useful framework from which to understand the evidence identifying higher physical abuse and, especially, neglect rates in single-mother households and households with unemployed fathers (Gillham et al., 1998; Paxson & Waldfogel, 1999, 2002; Pelton, 1994).

Still other theories such as psychodynamic, family systems and attachment theories have emphasized fathers' relational patterns within the family, their affective ties (and their origins) in the mother-father dyad, and in parent-child relationships. These theories have pointed out for example, tensions and potential alliances in the adult dyad that contribute to maltreating parenting or problems in the attachment between father and child (Rothbaum, Rosen, Ujiie, & Uchida, 2002). As an example, psychodynamic literature has emphasized the potential for a father or a mother to displace anger toward a child in response to felt abandonment or coercion by the other member of the parental dyad or the potential for role reversal, leading to physical abuse of a child in response to unfulfilled paternal expectations that the child behave like their parent (e.g., Steele, 1987).

An ecological model or framework has often been relied on to help integrate the varied theoretical lenses and empirical patterns that might explain the multiple levels of influence in risk for physical child abuse and neglect (Belsky, 1980; Garbarino, 1977). Although ecologically informed investigations have served, in particular, to underscore the importance of transactional patterns from micro- to macrosystem contexts, neither this broader ecological framework nor more specific theoretical traditions have yet offered a comprehensive explication of the range of biopsychosocial elements of fathering that might shape risk for physical child abuse and neglect. Given this, the present article seeks to examine, in a more multifaceted and comprehensive fashion, the many potential father pathways that may shape physical child abuse and neglect risk and the existing empirical evidence in relation to these pathways. Toward this end, we review prior research that discerns these possible pathways or provides support for their presence, and, at the same time, we offer unanswered questions and research recommendations that would assist in establishing a more complete and empirically grounded understanding of the role that fathers play in physical child abuse and neglect risk.

For this article, given the nascent state of the empirical findings on this subject, we examine the role of fathers in risk for both physical child abuse and physical child neglect. Although empirical investigations increasingly isolate findings with regard to the occurrence of physical child abuse in contrast with child neglect, a substantial proportion of the existing empirical studies specifically examining fathering elements combine findings for both these forms of child maltreatment. Given this, we identify when available, specific findings from the empirical base for each form of physical child maltreatment. We also specifically focus in this article on the role of fathers, per se, in physical abuse and neglect risk as a point of departure rather than also examining the role of the father figures of other adult males who may have involvement in the homes of families at risk. We recognize that these latter males present an additional set of specialized considerations for physical child abuse and neglect risk beyond the scope of this present review and for which, at present, there is preciously little extant empirical information available (cf. Holden & Barker, 2004, for further discussion). Nonetheless, a focused review of fathers' physical child abuse and neglect risk, per se, can aid in furthering a next generation of studies that can begin to examine more specialized questions in connection with father figures or other unrelated adult males in the home.

⊠ Fathers' Presence or Absence and Maltreatment Pathways

Among the earliest and most consistently reported findings implicating the important role of fathers in physical child abuse and neglect risk are those detecting higher maltreatment rates in single-parent (i.e., mother-headed) households (Gelles, 1989; Giovannoni & Billingsley, 1970; Seagull, 1987). In their examination of data from the Third National Incidence Study, Sedlak and Broadhurst (1996) reported that children in single-parent families experienced a 77% greater risk of being harmed by physical abuse and an 87% greater risk of being physically neglected. With respect to physical abuse, some researchers have posited that the absence of fathers contributes to fewer financial, child caring, and emotionally supportive resources available in the home, thus straining the mother's capacity to care for her child and heightening the likelihood she will act in a coercive and abusive fashion (Gelles, 1989; Seagull, 1987). Although some researchers have posited similar pathways explaining the links between physical child neglect and fathers' absence (e.g., Polansky, Chalmers, Buttenwieser, & Williams, 1979); other more recent research has indicated that fathers' absence, by itself, does not predict child-neglect risk (Dubowitz et al., 2000).

Prior studies have documented that father absence is often associated with familial economic deprivation (Black, Dubowitz, & Starr, 1999; McLoyd, 1990; Paxson & Waldfogel, 1999; Pelton, 1994), with, for example, 34% of single-mother-headed households living below the poverty level (as compared with 16% of single-father-headed households; Fields & Casper, 2001). Although poverty continues to be identified as among the most closely associated risk factors for both physical abuse (Straus, Gelles, & Steinmetz, 1980) and neglect, especially among the poorest of the poor (Brown, Cohen, Johnson, & Salzinger, 1998; National Research Council, 1993; Pelton, 1994), empirical research is still necessary to directly document the specific pathway from father absence, per se, to heightened physical child neglect and abuse risk via family impoverishment.

Complicating the picture, fathers' absence, most often assessed dichotomously in prior studies, does not necessarily denote an absence of fathers' involvement in the life of the child or family. Conversely, fathers' mere presence in the home may not necessarily mean a higher degree of their involvement when compared to those families assessed dichotomously as having absent fathers. Some have emphasized that single-mother homes, in fact, consist of a wide variety of patterns of father and/or other adult male involvement. For example, Radhakrishna et al. (2001) have noted that some single-mother homes involve a comparatively greater number of unrelated adult male figures, which may contribute to greater instability in the mother-child dyad, thereby heightening maltreatment risk.

Whether a father resides at home or not, he may play a variety of roles, economically related and otherwise, that shape a child's safety, risk, and well-being. Other aspects of fatherhood and fathers' involvement in family life may play significant roles in shaping child neglect and abuse risk such as their employment status and age, their own socialization experiences in childhood (particularly those that may be related to abuse or neglect themselves), the characteristics of their relationships with the mother and the child, and their potential use of psychoactive substances.

Fathers' Employment Status, Economic Hardship, and Pathways to Child Maltreatment

Beyond whether fathers are present or absent in the family, the specific role they play in a family's economic well-being has been linked with physical child abuse and neglect risk. It has long been documented that economic hardship is one of the most consistently identified risk factors for physical child abuse and neglect. For example, a number of prior studies have reported that severe or fatal injuries due to physical abuse and neglect are more likely to be found among families with low annual incomes (e.g., Gelles, 1992; Gill, 1970; Kruttschnitt, McLeod, & Dornfeld, 1994; Sedlak & Broadhurst, 1996; Wolock & Horowitz, 1979). As well, Kruttschnitt et al. (1994) reported that the risk of recurrent abuse appears to be related to the length of time a family has been in poverty.

Although the majority of such studies have not isolated the ways in which fathers' specific economic contribution to the family may shape maltreatment risk, several studies have identified father-specific aspects of economic hardship that are correlated with physical child abuse and neglect risk. For example, studies have reported that unemployed fathers are far more likely than employed fathers to physically abuse their children (Jones, 1990; Wolfner & Gelles, 1993). One study, using state-level aggregate data, found that states with higher proportions of nonworking fathers also report higher rates of maltreatment (Paxson & Waldfogel, 1999), although these researchers note that caution should be taken in generalizing such findings to individual-level behaviors.

Some researchers have hypothesized that unemployment can lower the male breadwinner's status within the family and that such loss in status might provoke a father to attempt to reassert his authority by engaging in physically abusive and violent behaviors toward the child and/or other family members (e.g., Madge, 1983; Straus, 1974). Studies have reported that fathers who have sustained heavy financial losses tend to become more irritable, tense, and explosive, which in turn increases their tendency to become more punitive toward their children. (cf. McLoyd, 1990). It may be that economic losses are perceived as stressful, especially in an uncontrollable way. According to stress theory, it is such uncontrollable stresses that appear to most directly contribute to the breakdown of personal coping capacities (cf. Lefcourt, 1992; Pearlin, 1999) and to thus elicit more negative psychosocial outcomes (e.g., Brosschot et al., 1998; Peeters, Buunk, & Schaufeli, 1995). In relation to parenting, Bugental and colleagues have specifically documented that when parents perceive a loss of control or power in their lives, they tend either to behave coercively toward their children in response, motivated by a desire to regain lost control, or, conversely, to behave in a tenuous or withdrawing fashion toward their children in response to their perceived precarious state of authority (Bugental, Lewis, Lin, Lyon, & Kopeikin, 1999). Such research is particularly instructive because it begins to suggest one potentially important mediating pathway explaining the mechanisms through which fathers' economic pressures may directly shape physical child abuse and neglect risk, thus outlining potential targets for intervention.

Economic hardship has also been closely related to greater transience in residence, lower educational attainment, higher rates of mental health disorders (including substance abuse), and less adequate social support (Sedlak & Broadhurst, 1996), each of which may also independently influence fathers' problematic parenting as a consequence. Thus, future research is necessary that examines the potentially varied direct and indirect pathways that fathers' economic hardship takes in shaping high-risk fathering behavior and in shaping mothers' own high-risk parenting behavior. Nonetheless, it appears that fathers' economic insecurity and job loss are likely to contribute both directly and indirectly to heightened physical child abuse and neglect risk via multiple pathways.

Young Fatherhood and Pathways to Child Maltreatment

As with mothers, the age at which a man becomes a parent may play an important role in the degree to which risk and protective elements may be in place, shaping the potential for physical child abuse and neglect, particularly if this age is a very young one. Prior research suggests that several characteristics of young fatherhood foreshadow the possibility of future child maltreatment. The transition to parenthood is a difficult passage for most. It has been suggested that adolescent fathers, in particular, begin to experience inordinate stress, fear, and negative emotions from the point at which they discover that a women they have had sexual relations with is pregnant (Elster & Panzarine, 1980; Westney, Cole, & Munford, 1986). Early fatherhood has been linked with more negative parenting attitudes and behaviors, sometimes leading to a withdrawal of involvement in the relationship with the mother and the baby, and to declining satisfaction with parenting more generally (cf. D. B. Miller, 1994).

Various studies have noted that younger fathers are particularly vulnerable to experiencing economic hardships because of their relatively lower educational status and the relatively fewer employment opportunities they face in comparison with older fathers (Bolton, 1987; Elster & Panzarine, 1983; Lamb & Elster, 1985;

Rhein et al., 1997; Rivara, Sweeney, & Henderson, 1986; Samuels, Stockdale, & Crase, 1994). However, what has not yet been empirically documented is whether young fatherhood may in some ways interact with or potentiate the influence that financial insecurity plays in shaping future physical child abuse and neglect risk. Like older fathers, younger fathers facing difficult economic circumstances may encounter common challenges in coping with seemingly uncontrollable stresses.

However, young fatherhood presents additional realities that might complicate fathers' relationships with their children and their partners, potentially heightening physical child abuse and neglect risk. For example, several studies of young fathers have reported their relative lack of preparedness for fatherhood, both cognitively and emotionally (Caparulo & Lonson, 1981; Rivara et al., 1986). Studies have documented that adolescent fathers' knowledge of infant development tends to be deficient and unrealistic (De Lissovoy, 1973). Furthermore, Vaz, Smolen, and Miller (1983) reported that psychological depression and social isolation were found to be present in almost one third of adolescent fathers studied, conditions closely associated with child maltreatment risk in mothers (e.g., Sidebotham, Golding, & the ALSPAC Study Team, 2001; Whipple & Wilson, 1996). Nonetheless, empirical evidence has yet to directly link young fathers' depression and physical child abuse and neglect risk.

Adolescent fathers have also been reported to be inordinately involved in illicit activities and drug use (Bolton, 1987; Fagot, Pears, Capaldi, Crosby, & Leve, 1998; Larson, Hussey, Gillmore, & Gilchrist, 1996; Rhein et al., 1997) and to have difficulty in controlling their tempers (Bolton, 1987). As with depression, although maternal substance abuse and involvement in illicit activity have been linked with greater physical child abuse and neglect risk, future research can establish whether and to what degree such a link exists for young fathers as well.

It is important to recognize that the role of young paternal age in physical child abuse and neglect risk is a little-understood area, and the existing evidence base rests largely on studies with a variety of methodological limitations, many of which derive from the challenges

researchers face in enrolling representative samples of young fathers in research studies. At present, the large majority of studies in this area relies on small numbers of study participants, the majority of whom have been limited to African American and low-income fathers or fathers already identified as at risk.

Fathers' Substance Abuse and Maltreatment Pathways

Notwithstanding limited empirical evidence concerning young fathers, prior studies have reported clear links between fathers' substance abuse (including alcohol) more generally and heightened risk for both physical abuse and physical neglect (Ammerman, Kolko, Kirisci, Blackson, & Dawes, 1999). Earlier work has reported parental substance abuse to be a strong predictor of risk for subsequent physical child abuse and neglect (Chaffin, Kelleher, & Hollenberg, 1996), and several studies have specifically reported fathers' substance abuse as correlated with physical abuse and neglect. For example, Moss, Mezzich, Yao, Gavaler, and Martin (1995) reported that substance abusing fathers exhibited more than twice the scores on the Child Abuse Potential Inventory than comparison group fathers where no substance abuse was present. Furthermore, drawing from protective service records, Murphy et al. (1991) reported that 59% of the maltreatment (both abuse and neglect) cases that involved substance abuse identified fathers as substance abusers.

Despite this clear correlational evidence identifying fathers' substance abuse as a risk factor, little is known with regard to exactly how fathers' substance abuse may serve to heighten child maltreatment risk (McMahon & Rounsaville, 2002). It appears likely that fathers' substance abuse may influence child maltreatment risk through multiple pathways and in ways that affect other risk and protective factors. Although little is known about what role fathers' substance abuse may play on a family's overall financial stability or its management of stressors, some evidence indicates that paternal substance abuse plays a key role in the functioning of the parental dyad. Paternal substance abuse has been associated, for example, with maternal substance abuse (Barnett & Fagan, 1993) and domestic violence (Bennett & Lawson, 1994), both of which have high comorbidity with physical child abuse and neglect (e.g., Edleson, 1999; Magura & Laudet, 1996). Male partners have been found to influence a woman's introduction to alcohol or substance use, including the use of harder drugs (e.g., Amaro & Hardy-Fanta, 1995). Once involved in drug or alcohol abuse, women appear to face far greater odds of experiencing partner violence than women not involved (B. Miller, 1998).

Fathers' substance abuse also conceivably influences their own direct provision of child care linked with both physical neglect and physical abuse. Studies by Eiden and colleagues have found that, in comparison with non-alcoholic fathers, alcoholic fathers are less sensitive and show higher levels of negative affect toward their infants, and their infants are less securely attached (e.g., Eiden, Chavez, & Leonard, 1999; Eiden, Edwards, & Leonard, 2002). It has also been found that fathers' alcoholism is associated with higher paternal irritation with the infant, greater aggression toward the parental partner, and greater maternal antisocial behavior and depression (Eiden & Leonard, 2000; Leonard et al., 2002). Substance and/or alcohol abusing fathers, similar to mothers, may also plausibly show impaired judgment about appropriate parenting expectations or a child's developmentally appropriate behavior, although this remains to be directly empirically documented.

Fathers' Own Upbringing and Pathways to Child Maltreatment

Although it is plausible that fathers' own socialization experiences may play an important part in determining fathering behavior deemed as abusive or neglectful, little direct evidence has examined the influence of father's childhood experiences on their parenting. A growing literature has sought to identify an intergenerational association between abuse experiences in childhood with current parenting behaviors, although a number of important methodological issues limit these findings, which are most frequently based on cross-sectional, retrospective research designs drawing on

parents' own memories (Egeland, 1993; Widom, 1989). Such retrospective studies have variously estimated intergenerational transmission rates of physical child abuse at approximately 30% (Kaufman & Zigler, 1993), with studies reporting rates ranging between 7% and 70% (Belsky, 1993).

Few studies have examined this pattern specific to fathers. In cross-sectional studies, Ferrari (2002) reported that fathers experiencing childhood abuse used physical punishment less frequently with their own children, whereas Merrill, Hervig, and Milner (1996) reported that fathers recalling parental violence toward them also reported significantly higher physical child abuse potential. More recent longitudinally executed studies have begun to more clearly outline a modest relationship between childhood abuse and fathers' physical abuse risk, although the present findings are far from conclusive. For example, Horwitz, Widom, McLaughlin, and White (2001) have reported that experiences of childhood abuse and neglect appear to increase the likelihood of subsequent antisocial personality disorder in men, a factor that has been found more prevalent in child abusing parents than nonabusing parents (e.g., Dinwiddie & Bucholz, 1993). In twin-study research controlling for genetic influences on intergenerational patterns of violence for boys, Jaffe, Moffitt, Caspi, Taylor, and Arseneault (2002) reported that domestic violence exposure in early childhood accounted for 5% of the variance in boys (as well as girls) externalizing behavior at 5 years of age. However, neither study examined males' parenting behavior as an outcome of experienced maltreatment in childhood.

More direct evidence linking fathers' childhood maltreatment experiences and their future risk for abuse or neglect of their own children remains sparse. Doumas, Margolin, and John (1994) reported that males' exposure to marital aggression in their families of origin predicts their own parental (as well as marital) aggression. Similarly, several studies (e.g., Halford, Sanders, & Behrens, 2000; Swinford, DeMaris, Cernkovich, & Giordano, 2000) have reported that male partners who recounted observing or experiencing violence in their family of origin exhibit greater propensity to perpetrate violence with their intimate partners. However, these studies have typically over-looked an examination of males' parenting behavior.

Although social learning theory posits the likely intergenerational link between male socialization experiences and fathering behavior, a highly underexplored area from an empirical standpoint includes ways in which fathers' childhood experiences directly predict their future risk for physically abusing and especially for neglecting their children. Studies examining this linkage will remain highly challenging to execute rigorously given the limitations attendant to retrospective reporting and the challenges of controlling for genetic and other environmental influences in teasing out the unique role of family-of-origin experiences on fathering behaviors (e.g., DiLalla & Gottesman, 1991; Horwitz et al., 2001).

Father, Mother, and Child Interactions and Maltreatment Pathways

Direct and indirect evidence indicates that fathers' interactions with mothers play an important role in physical child abuse and neglect risk. Early work by Belsky (1979) notably pointed out that relational qualities of the parental dyad have important correspondences with qualities of the parent-child relationship. Such mutually influential interactional patents in the father-mother-child triad have been demonstrated across numerous studies. For example, the quality of mothering provided to an infant has been linked with supports the mother receives from her partner, and the quality of the relationship between partners has been shown to predict how both mothers and fathers nurture and respond to their children's needs (Brunelli, Wasserman, Rauh, Alvarado, & Caraballo, 1995; Donovan & Leavitt, 1989; Parks, Lenz, Jenkins, 1992; Samuels et al., 1994).

Several prior studies have specifically identified a buffering role that fathers' support may play in maternal-child relationships at risk for future physical child abuse and neglect. For example, fathers' support can play a protective role in relation to mothers' depression, shielding infants from negative outcomes (Field, 1998), promoting greater maternal responses to their children (Jackson, 1999), and minimizing power-assertive

maternal child-rearing attitudes (Brunelli et al., 1995). Sunukarky, Unger, and Wandersaman (as cited in Samuels et al., 1994) have shown that teen mothers with positive partner support are less rejecting and punitive toward children. Conversely, some direct empirical evidence suggests that low father support toward the mother is intertwined with mothers' risk of both physical abuse and physical neglect (Kotch et al., 1995; Whipple & Webster-Stratton, 1991). For example, Coohey (1995) reported that, although three quarters of regretful mothers in her study stated having partners, they also recounted feeling less companionship, less instrumental support, and less exchange of resources from their partners in comparison with matched nonneglectful mothers.

Fathers' direct involvement in child care also predicts qualities of the home environment (Cutrona, Hessling, Bacon, & Russell, 1998) and child developmental outcomes (Feldman, Greenbaum, Mayes, & Erlich, 1997), including risk for child neglect (Dubowitz, 1999). Prior studies have noted that fathers often form independent attachments with infants that promote their security, and infants security of attachment with both mothers and fathers appears to be mutually influenced and interdependent (Field, 1998; Fox, Kimmerly, & Schafer, 1991; Hossain & Roopnarine, 1994). However, evidence suggests that the nature of fathers' involvement with their children likely plays a complex role in directly shaping risk, specifically for abuse and neglect risk.

For example, in one of the first studies directly examining fathers' involvement and child neglect risk, Dubowitz et al. (2000) reported that fathers' greater direct involvement with child care was positively linked with higher child neglect risk but that their involvement in other household domains was linked with lower child neglect risk. These researchers suggest that the fathers' greater participation in child care in this study may have been an indicator of mothers' relative unavailability in the child caring role, which itself may have led to heightened child neglect risk. Such empirical findings suggest a complicated and not yet well-understood picture of ways in which fathers' involvement in caring for their children in conjunction with other roles and interactions in the home may heighten and/or reduce physical child maltreatment risk. It is likely that future studies examining this complex interplay will not only require study of both mother and father behaviors simultaneously and in conjunction with one another, but the collection of data from both partners, given that empirical patterns have been reported differentially across informants (Phares & Compas, 1992).

Particularly in the early phases of parenting, partners must make major adjustments to parenting a new infant in the home, and they face increasing stresses that challenge their relationship and parenting behaviors. Fathers are frequently concerned about what role they will play in parenting, their increasing family burdens, and about what changes are being brought about by the infant's presence (e.g., Guterman, 2000). Mothers are often concerned about the evolving role their partner will play. As well, the relationship between mothers and their partners may pass through a substantial transition, raising the potential of increased conflict and declining satisfaction in the relationship (cf. Nitz, Ketterlinus, & Brandt, 1995; Osofsky et al., 1985).

Mounting evidence underscores that troubled or violent relationships between fathers and mothers appear particularly linked with physical child abuse and neglect risk, and, similarly, that coercive interactions between mothers and fathers appear linked with heightened coercive behaviors toward children (Appel & Holden, 1998; Corse, Schmid, & Trickett, 1990; Dumas, 1986; Edleson, 1999; Kirkham et al., 1986; O'Keefe, 1995; Salzinger et al., 2002; Straus & Gelles, 1986; Straus & Kantor, 1987). For example, Rosenbaum and O'Leary (1981) reported that parents who use physically aggressive tactics to resolve spousal disputes also tend to use similar tactics in disciplining their children. Similarly, according to reports from battered women, violent husbands are less involved in child rearing and use less induction and physical affection and more negative control techniques in their child-rearing practices (Holden & Ritchie, 1991).

In line with these findings, a number of studies have reported that the presence of domestic violence between adult partners is closely associated with physical child abuse and neglect, particularly in its most severe forms.

For example, domestic violence was found to be present in 41% of cases of critical and fatal child maltreatment in a state of Oregon review (Oregon Children's Services Division, 1993) and in 40% to 43% of child maltreatment fatalities in child fatality reviews from New York City and the state of Massachusetts (Child Fatality Review Panel, 1993; Felix & McCarthy, 1994). Similarly, studies conducted from hospital settings of children suspected of being maltreated have reported that between 45% and 59% of their mothers showed evidence of being battered by their partners (McKibben, De Vos, & Newberger, 1991; Stark & Filtcraft, 1988).

Aside from heightened risk of direct physical abuse and neglect against the child, interparental violence has been linked with other detrimental mental health sequelae in children, including increased aggression, depression, cognitive delays, and symptoms of posttraumatic stress disorder (Shipman, Rossman, & West, 1999). Given such consequences, children's mere exposure to partner violence (independent of their direct victimization) has been increasingly considered within legal and conceptual definitions of child neglect (e.g., Magen, 1999). The complex interplay among interparental violence and child maltreatment continues to require further empirical unraveling, for example, in tracing the temporal sequencing of varying forms of family violence and neglect when they are found to co-occur and why child and partner abuse co-occur in some families and not in others.

⬚ Future Directions for Research and Prevention

Recent research has shown that fathers play a key role in the well-being of family members and, particularly, in the quality of child rearing provided in the home by both mothers and fathers. In particular, findings continue to accumulate identifying an important set of father-related factors linked with risk for physical child abuse and neglect (see Table 1). These include key sociodemographic and economic factors such as a father's age, his presence or absence in the home, his employment status, and the degree of job insecurity that he faces. In addition, a number of psychosocial factors related to fathers' experiences and behaviors play a role in shaping families' risk for physical child abuse and neglect, including a father's potential involvement in substance or alcohol abuse, the degree to which he may have experienced maltreatment in his own family of origin, the nature of the direct child-caring activities he provides in the home, and the degree to which he supports or undermines the mother in her parental role, most especially if such parental interactions involve domestic violence. As Table 1 indicates, each of these father factors has some direct or indirect correlational evidence supporting its role in shaping a family's physical child abuse and neglect risk. Such an inventory of factors can begin to hint at the contours of an empirically guided risk and clinical assessment strategy for professionals considering the role of fathers in physical maltreatment risk.

At the same time, such findings raise numerous unanswered questions in relation to how and via what specific pathways such elements may directly operate to heighten or reduce maltreatment risk, as the existing empirical base is almost completely devoid of studies that employ casual modeling strategies (e.g., via the use of prospective research designs and such statistical strategies as structural equation modeling). As noted earlier, for example, fathers' substance abuse, as with the majority of other factors reviewed herein, may directly and indirectly influence pathways, and/or may be a comorbid outcome with child maltreatment, stemming from other causal factors. Similarly, although fathers' unemployment experiences are associated with heightened physical child abuse and neglect risk, it has not yet been clearly established as to which process mediates and moderates the influence of such experiences on maltreatment risk. Are there, for example, common paternal stress responses, maternal stress responses, or broader familiar responses to fathers' unemployment that ultimately lead to detrimental parenting and clear detriment with respect to the child's safety? Knowledge of such mediating and moderating processes would enable professionals to begin to develop prevention strategies that

Table 1   Empirical Evidence Supporting the Role of Specific Father Factors in the Risk for Physical Child Abuse and Neglect

Paternal Factor	Indirect Evidence	Direct Correlational Evidence	Causal Pathway Identified
Socioeconomic factors			
Absence		X	
Low income	X		
Unemployed		X	
Job or major financial loss	X		
Young age	X		
Psychosocial factors			
Substance abuse		X	
Family-of-origin maltreatment	X		
Support/undermining of mother		X	
Violence toward mother		X	
Direct provision of child care		X	

can assist fathers and other family members during periods of unemployment that might mitigate subsequent maltreatment risk.

With regard to some father factors, empirical evidence is available that specifies a role in one parent's (either a mother's or a father's) future abuse or neglect risk. For example, evidence is available indicating that fathers' support or undermining of the mother directly plays a significant role in mothers' parenting behavior. On the other hand, for many fathering factors, such as a father's young age, his use of substances, or his family-of-origin experiences, studies must still trace the differential roles these factors specifically play on each parent's at-risk parenting behavior separately. For example, how does a father's substance abuse shape his own parenting behavior and, differentially, play a role in shaping a mother's interactions with her children? Similarly, in what ways might a father's direct provision of child care both play a role in his own propensity toward physical child abuse or neglect and, separately, influence the mother's own maltreatment propensity?

Limiting the empirical base, the large majority of studies examining fathering and parent-child interaction derive their finding from fathers' self-reports or, more commonly, mothers' reports of fathering behavior rather than from direct observations of fathering behavior itself. However, comparisons of fathers' versus mothers' reports of parenting behavior indicate important differences across informants, emphasizing the need to triangulate the data collected in relation to fathering behavior (Phares & Compas, 1992).

Also limiting the present empirical base, the majority of studies that have examined the potential role that fathers may play in physical child abuse and neglect jointly have only recently begun to separate physical neglect from physical abuse outcome findings (e.g., Dubowitz et al., 2000). It is highly plausible, however, that different father-related factors may play unique roles across these forms of child maltreatment. For example, might socioeconomic enactors such as job loss or low income play different roles in shaping risk for physical child abuse as opposed to physical child neglect? Similarly, in what ways might fathers' direct

involvement in child caring predict physical child abuse versus physical child neglect risk? It is likely that differential father pathways are discernible across these forms of child maltreatment and across the maltreating perpetrator, mother or father.

A looming, unanswered question with respect to fathers' roles in physical maltreatment risk is the way that culture shapes fathers' roles in family life and how such variations may alter physical child abuse and neglect risk. The vast majority of studies reviewed in this article have examined either White or African American fathers and have not specifically examined the role of cultural factors in maltreatment risk. Studies examining ethnicity in the context of physical child abuse and neglect have most often reported findings on mothers alone (e.g., Coohey, 2001) or have not separately examined fathers from mothers in their analyses (e.g., Wissow, 2001). We are aware of only one study that has compared fathering behaviors and outcomes across cultures vis-à-vis physical child abuse and neglect risk (Ferrari, 2002), which preliminarily reported that Latino ethnicity and value placed on machismo and formalism played significant roles in predicting fathers' use of physical punishment. Several studies examining close proxies of physical child abuse and neglect have similarly reported important fathering differences across cultures. In a nationally representative sample of two-parent families, Hofferth (2003), for example, found African American fathers exhibiting less warmth, more control, and more responsibility toward their children than White fathers, and White fathers exhibiting more control and less responsibility for their children than Latino fathers.

These limitations in the existing empirical base prevent a more comprehensive and precise understanding of fathers' roles in physical child abuse and neglect risk. Given the present state of the empirical base, we offer the following set of recommendations for the next generation of studies aiming to advance the knowledge base examining fathers' roles in physical child maltreatment risk.

First, future studies should assess fathers' involvement in family life and child rearing in a multifaceted fashion beyond global indicators (such as presence versus absence or time spent with a child). The early and emerging evidence suggests that father involvement is a highly complex process and may operate in unexpected ways, shaping physical child abuse and neglect risk. Fortunately, the growing empirical base on fathers and fathering more generally is yielding more sophisticated and psychometrically sound assessment measures that can be employed to trace a more multidimensional picture of fathers' influence on family life and child rearing.

Second, given differing findings reported across informants in earlier studies, future studies should move beyond the mere use of maternal self-report, collecting data on behavioral patterns and hypothesized predictors from multiple informants, and especially from fathers themselves. Furthermore, parent-child observational coding systems (cf. Eyberg & Robinson, 1982; Roberts, 2001) can be readily adapted to track fathers' interactions with their children to cross-validate and track informant biases of either parent about fathering behavior.

Third, future studies should attempt to document father-relevant factors as predictors of each parent's (mother's versus father's) at-risk behavior separately, as influential processes may operate differently across parents and/or operate in an interactional pattern between parents.

Fourth, future studies should track physical child abuse and physical child neglect risk as separate outcomes rather than combining both as a single outcome of interest. Early evidence suggests that although each of these forms of maltreatment may share some overlapping etiological elements, fathering factors may likely operate differentially across types.

Fifth, future studies should attempt to more precisely tease out causal directionality in fathering factors under study, and the mediating and moderating processes that accompany these factors, so as to enable more relevant application to intervention and prevention purposes. Toward this end, prospective research designs drawing from population-based studies offer distinct methodological advantages over retrospective designs using clinically based samples

(Guterman, 2004; Widom, Raphael, & DuMont, 2004). In addition, the employment of multivariate techniques, especially the use of causal modeling strategies (such as structural equation modeling), will hold the greatest potential to identify causal pathways, their directionality, and important mediating and interactional processes.

Finally, early findings on cultural differences in fathering indicate that future research must begin to more explicitly trace cultural elements that might shape fathers' contribution to the risk and protective elements predicting child maltreatment risk. These may minimally include such culturally based elements as fathers' attitudes and behaviors concerning gender relations, power assertion, and child care, as well as culturally based expectations about fathers' roles as economic providers, especially given these factors' identified links with risk status for physical child abuse and neglect.

Clear opportunities exist to achieve major advances in developing a more comprehensive and detailed understanding of the father pathways that determine physical child abuse and neglect risk. Such knowledge, once attained, will provide direct guidance to those aimed at engaging and working with fathers and their families in ways that can reduce their risk for future physical child abuse and neglect. Although professionals working to avert and to reduce physical child abuse neglect risk must consider how best to address fathers' role in family life, they presently face precious little available empirically validated knowledge that can guide their practices. The evidence base has clearly established that fathers play an important role in physical child abuse and neglect, and, given this, it is essential to begin to develop intervention strategies and models that address fathers' needs and motivation for services, their risk profiles, their help-seeking patterns, as well as intervention strategies to engage and work with fathers in ways that help to enhance the protective elements and minimize the risk elements shaping physical child abuse and neglect. Given the inordinate role that fathers play in the most severe cases of physical child abuse and neglect risk, empirical advances that help develop an understanding of and an effective response toward at-risk fathers are likely to make a major contribution to protecting the lives and safety of vulnerable children.

⊠ References

Amaro, H., & Hardy-Fanta, F. (1995). Gender relation in addiction and recovery. *Journal of Psychoactive Drugs, 27*(4), 325–337.

Ammerman, R. T., Kolko, D. J., Kirisci, L., Blackson, T. C., & Dawes, M. A. (1999). Child abuse potential in parents with histories of substance use disorder. *Child Abuse & Neglect, 23*(12), 1225–1238.

Appel, A. E., & Holden, G. W. (1998). The co-occurrence of spouse and physical child abuse: A review and appraisal. *Journal of Family Psychology, 12*(4), 578–599.

Barnett, O. W., & Fagan, R. W. (1993). Alcohol use in male spouse abusers and their female partners. *Journal of Family Violence, 8*(1), 1–25.

Belsky, J. (1993). Etiology of child maltreatment: A developmental ecological analysis. *Psychological Bulletin, 114*(3), 413–434.

Belsky, J. (1979). The interrelation of parental and spousal behavior during infancy in traditional nuclear families: An exploratory analysis. *Journal of Marriage and the Family, 41*(4), 749–755.

Belsky, J. (1980). Child maltreatment: An ecological integration. *American Psychologist, 35*(4), 320–335.

Bennett, L., & Lawson, M. (1994). Barriers to cooperation between domestic-violence and substance-abuse programs. *Families in Society, 75*(5), 277–286.

Black, M. M., Dubowitz, H., & Starr, R. H., Jr. (1999). African American fathers in low-income, urban families: Development, behavior and home environment of their three-year-old children. *Child Development, 70*(4), 967–978.

Bolton, F. G., Jr. (1987). The father in the adolescent pregnancy at risk for child maltreatment. 1. Helpmate or hindrance? *Journal of Family Violence, 2*(1), 67–80.

Brewster, A. L., Nelson, J. P., Hymel, K. P., Colby, D. R., Lucas, D. R., McCanne, T. R., et al. (1998). Victim, perpetrator, family, and incident characteristics of 32 infant maltreatment deaths in the United States Air Force. *Child Abuse & Neglect, 22*(2), 91–101.

Brosschot, J. F., Godaert, G. L., Benschop, R. J., Olff, M., Ballieux, R. E., & Heijnen, G. J. (1998). Experimental stress and immunological reactivity: A closer look at perceived uncontrollability. *Psychosomatic Medicine, 60*(3), 359–361.

Brown, J., Cohen, P., Johnson, J. G., & Salzinger, S. (1998). A longitudinal analysis of risk factors for child maltreatment: Findings of a 17-year prospective study of officially recorded and self-reported child abuse and neglect. *Child Abuse & Neglect, 22*(11), 1065–1078.

Brunelli, S. A., Wasserman, G. A., Rauh, V. A., Alvarado, L. E., & Caraballo, L. R. (1995). Mothers' report of paternal support: Association with maternal child-rearing attitudes. *Merrill Palmer Quarterly, 41*(2), 152–171.

Bugental, D. B., Lewis, J. C., Lin, E., Lyon, J., & Kopeikin, H. (1999). In charge but not in control: The management of teaching relationship by adults with low perceived power. *Developmental Psychology, 35*(6), 1367–1378.

Cabrera, N. J., Tamis-LeMonda, C. S., Bradley, R. H., Hofferth, S., & Lamb, M. E. (2000). Fatherhood in the twenty-first century. *Child Development, 71*(1), 127–136.

Caparulo, F., & Lonson, K. (1981). Adolescent fathers: Adolescents first, fathers second. *Issues in Health Care of Women, 3*(1), 23–33

Chaffin, M., Kelleher, K., & Hollenberg, J. (1996). Onset of physical abuse and neglect: Psychiatric, substance abuse, and social risk factors from prospective community data. *Child Abuse & Neglect, 20*(3), 191–203.

Child Fatality Review Panel. (1993). *Child fatality review panel annual report for 1993.* New York City Human Resources Administration.

Coohey, C. (1995). Neglectful mothers, their mothers, and partners: The significance of mutual aid. *Child Abuse & Neglect, 19*(8), 885–895.

Coohey, C. (2001). The relationship between familism and child maltreatment in Latino and Anglo families. *Child Maltreatment, 6*(2), 130–142.

Corse, S. J., Schmid, K., & Trickett, P. K. (1990). Social network characteristics of mothers in abusing and nonabusing families and their relationships to parenting beliefs. *Journal of Community Psychology, 18*(1), 44–59.

Cutrona, C. E., Hessling, R. M., Bacon, P. L., & Russell, D. W. (1998). Predictors and correlates of continuing involvement with the baby's father among adolescent mothers. *Journal of Family Psychology, 12*(3), 369–387.

Danziger, S. K., & Radin, N. (1990). Absent does not equal uninvolved: Predictors of fathering in teen mother families. *Journal of Marriage and the Family, 52*(3), 636–642.

De Lissovoy, V. (1973). Child care by adolescent parents. *Children Today, 4*(1), 22–25.

DiLalla, L. F., & Gottesman, I. I. (1991). Biological and genetic contributors to violence: Widom's untold tale. *Psychological Bulletin, 190*(1), 125–129.

Dinwiddie, S. H., & Bucholz, K. K. (1993). Psychiatric diagnoses of self-reported child abusers. *Child Abuse & Neglect, 17*(4), 465–476.

Donovan, W. L., & Leavitt, L. A. (1989). Maternal self-efficacy and infant attachment: Integrating physiology, perceptions, and behavior. *Child Development, 6*(10), 43–56.

Doumas, S., Margolin, G., & John, R. S. (1994). The intergenerational transmission of aggression across three generations. *Journal of Family Violence, 9*(2), 157–175.

Dubowitz, H. (1999). The families of neglected children. In M. E. Lamb (Ed.), *Parenting and child development in "nontraditional" families* (pp. 327–345). Mahwah, NJ: Lawrence Erlbaum.

Dubowitz, H., Black, M. M., Kerr, M. A., Starr, R. H., Jr., & Harrington, D. (2000). Father and child neglect. *Archives of Pediatrics and Adolescent Medicine, 154*(2), 135–141.

Dubowitz, H., Hampton, R. L., Bithoney, W. G., & Newberger, E. H. (1987). Inflicted and noninflicted injuries: Differences in child and familial characteristics. *American Journal of Orthopsychiatry, 57*(4), 525–535.

Dumas, J. E. (1986). Indirect influence of maternal social contacts on mother-child interactions: A setting event analysis. *Journal of Abnormal Child Psychology, 14*(2), 205–216.

Edleson, J. L. (1999). The overlap between child maltreatment and women battering. *Violence Against Women, 5*(2), 134–154.

Egeland, B. (1993). A history of abuse is a major risk factor for abusing the next generation. In R. J. Gelles & D. R. Loseke (Eds.), *Current controversies in family violence* (pp. 197–208). Newbury Park, CA: Sage.

Eiden, R. D., Chavez, F., & Leonard, K. E. (1999). Parent-infant interactions among families with alcoholic fathers. *Development and Psychopathology, 11*(4), 745–762.

Eiden, R. D., Edwards, E. P., & Leonard, K. E. (2002). Mother-infant and father-infant attachment among alcoholic families. *Development and Psychopathology, 14*(2), 253–278.

Eiden, R. D., & Leonard, K. E. (2000). Paternal alcoholism, parental psychopathology, and aggravation with infants. *Journal of Substance Abuse, 11*(1), 17–29.

Elster, A. B., & Panzarine, S. (1980). Unwed teenage fathers: Emotional and health educational needs. *Journal of Adolescent Health Care, 1*(2), 116–120.

Elster, A. B., & Panzarine, S. (1983). Teenage fathers. *Clinical Pediatrics, 22*(10), 700–703.

Eyberg, S., & Robinson, E. (1982). Parent-child interaction training: Effects on family functioning. *Journal of Clinical Child Psychology, 11*(2), 130–137.

Fagot, B. I., Pears, K. C., Capaldi, D. M., Crosby, L., & Leve, C. S. (1998). Becoming an adolescent father: Precursors and parenting. *Developmental Psychology, 34*(6), 1209–1219.

Feldman, R., Greenbaum, C. W., Mayes, L. C., & Erlich, S. H. (1997). Change in mother-infant interactive behavior: Relations to change in the mother, the infant, and the social context. *Infant Behavior and Development, 20*(2), 151–163.

Felix, A. C., & McCarthy, K. F. (1994). *An analysis of child fatalities, 1992.* Boston: Commonwealth of Massachusetts Department of Social Services.

Ferrari, A. M. (2002). The impact of culture upon child rearing practices and definitions of maltreatment. *Child Abuse & Neglect, 26*(8), 793–813.

Field, T. (1998). Maternal depression effects on infants and early interventions. *Preventive Medicine, 27*(2), 200–203.

Fields, J., & Casper, L. M. (2001). *America's families and living arrangements: March 2000* (Current Population Reports, P20-537). Washington, DC: U.S. Census Bureau.

Fox, N. A., Kimmerly, N. L., & Schafer, W. D. (1991). Attachment to mothers/attachment to fathers: A meta-analysis. *Child Development, 62*(1), 210–225.

Garbarino, J. (1977). The human ecology of child maltreatment: A conceptual model for research. *Journal of Marriage and the Family, 39*(4), 721–735.

Gelles, R. J. (1989). Child abuse and violence in single-parent families: Parent absence and economic deprivation. *American Journal of Orthopsychiatry, 59*(4), 492–501.

Gelles, R. J. (1992). Poverty and violence toward children. *American Behavioral Scientist, 35*(3), 258–274.

Gill, D. (1970). *Violence against children: Physical abuse in the United States.* Cambridge, MA: Harvard University Press.

Gillham, G., Tanner, G., Cheyne, B., Freeman, I., Rooney, M., & Lambie, A. (1998). Unemployment rates, single parent density, and indices of child poverty: Their relationship to different categories of child abuse and neglect. *Child Abuse & Neglect, 22*(2), 79–90.

Giovannoni, J. M., & Billingsley, A. (1970). Child neglect among the poor: A study of parental adequacy in families of three ethnic groups. *Child Welfare, 49*(4), 196–204.

Guterman, N. B. (2000). *Stopping child maltreatment before it starts: Emerging horizons in early home visitation services.* Thousand Oaks, CA: Sage.

Guterman, N. B. (2004). Advancing prevention research on child abuse, youth violence, and domestic violence: Emerging strategies and issues. *Journal of Interpersonal Violence, 19*(3), 299–321.

Halford, W. K., Sanders, M. R., & Behrens, B. C. (2000). Repeating the errors of our parents? Family-of-origin spouse violence and observed conflict management in engaged couples. *Family Process, 39*(2), 219–235.

Hofferth, S. L. (2003). Race/ethnic differences in father involvement in two-parent families. *Journal of Family Issues, 24*(2), 185–216.

Holden, G. W., & Barker, T. (2004). Fathers in violent homes. In M. E. Lamb (Ed.), *The role of the father in child development* (4th ed., pp. 417–445). Hoboken, NJ: John Wiley & Sons.

Holden, G. W., & Ritchie, K. L. (1991). Linking extreme marital discord, child rearing, and child behavior problems: Evidence from battered women. *Child Development, 62*(2), 311–327.

Horwitz, A. V., Widom, C. S., McLaughlin, J., & White, H. R. (2001). The impact of childhood abuse and neglect on adult mental health: A prospective study. *Journal of Health and Social Behavior, 42*(2), 184–202.

Hossain, Z., Field, T., Pickens, J., Malphurs, J., & Del Valle, C. (1997). Fathers' caregiving in low-income African-American and Hispanic-American families. *Early Development and Parenting, 6*(2), 73–82.

Hossain, Z., & Roopnarine, J. L. (1994). African-American fathers' involvement with infants: Relationship to their functioning style, support, education and income. *Infant Behavior and Development, 17*(2), 175–184.

Jackson, A. P. (1999). The effects of nonresident father involvement on single Black mothers and their young children. *Social Work, 44*(2), 157–166.

Jaffe, S. R., Moffitt, T. T., Caspi, A., Taylor, A., & Arseneault, L. (2002). Influence of adult domestic violence on children's internalizing and externalizing problems: An environmentally informative twin study. *Journal of the American Academy of Child and Adolescent Psychiatry, 41*(9), 1095–1103.

Jones, L. (1990). Unemployment and child abuse. *Families in Society, 71*(10), 579–588.

Kaufman, J., & Zigler, E. (1993). The intergenerational transmission of abuse is overstated. In R. J. Gelles & D. R. Loseke (Eds.), *Current controversies in family violence* (pp. 209–221). Newbury Park, CA: Sage.

Kirkham, M. A., Schinke, S. P., Schilling, R. F., & Meltzer, N. J. (1986). Cognitive-behavioral skills, social supports, and child abuse potential among mothers of handicapped children. *Journal of Family Violence, 1*(3), 235–245.

Kotch, J. B., Browne, D. C., Ringwalt, C. L., Stewart, P. W., Ruina, E., Holt, K., et al. (1995). Risk of child abuse or neglect in a cohort of low-income children. *Child Abuse & Neglect, 19*(9), 1115–1130.

Krugman, R. D. (1985). Fatal child abuse: Analysis of 24 cases. *Pediatrician, 12*(1), 68–72.

Kruttschnitt, C., McLeod, J. D., & Dornfeld, M. (1994). The economic environment of child abuse. *Social Problems, 41*(2), 299–315.

Lamb, M. E., & Elster, A. B. (1985). Adolescent mother-infant-father relationships. *Developmental Psychology, 21*(5), 768–773.

Larson, N. C., Hussey, J. M., Gillmore, M. R., & Gilchrist, L. D. (1996). What about dad? Fathers of children born to school-age mothers. *Families in Society, 77*(5), 279–289.

Lefcourt, H. M. (1992). Perceived control, personal effectiveness, and emotional states. In B. N. Carpenter (Ed.), *Personal coping: Theory, research, and application* (pp. 111–131). Westport, CT: Praeger/Greenwood.

Leonard, K. E., Eiden, R. D., Wong, M. M., Zucker, R. A., Puttler, L. I., Fitzgerald, H. E., et al. (2002). Developmental perspectives on risk and vulnerability in alcoholic families. *Alcoholism: Clinical & Experimental Research, 24*(2), 238–240.

Madge, N. (1983). Unemployment and its effects on children. *Journal of Child Psychology & Psychiatry & Allied Disciplines, 24*(2), 311–319.

Magen, R. H. (1999). In the best interests of battered women: Reconceptualizing allegations of failure to protect. *Child Maltreatment, 4*(2), 127–135.

Magura, S., & Laudet, A. B. (1996). Parental substance abuse and child maltreatment: Review and implications for intervention. *Children & Youth Services Review, 18*(3), 193–220.

Malkin, C. M., & Lamb, M. E. (1994). Child maltreatment: A test of sociobiological theory. *Journal of Comparative Family Studies, 25*(1), 121–134.

Margolin, L. (1992). Child abuse by mothers' boyfriends: Why the overrepresentation? *Child Abuse & Neglect, 16*(4), 541–551.

Marsiglio, W., Amato, P., & Day, R. D. (2000). Scholarship on fatherhood in the 1990s and beyond. *Journal of Marriage and the Family, 62*(4), 1173–1191.

McKibben, L., De Vos, E., & Newberger, E. H. (1991). Victimization of mothers of abused children: A controlled study. In R. L. Hampton (Ed.), *Black family violence* (pp. 75–83). Lexington, MA: Lexington Books.

McLoyd, V. C. (1990). The impact of economic hardship on Black families and children: Psychological distress, parenting, and socioemotional development. *Child Development, 61*(2), 311–346.

McMahon, T. J., & Rounsaville, B. J. (2002). Substance abuse and fathering: Adding poppa to the research agenda. *Addiction, 97*(9), 1109–1115.

Merrill, L. L., Hervig, L. K., & Milner, J. S. (1996). Childhood parenting experiences, intimate partner conflict resolution, and adult risk for child physical abuse. *Child Abuse and Neglect, 20*(11), 1049–2065.

Miller, B. (1998). Partner violence experiences and women's drug use: Exploring the connection. In C. Wetherington & A. Roman (Eds.), *Drug addiction research and the health of women.* Washington, DC: National Institute on Drug Abuse.

Miller, D. B. (1994). Influences on parental involvement of African American adolescent fathers. *Child and Adolescent Social Work Journal, 11*(5), 363–378.

Moss, H. B., Mezzich, A., Yao, J. K., Gavaler, J., & Martin, C. S. (1995). Aggressivity among sons of substance-abusing fathers: Association with psychiatric disorder in the father and son, paternal personality, pubertal development, and socioeconomic status. *American Journal of Drug and Alcohol Abuse, 21*(2), 195–208.

Murphy, J. M., Jellinek, M., Quinn, D., Smith, G., Poitrast, F. G., & Goshko, M. (1991). Substance abuse and serious child mistreatment: Prevalence, risk, and outcome in a court sample. *Child Abuse & Neglect, 15*(3), 197–211.

National Research Council. (1993). *Understanding child abuse and neglect.* Washington, DC: National Academy Press.

Nitz, K., Ketterlinus, R. D., & Brandt, L. J. (1995). The role of stress, social support and family environment in adolescent mothers' parenting. *Journal of Adolescent Research, 10*(3), 358–383.

O'Keefe, M. (1995). Predictors of child abuse in martially violent families. *Journal of Interpersonal Violence, 10*(1), 3–25.

Oregon Children's Services Division. (1993). *Task force report on child fatalities and critical injuries due to abuse and neglect.* Salem, OR: Oregon Department of Human Resources.

Osofsky, H. J., Osofsky, J. D., Culp, R., Krantz, K., Litt, K., & Tobiasen, J. (1985). Transition to parenthood: Risk factors for parents and infants. *Journal of Psychosomatic Obstetrics and Gynecology, 4*(4), 303–315.

Parks, P. I., Lenz, E., & Jenkins, L. S. (1992). The role of social support and stressors for mothers and infants. *Child Care, Health and Development, 18*(3), 151–171.

Paxson, C., & Waldfogel, J. (1999). Parental resources and child abuse and neglect. *Child Welfare, 89*(2), 239–244.

Paxson, C., & Waldfogel, J. (2002). Work, welfare and child maltreatment. *Journal of Labor Economics, 20*(3), 435–474.

Pearlin, L. I. (1999). Stress and mental health: A conceptual overview. In A. Horwitz & T. L. Schied (Eds.), *A handbook for the study of mental health: Social contexts, theories, and system* (pp. 161–175). New York: Cambridge University Press.

Peeters, M. C., Buunk, B. P., & Schaufeli, W. B. (1995). A micro-analysis exploration of the cognitive appraisal of daily stressful events at work: The role of material factors in child abuse and neglect. In G. B. Melton & F. D. Barry (Eds.), *Protecting children from abuse and neglect: Foundation for a new national strategy* (pp. 131–181). New York: Guilford Press.

Pelton, L. H. (1994). The role of material factors in child abuse and neglect. In G. B. Melton & F. D. Barry (Eds.), *Protecting children from abuse and neglect: Foundations for a new national strategy* (pp. 131–181). New York: Guilford Press.

Phares, V., & Compas, B. E. (1992). The role of father in child and adolescent psychopathology: Make room for daddy. *Psychological Bulletin, 111*(3), 387–412.

Polansky, N. A., Chalmers, M. A., Buttenwieser, E., & Williams, D. P. (1979). The absent father in child neglect. *Social Service Review, 53*(2), 163–174.

Radhakrishna, A., Bou-Saada, I. E., Hunter, W. M., Catellier, D. J., & Kotch, J. B. (2001). Are father surrogates a risk factor for child maltreatment? *Child Maltreatment, 6*(4), 281–289.

Rhein, L. M., Ginsburg, K. R., Schwarz, D. F., Pinto-Martin, J. A., Zhao, H., Morgan, A. P., et al. (1997). Teen father participation in child rearing: Family perspectives. *Journal of Adolescent Health, 21*(4), 244–252.

Rivara, F. P., Sweeney, P. J., & Henderson, B. F. (1986). Black teenage fathers: What happens when the child is born? *Pediatrics, 78*(1), 151–158.

Roberts, M. W. (2001). Clinic observations of structured parent-child interaction designed to evaluate externalizing disorders. *Psychological Assessment, 13*(1), 46–58.

Rosenbaum, A., & O'Leary, K. D. (1981). Children: The unintended victims of marital violence. *American Journal of Orthopsychiatry, 51*(4), 692–699.

Rothbaum, F., Rosen, K., Ujiie, T., & Uchida, N. (2002). Family system theory, attachment theory, and culture. *Family Process, 41*(3), 328–350.

Salzinger, S., Feldman, R. S., Ng-Mak, D. S., Mojica, E., Stockhammer, T., & Rosario, M. (2002). Effects of partner violence and physical child abuse on child behavior: A study of abused and comparison children. *Journal of Family Violence, 17*(1), 23–52.

Samuels, V. J., Stockdale. D. F., & Crase, S. J. (1994). Adolescent mothers' adjustment to parenting. *Journal of Adolescence, 17*(5), 427–443.

Schloesser, P., Pierpont, J., & Poertner, J. (1992). Active surveillance of child abuse fatalities. *Child Abuse & Neglect, 16*(1), 3–10.

Seagull, E. A. W. (1987). Social support and child maltreatment: A review of the evidence. *Child Abuse & Neglect, 11*(1), 41–52.

Sedlak, A., & Broadhurst, D. D. (1996). *The third national incidence study of child abuse and neglect: Final report.* Washington, DC: U.S. Government Printing Office.

Shipman, K. L., Rossman, B. R., & West, J. C. (1999). Co-occurrence of spousal violence and child abuse: Conceptual implications. *Child Maltreatment, 4*(2), 93–102.

Sidebotham, P., Golding, J., & the ALSPAC study team. (2001). Child maltreatment in the "children of the nineties." *Child Abuse & Neglect, 25*(9), 1177–1200.

Sinal, S. H., Petree, A. R., Hermen-Giddens, M., Rogers, M. K., Enand, C., DuRant, R. H. (2000). Is race or ethnicity a predictive factor in shaken baby syndrome? *Child Abuse & Neglect, 24*(9), 1241–1246.

Stark, E., & Filtcraft, A. H. (1988). Women and children at risk: A feminist perspective on child abuse. *International Journal of Health Services, 18*(1), 97–118.

Steele, B. (1987). Psychodynamic factors in child abuse. In R. E. Helfer & R. S. Kempe (Eds.), *The battered child* (pp. 81–114). Chicago: University of Chicago Press.

Stiffman, M. N., Schnitzer, P. G., Adam, P., Kruse, R. L., & Ewigman, B. G. (2002). Household composition and risk of fatal child maltreatment. *Pediatrics, 109*(4), 615–621.

Straus, M. A. (1974). Leveling, civility, and violence in the family. *Journal of Marriage and the Family, 36*(1), 13–29.

Straus, M. A., & Gelles, R. J. (1986). Societal change and change in family violence from 1975 to 1985 as revealed by two national surveys. *Journal of Marriage and the Family, 48*(3), 465–479.

Straus, M. A., Gelles, R. J., & Steinmetz, S. K. (1980). *Behind closed doors: Violence in the American Family.* New York: Doubleday/Anchor.

Straus, M. A., & Kantor, G. (1987). Stress and child abuse. In R. Helfer & R. S. Kempe (Eds.), *The battered child* (pp. 42–59). Chicago: University of Chicago Press.

Swinford, S. P., DeMaris, A., Cernkovich, S. A., & Giordano, P. D. (2000). Harsh physical discipline in childhood and violence in later romantic involvement: The mediating role of problem behaviors. *Journal of Marriage and the Family, 62*(2), 508–519.

Vandell, D. L., Hyde, J. S., Plant, A., & Essex, M. J. (1997). Fathers and "others" as infant-care providers: Predictors of parents' emotional well-being and marital satisfaction. *Merrill-Palmer Quarterly, 43*(3), 361–385.

Vaz, R., Smolen, P., & Miller, C. (1983). Adolescent pregnancy: Involvement of the male partner. *Journal of Adolescent Health Care, 4*(4), 246–250.

Vogel, C. A., Boller, K., Farber, J., Shannon, J. D., & Tamis-LeMonda, C. S. (2003). *Understanding fathering: The Early Head Start study of fathers of newborns.* Princeton, NJ: Mathematica Policy Research.

Westney, O. E., Cole, O. J., & Munford, T. I. (1986). Adolescent unwed prospective fathers: Readiness for fatherhood and behaviors toward the mother and the expected infant. *Adolescence, 21*(84), 901–911.

Whipple, E. E., & Webster-Stratton, C. (1991). The role of parental stress in physically abusive families. *Child Abuse & Neglect, 15*(3), 279–291.

Whipple, E. E., & Wilson, S. R. (1996). Evaluation of a parent education and support program for families at risk of physical child abuse. *Families in Society, 77*(4), 227–239.

Widom, C. S. (1989). Does violence beget violence? A critical examination of the literature. *Psychological Bulletin, 106*(1), 3–28.

Widom, C. S., Raphael, K. G., & DuMont, K. A. (2004). The case for prospective longitudinal studies in child maltreatment research: Commentary of Dube, Williamson, Thompson, Felitti, and Anda (2004). *Child Abuse & Neglect, 28*(7), 715–722.

Wissow, L. S. (2001). Ethnicity, income, and parenting contexts of children. *Child Maltreatment, 6*(2), 118–129.

Wolfner, G. D., & Gelles, R. J. (1993). A profile of violence toward children: A national study. *Child Abuse & Neglect, 17*(2), 197–212.

Wolock, I., & Horowitz, B. (1979). Child maltreatment and material deprivation among AFDC-recipient families. *Social Service Review, 53*(2), 175–194.

Yeung, W. J., Sandberg, J. F., Davis-Kean, P. E., & Hofferth, S. L. (2001). Children's time with fathers in intact families. *Journal of Marriage & the Family, 63*(1), 136–154.

DISCUSSION QUESTIONS

1. How does stress experienced by a father impact the likelihood of child abuse and/or neglect occurring?

2. Explain how age and substance abuse play a role in child abuse and neglect.

3. How does violence get "passed down" from one generation to the next? What can be done to stop this intergenerational transmission of violence?

4. How do economic factors play a role in the etiology of child abuse and neglect? How are they tied to the traditional role of males as caregivers?

5. What are the policy implications that can be derived from this article? How can child abuse and neglect be reduced, given what we know about the role of fathers in child abuse and neglect?

Introduction to Reading 2

The cycle of violence occurs when persons who are abused in childhood perpetuate this abuse and become abusers themselves later in life. Gómez (2011) investigates the link between experiencing childhood abuse, adolescent dating violence victimization, and intimate partner perpetration and victimization in adulthood. She uses a sample of 4,191 respondents who were part of the National Longitudinal Study of Adolescent Health. She uses the first three waves of data from this study to examine the cycle of violence for both men and women.

Testing the Cycle of Violence Hypothesis

Child Abuse and Adolescent Dating Violence as Predictors of Intimate Partner Violence in Young Adulthood

Anu Manchikanti Gómez

Introduction

Intimate partner violence (IPV) is a pervasive public health problem in the United States. Nearly 29% of American women and 23% of men have experienced IPV—broadly defined as psychological, physical, or sexual violence perpetrated by a current or former spouse, partner, or lover—during their lifetimes (Coker et al., 2002). Of violent crimes committed by partners, 85% are against women (Rennison & Welchans, 2000), although men may also be victimized. Women between

SOURCE: Gómez (2010). Reprinted with permission.

the ages of 16 and 24 are at the greatest risk for nonfatal IPV, the time during life when they are most likely to be dating (Rennison & Welchans, 2000). Dating violence is alarmingly common among adolescents, with 32% of in-school adolescents reporting some form of psychological or physical abuse by heterosexual partners (Halpern, Oslak, Young, Martin, & Kupper, 2001). The sequalae of IPV include increased risk for sexually transmitted infections, unintended pregnancy, decreased condom health (Campbell et al., 2002; Coker et al., 2002; Goodwin et al., 2000; Hathaway et al., 2000; Plichta & Falik, 2001; Silverman, Raj, Mucci, & Hathaway, 2001; Smith, Thornton, DeVellis, Earp, & Coker, 2002; Weinbaum et al., 2001). Furthermore, women who are victims of child abuse face many of the same adverse health outcomes (Bensley, Van Eenwyk, & Simmons, 2003; Noll, Horowitz, Bonanno, Trickett, & Putnam, 2003; Noll, Trickett, & Putnam, 2003). Nearly all research studies on the health effects of violence in relationships focus on IPV and its effects on women (Chen & White, 2004). IPV is a serious human rights and health issue, and a better understanding of its risk factors is necessary for the development of effective public health interventions.

The cycle of violence hypothesis postulates that children who experience abuse and maltreatment are more likely to experience and perpetrate violence as they age (Heyman & Sleps, 2002). Abused children may often be rejected by their "normal" peers and seek friendships with deviant peer groups, choosing romantic partners from these peers during adolescence and young adulthood (Feiring & Furman, 2000). Indeed, many studies find a greater risk of adulthood violence victimization and perpetration among victims of child abuse (Bensley et al., 2003; Dunkle et al., 2004; Ehrensaft et al., 2003; Heyman & Sleps, 2002; Noll, Horowitz, et al., 2003). However, parental maltreatment of children may represent a constellation of other disadvantages, including sociodemographic, economic, cultural, and environmental influences that are risk factors for later aggression (Neugebauer, 2000). For example, in a prospective birth-cohort study in New Zealand, Fergusson and Lynskey (1997) find that participants who retrospectively report child maltreatment

at age 18 had a tendency to come from disadvantaged families and experienced more childhood adversity than those who were not maltreated. In interpreting these findings, the authors argue that interventions should not solely focus on individual-level factors but should consider context. There is a dearth of studies that recognize that social and contextual factors correlated with both child abuse and the risk of adult IPV may be reflected in the strong association between these two factors, rather than a singular, direct effect of violence (Mullen, Martin, Anderson, Romans, & Herbison, 1996). Similarly, much research on child abuse fails to consider the effect of adolescent dating violence and vice versa (Maker, Kemmelmeier, & Peterson, 2001). Women who experience dating violence during adolescence (broadly, ages 12–19) are found to be at risk of repeat violence in young adulthood (ages 20–26; Smith, White, & Holland, 2003).

Past research examining linkages between child abuse and adult IPV is limited by its focus on married individuals (Feiring & Furman, 2000; Kwong, Bartholomew, Henderson, & Trinke, 2003). As the greatest risk of IPV comes during a period prior to the average age of marriage, it is vital to examine violence in sexual and romantic relationships that are characterized as dating, cohabiting, and casual. In addition, many studies use cross-sectional, nonrepresentative samples and tend to only consider women as victims and men as perpetrators (Chen & White, 2004; Fergusson & Lynskey, 1997; Kimmel, 2002). Although the majority of reported IPV crimes in the United States are committed against women, men are also often the victim of women's psychological and physical abuse (Rennison & Welchans, 2000). Female IPV perpetration may be concurrent with their victimization; that is, women may perpetrate violence to protect themselves and/or their children or in retaliation to being victimized (Chen & White, 2004).

An understanding of the causes of violence against women is offered by Heise's (1998) adaptation of ecological systems theory (Bronfenbrenner, 1979), a framework for studying violence against women that incorporates individual, situational, and sociocultural factors. As relevant to this analysis, Heise suggests several layers of risk factors. Individual factors, or those that shape an individual's response to stressors from other levels, that put men at greater risk of perpetrating IPV include witnessing domestic violence as a child, being abused as a child, and having an absent or rejecting father. The exosystem refers to social structures that influence individual behaviors. Exosystem factors linked to violence against women include unemployment, low socioeconomic status, social isolation, and delinquent peer association. Although the framework had been specifically developed as a tool for organizing research on violence against women, many of the factors that Heise identifies are also germane to female perpetration of violence and male victimization.

In addition, the cycle of violence hypothesis may be explained by social learning theory (Bandura, 1977). According to social learning theory, behaviors are learned from observations. Thus, victims of child abuse enter adolescence and adulthood with the belief that aggression is a method for dealing with interpersonal conflict. Furthermore, victims of child abuse may respond to IPV with learned helplessness. Feelings of powerlessness or inability to cope with trauma may be initiated through experience of uncontrollability in the family of origin (Walker, 1983).

Social disorganization theory incorporates contextual factors that may precede child abuse and adult IPV. The theory suggests that a lack of neighborhood cohesiveness affects communities' ability to mobilize resources to address crime and violence. The confluence of poverty, racial heterogeneity, and residential instability affect collective efficacy by limiting the formation of lasting relationships, community attachment, and common goals (Sampson, Raudenbush, & Earls, 1997). Sampson and colleagues (1997) apply social disorganization theory to the study of collective efficacy and violent crimes in Chicago neighborhoods. The authors find that three dimensions of neighborhood social characteristics (concentrated disadvantage, immigrant concentration, and residential stability) explain 70% of neighborhood variation in collective efficacy. Browning (2002) extends the work of Sampson et al. to the study of community-level processes influencing IPV in Chicago neighborhoods. Both

Browning and Sampson et al. find that the influence of the three areas of social factors on violence is mediated by collective efficacy.

Little quantitative research incorporates parental and social factors in examining the association between child abuse and adult IPV. This study aims to fill a gap in the literature by using a longitudinal, nationally representative, school-based survey of adolescents to examine the impact of child abuse and adolescent dating violence victimization on the likelihood of IPV perpetration and victimization in young adulthood. While a previous study using these data has found links between child abuses, youth violence, and IPV (Fang & Corso, 2007), this analysis further contributes to the literature by examining dating violence and applying social disorganization theory.

Method

Data

Data from three waves of the National Longitudinal Study of Adolescent Health (Add Health) are used. Add Health is a nationally representative, school-based study of youth in Grades 7 to 12 conducted during the 1994–1995 school year. Respondents were interviewed again during Waves 2 (1996) and 3 (2001–2002). All interviews used in this study were conducted in the homes of the respondents. During Waves 1 and 2, audio computer-assisted self-interview technology was used for sensitive subjects (e.g., sexual activity and drug use). In addition, parents were interviewed during Wave 1, and 1990 census data has been linked to individual records. The methods are detailed elsewhere (Harris, 2005).

This analysis uses a subset of 4,191 Add Health respondents. Inclusion criteria for the study are as follows: (a) having completed Wave-2 and Wave-3 interviews, (b) having reported being in at least one romantic or sexual relationship after the age of 18 at Wave 3, and (c) being age 22 or older at Wave 3. While the key independent and dependent variables are drawn from Waves 2 and 3, the age truncation is necessary to allow for exposure to adult IPV. That is, it is believed that respondents who are 22 years or older, who have completed at least 4 years of their young adult lives, and who are either in or beyond the prime age group for IPV are the most valid sample for this analysis.

Measures

Both IPV perpetration and victimization are examined as outcome variables in this analysis. During the Wave-3 interview, respondents listed all romantic and sexual partners since the summer of 1995 in a relationship roster. Recent sexual relationships and relationships that respondents identified as important are selected for a more detailed relationship history. For each selected partner, respondents are asked a series of questions related to IPV, adapted from the revised Conflicts and Tactics Scale (CTS; Straus, Hamby, Boney-McCoy, & Sugarman, 1996). All IPV questions are asked in terms of both perpetration and victimization. The ordinal outcome variables for both perpetration and victimization include three categories: no IPV, less severe IPV, and more severe IPV in young adulthood. Less severe IPV includes responses to two questions about whether the respondent had been the victim or perpetrator of the following types of abuse: (a) threats of violence, pushing or shoving, throwing objects that could injure a partner and (b) kicking, slapping, or hitting. Most severe IPV includes two questionnaire items: (a) sexual abuse (insisting on or making a partner have sex when he or she did not want to, or having a partner insist or make the respondent have sex when he or she did not what to) and (b) physical abuse that led to an injury, such as sprains, bruises, and cuts. Each relationship that began after the age of 18 is examined for both IPV perpetration and victimization. Because only relationships initiated in adulthood are included in this analysis, there is no overlap between the young adult IPV and adolescent dating violence measures. On average, respondents have 2.4 sexual or romantic partners during young adulthood, ranging from 2.1 for those aged 25 or older to 2.6 for 24-year-olds.

The key independent variables are child abuse and adolescent dating violence victimization. Child abuse is measured retrospectively at Wave 3. Child abuse is a dichotomous variable, with respondents coded as "1" if they responded affirmatively to two survey items about behaviors perpetrated by a parent or caregiver before the sixth grade, including (a) being slapped, kicked, or hit and (b) forced to have sex.

Adolescent dating violence is measured at Wave 2 and captures only victimization. Items from the CTS assess psychological and physical abuse from a maximum of three sexual and three romantic partners (Straus et al., 1996). For each reported partnership, respondents are asked if the partner ever called them names, insulted them, or treated them disrespectfully in front of others (Item 1); swore at them (Item 2); threatened them with violence (Item 3); threw something at them that could hurt them (Item 4); or pushed or shoved them (Item 5). Adolescents reporting only Items 1 to 3 for any partner are considered to have experienced less severe dating violence, whereas those reporting Items 4 to 5 experienced more severe violence.

Family factors included in the final models were parent's income in 1995 and family structure. Parent's income is included as categorical variables for less than US$16,000 to US$29,999, US$30,000 to US$49,999, US$50,000 to US$79,999, and more than US$80,000. Eleven percent of the subsample did not have a parental interview; rather than dropping these observations from the analysis, a variable is included that indicates that no parental data were available. In addition, nearly 9% of observations are missing parental income despite having had a parental interview. A variable indicating whether income was not reported by interviewed parents is included in multivariate regression models. Family structure, as reported by the respondent in Wave 1, is included as categorical variables indicating whether the respondent lived with two biological parents, another two parents (including combinations of biological, step, and adoptive parents), a single father, a single mother, or another situation.

Three indices as a proxy for collective efficacy are used to examine social disorganization. Using principal components analysis, measures are created with 1990 Census data at the census tract level. The concentrated disadvantage index includes the proportion of the population living below the poverty line, receiving public assistance, below age 18, African American, and of households that are headed by women. The residential stability index reflects the proportion of the population that has not moved since 1985 and houses that are owner occupied. Immigrant concentration includes the proportion of population that are Latino and foreign born. Each index is entered in the models as a continuous variable.

Sociodemographic factors from the Wave-3 interview included in the models are as follows: age, the time of interview, gender, educational attainment (some high school or less, received high school diploma or GED, some postsecondary, received college degree, or higher), and relationship status (married, cohabitating, or neither). Immigrant status is also included, indicating whether the respondent was foreign born (first generation), U.S. born to foreign-born parents (second), or U.S.-born parents (third plus generation; Harris, 1999). Wave-1 characteristics include region (West, Midwest, South, Northeast) and race/ethnicity (mutually exclusive categories for Latino, non-Latino Black, non-Latino Asian, non-Latino Native American, other non-Latino, non-Latino White). In addition, the models include a variable that indicates whether the respondent reported a romantic or sexual partner of the same sex in the relationship history at Wave 3.

Results

Descriptive statistics for the entire sample and for adult IPV perpetrators and victims are presented in Table 1. Greater proportions of young adult IPV perpetrators and victims report experiencing child abuse and adolescent dating violence as compared to the entire sample. For example, although only 12% of the sample report child abuse, 19% of IPV perpetrators and 18% of victims report being abused as children. IPV perpetrators and victims are most likely to be female, Black, cohabitating, and have less than a college education. Their parents are more likely to have income in the lowest two categories, and there is a higher mean level of concentrated disadvantage and lower level of residential stability among the perpetrators and victims. The most

notable demographic difference between IPV victims and perpetrators is gender. Whereas female victims constitute 53%, female perpetrators constitute 61%.

For each set of multivariate models, individual sociodemographic characteristics are first entered, followed by separate models—including parental and social factors—and a final model with all three sets of variables. Each table presents the results of the model with only individual characteristics and with all sets of variables. Specific odds ratios (OR) and 95% confidence intervals refer to the results of the final model with individual, parental, and social factors, unless otherwise noted. The first set of ordered logistic regression models considers IPV perpetration and victimization (none, less severe, or more severe) as the outcome. Across the models, child abuse and both levels of severity of adolescent dating violence victimization are highly significant and predictive of young adult IPV perpetration. In the final model, victims of child abuse have 97% higher odds of perpetrating IPV as young adults compared to those who were not abused. More severe dating violence victimization is associated with an 82% increase in the odds of violence perpetration, whereas less severe dating violence increases the odds by 60%. Growing up with a single mother is protective (OR = 0.70, 95% CI = 0.52–0.95). None of the social disorganization factors have a statistically significant relationship with IPV perpetration. Across the models, women are significantly more likely to perpetrate violence as compared to men.

For young adult IPV victimization, child abuse and both forms of adolescent dating violence victimization are again highly predictive of the outcome. The effect is not attenuated with the addition of parental and social factors to the model. Respondents with parents in the highest two income categories are less likely to be victimized by sexual and romantic partners as young adults. The social factors are not statistically significant. Being female increases the risk of IPV victimization.

Because of the elevated likelihood of IPV perpetration and victimization for women, the models are stratified by gender. Although child abuse and adolescent dating violence victimization remain highly significant, less severe dating violence presents a slightly

Table 1 Selected Characteristics of the Study Population (Weighted Frequencies and Means)

	Entire Sample (N = 4,191)	Perpetrators of Young Adult IPV (N = 986)	Victims of Young Adult IPV (N = 1,039)
Child abuse	12.0	18.9	17.9
Adolescent dating violence victimization			
Less severe	17.4	21.6	21.8
More severe	10.2	14.0	14.7
Age			
22	40.2	37.8	37.9
23	36.2	36.1	35.5
24	18.3	21.2	20.9
25 and older	5.3	5.0	5.6
Female	47.8	61.1	52.7
Race			
White	65.0	56.8	59.6
Latino	12.4	13.4	12.3
Black	15.7	21.8	20.7
Asian	3.9	4.2	4.2
Native American	2.1	2.7	2.4
Other	0.9	0.2	0.6
Immigrant status			
First generation	6.1	7.6	7.2
Second generation	10.2	10.5	9.0
Third generation or more	82.7	80.4	82.5
Relationship status			
Married	19.3	20.4	19.1
Cohabitating	20.6	24.6	24.8
Neither	60.1	55.0	56.2
Ever had same-sex relationship	3.8	4.3	4.1
Region			
South	36.3	40.1	39.0
West	16.2	15.3	16.0

(Continued)

Table 1 (Continued)

	Entire Sample (N = 4,191)	Perpetrators of Young Adult IPV (N = 986)	Victims of Young Adult IPV (N = 1,039)
Midwest	34.4	29.9	30.3
Northeast	13.2	14.6	14.7
Educational attainment			
College graduate	18.1	11.5	9.9
Some college	37.6	38.6	42.1
High school graduate	35.2	38.1	35.8
Some high school	9.0	11.6	12.0
Parental factors			
Parental income			
Less than US$16,000	13.2	15.7	16.9
US$16,000–US$29,999	14.5	17.6	16.5
US$30,000–US$49,999	20.9	20.6	21.2
US$50,000–US$79,999	19.1	16.4	15.8
More than US$80,000	23.7	22.3	21.4
Missing	8.6	7.3	8.3
Family structure			
Two biological parents	53.5	48.8	48.9
Two other parents	17.2	20.6	19.6
Single mother	20.1	19.8	22.0
Single father	3.4	4.6	3.8
Other	5.8	6.3	5.7
Social disorganization			
Concentrated disadvantage	−0.035	0.157	0.084
Residential stability	−0.055	−0.098	−0.075
Immigrant concentration	−0.248	−0.198	−0.218

NOTE: Ns are unweighted. The following variables were missing data (weighted frequency from entire sample in parentheses): Child abuse (5.0), race (0.1), immigrant status (1.0), educational attainment (1.0), concentrated disadvantage (0.3), residential stability (0.2), and immigrant concentration (0.2).

greater risk of violence perpetration for men. Men who experienced only psychological abuse in dating relationships have an 80% increase in the odds of perpetrating IPV in young adulthood, whereas those who experienced physical abuse have a 75% increase in the odds of perpetration. Among parental factors, only single fatherhood is a significant risk factor for violence perpetration. Men who are in single-father households have greater odds of perpetrating IPV (OR = 2.83, 95% CI = 1.13–7.10). Social factors do not have a statistically significant impact.

In the case of male victimization, more severe adolescent dating violence is the most salient historical abuse factor. Men who had been physically abused by partners as adolescents are more likely to be victimized as adults than those who had not (OR = 2.80, 95% CI = 1.68–4.68). Less severe adolescent dating violence continues to have a strong and significant effect (OR = 1.94, 95% CI = 1.39–2.70), whereas men who had been abused as children have a 66% greater odds of being IPV victims as young adults. No parental or social factors have a statistically significant impact on male relationship victimization in young adulthood.

For female perpetration of violence in young adulthood, child abuse and more severe adolescent dating violence victimization are significant predictors. Women who experienced child abuse have a 94% greater odds of perpetrating adult IPV. The effect of severe adolescent abuse is similar (OR = 1.96, 95% CI = 1.28–2.99). Unlike the models for men, less severe dating violence is only marginally significant (p = .08) and associated with a 43% increase in the odds of violence perpetration. Among parental factors, women with parental income of US$50,000 to US$79,999 are significantly less likely to perpetrate IPV compared to those with parental income of less than US$16,000. No social factors are statistically significant.

For female IPV victimization in young adulthood, child abuse is highly significant, with victims of child abuse having a 210% increase in the odds of victimization as compared to women who were not abused. More severe adolescent dating violence also has a significant influence on the likelihood of IPV victimization (OR = 1.67, 95% CI = 1.14–2.46). Less severe adolescent

dating violence does not have a significant association with IPV victimization of women in young adulthood. Among social factors, concentrated disadvantage has a slightly protective effect (OR = 0.91, 95% CI = 0.84–0.99). Parental factors do not reach statistical significance in this model.

✖ Discussion

Taken together, the results of this analysis indicate that child abuse and adolescent dating violence victimization are highly predictive of young adult IPV, and the relationship is not attenuated by parental or social factors. In fact, in a number of the models, the effects of child abuse and adolescent dating violence grow stronger with the inclusion of parental and social factors. Though the models are not shown in this article, the interaction of child abuse and adolescent dating violence do not have a significant influence on the likelihood of young adult IPV. Thus, child abuse and adolescent dating violence appear to work independently of each other.

These results demonstrate that women have a significantly greater likelihood of reporting both IPV perpetration and victimization in young adult sexual and romantic relationships. The perpetration result was unforeseen, particularly because the IPV literature tends to focus on women as victims. There are several explanations for this association. First, women may be IPV perpetrators and victims concurrently. As rich as the Add Health data are, they do not provide information about the context of violence in relationships. The temporality of perpetration and victimization is often unknown in research studies. For example, Magdol et al. (1997) find a higher prevalence of IPV perpetration among women than men and that anxiety is a significant predictor of perpetration for women. However, with survey data, they are unable to tease out whether this anxiety puts women at greater risk of victimization or if victimization is causing the anxiety. In the present study, 65% of women who perpetrated IPV report being in adult relationships where they were both the victim and perpetrator, as compared to 59% of men. Female perpetration may be driven by female victimization, as women may fight back to protect themselves and their children (Chen & White, 2004; Dasgupta, 2002; Hamberger & Guse, 2002; Kimmel, 2002).

Second, there may be reporting differences in violence perpetration by gender. Some studies have found that women are more likely to report violence perpetration because the behavior is considered less socially acceptable and thus may be more memorable. Conversely, men may underreport violence perpetration because it demonstrates a lack of control over their partners (Dobash, Dobash, Cavanagh, & Lewis, 1998). Though this cannot be explored over the entire course of adult relationships in Add Health, disparities in reporting IPV could be examined using a sample of current relationships where both partners are interviewed and asked the same questions about violence perpetration and victimization (Harris, 2005). Moreover, qualitative research would be useful to flesh out the temporality of events.

Regardless of whether gender is a risk factor for perpetration, it is worth noting that even if women are more likely to perpetrate IPV than men, the biological ability of women to injure their partners is generally lower than their male counterparts. Injuries are less likely to be caused by pushing, shoving, and grabbing, the perpetration behaviors that are more common among women (Kimmel, 2005). Thus, the physical health implications of female IPV perpetration may not be as profound as male perpetration. For example, the rate of homicide of spouses or former spouses is much higher among men than women. The gender imbalance has increased over time: In the mid-1970s, women represented half of victims murdered by intimate partners, and this proportion increased to three quarters in the late 1990s (Rennison & Welchans, 2000). However, there is scant research on the health consequences of violence victimization among men. Considering the linkages to child abuse and adolescent dating violence, it is possible that being the victim of IPV may have mental health consequences for men or that poor mental health caused by historical abuse may drive both victimization and perpetration among men.

In the present study, historical abuse factors are found to operate somewhat differently for men and women. For example, child abuse is highly predictive of IPV perpetration and victimization for both men and women. Although child abuse is a strong predictor of female victimization (OR = 2.07, 95% CI = 1.40–3.05), it is slightly less predictive for men (OR = 1.65, 95% CI = 1.11–2.45). The results for adolescent dating violence victimization are more disparate. For male IPV perpetration, both levels of adolescent victimization are significant predictors. For female perpetration, more severe dating violence is statistically significant, whereas less severe abuse is marginally significant. For men, IPV perpetration effect estimates for less severe psychological dating violence are greater than those for more severe abuse, whereas for women, more severe adolescent victimization has a stronger impact. Considering IPV victimization, the adolescent dating violence variables have a much weaker effect for women than men. Although both adolescent dating violence measures are highly significant for men, less severe dating violence is not significant for women and more severe abuse is significant only at the .05 level. Though it appears that a cycle of violence is in play for both men and women, the mechanism varies by gender.

Parental income is occasionally a protective factor for young adult IPV. Family structure does not play a significant role, except for male perpetration models, where male youth who lived in single-father homes had substantially greater odds of perpetrating violence than those who lived with two biological parents. Results from a previous analysis of Add Health data examining the effect of growing up in a single-father home on adolescent well-being indicate that the strong effects of living with a single father may be less related to the specific family structure and more connected to instability of living arrangements that bring youth to live with fathers. Harris, Cavanagh, and Elder (2000) find that for many youth in single-father homes, the living arrangements are recent and involve a shift from living in a home with a single mother to a single father. Youth living with single fathers report lower levels of parental monitoring and supervision and are more likely to witness, perpetrate, or be the victim of a violent crime compared to youth living with two biological parents or a single mother. It appears that the measure of living with single fathers may reflect aggression that leads to the instability of living arrangements and that this aggression may translate into higher levels of involvement in violence throughout the life course.

Neither of the social disorganization factors have any significant effect on the likelihood of either young adult IPV perpetration or young adult IPV victimization nor do they attenuate the effects of historical abuse. Although the three indices used explain much of the variation in collective efficacy in previous research on violence (Sampson et al., 1997), they are only a proxy for this construct. Perhaps measuring social disorganization at Wave 1 or at multiple points of time may be a better approach to explore this mechanism in future research.

Although this study fills a gap in the literature, it does face a number of limitations. First, the retrospective measurement of child abuse is problematic. Measures of child abuse were not included in the Add Health survey until Wave 3, when respondents were between the ages of 18 and 26. In examining the Add Health data, Hussey, Chang, and Kotch (2006) contend that the level of child abuse is potentially underreported, based on comparisons to other research. However, researchers have also argued for the prospective rather than retrospective measurement of child abuse, finding that respondents tend to overestimate abuse in retrospective reports for a myriad of reasons (Tajima, Harrenkohl, Huang, & Whitney, 2004). If the results are biased due to measurement error, it is difficult to ascertain in which direction. In addition, respondents were only asked about three types of physical abuse—slapping, kicking, and hitting—in one questionnaire item. As these behaviors vary in severity, it would be useful to examine the occurrence of each type of abuse as well as psychological and other types of physical abuse not included in Add Health. Second, although I am able to examine violence in young adulthood across numerous relationships, there is a lack of information to contextualize violence. This is particularly important for understanding some of the gender differences in IPV and why women are at higher risk for perpetrating

violence in this analysis. Third, the measure of adolescent dating violence only includes victimization and captures limited severity. Although there is generally a strong relationship between adolescent victimization and adult IPV, a more complete portrait would examine adolescent perpetration and measures of abuse comparable to the adult IPV items as well. The context of the violence would be important to explore in this case as well, particularly whether violence occurred throughout the relationship or only during a breakup, and whether adolescents were both perpetrators and victims in the same relationship. Finally, this analysis does not examine the impact of witnessing parental IPV on the likelihood of young adult IPV, an important piece of Heise's (1998) framework for examining the risk factors for violence against women. Although interviewed parents are asked about the frequency of arguments with their current partner, there is no degree of specificity that would allow an understanding of the severity of the abuse (psychological, physical, or sexual). In addition, frequent arguments do not necessarily imply that abuse occurred, and respondents are not asked if they witnessed their parents' arguments.

These results have important implications for public health interventions and programs. First, child abuse prevention is paramount, as both the short- and long-term consequences of abuse are significant. Second, the long-term consequences of child abuse should be considered in counseling efforts. Though certainly not all children who are abused go on to be perpetrators or victims of relationship violence, the elevated likelihood of abuse makes it vital to consider the implications of child abuse for future transmission of violence. Third, nearly a third of respondents had been the victims of dating violence by their Wave-2 interview (Grades 10–12). As even less severe psychological abuse is generally strongly predictive of both young adult IPV perpetration and victimization, interventions to educate adolescents on healthy relationships may be an important opportunity to stop the cycle of violence (Foshee et al., 1998). If adolescents develop ideals and expectations about relationships during this precocious time when abuse is common, intervening may provide an opportunity to reduce the likelihood of relationship violence throughout the life course.

References

Bandura, A. (1977). *Social learning theory.* Englewood Cliffs, NJ: Prentice Hall.

Bensley, L., Van Eenwyk, J., & Simmons, K. W. (2003). Childhood family violence history and women's risk for intimate partner violence and poor health. *American Journal of Preventive Medicine, 25,* 38–44.

Bronfenbrenner, U. (1979). *The ecology of human development: Experiments by nature and design.* Cambridge, MA: Harvard University Press.

Browning, C. R. (2002). The span of collective efficacy: Extending social disorganization theory to partner violence. *Journal of Marriage and Family, 64,* 833–850.

Campbell, J., Jones, A. S., Dienemann, J., Kub, J., Schollenberger, J., O'Campo, P., et al. (2002). Intimate partner violence and physical health consequences. *Archives of Internal Medicine, 162,* 1157–1163.

Chen, P.-H., & White, H. R. (2004). Gender differences in adolescent and young adult predictors of later intimate partner violence. *Violence Against Women, 10,* 1283–1301.

Coker, A. L., Davis, K. E., Arias, I., Desai, S., Sanderson, M., Brandt, H. M., et al. (2002). Physical and mental health effects of intimate partner violence for men and women. *American Journal of Preventive Medicine, 23,* 260–268.

Dasgupta, S. D. (2002). A framework for understanding women's use of nonlethal violence in intimate heterosexual relationships. *Violence Against Women, 8,* 1364–1389.

Dobash, R. P., Dobash, R. E., Cavanagh, K., & Lewis, R. (1998). Separate and intersecting realities: A comparison of men's and women's accounts of violence against women. *Violence Against Women, 4,* 383–414.

Dunkle, K., Jewkes, R. K., Brown, H. C., Yoshihama, M., Gray, G. E., McIntyre, J. A., et al. (2004). Prevalence and patterns of gender-based violence and revictimization among women attending antenatal clinics in Soweto, South Africa. *American Journal of Epidemiology, 160,* 230–239.

Ehrensaft, M. K., Cohen, P., Brown, J., Smailes, E., Chen, H., & Johnson, J. G. (2003). Intergenerational transmission of partner violence: A 20-year prospective study. *Journal of Counseling and Clinical Psychology, 71,* 741–753.

Fang, X., & Corso, P. S. (2007). Child maltreatment, youth violence, and intimate partner violence: Developmental relationships. *American Journal of Preventive Medicine, 33,* 281–290.

Feiring, C., & Furman, W. C. (2000). When love is just a four-letter word: Victimization and romantic relationships in adolescence. *Child Maltreatment, 5,* 293–298.

Fergusson, D. M., & Lynskey, M. T. (1997). Physical punishment/maltreatment during childhood and adjustment in young adulthood. *Child Abuse & Neglect, 21,* 617–630.

Foshee, V. A., Bauman, K. E., Arriaga, X. B., Helms, R. W., Koch, G. G., & Linder, G. F. (1998). An evaluation of Safe Dates, an adolescent dating violence prevention program. *American Journal of Public Health, 88,* 45–50.

Goodwin, M. M., Gazmararian, J. A., Johnson, C. H., Gilbert, B. C., Saltzman, L. E., & PRAMS Working Group. (2000). Pregnancy intendedness and physical abuse around the time of pregnancy: Findings from the pregnancy risk assessment monitoring system, 1996–1997. *Maternal and Child Health Journal, 4,* 85–92.

Halpern, C. T., Oslak, S. G., Young, M. I., Martin, S. L., & Kupper, L. L. (2001). Partner violence in clinical samples. *Violence Against Women, 8,* 1301–1331.

Hamberger, L. K., & Guse, C. E. (2002). Men's and women's use of intimate partner violence in clinical samples. *Violence Against Women, 8,* 1301–1331.

Harris, K. M. (1999). The health status and risk behavior of adolescents in immigrant families. In D. J. Hernandez (Ed.), *Children of immigrants: Health, adjustment, and public assistance* (pp. 268–347). Washington, DC: National Academy Press.

Harris, K. M. (2005). *Design features of Add Health.* Chapel Hill: Carolina Population Center, University of North Carolina at Chapel Hill.

Harris, K. M., Cavanagh, S. E., & Elder, G. H. (2000, March). *The well-being of adolescents in single-father families.* Paper presented at the Annual Meeting of the Population Association of America, Los Angeles.

Hathaway, J. E., Mucci, L. A., Silverman, J. G., Brooks, D. R., Mathews, R., & Pavlos, C. A. (2000). Health status and health care use of Massachusetts women reporting partner abuse. *American Journal of Preventive Medicine, 19,* 302–307.

Heise, L. (1998). Violence against women: An integrated, ecological framework. *Violence Against Women, 4,* 262–290.

Heyman, R. E., & Sleps, A. M. S. (2002). Do child abuse and interparental violence lead to adulthood family violence? *Journal of Marriage and Family, 64,* 864–870.

Hussey, J. M., Chang, J. J., & Kotch, J. B. (2006). Child maltreatment in the United States: Prevalence, risk factors, and adolescent health consequences. *Pediatrics, 118,* 933–942.

Kimmel, M. S. (2002). "Gender symmetry" in domestic violence: A substantive and methodological review. *Violence Against Women, 8,* 1332–1363.

Kwong, M. J., Bartholomew, K., Henderson, A. J., & Trinke, S. J. (2003). The intergenerational transmission of relationship violence. *Journal of Family Psychology, 17,* 288–301.

Magdol, L., Moffitt, T. E., Caspi, A., Newman, D. L., Fagan, J., & Silva, P. A. (1997). Gender differences in partner violence in a birth cohort of 21-year-olds: Bridging the gap between clinical and epidemiological approaches. *Journal of Consulting and Clinical Psychology, 65,* 68–78.

Maker, A. H., Kemmelmeier, M., & Peterson, C. (2001). Child sexual abuse, peer sexual abuse, and sexual assault in adulthood: A multi-risk model of revictimization. *Journal of Traumatic Stress, 14,* 351–368.

Mullen, P. E., Martin, J. L., Anderson, J. C., Romans, S. E., & Herbison, G. P. (1996). The long-term impact of the physical, emotional, and sexual abuse of children: A community study. *Child Abuse & Neglect, 20,* 7–21.

Neugebauer, R. (2000). Research on intergenerational transmission of violence: The next generation. *Lancet, 355*, 1116–1117.

Noll, J. G., Horowitz, L. A., Bonanno, G. A., Trickett, P. K., & Putnam, F. W. (2003). Revictimization and self-harm in females who experienced childhood sexual abuse: Results from a prospective study. *Journal of Interpersonal Violence, 18*, 1452–1471.

Noll, J. G., Trickett, P. K., & Putnam, F. W. (2003). A prospective investigation of the impact of childhood sexual abuse on the development of sexuality. *Journal of Consulting and Clinical Psychology, 71*, 575–586.

Plichta, S. B., & Falik, M. (2001). Prevalence of violence and its implications for women's health. *Women's Health Issues, 11*, 244–258.

Rennison, C. M., & Welchans, S. (2000). *Intimate partner violence.* Washington, DC: U.S. Department of Justice.

Sampson, R. J., Raudenbush, S. W., & Earls, F. (1997). Neighborhoods and violent crime: A multilevel study of collective efficacy. *Science, 277*, 918–924.

Silverman, J. G., Raj, A., Mucci, L. A., & Hathaway, J. E. (2001). Dating violence against adolescent girls and associated substance use, unhealthy weight control, sexual risk behavior, pregnancy, and suicidality. *Journal of the American Medical Association, 286*, 572–579.

Smith, P. H., Thornton, G. E., DeVellis, R., Earp, J. A., & Coker, A. L. (2002). A population-based study of the prevalence and distinctiveness of battering, physical assault, and sexual assault in intimate relationships. *Violence Against Women, 8*, 1208–1232.

Smith, P. H., White, J. W., & Holland, L. J. (2003). A longitudinal perspective on dating violence among adolescent and college-age women. *American Journal of Public Health, 93*, 1104–1109.

Straus, M. A., Hamby, S. L., Boney-McCoy, S., & Sugarman, D. B. (1996). The revised Conflict Tactics Scales (CTS2): Development and preliminary psychometric data. *Journal of Family Issues, 17*, 283–316.

Tajima, E. A., Herrenkohl, T. I., Huang, B., & Whitney, S. D. (2004). Measuring child maltreatment: A comparison of prospective parent reports and retrospective adolescent reports. *American Journal of Orthopsychiatry, 74*, 424–435.

Walker, L. (1983). The battered women syndrome study. In D. Finkelhor, R. J. Gelles, G. T. Hotaling, & M. A. Straus (Eds.), *The dark side of families: Current family violence research* (pp. 31–48). Beverly Hills, CA: Sage.

Weinbaum, Z., Stratton, T. L., Chavez, G., Motylewski-Link, C., Barrera, N., & Courtney, J. G. (2001). Female victims of intimate partner physical domestic violence (IPP-DV), California 1998. *American Journal of Preventive Medicine, 21*, 313–319.

DISCUSSION QUESTIONS

1. Why would you expect that experiencing child abuse and/or adolescent dating violence would lead to intimate partner violence perpetration in young adulthood?

2. Why would you expect that experiencing child abuse and/or adolescent dating violence would lead to intimate partner violence victimization in young adulthood?

3. Why would you expect that experiencing child abuse and/or adolescent dating violence would differentially impact women and men in young adulthood?

4. How can we intervene to stop the cycle of violence? Should these interventions be gender-specific?

Introduction to Reading 3

Little attention has been given to the unique features of violent victimization perpetrated against the elderly. In their article, Bachman and Meloy (2008) use data from the Supplementary Homicide Reports and the National Crime Victimization Survey to compare the characteristics of homicides, robberies, and assaults of community-dwelling persons aged 65 and older to those of persons younger than 65. The authors also provide an overview of incidents of violence that occur in nursing homes, with special attention on the victim-offender relationship and risk factors for victimization.

The Epidemiology of Violence Against the Elderly

Implications for Primary and Secondary Prevention

Ronet Bachman and Michelle L. Meloy

Although elderly citizens generally have a much lower rate of victimization for virtually all types of crimes compared to their younger counterparts, the unique characteristics of their victimization and its aftermath illuminate the urgent need for informed prevention strategies. For example, we know that elderly violent crime victims, particularly elderly female victims, are more likely to require medical care for injuries sustained during a violent attack compared to their younger counterparts (Bachman, Dillaway, & Lachs, 1998; Bachman, Lachs, & Meloy, 2004). Another recent study found that a violent victimization for community dwelling elderly residents increased their risk of nursing home placement even after controlling for other variables traditionally thought to be predictive of nursing home placement such as functional and cognitive impairment, social networks, and age (Lachs et al., 2006). And finally, after examining data from the National Incident-Based Reporting System, Chu and Kraus (2004) concluded that elderly victims had a higher risk of death from assault than younger age groups.

Thus, despite their decreased risk of violent victimization, the more severe consequences associated with their victimizations suggest an urgent need to more fully understand both the epidemiology and etiology of elderly victimization. This is particularly important because those older than 65 are the fastest growing age group in the United States. For example, in 1990, one in eight persons were older than the age of 65; by 2030, this ratio will decline to one in five (Wan, Sengupta, Velkoff, & Debarros, 2005).

Before we begin, we will highlight a few methodological issues. One of the problems with the extant research studying elderly victimization is that there is wide variability in the definitions of "elderly victimization" and the corresponding instruments used to measure these victimizations. Although this is true of the literature measuring violence in general, these methodological issues are particularly problematic here because of the global definition of "elder abuse," which may include a myriad of victimizations from financial abuse and neglect to aggravated assault and murder. Clearly, research that provides one estimate of victimization for such extremely different types of crime precludes making comparisons with research using more refined categories of victimization. Moreover, many estimates of elderly victimization rely solely on reports that come to the attention of authorities such as the police or Adult Protective Services (APS). For example, the National Elder Abuse Incidence Study relied on reports that had been substantiated by APS and other "sentinels" within county government (Administration on Aging, 1998). Obviously, these estimates are extreme under-reports because we know that the majority of violent victimizations against the elderly are never reported to anyone (Bachman et al., 1998).

In this article, we divide victimizations into three more homogeneous categories: murder, robbery, and assault. Although elders do become victims of rapes and sexual assaults, these cases are rare and sample sizes from even the largest surveys, such as the National Crime Victimization Survey (NCVS), are too few from which to make reliable generalizations to the larger elderly population (Klaus, 2005). As such, we will not discuss sexual assault victimizations here. We also examine what is known about violence against elders living in the

SOURCE: Bachman and Meloy (2008). Reprinted with permission.

community compared to those residing in nursing homes. And finally, although there is no satisfactory or universal age at which individuals suddenly become "elderly," we operationalize individuals 65 years of age or older as elderly to be consistent with past research.

✉ Elderly Homicide Victimization

According to the supplementary homicide reports from the Federal Bureau of Investigation, those aged 65 and older have the lowest rates of murder compared to other age groups. Since declining from their highs in the early 1990s, these elderly homicide rates have been relatively constant since 1998 (Klaus, 2005). However, when the context of murder is examined, victimizations against the elderly look very different from those of younger cohorts.

Table 1 displays the percentage distribution for weapon used, the victim and offender relationship, and the precipitating circumstances of the murder for those younger than 65 and for those aged 65 and older. Although the most common method of killing for all homicide victims is firearms, elderly victims are more likely to be stabbed or bludgeoned to death compared to their younger counterparts. In fact, nearly one quarter of elderly victims are bludgeoned to death. All homicide victims were more likely to be killed by known offenders compared to strangers; however, elderly individuals had an increased risk of being killed by other family members and by strangers. This latter vulnerability is related to the difference in precipitating circumstances across age groups; elderly victims were more likely to be killed in the context of another felony, primarily robberies, compared to their younger counterparts who were more likely to be killed as a result of a conflict situation like an argument or fight.

This vulnerability to robbery-related death has been found in city-level data as well. For example, homicide in the Chicago homicide file indicates that for younger age groups, assault-related homicides (e.g., argument) outnumber robbery-related homicides by an average of six to one, but this differential completely

Table 1 Percentage of National Sup... Homicide Reports Characteri... 1976 to 2004

	Victims 64 and Younger (%)	Victims 65 and Older (%)
Weapon used		
Gun	66	40
Knife	16	19
Blunt object	10	24
Other	7	17
Victim-offender relationship		
Intimate	17	16
Other family	12	19
Other known	50	38
Stranger	21	26
Precipitating circumstance		
Other felony	19	35
Conflict or argument	39	21
Other	41	44

reverses by the age of 65 where robbery related homicides become more prevalent than conflict-related homicides (Bachman, Meloy, & Block, 2005). As you will see in the next section, this vulnerability to economic predation exists for nonfatal forms of violence against the elderly as well.

✉ Nonlethal Violent Victimization of the Elderly

The most reliable nationally representative data on nonfatal forms of violent victimization come from the annual NCVS that is sponsored by the Bureau of Justice Statistics. Those eligible for the sample include individuals age 12 or older living in the United States, including persons living in group quarters such as rooming houses or dormitories. The survey does *not* interview members of any type of institutionalized population, however, including persons in nursing homes. Of importance, because the NCVS is a random sample of U.S. households, it uncovers victimizations that were both reported and not reported to police. Unless otherwise noted, the findings discussed below were obtained from the NCVS.

Robbery

Similar to lethal violence, the probability of becoming a robbery victim significantly decreases as individuals move through the life course. However, there are contextual differences among robbery victimizations against the elderly compared to their younger counterparts. To begin, the gender differentials in victimization that exist for younger age groups essentially disappear for those older than 65. For example, younger males are significantly more likely to become robbery victims compared to their female counterparts. However, elderly women are just as likely to be robbed as their male counterparts (Bachman et al., 1998). This is like no other time in the life course. The lack of gender symmetry with regard to victimization is also true for personal larcenies with contact such as purse snatching and pocket picking (Klaus, 2005). These findings are consistent with other victimization research from a cohort of community-dwelling elders that found there was no difference in the risk of victimization for male and female elders (Lachs et al., 2004).

Table 2 displays the percentage distribution for robbery victims by age, the victim and offender relationship, and the time and place of occurrence. Although males of all ages and elderly women are more likely to be robbed by strangers, younger women are equally likely to be robbed by known offenders.

Of importance, this vulnerability to strangers for the elderly does not simply exist out on the street while conducting the routine activities of the day, but in their homes as well. That is, elderly women are just as vulnerable in their homes to economic predation by strangers as they are on the street (Bachman et al., 1998).

Although elderly robbery victims are slightly more likely to sustain injuries as the result of being robbed compared to younger victims, elders are much more likely than their younger counterparts to require medical care for their injuries, particularly elderly women. In fact, more than two thirds of elderly female robbery victims who were injured required medical care for these injuries.

Table 2   Percentage of Robbery Victimization by Victim-Offender Relationship, Time, and Location of Occurrence by Gender and Age, National Crime Victimization Survey 1992 to 2003

	Victims 64 and Younger (%)	Victims 65 and Older (%)
Male robbery victims		
Victim-offender relationship		
Intimate	2	0
Other relative	2	6
Acquaintance or friend	27	21
Stranger	69	73
Occurred in private	40	70
Occurred in daytime	21	37
Victim injured	31	35
Required medical care	56	64
Female robbery victims		
Victim-offender relationship		
Intimate	17	2
Other relative	6	2
Acquaintance or friend	27	20
Stranger	50	76
Occurred in private	60	81
Occurred in daytime	30	48
Victim injured	33	37
Required medical care	46	67

These findings coupled with the homicide patterns in the previous section underscore the vulnerability elderly individuals have to economic predation related to violent crimes. Violence against the elderly is more likely to be predicated on the possibility of economic gain compared to victimizations against younger people. Of importance, this vulnerability affects both elderly men and women equally.

Assault Victimizations

Community-Dwelling Elderly

Similar to rates of other types of violence, elderly rates of assault victimization are lower compared to their younger cohorts (Klaus, 2005). Unlike robbery and homicide victimization, there are actually more similarities for assault victimizations when gender-specific data are examined. Table 3 displays the percentage distribution for assault victim by age, the victim and offender relationship, and the time and place of occurrence. Assaults for both males and females of all ages are more likely to be committed by known offenders compared to strangers, particularly female victims. Elderly women, like women of all ages, share a unique vulnerability from attacks by their intimate partners and by other family members. However, this vulnerability to intimate partners actually decreases as women move through the life course, whereas the percentage of assaults perpetrated by other family members remains relatively constant. This may be related to life situation changing and other family members becoming the caretakers of their elderly parents or other elderly relative. It may also be related to perpetrators of intimate partner violence aging out of offending as they become older.

Elderly assault victims are more likely to be victimized in private locations and in the daytime compared to their younger counterparts. Data indicate that the elderly are more likely to sustain injuries as the result of an assault; however, elderly women are more likely to require medical care for their injuries when they occur compared to other assault victims. And similar to robbery victimizations, elderly victims of assault are more likely to be assaulted at their private residences and in the daytime compared to younger assault victims. It is quite possible that these risks are linked to the routine activities of elders who are generally more often at home than younger cohorts because of retirement and a decrease in nightly activities away from the home.

Institutionalized Assaults Against the Elderly

When cases of substandard nursing home environments and heinous acts of assault therein make it to the news media, people are justifiably outraged. Unfortunately, we know relatively little about the magnitude of violence against the elderly residing in nursing homes. Furthermore, it is hard to make comparisons across the few studies that have been done because of the methodological issues already highlighted in this article regarding the measurement of violence. In addition, there is the added difficulty of defining what institutes a "nursing home" because there are a wide variety of institutions ranging from short- to long-term care, from assisted living facilities to skilled nursing facilities, and so on. The few studies that have attempted to measure the incidence of victimization in nursing homes have primarily relied on convenience samples of nursing home staff (Douglas, Hickey, & Noel, 1980; Pillemer & Bachman, 1991; Pillemer & Hudson, 1993). Unfortunately the majority of these studies have reported incidents based on a global category of "elder abuse," in which incidents of physical violence cannot be disentangled from incidents of emotional or psychological abuse or neglect. We will highlight a few studies that did measure incidents of violence separately.

In a study conducted on the nursing aid staff of 10 Philadelphia-based nursing homes, a little more than half (51%) of the respondents stated they had yelled at a resident in anger; nearly one fourth of the respondents (23%) stated that they insulted or called a resident names; 17% used excessive physical restraint to hold a resident; and 1 in 10 nurses' aides stated they pushed, grabbed, or shoved a resident within the last 30 days (Pillemer & Hudson, 1993). In another study relying on surveys of both nurses and nursing aides

Table 3   Percentage of Assault Victimizations by Victim-Offender Relationship, Time, and Location of Occurrence by Gender and Age, National Crime Victimization Survey 1992 to 2003

	Victims 64 and Younger (%)	Victims 65 and Older (%)
Male assault victims		
Victim-offender relationship		
Intimate	3	2
Other relative	4	4
Acquaintance or friend	51	60
Stranger	42	34
Occurred in private	28	47
Occurred in daytime	58	76
Victim injured	19	14
Required medical care	48	50
Female assault victims		
Victim-offender relationship		
Intimate	21	5
Other relative	9	10
Acquaintance or friend	48	62
Stranger	22	24
Occurred in private	47	63
Occurred in daytime	57	74
Victim injured	42	39
Required medical care	41	50

employed at skilled nursing home facilities, less than 3% reported they had engaged in some form of violence in the last year such as pushing, grabbing, slapping, or hitting (Pillemer & Bachman, 1991).

Although these studies have examined incidents of staff-to-resident abuse, a recent study investigating calls to police from nursing homes revealed that the most common incidents reported were cases of resident-to-resident assault, not cases of staff-to-resident assault, even though facilities were mandated to report all acts of assault (Lachs, Bachman, Williams, & O'Leary, 2007). After examing the narratives of the police reports, Lachs and his colleagues (2007) developed the following typologies of resident-to-resident assault: (a) unprovoked assault in which a resident with dementia-related behavioral problems assaults other residents without provocation, (b) invasion of space assaults in which residents assault others who enter their rooms uninvited, (c) male-on-male assaults similar to the most common types of assaults in the general population that are precipitated over issues that appear insubstantial (e.g., who enters the elevator first), (d) competition for resources in which assaults occur over things like chairs in the television room or in the dining room, and (e) breaking point assaults when residents assault other residents after their prolonged or repetitive disruptive behaviors like talking, screaming, or unwanted touching. This study illuminates the problematic nature of assault in nursing home settings. Unlike other assaults, resident-to-resident assaults are unique in that dementia often plays a role in both acts of victimization and offending. Lachs and his colleagues conclude,

> "Victim," "perpetrator," or both may be "blameless" in many of these episodes, in that they are facilitated by brain disease and not volitional ill will. Indeed, even the label "elder mistreatment" suggests, often incorrectly, that malevolence is a motivator in these situations, although it may promulgate more work in this sorely ignored area. (p. 844)

Not surprising, institutional-level risk factors for elder abuse include nursing homes with high percentages of residents with dementia (Talerico, Evans, & Strumpf, 200) or other types of highly dependent residents (National Center on Elder Abuse, 2000, 2005), staffing issues such as training (National Center on Elder Abuse, 2000), low staff-to-patient ratios and frequent staff turnover, stress, and "burn-out" (Pillemer & Bachman, 1991; Pillemer & Moore, 1990; Shaw, 1998), institutional culture of the nursing home (National Center on Elder Abuse, 2000), and facility infrastructure (long hallways, poor lighting, many floors and stairwells, far away nurses stations; National Center on Elder Abuse, 2000). Perpetrators of staff abuse are often nurses' aides as opposed to nurses, doctors, or administrative staff. Nursing aides have the most direct contact with residents so, in part, their overrepresentation among offenders is to be expected. Other perpetrator risk factors are being young, having less education, having less experience working with elder populations, and an overall dissatisfaction with the job (Pillemer & Bachman, 1991; Pillemer & Moore, 1990).

As noted earlier, resident-level risk factors for nursing home abuse include cognitive conditions associated with aggressive behavior such as dementia and Alzheimer's disease (National Center on Elder Abuse, 2005). Residents living in tight quarters or subjected to other institutional conditions where their lack of autonomy is exacerbated are more likely to lash out, especially at other residents (National Center on Elder Abuse, 2005). Finally, residents who rarely receive visitors experience an increased risk in abuse and mistreatment by staff and other residents (Menio, 1996), undoubtedly related to the lack of guardianship for these residents.

Added to these risks is the fact that state judges are increasingly ordering older felons to nursing facilities (National Center on Elder Abuse, 2000) likely because of the fact that the correctional system is ill equipped to deal with its aging prison population. Although most criminal offenders do age out of crime, increasing the number of convicted felons in the nursing home system may result in negative safety consequences for staff and residents. Moreover, the increasing presence of younger mentally ill residents in long-term care facilities further increases the likelihood of resident-to-resident abuse as these individuals have a greater propensity toward anger, hostility, and aggressive behavior (National Center on Elder Abuse, 2000).

⊠ Discussion and Recommendations

Based on the epidemiological evidence presented in this article, it is clear that both male and female elders experience unique vulnerabilities to victimization that are unlike any other time in the life course, despite their decreased rates of victimization. These findings coupled with the extant literature illustrate the often deleterious consequences victimization has for the elderly and the urgent need for primary and secondary prevention.

Future research is needed to investigate the etiology of elderly victimization. What factors increase the probability of an elder becoming the victim of violence? Are these factors different across gender and ethnic groups? Are there community level factors related to the risk of victimization against the elderly? Although we know very little about the answers to these questions, the epidemiological patterns of victimization against the elderly do illuminate specific policy recommendations.

To prevent predatory victimizations against the elderly, policies should concentrate on placing guardianship where none exists through such initiatives as neighborhood watches, transportation and home security assistance, and block clubs. One innovative effort in this area is called the Triad program, which couples local police and sheriffs' department with senior citizens themselves to prevent victimization in their community. Like other community policing initiatives, the goal of Triad programs is to share ideas and resources to provide programs and training directed at increasing protection for vulnerable elderly populations (Cantrell, 1994).

The murders and assault against the elderly committed by other family members are likely within the context of a caregiving situation for an elderly parent or relative. As such, policies aimed at alleviating the stress of such situations and providing respite services to caregivers are extremely important. Detecting these victimizations, of course, is of paramount importance. All states and the District of Columbia have authorized the provision of APS, which establishes a system for the reporting and investigation of elderly victimization, and the provision of social services to help the victim and ameliorate the abuse. Laws in most states require professionals who have direct contact with elders such as doctors and home health providers, to be mandated reporters when it comes to cases of suspected victimization. But what services are available for elders who find themselves in an abusive home environment?

Elderly victims of familial abuse face many unique obstacles different from those of their younger counterparts. As such, services and safe havens that can serve elderly victims are an urgent priority. The culture of traditional nursing home care is generally geared toward involving the family in an elder's care and decision making. However, when a family member is an abuser, preventing family contact becomes a priority. Despite these unique needs, there are only a handful of organizations that are designed to serve elderly victims of domestic abuse. One shelter that caters to the needs of elderly victims of familial abuse is the Hebrew Home for the Aged in the Bronx of New York City (Leland, 2005). The Hebrew Home, in conjunction with the Jeanette Weinberg Center for Elder Abuse Prevention, opened this free program in January 2005 to provide temporary emergency shelter for up to 31 elders. They also provide services to find long-term housing if necessary along with important counseling and advocacy (Leland, 2005).

Finally, because violence against the elderly has such severe physical consequences in terms of increased risk of injury, requiring medical care, nursing home placement, and injuries resulting in death, secondary prevention efforts following nonfatal victimizations against the elderly are extremely important. This is true as much for physical injuries as it is for the emotional trauma that undoubtedly accompanies such brutality. Health care providers need to be educated on how to aggressively respond to the unique needs of elderly victims of crime to prevent any spiral decline of health for elders.

Protection of Institutionalized Elderly

Prevention strategies for reducing staff-to-resident victimization in nursing homes entail extensive staff training, staff employment screening, incorporation of abuse prevention policies and reporting procedures, implementation of measures to decrease staff attrition, decrease of staff stress levels, and decrease of staff "burn-out" (Clough, 1999; Menio, 1996; National Center on Elder Abuse, 2000; Payne & Cikovic, 1995; Pillemer & Bachman, 1991). To address some of these issues it is suggested that institutions increase their pay scale for the positions most vulnerable to committing abuse and to being the direct target of abuse (nurse's aides in particular), keep staffing levels high at all times, and implement career trajectories and upward mobility tracks within the organization (National Center on Elder Abuse, 2005; Pillemer & Bachman, 1991; Shaw, 1998). Oversight agencies, such as Ombudsman Programs, should also increase their vigilance at investigating and adjudicating all credible allegations.

Regarding the more prevalent resident-to-resident assaults, institutions should be proactive in providing staff training for managing dementia-related behaviors, particularly the monitoring and management of aggressive behavior. General training in conflict resolution should also be provided. Unfortunately, when training protocols were examined, specific guidelines or training materials for nursing home staff on how to interdict in cases of resident-to-resident assault could not be found (Lachs et al., 2004).

In sum, during the next 20 years, the United States will become a nation in which those 65 and older will represent more than 20% of the total population. The government has spent a great deal of time and energy considering what this growth will mean for such programs as Social Security and Medicare. However, very little attention has been given to issues regarding the quality of life older Americans can come to expect, including feeling safe in their own homes, or having safety nets available when their own homes become domains of abuse. The time is now to develop well-informed policies aimed at both preventing violence against the elderly and ameliorating the consequences of this violence when it does occur. We cannot afford to ignore victimization against the elderly.

References

Administration on Aging. (1998). *The National Elder Abuse Incidence Study: Final report.* Available from the Administration on Aging website, http://www.aoa.gov/eldfam/Elder_Rights/Elder_Abuse/AbuseReport_Full.pdf

Bachman, R., Dilloway, H., & Lachs, M. S. (1998). Violence against the elderly: A comparative analysis of robbery and assault across age and gender groups. *Research on Aging, 20*(2), 183–198.

Bachman, R., Lachs, M. S., & Meloy, M. (2004). Reducing injury through self-protection by elderly victims of violence: The interaction effects of gender of victims and the victim/offender relationship. *Journal of Elder Abuse & Neglect, 16*(4), 1–24.

Bachman, R., Meloy, M., & Block, R. (2005, November). *Examining violence against the elderly: Lethal and nonfatal patterns of robbery and assault.* Paper presented at the annual conference of the American Society of Criminology, Toronto, Ontario, Canada.

Cantrell, B. (1994, February). Triad: Reducing criminal victimization of the elderly. *FBI Law Enforcement Bulletin.* Retrieved January 12, 2007, from http://www.findarticles.com/p/articles/mi_m2194/is_n2_v63/ai_15267890

Chu. L. D., & Kraus, J. F. (2004). Predicting fatal assault among the elderly using the National Incident-Based Reporting System crime data. *Homicide Studies, 8*, 71–95.

Clough, R. (1999). "Scandalous care": Interpreting public inquiry reports of scandals in residential care. *Journal of Elder Abuse & Neglect, 19*, 13–27.

Douglas, R. L., Hickey, T., & Noel, C. (1980). *A study of maltreatment of the elderly and other vulnerable adults: Final report to the United States Administration on Aging and the Michigan Department of Social Sciences.* Ann Arbor: University of Michigan.

Klaus, P. (2005). *Crime against persons age 65 or older, 1993–2002* (NCJ 206154). Washington, DC: U.S. Department of Justice, Bureau of Justice Statistics.

Lachs, M., Bachman, R., Williams, C., Kossack, A., Bove, C., & O'Leary, J. (2004). Older adults as crime victims, perpetrators, witnesses, and complainants: A population-based study of police interactions. *Journal of Elder Abuse, 16*(4), 25–40.

Lachs, M., Bachman, R., Williams, C., Kossack, A., Bove, C., & O'Leary, J. R. (2006). Violent crime victimization increases the risk of nursing home placement in older adults. *The Gerontologist, 46*, 583–589.

Lachs, M., Bachman, R., Williams, C. S., & O'Leary, J. R. (2007). Resident-to-resident elder mistreatment and police contact in nursing homes: Findings from a population-based cohort. *Journal of the American Geriatrics Society, 55*, 840–845.

Leland, J. (2005, November 8). For the elderly, a place to turn to when abuse comes from home. *New York Times.* Retrieved October 3, 2007, from http://www.nytimes.com/2005/11/08/nyregion/08elder.html?_r=1&oref=slogin&emc=etal& pagewanted

Menio, D. A. (1996). Advocating for the rights of vulnerable nursing home residents: Creative strategies. *Journal of Elder Abuse & Neglect, 8*(3), 59–72.

National Center of Elder Abuse. (2000). *Adult protective services role in the prevention of nursing home abuse: Report of a national teleconference.* Washington, DC: Author.

National Center on Elder Abuse. (2005). *Nursing home abuse: Risk prevention profile and checklist.* Washington, DC: National Center on Elder Abuse and U.S. Department of Health and Human Resources, Administration on Aging.

Payne, B., & Cikovic, R. (1995). An empirical examination of the characteristics, consequences, and causes of elder abuse in nursing homes. *Journal of Elder Abuse & Neglect, 4*, 61–74.

Pillemer, K., & Bachman, R. (1991). Helping and hurting: Predictors of maltreatment of patients in nursing homes. *Research on Aging, 13*(1), 74–95.

Pillemer, K., & Hudson, B. (1993). A model abuse prevention program for nursing assistants. *The Gerontologist, 33*, 128–131.

Shaw, M. M. C. (1998). Nursing home resident abuse by staff: Exploring the dynamics. *Journal of Elder Abuse & Neglect, 9*(4), 1–22.

Talerico, K., Evans, L., & Strumpf, N. (2002). Mental health correlates of aggression in nursing home residents with dementia. *The Gerontologist, 42*, 169–177.

Wan, H., Sengupta, M., Velkoff, V. A., & DeBarros, K. A. (2005). *65+ in the United States, 2005: Current population reports.* Washington, DC: U.S. Department of Health and Human Services.

DISCUSSION QUESTIONS

1. What are the unique vulnerabilities to violent victimization that elderly persons face as compared with younger persons?

2. Are the consequences of being victimized different for elderly persons? Why or why not? Explain.

3. What are the policy implications of these findings?

4. Are community-dwelling elderly persons at greater risk of being victimized? Why or why not? Be sure to use information from the reading to support your answer.

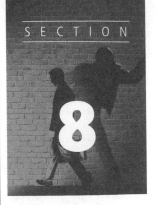

SECTION

8

Victimization of Special Populations

On February 1, 2011, it was reported that a Swiss Social Worker had confessed to sexually assaulting 114 disabled children and adults. The man, working as a therapist, assaulted these people in nine different institutions over a 28-year span. He confessed to sexually abusing these people, all of whom were mentally disabled—72 of them were under the age of 18, and one was only 1 year old! He went so far as to videotape or photograph 18 of the incidents (CNN Wire Staff, 2011). You probably are wondering what could possess a person to do this to individuals under his care. But consider the victims here—these are mentally disabled individuals in institutions. These people may have been especially vulnerable to victimization given their reliance on others for care and their diminished mental capacity.

This chapter is dedicated to such vulnerable victims. It begins with a discussion of persons who are mentally disabled and then turns to a discussion of persons who are mentally ill. Victimization of persons who are incarcerated is then covered. You may wonder why incarcerated victims are included in the same chapter as mentally disabled and mentally ill victims, but as you will read below, these three groups are at particular risk of being victimized, given their status and unique vulnerability.

⬛ Victimization of Persons With Disabilities

Another group of individuals who are especially vulnerable to victimization are people who have disabilities. It may seem odd, at first, to discuss persons with disabilities in the same section as those who are incarcerated. But let's consider the fact that both groups of people, given their status, are likely targets of would-be offenders. Given what we know about lifestyle and routine activities theory, being a vulnerable target is a key factor in a person's risk for victimization. For some people, their personal characteristics will place them at particular risk of being victimized while incarcerated. Predators and offenders in prison will target them. The same is true for people with disabilities. Predators and offenders will likely see them as easy marks for a variety of reasons. Because of this, persons with disabilities who are victimized deserve special attention.

342

Defining Persons With Disabilities

When you first began reading this section, an image of a person with a disability may have come to you. This image may have been of a person with a physical disability or an intellectual disability. Both of these are, in fact, examples of persons with disabilities. But what do these terms mean? A physical disability is present when a person is unable to perform daily activities such as walking, climbing stairs, bathing, getting dressed, and taking care of himself or herself. An intellectual disability falls under the broader category of **developmental disability**. Developmental disabilities include such things as cerebral palsy, epilepsy, severe learning disabilities, mental retardation, and autism (Petersilia, 2001). See Table 8.1 for descriptions of common developmental disabilities. A developmental disability manifests before the age of 22, is long-lasting, and causes severe impairment in at least three of five areas of activity—self-care, language, learning, mobility, and capacity for independent living (Centers for Disease Control and Prevention, 2010). In this section, we will mainly discuss persons with physical disabilities, which can occur at any point in life, and persons with developmental disabilities. If the research or study being referenced does not distinguish between developmental and physical disability, the term *disability* will be used.

Extent of Victimization of Persons With Disabilities

As with most types of victimization, the first place we turn for a snapshot of the extent of victimization is the National Crime Victimization Survey (NCVS). For persons with disabilities, this is no different. The **Crime Victims with Disabilities Awareness Act of 1998** mandated that the NCVS begin to collect information about the victimization of people with disabilities. To do so, the NCVS incorporated questions asking people if they had a sensory, physical, mental, or emotional condition (Harrell & Rand, 2010). If a person reported that he or she did have such a condition, that it had lasted 6 months or longer, and that it had made performing daily life activities difficult, that person would be considered to suffer from a disability (Harrell & Rand, 2010). Notice that the NCVS's definition of disability includes a wide range of limitations on behavior and interaction.

Based on this definition, according to findings from the 2008 NCVS, there were about 730,000 violent crime victimizations and 1,793,490 property crime victimizations perpetrated against persons with disabilities. The most

Table 8.1   Description of Common Developmental Disabilities

Cerebral palsy	A condition that involves the brain and nervous system functions. Persons with cerebral palsy have damage to the area of the brain responsible for muscle tone, resulting in difficulty in movement, balance, and posture.
Epilepsy	A disorder in the brain that typically causes various spontaneous seizures.
Severe learning or intellectual disability	Limited ability to perform daily functions (i.e., communication, self-care, and getting through basic and often very simple learning/education courses) and below-average score on mental ability or intelligence.
Mental retardation	Intellectual level is significantly below average; person lacks basic skills for everyday living. This condition is generally diagnosed before a person reaches age 18.
Autism	Within the first 3 years of a child's life, this developmental disorder appears and affects brain development (usually disrupting the child's social and communication skills).

SOURCE: Centers for Disease Control and Prevention (2010).

common types of victimization were simple assault and theft, the same types most commonly experienced by persons without disabilities. Comparing victimization rates of persons with disabilities with those of persons without disabilities in the NCVS shows that people with disabilities have higher victimization rates. The victimization rate of people with disabilities is 40 per 1,000 compared with 20 per 1,000 people without disabilities (Harrell & Rand, 2010).

The NCVS is, of course, not the only source of victimization statistics for persons with disabilities. It is the only source for national-level survey-derived estimates in the United States, but other research has been conducted on smaller samples and in other countries. Much of the research was conducted in the 1980s and early 1990s. Not to say that these estimates are unreliable, but it should be noted that this area of research has not received as much attention as others. Also of importance, because most of these studies have been conducted on nonrepresentative samples, it is hard to say how generalizable the findings are. In fact, there is actually some debate in terms of the extent to which people with disabilities are more likely to be victimized than persons without, given the methodological weaknesses of the research studies (Marge, 2003).

From these studies, other pictures of the extent of victimization against persons with disabilities are revealed. In Australia, the Bureau of Statistics administered a version of their Victims of Crime Survey to adults with intellectual handicaps. From this survey, it was found that these individuals had higher rates of victimization than persons without intellectual handicaps (Petersilia, 2010). An oft-cited study by Wilson and Brewer (1992), in which they studied 174 adults with intellectual disabilities in South Australia, found that the relative risk of victimization was highest for violent victimizations—persons with intellectual disabilities were 12.8 times more likely to be robbed, 10.7 times more likely to be sexually assaulted, and 2.8 times more likely to be assaulted (as cited in Petersilia, 2001). Other research has found that persons with disabilities are at risk of being victimized by their personal care attendants (Ulicyn, White, Bradford, & Matthews, 1990).

Official data sources also can be used to glean information about abuse of persons with disabilities. In one study, McCartney and Campbell (1998) reviewed records of 9,400 men and women with mental retardation living in 23 institutional facilities in six states. They found that 5% of them had experienced abuse by a staff member over a 22-month period. Most typically, the victim had been neglected or physically abused.

◪ Who Is Victimized?

Among those persons with disabilities surveyed in the NCVS, women were more likely to be the victims of a violent crime than were men (Harrell & Rand, 2010). This is unlike the patterns we see for violent crime victimization in general, in which men are more likely than women to be violently victimized for all types of victimization except rape and sexual assault. Black persons and White persons with disabilities are equally likely to experience a violent victimization, another finding that is contrary to violent victimization patterns. Black persons have higher rates of violent victimization overall (Truman & Rand, 2010). All in all, even when disaggregating different racial groups (White, Black, other race) and examining Hispanics and non-Hispanics, persons with disabilities have higher rates of violent victimization than do persons without disabilities.

When examining the types of disabilities separately, it does seem that not all disabilities impact victimization risk similarly. Persons with cognitive disabilities have the highest rates of violent victimization, compared with the other disabilities measured in the NCVS (Harrell & Rand, 2010). Rates of robbery and simple assault were significantly higher for persons with cognitive disabilities than for persons with other types of disabilities, but no differences were found for rape and sexual assault and aggravated assault (Harrell & Rand,

2010). Having a hearing disability elevated the risk of violent crime more so for females than for males (Harrell & Rand, 2010).

Violence Against Women With Disabilities

Special attention has been given to studying women with disabilities. In one study of physically disabled women who used personal assistants, Powers, Curry, Oschwald, and Maley (2002) discovered these women had experienced a wide range of abusive behavior. Many women reported that their personal assistants had stolen money or items from them (35.5%) and that their checks had been misused or forged (30%). Neglect and physical abuse were also experienced by the women—19% had their physical needs neglected, 14% had been physically abused, 14% had been sexually abused, and 11% reported unwanted sexual touching. Other studies estimate that 2% of women with disabilities (physical, mental, or emotional) have been physically abused in the past year (Martin et al., 2006), 10% of women with physical disabilities have experienced abuse within the past year (McFarlane et al., 2001), and 36% of women with disabilities have experienced physical assault abuse during their lifetimes (Young, Nosek, Howland, Chanpong, & Rintala, 1997).

When we discuss victimization against women, we must, of course, give attention to rape and sexual victimization. Like other women, women with disabilities are at risk of being sexually victimized. Women with developmental disabilities are 4 to 10 times more likely to be sexual assault victims than women in the general population (National Council on Disability, 2007). When only rape is examined, between 15,000 to 19,000 people with developmental disabilities are raped each year in the United States (Subsey, 1994). It is estimated that more than 80% of women with a disability will be sexually assaulted over their lifetimes (Wisconsin Coalition Against Sexual Assault, 2003). Research on women who use personal assistance services has revealed that women may be abused by people employed by these services. In fact, Powers et al. (2002) found that in their study, women with physical disabilities and those with both physical and intellectual disabilities were more likely to report being sexually abused in general by persons employed by personal assistance services.

Most of the research on sexual victimization of women with disabilities has been largely descriptive—identifying the extent to which it occurs. Little research has identified the characteristics that place certain women with disabilities at risk. It does appear, though, that women who are younger, non-White, nonmarried, employed, and less-educated, are more likely than other women with disabilities (physical, mental, or emotional problems that limit activity) to be sexually assaulted (Martin et al., 2006). Also important, research on women with disabilities shows that repeat sexual victimization is also quite common. Some estimate that almost half of women with developmental disabilities who are victims of sexual victimization will experience at least 10 incidents during their lives (Valenti-Hein & Schwartz, 1995). Past sexual victimization also has been shown to increase the likelihood of being sexually assaulted 2.5 times for this population (Nannini, 2006). Although this section refers to female victims, males with disabilities are also at risk for experiencing sexual assault. In one study, 32% of males with developmental disabilities reported being sexually assaulted by an intimate partner (Johnson & Sigler, 2000).

Along with sexual victimization, women with disabilities are at risk of being violently victimized by their intimate partners. They have indicated through surveys a particular worry about intimate partner violence, naming it one of their top five concerns (Grothaus, 1985). Given the prevalence of intimate partner violence among women with disabilities, this concern does not appear to be unfounded. One study of Canadian women found that women with disabilities (long-term physical, mental, or health problems that limit home, school, work, or other activities) were 40% more likely to experience partner violence within the previous 5 years than were women without disabilities (Brownridge, 2006). For more explanations regarding why women with disabilities are likely to be victimized by their intimate partners, see the article by Douglas A. Brownridge (2006) included in this section.

Victimization of Youth With Disabilities

Children with disabilities may be especially vulnerable to victimization. Young people in general are at risk of victimization because they may be unable to defend themselves, thus making them attractive targets. Children with disabilities may be around caretakers who become frustrated with them and are less emotionally invested in them than their parents. They also may be targets of peers at school. For these reasons, their victimization patterns have been examined. Professionals working with youth with disabilities perceive that children with intellectual disabilities have higher rates of maltreatment than children without them (Verdugo, Bermejo, & Fuertes, 1995). Formal school records also support this contention (Sullivan & Knutson, 2000).

Bullying, although not necessarily reaching levels that would be considered criminal, may also be a problem facing youth with disabilities. Bullying—discussed in detail in Section 9—is harmful behavior directed at a youth that occurs more than once and is characterized by a power imbalance (Olweus, 1993).

A study of children with disabilities apparent to other children found that they were more likely than other children to be bullied at school—half of them reported being bullied during the school term (Dawkins, 1996). Similarly, when mothers were asked whether their children with Asperger's syndrome or other nonverbal learning disabilities had experienced victimization by their peers, more than 90% responded "yes" (Little, 2002). We have already learned that many children are bullied and victimized by their peers, but it seems that children with disabilities are particularly vulnerable.

Another type of victimization of children with disabilities that appears to be especially problematic is sexual abuse. Children with disabilities are estimated to be 4 to 10 times more likely than the general population to be sexually abused. As noted by Baladerian (1991), an estimated 25% of the general population of females is sexually victimized before reaching adulthood; thus, girls with disabilities are particularly vulnerable (it is estimated that almost all of them will experience a sexually abusive event). Also, between 16% and 32% of boys with developmental disabilities are believed to be at risk for sexual abuse before the age of 18 (Badgley et al., 1984; Hard, 1986).

◤ Patterns of Victimization

The NCVS provides interesting information about each incident of victimization that occurred against persons with disabilities. Individuals were asked if they believed they were victimized because of their disability; about 15% of violent crime victims with disabilities indicated that they thought so (Harrell & Rand, 2010). About one third of victims with disabilities noted they were unsure if they were victimized because of their disability (Harrell & Rand, 2010). Victims were also asked if they did anything during the victimization to resist or try to stop it. Persons with disabilities were less likely than persons without disabilities to attempt to employ any measure of resistance during a violent victimization (Harrell & Rand, 2010). We will return to the use of resistance below, but for now, think about the use of resistance, or lack thereof, and how it may factor into victim selection by offenders.

Violent victimizations against persons with disabilities most commonly do not involve a weapon (only 20% do), and about one fourth of victims with disabilities suffer injury as a result (Harrell & Rand, 2010). About half of violent crime victims with disabilities reported their experiences to the police, a finding similar to violent crime victims overall. They were, however, more likely than violent crime victims without disabilities to seek assistance from victim advocacy agencies (Harrell & Rand, 2010). Property crime victims with disabilities were *less likely* to report their victimizations to police (Harrell & Rand, 2010).

Who is likely to perpetrate these victimizations? One of the unique features of victimization of persons with disabilities is that they often rely on others for support and care. Adults may receive care from friends and family and/or formal personal care assistance services. The victim-offender relationship, then, is worthy of discussion, because for this type of victimization, exposure to motivated offenders is likely to be different than for other potential victims. It is difficult to say exactly who is most likely to offend against victims with disabilities because many studies have exclusively asked about certain types of offenders. That is, some surveys have explicitly asked victims only about whether care providers have victimized them and not asked about other types of offenders. Nonetheless, these studies have revealed a startling picture of persons who are charged with caring for individuals actually harming them. For example, one study compared individuals with severe cognitive impairment receiving in-home supportive services in California with individuals having no severe cognitive impairment. Those with severe impairment had significantly higher rates of neglect, injury, being yelled at, suspected theft, threats by their providers, physical abuse, and sexual advances (Matthias & Benjamin, 2003). In another study, researchers reviewed the records of 9,400 institutionalized women and men with mental retardation living in 23 residential facilities in six U.S. states. They found that about 5% had experienced some form of abuse by a member of the facility staff during a 22-month period (McCartney & Campbell, 1998). Other studies have included questions about who the perpetrator is or have located in official records who the offender was. These studies revealed that persons are most likely to be victimized by family members, intimate partners, and personal assistance/health care providers (McFarlane, Hughes, Nosek, Groff, Swedlend, & Mullen, 2001; National Council on Disability, 2007).

⊠ Risk Factors for Victimization for Persons With Disabilities

As noted, there are a number of reasons why persons with disabilities are at risk of being victimized. Let's discuss these factors in a bit more detail so that the risk of victimization can be better understood. First, persons with disabilities are likely seen as vulnerable, especially if they rely on others for care. They may, by necessity, allow other people access to their money, bank accounts, checking accounts, and other financial records. As you may imagine, doing so places them at risk for theft. Having home health care and personal assistants requires allowing strangers access to their homes and personal effects. According to routine activities theory, these people could be providing capable guardianship, but they may be motivated offenders instead.

Second, people with intellectual or developmental disabilities may not be able to recognize and process risk. They may not, despite doing all the "right" things to protect themselves, realize that they are in the wrong place or with the wrong people. Unfortunately, they also may not realize that a victimization has occurred. Even when they do, some research shows that they are less likely to report their victimization to police than are other victims. The decision not to report could be a result of several factors. They may have limited communication skills, which could make reporting difficult and frightening (Nettelbeck & Wilson, 2002). They may be unaware that money has been taken if they typically do not control their own finances. They may not know that sexual activity that has occurred is inappropriate and illegal if they have never been exposed to sex education (Petersilia, 2001). Reporting may also be unlikely if the offender is a family member or someone else who is caring for them. Victims may fear retaliation for reporting (National Council on Disability, 2007). Reporting may mean that the victim will have to move into a more restrictive living situation, an option that may be worse than saying nothing (Petersilia, 2001). For victims of intimate partner violence, they may not have access to shelters that can accommodate persons with disabilities, or similar to other victims of intimate partner violence, they may feel they lack the financial resources to leave (Petersilia, 2001).

Third, persons with disabilities are more likely to live in poverty than people without disabilities. At least one in every three adults with disabilities live in households with an income totaling less than $15,000 per year. Overall, only 12% of those without disabilities live in households in this income range (as cited in Petersilia, 2001). Poverty is a risk factor for victimization in that people living in poverty are likely to live in crime-ridden neighborhoods in areas that are unlikely to be diligently patrolled by law enforcement.

Fourth, victimization of persons with disabilities has been linked to a **dependency-stress model**, especially for victimization of younger persons with disabilities. That is, because children with disabilities tend to be more dependent than other children on their parents and caregivers, it creates more stress, which may result in abuse if their parents and caregivers cannot appropriately deal with the stress of their responsibilities (Petersilia, 2001). Some have argued that persons with disabilities display behaviors that may promote victimization—without suggesting a victim-blaming model. Persons with disabilities may misinterpret social cues, want to please others, and think that people want to be their friends (Petersilia, 2001). In this way, they may comply with inappropriate requests, resulting in victimization (Nettelbeck & Wilson, 2002). In addition, victims with disabilities may evoke aggressive responses with their behaviors. In one study that compared nonvictims to victims matched on age, IQ, and adaptive behavior, Wilson, Seaman, and Nettelbeck (1996) found that victims were more likely to react with inappropriately angry or aggressive responses in everyday situations with strangers, indicating that poor social competence (anger and aggression in interpersonal interactions) rather than intellectual disability was a contributing factor in victimization.

⊠ Responses to Victims With Disabilities

Remember that much research shows that crime victims with disabilities are unlikely to seek assistance from the police (Focht-New, Clements, Barol, Faulkner, & Service, 2008). This low level of reporting may be due to several reasons. They may be unaware, especially in cases of property victimization (Petersilia, 2001) or because of not receiving appropriate sex education, that a crime ever occurred (Nosek, 1996) In addition, persons with disabilities may be particularly dependent on the person who perpetrated the incident; thus, reporting a crime may make their day-to-day living difficult (Nosek, 1996). Even when crime victims with disabilities do come forward, there is evidence that police may be hesitant to treat the cases the same as they do when other victims come forward (Petersilia, 2010). These victims may be seen as less legitimate and as unreliable witnesses due to their diminished intellectual capacity. Law enforcement may not realize that persons with mental retardation and autism have good memories and can be reliable witnesses (Petersilia, 2010). Indeed, in Petersilia's (2001) review of research on crime victims with developmental disorders, she notes that much research shows that people with mental retardation forget information at rates similar to those of persons without mental retardation and that they recall with similar accuracy to others their age (in studies of children). Children with mental retardation are, however, highly impressionable and eager to please; thus, interviewers and attorneys should take special precautions in criminal justice settings (Petersilia, 2001). Prosecutors also may be hesitant to pursue these cases for similar reasons.

Research on people with disabilities who have been sexually abused is particularly instructive. It has shown that victims rarely report (Wacker, Parish, & Macy, 2008). Even when victims do report, offenders are rarely charged—this may be due to a number of reasons, one of which is that perpetrators are commonly acquaintances and intimate partners (Wacker et al., 2008). In one study, of the 22% of alleged offenders charged with sexual abuse, only 38% were ultimately convicted (Sobsey & Doe, 1991). This lack of reporting, charging, and conviction is not unique to victims with disabilities, but the rates do appear to be particularly low for female victims of sexual abuse/violence with disabilities.

Recognizing the difficulties victims with disabilities face, the **Federal Crime Victims with Disabilities Awareness Act of 1998** required the collection of statistics on the victimization of persons with disabilities (Pub. L. 105-301). Also proposed was the **Crime Victims with Disabilities Act of 1998**, which would require reporting of abuse to law enforcement; provide training for law enforcement, prosecutors, rape crisis counselors, and health care providers; require victim-witness assistance programs to cover victims with disabilities; and develop continued reform for working with victims with disabilities (Assembly Bill 2038). This act was proposed by then Senator Biden but has not been reintroduced as of January 2011.

Without federal legislation requiring reporting and victim assistance, some states have specifically addressed victims with disabilities. For example, Ohio has a mandatory reporting law requiring that a mandatory reporter who has a "reason to believe that a person with mental retardation or a developmental disability has suffered or faces a substantial risk of suffering any wound, injury, disability, or condition of such a nature as to reasonably indicate abuse or neglect of that person" (RAINN, 2009) must report. Table 8.2 provides the specifics of Ohio's law. Other states have amended their criminal codes so that persons with cognitive disabilities are afforded similar protections as those afforded to children when testifying in court, such as videotaped testimony, testimony via closed-circuit television, the use of breaks during testimony, allowing the presence of support persons in the courtroom, and adjusting the seating arrangements in the courtroom to be less intimidating (Petersilia, 2001).

Even without legislation, victim service providers are already faced with addressing the needs of victims with disabilities, although some jurisdictions are likely better able to do this than others. Nonetheless, victim service providers should be able to assist victims with disabilities in finding suitable housing, applying for victim compensation, receiving counseling, getting legal services, having someone attend court with them, and navigating the criminal justice system (National Council on Disability, 2007). Beyond victim services, there are training videos,

Table 8.2   Ohio's Mandatory Reporting Law for Crime Victims With Mental Disabilities

Who is disabled?	Persons with mental retardation and persons with developmental disabilities
Who is the mandatory reporter?	Any physician (including a hospital intern or resident); dentist; podiatrist; chiropractor; practitioner of a limited branch of medicine; hospital administrator or employee; licensed nurse; ambulatory health facility employee; home health agency employee; adult care facility employee; community mental health facility employee; school teacher or authority; social worker; psychologist; attorney; peace officer; coroner; residents' rights advocate; superintendent, board member, or employee of a county board of developmental disabilities; administrator, board member, or employee of a residential facility; administrator, board member, or employee or any other public or private provider of services to a person with mental retardation or a developmental disability; MR/DD employee; member of a citizen's advisory council established at an institution or branch institution of the department of developmental disabilities; a person who is employed in a position that includes providing specialized services to persons with disabilities, and a clergyman who is employed in a position that includes providing specialized services to persons with disabilities and who renders spiritual treatment through prayer in accordance with the tenets of an organized religion
Who is reported to?	• Law enforcement agency or the county board of developmental disabilities • State highway patrol if the individual is an inmate in a state correctional institution • Law enforcement agency or the department of developmental disabilities if the report concerns a resident of a facility operated by the department
Standard of knowledge	Reason to believe that person with disability has suffered or faces substantial risk of suffering wound, injury, disability, or condition as to reasonably indicate abuse or neglect of that person

SOURCE: RAINN (2009). Reprinted by permission of RAINN, Rape, Abuse and Incest National Network.

classes, and pamphlets available for law enforcement and other criminal justice professionals preparing to respond to crime victims with disabilities. Most of the training, however, has primarily focused on victims with mental illness, discussed in the next section.

Victimization of Persons With Mental Illness

Mental illness may also put individuals at risk of being victimized, for many of the same reasons that people with disabilities find themselves at risk. What, though, constitutes mental illness? You may think of mental illness as being debilitating, and sometimes it is. But many people live with mental illness—they go to work, they go to school, they are, by all accounts, productive members of society. Others do, however, face severe challenges. For many, one of these challenges is an increased chance of being victimized. Before we begin a discussion of this risk, let's first discuss what mental illness is.

Defining Mental Illness

Much of the research examining the link between mental illness and victimization has focused on persons with severe mental illness. Persons with severe mental illness often have symptoms of "impaired reality testing, disorganized thought processes, impulsivity, poor planning, and poor problem solving" (Teplin, McClelland, Abram, & Weiner, 2005, p. 911). In terms of operationalization, severe mental illness is often specified when persons have taken psychoactive medications for extended periods of time (e.g., 2 years) or have been hospitalized for psychiatric reasons (Teplin et al., 2005). But people can suffer from mental illness that may not meet these criteria. In addition, classifying severe mental illness as such does not identify the type of mental disorder diagnosis. There are, of course, a range of mental illnesses. Some of the most common ones that have been studied and linked to victimization are found in Table 8.3. Substance abuse issues such as alcohol and marijuana dependence are also mental disorders identified by the *Diagnostic and Statistical Manual* (4th edition, textual revision), which is a diagnostic manual published by the American Psychiatric Association. It is used by mental health professionals to diagnose and treat patients.

Extent and Type of Victimization of Persons With Mental Illness

When persons in criminal justice first paid attention to persons with mental illness, it was not out of concern for their victimization. Instead, persons with mental illness were perceived to be violent offenders rather than victims (Silver, Arseneault, Langley, Caspi, & Moffitt, 2005). What research has found, however, is that persons with mental illness are more likely to be victims of violence than perpetrators (Levin, 2005). In terms of the extent to which persons with mental illness are victimized, it is important to note the methodology of the study. There are some studies that have examined the victimization of persons who are or have been institutionalized, while others have used samples drawn from those living in the community. We would expect to find different rates of victimization for persons living in the community than for those who are institutionalized.

When comparing persons with mental disorders in the community to persons in the community without mental disorders, it becomes clear that mental disorder is a risk factor for victimization. When comparing persons with severe mental illness to those in the general population, those with mental illness are 2.5 times more likely to be violently victimized (Hiday, Swartz, Swanson, Borum, & Wagner, 1999). Indeed, studies on persons with mental disorders have consistently found high rates of victimization—studies have found estimates of victimization from 8% (sexual assault victimization in the past year for males with severe mental illness; Goodman et al., 2001) to more than 50% (physical abuse in the past year for females with severe mental illness; Goodman et al., 1999).

Table 8.3   Common Mental Illnesses, *DSM-IV-TR* Definitions and Diagnostic Criteria

Disorder	Definition	Diagnostic Criteria
Depressive	Five or more of the following symptoms have been present during the same 2-week period or represent a change from previous functioning; at least one of the symptoms is either (1) depressed mood or (2) loss of interest or pleasure. • Depressed mood most of the day, nearly every day • Markedly diminished interest or pleasure in all, or almost all, activities most of the day, nearly all day • Significant weight loss when not dieting or weight gain • Insomnia or hypersomnia nearly every day • Psychomotor agitation or retardation nearly every day • Fatigue or loss of energy nearly every day • Feelings of worthlessness or excessive or inappropriate guilt (which may be delusional) nearly every day • Diminished ability to think or concentrate, or indecisiveness, nearly every day • Recurrent thoughts of death, recurrent suicidal ideation without a specific plan, or a suicide attempt or a specific plan for committing suicide	• Cause clinically significant distress or impairment in social, occupational or other important areas of functioning • Are not due to the direct physiological effects of a substance or a general medical condition • Are not better accounted for by bereavement • Persist for longer than 2 months or are characterized by marked functional impairment, morbid preoccupation with worthlessness, suicidal ideation, psychotic symptoms, or psychomotor retardation
Anxiety	Characterized by excessive worry, but such worries are distinguished from obsessions by the fact that the person experiences them as excessive concerns about real-life circumstances.	Ranging from worrying about losing one's job to the intrusive distressing idea that "God" is "dog" spelled backward
Schizophreniform	The essential features of Schizophreniform disorder are identical to those of Schizophrenia except for two differences: the total duration of the illness (including prodromal, active, and residual phases) is at least 1 month but less than 6 months and impaired social or occupational functioning during some part of the illness is not required (although it may occur).	• Onset of prominent psychotic symptoms within 4 weeks of the first noticeable change in usual behavior or functioning • Confusion or perplexity at the height of the psychotic episode • Good premorbid social and occupational functioning • Absence of blunted or flat affect

SOURCE: Data from American Psychiatric Association (2000).

One way to determine if persons with mental illness are in fact more prone to victimization is to compare their victimization rates with those of the general population. One rich national-level data source, is, as you know, the NCVS. It provides data with which researchers can compare their estimates—with national data or with data for smaller areas such as central cities. Doing just this, Teplin and colleagues (2005) used the NCVS instrument to measure victimization experiences of persons with severe mental illness who were receiving psychiatric services in Chicago from January 31, 1997, to October 4, 1999, and to compare victimization incidence and prevalence to the NCVS. They discovered that more than one fourth of those with severe mental illness in the sample had experienced either a completed or attempted violent crime in the past year. This was 11.8 times higher than the prevalence rate in the NCVS. If this finding is extrapolated to the population of persons with severe mental illness in the United

States, almost 3 million persons with severe mental illness are violently victimized each year! In addition, the researchers found that almost 28% had experienced a property crime, rates that were about 4 times higher than the NCVS. Also important, they found that females with severe mental illness were more likely to experience completed violent victimization, rape/sexual assault, personal theft, and motor vehicle theft than were males with severe mental illness. Males with severe mental illness experienced robbery at higher rates than did females with severe mental illness. For some types of victimization, African Americans had higher rates, but rates were generally similar and high for all racial and ethnic groups.

These findings that victimization risk is higher for persons with mental illness are not specific to the United States. In a study examining a birth cohort of males and females born in Dunedin, New Zealand, Silver et al. (2005) found that 34% of persons with any mental disorder had been physically assaulted during the previous year, compared with 21% of persons without a mental disorder. They also examined specific types of mental disorders—persons with schizophreniform disorders had the highest odds of completed physical assault and threats of physical assault. Persons with anxiety disorders were the most at risk for sexual assault victimization.

Why Are Persons With Mental Illness at Risk for Victimization?

Beginning in the 1950s, the United States implemented policies of **deinstitutionalization** that closed many institutions for mentally disordered persons, releasing them into the community (Silver, 2002). Intended to be a more humane approach to care as well as a cost-saving device, deinstitutionalization has had negative consequences. For example, even for people who may benefit from in-patient care, there may not be beds available. In 1970, there were 200 beds available for every 100,000 civilians, but that number had declined to less than 50 by 1992 ("Deinstitutionalization," 2011). The lack of beds has led to shorter stays when in-patient care is received—on average less than 10 days (Teplin et al., 2005). Consequently, persons with mental disorders, even severe mental disorders, may not receive necessary treatment or medicine. Without treatment or medicine, victimization risk actually increases. In fact, research shows that when individuals who have been hospitalized are released under outpatient commitment rather than without any outpatient commitment, they are less likely to be criminally victimized (Hiday, Swartz, Swanson, Borum, & Wagner, 2002).

Persons with mental disorders also have high rates of homelessness (Teasdale, 2009), poverty (Hudson, 2005), and substance abuse (Teasdale, 2009), all of which are risk factors for victimization. These factors place persons with mental disorders at risk for victimization much in the same manner as they do for others—they become easy targets, without capable guardianship. It has been argued that persons with mental illness also may be less likely to report their victimizations to law enforcement because they fear not being believed or taken seriously (Silver et al., 2005). If this is the case, then motivated offenders may target them if the offenders believe they can commit crime with a lower probability of detection (Silver et al., 2005).

Along with lacking capable guardians, persons with mental disorders may be less diligent in their environment and less able to engage in meaningful and effective self-protection (Silver et al., 2005). Also, would-be offenders may perceive that the mentally ill are less able to defend themselves (Silver et al., 2005). This may explain why persons with mental illness who are institutionalized are also at risk. Thus

▲ **Photo 8.1** A mentally ill man sleeps on a city bench, clearly vulnerable to victimization.

Photo credit: ©Thinkstock/BananaStock

far, we have mainly discussed the victimization of persons with mental illness who reside in the community, but being in a care facility or hospital may also place people at risk. The rates of violent victimization in institutions are "often as high or higher than those in the community at large" (Petersilia, 2010, p. 2). Given their unique vulnerability and perceived or real inability to defend themselves, institutionalized persons may be at a great risk for victimization. Also, there may little oversight as to the care provided in these institutions. In addition, patient-on-patient abuse may occur.

Exposure to motivated offenders also places persons at risk of being victimized. Persons with mental disorders may very well be exposed to motivated offenders since they engage in violence at higher rates than people without mental disorders (Arseneault, Moffitt, Caspi, Taylor, & Silva, 2000). This is similar to what we expect generally for people who engage in more violent lifestyles, remembering that these individuals are more likely to be around people who are also violent. Also, people may find that their aggressive and violent behavior evokes aggressive and violent responses.

Why are persons with mental illness more violent? Are they naturally aggressive? Research suggests that violence and victimization are linked to **symptomology**. Exhibiting psychological symptomology, such as delusions and hallucinations, is linked to violent behavior (Appelbaum, Robbins, & Monahan, 2000; Link, Monahan, Stueve, & Cullen, 1999). When persons are exhibiting these symptoms, they are also likely to be victimized because they may cope with their symptoms by becoming aggressive and violent. They are hence likely to evoke similar responses in others (Teasdale, 2009).

Others have argued that when people are exhibiting heightened symptomology, they may act in bizarre ways to which others may not know how to respond (Hiday, 1997). Others may act to control a person who is mentally disordered, and this action may be victimizing. In this way, persons with mental disorders may find themselves in conflicted relationships. Consider how stressful it can be to care for persons who are mentally disordered! Although it is not excusable to abuse anyone, caring for a mentally ill person may at times be overwhelming. In support of this, some research has found that involvement in conflicted relationships drives the relationship between mental disorder and violent victimization (Silver, 2002). This suggests that it may not be the mental disorder per se that places a person at risk but the fact that persons with mental disorders are often involved in relationships characterized by conflict that produces victimization. To read more about this link between mental disorder, conflicted relationships, and violent victimization, see the article in this section by Eric Silver (2002).

Responses to Victims With Mental Illness

It is likely that victims with mental illness face particular challenges. As previously noted, they may be less likely than others to report their victimizations to police for fear that they will not be believed or will not be taken seriously. For women, this risk may be especially germane. Females with mental illness may be unlikely to be taken seriously and may be blamed for being complicit in their own victimization (Salasin & Rich, 1993).

This lack of reporting is especially problematic since victimization is likely to bear psychological costs that may be particularly traumatizing for persons with mental illness. Mental illness severity and symptomology may be exacerbated by victimization—in this way, mental illness may be a contributory factor and a consequence of victimization (Goodman et al., 2001). The correlation between victimization and substance abuse among the mentally ill may indicate the need to use these substances as a coping mechanism (Goodman, et al., 2001). Again, it is likely both a risk factor for and a consequence of victimization. As previously noted, treatment providers and victim service agencies should be aware that victims with mental illnesses may need special care, treatment, and services to ensure that they can reduce their chances of being revictimized.

Also interesting to note is that a wide range of programs are geared specifically toward addressing the needs of offenders with mental health issues and disorders, such as mental health and drug courts, crisis-intervention training for law enforcement officers, and limited mental health treatment in programs. Of course, participation in and funding for these programs is mandated, which may explain why they abound, but similar attention to the needs of crime victims with mental illness has not been given.

⊠ Victimization of the Incarcerated

Incarceration is, by definition, punishment. It includes loss of freedom, loss of dignity, loss of heterosexual relationships, and loss of liberty. Despite what you may have heard, what happens to prisoners while they are incarcerated is not just part of being in prison. Being victimized is not part of the punishment. In fact, the state or federal government, depending on the type of facility in which an inmate is housed, is supposed to protect inmates from any type of victimization—those perpetrated by staff or other inmates. It is actually part of the correctional staff's and administration's jobs to make sure that the jail or prison environment is safe. Why, then, is there a whole section in this text dedicated to victimization of persons who are incarcerated? Because, despite this charge to protect and to keep a secure and safe environment, prisons are, in fact, dangerous places. This section discusses the extent to which victimization occurs to inmates, the types of victimizations that are most prevalent, the risk factors for victimization, and responses to victimizations that occur.

The Extent of Victimization of People in Jail and Prison

As you may imagine, determining how much victimization occurs in jails and prisons is difficult. Much victimization goes unreported—inmates may be fearful of retaliation if they report, they may be scared to report if the perpetrator was a staff member, and they may feel that nothing will be done if they report. As a result, official data sources may underrepresent the true amount of victimization that occurs. Instead, survey data may be a better source of information to reveal the "**dark figure of crime.**"

Results from surveys indicate that victimization is more prevalent in prisons than in the general population (Wolff, Shi, & Siegel, 2009b). In one study of inmates housed in three prisons in Ohio, it was found that half of all inmates had experienced some type of victimization during the previous 6 months (Wooldredge, 1998). Different types of victimization can also be examined separately. In this same study of inmates in Ohio prisons, 10% of inmates had been physically assaulted (Wooldredge, 1998). In an even larger study of about 7,500 prison inmates in 12 adult male and 1 adult female prison, Wolff, Shi, and Siegel (2009a) found that 34% had experienced a physical victimization during the previous 6 months. Of course, the most serious type of physical victimization that can occur behind bars is homicide. In 2007, three inmates in local jails (Noonan, 2010a) and 57 inmates in state prisons were the victims of homicides (Noonan, 2010b). For jail inmates, homicide is most likely to occur in the largest jails, to the most violent offenders, and within 7 days of admission (Noonan, 2010a). For prison inmates, homicide rates are greatest for males (99% of victims) and for Whites (46%) (Noonan, 2010b). Physical assault, however, is not the only type of victimization that inmates experience. Inmates may also have their personal effects taken. In the same study of Ohio inmates introduced above, it was found that 20% had reported a theft in the previous 6 months (Wooldredge, 1998), while other studies of property victimization indicate that one fourth of inmates experienced a property victimization in the past 12 months (Lahm, 2009). Sexual victimization also occurs in prison, but we will discuss that separately in a later section.

Who Is Victimized?

After researchers recognized that inmates are, in fact, at risk of being victimized while incarcerated, attention was given to identifying who is most at risk. Although little research has been done comparing males and females in terms of risk, it does appear that a greater percentage of male inmates experience physical victimization compared with females (Wolff et al., 2009a). A gendered difference discovered in physical victimization is that when females are physically victimized, they are more likely to be victimized by another inmate than are males; males are more likely to be victimized by a staff member (Wolff et al., 2009a). In addition, younger, White inmates are more likely to experience physical victimization perpetrated by other inmates than by staff (Wolff et al., 2009b). Being non-White, however, is associated with an increased risk of physical assault perpetrated by a staff member (Lahm, 2009; Wolff et al., 2009b). Other research has shown that Mexican American inmates are more likely to be victims of both personal and property crimes compared with White or Black inmates (Wooldredge, 1998). Education and income also have been examined in terms of risk for victimization. Higher-income, more-educated inmates are more at risk for being victimized by theft (Wooldredge, 1998), although lower-income inmates face greater risk of physical assault victimization (Wooldredge, 1998).

▲ **Photo 8.2**   An inmate involved in a fight is pictured here being taken to a hospital.

Photo credit: © Elizabeth Ruiz/epa/Corbis

Risk Factors for Victimization While Incarcerated

Although the demographic characteristics of gender, age, and race give us a picture of the typical victim of prison or jail assaults, they do not identify additional risk factors that can be used to help staff and administrators identify who may be at risk in their jails and prisons.

Previous History of Victimization

Remember from Section 2 that one of the risk factors for victimization is having been victimized in the past and that many victims are actually revictimized. This also is true for prisoners. Incarcerated individuals are more likely to be physically victimized if they have been physically victimized in the past (Wolff et al., 2009b). About two thirds of all inmates in Wolff and colleagues' study (2009a) who had been victimized in prison also had been physically victimized before the age of 18. Of those who had been victimized before the age of 18, about half had been victimized in prison during the past 6 months, suggesting a relationship between past victimization and victimization in prison. A relationship between sexual victimization in prison and prior sexual victimization history also was found (Wolff et al., 2009a). To read more about the link between past victimization and victimization in prison, read the article in this section by Wolff and colleagues (2009a).

Mental Illness

Those who are vulnerable may be seen as easy targets in prison. As such, having a serious mental illness may impact people's cognitive abilities and behavior, which may make them prone to victimization (Baskin, Sommers, & Steadman, 1991). In addition, if they have affective disorders such as depression, anxiety, or posttraumatic stress disorder, they may be withdrawn or act in ways that suggest defenselessness, hence producing an increased risk for victimization (Wolff et al., 2009b). Some prisons separate persons with severe mental illness so that they can be treated for their disorders. One of the benefits of doing so is that they can be isolated from other prisoners who may otherwise harm them.

Risk Taking/Self-Control

One criminological theory that has been used to explain victimization is Gottfredson and Hirschi's (1990) **theory of low self-control**. In this theory, they propose that persons with low self-control are likely to become involved in delinquency, crime, and other analogous behaviors. In Section 2, you read about how Schreck (1999) utilized self-control to explain victimization. Building on this work, Kerley, Hochstetler, and Copes (2009) used the theory of self-control to explain prison victimization. In partial support of the theory of low self-control (remember that self-control encompasses six constructs), they found that inmates who like to take risks are more likely to be victimized than other inmates.

Institutional Factors

Aspects of institutional life such as sentence length, type of crime for which an inmate is in prison, security level, proportion of non-White inmates, and lifestyles also play a role in victimization risk. Inmates closer to the beginning of their sentence are more likely to be property crime victims than other inmates (Wooldredge, 1998). Inmates serving time for sexual offenses find themselves at increased risk of being victimized by fellow prisoners, while inmates serving time for violent offenses are especially at risk of being physically assaulted by staff (Wolff et al., 2009b). Security level also matters. Although some of the most dangerous inmates are housed in maximum-security facilities due to the seriousness of their offenses and sentence length, inmates housed in these facilities are less likely to experience both physical and property victimization (Lahm, 2009). Racial composition also is impor-tant. As the proportion of non-White inmates increases in prisons, the likelihood of experiencing property and personal victimization also increases (Lahm, 2009). Although speculative, this finding could be due to a lack of prison management keeping housing units' racial composition in balance.

The lifestyles and routine activities perspective also has been used to explain victimization in prison. Based on the activities in which inmates routinely engage in prison and the social distance inmates feel in relation to other inmates, it may be possible to identify who is more likely to be victimized. In one study, inmates who spent more hours in education programs and more time studying were less likely to be violently victimized but were more likely to become the victims of theft (Wooldredge, 1998). Spending time in recreation hours and the more visits an inmate had per month also increased physical assault risk. Time spent in recreation was also positively related to theft victimization (Wooldredge, 1998). Why does spending time in these structured activities increase risk? If you think about lifestyles and routine activities theory, it may make sense. Time spent in these activities may leave a person's things vulnerable and without capable guardianship, while placing them in the presence of potential offenders.

Finally, feelings that inmates have about other inmates and staff may also impact their risk of being victimized. **Social distance** also increased physical and theft victimization risk (Wooldredge, 1998), while the number of friends at the facility decreased personal crime victimization risk (Wooldredge, 1998). Social distance may indicate an unwill-ingness to incorporate oneself into the prison culture, which may signal to other inmates that others will not protect that person or that person's property (Wooldredge, 1998). Institutions in which inmates report higher levels of dissat-isfaction with officers had higher rates of physical victimization than other facilities (Wolff et al., 2009b).

Special Case: Sexual Victimization of Incarcerated Persons

With the passage of the **Prison Rape Elimination Act of 2003** (Pub. L. 108-79), even more attention has been given to the sexual assault and rape of prisoners than to physical victimization or property victimization. This piece of legislation requires that the Bureau of Justice Statistics, the research arm of the Department of Justice, annually statistically analyze the incidence and effects of prison rape in at least 10% of all state, county, and federal prisons and a representative sample of municipal prisons. This means that data about the occurrence and outcomes of prison rape must be collected and analyzed each year from more than 8,700 of our nation's prisons. From this data, a report is submitted annually that lists institutions and identifies them, in order, in terms of incidence of prison rape. This is one list an institution would not want to be at the top of!

Since this act has required data collection, we now have a better idea as to how much sexual victimization occurs in prisons and jails. The data collected for the **National Inmate Survey** on adult inmates by the Bureau of Justice Statistics was done through self-report surveys, so it did not rely on inmates to make formal reports of their victimizations to institutional authorities. They did, however, have to admit to the victimization during the interview process, which was conducted via an audio computer-assisted self-interview using a touch screen and instructions delivered through headphones (Beck & Harrison, 2010).

From this data, it was found that 4.4% of inmates in prison and 3.1% of jail inmates experienced a sexual victimization by another inmate or someone working at the facility during the past 12 months or since they were admitted (Beck & Harrison, 2010). A slightly greater percentage of inmates in both prisons and jails reported experiencing sexual victimization perpetrated by staff rather than by other inmates (Beck & Harrison, 2010).

Who Is Sexually Victimized?

Similar to sexual victimization outside prison, female inmates are more likely to be sexually victimized than male inmates. According to findings from the National Inmate Survey, 4.7% of female prison inmates and 3.1% of female jail inmates had been sexually victimized by another inmate during the previous 12 months or since admission (Beck & Harrison, 2010). This rate is more than twice that of inmate-on-inmate sexual victimization experienced by male inmates. When males are sexually victimized, they are more likely to be victimized by staff. Almost 3% of male prisoners and 2.1% of male jail inmates reported staff-perpetrated sexual victimization (note that consensual sexual activity with staff is considered victimization) compared with 2.1% of female prisoners and 1.5% of female jail inmates (Beck & Harrison, 2010). White prison and jail inmates were more likely to be sexually victimized by other inmates than were Black or multiracial inmates, although Black inmates were more likely to experience staff-perpetrated sexual victimization (Beck & Harrison, 2010). In addition to gender and race, education and age are factors related to sexual victimization experience. More-educated inmates report higher levels of inmate-on-inmate and staff-on-inmate sexual victimization than do less-educated inmates (Beck & Harrison, 2010). Inmates aged 20 to 24 years are more at risk of being sexually victimized by staff than are inmates aged 25 years and older (Beck & Harrison, 2010).

Adults, however, are not the only inmates we house. Juveniles are also kept in correctional facilities in the United States. Their potential sexual victimization in these facilities is also being studied as part of the requirements of the Prison Rape Elimination Act. Between 2008 and 2009, the Bureau of Justice Statistics completed the first **National Survey of Youth in Custody** (Beck, Harrison, & Guerino, 2010). This survey covered 166 state-owned or state-operated and 29 locally or privately operated facilities that house adjudicated juveniles for at least 90 days. From this study, it was discovered that 12% of youth in these facilities had experienced at least one incident of sexual victimization in the past 12 months or since admission (Beck et al., 2010). Similar to the patterns seen with adult inmates, youth were more likely to be victimized by staff than by other inmates. More often than not, however, the sexual contact was achieved without any force, threat, or explicit form of coercion, although it was still illegal (Beck et al., 2010). Almost all youth (95%) who said they had been sexually victimized said that a female staff member was the perpetrator. Males were more likely than females to report being victimized by staff (10.8% vs. 4.7%), while a greater percentage of females than males indicated they had unwanted sexual activity with other inmates (9.1% vs. 2.0%) (Beck et al., 2010).

Risk Factors for Sexual Victimization in Prison and Jail

In addition to demographic characteristics, other qualities seem to place inmates at an increased risk for being targeted for sexual victimization. One of these risk factors is having experienced a sexual victimization before the current incarceration. In the National Inmate Survey, those inmates who had experienced a sexual victimization prior to coming to the facility were more likely to be sexually victimized by staff and other inmates than were inmates without a previous sexual victimization history (Beck & Harrison, 2010). Similar links between previous sexual victimization history and in-prison sexual victimization have been found (Wolff et al., 2009a). These findings are similar to what other research on the risk of revictimization has found—being victimized at one point in a person's life places him or her at risk of later revictimization.

Another risk factor for sexual victimization is sexual orientation. Specifically, inmates who have a sexual orientation other than heterosexual are more often targets for sexual victimization than are heterosexual inmates (Beck & Harrison, 2010; Hensley, Koscheski, & Tewksbury, 2005). Being vulnerable may also place inmates at risk. For example, being small in stature has been identified in some research as a risk factor for sexual victimization in prisons (Chonco, 1989; Smith & Batiuk, 1989; Toch, 1977). Other research has found that body types, such as having a large build, may also put inmates at risk, perhaps because they cannot defend themselves (Hensley, Tewksbury, & Castle, 2003). Inmates who suffer from mental disorders are also at an elevated risk of experiencing sexual victimization while incarcerated, and females with mental disorders are 3 times as likely to be sexually victimized as males with mental disorders (Wolff, Blitz, & Shi, 2007).

Responses to Victimization in Prison

Inmate Response

The very real likelihood of being victimized while in prison carries with it consequences. Many inmates spend their days in prison fearful of this potential threat. This fear leads some to request placement in **protective custody**, which involves placement in secure housing where an inmate is often kept alone in a cell for 24 hours a day, maybe only getting out for an hour for recreation or showering. Even then, the inmate will be isolated (McCorkle, 1992). Protective custody further impedes inmates from participating in programming such as education, treatment, or vocation training. It hardly seems fair, then, that a common response, albeit for the inmate's "own good," is one that is also used for punishment (solitary confinement) for inmates who have violated institutional policy.

Other inmates may use avoidance strategies such as not going to areas in the prison that they perceive to be dangerous (Irwin, 1980). Research shows that inmates who have been robbed and who are fearful are, indeed, more likely to spend more time in their cells, to avoid areas of the prison, and to avoid activities (McCorkle, 1992). Even more risky, inmates may carry "shanks"—crude, often handmade weapons—as a way to protect themselves should other inmates try to harm them or to take on a tough persona so that others will not target them (McCorkle, 1992). In addition, having been assaulted, threatened, or robbed and being fearful increase the likelihood of using aggressive precautionary behaviors such as using a "get-tough attitude," lifting weights, and keeping a weapon nearby (McCorkle, 1992).

Institutional Response

Given the factors that place inmates at risk of being victimized while in prison, there are many things that institutions do to increase safety and security. Some of these things are done out of response to judicial mandates and constitutional requirements; others are out of recognition of the dangers inherent in prison. Let's first discuss legal and constitutional requirements.

The Eighth Amendment to the United States Constitution, which prohibits cruel and unusual punishment, has been interpreted by the Supreme Court to prohibit inhumane treatment of prisoners (*Farmer v. Brennan*, 1994).

As the court noted in *Farmer v. Brennan* (1994), not only do prison officials have to provide a humane environment, but failing to do so can make a prison official liable if he or she acts with **deliberate indifference** to inmate health or safety or if he or she knows that the inmate faces a substantial risk of serious harm and disregards that risk by failing to take reasonable measures to halt it (for a discussion of the facts of this case, see box).

THE CASE OF *FARMER V. BRENNAN*

Farmer was a preoperative transsexual with feminine characteristics who was incarcerated with other males in an all-male federal prison. During his imprisonment, he was sometimes in the general population but was more often in segregation. He was transferred from a correctional institute to a penitentiary, which is generally a higher-security facility with more serious offenders. After the transfer, he was placed in the general population. He was subsequently beaten and raped by other prisoners. He sought damages and an injunction against further confinement. The Supreme Court eventually heard the case and ruled that "a prison official's 'deliberate indifference' to a substantial risk of serious harm to an inmate violates the Eighth Amendment" (*Farmer v. Brennan*). It is the prison officials' duty under the Eighth Amendment to provide humane conditions for all prisoners; this means adequate food, clothing, shelter, medical care, and protection from violence at the hands of other prisoners.

SOURCE: Retrieved from http://caselaw.lp.findlaw.com/cgi-bin/getcase.pl?court=US&vol=000&invol=U10394

In addition to these legal requirements, prisons have also put into place policies to reduce victimization. Because physical victimization cannot occur if inmates are not in contact with one another, one way to reduce victimization is through effective classification. **Classification** is the process through which offenders are screened and placed in facilities that match both their needs and characteristics, both good and bad. For instance, a prisoner who has a substance abuse problem should, theoretically, be placed in a facility that offers substance abuse treatment. Likewise, a prisoner who has anger management issues, who has demonstrated an inability to control impulses, and who has victimized staff and inmates in previous incarcerations should be placed in a maximum-security facility or one that can effectively respond to violent inmates.

Once in prison, administrators and staff may reduce movement in prisons and reduce prisoner interaction, thus reducing opportunity for victimization to occur. Moreover, because race appears to be a risk factor for victimization and the racial composition of the prison is related to victimization rates, some prisons track the number of White and non-White inmates living in housing units and try to keep a particular balance (Lahm, 2009).

One way prisons have responded to the deprivation of heterosexual relationships that inmates face is through the use of conjugal visits. **Conjugal visits** are those in which a married inmate is allowed to spend time with his or her partner, usually in prison quarters—such as in a trailer or separate housing unit—overnight or for a weekend. During these visits, the couple is afforded more privacy than the inmate normally receives, although they are regularly checked on. Currently, only six states allow conjugal visits, however, and these visits serve to reduce only consensual sexual activity and forced sexual activity borne out of the lack of access to heterosexual relationships, not sexual activity motivated by a desire to exercise power and control

(Knowles, 1999). In this way, conjugal visits are unlikely to be an effective strategy to reduce rape and sexual victimization. In addition, remember that many sexual victimizations are perpetrated by prison staff, not other inmates. As such, prison staff need to be screened, rotated on different assignments throughout the prison, and trained in ethics and law. One study on correctional officers revealed that they are likely to engage in victim blaming in regard to the sexual victimization of inmates (Eigenberg, 1989), thus suggesting a need for continued education. These are but a few of the ways that prison staff and administrators may work to reduce victimization in prisons. As you may imagine, given the unique culture and environment in prisons, seriously curtailing victimization may mean that the prison culture itself—one that involves not just inmates but the relationship between inmates and staff—needs to be addressed. This is not a task so easily undertaken.

Many people are at risk of being victimized, but certain people are more vulnerable to victimization because of qualities or characteristics they possess. In this section, persons with disabilities, persons with mental illness, and persons who are incarcerated were discussed because they are all vulnerable to victimization, given their status. You should consider what responsibility we have to people who are uniquely vulnerable to victimization. What should we do to protect these people?

SUMMARY

- In order to get an accurate measure of the victimization of people with disabilities, the Crime Victims with Disabilities Awareness Act of 1998 mandated that the National Crime Victimization Survey (NCVS) begin to collect information about the victimization of people with disabilities.

- Research suggests that persons with disabilities are more likely than persons without disabilities to be victimized.

- The fact that victims with disabilities and disorders are often in the care of various people who recognize their vulnerability makes them targets for violence.

- Women with disabilities face particular risks of being victimized. According to the NCVS, women with disabilities are more likely to be victimized by their intimate partners than are men. Sexual victimization is also prevalent. It is estimated that women with developmental disorders are between 4 and 10 times more likely to be sexually assaulted than women in the general population (National Council on Disability, 2007).

- Often, victims with mental disorders and/or disabilities are unsure of their victimizations and, therefore, may not receive assistance.

- Children and adolescents with disabilities are also vulnerable to victimization. They face increased risks for bullying, maltreatment, and sexual abuse.

- Risk factors for victimization for persons with disabilities include relying on others for assistance (routine activities/lifestyles theory); inability to recognize, process, and/or respond to risk; living in poverty; and the dependency-stress model.

- When mental disorders were first considered in the criminal justice system, persons with mental disorders were expected to be the offenders and not generally expected to be victims.

- A testable way to find out if victims with mental illness are more prone to victimization is to compare them with victimizations of the general population. Findings from such research suggest that mental illness is linked to an increased risk for violent and property victimization.

- The severity of one's mental illness could make it more likely that he or she will be victimized and revictimized.

- Deinstitutionalization, homelessness, poverty, and substance abuse contribute to victimization of persons with mental illness. Symptomology also has been linked to an increase in victimization risk for persons with mental illness.

- Some estimate that up to half of all prisoners are victimized while incarcerated.
- Male, young, non-White inmates appear to be most at risk for victimization in prison.
- Previous victimization, suffering from mental illness, and having low self-control increase risk for victimization in prison. Institutional factors also play a role in prison victimization.
- Quite a bit of prison victimization goes unreported due to the inmate's fear of being caught reporting.
- Prisoners' sentence length, sexual orientation, physical size, age, lifestyle while incarcerated, and even what they were convicted for may or may not make them suitable targets for sexual victimization.
- Inmates serving time for sexual offenses find themselves at increased risk of being victimized by fellow prisoners, while inmates serving time for violent offenses are especially at risk of being physically assaulted by staff.
- A response to victimization in prison and being fearful is increased likelihood of using aggressive precautionary behaviors such as a "get-tough" attitude, lifting weights, and keeping a weapon nearby.
- It is the prison officials' duty to separate those prisoners who are more vulnerable and likely to be victimized from the general population. In this way, crime within the prison should be reduced.

DISCUSSION QUESTIONS

1. Given that exposure to motivated offenders is a direct cause of victimization of disabled persons and those with mental disorders, do you think that offenders are initially motivated or that they become motivated once they recognize how vulnerable these special populations are? What would Cohen and Felson say?

2. What special services do victims with mental illness and victims with disabilities need from criminal justice professionals and agencies? Do we currently have enough resources in place to address their special needs? What about other social service agencies that help crime victims?

3. Think about the *Farmer v. Brennan* (1994) case. Do you feel special prison conditions should be available for members of the LGBT community who are more likely to be victimized in male or female prisons?

4. As the laws for people in the United States change, do our prison communities also need to evolve so we are providing humane living conditions as defined in our U.S. Constitution?

5. Put yourself in one of the incarcerated victim's position. Would being victimized in prison make you more likely to continue being a victim or would you use other coping strategies? What are the consequences of prison victimization?

KEY TERMS

developmental disability

dependency-stress model

Federal Crime Victims with Disabilities Awareness Act of 1998

Crime Victims with Disabilities Act of 1998

deinstitutionalization

symptomology

dark figure of crime

theory of low self-control

social distance

Prison Rape Elimination Act of 2003

National Inmate Survey

National Survey of Youth in Custody

protective custody

deliberate indifference

classification

conjugal visits

INTERNET RESOURCES

Centers for Disease Control and Prevention (http://www.cdc.gov/)

This website offers full definitions of the developmental diseases as well as their symptoms, all of which could make a person a more suitable target for victimization. The site provides some safety and health tips for parents and caregivers of people who have developmental disorders. Most helpful are the emergency preparedness and response topics listed on the website.

State of the USA (http://www.stateoftheusa.org/content/44-percent-of-prison-inmates-s.php)

This website provides a report about prison inmate victimization, with statistics from the Bureau of Justice Statistics. The charts on the website provide an in-depth breakdown of what types of prison victimization occur based on what incidents are actually reported. This page also offers data on variations and circumstances surrounding prison inmates' victimization. Prison facilities are ranked based on inmates' reports of victimization.

"Crime and Disability" (http://www.ovc.gov/publications/infores/ServingVictimsWithDisabilities_bulletin/crime.html)

This page on the Office for Victims of Crime website outlines patterns and responses of victimization. Resources are listed for caregivers or the victims themselves to prevent victimization. There are statistics for how often this type of victimization occurs as well as recommendations for coordinating with law enforcement to stop victimization of persons with disabilities.

"Violence and Mental Illness: The Facts" (http://www.stopstigma.samhsa.gov/topic/facts.aspx)

This webpage is a very specific information source on the common stigmas and stereotypes associated with people who have mental health issues as they pertain to violence. This page does well explaining why certain discriminations exist and why people typically have a hard time being caregivers for them. Believing that persons with mental illness are always the suspects in violent situations does not help prevent their victimization.

Introduction to Reading 1

In his article, Brownridge (2006) discusses the extent to which women with disabilities (having a physical, mental, or health condition that curtails their ability to participate in the amount or kind of activities they can do at home, work, school, or otherwise) who are married or in a common-law union experience intimate partner violence. This study was conducted using a sample of 7,027 Canadian women who participated in the 13th cycle of the General Social Survey, which is a random sample of adults over the age of 15. Brownridge's findings comport with the more general research on persons with disabilities in that women with disabilities were more likely than others to report intimate partner violence victimization. The risk factors are discussed in this paper.

Partner Violence Against Women With Disabilities

Prevalence, Risk, and Explanations

Douglas A. Brownridge

Women with disabilities rank issues of violence as their most important research and health priorities (Curry, Hassounch-Phillips, & Johnston-Silverberg, 2001). Despite an apparent consensus on the importance of and need for research on violence against women with disabilities, the issue remains an understudied social problem. A review by Curry et al. (2001) found that "there is practically no literature regarding the risk of abuse, women's experiences of abuse, and barriers to seeking help among women with disabilities" (p. 60), and that "the absence of attention to this issue from both disability and violence researchers has contributed to the 'invisibility' of the victimization of women with disabilities" (p. 68).

The small body of existing research on violence against persons with disabilities has identified a wide range of prevalence rates. Based on a review of research, Chappell (2003) concluded that "women with disabilities

face an epidemic of monumental proportions" (p. 12). Indeed, it is common in the literature to see very high estimates of violence against persons with disabilities, such as being 50% more likely to encounter abuse than the rest of the population (Hightower & Smith, 2003) or having 2 to 5 times the likelihood of abuse compared to nondisabled persons (Melcombe, 2003). Other research suggests less extreme disparities in risk between those with and without disabilities. A study for the Ontario Ministry of Community and Social Services in which 62 Toronto women were sampled found that 33% of those with disabilities were physically abused compared to 22% of those without disabilities. In the same study, however, women with disabilities were less likely to report having been sexually abused (23% vs. 31%). Using data from a national survey on sexuality in the United States, Young and colleagues (Young, Nosek, Howland, Chapong, & Rintala, 1997) examined violence

SOURCE: Brownridge (2006). Reprinted with permission.

against women with physical disabilities. Based on a comparison of a nonrandomly selected sample of 439 women with physical disabilities and 421 women without physical disabilities, these researchers found that both groups of women had an equally high lifetime prevalence of emotional, physical, and sexual abuse (62%). Given that most research has focused on persons with developmental disabilities, typically examining sexual abuse (Curry et al., 2001), one would expect relatively little debate about the prevalence of violence against persons with developmental disabilities. However, Newman, Christopher, and Berry (2000) argued that there is little evidence for the common assumption that persons with developmental disabilities are at greater risk for violence than persons without developmental disabilities, and they concluded that "until that notion is supported, it may be prejudicial to assume that people with developmental disabilities are especially vulnerable" (p. 165). Clearly, current knowledge provides an insufficient basis on which to identify the prevalence of violence against persons with disabilities.

Although there is a dearth of research on violence against persons with disabilities, there is an even greater paucity of research on women with disabilities who experience violence by an intimate partner. One possible explanation for a lack of attention to partner violence against women with disabilities concerns societal myths of these women as single and asexual (Barnett, Miller-Perrin, & Perrin, 2005). It is ironic to note, research suggests that the most common perpetrators of violence against women with disabilities are their male partners. Milberger and colleagues (2003) found that 56% of a nonrandom sample of 177 women with disabilities reported abuse, and the abusers were typically their male partners. The Disabled Women's Network of Canada (DAWN) surveyed women with disabilities and found that the most common perpetrators of violence were current or former intimate partners (37%; Ridington, 1989). In Young et al.'s (1997) study of violence against women with physical disabilities, intimate partners were most likely to be the perpetrators of physical and emotional abuse. In fact, the sampled women with disabilities were equally as likely as those

without disabilities to report having experienced violence by husbands and/or live-in partners, with 30% of each group reporting any experience of partner violence in their lifetime. However, Young et al. (1997) used a sample that was representative neither of women with nor without disabilities. As a result, it is not possible to comment on the risk for partner violence against women with disabilities relative to women without disabilities with any measure of confidence.

The purposes of the current study were (a) to identify whether Canadian women with disabilities report an elevated risk for partner violence compared to their counterparts without disabilities and, if so; (b) to examine the extent to which disabled women's risk is elevated; and (c) to examine risk markers derived from potential explanations in terms of their impact on, and the extent to which they account for an elevated risk of, partner violence against women with disabilities.

◪ Explanatory Framework

The diversity of disabled person's situations and potential variables affecting their likelihood of violence impedes the development of a single theoretical explanation to account for their experiences of violence. Instead, the state of the art is to recognize an array of potential reasons for an increased risk for violence against persons with disabilities and to organize them based on where they fit within an ecological context (Curry et al., 2001; Petersilia, 2001; Sobsey, 1994; Sobsey & Calder, 1999). In the current study, available risk markers for violence against women with disabilities were organized into a framework based on whether they related primarily to the context of the relationship, to the victim, or to the perpetrator.

Relationship Factors

A commonly cited risk marker for violence against women with disabilities is dependence. To the extent that persons with disabilities are dependent, differences in power may result that, in turn, could lead to abuse (Petersilia, 2001). One indication of potential dependence on a relationship is the couples' education compatibility (Anderson, 1997). The more education resources a woman with disabilities has relative to her partner, the more power she should have in the relationship. Disabled women with fewer relative education resources may be more dependent, less powerful, and thus more prone to violent victimization.

Another relationship factor that is known to affect partner violence against women is duration of relationship. The rate of disability in Canada increases with age (Statistics Canada, 2002). Because duration of relationship also tends to increase with age (Brownridge & Halli, 2001), one would expect women with disabilities on average to have longer duration unions. If disability increase the likelihood of violence and disability is more likely among longer duration unions, one might expect that union duration will be positively related to partner violence against women with disabilities. On the other hand, research shows that duration of relationship is negatively related to violence against women in the general population (Brownridge & Halli, 1999). Based on this research, one would expect partner violence to be more likely among women with disabilities who have been in their relationship for a relatively short duration. In the absence of empirical evidence, it was not possible to choose one hypothesized direction of the relationship between duration and partner violence against women with disabilities over the other.

Victim-Related Characteristics

At the outset, it must be understood that an examination of victim-related characteristics does not equate to a victim-blaming approach. Indeed, the importance of including victim-related characteristics is widely acknowledged in the literature on violence against persons with disabilities (Curry et al., 2001; Nettelbeck & Wilson, 2002; Nettelbeck, Wilson, Potter, & Perry, 2000; Petersilia, 2001; Sobsey & Calder, 1999). For instance, Nettelbeck et al. (2000) wrote that "although any instance of victimization is dependent on offender attributes and behaviors, victim attributes and behavior will also contribute" (p. 47). Proponents of including victim-related characteristics

typically argue that denial of the potential for victim-related characteristics to affect violence suggests that individuals do not have the power to reduce their risk (Nettelbeck et al., 2000; Petersilia, 2001; Sobsey & Calder, 1999). Although some victim-related characteristics may be amenable to change on an individual level, there are others that are less easily changed by the individual. The latter are typically "the result of a failure of health and social policy to adequately address the needs of persons with disabilities" (Curry et al., 2001, p. 72). Empirical evidence for the impact of these victim-related characteristics would direct prevention efforts toward formative changes in health and social policy.

Victim-related characteristics stemming from socioeconomic status (SES) have been identified as potentially affecting the likelihood of violence. Women with disabilities are more likely than nondisabled women to be of low SES. They tend to have lower educational attainment than women without disabilities, and disability and lower educational attainment are barriers to employment (Nosek, Howland, & Hughes, 2001). The unemployment rate among women with disabilities has been identified as being as high as 75% (Melcombe, 2003). This results in women with disabilities being more likely to live in poverty (Curry et al., 2001). Low SES has generally been associated with violence against women (Barnett et al., 2005). Hence, it is expected that women with disabilities will score lower on SES indicators and that this will increase their likelihood of experiencing partner violence.

As noted above, the rate of disability increases with age. It has been suggested that "older age greatly increases the risk of disabilities, particularly for women, with concomitant risks of violence and abuse" (Hightower & Smith, 2003, p. 18). A woman's age is associated with her risk for partner violence; though the relationship is usually negative (Brownridge & Halli, 2001). As well, the prevalence of disability is higher among Aboriginal than non-Aboriginal Canadians (Melcombe, 2003) and Aboriginal status is strongly associated with an increased risk for violence against women in Canada (Brownridge, 2003). Given an association between each of these victim-related characteristics (age and Aboriginal status) and violence against women, these variables needed to be controlled in the current study.

Perpetrator-Related Characteristics

Feminist disability theorists essentially view women with disabilities as being in a position of double vulnerability. This approach directs attention toward the fact that women with disabilities live in a society that is at once disablist and patriarchal (Curry et al., 2001; Thomson, 1994). Research has found an association between violence against women and male patriarchal domination (Brownridge, 2002) and male espousal of patriarchal ideology and beliefs (Smith, 1990). It is possible that women with disabilities are perceived by men who espouse a patriarchal ideology as being less difficult to dominate, which may include domination through violence.

Although women with disabilities are in a position of double vulnerability, Petersilia (2001) noted that "vulnerability by itself is rarely, if ever, sufficient to motivate a crime. The potential victim must have something the offender wants or have the ability to produce an event the offender finds desirable" (p. 676). One key factor identified by Petersilia (2001) as motivating many crimes against persons with disabilities is the effort to gain control over the victim's behavior. Evolutionary psychology directs attention to men's need to maintain control over "their sexual property" (Wilson & Daly, 1998). Sexual proprietariness, in terms of male sexual jealousy and possessive behavior, has been linked to violence against women (Brownridge, 2004). Men who are sexually proprietary may be more likely to act violently toward women with disabilities to gain or maintain control over "their sexual property."

A final perpetrator-related characteristic is substance abuse. Men, who feel affected by their partner's disability, particularly if they are the primary caregiver, may experience stress (Milberger et al., 2003). Such dependency-stress models suggest that caregivers who cannot cope with the stress of caregiving abuse their charges (Petersilia, 2001). One indirect indication of stress that has also been linked to partner violence

against women is heavy alcohol consumption (Johnson, 2001). Li, Ford, and Moore (2000) investigated the role of substance abuse in violence among a sample of 1,876 persons with disabilities. The study found that disabled women were more likely to be victims of violence related to alcohol or drug use than were disabled men. Although the study demonstrated that substance use can play a role in violence against persons with disabilities, and particularly women, the fact that the sample included only persons with disabilities rendered the research unable to speak to the extent to which substance abuse is responsible for violence against women with disabilities relative to nondisabled women.

⊠ Materials and Method

Any one of three definitions of *disability* is typically cited in the literature. These are the definitions of the World Health Organization (WHO), the United Nations (UN), and the Americans with Disabilities Act (ADA) of 1990. A comparison of the WHO and UN definitions, which have been reprinted in Howe (2000) and Curry et al. (2001), respectively, showed that they are identical; that is, *disability* is defined as "any restriction or lack (resulting from an impairment) of ability to perform an activity in the manner or within the range considered normal for a human being" (Curry et al., 2001, p. 62). Although this definition is not perfect, particularly in terms of its emphasis on the individual's deviation from "normality," it is widely accepted around the world and is an inclusive definition that is well suited for establishing the population of persons with disabilities. In the ADA definition, disability is determined based on whether a person

> (a) has a physical or mental impairment that substantially limits one or more major life activities, (b) has a record of physical or mental impairment, or (c) is regarded as having a physical or mental impairment that substantially limits a major life activity. (Gilson, DePoy, & Cramer, 2001, pp. 427–428)

The main difference between the ADA and WHO and/or UN definitions appears to be the ADA's inclusion of persons who have had disabilities in the past and the possibility for disability to be defined based on the perceptions of others.

The data in the current study identify persons with disabilities based on the WHO and/or UN definition. In its 2001 Participation and Activity Limitation Survey (PALS) and its 1999 GSS, Statistics Canada (2000, 2002) identified persons with disabilities as those who reported that a long-term physical or mental condition or health problem reduced the amount or the kind of activities they could do at home, school, work, or in other activities. Of the women included in the current study, 1,092 were disabled, and 5,935 were not disabled. In other words, the prevalence of disability in the sample was 15.5%. In all analysis, the weighting scheme suggested by Statistics Canada (2002) was followed. With the data weighted, the prevalence of disability was 15.7%, which was identical to the prevalence of disability among noninstitutionalized women age 15 years and older found in the PALS (Statistics Canada, 2002). This is also virtually the same of disability found in the population of non-institutionalized women in the United States (15.4%; Curry et al., 2001).

Measurement

Independent Variables

Education compatibility was obtained by calculating the ratio of the respondent's years of education to the couple's total years of education. For the analyses, education compatibility was grouped into the following categories: the woman had much less education (ratio < .46), the woman had less education (ratio = .46 to .49), the woman had the same years of education as her partner (ratio = .50), the woman had more education (ratio = .51 to .54), and the woman had much more education (ratio > .54). Duration of relationship was measured with a variable derived from the respondent's report of the year in which she became married or began living with her common-law partner. Education referred to the respondent's education in years. Employment was measured in terms of whether the respondent's main activity in the 12 months prior to the interview was looking for work, caring for children or housework (unemployed), or working at a paid job or business (employed). Age referred to the respondent's age at the time of the interview. Aboriginal status was measured in terms of whether the respondent identified her cultural or racial background as being Aboriginal (North American Indian, Metis, or Inuit) or some other background (non-Aboriginal). Patriarchal dominance was measured with an item that asked the respondent if her partner prevented her from knowing about or having access to the family income, even if she asked. Sexual possessiveness was measured with an item asking the respondent if her partner demanded to know who she was with and where she was at all times. The measure of sexual jealousy was based on a question asking the respondent if her partner was jealous and did not want her to talk to other men. Heavy alcohol consumption was measured with a question that asked the respondent how many times in the month prior to the interview her partner had five or more drinks on one occasion.

Dependent Variable

The current study employed a modified version of the Conflict Tactics Scales (CTS; Statistics Canada, 2000). *Male partner violence against women* was defined as acts of physical assault (being pushed, grabbed, or shoved in a way that could hurt; being slapped; being choked; having something thrown that could hurt; being hit with something that could hurt; being threatened with or having a knife or gun used; being kicked, bit, or hit with a fist; being beaten), physical threat (being threatened to be hit with a fist or anything else that could hurt), and sexual assault (being forced into any sexual activity by being threatened, held down, or hurt in some way) perpetrated by a woman's current marital or common law partner within a specified time frame preceding the interview. Hence, if respondents reported having experienced any of the aforementioned forms of violence within the specified time frame preceding the interview, they were coded as having experienced violence. Two reference periods were employed in the current study, namely, 1-year and 5-year frames.

Method of Data Analysis

The analysis was conducted in two stages. To document the prevalence of violence among women with and without disabilities and investigate risk markers, the first stage consisted of descriptive analyses in which bivariate relationships were examined using cross-tabulations with chi-square tests of significance. In the second stage, more elaborate analyses were conducted using multivariate statistical techniques. These analyses allowed an assessment of the impact of the independent variables and determination of the importance of the independent variables for understanding and elevated risk for violence against women with disabilities. The multivariate technique used for this purpose was logistic multiple regression. Logistic regression is an appropriate technique for predicting a dichotomous dependent variable from a set of independent variables. This technique also has a very simple interpretation. For a given variable it simply provides a ratio of the odds of violence occurring. If the value of the odds is greater than 1, the variable is positively related to violence. If it is less than 1, the variable is negatively related to violence.

▧ Result

Descriptive Analysis

Violence by Disability Status

Although women with disabilities reported a higher 1-year prevalence of violence than women without disabilities (2.0% vs. 1.7%), the difference was not statistcally significant. However, when recalling violence over the 5 years prior to the interview, women with disabilities reported a significantly higher prevalence of violence (4.9% vs. 3.5%, p < .05).[1]

Table 1 contains the results of each component of the 5-year prevalence of violence cross-tabulated with disability status. Women with disabilities were significantly more likely to report experiencing 6 of the 10 items measuring violence. They were 1.4 to 1.9 times more likely than women without disabilities to report being threatened to be hit with a fist or anything else that could hurt; to be pushed, grabbed, or shoved in a way that could hurt; and to be slapped. The greatest disparities occurred on some of the more severe forms of violence. Women with disabilities were twice as likely to report being beaten and kicked, bit, or hit with a fist. They were also 3 times more likely to report being forced into any sexual activity by being threatened, held down, or hurt in some way.

Independent Variables by Disability Status

Table 2 provides the results of the cross-tabulations of the independent variables by disability status. In terms of relationship factors, Table 2 shows that women with disabilities were more likely than women without disabilities to be in relationships that were educationally incompatible at both ends of the continuum; that is, they were more likely to report having either much less education or much more education than their partner. Women with disabilities were more

---

Table 1 Five-Year Prevalence of Each Component of Violence by Disability Status (%)

	Physical Assault									Sexual Assault
	Physical Threat									
	Threaten	Push	Slap	Choke	Throw	Hit	Threaten and/or Use Gun and/or Knife	Kick	Beat	Sex
Disabled	3.0	3.5	1.9	0.6	1.5	0.6	0.2	1.2	0.8	0.6
Nondisabled	1.9**	2.5*	1.0**	0.3	1.2	0.4	0.1	0.6**	0.4*	0.2**

*p < .10. **p < .05.

likely to have had longer duration relationships. With respect to victim-related characteristics, women with disabilities were more likely to have less than high school education and less likely to have a university degree. They were also more likely to be unemployed, older, and Aboriginal. In terms of perpetrator-related characteristics, partners of women with disabilities were more likely to engage in patriarchal dominance, sexual possessiveness, and sexual jealousy. However, they were less likely to have consumed alcohol heavily

and no more likely to have drank heavily 5 or more times in the month prior to the survey.

Multivariate Analysis

Separate Logistic Regressions for Disabled and Nondisabled Women

Table 3 provides the results of the logistic regressions on the 5-year prevalence of violence for women with and without disabilities. In terms of relationship

Table 2   Independent Variables by Disability Status (%)

Independent Variables	Disabled	Nondisabled
Relationship factors		
Education compatibility		
Woman has much less education	18.3	15.2
Woman has less education	10.5	14.2
Woman has same education	33.8	33.3
Woman has more education	14.9	18.4
Woman has much more education	22.5	18.9***
Duration of relationship		
Less than 4 years	8.4	12.3
4 to 9 years	11.4	18.4
10 or more years	80.1	69.2***
Victim-related characteristics		
Education		
Less than high school	35.0	17.7
High school	13.7	17.7
Some postsecondary	13.6	13.8
Community college diploma and/or certificate	24.1	28.6
University degree	13.5	22.3***

Employment			
Unemployed		42.5	27.8
Employed		57.7	72.2***
Age			
15 to 34 years		12.0	27.7
35 to 54 years		39.4	50.8
55 years and older		48.6	21.5***
Aboriginal status			
Aboriginal		2.9	1.8
Non-Aboriginal		97.1	98.2**
Perpetrator-related characteristics			
Patriarchal dominance			
Yes		2.1	0.8
No		97.9	99.2***
Know whereabouts			
Yes		5.6	3.7
No		94.4	96.3***
Jealousy			
Yes		7.1	4.1
No		92.9	95.9***
Heavy drinking (past month)			
None		83.3	79.1
Once		6.6	9.4
2 to 4 times		6.9	8.5
5 or more times		3.2	3.1**

**p < .05. ***p < .01.

factors, the results show that, controlling for all other variables in the models, education compatibility was not significantly linked to the odds of violence for either group of women. Duration of relationship was negatively related to the odds of violence for women without disabilities. For each additional year in which a nondisabled woman had been with her current partner, her odds of violence decreased by 2.5%. For each additional year that a disabled woman had been with her partner, her odds of violence decreased by 1.9%, a difference that was not statistically significant. With respect to victim-related characteristics, education was not significantly linked to violence for either group of women. Although unemployed women with disabilities

had 43% increased odds of abuse, this increase was not statistically significant. Age was negatively related to violence for women with disabilities. Each year of increase in age for women with disabilities was associated with a 4% reduction in odds of violence. Aboriginal status was not linked to significantly increased odds of violence for women with disabilities but was for those without disabilities. Aboriginal women without disabilities faced 180% increased odds of violence compared to non-Aboriginal women without disabilities. In terms of perpetrator-related characteristics, patriarchal dominance, sexual possessiveness, and sexual jealousy were linked to significantly increased odds of violence for women with and without disabilities.

Table 3 Results of Logistic Regressions on 5-Year Prevalence of Violence for Women With and Without Disabilities

Covariates	Model 1 Disabled $n = 966$ Odds Ratio	Model 2 Nondisabled $n = 5,338$ Odds Ratio
Relationship factors		
Education compatibility		
Woman has much less education	0.670	0.834
Woman has less education	0.892	1.023
Woman has same education	1.000	1.000
Woman has more education	0.454	1.303
Woman has much more education	1.360	0.867
Duration of relationship	0.981	0.975*
Victim-related characteristics		
Education	1.033	1.002
Employment		
Unemployed	1.433	0.925
Employed	1.000	1.000
Age	0.957**	0.981
Aboriginal status		
Aboriginal	1.136	2.795**
Non-Aboriginal	1.000	1.000
Perpetrator-related characteristics		
Patriarchal dominance		
Yes	3.689**	12.602***
No	1.000	1.000
Know whereabouts		
Yes	3.412**	4.521***
No	1.000	1.000
Jealousy		
Yes	2.293**	4.975***
No	1.000	1.000
Heavy drinking	1.044	1.106***
Constant	0.237	0.067***
−2 log likelihood	305	1,310
χ^2	79	332

*$p < .10.$ **$p < .05.$ ***$p < .01.$

Of all the variables included in the current study, these had the strongest impact on the odds of violence for both groups of women. On the other hand, heavy alcohol consumption was significantly linked to increased odds of violence only for women without disabilities, who faced 11% increased odds of violence for each additional occasion in the month prior to the survey that their partner drank heavily.

Logistic Regression for Disabled and Nondisabled Women Combined

The next stage of the analysis was to enter the independent variables into the logistic regression model sequentially based on the explanatory framework. This allowed identification of whether variables derived from any particular level of the explanatory framework, or in combination, accounted for the elevated risk for violence against women with disabilities.

Table 4 provides the results of the sequential logistic regressions. The first model in Table 4 contains the results of the disability status variable without any controls. The difference in prevalence of violence between the two groups translated to women with disabilities having 39% higher odds of violence in the 5 years preceding the survey compared to women without disabilities. The second model in Table 4 controlled for the relationship factors. Controlling for relationship factors did not decrease the difference in odds between the two groups of women and, in fact, significantly increased the difference in odds. Similarly, when controlling for victim-related characteristics in the third model, the difference in odds was not decreased. The fourth model in Table 4 controlled for the perpetrator-related characteristics. When the perpetrator-related characteristics were controlled, the odds of violence for women with disabilities became only 3% higher compared to women without disabilities, a difference for all of the variables from the explanatory framework. With all variables

Table 4   Results of Sequential Logistic Regressions of 5-Year Prevalence of Violence

Covariates	Model 1 Disabled or Nondisabled $n = 6,912$ Odds Ratio	Model 2 Relationship Factors $n = 6,610$ Odds Ratio	Model 3 Victim-Related Characteristics $n = 6,806$ Odds Ratio	Model 4 Perpetrator-Related Characteristics $n = 6,544$ Odds Ratio	Model 5 Full Model $n = 6,304$ Odds Ratio
Disability status					
Disabled	1.390**	1.867***	2.008***	1.025	1.409*
Nondisabled	1.000	1.000	1.000	1.000	1.000
Relationship factors		BLOCK			BLOCK
Victim-related characteristics			BLOCK		BLOCK
Perpetrator-related characteristics				BLOCK	BLOCK
Constant	.037***	.079***	.331**	.023***	.085***
−2 Log-likelihood	2.196	2,048	2,061	1,779	1,641
χ^2	4	86	110	320	390

*$p < .10$. **$p < .05$. ***$p < .01$.

controlled, the odds of violence for women with disabilities relative to women without disabilities were not reduced. In short, the perpetrator-related characteristics accounted for disabled women's significantly higher odds of violence relative to women without disabilities.

Discussion

Sobsey (2000) cautioned that research on partner violence against women with disabilities that draws comparisons with the general population may yield misleading results because women who have severe developmental disabilities are underrepresented among women in intimate relationship. Although the 1999 GSS is representative of the population of Canada, thereby including some women with severe disabilities,[2] it is nevertheless the case that women with the most severe developmental disabilities may be underrepresented among women with partners. Of course, results from an examination of partner violence against women with and without disabilities would be misleading if one were to generalize to all women with disabilities. It behooves the reader to bear in mind that this research focused on violence by partner, who, as shown in the review of the literature, have nevertheless been identified as the most common perpetrators of violence against women with disabilities.

Based on the wide variation in the literature on prevalence rates, the prevalence of violence against women with disabilities is generally described as being "equal to or greater than their nondisabled peers" (Gilson et al., 2001, p. 419). The current study shows that partner violence against women with disabilities in Canada is no exception. Although women with disabilities reported 1.2 times the rate of violence compared to women without disabilities in the year prior to the interview, this difference was not statistically significant. When examining a longer time frame that provided a more accurate account of all women in the sample who had experienced partner violence, women with disabilities reported a significantly elevated risk for violence. The disparity in the rate of violence for women with disabilities compared to those without disabilities grew to 1.4 times in the 5 years preceding the interview. In other words, depending on the time

frame of the dependent variable, women with disabilities in Canada reported experiencing violence by intimate partners at rates that either equaled or surpassed those of women without disabilities. Moreover, the results showed that women with disabilities were particularly vulnerable to severe acts of violence by their partners in the 5 years preceding the survey. Barnett et al. (2005) commented that "investigators rarely assume that disabled women have intimate partners, so IPV (intimate partner violence) often goes undetected" (pp. 353–354). To effectively prevent and address violence against women with disabilities, stakeholders need to be aware of this false assumption and of the reality of partner violence against women with disabilities. In addition, they must also have an understanding of the dynamics underlying an elevated risk of violence against women with disabilities. To provide insights into these dynamics, an explanatory framework that organized risk markers in terms of relationship factors, victim-related characteristics, and perpetrator-related characteristics was examined.

In terms of relationship factors, although women with and without disabilities significantly differed on education compatibility and duration of union, the multivariate analyses showed that these variables did not significantly affect the odds of violence against women with disabilities, and they did not contribute to the elevated risk of violence against women with disabilities. These relationship factors were not important for understanding the elevated risk for violence against women with disabilities.

The results of the descriptive analysis were consistent with expectations on all of the victim-related characteristics. Women with disabilities were less likely to be well educated, and more likely to be unemployed, older, and of Aboriginal ancestry. However, controlling for all other independent variables in the current study, only age significantly affected the odds of violence against women with disabilities. Although disability was positively related to age, the negative relationship between age and violence generally found in the literature also held true for partner violence against women with disabilities; that is, it was young women with disabilities who were most vulnerable to violence by their partners. Given that it is conventional in Canadian society for partners within a couple to be of the same age or for the male partner to be slightly older (Eshleman & Wilson, 1998; Martin-Matthews, 2000), it is reasonable to extrapolate that young males are most likely to perpetrate partner violence against women with disabilities. Despite a significant impact of age on the odds of violence against women with disabilities, victim-related characteristics did not contribute to understanding the elevated risk for violence against women with disabilities.

With respect to perpetrator-related characteristics, contrary to expectations, male partners of women with disabilities were not more likely to abuse alcohol, and alcohol abuse did not have a significant impact on the odds of violence against women with disabilities. These results did not lend support to the dependency-stress model as an explanation for the elevated risk of partner violence against women with disabilities. On the other hand, male partners of women with disabilities were about 2.5 times more likely to behave in a patriarchal dominating manner and about 1.5 times more likely to engage in sexually proprietary behaviors than were partners of women without disabilities. The results of the multivariate analysis showed that patriarchal dominance and sexually proprietary behaviors were strongly linked to increased odds of violence for women with and without disabilities. Moreover, controlling for these perpetrator-related characteristics accounted for the elevated risk for violence against women with disabilities. These results did lend support to the applications of feminist disability and evolutionary psychology theories. Hence, it is important for stakeholders to recognize not only that women with disabilities have partners and are susceptible to partner violence but also that ideologies of patriarchy and male sexual proprietariness also affect these women's lives and, in fact, appear to account for their elevated risk for partner violence.

At this juncture it is important to note some limitations of the current research. It has been argued that a tendency for women with disabilities to have lower education levels, low self-esteem, and less assertiveness may render these women less likely to interpret violent acts as violence (Curry et al., 2001). This suggested that behavioral indicators were appropriate for measuring violence against the sampled women with disabilities.

However, it also bears adding that women with disabilities have been found to experience forms of abuse that were not measured with the modified version of the CTS used in the current study. For example, Gilson et al. (2001) identified several forms of abuse that are unique to women with disabilities including removal of an accessibility device, withholding medication, and threatening institutionalization. Exclusion of such acts results in underestimating the prevalence of partner abuse against women with disabilities. This was illustrated in a study by McFarlane et al. (2001), who employed the Abuse Assessment Screen-Disability (AAS-D) instrument with a clinical sample of 51 women with physical disabilities. The AAS-D included two standard questions (one on physical and one on sexual assault) and two disability-related questions on abuse (e.g., being prevented from using a wheelchair, cane, respirator, or other assistive devices). Of the sampled women, 9.8% reported at least one form of abuse having occurred in the past year. Without the two disability-related questions, the prevalence would have been 7.8%. Of course, the disability-related questions would not be applicable to women without disabilities and so would not have been appropriate for use in the current study because the analysis required use of the same dependent variable for women with disabilities and women without disabilities. Nevertheless, readers must recognize that the prevalence rates of violence in the current study likely underestimate the rate of partner abuse experienced by women with disabilities in Canada. A second limitation concerned the survey method. Women with disabilities that would prevent them from understanding the survey were excluded, and so these women were not represented. As well, women who would have required assistance to complete the survey by phone were excluded. Although the research conducted by Young et al. (1997) suffered from its own limitations, in addition to completion of the survey over the phone they were able to offer respondents hard-copy, computerized, or audio cassette versions to allow women with severe disabilities to complete the survey in privacy. On the other hand, if offering such options in the GSS would have required these women to lose their anonymity, then new compromises would have been introduced into the survey.

Although this research was not without limitations, it has filled an important gap in our knowledge of partner violence against women with disabilities. Extant research was widely acknowledged as "acutely deficient" (Barnett et al., 2005, p. 352). Nosek et al. (2001) criticized past studies of abuse and disability based on seven principles of valid research: (a) failure to operationally define variables and distinguish between different forms of violence, such as physical and sexual violence; (b) use of no standardized measures; (c) failure to categorize specific incidents; (d) heterogeneity in terms of gender, age, and type of disability; (e) use of convenience sampling; (f) failure to use comparison groups; and (g) lack of multivariate statistical analyses. Although the data did not allow a meaningful categorization of women by type of disability, the current study largely addressed these weaknesses of past research (variables were operationalized, standardized measures of violence were employed, distinctions were made between types of violence, there was a focus on gender, age was controlled, advanced sampling methods were used, women with disabilities were compared to the population of women without disabilities, and multivariate statistical analyses were employed).

In conclusion, partner violence against women with disabilities is a social problem in need of recognition and attention by stakeholders. Although women with disabilities need supports to reduce their vulnerability, the apparent importance of perpetrator-related characteristics derived from feminist disability and evolutionary psychology theories suggested that more attention should also be directed toward perpetrators. Efforts to reduce patriarchal dominating and male sexually proprietary behaviors in the general population will also help to reduce partner violence against women with disabilities. In addition, given that partners of women with disabilities are more likely to express these behaviors, efforts specifically directed to these men are needed. For example, community awareness campaigns for violence prevention could specifically address patriarchal domination, sexual proprietariness, and violent behavior in the context of relationships involving women with disabilities. Men who espouse patriarchy and sexual proprietariness need to receive the message that such ideologies are inappropriate and, along with violence, such behaviors toward women, including women with disabilities, will not be tolerated.

Notes

1. Given that the purpose of the remaining analyses was to examine an elevated risk for violence against women with disabilities, the remaining analyses were conducted with the 5-year prevalence rate.

2. Given that the Participation and Activity Limitations Survey (PALS) and GSS were representative of the Canadian population and used the same definition of disabilities that spanned the continuum in terms of severity. Of the women reporting disabilities in the PALS, 32.2% reported mild disabilities, 25.3% moderate disabilities, 28.3% severe disabilities, and 14.1% very severe disabilities (Statistics Canada, 2002).

References

Anderson, K. L. (1997). Gender, status, and domestic violence: An integration of feminist and family violence approaches. *Journal of Marriage and the Family, 59*, 655–669.

Barnett, O., Miller-Perrin, C. L., & Perrin, R. D. (2005). *Family violence across the lifespan: An introduction* (2nd ed.). Thousand Oaks, CA: Sage.

Brownridge, D. A. (2002). Cultural variation in male partner violence against women: A comparison of Quebec with the rest of Canada. *Violence Against Woman, 8*, 87–115.

Brownridge, D. A. (2003). Male partner violence against Aboriginal women in Canada: An empirical analysis. *Journal of Interpersonal Violence, 18*, 65, 83.

Brownridge, D. A. (2004). Male partner violence against women in stepfamilies: An analysis of risk and explanations in the Canadian milieu. *Violence and Victims, 19*, 17–36.

Brownridge, D. A., & Halli, S. S. (1999). Measuring family violence: The conceptualization and utilization of prevalence and incidence rates. *Journal of Family Violence, 14*, 333–350.

Brownridge, D. A., & Halli, S. S. (2001). *Explaining violence against women in Canada.* Lanham, MD: Lexington.

Chappell, M. (2003). Violence against women with disabilities: A research overview of the last decade. *AWARE: The Newsletter of the BC Institute Against Family Violence, 10*(1), 11–16.

Curry, M. A., Hassounch-Phillips, D., & Johnston-Silverberg, A. (2001). Abuse of women with disabilities: An ecological model and review. *Violence Against Women, 7*, 60–79.

Eshleman, J. R., & Wilson, S. J. (1998). The family (2nd Canadian ed.). Scarborough, Canada: Allyn & Bacon.

Gilson, S. R., DePoy, E., & Cramer, E. P. (2001). Linking the assessment of self-reported functional capacity with abuse experiences of women with disabilities. *Violence Against Women, 7*, 418–413.

Hightower, J., & Smith, G. (2003). Aging, disabilities, and abuse. *AWARE: The Newsletter of the BC Institute Against Family Violence, 10*(1), 17–18.

Howe, K. (2000). *Violence against women with disabilities: An overview of the literature.* Retrieved April 28, 2003, from www.wwda.org.au/keran.htm

Johnson, H. (2001). Contrasting views of the role of alcohol in cases of wife assault. *Journal of Interpersonal Violence, 16,* 54–72.

Li, L., Ford, J. A., & Moore, D. (2000). An exploratory study of violence, substance abuse, disability, and gender. *Social Behavior and Personality, 28,* 61–72.

Martin-Matthews, A. (2000). Change and diversity in aging families and intergenerational relations. In N. Mandell & A. Duffy (Eds.), Canadian families: Diversity, conflict, and change (2nd ed., pp. 323–360). Toronto, Canada: Harcourt Brace.

McFarlane, J., Hughes, R. B., Nosek, M. A., Groff, J. Y., Swedlend, N., & Mullen, P. D. (2001). Abuse Assessment Screen-Disability (AAS-D): Measuring frequency, type, and perpetrator of abuse toward women with physical disabilities. *Journal of Women's Health and Gender-Based Medicine, 10,* 861–866.

Melcombe, L. (2003). Facing up facts. *AWARE: The Newsletter of the BC Institute Against Family violence, 10*(1), 8–10.

Milberger, S., Israel, N., LeRoy, B., Martin, A., Potter, L., & Patchak-Schuster, P. (2003). Violence against women with physical disabilities. *Violence and Victims, 18,* 581 591.

Nettelbeck, T., & Wilson, C. (2002). Personal vulnerability to victimization of people with mental retardation. *Traumas, Violence, & Abuse, 3,* 289–306.

Nettelbeck, T., Wilson, C., Potter, R., & Perry, C. (2000). The influence of interpersonal competence on personal vulnerability of persons with mental retardation. *Journal of Interpersonal Violence, 15,* 46–62.

Newman, E., Christopher, S. R., & Berry, J. O. (2000). Developmental disabilities, trauma exposure and post-traumatic stress disorder. *Trauma, Violence, & Abuse, 1,* 154–170.

Nosek, M. A., Howland, C. A., & Hughes, R. B. (2001). The investigation of abuse and women with disabilities: Going beyond assumptions. *Violence Against Women, 7,* 477–499.

Petersilia, J. R. (2001). Crime victims with developmental disabilities: A review essay. *Criminal Justice and Behavior, 28,* 655–694.

Ridington, J. (1989). *Beating the "odds": Violence and women with disabilities* (Position Paper 2). Vancouver: Disabled Women's Network of Canada.

Smith, M. D. (1990). Patriarchal ideology and wife beating: A test of a feminist hypothesis. *Violence and Victims, 5,* 257–273.

Sobsey, D. (1994). *Violence and abuse in the lives of people with disabilities: The end of silent acceptance?* Baltimore: Paul H. Brookes.

Sobsey, D. (2000). Faces of violence against women with developmental disabilities. *Impact, 13*(3), 2–3.

Sobsey, D., & Calder, P. (1999). *Violence against people with disabilities: A conceptual analysis.* Washington, DC: National Research Council.

Statistics Canada. (2000). *1999 General Social Survey, Cycle 13 victimization: Public use microdata file documentation and user's guide.* Ottawa, Canada: Minister of Industry.

Statistics Canada. (2002, December 3). Participation and activity limitation survey: A profile of disability in Canada. *Daily,* pp. 1–8.

Thomson, R. G. (1994). Redrawing the boundaries of feminist disability studies. *Feminist Studies, 20,* 582–598.

Wilson, M., & Daly, M. (1998). Lethal and nonlethal violence against wives and the evolutionary psychology of male sexual proprietariness. In R. E. Dobash & R. P. Dobash (Eds.), *Rethinking violence against women* (pp. 224–230). Thousand Oaks, CA: Sage.

Young, M. E., Nosek, M. A., Howland, C., Chanpong, G., & Rintala, D. H. (1997). Prevalence of abuse of women with physical disabilities. *Archives of Physical Medicine & Rehabilitation, 78*(12, Suppl. 5): S34–S38.

DISCUSSION QUESTIONS

1. The author includes a quote from Chappell (2003) stating that "women with disabilities face an epidemic of monumental proportions" (as cited by Brownridge, 2006, p. 805). Based on the findings presented in this article, do you agree? Why or why not?

2. This study presents findings from Canada's General Social Survey. Do you think that these results would be similar if the study was conducted on data collected from persons living in the United States? Why or why not?

3. Many have argued that individual characteristics can explain why women with disabilities are more likely to be victimized than others, such as living in poverty, having lower educational attainment, and experiencing barriers to employment; however, Brownridge (2006) finds that perpetrator-related characteristics explain intimate partner violence against women with disabilities more readily than victim characteristics. How can you reconcile these findings?

4. Given the risk factors for intimate partner violence for women with disabilities, what are the policy implications for reducing this type of victimization?

Introduction to Reading 2

Historically, research on persons with mental disorders and crime focused on their criminal behavior rather than on their criminal victimization. More recently, the victimization experiences of mentally disordered individuals have come to light. Beyond just identifying that mentally disordered persons are at risk for victimization, Silver (2002) uses theory to examine explanations for their increased victimization risk. In doing so, he uses a sample of 270 discharged psychiatric patients and a sample of nonpatients drawn from the same neighborhood. He finds that involvement in conflicted social relationships influences victimization risk for persons with mental disorders.

Mental Disorder and Violent Victimization

The Mediating Role of Involvement in Conflicted Social Relationships

Eric Silver

Most prior studies of the relationship between mental disorder and violence depict mentally disordered people as violent actors (Link et al., 1992, 1999; Monahan et al., 2000; Quinsey et al., 1998; Silver, 2000a, 2000b; Silver et al., 1999; Steadman et al., 1998, 2000; Swanson et al., 1999). Few studies, however, have attempted to measure the amount of violence that is committed against people with mental disorder (but see Hiday et al., 1999; Lehman and Linn, 1984). Thus, we know relatively little about the risk of violent victimization that is associated with mental disorder and even less about why such victimizations occur.

As a result of deinstitutionalization policies that began in the 1950s and 1960s, most mentally disordered people currently reside in communities where they are engaged in social relationships with a range of others, including family members, friends, and other community residents (Johnson, 1990; Silver, 2000a). Further, several studies suggest that these social relationships may often become strained as family members and others seek to manage and control the behavior of their mentally disordered relatives and acquaintances (Cascardi et al., 1996; Estroff and Zimmer, 1994; Hiday, 1997; see also Felson, 1992). According to Hiday (1997:400), symptoms of serious mental disorder can "create tense situations between a mentally ill person and others," particularly when others attempt to exert social control by "persuading the person with mental disorder to desist in disturbing or annoying behavior or comply with treatment." Hiday further suggests that "these tense situations can escalate to accusations, anger, and eventually pushing, hitting, and fighting." Thus, it appears that

SOURCE: Silver (2002). Reprinted with permission of Wiley-Blackwell.

one important reason mentally disordered people may become victims of violence is that they are at greater risk of becoming involved in conflicted social relationships in which others are attempting to control their behavior.

The purpose of the current study is to determine whether the increased risk of violent victimization that is associated with mental disorder can be explained by the disproportionate involvement of mentally disordered people in conflicted social relationships. To the extent that mental disorder leads people to behave in ways that elicit grievances (or produce negative emotions) in others, mentally disordered people may be more likely to become the targets of social control efforts that may include violence. This may be particularly true for mentally disordered people who use illegal substances because the behaviors associated with acquiring and using substances may elicit additional social control responses (Hiday, 1997).[1] Indeed, research on violence committed by discharged psychiatric patients has shown that those who use drugs are more likely to engage in violence (Steadman et al., 1998; Swanson et al., 1990), which suggests that they also may be more likely to engage in other forms of behavior that people construe as unwanted or offensive.

To examine the link between involvement in conflicted social relationships and violent victimization, this study draws on Felson's (1992) social interactionist theory of violence and Agnew's (1992) general strain theory of crime and delinquency, both of which emphasize the situational dynamics surrounding violent interactions. From Felson's (1992) perspective, people use violence in social interactions with others as part of a strategy of informal social control (see also Felson and Steadman, 1983). From Agnew's (1992) perspective, violent behavior is one of several negative outcomes that result from exposure to stressful events that produce negative emotions, such as anger, rage, or fear. Both perspectives, therefore, support the notion that mentally disordered people who are involved in conflicted social relationships—that is, relationships in which their behavior elicits grievances (or negative

emotions) in others—will be at greater risk of being victimized by violence.

Mental Disorder and Violent Victimization

In the only recent study to directly compare the violent victimization rates of mentally disordered and nonmentally disordered people, Hiday et al. (1999) found that recently involuntarily hospitalized psychiatric patients in North Carolina were approximately 2.5 times more likely to be victimized by violence than were members of the U.S. general population (8.2% compared with 3.1%). However, this comparison remains tentative because it was based on victimization data from very different sources (i.e., North Carolina versus the entire United States) and because it was made without statistical controls for other known correlates of violent victimization (i.e., race, gender, SES, and neighborhood disadvantage). Thus, we do not know the extent to which the observed association between mental disorder and violent victimization is due to spurious associations with other known correlates of violent victimization. An important contribution of the current study is that it controls for both individual- and neighborhood-level correlates of violent victimization in examining this association.

Moreover, the study by Hiday et al. (1999) was not designed to provide information on the types of social relationships that give rise to violent victimization. Fortunately, however, qualitative data bearing on this question are available. In a study of 143 in-depth interviews with seriously mentally disordered people and their relatives, Estroff and Zimmer (1994) observed a significant amount of reciprocal antagonism and tension in the social relationship of mentally disordered people. They concluded that "violence seldom happens unilaterally, and that hostility and violence from others is not uncommon in the personal histories and current social networks of the respondents" (p. 271). This conclusion points to the importance of examining involvement in conflicted relationships with others as a possible explanation for the violent victimization of mentally disordered people.

However, because most studies of victimization have focused primarily on predatory street crime (Meier and Miethe, 1993), little is known about the risk of violent victimization that may result from involvement in conflicted relationships with others. To fill this gap, the current study addresses the following research questions: (1) Are people with mental disorder more likely than those without mental disorder to be victimized by violence, after controlling for known individual- and community-level correlates of victimization? (2) Is the increased risk of violent victimization that is associated with mental disorder mediated, in part, by the greater involvement of mentally disordered people in conflicted social relationship with others? (3) Is the increased risk of violent victimization that is associated with mental disorder and illegal drug use also, in part, mediated by involvement in conflicted social relationships?

To address these questions, this study uses data from the Pittsburgh site of the MacArthur Foundation's Violence Risk Assessment Study (Silver, 2000a, 2000b; Silver et al., 1999; Steadman et al., 1998). The data consist of a sample of discharged psychiatric patients ($N = 270$) and a sample of nonpatients ($N = 477$) drawn from the same neighborhoods as the patients, using the same data collection and measurement procedures. Both the patient and the nonpatient data contain measures of subjects' recent violent victimization and the quality of their social relationships with others, among other relevant factors. Thus, the MacArthur data provide a unique opportunity to measure the association between violent victimization and serious mental disorder and to examine the contribution that involvement in conflicted social relationships makes to this association.

✉ Data, Measures, and Statistical Procedures

Psychiatric Patient Sample

Between 1992 and 1995, the MacArthur Foundation's Violence Risk Assessment Study sampled admissions from the Western Psychiatric Institute and Clinic (WPIC), a university-based acute psychiatric inpatient facility located in Pittsburgh, Pennsylvania (Silver, 2000a, 2000b; Silver et al., 1999; Steadman et al., 1998). To be eligible for participation in the study, patients had to be acute (i.e., short-stay) civil admissions, between the age of 18 and 40, English-speaking, White or African-American, and have at least one of the following types of major mental disorder diagnoses: schizophrenia, depression, mania, psychosis, delusions, substance abuse disorder, or personality disorder.

The mean length of hospitalization for subjects treated at WPIC was 20 days (the median was 15), consistent with the lengths of hospitalization for patients admitted to psychiatric facilities in the United States (Rouse, 1998). The mean length of stay of 20 days was similar to the mean length of stay of 23 days for all patients admitted to WPIC during the three-year period that these data were gathered. Most patients (72.6%) were admitted to the hospital voluntarily. The fact that the patient sample was drawn from a population of acute (i.e., short-stay) inpatients must not be taken as an indication that they were not seriously mentally ill. Although subjects tended to be released quickly after admission, they also tended to require repeated admissions for psychiatric problems. Of the 270 patients in this study, 76% had experienced at least one prior admission to an inpatient psychiatric facility, and 30% had experienced four or more prior admissions. Hospital data collection was conducted in two parts: an initial interview to obtain a wide array of background data, and a second interview by the research clinician to confirm the subject's hospital chart diagnosis and to administer several other clinical scales. All subjects were asked for written informed consent to participate in the study. Five attempts were made to reinterview subjects in the community (approximately every 10 weeks) over the one-year period after discharge from WPIC.

Using the same interview schedule, interviewers also spoke with a collateral informant, i.e., a person who was familiar with the subject's behavior and functioning in the community. Collateral informants were nominated by subjects during each follow-up interview. If the collateral nominee did not have at least weekly contact with the subject, the interviewer suggested a more appropriate person based on a review of the subject's social network data. Collateral informants were most often family members (47.1%), but also were friends (22.0%), professionals (16.1%), significant others (11.8%), or others (e.g., coworkers, 3.0%). All subjects and collateral informants were paid for each completed interview.

Official records provided a third source of information about the patient's behavior in the community. If a patient was readmitted to a psychiatric hospital, the hospital was contacted to obtain information regarding the rehospitalization, including dates of hospitalization, diagnosis, and reasons for admission. Arrest records for all patients were requested at the end of the one-year follow-up.

Of the 4,069 patients admitted to WPIC between 1992 and 1995, 2,532 met the study's eligibility criteria. A total of 629 patients were approached to participate in the study. The refusal rate was 33.2%. Twenty-nine patients who agreed to participated were discharged before the hospital interview could be completed. The final sample given a hospital interview was 391, of whom 312 (80%) completed the first follow-up interview. Postdischarge census tract locations were identified for 270 of these 312 subjects (87%). There were no significant differences in terms of sex, age, race, or admission diagnosis between enrolled subjects who did not complete the first follow-up interview ($N = 79$) and those who did ($N = 321$), nor were there any significant differences between subjects for whom the postdischarge census tract location was known ($N = 270$) and those for whom it was not known ($N = 42$).

Community Comparison Sample

A detailed description of the methods used to gather the community comparison is provided by Steadman et al. (1998). Briefly, the University of Pittsburgh's Center for Social and Urban Research identified a community sample in Pittsburgh that was drawn from the same census tracts in which the patients lived during the one-year period after their discharge from WPIC.

The subjects in the community sample were interviewed only once. They and their collateral informants were questioned about the subject's behavior in the past ten weeks, the same time frame used for the patients. Official arrests recorded also were obtained. In addition to being selected from the same census tracts as the patients, the nonpatient community respondents had to have lived at the current address for at least two months, be between the age of 18 and 40, and be White or African-American (i.e., the same criteria used to gather the patient sample).

Measures

Violent Victimization

Violent victimization was measured based on the following questions about subjects' victimization experiences in the ten weeks after discharge from WPIC: (1) Has anyone hit you with a fist or object or beaten you up? (2) Has anyone tried to physically force you to have sex against your will? (3) Has anyone used a knife or fired a gun at you (or threatened to do so with the weapon in hand)? Seven percent of subjects reported having been hit or beat up during the past ten weeks; 3% reported having been forced to have sex; and 2% reported having been threatened or attacked with a knife or gun. Because affirmative responses to these items were relatively rare, they were combined into a dichotomous indicator coded 1 for subjects who answered yes to any of the questions (10%) and 0 for all others.

Involvement in Conflicted Social Relationships

Involvement in conflicted social relationships was measured by summing two items that were embedded in a larger instrument designed to elicit information on subjects' social networks during the past ten weeks. Specifically, subjects were asked to identify from their social networks anyone "with whom you really don't get along, or don't like, or who really upsets you," and anyone "who really doesn't seem to like you or who you seem to upset."[2] The distribution of responses to these items was skewed: Most subjects (78.3%) identified no one, 13% named one person, and 5% named two people. Because less than 3% of the sample named three or more people in response to these items, responses of three or more were recoded to 3. Although this measure gauges the extent to which subjects were involved in conflicted relationships with others, it does not differentiate between conflicts that resulted from social control efforts aimed at redressing grievances (Felson, 1992) and conflicts that resulted from the release of negative affect, more generally (Agnew, 1992).

Patient Status

Patient status was measured as a dichotomous variable coded 1 for subjects in the patient sample and 0 for subjects in the community comparison sample.

Individual-Level Control Variables

A large number of individual-level control variables are available in the MacArthur Violence Risk Assessment data (Monahan et al., 2000). For the purpose of this analysis, control variables were selected if they had been found consistently in previous research to be significant predictors of violent victimization (for a review, see Sampson and Lauritsen, 1994). These measures include gender, age, socioeconomic status, employment status, and race. Gender was measured as a dichotomous variable coded 1 for males. Age was measured as a dichotomous variable coded 1 for subjects who were under 25 at the time of the interview. Socioeconomic status was measured as a combination of subjects' education, income, and occupational status based on the Hollingshead and Redlich five-factor scale. From this scale, a dichotomous variable was created and coded 1 for subjects in the lowest SES category and 0 for all others. Employment status was measured by a dichotomous variable coded 1 for subjects who were unemployed for the entire ten-week follow-up period and 0 for all others. Race was measured by a dichotomous variable coded 1 for African Americans and 0 for Whites.

In addition, the analyses reported below control for subject violence committed during the same ten-week period covered by the violent victimization measure. This is a crucial control variable because an unknown proportion of the victimization experiences reported by subjects may have grown out of conflicts that were initiated by the subject's violent behavior. Unfortunately, the data do not allow the victimization incidents to be disaggregated into those that were and were not initiated by subject violence. Therefore, it is necessary to control statistically for subject violence.

Subject violence was measured using questions adapted from the Conflict Tactics Scale. Subjects and collaterals each were asked whether the subjects had been involved in a violent incident during the past ten weeks. Violent acts included acts of battery that resulted in physical injury, sexual assaults, assaultive acts that involved the use of a weapon, or threats made with a weapon in hand. Only the most serious act for each violent incident was coded. Incidents of child discipline without injury were excluded. Violent acts reported by any of the three data sources—subject self-report, collateral report, or official records—were reviewed by a team of trained coders to obtain a single reconciled report of violence. A high degree of inter-rater reliability was achieved (Cohens's kappa = .93; for additional details on coding and reliability, see Silver, 2000a). From these reconciled reports, a dichotomous violence measure was created and coded 1 for subjects who engaged in a violent act during the ten-week follow-up period, and 0 for those who did not.

Neighborhood-Level Control Variables

Studies of victimization that include individual- and community-level risk factors are rare in criminology. However, those that have been conducted have found that both community- and individual-level factors are important predictors of victimization risk (Lauritsen, 2001; Rountree et al., 1994; Sampson and Lauritsen, 1990; Sampson and Wooldredge, 1987; Sampson et al., 1997). Thus, all of the models reported below are estimated with controls for neighborhood characteristics, operationalized here as the census tract in which subjects lived.[3] Using census tract identifiers, data from the patient and community comparison samples were matched to census tract records from the 1990 U.S. census to create a multilevel analysis file consisting of both neighborhood- and individual-level characteristics.

The neighborhood structural characteristics examined here include neighborhood socioeconomic disadvantage and neighborhood residential mobility. These measures were derived from a principle components factor analysis (using oblique rotation), in which the following variables were found to load on the disadvantage factor (factor loadings are provided in parentheses): percent public assistance (.92), percent female-headed households (.91), poverty rate (.87), unemployment rate (.87), percent African-American (.81), high income (–.80), percent vacant dwellings (.76), percent managerial employment (–.75), and mean household wage (–.73). The eigenvalue for this factor was 6.2; the explained variance was 56.6%. A second factor, labeled "neighborhood mobility," consisted of the proportion of individuals residing in the same home as five years ago (–.90) and the percent foreign born (.86). The eigenvalue for this factor was 1.8; explained variance was 16.4%.

Results

Table 1 compares the patient and community samples on each of the variables included in this study. Although the community sample was drawn from the same neighborhoods as those of the patients, significant differences were observed in the characteristics of the individuals in each sample. As shown, the patient sample contained a higher percentage of males, a lower percentage of African Americans, a higher proportion of low SES individuals, a lower proportion of employed individuals, a higher proportion of individuals who had engaged in violence toward others, and a higher concentration of individuals in highly residentially mobile neighborhoods. There were no significant differences between the samples in terms of age or neighborhood disadvantage. Moreover, Table 1 shows that patients were significantly more likely than were nonpatients to have been involved in conflicted social relationships

Table 1 Comparison of the Patient and Community Samples

	Patient Sample (*N* = 270)	Nonpatient Sample (*N* = 477)
% Male***	55.2	36.7
% African-American*	33.7	41.3
% Under Age 25	24.8	22.4
% Low SES***	58.9	19.3
% Unemployed During Prior 10 Weeks***	70.4	41.5
% Violent Toward Others During Prior 10 Weeks***	10.4	3.4
% Living in Highly Mobile Neighborhoods[a] **	40.4	29.1
% Living in Highly Disadvantaged Neighborhoods[b]	29.3	35.6
% Involved in Any Conflicted Social Relationships***	37.8	20.8
% Victimized by Violence During Prior 10 Weeks***	15.2	6.9

*$p < .05$; ** $p < .01$; *** $p < .001$ (based on Pearson's chi-square).

[a]Includes subjects in the upper third of the neighborhood mobility factor score.

[b]Includes subjects in the upper third of the neighborhood disadvantage factor score.

with others. In addition, patients were more likely to have been victims of violence, although a more rigorous test to this relationship (controlling for demographic differences between the samples) is provided in the multivariate analyses below.

As a backdrop to the multivariate logistic regression analyses that follow, Table 2 presents bivariate associations with violent victimization for each of the study variables. Column 1 of Table 2 shows the unstandardized logistic regression coefficients and standard errors for each variable, and column 2 shows the odds ratios, which represent the multiplicative change in the odds of violent victimization that is associated with a one-unit change in each predictor variable. Consistent with prior research on victimization (Lauritsen, 2001; for a review, see Sampson and Lauritsen, 1994), the risk of violent victimization was found to be higher for males, African Americans, younger people, people of low SES, the unemployed, people who have been violent

toward others, and those who live in more disadvantaged neighborhoods. Moreover, Table 2 shows that involvement in conflicted social relationships was significantly and positively related to the likelihood of violent victimization at the bivariate level.

The multivariate analyses are presented in Table 3. Model 1 of Table 3 shows the significant bivariate relationship between patient status and violent victimization reported previously in Table 1. Model 2 adds the individual- and neighborhood-level control variables to the equation. Not surprisingly, violence toward others in the past ten weeks was strongly and significantly related to violent victimization, net of the other control variables, as was neighborhood disadvantage. The strong effect of violence toward others on victimization suggests that a significant proportion of violent victimizations may grow out of situations in which violence is mutually inflicted.

Model 2 also shows that the effect of patient status on violent victimization was reduced when the control variables were introduced to the equation, which suggests that some of the effect of patient status on violent victimization was spuriously related to the control variables.[4] However, the effects of patient status on violent victimization remained significant, net of the control variables. Thus, patient status contributed significantly to the probability of violent victimization, over and above the effects of the individual- and neighborhood-level control variables.

Model 3 of Table 3 shows that involvement in conflicted social relationships was significantly and positively related to violent victimization, net of the individual and neighborhood control variables. Moreover, controlling for involvement in conflicted relationships resulted in a 22% reduction in the unstandardized coefficient for patient status [(.60–.47)/.60], and it rendered the relationship between patient status and violent victimization nonsignificant. This result suggests that a substantial amount of the increased risk of violence victimization that was associated with patient status was due to the disproportionate involvement of patients in conflicted social relationships with others. In contrast, the coefficient for neighborhood disadvantage remained significant in Model 3, which suggests that the risk of victimization associated with

Table 2 Bivariate Associations With Violent Victimization From Logistic Regression

	UC (S.E.)[a]	Odds Ratio
Male	0.54 (.25)	1.71*
African-American	0.64 (.25)	1.90**
Under Age 25	0.51 (.25)	1.67*
Low SES	0.82 (.25)	2.28***
Unemployed During Prior 10 Weeks	0.75 (.25)	2.12**
Violent Toward Others During Prior 10 Weeks	1.97 (.25)	7.12***
Neighborhood Residential Mobility[b]	−0.09 (.93)	0.92
Neighborhood Disadvantage[b]	0.33 (0.9)	1.40***
Involvement in Conflicted Social Relationships	0.48 (.11)	1.62***

*$p < .05$; ** $p < .01$; *** $p < .001$.

[a]UC = unstandardized coefficient; S.E. = standard error.

[b]Unrecoded factor score.

Table 3 Logistic Regression Predicting Violent Victimization in the Past 10 Weeks

	Model 1		Model 2		Model 3	
	UC (S.E.)[a]	Odds Ratio	UC (S.E.)[a]	Odds Ratio	UC (S.E.)[a]	Odds Ratio
Patient Status	0.88 (0.25)	2.42***	0.60 (.30)	1.82*	0.47 (.31)	1.60
Male	—	—	0.38 (.27)	1.46	0.37 (.28)	1.45
African-American	—	—	0.02 (.37)	1.02	−0.02 (.37)	0.98
Under Age 25	—	—	0.46 (.29)	1.59	0.52 (.29)	1.67
Low SES	—	—	0.19 (.29)	1.21	0.16 (.29)	1.18
Unemployed During Prior 10 Weeks	—	—	0.40 (.29)	1.49	0.37 (.29)	1.45
Violent Toward Others During Prior 10 Weeks	—	—	1.47 (.36)	4.35***	1.42 (.37)	4.14*
Neighborhood Residential Mobility[b]	—	—	−0.12 (.15)	0.89	−0.15 (.15)	0.86
Neighborhood Disadvantage[b]	—	—	0.29 (.14)	1.34*	0.32 (.15)	1.37*
Involvement in Conflicted Social Relationships	—	—	—	—	0.42 (.12)	1.52***
Nagelkerke (pseudo) R^2	.04		.15		.17	
χ^2 Model Improvement(df)	12.65(1)***		41.12(8)***		10.51(1)***	

*$p < .05$; ** $p < .01$; *** $p < .001$.

[a]UC = unstandardized coefficient; S.E. = standard error.

[b]Unrecoded factor score.

living in a disadvantaged neighborhood involves processes other than relationship conflict. Such processes may include proximity to a larger pool of motivated offenders (see below).

Because research on violence committed by discharged psychiatric patients has shown that those with drug abuse problems are at greater risk of engaging in violence than are patients without such problems (Steadman et al., 1998; Swanson et al., 1990), and because substance use may further lead patients to behave in ways that elicit social control efforts from others, the analyses in Table 3 were replicated after disaggregating the patients into those who reported using illegal drugs in the ten-week period prior to hospital admission and those who did not.

Patients who used illegal drugs were indeed more likely than were control subjects to be victims of violence, as were patients who did not use illegal drugs. However, the increased risk of victimization for patients who did not use illegal drugs disappeared when the control variables were introduced. In contrast, the effect for patients who used illegal drugs was reduced but remained significant. These results suggest that it is the combination of mental disorder and substance use that may be particularly important in increasing the risk of violent victimization among patients. More important, however, is that we see that the likelihood of violent victimization among patients who used illegal drugs was reduced by 23% and rendered nonsignificant when involvement in conflicted social relationships was introduced to the equation. Thus, involvement in conflicted social relationships appears to explain the increased risk of violent victimization that is associated with drug use among the patient sample. Once again, the effect of neighborhood disadvantage remained significant after controlling for relationship conflict.

Discussion

This study showed that mentally disordered people were more likely than were nonmentally disordered people to be victimized by violence, after controlling for known individual- and community-level correlates of violent victimization. Moreover, the risk of victimization was found to be particularly strong when mental disorder co-occurred with illegal drug use. Results also showed that mentally disordered people were more likely than were nonmentally disordered people to be involved in conflicted social relationships with others. More importantly, however, this study showed that the significant relationship between mental disorder and violent victimization was rendered nonsignificant when involvement in conflicted social relationships was controlled statistically, a result that held after illegal drug use by the patients was taken into account. Together, these results suggest that one important reason mentally disordered people are more likely than are nonmentally disordered to become victims of violence is that their relationships with others are more likely to involve conflict.

The theoretical implications of this study lie in highlighting the potentially important contributions that situational theories such as those proposed by Felson (1992) and Agnew (1992) can make to the study of violent victimization. Felson (1992) views violence as an informal social control strategy aimed at deterring the behavior of others or obtaining "justice" as defined by the actor. As a subclass of what Felson (1992:4) refers to as "distressed people," people who are mentally disordered (and who abuse drugs) may find it difficult to feign positive emotions, and they may be less able to show deference, thereby resulting in behavior that others may consider inappropriate or provocative. To the extent that people with mental disorder (and drug use problems) perform less competently in

this regard, their behavior is likely to produce grievances in others. Such grievances, in turn, may elicit aggression from others as a form of informal social control or retaliation.

Similarly, Agnew (1992) argues that exposure to negative stimuli, including (but not limited to) behaviors or verbalizations that the individual experiences as noxious, may lead to violence. People with serious mental disorders, particularly those experiencing delusional beliefs or hallucinations, or those with substance abuse disorders, may introduce a variety of negative stimuli into their relationships with others (Hiday, 1997). Such negative stimuli may, in turn, increase the likelihood that the mentally disordered person will become the victim of violence as others attempt to apply "corrective actions" in the situation.

In emphasizing motivational factors related to relationship conflict, the current study departs from the more familiar lifestyle/routine activity perspective that is typically invoked in criminological studies of victimization (Meier and Miethe, 1993). Nonetheless, the finding that involvement in conflicted social relationships contributes to the violent victimization of mentally disordered people does not imply that the lifestyle/routine activity perspective is irrelevant. First, as noted above, the effect of neighborhood disadvantage on violent victimization remained significant after controlling for involvement in conflicted social relationships and other demographic variables, suggesting that proximity to motivated offenders may be an important factor. In addition, as one of the reviewers of this paper pointed out, it may be that involvement in conflicted social relationships leads to rejection and avoidance of the mentally disordered person, thereby decreasing his or her access to capable guardianship.

Thus, an important implication of this study may lie in suggesting that the situational theories of Felson and Agnew can complement the more traditional routine activity approaches by enabling an examination of the role of motivation in the violent encounter. Whereas the traditional routine activity approach to predatory crime treats offender motivation as constant (Cohen and Felson, 1979), the situational approach described here treats motivation as central by arguing that the grievances (or negative affect) produced by conflicted relationships may motivate situational participants to behave violently (Felson, 1993). Thus, we may expect the specification of motivational factors rooted in relationship conflict to increase the explanatory power of routine activity models of violent victimization.

A similar argument was recently put forth by Finkelhor and Asdigian (1996:6) in the context of youth victimization. They argued that "in addition to environmental conditions highlighted by the lifestyle theory ... more attention needs to be given to the risk-increasing potential of individual characteristics and attributes." Of particular relevance is their notion of "target antagonism," which suggests that individual characteristics, including possessions, skills, or attributes, increase the risk of victimization by arousing anger, jealousy, or destructive impulses in others. The results reported here suggest that an important source of target antagonism leading to the violent victimization of mentally disordered people may include the extent to which the psychological, emotional, and behavioral manifestations of mental disorder (and substances use) lead them to become involved in conflicted relationships with others.

In addition to explaining violent victimization, grievances (and negative affect) may also help to explain the increased rates of violent behavior that have been observed among people with mental disorder. For example, a recent study of the perception of coercion among hospitalized psychiatric patients found that approximately two-thirds of patients who felt coerced into being hospitalized also reported feeling angry and fearful as a result (Monahan et al., 1999). This suggests that the conflict associated with efforts at informal social control by caretakers may inadvertently contribute to violence among psychiatric patients by eliciting grievances among them, particularly when social control efforts include involuntary treatments. Similarly, in a study of a combined sample of former psychiatric patients, ex-offenders, and members of the general population, Felson (1992) found that the effect of stressful life events on violent behavior was mediated by the victimization experiences of the respondents. In other words, the reason respondents (including former patients) committed more violence was that they had been victimized.

Similarly, a recent study by Hiday and colleagues (2000) found that psychiatric patients who had been victimized by violence were significantly more likely to engage in violence during the four-month period prior to their hospitalization for psychiatric treatment. Further, several recent studies have found that people with mental disorders who have behaved, or threatened to behave, violently in the recent past also report feeling threatened and, in fact, often were victims of violence (Estroff and Zimmer, 1994; Swanson et al., 1998, 1999). Thus, measuring the victimization experiences of mentally disordered people may ultimately help to explain their greater involvement in violent behavior.

Before concluding, some caveats are in order. First, because the MacArthur Violence Risk Assessment Study was not designed to measure victimization, but was designed as a study of patient violence, the measure of victimization used here was limited. Specifically, I could not determine the number of victimizations suffered by subjects, nor could I determine the extent of injury that occurred. Moreover, the data did not allow me to identify the perpetrators of the victimizations. Therefore, the current study could not examine whether involvement in conflicted relationships had similar effects on victimizations that ranged in seriousness and frequency, or those that were committed by family members, friends, or acquaintances.

Second, as mentioned earlier, the measure of relationship conflict used here was limited in that it did not differentiate between conflict that resulted from social control efforts and conflict that resulted from the release of negative emotions, more generally. Thus, this study was unable to assess the relative merits of Felson's (1992) social interactionist perspective and Agnew's (1992) theory of general strain as explanations for the violent victimization experiences of people with mental disorder. Distinguishing between these mechanisms empirically is the next logical step in understanding the social psychological processes that lead to violent victimization. Finally, the current study focused only on drug use by the patients because of the prominence of this variable in the literature on mental disorder and violence (Silver, 2000a; Silver et al., 1999). Nonetheless, similar hypotheses to those examined here may also be tested with respect to alcohol abuse.

Clearly, a more detailed study of the nature of the relationship between mental disorder and violent victimization should be pursued in the future. Toward this end, the importance of the current study lies in suggesting that one important direction for future research should be to collect more detailed information on the situational and interpersonal dynamics that surround the victimization experiences of mentally disordered people, with a particular emphasis on the social, contextual, and clinical factors that may lead to involvement in conflicted social relationships.

Notes

1. Although the current study focuses on the effect of substance use on violent victimization, similar hypotheses may be formulated with respect to alcohol abuse.

2. For the patient sample, in which multiple follow-up interviews were conducted, the items refer to the ten-week period prior to hospital admission and therefore occur before the period in which violent victimization was measured. However, for the community comparison sample, in which only one interview was conducted, the items refer to the same ten-week period as the violent victimization measure. Thus, the likelihood of confounding between the violence measure and the measure of relationship conflict is potentially greater for the community comparison sample than for the patient sample.

3. Most people regard their neighborhood as larger than one block but smaller than an entire sector of a city. Census tract boundaries lie between these two extremes and are drawn to encapsulate relatively homogeneous populations in terms of demographic and economic characteristics. Although census tracts are imperfect operationalizations of neighborhoods, they come closer than any commonly available geographic entity in approximating the usual conception of a neighborhood.

4. Interestingly, the positive bivariate relationship between African-American status and violent victimization that was reported in Table 2 was not observed in Model 2 of Table 3. This is due to the high correlation between African-American status and neighborhood disadvantage (Pearson's $r = .69, p < .001$). In analyses not shown, in which the neighborhood disadvantage variable was not included in Model 2, the effect of African-American status remained significant (odds ratio = 1.70, $p < .05$). (Note that the same finding was reported by Silver (2000b) with regard to the relationship between African-American racial status and violence committed by the patient sample.)

References

Agnew, Robert. 1992. Foundation for a general strain theory of crime and delinquency. Criminology 30:47–87.

Cascardi, Michele, Kim T. Mueser, Joanne DeGiralomo, and Mary Murrin. 1996. Physical aggression against psychiatric inpatients by family members and partners. Psychiatric Services 47:531–533.

Cohen, Lawrence E. and Marcus Felson. 1979. Social change and crime rate trends: A routine activity approach. American Sociological Review 44:588–608.

Estroff, Sue E. and Catherine Zimmer. 1994. Social networks, social support, and violence among persons with severe, persistent mental illness. In John Monahan and Henry J. Steadman (eds.), Violence and Mental Disorder. Chicago, Ill.: University of Chicago Press.

Felson, Richard B. 1992. "Kick 'em when they're down": Explanations of the relationship between stress and interpersonal aggression and violence. The Sociological Quarterly 33:1–16.

———1993. Predatory and dispute-related violence: A social interactionist approach. In Ronald V. Clarke and Marcus Felson (eds.), Routine Activity and Rational Choice: Advances in Criminological Theory (Vol. 5.). New Brunswick, N.J.: Transaction.

Felson, Richard B. and Henry J. Steadman. 1983. Situational factors in disputes leading to criminal violence. Criminology 21:59–74.

Finkelhor, David and Nancy L. Asdigian. 1996. Risk factors for youth victimization: Beyond a lifestyle/routine activities theory approach. Violence and Victims 11:3–19.

Hiday, Virginia A. 1997. Understanding the connection between mental illness and violence. International Journal of Law and Psychiatry 20:399–417.

Hiday, Virginia A., Jeffrey W. Swanson, Marvin S. Swartz, Randy Borum, and H. Ryan Wagner. 2000. Victimization: A link between mental illness and violence? International Journal of Law and Psychiatry. In press.

Hiday, Virginia A., Marvin S. Swartz, Jeffrey W. Swanson, Randy Borum, and H. Ryan Wagner. 1999. Criminal Victimization of Persons with Severe Mental Illness. Psychiatric Services 50:62–68.

Johnson, Ann Braden. 1990. Out of Bedlam: The Truth About Deinstitutionalization. New York: Basic Books.

Lauritsen, Janet L. 2001. The social ecology of violent victimization: Individual and contextual effects in the NCVS. Journal of Quantitative Criminology 17:3–32.

Lehman, Anthony F. and Lawrence S. Linn. 1984. Crimes against discharged mental patients in board and care homes. American Journal of Psychiatry 141:271–274.

Link, Bruce G., Howard Andrews, and Francis T. Cullen. 1992. The violent and illegal behavior of mental patients reconsidered. American Sociological Review 57:275–292.

Link, Bruce G., John Monahan, Ann Steve, and Francis T. Cullen. 1999. Real in their consequences: A sociological approach to understanding the association between psychotic symptoms and violence. American Sociological Review 64:316–332.

Meier, Robert F. and Terance D. Miethe. 1993. Understanding theories of criminal victimization. In Michael Tonry and Norval Morries (eds.), Crime and Justice: A Review of Research, Vol. 17. Chicago, Ill.: University of Chicago Press.

Monahan, John Charles W. Lidz, Steven K. Hoge, Edward P. Mulvey, Marlene M. Eisenberg, Loren H. Roth, William P. Gardner, and Nancy Bennett. 1999. Coercion in the provision of mental health services: The MacArthur studies. In J. Morrissey and J. Monahan (eds.), Research in Community and Mental Health, Vol. 10: Coercion in Mental Health Services—International perspectives. Stamford, Conn.: JAI Press.

Monahan, John, Henry J. Steadman, Paul S. Appelbaum, Pamela C. Robbins, Edward P. Mulvey, Eric Silver, Loren H. Roth, and Thomas Grisso. 2000. Developing a clinically useful actuarial tool for assessing violence risk. British Journal of Psychiatry 176:312–319.

Quinsey, Vernon L., Grant T. Harris, Marnie E. Rice, and Catherine A. Cormier. 1998. Violent Offenders: Appraising and Managing Risk. Washington, D.C.: American Psychological Association.

Rountree, Pamela Wilcox, Kenneth Land, and Terance Miethe. 1994. Macro-micro integration in the study of victimization: A hierarchical logistic model analysis across Seattle neighborhoods. Criminology 32:387–414.

Rouse, Beatrice A. 1998. Substance Abuse and Mental Health Statistics Source Book. Washington, D.C.: Department of Health and Human Services. Publication 98-3170.

Sampson, Robert J. and Janet L. Lauritsen. 1990. Deviant lifestyles, proximity to crime, and the offender-victim link in personal violence. Journal of Research in Crime and Delinquency 27:110–139.

———1994. Violent victimization and offending: Individual-, situational-, and community-level risk factors. In Albert J. Reiss and Jeffrey A. Roth (eds.), Understanding and Preventing Violence, Vol. 3. Washington, D.C.: National Academy Press.

Sampson, Robert J. and John Wooldredge. 1987. Linking the micro- and macro-dimensions of lifestyle-routine activity and opportunity models of predatory victimization. Journal of Quantitative Criminology 3:371–393.

Sampson, Robert J., Stephen W. Raudenbush, and Felton Earls. 1997. Neighborhoods and violent crime: A multilevel study of collective efficacy. Science 277:918–924.

Silver, Eric. 2000a. Extending social disorganization theory: A multilevel approach to the study of violence among persons with mental illnesses. Criminology 38:301–332.

———2000b. Race, neighborhood disadvantage, and violence among persons with mental disorders: The importance of contextual measurement. Law and Human Behavior 24:449–456.

Silver, Eric, Edward P. Mulvey, and John Monahan. 1999. Assessing violence risk among discharged psychiatric patients: Toward an ecological approach. Law and Human Behavior 23:235–253.

Steadman, Henry J., Edward P. Mulvey, John Monahan, Pamela C. Robbins, Paul S. Appelbaum, Thomas Grisso, Loren H. Roth, and Eric Silver. 1998. Violence by people discharged from acute psychiatric inpatient facilities and by others in the same neighborhoods. Archives of General Psychiatry 55:393–401.

Steadman, Henry J., Eric Silver, John Monahan, Paul S. Appelbaum, Pamela Clark Robbins, Edward P. Mulvey, Thomas Grisso, Loren Roth, and Steven Banks. 2000. A classification tree approach to the development of actuarial violence risk assessment tools. Law and Human Behavior 24:83–100.

Swanson, Jeffrey W., Randy Borum, Marvin Swartz, and Virginia A. Hiday. 1998. Violent behavior preceding hospitalization among persons with severe mental illness. Law and Human Behavior 23:185–204.

Swanson, Jeffrey W., Charles E. Holzer, Vijay K. Ganju, and Robert T. Jono. 1990. Violence and psychiatric disorders in the community: Epidemiologic Catchment Area Surveys. Hospital and Community Psychiatry 41:761–770.

Swanson, Jeffrey W., Marvin Swartz, Sue Estroff, Randy Borum, H. Ryan Wagner, and Virginia A. Hiday. 1999. Psychiatric impairment, social contact, and violent behavior: Evidence from a study of outpatient-committed persons with severe mental disorder. Social Psychiatry and Psychiatric Epidemiology 33:86–94.

DISCUSSION QUESTIONS

1. Do you think that involvement in conflicted social relationships would mediate the effect of mental disorder for persons with less severe mental disorder? Why or why not?

2. What are some alternate explanations for why persons with mental disorder are victimized?

3. This study showed that it is not mental disorder per se that increases victimization risk but that conflicted social relationships mediate the effects of mental disorder. Discuss the reasons that mentally disordered individuals are likely to be involved in conflicted social relationships. How can this likelihood be reduced?

4. How might the theoretical ideas presented in this article apply to the victimization of persons with disabilities? What other theories could apply to the victimization of persons with mental disorders?

Introduction to Reading 3

One of the pervasive notions about prisons is that they are violent places where inmates face an almost daily risk of victimization. This notion has only recently been empirically tested. Wolff and colleagues (2009a), in one of the most comprehensive examinations of inmates' victimization experiences (6,964 male and 564 female inmates from 13 prisons in a single state), examine the victimization experiences of both male and female inmates. Not only do they discover that victimization is prevalent, but they also discover that inmates who come into prison with a victimization history are more likely than other inmates to be victimized in prison.

Patterns of Victimization Among Male and Female Inmates

Evidence of an Enduring Legacy

Nancy Wolff, Jing Shi, and Jane A. Siegel

People inside prison are different in many ways from people without criminal histories. One difference, health disparities, has received increasing attention over the past dozen years (Baillargeon, Black, Pulvino, & Dunn, 2000; Freudenberg, 2002; Goff, Rose, Rose, & Purves, 2007; Hammet, Harmon, & Rhodes, 2002; Hammet, Roberts, & Kennedy, 2001; Teplin, Abram, & McClellan, 1996). According to this research, incarcerated people have higher rates of particular chronic and infectious disease (e.g., HIV/AIDS, hepatitis C, heart disease) and behavioral disorders (e.g., substance abuse disorders, depression schizophrenia, posttraumatic stress disorder). Another disparity receiving less but growing attention is their elevated rates of victimization both before and during incarceration.

Both men and women in prison have histories of interpersonal violence. Extant estimates suggest that at least half of incarcerated women have experienced at least one traumatic event in their lifetime (Browne, Miller, & Maguin, 1999; Sacks, 2004). Rates reported by men are lower by comparison but significant nonetheless (McClellan, Farabee, & Crouch, 1997). Childhood abuse is reported by 25% to 50% of incarcerated women (Bloom, Owen, & Covington, 2003; Bureau of Justice Statistics, 1999; Fletcher, Shaver, & Moon, 1993; Greenfeld & Snell, 1999) and by 6% to 24% of their male counterparts (Bureau of Justice Statistics, 1999; McClellan et al., 1997). Prior to age 18, physical abuse is more likely than sexual abuse for males, but the two forms of maltreatment occur at equal rates for females (Bureau of Justice Statistics, 1999; McClellan et al., 1997). Abuse in childhood is strongly correlated with adult victimization, substance abuse, and criminality (Browne et al., 1999; Chesney-Lind, 1997; Dutton & Hart, 1992; Goodman et al., 2001; Ireland & Widom, 1994; McClellan et al., 1997; Siegel & Williams, 2003a, 2003b; Smith & Thornberry, 1995; Widom, 1989a).

SOURCE: Wolff et al. (2009a). Reprinted with permission from Springer Publishing Company.

Victimization continues inside prison for many of these individuals. Correctional settings are known for their violence between inmates and between inmates and staff. The research evidence here also shows that rates of victimization are higher in prison settings than in the general community. Violent victimization rates, inclusive of robbery and sexual and physical assault, are estimated at approximately 21 per 1,000 in the community (Bureau of Justice Statistics, 2006). Rates of victimization for an incarcerated population are considerably higher. Using a sample of 581 male inmates drawn from three Ohio prisons, Wooldredge (1998) found that approximately one of every 10 inmates reported being physically assaulted in the previous 6 months, while one of every five inmates reported being a victim of theft during that same time frame. Aggregating all crime, one of every two inmates surveyed reported being a victim of crime in the previous 6 months. More recently, Wolff, Blitz, Shi, Siegel, and Bachman (2007), based on a sample of more than 7,000 inmates, reported 6-month inmate-on-inmate physical victimization rates at 21% for both female and male inmates—a rate 10 times higher than the overall victimization rate in the community.

Since passage of the Prison Rape Elimination Act (Bureau of Justice Statistics, 2004), sexual assault in prison has received more research attention. Rates of sexual victimization in America's prisons vary greatly, ranging from 41% to less than 1%. Based on a meta-analysis by Gaes and Goldberg (2004), a conservative "average" prevalence estimate of prison sexual assault was estimated at 1.9%. Wolff, Blitz, Shi, Bachman, and Siegel (2006) estimated sexual inmate-on-inmate victimization rates over a 6-month period at 3.2% for female inmates and 1.5% for male inmates. The risk of victimization doubled for female inmates who experienced sexual abuse prior to age 18. For male inmates, those who experienced sexual victimization prior to age 18 were approximately two to five times more likely to report sexual victimization inside prison during a 6-month time period than their counterparts who had no sexual victimization prior to age 18 (Wolff et al., 2007).

Previous research has focused on the experience of either physical or sexual victimization inside prison.

No attention, however, has focused on the relationship between the two forms of victimization inside prison or to specific types of childhood victimization. The literature shows a strong association among childhood victimization, delinquency, and criminality (Dutton & Hart, 1992; Ireland & Widom, 1994; McClellan et al., 1997; Siegel, 2000; Siegel & Williams, 2003a; Smith & Thornberry, 1995; Widom, 1995) and between childhood and adult victimization (Siegel & Williams, 2003b; Widom, 1989b) as well as gender patterns in the experiences of and reactions to victimization over the life cycle (Cutler & Nolen-Hoeksema, 1991; Nolen-Hoeksema, 1987, 1990; Toray, Coughlin, Vuchinich, & Patraicelli, 1991; Widom, 1989b, 1995).

This gap in the literature reflects in part the dearth of data on prisoners, particularly victimization of people inside prison, and in part that researchers specialize in types of victimization, that is, physical or sexual, not their conjointness. This article estimates rates of victimization for male and female inmates by type of perpetrator (inmate, staff, or either) and form of victimization (sexual, physical). Data for these estimates are based on a random sample of approximately 7,500 inmates housed in 12 adult male prisons and one adult female prison in a single state. The significance of the findings for practice and policy are discussed along with recommendations to improve the health and welfare of people inside prison.

▧ Participants

Of the 19,788 eligible to participate, a total of 6,964 men (mean age = 34.0 years, $SD = 7.9$) from male general population prisons and 564 woman (mean age 35.5 years, $SD = 6.8$) aged 18 or older were recruited and completed the survey. Sample bias was tested using demographic data from the information system of the prison system. Of the male general population sample, 80% were classified as non-White (with a mean age of 33.3), which is roughly equivalent to the general prison population (80% of males are non-White with a mean age of 34.3). Among female inmates, 67% were non-White (with a mean age of 35.5), which is also equivalent to the population data (67% of females are non-White with a mean age of 35.4). The percentage of the survey sample that was Hispanic (inclusive of White and non-White) was similar to the prison population as a whole (males: 16% vs. 15%; females: 9% vs. 10%), and the ages of that portion of the sample were comparable as well (males: 32.0 vs. 32.5; females: 35.0 vs. 33.7).

Instruments

The questions regarding sexual victimization were adapted from the National Violence Against Women and Men Survey (Tjaden & Thoennes, 2000). Physical violence was measured in the survey through the use of two general questions for two categories of perpetrator: inmates or staff members. Specifically, the question was "Have you been physically assaulted by an inmate (or staff member) within the past 6 months?" Behavior-specific questions about physical victimization were asked as well (e.g., "During the past 6 months, has another inmate [or staff member] slapped, hit, kicked, or bit you; choked or attempted to drown you; hit you with an object with the intent to do harm; beat you up; or threatened or harmed you with a knife or shank?"). Respondents who responded affirmatively to any one of the six questions were classified as experiencing physical victimization.

Sexual victimization was measured using one general question for each of two types of perpetrator (inmate or staff member): "Have you been sexually assaulted by (an inmate or staff member) within the past 6 months?" Then additional questions about specific types of sexual victimization were used (e.g., "During the past 6 months, has [another inmate or staff member] ever . . . touched you, felt you, or grabbed you in a way that you felt was sexually threatening or made you have sex by using force or threatening to harm you or someone close to you?"). Seven of the specific questions involving penetration or sexual acts were included in the category for nonconsensual sexual acts (e.g., "Has [another inmate or staff member] ever . . . made you have oral sex by using force . . . ?"). Three questions focused on abusive sexual contacts (e.g., "Has [another inmate or staff member] ever touched you, felt you, or grabbed you in a way that felt sexually threatening?"). Respondents who responded affirmatively to any one of the 11 questions were considered victims of sexual abuse.

At the end of the survey instrument, respondents were asked, sequentially, if prior to age 18 anyone, including a relative or friend, ever "choked or attempted to drown you, hit you with some object that left welts or caused bleeding, burned you with a match, cigarette, hot liquid, or any other hot object, threatened or harmed you with a knife or gun." Responses to these questions were used to measure physical victimization prior to age 18. In addition, they were asked another series of questions about sexual victimization prior to age 18, which included any experiences where anyone "touched, felt, or grabbed you in a way that you felt was sexually threatening, tried or succeeded in getting you to touch their genitals when you didn't want to, made you have sex by using force or threat of force, made you have oral or anal sex by using force or threat of force."

Analyses

Weighted analyses were conducted. Both sexual and physical victimization were measured by combining any positive response to either the general or the specific questions. Prevalence (expressed as a percentage) of sexual (physical) victimization measures the percentage of people in the population experiencing sexual (physical) victimization within a 6-month period. The 95% confidence intervals presented in Table 1 are equivalent to two-sided t tests for means or proportions based on Taylor expansion. The overlap of the confidence intervals between comparison groups suggests that the null hypothesis that the means or proportions are the same between comparison groups at a significance level of 0.05 cannot be rejected.

Results

In this section, the victimization experiences of male and female inmate respondents are described for a 6-month time period. Respondents may have experienced in-prison physical victimization, sexual victimization, or both. They may have been victimized by other inmates (inmate on inmate), staff (staff on inmate), or either (inmate or staff).

Prevalence of Victimization

Six-month prevalence rates (expressed as a percentage) by type of perpetrator and victimization appear in Table 1. Several patterns are noteworthy. First, while percentages of inmate-on-inmate physical victimization are equal for male and female inmates (21%), male inmates report a significantly higher percentage of physical victimization by staff than do females (25% vs. 8%). Second,

female inmates report roughly equal percentages of sexual (24%) and physical (24%) victimization independent of type of perpetrator. By contrast, males are far more likely to experience physical than sexual victimization (35% vs. 10%). Third, sexual victimization is more common between female inmates (inmate on inmate) than between female inmates and staff (staff on inmate), but the reverse is true for male inmates. Fourth, nearly

Table 1 Six-Month Prevalence Estimates (Expressed as a Percent) for Physical and Sexual Victimization by Gender

Source of In-Prison Victimization	Male Facilities (*N* = 6,964)	Female Facility (*N* = 564)
Physical victimization, 6 month		
Inmate on inmate[a]	20.7 (19.6–21.8)	20.7 (17.2–24.1)
Staff on inmate[b]	25.2 (24.0–26.3)*	8.3 (6.0–10.7)
Either inmate on inmate or staff on inmate	35.3 (34.0–36.5)*	24.0 (20.4–27.6)
Sexual victimization, 6 month		
Inmate on inmate	4.3 (3.8–4.8)*	21.3 (17.8–24.8)
Staff on inmate	7.6 (7.0–8.3)	7.7 (5.5–9.9)
Either inmate on inmate or staff on inmate	10.3 (9.6–11.1)*	24.5 (20.9–28.2)
Physical or sexual victimization, 6 month[d]		
Inmate on inmate	22.3 (21.2–23.4)*	32.3 (28.3–36.3)
Staff on inmate	28.0 (26.8–29.1)*	12.8 (10.1–15.6)
Either inmate on inmate or staff on inmate	38.4 (37.1–39.7)	36.9 (32.8–41.1)
Physical and sexual victimization, 6 month[d]		
Inmate on inmate	2.6 (2.2–2.9)*	9.5 (7.1–12.0)
Staff on inmate	4.7 (4.2–5.2)	3.1 (1.7–4.6)
Either inmate on inmate or staff on inmate	7.0 (6.4–7.6)*	11.4 (8.7–14.1)

NOTE. Estimates are based on a weighted valid percentage.

[a]Denotes inmate perpetrator and inmate victim.

[b]Denotes staff perpetrator and inmate victim.

[c]Denotes either inmate or staff perpetrator and inmate victim.

[d]Physical or sexual victimization includes unduplicated counts of inmates who reported either sexual victimization or physical victimization during the 6-month period. Physical and sexual victimization includes unduplicated counts of inmates reporting both physical and sexual victimization during the 6-month period. Together, they equal the number of unduplicated counts of inmates reporting victimization, either physical or sexual. Consequently, percentages for the specific forms of victimization (physical, sexual) add to the percentages of conjoint forms of victimization (physical or sexual and physical and sexual). Deviations are due to rounding error.

*Statistically significant difference between males and females ($p < .05$).

40% of male and female inmates experience some form of victimization during a 6-month period. Finally, while it is relatively uncommon for either male or female inmates to experience both physical and sexual victimization, still 7% of males (18% of those reporting any victimization) and 11% of females (31% of those reporting any victimization) report experiencing both forms of victimization by other inmates, staff, or both inmates and staff over a 6-month period.

Patterns of Victimization

Patterns of victimization for male and female inmates reporting victimization during a 6-month period are shown in Table 2. First, looking at the male facilities, approximately one-fifth (22%) of the male inmates reporting at least one incident of physical victimization also reported an incident of sexual victimization either by staff or another inmate. Half the male inmates (15%) physically victimized by other inmates (inmate on inmate) also reported being physically victimized by staff (staff on inmate). Nearly three-quarters of male inmates reporting either inmate-on-inmate or staff-on-inmate sexual victimization also report experiencing physical victimization by another inmate or staff.

Conjoint victimization is also common in the female facility. Roughly half the female inmates reporting either inmate-on-inmate or staff-on-inmate physical

Table 2   Percentage With Overlapping Physical and Sexual Victimization, Male and Female Facilities

Type of In-Prison Victimization	Victimization Experience Over a 6-Month Period					
	Percentage Also Experiencing Sexual Harm			Percentage Also Experiencing Physical Harm		
	Inmate on Inmate[a]	Staff on Inmate[b]	Either Inmate on Inmate or Staff on Inmate[c]	Inmate on Inmate	Staff on Inmate	Either Inmate on Inmate or Staff on Inmate
Male facilities (N = 12)						
Physical victimization, 6 month						
Inmate on inmate (N = 1,419)	12.6	15.0	22.2	—	50.8	—
Staff on inmate (N = 1,725)	9.5	19.0	23.2	41.8	—	—
Sexual victimization, 6 month						
Inmate on inmate (N = 295)	—	37.1	—	60.2	55.2	72.7
Staff on inmate (N = 519)	20.9	—	—	40.4	62.2	70.4
Female facility (N = 1)						
Physical victimization, 6 month						
Inmate on inmate (N = 116)	46.5	16.2	48.3	—	24.3	—
Staff on inmate (N = 47)	51.6	37.5	63.3	60.2	—	—
Sexual victimization, 6 month						
Inmate on inmate (N = 119)	—	21.1	—	44.9	20.3	49.3
Staff on inmate (N = 43)	57.9	—	—	43.6	41.3	53.7

NOTE. Estimates are based on a weighted valid percentage.

[a]Denotes inmate perpetrator and inmate victim.

[b]Denotes staff perpetrator and inmate victim.

[c]Denotes either inmate or staff perpetrator and inmate victim.

victimization also report experiencing sexual victimization by other inmates. More than half the female inmates (60%) who report staff-on-inmate physical victimization reported inmate-on-inmate physical victimization as well. Similarly, of those who reported sexual victimization by either another inmate or staff, half reported experiencing physical victimization during the same 6 months.

Victimization Histories

Table 3 shows the pattern between in-prison victimization and childhood victimization by type of

perpetrator and victimization. Before discussing differences between in-prison and childhood victimization, it is noteworthy that half or more of all male and female inmates reported childhood physical victimization. More specifically, 56% of all male inmates experienced physical abuse as children, as did 54% of all female inmates. By contrast, slightly less than 10% of all male inmates, compared to 47% of all female inmates, reported childhood sexual victimization. In general, inmates, both male and female, who experienced victimization inside prison, independent of type of victimization and perpetrator, were significantly more likely to report having experienced victimization in

Table 3   Percentage With Current and Previous Victimization by Type of Victimization and Gender

Source of Victimization	Male Facilities		Female Facility	
	In-Prison Physical Victimization Experienced at Least Once in Past 6 Months			
	Yes	No	Yes	No
Inmate on inmate[a]	N = 1,419	N = 5,440	N = 116	N = 446
Prior physical victimization < 18	67.4*	54.0	65.7*	49.5
Prior sexual victimization < 18	14.0*	8.7	56.1*	43.7
Staff on inmate[b]	N = 1,725	N = 5,131	N = 47	N = 516
Prior physical victimization < 18	63.2*	54.5	61.9	52.0
Prior sexual victimization < 18	10.8	9.5	53.2	45.6
	In-Prison Sexual Victimization Experienced at Least Once in Past 6 Months			
Inmate on inmate	N = 295	N = 6,545	N = 119	N = 441
Prior physical victimization < 18	64.0*	56.3	69.1*	48.4
Prior sexual victimization < 18	27.2*	9.0	61.8*	41.9
Staff on inmate	N = 519	N = 6,300	N = 43	N = 517
Prior physical victimization < 18	66.6*	55.9	65.4	51.7
Prior sexual victimization < 18	15.6*	9.3	56.8	45.1

NOTE. Estimates are based on a weighted valid percentage.

[a]Denotes inmate perpetrator and inmate victim.

[b]Denotes staff perpetrator and inmate victim.

*Statistically significant difference between victimization group and no victimization group with chi-square test ($p < .05$).

childhood. Specifically, roughly two-thirds of male and female inmates experiencing victimizations in prison reported being physically victimized prior to age 18 compared to roughly half of those who did not report experiencing victimization during the past 6 months. Similarly, percentages of childhood sexual victimization were also higher among those reporting in-prison physical victimization (but differences were not always significant in part because of small numbers in the comparison groups), although the prevalence of childhood sexual victimization was lower than childhood physical victimization.

Patterns for in-prison sexual victimization are similar. Childhood sexual victimization is slightly but significantly more prevalent in inmates experiencing in-prison sexual victimization than in inmates experiencing in-prison physical victimization. Male and female inmates experiencing in-prison sexual victimization were statistically significantly more likely to have experienced both childhood sexual and physical victimization than those not victimized in prison, except in the case of staff-perpetrated sexual acts in the female facility.

Discussion

The patterns presented herein show, in general, that prison is a harmful place for many of the people residing there, that people inside prison have high percentages of childhood victimization (both of these patterns are consistent with extant research), that people who were victimized inside prison by other inmates and/or staff are more likely to have experienced victimization prior to age 18, that people who experience sexual victimization inside prison are also very likely to experience physical victimization, that males who experience physical victimization by other inmates are also likely to experience similar victimization by correctional staff, and that females who experience physical victimization by staff also report physical victimization by other inmates (but not vice versa).

Some very distinct differences were found in the patterns of victimization among male and female inmates. Male inmates, relative to their female counterparts, reported significantly higher percentages of physical victimization perpetrated by staff, although percentages of inmate-on-inmate physical victimization were equal for male and female inmates. This suggests gender-patterned interactions between inmate and staff in which (a) male inmates, compared to female inmates, are more aggressive against authority figures, resulting in physical altercations with staff; (b) staff is more willing to use physical force against male inmates than female inmates; or (c) some combination of both. This warrants further investigation. In contrast, sexual victimization between inmates was more common among female than male inmates, but their percentages of sexual victimization perpetrated by staff were roughly equal. This gender difference in inmate-on-inmate victimization percentages is explained by differences in inappropriate touching, not sexual assault (i.e., rape). Compared to male inmates, female inmates were significantly more likely to report that other inmates touched, felt, or grabbed them in sexually threatening ways.

Before discussing the significance of our results, it is important to note their limitations. First, we measured prevalence of victimization within a 6-month period. Individuals who did not report physical or sexual victimization inside prison may have experienced such trauma prior to the 6-month period. A 6-month reflection period was adopted for several reasons. First, one of our goals was to test for interfacility variation in rates of victimization. Because it is common for inmates to move between and among facilities, a 6-month reflection period, compared to a 12-month period, was expected to yield more stability within the denominator of the prevalence rates for each facility. Relatedly, the literature suggests that the likelihood of victimization is greatest in the first 6 months at a facility (see Hensley, Castle, & Tewksbury, 2003; Hensley, Koscheski, & Tewksbury, 2005). By limiting the reflection period to 6 months, we had the ability to test this as a hypothesis. (Testing for interfacility variation and the effect of time in the facility is outside the scope of this article.) In addition, given the frequency of violence expected inside prison, especially in terms of physical victimization, more reliable reporting is expected from shorter reflection periods. For these reasons, we used a conservative test (6-month exposure window at the current facility) to explore victimization inside prisons.

Second, sample bias is possible. Our samples ranged from 26% to 53% of the general population among 13 facilities. Nonrepresentativeness was tested in terms of age, race/ethnicity, and length of incarceration and adjusted for in the weighting strategy, yet these characteristics may not fully predict variation in victimization within or across facilities. We account for such uncertainty by estimating confidence intervals that provide a reasonable (95%) approximation of the range of variation in percentage of victimization. Third, biased reporting may have occurred. Audio-CASI is the most reliable method for collecting information about activities or events that are shaming or stigmatizing. Bias also may arise from the intent to make the facility and its staff members look bad. To improve accuracy and reliability, the consent process stressed the importance of accurate reporting and its impact on the legitimacy of the data and survey; we surveyed respondents by units and rapidly over a 2- to 4-day period, staff did not have access to the survey questions, and the victimization questions were nested deep into a general survey focusing on quality of prison life. Systematic false reporting of victimization would have yielded much higher and clustered rates than those reported here. Our data indicated systematic enterprising differences in victimization rates as well as in prison conditions that were consistent with the "reputations" of the individual facilities. Variation in inmate responses across a variety of survey questions both within and across prisons lends credibility to the data (Camp, 1999). However, research has found that many known victims of child sexual abuse fail to report their victimization when asked about childhood experiences as adults (Williams, 1994), so the rates reported here actually may be conservative.

Fourth, our estimates are based on a single state correctional system. There was evidence of variation in physical victimization rates by facility, with rates highest for facilities with a younger population. It is only by collecting data on all prisons within a correctional system that we could have identified this pattern. However, whether our rates are representative of other state correctional systems is an empirical issue that warrants further research.

Our findings suggest that victimization inside prison warrants an increase in research, clinical, and administrative attention. Prison is a breeding ground for traumatization and retraumatization. Most people come to prison with a legacy of victimization, which, as the evidence clearly shows, elevates their risk for drug and/or alcohol abuse, posttraumatic stress disorder, depression, low self-esteem, and criminality *before* the experience of incarceration (Goff et al., 2007; McClellan et al., 1997; Mullings, Hartley, & Marquart, 2004).

The prison experience itself is likely to activate and exacerbate past trauma. The culture and climate of the prison environment may itself trigger unwelcome memories of prior victimization and provoke symptoms and create opportunities for (re)victimization (Wortley, 2002). More specifically, the ecology of prison environments may produce conditions that support or encourage victimization. Camp, Gaes, Langan, and Saylor (2003) argue that "it would be naïve to assume that inmate behavior is independent of other behavior occurring in the same institution" (p. 504). Bottoms (1999), using an interactionist framework, advances a view that each prison is a microsocial organization with a unique culture that influences the prison's operation, particularly how social control is wielded inside the facility. Using a situational perspective of prisons, Wortley (2002) elaborates on the interactionist framework, arguing that "behavior can only be understood in terms of an interaction between the characteristics of the actor and the characteristics of the environment in which an act is performed" (p. 3). Together, this suggests that to understand and prevent victimization inside prison, the incident must be contextualized to and remedied within the specific environment. Additional research is needed that explores how social disorder factors (such as incivilities between inmates and inmates and officers) within prisons affect fear and rates of victimization (Edgar, O'Donnell, & Martin, 2003).

Overall, the potential number of individuals liable to suffer harmful consequences as a result of victimization inside prisons can be expected to be large for several reasons. The first concerns the psychological impact of prison conditions on inmates. Haney (2003), in describing the psychological impact of incarceration, notes "that the harsh, punitive, and uncaring nature of prison life may represent a kind of 're-traumatization' experience for many ... [such that] time spent in prison may rekindle not only the memories but the disabling psychological reactions and consequences of these earlier damaging experiences" (p. 52). Added to this is the fact that a large minority of people in prison will be physically victimized by other inmates and by staff (an abusing authority figure) within a 6-month period and that a smaller but significant minority will experience sexual victimization, some even in addition to physical assault. Furthermore, even those who are not directly victimized may well be witnesses to the violence perpetrated inside prisons, and research has found that even such passive activity is associated with emotional and behavioral effects similar to those found among direct victims of violence (Buka, Stichick, Birdthistle & Earls, 2001; Fitzpatrick & Boldizar, 1993; Kitzmann, Gaylord, Holt, & Kenny, 2003; Nofziger & Kurtz, 2005).

Preventing victimization requires a two-pronged approach. First, it requires changing the prison environment in ways that reduce the opportunities for victimization and eliminate the conditions that encourage predatory and traumatizing behavior. Understanding the social dynamics inside prison as well as the routines and architectural conditions that promote hostile and harmful environments is of paramount importance (Bottoms, 1999; Edgar et al., 2003; Wortley, 2002). Second, effective diagnosis and treatment of trauma among inmates is required. Researchers have noted the failure of correctional facilities for women to take women's victimization experiences into account when considering their programming needs (Bloom et al., 2003; Morash, Bynum, & Koons, 1998). Results from this study suggest that attention should be paid as well to men's victimization because both men and women are susceptible to posttraumatic stress disorder as a result of being victimized.

The prevalence of full or subthreshold posttraumatic stress disorder (Grubaugh et al., 2005) inside prison has received very little research attention (Goff et al., 2007), and its impact on inmate health and the quality of life inside prison has been largely ignored. This is problematic in part because in a social psychological sense, prison is a stressor, like any chronic strain, that is very likely to cause stress reactions within an individual. Stressors, according to an abundance of health research, are connected to a wide variety of behavioral and health-related problems requiring treatment (Thoits, 1995). Furthermore, individuals generally manage stress reactions through learned coping styles, which may include self-harm, self-medicating with drugs or alcohol, withdrawal, or externalized aggression—all of which make managing the prison more difficult, expensive, and risky (Haney, 2006; McClellan et al., 1997).

While victims of trauma internalize it differently, some of the more common responses are dissociation, affect dysregulation, chronic characterological changes, somatization, and hyperarousal (Harris & Fallot, 2001; Kluft, Bloom, & Kinzie, 2000; Rosenberg, Mueser, Friedman, & Gorman, 2001; Sacks, 2004). Trauma-related psychological difficulties are amenable to intervention, and an array of interventions exist (Harris & Fallot, 2001). In considering interventions most suitable for correctional settings, some general guidance can be gleaned from the literature. First, integrated treatment for comorbid conditions is considered optimal, compared to parallel, sequential, or single treatment models (Harris & Fallot, 2001; Mueser, Drake, Sigmon, & Brunette, 2005). Second, trauma-related difficulties are best treated in stages (Herman, 1992), with the first stage focusing on safety through recognition, education, and skills to replace the use of drugs or alcohol or other self-harming behaviors. Later stages of trauma recovery focus on processing the trauma directly after the person has achieved stable functioning. Third, trauma interventions must be sensitive to environment (Harris & Fallot, 2001). Trauma processing therapies (e.g., exposure therapy, cognitive restructuring), while efficacious, require environments that are healing and supportive (Bradley, Greene, Russ, Durta, & Westen, 2005; Van Etten & Taylor, 1998). Correctional settings are not healing. To the contrary, they often evoke memories of environments where trauma occurred in the past.

A variety of integrated, skill-based, gender-sensitive (first stage) approaches have been developed to promote trauma recovery and to treat substance abuse disorder in people with and without co-occurring mental illness (Hien, Cohen, Miele, Litt, & Capstick, 2004; Najavits, Sonn, Walsh, & Weiss, 2004; Rosenberg et al., 2001). These approaches have been found to be generally effective in quasi-experimental or small pilot studies, often without randomization (Finkelstein et al., 2004; Jennings, 2004). Some trauma recovery treatments have been pilot-tested in correctional settings and were found to be generally effective (Zlotnick, Johnson, & Najavits, in press; Zlotnick, Najavits, Rohsenow, & Johnson, 2003).

Trauma and retraumatization are part of prison reality. The issue remaining is how best to respond. Ignoring the evidence (i.e., doing nothing) courts higher costs, greater safety and health risks, and, in view of the evidence documenting a relationship between the experience of being victimized and subsequent violence perpetration (Siegel, 2000; Widom, 1989c), perpetuates behaviors that elevate recidivism on release. Treating the symptoms of new incidents of inmate-on-inmate or staff-on-inmate victimization without changing the prison culture and climate is likely to be better than nothing but probably as effective as treating combat trauma in a war zone. Real value for the treatment dollar would require changing the prison environment in ways that make it more humanizing, more healthy, more habitable, and hence, more in keeping with the long-term societal goals and expectations of public safety and rehabilitation.

◼ References

Baillargeon, J., Black, S. A., Pulvino, J., & Dunn, K. (2000). The disease profile of Texas prison inmates. *Annals of Epidemiology, 10*(2), 74–80.

Bloom, B., Owen, B., & Covington, S. (2003). *Gender-responsive strategies: Research, practice, and guiding principles for women offenders.* Washington, DC: National Institute of Corrections, U.S. Department of Justice.

Bottoms, A. E. (1999). Interpersonal violence and social order in prisons. In M. Tonry & J. Petersilia (Eds.), *Prisons* (pp. 205–282). Chicago: University of Chicago Press.

Bradley, R., Greene, J., Russ, E., Durta, L., & Westen, D. (2005). A multidimensional meta-analysis of psychotherapy for PTSD. *American Journal of Psychiatry, 162*, 214–227.

Browne, A., Miller, A., & Maguin, E. (1999). Prevalence and severity of lifetime physical and sexual victimization among incarcerated women. *International Journal of Law Psychiatry, 22*, 301–322.

Buka, S. L., Stichick, T. L., Birdthistle, I., & Earls, F. J. (2001). Youth exposure to violence: Prevalence, risks and consequences. *American Journal of Orthopsychiatry, 71*, 298–310.

Bureau of Justice Statistics. (1999). *Prior abuse reported by inmates and probationers* (NCJ 172879). Washington, DC: U.S. Department of Justice.

Bureau of Justice Statistics. (2004). *Data collections for the Prison Rape Elimination Act of 2003.* Washington, DC: U.S. Department of Justice.

Bureau of Justice Statistics. (2006). *Criminal victimization, 2003* (NCJ 214644). Washington, DC: U.S. Department of Justice.

Camp, S. D. (1999). Does inmate survey data reflect inmate conditions? Using surveys to assess prison conditions of confinement. *The Prison Journal, 79*, 250–268.

Camp, S. D., Gaes, G. G., Langan, N. P., & Saylor, W. G. (2003). The influence of prisons on inmate misconduct: A multilevel investigation. *Justice Quarterly, 20*, 501–533.

Chesney-Lind, M. (1997). *The female offender: Girls, women, and crime.* Thousand Oaks, CA: Sage.

Cutler, S. E., & Nolen-Hoeksema, S. (1991). Accounting for sex difference in depression through female victimization: Childhood sexual abuse. *Sex Roles, 24*, 425–438.

Dutton, D., & Hart, S. (1992). Evidence for long-term, specific effects of childhood abuse and neglect on criminal behavior in men. *International Journal of Offender Therapy and Comparative Criminology, 36*, 129–137.

Edgar, K., O'Donnell, I., & Martin, C. (2003). *Prison violence: The dynamics of conflict, fear, and power.* Devon: Willan Publishing.

Finkelstein, N., VandeMark, N., Fallot, R., Brown, V., Cadiz, S., & Heckman, J. (2004). *Enhancing substance abuse recovery through integrated trauma treatment.* Saratoga, FL: National Trauma Consortium.

Fitzpatrick, K. M., & Boldizar, J. P. (1993). The prevalence and consequences of exposure to violence among African-American youth. *Journal of the American Academy of Child and Adolescent Psychiatry, 32*, 424–430.

Fletcher, B. R., Shaver, L. D., & Moon, D. G. (1993). *Women prisoners: A forgotten population.* Portsmouth, NH: Greenwood Press.

Freudenberg, N. (2002). Adverse effects of U.S. jail and prison policies on the health and well-being of women of color. *American Journal of Public Health, 92*, 1895–1899.

Gaes, G. G., & Goldberg, A. L. (2004). *Prison rape: A critical review of the literature.* Washington, DC: National Institute of Justice.

Goff, A., Rose, E., Rose, S., & Purves, D. (2007). Does PTSD occur in sentenced prison population? A systematic literature review. *Criminal Behavior and Mental Health, 17,* 152–162.

Goodman, L. A., Salyers, M. P., Mueser, K. T., Rosenberg, S. D., Swartz, M., Essock, S. M., et al. (2001). Recent victimization in women and men with severe mental illness: Prevalence and correlates. *Journal of Trauma Stress, 14,* 615–632.

Grubaugh, A. L., Magruder, K. M., Waldrop, A. E., Elhai, J. D., Knapp, R. G., & Frueh, B. C. (2005). Subthreshold PTSD in primary care: Prevalence, psychiatric disorders, healthcare use, and functional status. *Journal of Nervous Mental Disorders, 193,* 658–664.

Greenfeld, L. A., & Snell, T. L. (1999). *Women offenders.* Washington, DC: U.S. Department of Justice, Office of Justice Programs, Bureau of Justice Statistics.

Hammet, T. M., Harmon, M. P., & Rhodes, W. (2002). The burden of infectious disease among inmates of and releasees from U.S. correctional facilities, 1997. *American Journal of Public Health, 92,* 1789–1794.

Hammet, T. M., Roberts, C., & Kennedy, S. (2001). Health-related issues in prisoner reentry. *Crime & Delinquency, 47,* 390–409.

Haney, C. (2003). The psychological impact of incarceration: Implications for post-prison adjustment. In J. Travis & M. Waus (Eds.), *Prisoners once removed: The impact of incarceration and reentry on children, families, and communities* (pp. 33–66). Washington, DC: Urban Institute Press.

Haney, C. (2006). *Reforming punishment: Psychological limits to the pains of imprisonment.* Washington, DC: American Psychological Association Books.

Harris, M., & Fallot, R. D. (2001). Designing trauma-informed addictions services. *New Directions in Mental Health Services, 89,* 57–73.

Hensley, C., Castle, T., & Tewksbury, R. (2003). Inmate-to-inmate sexual coercion in a prison for women. *Journal of Offender Rehabilitation, 37,* 77–87.

Hensley, C., Koscheski, M., & Tewksbury, R. (2005). Examining the characteristics of male sexual assault targets in a southern maximum-security prison. *Journal of Interpersonal Violence, 20,* 667–679.

Herman, J. L. (1992). *Trauma and recovery: The aftermath of violence— From domestic abuse to political terror.* New York: Basic Books.

Hien, D. A., Cohen, L. R., Miele, G. M., Litt, L. C., & Capstick, C. (2004). Promising treatments for women with comorbid PTSD and substance abuse. *American Journal of Psychiatry, 161,* 1426–1432.

Ireland, T., & Widom, C. S. (1994). Childhood victimization and risk for alcohol and drug arrests. *International Journal of the Addictions, 29,* 235–274.

Jennings, A. (2004). *Models for developing trauma-informed behavioral health systems and trauma-specific services.* Alexandria, VA: National Technical Assistance Center for State Mental Health Program Directors.

Kitzmann, K. M., Gaylord, N. K., Holt, A. R., & Kenny, E. D. (2003). Child witnesses to domestic violence: A meta-analytic review. *Journal of Consulting and Clinical Psychology, 71,* 339–352.

Kluft, R. P., Bloom, S. L., & Kinzie, J. D. (2000). Treating traumatized patients and victims of violence. *New Directions in Mental Health Service, 86,* 79–102.

McClellan, D. S., Farabee, D., & Crouch, B. M. (1997). Early victimization, drug use, and criminality: A comparison of male and female prisoners. *Criminal Justice and Behavior, 24,* 455–467.

Morash, M., Bynum, T., & Koons, B. A. (1998). Women offenders: Programming needs and promising approaches. *Research In Brief.* Washington, DC: National Institute of Justice.

Mueser, K. T., Drake, R. E., Sigmon, S. C., & Brunette, M. R. (2005). Psychosocial interventions for adults with severe mental illness and co-occurring substance use disorders: A review of specific interventions. *Journal of Dual Disorders, 1,* 57–82.

Mullings, J. L., Hartley, D. J., & Marquart, J. W. (2004). Exploring the relationship between alcohol use, childhood maltreatment, and treatment needs among female prisoners. *Substance Use and Misuse, 39,* 277–305.

Najavits, L. M., Sonn, J., Walsh, M., & Weiss, R. D. (2004). Domestic violence in women with PTSD and substance abuse. *Addiction Behavior, 29,* 707–715.

Nofziger, S., & Kurtz, D. (2005). Violent lives: A lifestyle model linking exposure to violence to juvenile violent offending. *Journal of Research in Crime and Delinquency, 42,* 3–26.

Nolen-Hoeksema, S. (1987). Sex differences in unipolar depression: Evidence and theory. *Psychological Bulletin, 201,* 259–282.

Nolen-Hoeksema, S. (1990). *Sex differences in depression.* Stanford, CA: Stanford University Press.

Rosenberg, S. D., Mueser, K. T., Friedman, M. J., & Gorman, P. G. (2001). Developing effective treatments for posttraumatic disorder among people with severe mental illness. *Psychiatric Services, 52,* 1453–1461.

Sacks, J. Y. (2004). Women with co-occurring substance use and mental disorders (COD) in the criminal justice system: A research review. *Behavior Sciences and the Law, 22,* 449–466.

Siegel, J. A. (2000). Aggressive behavior among women sexually abused as children. *Violence and Victims, 15,* 235–255.

Siegel, J. A., & Williams, L. M. (2003a). The relationship between child sexual abuse and female delinquency and crime: A prospective study. *Journal of Research in Crime and Delinquency, 40,* 71–94.

Siegel, J. A., & Williams, L. M. (2003b). Risk factors for sexual victimization of women: Results from a prospective study. *Violence Against Women, 9,* 902–930.

Smith, C., & Thornberry, T. P. (1995). The relationship between childhood maltreatment and adolescent involvement in delinquency. *Criminology, 33,* 451–481.

Teplin, L., Abram, J. M., McClellan, G. M. (1996). Prevalence of psychiatric disorders among incarcerated women: Pretrial jail detainees. *Archives of General Psychiatry, 53,* 505–512.

Thoits, P. A. (1995). Stress, coping, and social support processes: Where are we? What next? *Journal of Health and Social Support, 35,* 53–79.

Tjaden, P., & Thoennes, N. (2000). *Full report of the prevalence, incidence, and consequences of violence against women: Findings from the National Violence Against Women Survey* (NCJ 183781).

Washington, DC: National Institute of Justice and Centers for Disease Control and Prevention.

Toray, T., Coughlin, C., Vuchinich, S., & Patraicelli, P. (1991). Gender differences associated with adolescent substance abuse: Comparisons and implications for treatment. *Family Relations, 40,* 338–344.

Van Etten, M. L., & Taylor, S. (1998). Comparative efficacy of treatments for posttraumatic stress disorder: A meta-analysis. *Clinical Psychological Review, 5,* 126–144.

Widom, C. S. (1989a). Child abuse, neglect, and violent criminal behavior. *Criminology, 27,* 251–271.

Widom, C. S. (1989b). Does violence beget violence? A critical examination of the literature. *Psychological Bulletin, 106,* 3–28.

Widom, C. S. (1989c). The cycle of violence. *Science, 244,* 160–166.

Widom, C. S. (1995). *Victims of childhood sexual abuse: Later criminal consequences.* Washington, DC: U.S. Department of Justice.

Williams, L. M. (1994). Recall of childhood trauma: A prospective study of women's memories of child sexual abuse. *Journal of Urban Health, 83,* 835–848.

Wolff, N., Blitz, C., Shi, J., Bachman, R., & Siegel, J. (2006). Sexual violence inside prison: Rates of victimization. *Journal of Urban Health, 83,* 835–848.

Wolff, N., Blitz, C., Shi, J., Siegel, J., & Bachman, R. (2007). Physical violence inside prison: Rates of victimization. *Criminal Justice and Behavior, 34,* 588–599.

Wooldredge, J. D. (1998). Inmate lifestyles and opportunities for victimization. *Journal of Research in Crime and Delinquency, 35,* 480–502.

Wortley, R. (2002). *Situational prison control: Crime prevention in correctional institutions.* Cambridge: Cambridge University Press.

Zlotnick, C., Johnson, J., & Najavits, L. M. (in press). Randomized controlled pilot study of incarcerated women with substance use.

Zlotnick, C., Najavits, L. M., Rohsenow, D. J., & Johnson, D. M. (2003). A cognitive-behavioral treatment for incarcerated women with substance abuse disorder and posttraumatic stress disorder: Findings from a pilot study. *Journal of Substance Abuse Treatment, 25,* 99–105.

DISCUSSION QUESTIONS

1. There are gender differences in terms of who is likely to perpetrate physical and sexual violence against female and male inmates. Why do you think these gender differences exist?

2. How does previous victimization impact risk of victimization in prison? Is the explanation for the link between past victimization and current victimization for prisoners similar to that for persons outside prison? Explain.

3. What are the implications of this article's findings for prison administrators? For persons working in the criminal justice system more generally? For the community when these inmates are released from prison?

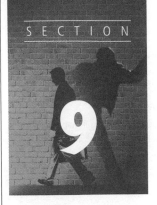

Victimization at School and Work

P olly, the young woman first introduced in Section 2, was walking home from a local bar when she was accosted by two men. She was shoved by one of the men, and they were able to take her bag and its contents. Polly suffered several negative consequences from her victimization—most obviously, the 10 stitches in her head. But she also had a hard time coping with what happened. She found herself staying in bed when she should have been going to class. As a college student, this negative consequence could have prevented her from successfully completing the semester. So far, you have considered Polly's victimization in terms of why she was victimized, whether she meets the criteria for being a "typical" victim, and the types of services to which Polly should have access. What you should also consider, however, is whether Polly's victimization would be included in school victimization statistics. She is, after all, a college student. Is this designation enough for her victimization to be classified or counted as a victimization at school? Does it matter that she was not in class at the time? Would she need to be on campus for it to count as a school victimization? What would the college have to do if this was considered a school victimization? These issues will be discussed in this section along with another special case of victimization, those that occur while people are at work or on duty.

Victimization at School

We often think of school as being a safe place. Schools are designed as places where young people come together to learn and grow. Attendance is not voluntary; instead, children are required to go to school, and it is the place where they spend a great deal of their lives. Parents assume that when they send their children off to school they will willingly go and that, even if they do not love school, they are at least safe. For the most part, this is true, but not all students matriculate through school without experiencing some type of victimization. When they are victimized in the school building, on school grounds, while riding a school bus, or while attending or participating in a school function, it is termed **school victimization**. This type of victimization can take the form of any other victimization—property victimization or personal victimization that can include theft, robbery, simple assault, aggravated assault, rape, and murder. Other victimizations may include harassment or bullying—a focus of recent research that is discussed in more detail below.

▲ **Photo 9.1** On April 20, 1999, Columbine High School experienced one of the worst school shootings in U.S. history. Twelve students and one teacher were shot dead by two students who also took their own lives. Although the reasons behind the shootings cannot be known for certain, both boys experienced bullying at school.

Photo credit: Seraphimblade

▧ Victimization at School: Grades K–12

The type of victimization that likely comes to mind when you think about children being harmed at school is a school shooting. When a school shooting—such as the one at Columbine in 1999—does occur, it is difficult to watch the news or read the newspaper without hearing about the incident. Indeed, the media pay close attention when young people are shot at school. Although they get heightened media attention, fortunately, school shootings are very rare. But how do we know this? The information on victimizations that occur at school comes from a variety of resources. Remember that the National Crime Victimization Survey (NCVS) asks individuals aged 12 and over about their victimization experiences during the previous 6 months. If a person has been victimized, he or she is then asked detailed questions about the victimization, including where it occurred. From the NCVS, then, it is possible to generate estimates of the extent of victimization occurring at school for people at least 12 years of age. To supplement the NCVS, the School Crime Supplement Survey is a national survey of about 6,500 students aged 12 to 18 enrolled in schools in the United States. Students are asked about victimizations they experienced in the previous 6 months (Robers, Zhang, Truman, & Snyder, 2010). Keep in mind, though, that students younger than 12 are not included in this survey. In addition, other nationally representative surveys are commonly used to determine the extent to which students, staff, and teachers are victimized at school. For example, a survey has been implemented to assess safety and crime at schools and is filled out by school principals or those knowledgeable about discipline (School Survey on Crime and Safety 1999–2000, 2003–2004, 2005–2006, and 2007–2008 school years); another is filled out by teachers in elementary, middle, junior, and high schools (Schools and Staffing Survey 1993–1994, 1999–2000, 2003–2004, and 2007–2008 school years); and another is completed by students enrolled in grades 9 through 12 in public and private schools (Youth Risk Behavior Surveillance System, 1993–2009 biennially).

From these surveys and other official data sources, we have a fairly good idea as to what types of victimization experiences occur at school. In 2008, about 1.2 million nonfatal violent crimes occurred against children aged 12 to 18. Importantly, more nonfatal violent crimes occurred against this age group *at* school then *away* from school (Robers et al., 2010). It is estimated that 4% of students experienced some type of victimization in the previous 6 months. The most common type of victimization they experienced was theft—less than 0.5% experienced a violent victimization (Robers et al., 2010). For students enrolled in grades 9 through 12, violent victimization was more common—in 2009, 11% reported being in a physical fight on school property, and 8% reported being threatened or injured with a weapon on school property (Centers for Disease Control and Prevention, 2010). Although relatively uncommon, violent death also can occur at school. Between July 1, 2008, and June 30, 2009, 15 students were victims of homicide and 7 were victims of suicide at school. Violent deaths at schools are tracked through the School-Associated Violent Deaths Surveillance Study, the Supplementary Homicide Reports, and the Web-Based Injury Statistics Query and Reporting System.

Who Is Victimized?

Similar to most other types of victimization, except for sexual assault and rape, male students are more likely to be victimized at school than are female students. Ten percent of male students in grades 9 through 12 reported being threatened or injured with a weapon on school property during the previous year, compared to only 5% of female students (Centers for Disease Control and Prevention, 2010). Of all nonfatal crimes against students ages 12 to 18, 59% were committed against males. Males experienced a greater percentage of all violent events (61%) than thefts (57%) that occurred at school (Robers et al., 2010). Age is another correlate of school victimization that we can examine. Violent victimization appears to be most common among younger school children than those in high school. For example, in the Youth Risk Behavior Surveillance System survey, it was found that students in 9th grade had higher rates of fighting than did students in 10th through 12th grade. In addition, more than half of violent victimizations that occurred at school were experienced by children aged 12 to 14 (Robers et al., 2010). Race/ethnicity is another important correlate in examining any type of victimization. More than half of victimizations at school are experienced by White youth, although the rates of both violent and theft victimization are highest for Black and Hispanic youth (Robers et al., 2010). A final factor to consider is household income of the student. For theft, there is little difference in victimization rates across household income levels; however, violent victimization impacts children who reside in households with annual incomes less than $15,000 at the highest rate (Robers et al., 2010).

Not only are students at risk of being victimized at school, but teachers, administrators, and staff may also become victims. Technically, this type of victimization would be victimization at work, but since it occurs at schools, we will discuss it here. Fortunately for them, teachers, administrators, and staff are fairly unlikely to be victimized while working at school. As discussed later in the section on workplace victimization, persons working in service and retail occupations, law enforcement/corrections, and mental health have much higher rates of workplace victimization than do educators and persons employed at schools. Nonetheless, victimization does occur, but risk is not constant for all teachers. Special education teachers are more likely to be violently victimized than are other teachers (May, 2010). Data from the Schools and Staffing Survey show that a larger percentage of secondary school teachers reported being threatened with injury by a student than did elementary school teachers, but elementary school teachers are more likely to be physically attacked (Robers et al., 2010). Risk of victimization also varies depending on the location of the school. Teachers employed in public schools and those who work in city schools are at greater risk of being threatened with injury and being physically attacked than those working in private schools or town, suburban, or rural schools (Robers et al., 2010). As with students, teachers are more likely to be victims of theft than of violent offenses (Robers et al., 2010).

Risk Factors for School Victimization

Much like with other types of victimization, victimologists and others have attempted to discover what causes school victimization. Why are some school children victimized while others are not? Why are some schools safe and others riddled with crime? It is difficult to determine the exact reasons why school victimization occurs, but it is likely a combination of structural forces as well as individual factors. For example, much attention has been paid to the school's location and its relationship to the amount of victimization that occurs there. As you may expect, schools located in crime-ridden neighborhoods are often likely to have high levels of violence and other types of victimization (Laub & Lauritsen, 1998). But there are high-performing and safe schools in these same neighborhoods, so it is not enough to locate "bad" neighborhoods to identify unsafe schools. It is also possible that the factors that place youth at risk for victimization outside school are similar to the ones that place them at risk of being victimized in school—such as low self-control, lack of capable guardianship, and having deviant or delinquent peers. In addition, adolescence is a period marked by biological changes; hormones run amuck and bodies change. Both males and females go through transitions, both physically and emotionally, and they navigate new social situations. This time can be pressure filled and stressful, which may lead to outbursts, aggression, bullying, and other maladaptive behaviors.

Consequences

We require youth to attend school in the United States, and in turn, we should be providing them a safe, productive learning environment. When this does not occur and students are victimized, many negative outcomes may arise. Results from the 2007 School Crime Supplement to the NCVS show that 13% of victims of any crime and 23% of violent crime victims reported being afraid of harm or being attacked (Devoe, Bauer, & Hill, 2010). Victims reported higher levels of fear than nonvictims, which shows the powerful effects that victimization has on school-aged youth. Victims are also likely to skip or avoid school. In a study conducted by the Centers for Disease Control and Prevention (2004), it was discovered that more than 5% of students felt unsafe at school and stayed away from school for at least 1 day during the 30 days prior to the survey. Victimized students also indicated that they avoided school activities and specific places inside school buildings, such as certain hallways, the school entrance, parts of the cafeteria, and restrooms, at higher levels than did nonvictims (Devoe et al., 2010).

Bullying

One specific type of victimization that can occur at school and has garnered recent attention is bullying. **Bullying** is the intentional infliction of injury or discomfort (or the attempt to do so) on another person repeatedly over time when there is a power imbalance between the perpetrator and the victim (Olweus, 2007). Bullying can be both direct and indirect. **Direct bullying** involves both physical and verbal actions in the presence of the victim. **Physical bullying** can include hitting, punching, shoving, pulling the chair out from under another person, tripping, and other physical actions. **Verbal bullying** includes direct name calling and threatening. **Indirect bullying** can be more subtle and harder to detect. It is often referred to as social bullying and includes actions such as isolating individuals, making obscene gestures, excluding from activities, and manipulation. Even though it is often said that "kids will be kids" and bullying is simply a natural part of children's interactions, research suggests that it can have pernicious effects and that ignoring bullying in schools may be dangerous.

Before discussing the consequences of bullying, let's first uncover the extent to which bullying occurs in schools. Bullying appears to be more common than other types of victimization. In 2007, almost one third of students aged 12 to 18 said they had been bullied at school during the school year (Devoe et al., 2010). A report published by the National Institute of Child Health and Human Development found that 1.6 million children are bullied at least weekly and that 17% of children in grades 6 through 10 have been bullied (Ericson, 2001). Most commonly, students reported being made fun of (21%), but 11% reported having been pushed, shoved, tripped, or spit on, and 6% said they had been threatened with harm (Robers et al., 2010).

Like school victimization more generally, bullying differentially impacts some youth. In the 2001 World Health Organization's Health Behavior in School-Aged Children Study, it was found that Black youth are less likely to be bullied than are White or Hispanic children (Nansel et al., 2001). Males were more likely than females to be bullied (Nansel et al., 2001). Other recent research has shown that certain groups are more at risk of being bullied than others. Youth who have learning disabilities, those who have attention-deficit hyperactivity disorder, those with physical disabilities, obese children, and those who stutter experience higher rates of bullying victimization (Miller & Miller, 2010). Recently, attention has been given to the fact that gay youth are more likely to be targeted by bullies than are other youth and that this bullying often occurs on a daily basis (Miller & Miller, 2010). The largest group of victims tends to be slight or frail. They tend to be average or poor students and are generally passive socially (Olweus, 1993). This is not to say, however, that all bullied victims "look" the same. Other victims tend to be more assertive and hot-tempered; they react aggressively when they are bullied. They start fights in addition to being picked on (Pellegrini, 1998). Both types of bullied youth are unlikely to be in the "popular" groups in school.

As with other forms of victimization, research in bullying has uncovered the fact that there is a subset of individuals who are bully/victims—they bully and are bullied (Haynie et al., 2001). What is important about this group of youth is that they seem to fare even worse in terms of psychosocial or behavioral functioning than do those who are either bullies or victims.

Psychosocial Effects of Bullying Victimization

Although, as mentioned, bullying sometimes has been treated as "normal" behavior for children, its effects can be quite serious. It has been linked to poor psychosocial adjustment—students who are bullied more often report greater levels of unhappiness (Arseneault et al., 2006) and lower self-worth (Egan & Perry, 1998). Bullying during adolescence has been linked to anxiety and depression both contemporaneously and later in life (Bond, Carlin, Thomas, Ruin, & Patton, 2001; Olweus, 1993). Being bullied also has been shown to be linked to health symptoms in children. Children who reported being bullied were more likely to report not sleeping well, bed wetting, and getting occasional headaches and tummy aches (Williams, Chambers, Logan, & Robinson, 1996). In addition to these consequences, being bullied also has negative outcomes on school adjustment and performance. Bullied youth are more likely to say they dislike school than are nonbullied youth (Kochenderfer & Ladd, 1996; Rigby & Slee, 1993), are more likely to report absenteeism from school (Rigby, 1997; Zubrick et al., 1997), and have higher levels of school avoidance (Kochenderfer & Ladd, 1996).

Violent Effects of Bullying Victimization

Perhaps the most serious outcome of being bullied is acting out in response. Bullying victimization has been linked to violent behavior by the *victim*. A report by the Secret Service revealed that 71% of school shooters whose friends, families, and neighbors were interviewed had been the targets of a bully (Espelage & Swearer, 2003). It should be noted, however, that if bullying caused school shootings, we would see many more than we do. As such, it may be a contributing factor in some instances but by no means can be considered a cause. Recently, the media have given widespread attention to several young people who have committed suicide after being bullied by their peers in various ways. With the widespread use of the Internet and cell phones, bullying methods have expanded to include what is known as cyberbullying. **Cyberbullying** is bullying behavior that takes place via mobile phones, the Internet, and/or digital technologies. It can involve threats or harassment sent over the phone, threatening and insulting comments posted on a social networking site, and vulgar or scary text messages. Children can do the bullying themselves or enlist their friends or family. Cyberbullying can be particularly harmful given its ease of use and the fact that people can bully without having to be in the presence of the victim. They can even do it anonymously. As noted by Aftab, "The schoolyard bullies beat you up and then go home. The cyberbullies beat you up at home, at grandma's house, wherever you're connected to technology" (as quoted in Nies, James, & Netter, 2010). In the following box read the heartbreaking story of Phoebe Prince, a young teen who committed suicide after being cyberbullied. The School Crime Supplement to the NCVS began including questions regarding bullying behavior through electronic means in 2007. Almost 4% of students surveyed reported experiencing electronic bullying during the previous school year (Devoe et al., 2010). To learn more about the extent and consequences of cyberbullying, read the article by Carter Hay, Ryan Meldrum, and Karen Mann (2010) included in this section.

THE STORY OF PHOEBE PRINCE

Phoebe Prince, a recent Irish immigrant, hanged herself January 14, 2010 after nearly 3 months of routine torment via text message and through Facebook by students at South Hadley High School. Police believe she was the victim of cyberbullying from multiple "girls at the school who had an unspecified beef with her over who she was dating" (Kotz, 2010). Her case has been called "the culmination of a nearly 3-month campaign of verbally assaultive behavior and threats of physical harm" (as quoted in Goldman, 2010). In at least one instance, she was physically attacked when a girl pelted her with a soft drink can. As of January 28, 2010, "nine students have been indicted on charges ranging from statutory rape to civil rights violations and stalking. It appears that Phoebe may finally get her justice" (Kotz, 2010). As a result of the attention that bullying has garnered through cases like Phoebe's and others, "forty-five states now have anti-bullying laws; in Massachusetts, which has one of the strictest, anti-bullying programs are mandated in schools" (Bennett, 2010).

Responses to School Victimization

In response to victimization and bullying at schools, many schools have instituted security measures. Most commonly, schools have hired law enforcement officers, installed metal detectors, installed security cameras, begun to lock entrances and exits during school hours, and implemented supervision of hallways during the school day (Devoe et al., 2010). According to a survey of principals, almost half of public schools in 2008 had paid law enforcement or security staff, more than half used security cameras, and 5% used metal detectors (Devoe et al., 2010). The School Crime Supplement Survey also indicates that many schools are implementing security measures. Almost one fourth of students reported that they were required to wear picture identification at school, and 94% indicated that visitors were required to sign in at their school (Devoe et al., 2010).

In addition to school security measures, laws and policies are in place to address violence within schools. Current federal law, under the **Gun-Free Schools Act,** mandates that each state that receives federal funding must suspend for at least 1 year any student who brings a firearm to school. As such, most states have laws to address bullying, harassment, and hazing that occurs at school (Olweus Bullying Prevention Program, 2011)—some of which are **zero-tolerance policies** that mandate specific punishments for fighting and violence along with bringing weapons to school—which may serve to limit school victimization and bullying. Fortunately, most states have instituted a broad range of policies that attempt to address school victimization more holistically. For example, see next box, which provides a description of Florida's laws. Although it remains to be seen how effective these laws are at reducing the amount of victimization that occurs at school, many of the laws also require mandatory reporting of suspected victimization, mandate that schools have programs and resources to reduce school victimization, and require that people are in place to oversee these programs (Limber & Small, 2003).

FLORIDA'S BULLYING/HARASSMENT, CYBERBULLYING, AND HAZING LAWS

Bullying/Harassment

Statute 1006.147 (2008) prohibits bullying or harassment of any student or employee of a public K–12 educational institution, during any program or activity conducted by a public K–12 educational institution, during any school-related or school-sponsored program or activity, or through the use of data or computer software accessed through a computer, computer system, or network of a public K–12 educational institution. Specific definitions of bullying and harassment are outlined in the statute.

Statute 1006.147 (2008) provides immunity from a cause of action to a school employee, school volunteer, student, or parent who promptly reports in good faith an act of bullying or harassment to the appropriate school official.

Statute 1006.147 (2008) requires school districts to adopt a policy prohibiting bullying and harassment of any student or employee of a public K–12 educational institution. The policy must substantially conform to the model policy of the state Department of Education, and must afford all students the same protection regardless of their status under the law. Requirements of the policy are outlined in the statute.

Statute 1006.07(6) requires district school boards to provide for the welfare of students by using the Safety and Security Best Practices to conduct a self-assessment of the district's current safety and security practices. The self-assessment includes indicators for districts to develop and enforce policies regarding anti-bullying, anti-harassment, and due process rights in accordance with state and federal laws. The assessment also includes indicators of schools surveying students on school climate questions related to discipline, bullying, threats perceived by students, and other safety or security related issues.

Statute 1006.07(2) also requires a student to be subject to in-school suspension, out-of-school suspension, expulsion, or imposition of other disciplinary action by the school and possibly criminal penalties for violating the district's sexual harassment policy.

State Board of Education Administrative Rule 6A-19.008 (1985) requires schools to have environments that are free of harassment and prohibit any slurs, innuendos, or other verbal or physical conduct reflecting on one's race, ethnic background, gender, or handicapping condition, which creates an intimidating, hostile, or offensive educational environment, or interferes with students' school performance or participation or other educational opportunities.

Cyberbullying

Statute 1006.147 (2008) prohibits bullying and harassment of any student or employee of a public K–12 educational institution through the use of data or computer software that is accessed through a computer, computer system, or computer network of a public K–12 educational institution. The definition of "harassment" in the statute includes any threatening, insulting, or dehumanizing gesture, use of data or computer software, or written, verbal, or physical conduct directed against a student or school employee that does one of the following: (1) places them in reasonable fear of harm to his or her person or damage to his or her property, (2) substantially interferes with a student's educational performance, opportunities, or benefits, or (3) substantially disrupts the orderly operation of a school. The definition of "bullying and harassment" includes perpetuation of actions by an individual or group with intent to demean, dehumanize, embarrass, or cause physical harm to a student or school employee by accessing or knowingly causing or providing access to data or computer software through a computer, computer system, or computer network within the scope of the district school system.

Hazing

Statute 1006.135 (2005), referred to as the Chad Meredith Act, defines hazing and makes the hazing of students at a high school with grades 9–12 a criminal offense as defined within the statute. The bill prohibits the following defenses to a charge of hazing: obtained consent of the victim; the conduct or activity that resulted in the death or injury was not part of an official organizational event or was not otherwise sanctioned or approved by the organization; the conduct or activity that resulted in the death or injury was not done as a condition of membership in an organization.

SOURCE: National Association of State Boards of Education (2010). The NASBE Healthy Schools Policy Database is a comprehensive set of state-level laws and policies from 50 states on more than 40 school health topics that includes hyperlinks to the actual policies whenever possible.

Most school-based programs are targeted at reducing school violence and/or bullying specifically. The most effective of these programs are proactive and involve parents, students, and the community (Ricketts, 2010). A common type of a violence-reduction program is peer mediation. Peer mediation programs train a group of students in interest-based negotiation skills, communication skills, and problem-solving strategies so they can help their peers settle disagreements peacefully and without violence (Ricketts, 2010). Findings from evaluations of peer mediation programs show that they can change the school climate over time (Ricketts, 2010). To specifically attack bullying, some schools have adopted bullying prevention programs. One of the most widely adopted of these programs, the Olweus Bullying Prevention Program, has shown promising reductions in bullying perpetration and victimization in both the United States and Norway (Olweus, 1991).

Victimization at School: College

You may have been wondering whether college students are similarly at risk for victimization while attending school. Most parents send their children to college feeling pretty confident that they will be safe—most college students report feeling safe at school. Are these feelings justified?

Who Is Victimized?

You will be happy to learn that college students actually have lower average annual rates of victimization than their similarly aged nonstudent counterparts (Baum & Klaus, 2005). The difference for students and nonstudents was not significant, however, for rape and sexual assault. When college students are victimized, they are most likely to experience a nonviolent victimization, such as theft. In a study of college students enrolled in 12 institutions of higher learning, Fisher, Sloan, Cullen, and Lu (1998) found 169.9 theft victimizations per 1,000 students. Of those who are violently victimized, the most common victimization is simple assault (63% of all violent victimizations) (Baum & Klaus, 2005). Most violent victimizations are committed by strangers, at night, without a weapon, and off-campus (Baum & Klaus, 2005). In fact, college students experience violent victimizations off-campus at 20 times the rate they experience them on campus (Hart, 2007). Most violent victimizations of college students do not result in physical injury (75%) (Baum & Klaus, 2005).

In addition to these incident-level characteristics, college students who are victimized share common characteristics. White college students have higher violent victimization rates than do students of other races (Baum & Klaus, 2005). White students and students of other races have lower violent victimization rates than nonstudents, while Hispanic students have similar violent victimization rates to Hispanic nonstudents (Baum & Klaus, 2005). Violently victimized college students also tend to be male. Male college students are twice as likely as female college students to be violently victimized (Baum & Klaus, 2005). The only type of violent victimization that female college students experience at higher rates than males is sexual victimization. Female college students have an average annual rape/sexual assault rate of 6 per 1,000 persons ages 18 to 24, compared with a rate of 1.4 for males. Male college students, on the other hand, have an average annual rate of violent victimization of 80.2, compared with 42.7 for female college students (Baum & Klaus, 2005). For a detailed account of sexual victimization of college students, see Section 5 on sexual victimization.

Less attention has been given to property victimization of college students; however, we do have an idea as to who the "typical" college property victim is. Males tend to be property victims more so than female college students (Fisher et al., 1998). In addition, younger students, aged 17 to 20, are at greater risk of property victimization than are older college students (Fisher et al., 1998). Employed students report higher levels of property victimization than do unemployed students or those who work part-time (Johnson & Kercher, 2009).

Risk Factors for Victimization at College

Although college students are not at a greater risk than nonstudents of being violently victimized, they do still experience a great deal of violent victimization. With an average of 7.9 million 18- to 24-year-olds enrolled annually in college between 1995 and 2002, a violent crime rate of 60.7 per 1,000 persons equates to 479,530 violent victimization incidents each year. So, why do this many victimizations occur?

Lifestyle/Routine Activities

Recall from Section 2 that routine activities theory is one of the hallmark theories of victimization. According to this theoretical perspective, risky lifestyles and daily routines place individuals at risk of being victimized. When persons engage in risky lifestyles and routines that bring them together in time and space with would-be offenders and they are without capable guardians—thus making them suitable targets—they are likely to be victimized. Let's consider how college students may engage in routine activities and risky lifestyles that could place them at risk.

Spending time in the presence of potential offenders increases the risk of being victimized. It is instructive that, for college students, the most common offender is another college student (Fisher et al., 1998). Students who spend more time away from home in the evening are at risk of being victimized (Mustaine & Tewksbury, 2007). Spending many nights on campus during the week has been found to increase theft victimization for college students (Fisher et al., 1998). For college females, spending time in places where men are seems to increase risk, particularly for sexual victimization. That is, women who spend more time in fraternity houses are more likely than other women to be sexually victimized (Stombler, 1994). Being a member of a fraternity or sorority also has been linked to an increased risk of property victimization (Johnson & Kercher, 2009). Another way that college students may be exposed to potential offenders is through engaging in victimizing behavior. Those who do are more likely to be victimized than others (Tewksbury & Mustaine, 2000).

Being a suitable target also increases risk of victimization for college students. A person can be deemed "suitable" for a variety of reasons. A person may have items that are valuable and easy to steal, and if they carry these items in public, they are at greater risk of being victimized than others (Johnson & Kercher, 2009). A person may be walking alone and not seem as though he or she is able to protect himself or herself. Or, as is quite relevant for college students, a person may be visibly intoxicated and, thus, may be seen as easy to victimize and unable to resist an attack; the offender may not fear that the person will fight back or even recall enough details about the event to make a reliable police report. We will return to alcohol and its role in college students' victimization below.

The last element relevant to routine activities theory is lack of capable guardianship. College students who live in settings with high levels of transience and low levels of cohesion—for example, student apartments where residents move year to year and change roommates frequently—are at greater risk of victimization than students who live at home or in campus dorms (Mustaine & Tewksbury, 2007). Guardianship can also be created through physical means such as carrying a weapon, mace, or pepper spray or by attending prevention or crime awareness seminars. Attending crime prevention or crime awareness seminars has been found to reduce the risk of violent victimization among college students (Fisher et al., 1998). On the other hand, some research shows that the use of physical guardianship is actually related to an increase in victimization, but this may be due to people purchasing and carrying these items after being victimized (see Fisher, Daigle, & Cullen, 2010).

Alcohol

In addition to the three elements of routine activities theory, college students also often engage in behaviors that likely increase their risk of victimization. The first risky behavior we should consider is the use of alcohol and drugs. As you are well aware, the use of alcohol is pervasive among college students. Research on the use of alcohol by college students shows that between 75% and 96% of college students consume alcohol (National Institute on Drug Abuse, 1995, 1998). One of the most serious consequences of alcohol use among college students is criminal victimization. In fact, "college students who drink heavily are more likely to be both criminal offenders and victims of crimes" (Tewksbury & Pedro, 2003, p. 32).

Why does alcohol increase college students' risk of being victimized? Well, as noted in Section 2, alcohol impairs cognitive functioning and reduces a person's ability to assess situations as risky. Even if a person can see that a situation is dangerous, he or she may not be physically capable of warding off a potential attacker if inebriated. A person may also have lowered inhibitions and say and do things that he or she would not normally do; thereby, a person can find herself or himself in a situation with a potential offender. Given the effects of alcohol, a person may say or do things to anger others, thus unintentionally getting into a fight or argument. Not to blame the victim, but when alcohol is involved, the victim's actions or words may set in motion a series of events that lead to victimization.

Responses to Campus Victimization

Legislation

No parents ever want to get a phone call informing them that the child they dropped off at college was harmed, but that is just what the parents of Jeanne Clery experienced. They received the worst phone call of all: Their daughter had been raped and murdered while attending Lehigh University. She was sleeping in her dorm room when Joseph Henry entered through propped-open and unlocked doors (Clery & Clery, 2008). In response to their daughter's death, the Clerys sued the university after they found out that the university knew about the propped-open and unlocked doors and did not tell the students about the potential dangers lurking on their campus. In addition to filing a lawsuit, they pushed for legislation that would require college campuses to better inform students about crime. As a result, the Student Right-to-Know and Campus Security Act of 1990, renamed in 1998 the **Jeanne Clery Disclosure of Campus Security Policy and Campus Crime Statistics Act** [20 USC 1092 (f)], was passed (hereafter referred to as the Clery Act).

The Clery Act applies to all institutions of higher learning that are eligible to participate in student aid under Title IV of the Higher Education Act of 1965. Enforced by the Department of Education, the act has three main requirements. First, by October 1 of each year, schools must publish an annual campus security report with crime statistics and a security policy that includes information about sexual assault policies, the authority of campus security officers, and where students should go to report a crime. The security report must include information about the three most recent calendar years and must be made available to all current students and employees. The crimes that must be included in the report are homicide, sex offenses, robbery, aggravated assault, burglary, motor vehicle theft, and arson. If an arrest or disciplinary referral was made, the report must also include liquor law violations, drug law violations, and illegal weapons possession. The location of the incident must be provided in terms of occurring on campus, in residential facilities for students on campus, in noncampus buildings, or on public property immediately adjacent or running through the campus.

Second, colleges and universities are required to disclose incidents in a timely fashion through crime logs and warnings about ongoing threats. A crime log must be kept by the campus police or security department. Within two days of being made aware of a crime, the police or security must include the offense in the crime log, including the nature, date, time, and location of each crime. This crime log is public record—it is to be made publicly available during normal business hours and kept open for 60 days. Along with the crime log, warnings must be provided about those crimes that are required to be disclosed or those believed to represent an ongoing threat to students and employees. Warnings are commonly delivered through e-mails, phone calls, and text messages.

Third, the Clery Act requires that certain rights for both accusers and victims are protected in cases of sexual assault handled on campus. Both parties are given the same opportunity to have other people present at campus disciplinary hearings. Both parties have the right to be informed of the outcome of disciplinary hearings. Victims have the right to notify law enforcement and also have the right to be notified about counseling services available to them and options for changing academic and living situations.

▲ **Photo 9.2** Howard and Connie Clery, sitting on the bed of their daughter, Jeanne Clery, who was brutally murdered at Lehigh University. Howard and Connie are the heads of Security on Campus, Inc.

Photo credit: © Time & Life Pictures/Getty Images

Despite these requirements, there are some limitations to the Clery Act. Not all crimes that occur are required to be disclosed in the security report, not all crimes are reported to the police, and only those crimes that occur on campus property or public property adjacent to the campus are required to be in the report; thus, individuals who look to the security report for guidance about their campus safety may be getting only a partial picture of the true amount of crime that occurs. In addition, although the Clery Act mandates that colleges and universities comply with its requirements, research shows that not all schools are doing so (Fisher et al., 2007).

In addition to the Clery Act, other federal legislation has been passed that allows for the tracking of sex offenders who have been convicted and who are required to register if they are enrolled in an institution of higher learning or volunteering on campus (Carter & Bath, 2007). In addition, at least 19 states have Clery-type legislation (Sloan & Shoemaker, 2007). To read about how college students—in particular, college women—respond to actual victimization and perceived danger on campus and how the Clery Act's requirements impact this response, read Pamela Wilcox, Carol E. Jordan, and Adam J. Pritchard's (2007) article included in this section.

Campus Police and Security Measures

Of those 4-year colleges and universities that have 2,500 students or more, almost 75% have sworn law enforcement officers (Reaves, 2008). Almost all public institutions have sworn personnel, while less than half of private campuses have sworn officers (Reaves, 2008). Instead, campuses that do not have sworn personnel have nonsworn security officers. Those that do have sworn personnel grant these officers full arrest powers. Campuses differ greatly in terms of the number of officers they employ. As you may imagine, institutions of higher learning that have a large student body and are located in large, urban centers tend to have larger police and security forces than do smaller and more traditional rural institutions (Reaves, 2008). For a list of the campuses with the largest number of sworn law enforcement officers, see Table 9.1.

Almost all college campuses have a three-digit emergency telephone number that persons can call for assistance or to report a crime (Reaves, 2008). Those who are on campus and need assistance can use blue-light emergency campus phones that provide direct access to campus law enforcement. More than 90% of 4-year institutions of higher learning that enroll at least 2,500 students have these phones (Reaves, 2008). In addition to these security measures, most institutions of higher learning have written terrorism plans, written emergency response plans, and provide students with access to crime prevention programs (Reaves, 2008).

⬚ Victimization at Work

So far, we have discussed victimization that occurs when people are attending school, despite the fact that schools should provide a safe environment to encourage active learning. The work environment, you would think, should be similarly safe—and there are strict rules and laws in place, discussed below, that attempt to ensure it is just that. Nonetheless, people are not

Table 9.1 Largest Full-Time Sworn Officers for Colleges, 2004–2005 School Year

Campus Served	Full-Time Sworn Officers
Howard University	166
Temple University	119
University of Pennsylvania	100
University of Medicine and Dentistry of New Jersey	97
George Washington University	95
University of Florida	86
Georgia State University	79
Yale University	78
University of Maryland—College Park	76
Vanderbilt University	76

SOURCE: Reaves (2008).

always safe from harm while they try to earn a living. But just how often are people victimized on the job? Why does it happen, and how can it be prevented? These questions and more will be addressed in this section.

Definition of Workplace Victimization

Before we address the extent to which victimization occurs to people at work, let's address what victimization at work encompasses. Obviously, if a person is victimized while physically at work—say, in his or her office—that constitutes a **workplace victimization**. But what if a person is victimized in the parking garage at work? Would this too be "counted" as a workplace victimization? What if a victimization occurs when a person is traveling for work or doing official work business, such as making a delivery? In short, if a person is working or on duty and victimized, the incident is considered a workplace victimization. Workplace victimizations can be violent—ranging from threats and simple assaults to homicide—or can be nonviolent—such as theft.

Much of the research that has examined workplace victimization has focused on violence that occurs in the workplace, as opposed to nonviolent victimizations. You may be surprised to learn that the concept of workplace violence has been studied for many years. In fact, the first discussions of workplace violence in the literature occurred in 1892 (Jenkins, 2010)! Data on workplace violence though were not collected and criminologists did not begin to consider seriously the etiology and causes of workplace victimization until the 1970s and 1980s (Jenkins, 2010). Since then, however, widespread attention has been given to violence that occurs in the workplace. One useful tool that has been developed to understand workplace violence is a typology of workplace violence (Jenkins, 2010). The first type is **criminal intent incidents,** which include incidents in which the perpetrator has no legitimate relationship to the business at which the crime occurs. Most commonly, the perpetrator in this type commits a crime in conjunction with the violence, such as a person who robs a gas station and shoots the attendant. The second type is **customer/client incidents**. These incidents occur when the perpetrator has a legitimate relationship with the business and becomes violent when receiving services from the business. An example of this second type would be a person at a doctor's office who is quite agitated and begins punching the doctor. The third type is **worker-on-worker incidents**. In these incidents, the perpetrator is a current or former employee of the business and aggresses against another employee. The fourth, and last, type of incident is **personal relationship incidents,** in which the perpetrator has a personal relationship with the intended victim, who is targeted while at work. An example of this type would be a domestic violence incident in which a man shows up at his ex-wife's place of employment and shoots her. Now that you have an idea of the types of victimization that constitute workplace victimization, let's find out how much of this occurs each year.

Extent of Workplace Victimization

As with victimization at school, most people can safely go to work each day and be free from victimization. But some people do, in fact, experience victimization. We can get an idea of the extent of workplace victimization from a variety of data sources. As with most types of victimization, one rich source of data is the NCVS. Recall that when a person indicates in the NCVS that he or she experienced a victimization, an incident report is then completed. In the incident report, if persons indicate they were at work or on duty at the time of the victimization, the incident is counted as a workplace victimization incident. Between 1993 and 1999, an average of 1.7 million violent victimizations occurred at work or while people were on duty, accounting for 18% of all violent crime during this time period (Duhart, 2001).

According to findings from the NCVS, the most common type of victimization experienced by people at work, however, is personal theft—between 1987 and 1992, more than 2 million personal thefts, on average, were experienced at work each year (Bachman, 1994). Almost a quarter of all thefts reported in the NCVS occurred at work or while persons were on duty (Bachman, 1994). The second most common type of victimization measured in the NCVS is simple assault, a violent victimization. Between 1993 and 1999, an average of 1,311,700 simple assaults occurred against people at work or while on duty (Duhart, 2001). When aggravated assaults are included, almost 19 of every 20 workplace violent incidents were aggravated or simple assaults, which shows that other forms of violent workplace victimization are relatively uncommon (Duhart, 2001). In fact, during this same time period, only 6% of all workplace violent crime was rape/sexual assault, robbery, or homicide (Duhart, 2001).

Who Is Victimized at Work?

Demographic Characteristics of Victims

By now, you could probably guess that males are more likely than females to be violently victimized at work or while on duty, except for rape and sexual assault (Duhart, 2001). There is no difference, however, in the rates of theft victimization for males and females—they are equally likely to be the victims of theft while working (Bachman, 1994). Although the patterns for workplace violent victimization for gender follow the trends for victimization more generally (i.e., males have higher rates than do females), the trends for race do not. Rates of workplace violent victimization are highest for Whites. In fact, from 1993 to 1999, the average workplace violent victimization rate was 25% higher for White workers than for Black workers (Duhart, 2001). The rate for workplace violent victimization was similar for Black and Hispanic workers (Duhart, 2001).

The last two demographic characteristics to consider are age and marital status. Young adults aged 20 to 34 had the highest rates of workplace violent victimization. Persons who were married or widowed had lower workplace violent victimization rates than those who were never married, divorced, or separated. The last three groups all have similar rates of victimization (Duhart, 2001).

Occupations With Greatest Risk

Knowing what demographic characteristics are correlated with workplace victimization does little to inform us what jobs are most risky. You would be right if you thought that some jobs are replete with danger. Data from the NCVS show that law enforcement jobs are the most dangerous. Persons working in this field, which includes police, corrections, and private security officers, experience 11% of all workplace violent victimizations (Duhart, 2001). Fortunately for your professor, college and university teachers have the lowest rates. Other occupational fields that face the highest risk (in order) are mental health, retail sales, teaching, transportation, and medicine. Within these broad categories, police, taxicab drivers, corrections workers, private security, bartenders, custodians in mental health settings, professionals in mental health settings, special education teachers, and gas station attendants have some of the highest violent victimization rates at work (Duhart, 2001). As you might imagine, persons in retail sales have the highest robbery victimization rates, while people in law enforcement have the highest assault rates (Duhart, 2001). Persons working in government jobs are also at risk of workplace violence. Vivian Lord's (1998) article, included in this section, addresses these workers' risk of being victimized on the job.

Special Case: Fatal Workplace Victimization

Of course, the most serious outcome of victimization is when a person loses his or her life. Unfortunately, this sometimes does occur when a person is at work—and it is not just caused by people "going postal"! In fact, homicide is not the leading cause of occupational injury deaths of postal workers (Jenkins, 2010). How do we know this? To track the number of fatal injuries for various occupations, the National Census of Fatal Occupational Injuries program was initiated in 1992. Each year, the Bureau of Labor Statistics publishes its findings about the extent of fatal injuries from this program so that a picture of the extent and types of fatal occupation injuries can be drawn (Bureau of Labor Statistics, 2011).

In 2010, 808 persons died in the workplace or while on duty as a result of violence or assaultive behavior (Bureau of Labor Statistics, 2011). Of these 808 people, 506 were homicide victims and 258 died as a result of self-inflicted wounds. Of the 506 homicide victims, most (401) were the victims of shootings in the workplace and 34 were the victims of stabbings. The most common type of workplace homicide is criminal incident, with robberies alone accounting for 40% of the cases in 2008 (Bureau of Labor Statistics, 2010). Coworkers and former coworkers (worker-on-worker type) accounted for only 12% of incidents in the same year (Bureau of Labor Statistics, 2010). Another special case of workplace victimization is sexual harassment. See the article by Anne M. O'Leary-Kelly, Lynn Bowes-Sperry, Collette Arens Bates, and Emily R. Lean (2009), included in this section, for an in-depth analysis of the state of the research on sexual harassment victimization at work.

Demographic Characteristics of Victims

As with workplace violence in general, males are more likely to be murdered at work than are females. One interesting difference between men and women, however, is who the perpetrator is when homicide at work does occur. Although males and females are equally likely to be the victims of workplace robbery and have other assailants, females are more likely to be murdered at work by a relative or personal acquaintance than are males (Bureau of Labor Statistics, 2010). In addition to males having higher rates of workplace homicide, minorities face a greater risk of becoming victims of this specific type of workplace victimization than do others (Sygnatur & Toscano, 2000). Adults aged 25 to 44 account for the greatest percentage of workplace homicides, while persons under the age of 18 have the lowest rates (Sygnatur & Toscano, 2000).

Occupations and Workplaces With Greatest Risk

You have already learned that workers in some occupations face greater risk of being victimized than do others. Do you think that the same types of occupations pose the same dangers in terms of fatal workplace violence? You may be surprised to learn that, although law enforcement jobs are "risky" in terms of nonfatal violence, they are not the most risky in terms of fatal workplace violence. Taxicab drivers and chauffeurs face the greatest risk of being murdered of any type of worker in the United States (Sygnatur & Toscano, 2000). They face 36 times the risk of all employed individuals! Consider this fact: Taxicab drivers and chauffeurs compose only 0.2% of employed workers but account for 7% of all work-related homicides (Sygnatur & Toscano, 2000). They account for a disproportionate share of workplace homicide. Law enforcement officers have the next highest rate of workplace homicide.

It is also instructive to consider workplaces that have high rates of occupational homicide. Retail trades have the highest rates of workplace homicide, while services, public administration, and transportation also have high rates. Retail establishments that have the most homicides include liquor stores, gas stations, grocery stores, jewelry stores, and eating/drinking establishments. Services include hotels and motels, for example. Public administration includes detective and protective order services as well as justice and public order establishments. Finally, transportation includes taxicab establishments, which, as previously mentioned, contains the most risky occupation for workplace homicide (National Institute for Occupational Safety and Health, 1995).

Risk Factors for Victimization at Work

Now that you know what types of occupations and workplaces have the highest risks of workplace victimization and workplace homicide, the next thing to consider is why these jobs and places are so dangerous. What is it about working in retail or law enforcement, for instance, that poses a risk? Generally, a number of characteristics of certain jobs place workers at risk of being victimized (Jenkins, 2010; National Institute for Occupational Safety and Health, 1995; Sygnatur & Toscano, 2000).

- Working in contexts that involve the exchange of money with the public
- Working with few people or alone
- Working late at night or during the early morning
- Working in high-crime areas
- Working in the community (such as police or taxicab drivers)
- Working with criminal, unstable, or volatile persons
- Having a mobile workplace
- Working in delivery of goods, passengers, or services
- Guarding valuables or property

▲ **Photo 9.3** A day at work can prove to be very dangerous for taxicab drivers, who face some of the greatest risks of violence at work.

Photo credit: ©Thinkstock/Creatas

Consequences of Workplace Victimization

One of the most obvious consequences of workplace victimization is that people may not be able to go to work. They may be injured and need to receive medical attention and, as a result, may need to take time off from work. They may be fearful and scared to return to work. Persons who participated in the NCVS indicated that they missed some 1,751,100 days of work each year as a result of workplace victimization (Bachman, 1994). On average, each workplace victimization incident cost 3.5 days of missed work, which resulted in more than $55 million in lost wages annually, not including days missed that were covered by sick and annual leave (Bachman, 1994).

Workplace homicide also is associated with a whole host of costs. When examining medical expenses, future earnings lost, and household production losses that include child care and housework, workplace homicides between 1992 and 2001 cost $6.5 billion (Hartley, Biddle, & Jenkins, 2005). Each workplace homicide cost an average of $800,000 (Hartley et al., 2005).

Responses to Workplace Victimization

We know the extent to which workplace victimization occurs, the occupations and workplaces at greatest risk of experiencing workplace victimization, and the risk factors for experiencing workplace victimization. But what have places of business, the government, and the legislature done to prevent victimization at work?

Prevention Strategies

For prevention strategies to be most effective, they should be tied to risk factors. That is, to prevent workplace victimization, the factors that place particular occupations or places of work at risk should be targeted and altered to effect change. In doing so, preventing workplace victimization has typically occurred in three main areas. The first prevention strategy is targeting **environmental design**. These strategies focus on ways to make a workplace more secure and a less attractive target. See Table 9.2 for examples of this type of prevention. The second type of prevention strategy, **organizational and administrative controls**, focuses on strategies that administrators and agencies can implement to reduce the risk of workplace victimization in their organizations. Although these controls can be varied, common strategies are identified in Table 9.2. The third type of prevention strategy is behavioral. **Behavioral strategies** are actions that workers can take to reduce their risk of workplace victimization. These include the behaviors identified in Table 9.2.

Employers can also do their part to reduce workplace violence perpetrated by current or former workers by being careful about who they hire and by being watchful of suspicious behavior from their employees. One thing employers should do is carefully screen their employees to uncover any issues with alcohol or drugs, violence, or issues with coworkers in the past (Morgan, 2010). They should also try to identify persons who tend to externalize blame for their problems, who are hostile, and who frequently change jobs, as all these characteristics are indicators of a person who may become violent at work (Morgan, 2010). Employers should also be watchful of warning signs of violence such as when a person is obsessed with weapons or brings weapons to work, when a person recently has been written-up or fired, when a person has made a threat, when a person is intimidating or has made others fearful, when a person has demonstrated romantic interest in a coworker that is not reciprocated, when a person is paranoid, when a person cannot accept criticism, when a person has experienced personal problems, or when a person begins changing work habits (e.g., showing up to work late when usually on time) (Morgan, 2010).

Table 9.2   Types of Workplace Prevention Strategies

Strategy	Definition	Examples
Environmental	Focuses on ways to make a workplace more secure and a less attractive target	Install better lighting, install security cameras and bullet-proof barriers or enclosures, post signs stating that only small amounts of cash are on hand, make high-risk areas visible to more people, install silent alarms, have police check on workers, etc.
Organizational and administrative	Focuses on strategies that administrators and agencies can implement to reduce the risk of workplace victimization in their organizations	Providing training on maintaining a safe work environment; instituting ban on working alone; recording verbal abuse accidents and suspicious behavior; policies that define what is considered workplace victimization/violence and methods for defusing volatile situations; training on how to use security equipment; access to psychological counseling and/or support to reduce likelihood of acting out at work; and access to services following acts of workplace victimization
Behavioral	Actions that workers can take to reduce their risk of workplace victimization	Training in nonviolent response and conflict resolution, training on how to anticipate and respond to potential violence, training on how to resist during a robbery

SOURCE: U.S. Department of Health and Human Services (1996).

Legislation and Regulation

You may be surprised to learn that there is not a national set of standards specific to the prevention of workplace violence (Jenkins, 2010; OSHA, n.d.). There are, however, federal agencies that provide legislative guidance for workplace safety and health. The **Occupational Safety and Health Administration (OSHA)** in the U.S. Department of Labor has occupational safety and health legislative responsibility, while the National Institute for Occupational Safety and Health in the U.S. Department of Health and Human Services is responsible for research in this area (Jenkins, 2010). Both agencies, along with others in the federal government, provide publications and recommendations regarding workplace violence prevention. In addition to these federal guidelines, 25 states, Puerto Rico, and the Virgin Islands have standards and enforcement policies to address workplace violence as well as plans that have been approved by OSHA (n.d.). Some states have standards or policies that are different from the guidelines put forth by OSHA. More generally, all employers are required to provide a safe work environment that is free of recognizable hazards likely to cause death or serious bodily harm (Occupational Safety and Health Act of 1970, Pub. L. 91-596). If workers are harmed while on duty, employers may be liable if they failed to disclose the dangers workers face or if they negligently ignored the threat of workplace violence (Smith, Gambrell, & Russell, LLP, 2005).

SUMMARY

- Males are the most likely targets for victimization in schools, much the same as research shows them to be for most other crimes.
- The most commonly occurring type of school victimization is theft.
- More violent acts of victimization typically happen in lower grade levels than in grades 10 through 12.
- Individual, school, and structural forces all play a role in school victimization.
- Though extreme cases of school victimization, such as school shootings, that would get increased media attention have occurred, this type of victimization is very rare.
- When looking at school victimization, it is important to note that some teachers are also the targets of aggressive or criminal behavior. The typical teacher-victim is the instructor of a special-education class.
- Though bullying can have some serious psychological effects on people, it often is considered a normative behavior in the socialization process among students.
- Bullying can be indirect or direct, and can be social, physical, or verbal. Cyberbullying—in which people use the Internet, mobile phones, and other digital technologies to bully others—has received recent attention in the news.
- The most typical victimization a college student experiences is theft.
- College students who experience victimization, specifically violent victimization, often engage in risky lifestyles and routine activities (i.e., participating heavily in the college partying culture, binge drinking, and going to events alone and late at night).
- The Jeanne Clery Disclosure of Campus Security Policy and Campus Crime Statistics Act requires most universities to have widely accessible campus crime reports, to keep crime logs and deliver warnings about crime threats, and to protect the rights of both victims and those who are accused of sexual assaults handled on campus.
- Campuses have also responded to the threat of crime by employing campus police or security officers and providing other safety precautions and programming for students.
- In the workplace, younger adults and children are, of course, at less risk for being victimized, while the age range at greatest risk for violent and even fatal victimization is 25 to 44.
- The typology of workplace violence classifies incidents into criminal, customer/client, worker-on-worker, and personal relationship incidents.

- Thefts are the most common type of workplace victimization. Males and White workers have higher rates of workplace victimization than do females and non-White workers.
- The Census of Fatal Occupation Injuries program tracks all cases of fatality in the workplace.
- Taxicab drivers have the highest rates of workplace homicide; law enforcement officers have the highest rates of nonfatal workplace victimization.
- Working at night, alone, with money, with the public, in high-crime areas, in a mobile workplace, or in the community places people at risk for workplace victimization.
- Though there is no standard set of workplace prevention guidelines, it is the employer's responsibility to provide a safe work environment for all employees (i.e., screening potential employees before hiring, making sure the physical work space is free from hazardous material or anything that could cause bodily harm, etc.).
- Much like school victimization, the most extreme cases of fatality in the workplace, which would get media attention, are not the most common. Personal theft of belongings is more likely to occur than a workplace shooting or any other physically violent act.

DISCUSSION QUESTIONS

1. Assess your risk of being victimized at college. Are you at high risk? Why or why not? How could you reduce your risk of being victimized?

2. Based on what you have read about workplace violence, assess your risk of being a victim at your current or former place of work. If you do not work, consider your parent's place of employment. Does your job have risk factors for violence? If yes, why? How could your place of employment reduce its risk?

3. Why do you think males and minorities have higher rates of workplace victimization than do females and nonminority workers?

4. In the workplace, do you think you would be at less risk being employed as a police officer or as a taxicab driver? Why? What does the research say?

5. How important is it for you personally to protect yourself and your belongings at work if personal theft is the most common form of victimization in the workplace? How can your own routine activities in the workplace contribute to or prevent you from being a victim of theft?

KEY TERMS

school victimization

bullying

direct bullying

physical bullying

verbal bullying

indirect bullying

cyberbullying

Gun-Free Schools Act

zero-tolerance policies

Jeanne Clery Disclosure of Campus Security Policy and Campus Crime Statistics Act

workplace victimization

criminal intent incidents

customer/client incidents

worker-on-worker incidents

personal relationship incidents

environmental design

organizational and administrative controls

behavioral strategies

Occupational Safety and Health Administration (OSHA)

INTERNET RESOURCES

"Fact Sheet: Workplace Shootings" (http://www.bls.gov/iif/oshwc/cfoi/osar0014.htm)

The Bureau of Labor Statistics website provides a workplace shootings fact sheet listing statistics on the year's fatalities, injuries, and illnesses. Charts separate the shootings by workplace industry. The links on the page offer information to victims of workplace violence about how to receive compensation and benefits for their losses.

Olweus Bullying Prevention Program (http://www.olweus.org/public/index.page)

This website is a helpful tool for understanding the types of bullying that exist as more technology develops. There is a host of information for teachers, parents, and students who may be affected by some form of school victimization, and headlining news videos about bullying that has occurred in various states. This is a website with the facts and harsh realities about school victimization but also with tips and testimonials that could prevent it in the future.

"Workplace Violence" (www.osha.gov/SLTC/workplaceviolence/)

The U.S. Department of Labor's Occupational Safety and Health Administration webpage defines and outlines the standards, rules, and regulations for workplace/office violence. The key resource this website offers is a detailed list of references for workplace/hazard awareness. A lengthy list of PDFs provides information on preventing victimization and being more aware of routines at work that could make you a suitable target for violence.

Stop Cyberbullying (http://www.stopcyberbullying.org/index2.html)

This website offers valuable information on how serious this technologically advanced form of bullying is and explains what the law says and does when this type of bullying is reported. This is more of a "take action" website that explains to victims and people who want to help exactly what can be done. The site also explains how cyberbullying works and why people choose to do it so readily.

"Harassment-Free Hallways" (http://www.aauw.org/learn/research/upload/completeguide.pdf)

This article discusses ways in which sexual harassment can be prevented in schools. It provides information for students, parents, and school personnel. It also includes a brief survey that you can take to learn whether you have been a victim of sexual harassment.

Introduction to Reading 1

One of the key considerations of victimization research is the effect victimization has on victims. For bullying, the effects can be dramatic. Hay and colleagues (2010) investigate the effects of bullying and cyberbullying among a sample of about 400 students enrolled in two schools (one high school and one middle school) in a Southeastern state. Data were collected from surveys completed by the students in the sample. This article makes several key contributions. First, it examines bullying and cyberbullying, a relatively new form of bullying. Second, it examines bullying of both male and female adolescents. Third, it uses general strain theory to understand the effects that bullying has on youth. Fourth, both internalizing and externalizing responses to bullying are considered. In this way, a theoretically driven analysis of the consequences of bullying is presented.

Traditional Bullying, Cyber Bullying, and Deviance

A General Strain Theory Approach

Carter Hay, Ryan Meldrum, and Karen Mann

Agnew's (1992, 2001) general strain theory (GST) has received significant empirical scrutiny, with much of it supporting the theory. Most tests confirm its central hypothesis that strainful events and relationships are positively related to involvement in delinquency (e.g., Broidy, 2001; Capowich, Mazerolle, & Piquero, 2001; Hoffmann & Miller, 1998; Paternoster & Mazerolle, 1994). Moreover, the effects of strain on delinquency appear to be partially explained by heightened levels of anger and frustration (e.g., Brezina, 1998; Mazerolle & Piquero, 1997; Piquero & Sealock, 2000). With these favorable results, it is not surprising that GST has moved to the "forefront of criminological theory" (Hoffmann & Miller, 1998, p. 83).

And yet, important empirical issues remain unresolved or even largely unexplored in GST research. The purpose of this study is to address three issues that we

see as neglected, substantively important, and logically linked to one another. The first of these involves the need to learn more about the criminogenic effects of bullying. In a significant elaboration of GST, Agnew (2001) identified bullying—or "peer abuse"—as a strain that should be especially consequential for delinquency. Yet even as bullying has received continued attention as an adolescent social problem (e.g., White & Loeber, 2008) GST research has neglected its effects on crime and delinquency. This study addresses this void, and we do so with data that allow us to examine not only traditional notions of bullying (e.g., physical and verbal harassment) but also cyber bullying, which has garnered significant recent attention (Hinduja & Patchin, 2009; Wang, Iannoti, & Nansel, 2009).

A second goal of our study is to examine the effects of bullying not just on crime but also on noncriminal,

SOURCE: Hay et al. (2010). Reprinted with permission.

internalizing forms of deviance. GST tests often assess the effects of strain on "externalizing" crimes—acts committed against others or their property. Some individuals, however, may respond with "internalizing" acts that harm themselves. Although neglected, this possibility is consistent with GST's position that individuals cope in different ways and deviant adaptations can come in many forms (Agnew, 1992). We consider this issue by examining the effects of bullying not just on crimes committed against others but on internalizing deviance, including suicidal ideation and acts of self-harm such as "cutting" or burning oneself. These outcomes are especially important to consider when examining the effects of bullying—those who are bullied may be socially isolated and ostracized, and this may lead to self-directed responses to strain (Moon, Morash, McCluskey, & Hwang, 2009).

Our final focus is on the possibility that these relationships vary across males and females. Broidy and Agnew (1997) identified important ways in which strain crime relationships may be moderated by sex, and a number of empirical studies support their arguments (e.g., Piquero & Sealock, 2004). This issue has not, however, been examined with respect to the above two issues—issues that may especially call for a consideration of sex-specific patterns. There still is uncertainty about whether males and females experience bullying to the same degree and whether they react to it in similar ways (see Espelage, Mebane, & Swearer, 2004). Sex differences also are important to consider when studying the effects of strain on internalizing deviance. As Broidy and Agnew noted, when confronted with strain, males may resort to externalizing responses, whereas females may be more susceptible to internalizing responses. This hypothesis rarely, however, has been tested in GST research, and it has not been examined with respect to bullying. Also, recent evidence contradicts the conventional wisdom that internalizing deviance is largely a problem among females—rates of deliberate self-harm and suicidal ideation are far from trivial among males in industrialized nations (Kerr, Owen, Pears, & Capaldi, 2008; Patton et al., 2007).

This study examines these three issues in conjunction with one another: Bullying is assessed in terms of its effects on both externalizing and internalizing deviance, and these relationships are examined separately for males and females. This is done with survey data collected from a sample of students in a nonmetropolitan county of a Southeastern state. First, however, we consider in more detail both GST's positions on these various issues and the findings from prior research.

Prior Theory and Research

Bullying as a Source of Strain

Although being the victim of bullying has always fit within GST's broad conception of strain (Agnew, 1992), attention to it emerged most notably from Agnew's (2001) elaboration of GST in which he identified the strains that should be most consequential for crime. One of these was bullying (or "peer abuse"), which, unlike such strains as parental rejection and negative experiences at school, "has been neglected as a type of strain" (Agnew, 2001, p. 346). Agnew (2001) contended that bullying should be consequential because it satisfies four conditions that should characterize consequential strains: (1) It should be perceived as unjust (because bullying often will violate basic norms of justice), (2) it should be perceived as high in magnitude (because peer relations often are central in the lives of adolescents), (3) it should not be associated with conventional social control (because bullying often will occur away from adult authority), and (4) it should expose the strained individual to others—the bullies themselves—who model aggressive behavior.

Despite these suggestions of an important effect, exposure to bullying largely has been ignored in GST research and, more broadly, in research on the causes of crime. Some exceptions to that pattern suggest that bullying—or the related concept of criminal victimization—is important for delinquency (Agnew, 2002; Agnew, Brezina, Wright, & Cullen, 2002; Baron, 2004; Hay & Evans, 2006). Baron's study of homeless youths, for example, revealed that being a property victim significantly increased delinquency, even after controlling for a wide array of alternative explanatory variables. Of more direct relevance to the effects of bullying, Agnew et al. (2002) found that subjects who were picked on by neighborhood peers were more involved in delinquency, although this was true only for those with personality characteristics conducive to delinquency. More recently, findings from Moon et al. (2009) contradicted GST's position: Bullying victimization was not associated with general delinquency in their study of Korean youths.

This dearth of criminological research stands in contrast to the extensive scholarship that reveals effects of bullying on many forms of social psychological maladjustment, including low self-esteem, loneliness, and depression (see Hawker & Boulton, 2000). When behavioral outcomes have been considered, however, the focus generally has been limited to school absenteeism or antisocial behavior in the preadolescent years, prior to the point in which serious crime and delinquency is pervasive. A few recent studies (e.g., Hinduja & Patchin, 2007; Ybarra & Mitchell, 2004) provide important exceptions to that pattern and find that bullying increases adolescent crime or deviance. Thus, there is potentially much to be gained from giving further attention to Agnew's (2001) hypothesis about the important effects of bullying on adolescent behavior.

It also is important for such research to focus on the newly emerging issue of cyber bullying (Hinduja & Patchin, 2009), which involves using the Internet or cell phones to mistreat others. This includes abusive e-mails or text messages, insulting messages or pictures on online message boards, and Web sites that disseminate degrading content. Recent surveys of adolescents indicate their potential exposure to cyber bullying—nearly 90% frequently use the Internet and 50% have their own cell phones (Lenhart, Madden, & Hitlin, 2005; Ybarra & Mitchell, 2004). These media often are used to harass or embarrass others (Ybarra & Mitchell, 2004), but little is known about how this victimization may affect behavior. There is reason for concern because, unlike face-to-face bullying, cyber bullying may be especially difficult to escape from. Its electronic nature may allow it to occur without attracting the attention of teachers or parents. Also, because many adolescents—for legitimate reasons—carry their cell phones at all times and frequently use the Internet, they can be exposed to cyber bullying even when physically removed from bullies. And once information is posted to the Web, it may be difficult for the bullying victim to have it removed from all of the sites in which it may have appeared. It also can reach a much wider audience than what may be possible with traditional bullying. Mason (2008, p. 324) comments on the relentless nature of cyber bullying, noting that it "can harass individuals even when [they are] not at or around school. . . . [U]nlike traditional forms of bullying, home may no longer be a place of refuge." Thus, if bullying is to be examined as an important source of strain, attention to cyber bullying should be a key consideration.

Internalizing Responses to Strain

GST predicts that individuals will respond to strain in different ways (Agnew, 1992). Tests of GST, however, have disproportionately focused on criminal responses, especially acts that harm another person either through violence or through the theft or damage of their property. When internalizing responses have been considered, the focus generally has been limited to substance use. Thus, an entire class of internalizing acts—aggression against oneself with acts of deliberate self-harm—has been ignored. This type of behavior includes such things as cutting or burning oneself, jumping from heights, running into traffic, poisoning, hanging, and self-battery, with each of these acts sometimes resulting in suicide (Hawton, Rodham, & Evans, 2006).

Ignoring this type of response to strain could be a significant omission in GST research. First, by not considering a broader array of responses to strain, prior studies may have misclassified some subjects. Specifically, some adolescents may have been seen as coping with strain in prosocial ways simply because their harmful responses were not captured by a study that focused only on criminal outcomes. Insight on this comes from a study by Sharp and her colleagues (Sharp, Terling-Watt, Atkins, Gilliam, & Sanders, 2001), who found that 23% of a sample of college females reported some type of eating disorder, and this outcome was affected by strain. Given that these individuals may have committed few crimes against others, their responses to strain might have been overlooked in a more conventional test of GST.

A second reason to consider acts of deliberate self-harm is that they may be more common than is typically recognized. This issue has been studied extensively outside the United States since the 1960s, with many studies finding higher than expected rates of self-harm. For example, in the United Kingdom, more than 20,000 adolescent hospital admissions occur each year because of self-inflicted overdoses, poisonings, or injuries (Hawton et al., 2006). Comparable prevalence rates were observed in such countries as France, Ireland, and Australia (Hawton et al., 2006). This issue has garnered recent attention in the United States from the Centers for Disease Control and Prevention (2008), who found that nearly 14% of high school students in a national survey seriously contemplated suicide in the prior year.[1]

The neglect of self-harm in criminological research is understandable, given our focus toward general theories that can explain a wide range of deviant or harmful outcomes. Gottfredson and Hirschi's (1990) self-control theory, for example, often is lauded for its ability to explain many noncriminal behaviors that are rewarding in the short-term but carry long-term costs. GST may be a similarly general theory—in addition to explaining criminal outcomes, it may also explain involvement in many harmful, self-directed actions that are used to cope with intense feelings of stress.

One final point should be emphasized: Self-harm may be especially important to consider when studying the effects of bullying. As many studies in psychology reveal, bullying victims suffer from a wide array of social psychological maladjustments and tend to be socially isolated. Thus, rather than responding to this strain with normal delinquent or criminal acts (many

of which are committed in the context of adolescent peer groups), bullied individuals may respond with acts committed against themselves.

Male-Female Differences in the Response to Strain

Broidy and Agnew (1997) systematically introduced to criminology the idea that males and females may differ in their levels of exposure and responses to strain. They offered several relevant hypotheses. First, relative to females, males should be exposed to criminogenic strains at higher levels. Second, because of sex differences in stress coping, males should have emotional reactions to strain that are conducive to externalizing responses (crime), whereas females should have emotional reactions more conducive to internalizing responses. For example, although both males and females may get angry in response to strain, females may also experience self-directed emotions like guilt, shame, and depression. And last, net of any differences in emotions, males should have behavioral reactions to strain that are more criminal, in part because they have lower personal coping resources and fewer social constraints to criminal coping.

These arguments have received at least moderate support, suggesting that higher male crime is partially explained by GST processes. For example, males are exposed to some criminogenic strains at higher levels than females (Baron, 2004; Hay, 2003), and they are more likely to respond to given strain with crime (Agnew & Brezina, 1997; Hay, 2003; Piquero & Sealock, 2004).[2] Sex differences have not yet been considered, however, in the few GST studies of bullying and peer victimization. Moreover, questions about sex differences remain in the larger study of bullying. The belief that bullying is largely a male problem disappeared when conceptions of bullying broadened to include "relational" forms of bullying (Espelage et al., 2004) like gossip, ridicule, and friendship withdrawal. All of these appear to be common among both male and female adolescents (see Crick & Grotpeter, 1995). And with respect to cyber bullying, strong conclusions on sex differences have yet to emerge, although it may be more common among females (Hinduja & Patchin, 2009).

These patterns leave open the question of whether males or females are exposed to bullying at higher levels, and they suggest that the answer to this question may depend on the type of bullying being considered. Just as important, the research on bullying has yet to examine Broidy and Agnew's (1997) hypotheses regarding the ways in which male and female adolescents differ in their responses to bullying. Thus, there is a clear need for research that examines sex-specific responses to bullying and does so for multiple outcomes that include both externalizing and internalizing forms of deviance.

⬙ The Present Study

The purpose of this study is to examine the effects of bullying on externalizing and internalizing forms of deviance and to assess whether these relationships vary across males and females. Our central hypothesis is that, consistent with GST, bullying is significantly related to both types of deviance. Testing this hypothesis reveals insight on the accuracy of Agnew's (2001) claims regarding the importance of exposure to bullying, and it clarifies whether GST can be extended to explain aggression directed against the self. With respect to sex differences, two hypotheses are tested, both of which draw from Broidy and Agnew (1997) and the related research. We predict that the effects of bullying should be greater for males than females when the dependent variable is externalizing behavior. Conversely, the effects of bullying should be greater for females for dependent variables that involve internalizing deviance. We should emphasize, however, that these hypotheses are offered tentatively. The stress and coping literature upon which Broidy and Agnew based their arguments has not focused on adolescent stressors like bullying. Moreover, its concern with internalizing deviance often has emphasized emotional (e.g., depression) rather than behavioral outcomes. Thus, our test of how these relationships vary according to sex is exploratory to some degree.

Data

These issues are considered with data collected from a sample of roughly 400 adolescents in a Southeastern state of the United States in the spring of 2008. Respondents were sampled from two participating schools—one high school and one middle school—located in a rural and relatively poor county. Using the standards set by the school district, a passive consent procedure was followed. Permission forms were distributed to all students 1 week prior to the survey administration, and students were excluded from the study if parents returned the form asking that their child be excluded. Each participating student then completed an anonymous, self-administered questionnaire during normal school hours and was given a small reward (a candy bar) for completion. This procedure allowed for a near complete census of the two schools' populations, with 93% of attending students participating in the study. This produced a fairly diverse sample. The average age of participants was 15 but ranged from 10 to 21. The sample was split evenly between males and females, and non-Whites represented 34 of the sample. Additionally, family disruption was common, with only 50% of respondents living in a household with both biological parents.

Measures

The survey allowed for multiple-item scales for most variables, and there are two features common to the measures that we used. First, all items in multiple-item scales included ordinal response categories, with almost all using a 4-point scale. For measures that assess frequencies, responses ranged from 1 = *never* to 4 = *often*. For items asking respondents to rate themselves on some characteristic, responses ranged from 1 = *strongly disagree* to 4 = *strongly agree*. Second, with respect to scale construction, each scale was computed by averaging its constituent items.

Independent Variables

To assess the effects of bullying, two measures were used. The first is a 6-item measure ($\alpha = .85$) that captures the traditional emphasis on physical and verbal harassment. Respondents indicated how often during the prior 12 months they were (a) the target of lies or rumors; (b) the target of attempts to get others to dislike them; (c) called names, made fun of, or teased in a hurtful way; (d) hit, kicked, or pushed by another student; (e) physically threatened by other students; and (f) picked on by others. Our second bullying measure is a 3-item scale ($\alpha = .80$) that captures the more recent interest in cyber bullying. Respondents were asked to indicate how frequently during the previous 12 months they were (a) the target of "mean" text messages; (b) sent threatening or hurtful statements or pictures in an e-mail or text message; and (c) made fun of on the Internet.

Dependent Variables

Externalizing delinquency was measured with a 5-item scale ($\alpha = .86$) of self-reported offending during the prior 12 months. Respondents indicated how often they had (a) stolen something worth less than $50; (b) stolen something worth more than $50; (c) damaged, destroyed, or tagged property that did not belong to them; (d) entered a building or house without permission from the owner; and (e) hit, kicked, or struck someone with the idea of seriously hurting them. Two measures of internalizing behavior were used, with both measured with a single item. The first is a measure of suicidal ideation in which respondents were asked how often "you think about killing yourself." Self-harm was measured by asking respondents how often "you purposely hurt yourself without wanting to die," with "cutting or burning" offered as examples.

Control Variables

A number of demographic control variables were included in the analyses to protect against concerns about spuriousness. These included five demographic variables: age (measure in years), sex (*male* = 1, *female* = 0), race (*non-White* = 1, *White* = 0), nonintact family structure (*nonintact* = 1, *living with both biological parents* = 0), and place of birth (*foreign-born* = 1, *native born* = 0). Also, to better isolate the independent relationship between exposure to bullying and the outcomes of interest, controls

were included to capture key aspects of respondents' school, family, peer, and personal characteristics. This included measures of school grades (as indicated by self-reported grades on the most recent report card); parental control, as indicated by a 10-item scale ($\alpha = .92$) of parental monitoring and discipline; and unstructured time spent with peers, as indicated by a 2-item scale ($r = .56$) measuring time spent with friends with no adults present and time spent with friends at a mall, restaurant, or street corner. And last, all analyses included an 8-item measure ($\alpha = .85$) of self-control, which included the 8 items used in the Grasmick et al. scale (Grasmick, Tittle, Bursik, & Arneklev, 1993) to measure impulsivity and risk seeking.[3]

Results

The first step in the analysis was to consider the effects of bullying on our externalizing and internalizing outcomes. Given the high correlation ($r = .67$) between traditional and cyber bullying, the effects of the two were estimated in separate equations. Thus, with two measures of bullying and three outcomes of interest (delinquency, self-harm, and suicidal ideation), we estimated six ordinary least squares (OLS) equations, each of which included all of the controls.

The results for these equations are shown in Table 2, which reveals a consistent effect of bullying—the effects of bullying are statistically significant and relatively large in all six equations (with betas ranging from .22 to .41). Cyber bullying has modestly higher effects than traditional bullying—standardized effects of .33 for delinquency, .39 for self-harm, and .41 for suicidal ideation, which compares to effects of .22, .33, and .39 for traditional bullying. Also, both types of bullying have greater effects on self-harm and suicidal ideation than on delinquency. For traditional bullying, for example, the effect on suicidal ideation ($B = .39$) is nearly 80% higher than the effect on delinquency ($B = .22$). The pattern is less extreme but still true for cyber bullying, which has an effect on suicidal ideation ($B = .41$) that is 24% higher than its effect on delinquency ($B = .33$). Thus, bullying has a consistent, relatively strong association with delinquency, self-harm, and suicidal ideation,

Table 1 Descriptive Statistics

Total Sample	N	M	SD	Minimum	Maximum
Age	424	14.99	2.18	10.00	21.00
Male	420	0.50	0.50	0.00	1.00
Non-White	422	0.34	0.48	0.00	1.00
Foreign-born	423	0.07	0.26	0.00	1.00
Nonintact family	407	0.50	0.50	0.00	1.00
Poor school grades	391	2.02	1.04	1.00	5.00
Parental control	416	3.08	0.76	1.00	4.00
Time spent with peers	407	2.65	1.78	0.00	5.00
Self-control	422	2.79	0.64	1.00	4.00
Cyber bullying victimization	417	1.33	0.64	1.00	4.00
Traditional bullying victimization	419	1.74	0.72	1.00	4.00
Delinquency	415	1.23	0.51	1.00	4.00
Self-harm	418	1.31	0.75	1.00	4.00
Suicidal ideation	417	1.33	0.76	1.00	4.00
Male Sample					
Cyber bullying victimization	204	1.30	0.62	1.00	4.00
Traditional bullying victimization	205	1.76	0.72	1.00	4.00
Delinquency	205	1.28	0.54	1.00	4.00
Self-harm	205	1.25	0.70	1.00	4.00
Suicidal ideation	204	1.31	0.74	1.00	4.00
Female Sample					
Cyber bullying victimization	209	1.37	0.65	1.00	4.00
Traditional bullying victimization	209	1.72	0.72	1.00	4.00
Delinquency	205	1.16	0.43	1.00	4.00
Self-harm	210	1.37	0.80	1.00	4.00
Suicidal ideation	210	1.35	0.79	1.00	4.00

Table 2   OLS Regressions of Dependent Variables on Traditional and Cyber Bullying Victimization

Predictor	Delinquency (N = 363)			Self-Harm (N = 365)			Suicidal Ideation (N = 364)		
	b	SE	B	b	SE	B	b	SE	B
Age	0.02	0.012	0.07	0.00	0.017	0.01	0.02	0.017	0.05
Male	0.09	0.049	0.09	−0.18	0.070	−0.12**	−0.08	0.071	−0.05
Non-White	0.14	0.055	0.13*	0.21	0.079	0.13**	0.19	0.080	0.12*
Foreign-born	0.01	0.098	0.00	0.06	0.141	0.02	0.27	0.142	0.10
Nonintact family	0.10	0.051	0.10*	0.09	0.073	0.06	0.10	0.073	0.07
Poor school grades	0.00	0.025	0.01	−0.01	0.036	−0.02	−0.06	0.037	−0.09
Parental control	−0.16	0.035	−0.24**	−0.14	0.050	−0.14**	−0.17	0.051	−0.17**
Time with peers	0.04	0.014	0.13**	0.03	0.020	0.06	0.00	0.020	0.00
Self-control	−0.04	0.042	−0.05	−0.21	0.061	−0.18**	−0.13	0.062	−0.11*
Traditional bullying	0.15	0.035	0.22**	0.34	0.050	0.33**	0.41	0.051	0.39**
Constant	1.10	0.264	—	1.62	0.379	—	1.29	0.382	—
Adjusted R^2	—	0.21	—	—	0.24	—	—	0.26	—
Age	0.01	0.011	0.02	−0.02	0.016	−0.06	−0.01	0.017	−0.03
Male	0.12	0.047	0.12*	−0.14	0.069	−0.10*	−0.04	0.070	−0.03
Non-White	0.13	0.053	0.12*	0.18	0.077	0.11*	0.15	0.078	0.10
Foreign-born	0.02	0.094	0.01	0.09	0.137	0.03	0.31	0.140	0.11*
Nonintact family	0.12	0.048	0.12*	0.14	0.070	0.09	0.16	0.072	0.11*
Poor school grades	0.01	0.024	0.01	−0.01	0.035	−0.01	−0.05	0.036	−0.07
Parental control	−0.17	0.034	−0.24**	−0.15	0.049	−0.15**	−0.18	0.050	−0.18**
Time with peers	0.02	0.014	0.08	0.00	0.020	0.01	−0.02	0.020	−0.05
Self-control	−0.05	0.040	−0.06	−0.24	0.058	−.21**	−.018	0.059	−0.16**
Cyber bullying	0.26	0.038	0.33**	0.45	0.054	0.39**	0.49	0.055	0.41**
Constant	1.21	0.231	—	2.08	0.336	—	1.94	0.344	—
Adjusted R^2	—	0.27	—	—	0.29	—	—	0.28	—

*$p < .05.$ **$p < .01.$

but this is especially true for cyber bullying in particular and for outcomes that involve internalizing rather than externalizing deviance.

Our next step in the analysis was to examine whether these relationships vary across males and females; in short, do males and females differ in their response to traditional and cyber bullying? It is first useful to consider whether there were sex differences in exposure to these forms of bullying. The descriptives provided in Table 1 reveal that there were not. On scales that ranged from 1 to 4 (indicating exposure as 1 = *never*, 2 = *rarely*, 3 = *sometimes*, or 4 = *often*), both males and females had average values of approximately 1.75 for traditional bullying and 1.35 for cyber bullying. (The differences between males and females were not significant.) Thus, if bullying victimization is to produce divergent outcomes for males and females, it will result not from their differing extent of exposure, but instead, from their differing reactions.

To consider this possibility, we estimated OLS regression equations identical to those presented in Table 2, except that they were estimated separately for males and females. Table 3 provides a summary of the key results from these equations. For each male-female comparison, we provide the unstandardized

coefficient and standard error for the bullying measure in question. Also, we provide the z-score statistic used to determine whether the coefficients for males and females significantly differed. We used the formula recommended by Paternoster and his colleagues (Paternoster, Brame, Mazerolle, & Piquero, 1998) that takes b1–b2 (the difference between the two coefficients) as the numerator and the square root of $SEb1^2 + SEb2^2$ (the estimated standard error of the differences) as the denominator. If this formula yields a value for z that exceeds 1.96, the null hypothesis that b1 = b2 is rejected for a two-tailed test with an alpha level of .05.

The figures in Table 3 reveal that in four of the six bullying-deviance combinations, there are no significant differences in effects between males and females. Traditional bullying has effects (shown in the top pane of Table 3) that are similar for males and females across all dependent variables—each effect is significant, and the differences between the coefficients for males and females are negligible and insignificant. This pattern also is true for cyber bullying (shown in the bottom panel of Table 3) when delinquency is the dependent variable—the effect of cyber bullying on delinquency is almost identical for males and females.

Table 3 Z-Score Test for Differences in Effects of Bullying for Males and Females

Type	Delinquency			Self-Harm			Suicidal Ideation		
	b	SE	Z	b	SE	Z	b	SE	Z
Traditional bullying									
Males	0.135	0.057	−0.54	0.366	0.067	0.46	0.454	0.074	0.63
Females	0.174	0.044	—	0.319	0.077	—	0.389	0.071	—
Cyber bullying									
Males	0.274	0.062	0.12	0.578	0.069	2.20*	0.648	0.078	2.41*
Females	0.265	0.045	—	0.342	0.082	—	0.382	0.078	—

NOTE: All equations included controls for age, non-White, foreign-born, nonintact family, poor school grades, parental control, time spent with peers, and self-control.

*$p < .05$.

A different pattern emerges, however, for the effects of cyber bullying on self-harm and suicidal ideation—these effects are significantly greater for males. To be clear, exposure to cyber bullying is associated with heightened internalizing deviance for both males and females. However, the effects on these two outcomes are about 70% higher than what is observed for females, and these effects for males are quite large in absolute terms, with standardized effects (not shown) of .52 on self-harm and .54 on suicidal ideation. Indeed, these effects of cyber bullying on males' self-harm and suicidal ideation are nearly double the standardized effect of cyber bullying on male delinquency ($B = .29$).

Taken as a whole, these results are consistent with the possibility that males and females sometimes differ in their responses to strain. However, support for this idea emerged only when considering the effects of cyber bullying on internalizing deviance. Moreover, the exact pattern of differences was unexpected—internalizing responses to strain were higher among males rather than females.

Discussion and Conclusion

This study used Agnew's GST as the theoretical foundation for studying the effects of bullying on both externalizing and internalizing forms of deviance. Moreover, given prior theory and research of sex differences in stress coping, we were interested in examining how the effects of bullying would vary across males and females. Three key conclusions emerged from the analysis.

The first is that both forms of bullying victimization—a "traditional" measure based on physical and verbal harassment and a "cyber" measure based on online or electronic harassment—were significantly related to delinquency. This finding confirms the conclusions from other recent studies (e.g., Hinduja & Patchin, 2007) that bullying has important effects on delinquency. Also, although the differences were not extreme, we found that effects were greater for cyber bullying. It will be interesting to observe whether this pattern emerges in other studies. As noted previously, there are reasons to suspect that cyber bullying could indeed be

the more problematic form of bullying. Its electronic form allows it to occur in ways that are less visible and overt; it, therefore, may not attract the attention of parents or teachers. Moreover, cyber-bullied youths may find it more difficult to gain relief than those who are bullied in more traditional ways, because being physically removed from bullies offers little relief, and the bullying may reach a wider audience. Thus, although the significant effects of both types of bullying support Agnew's (2001) position that bullying is a more consequential form of strain than earlier believed, our findings especially suggest the importance of moving "beyond the schoolyard" to consider bullying that is linked to adolescents' growing use of the Internet and cell phones (Hinduja & Patchin, 2009).

Our second key finding is that both forms of bullying affected not simply delinquency (which was measured in terms of externalizing acts against other people or their property) but also internalizing forms of deviance like intentional self-harm and suicidal ideation. Indeed, these relationships were of greater magnitude than those observed between bullying and delinquency. One possible explanation for this involves the way in which bullying may socially isolate its victims—if victims are rejected by others or voluntarily withdraw from social interactions, this may encourage internalizing rather than externalizing emotional and behavioral responses. Again, this pattern needs to be confirmed in future studies. This is an important issue to consider; however, given GST research rarely has considered internalizing behavioral responses to strain. The findings in this study suggest that GST processes are relevant to a potentially wide array of harmful, self-defeating action not typically evaluated by criminologists.

Our final key finding is that these relationships were moderated by sex in some instances. Specifically, the effects of bullying on self-harm and suicidal ideation were greater for males, and this difference was large and statistically significant with respect to cyber bullying in particular—its effects on self-harm and suicidal ideation were approximately 70% greater for males. This finding contradicts the specific GST arguments that have been made (Broidy & Agnew, 1997)—compared to females, males were expected to respond to strain with externalizing rather than internalizing deviance.

This unexpected pattern calls for speculation on what may explain it. Given that rates of self-harm and suicidal ideation are almost always higher among females (Hawton et al., 2006), males do not seem predisposed to respond to stress with internalizing deviance. Thus, a more plausible explanation for our finding could involve the focus on bullying in particular. Indeed, Hawton and colleagues reached a similar conclusion in their study of bullying—exposure to bullying increased the odds of internalizing deviance more for males than females. There may be two explanations for why bullying generated greater internalizing deviance for males. Both follow from the premise that while externalizing deviance often is facilitated by social engagement with peers (Warr, 1996), internalizing deviance often is the opposite—it is especially likely when a person is socially isolated (Hawton et al., 2006). Thus, one explanation may be that bullying victimization socially isolates males to a greater degree than what is observed for females. In connection, some have observed that relational forms of bullying are normative to some degree in female peer groups (Crick & Grotpeter, 1995); thus, rather than severing the victim's ties to the social network, some degree of bullying may be part and parcel of female network membership. With males, however, the victimization may be less consistent with one's membership in the group and may denigrate or emasculate the victim in ways that sever his or her ties with it (thus prompting internalizing deviance). A second and related possibility is that the bullying that males experience may be notably more severe or threatening than what is experienced by females, and this may prompt greater social withdrawal. Physical bullying is in fact more common among males (Espelage et al., 2004). Moreover, Hinduja and Patchin (2009) provided evidence on sex-differentiated emotional responses to bullying that may encourage greater social avoidance from males—males and females both expressed anger and frustration in response to bullying, but males were twice as likely to report being scared.

Taken as a whole, these conclusions can potentially advance and redirect future GST research, but they also should be viewed in the context of our study's limitations. First, our analyses were based on cross-sectional data gathered at one point in time. Thus, rather than assessing acts of deviance that necessarily followed exposure to bullying, our incidents of deviance and bullying occurred during the same time period (the prior 12 months). Our study, therefore, offers no guarantee of capturing the appropriate causal order (Lauritsen, Sampson, & Laub, 1991). To correct for this, future studies may use longitudinal data to examine lagged effects of bullying often amounting to 1 year or more. This approach provides a less than ideal match with theoretical arguments about the relatively short-term or instantaneous effects of strain (Agnew, 2001). A second limitation of our study involves the sample, which came from students from just two schools in a nonmetropolitan county. Although this sample does not appear to bias the results in favor of the observed findings, different results could emerge with samples that are more representative of the national population of adolescents. Last, and similar to other studies (although see Moon et al., 2009, for a recent exception), we were not able to confirm that those who experienced bullying perceived this to be strainful. Instead, we inferred the presence of strain from the significant positive relationships between bullying victimization and deviance. A fuller test of GST could consider this issue in a more direct way.

In concluding, it bears emphasizing that our findings suggest the notable gains in knowledge that may come from greater attention to the effects of bullying on adolescent behavior. Moreover, and in the spirit of GST's attention to general rather than narrow social dynamics, bullying should be examined in ways that emphasize the variety of forms in which it comes and the variety of consequences it may have.

⊠ Notes

1. It also should be noted that acts of deliberate self-harm often are quite serious. Even nonfatal incidents are injurious and suggest the emotional and mental suffering of those involved

(Vajani, Annest, Crosby, Alexander, & Millet, 2007). Moreover, non-fatal incidents often use the same techniques (especially poisoning, cutting and piercing, and suffocation) that are common in suicide attempts. Indeed, more than 50% of those admitted to a hospital for nonfatal self-harm have a history of suicidal behavior or ideation (Vajani et al., 2007).

2. See Hoffmann and Cerbone (1999) and Mazerolle (1998) for contrary evidence.

3. Table 1 provides descriptive statistics for the full sample and for the male and female subsamples (for key variables of interest).

References

Agnew, R. (1992). Foundation for a general strain theory of crime and delinquency. *Criminology, 30*, 47–87.

Agnew, R. (2001). Building on the foundation of general strain theory: Specifying the types of strain most likely to lead to crime and delinquency. *Journal of Research in Crime and Delinquency, 38*, 319–361.

Agnew, R. (2002). Experienced, vicarious, and anticipated strain: An exploratory study on physical victimization and delinquency. *Justice Quarterly, 19*, 603–632.

Agnew, R., & Brezina, T. (1997). Relational problems with peers, gender, and delinquency. *Youth & Society, 29*, 84–111.

Agnew, R., Brezina, T., Wright, J. P., & Cullen, F. T. (2002). Strain, traits, and delinquency: Extending general strain theory. *Criminology, 40*, 43–72.

Baron, S. W. (2004). General strain, strict youth and crime: A test of Agnew's revised theory. *Criminology, 42*, 457–483.

Brezina, T. (1998). Adolescent maltreatment and delinquency: The question of intervening processes. *Journal of Research Crime and Delinquency, 35*, 71–99.

Broidy, L. (2001). A test of general strain theory. *Criminology, 39*, 9–36.

Broidy, L., & Agnew, R. (1997). Gender and crime: A general strain theory perspective. *Journal of Research in Crime and Delinquency, 34*, 275–306.

Capowich, G. E., Mazerolle, P., & Piquero, A. (2001). General strain theory, situational anger, and social networks: An assessment of conditioning influences. *Journal of Criminal Justice, 29*, 445–461.

Centers for Disease Control and Prevention. (2008). Youth risk behavior surveillance—United States, 2007. *Surveillance summaries, June 6, 2008. Morbidity and Mortality Weekly Report 2008, 57* (No. SS-1).

Crick, N. R., & Grotpeter, J. F. (1995). Relational aggression, gender, and social-psychological adjustment. *Child Development, 66*, 710–722.

Espelage, D. L., Mebane, S. E., & Swearer, S. N. (2004). Gender differences in bullying: Moving beyond mean level differences. In D. L. Espelage & S. M. Swearer (Eds.), *Bullying in American schools: A social-ecological perspective on prevention and intervention* (pp. 15–35). Mahwah, NJ: Lawrence Erlbaum.

Gottfredson, M., & Hirschi, T. (1990). *A general theory of crime*. Palo Alto, CA: Stanford University Press.

Grasmick, H. G., Tittle, C. R., Bursik, R. J., & Arneklev, B. (1993). Testing the core empirical implications of Gottfredson and Hirschi's general theory of crime. *Journal of Research in Crime and Delinquency, 30*, 5–29.

Hawker, D. S. J., & Boulton, M. J. (2000). Twenty years' research on peer victimization and psychosocial maladjustment: A meta-analytic review of cross-sectional studies. *The Journal of Child Psychology and Psychiatry, 41*, 441–445.

Hawton, K., Rodham, K., & Evans, E. (2006). *By their own hand*. Philadelphia: Jessica Kingsley Publishers.

Hay, C. (2003). Family strain, gender, and delinquency. *Sociological Perspectives, 46*, 107–135.

Hay, C., & Evans, M. (2006). Violent victimization and involvement in delinquency: Examining predictions from general strain theory. *Journal of Criminal Justice, 34*, 261–274.

Hinduja, S., & Patchin, L. W. (2007). Offline consequences of online victimization: School violence and delinquency. *Journal of School Violence, 6*, 89–112.

Hinduja, S., & Patchin, J. W. (2009). *Bullying beyond the schoolyard: Preventing and responding to cyberbullying*. Thousand Oaks, CA: Corwin Press.

Hoffmann, J. P., & Cerbone, F. G. (1999). Stressful life events and delinquency escalation in early adolescence. *Criminology, 37*, 343–373.

Hoffmann, J. P., & Miller, A. S. (1998). A latent variable analysis of general strain theory. *Journal of Quantitative Criminology, 14*, 63–110.

Kerr, D. C. R., Owen, L. D., Pears, K. C., & Capaldi, D. M. (2008). Prevalence of suicidal ideation among boys and males assessed annually from ages 9 to 29 years. *Suicide and Life-Threatening Behavior, 38*, 390–402.

Lauritsen, J. L., Sampson, R. J., & Laub, J. H. (1991). The link between offending and victimization among adolescents. *Criminology, 29*, 265–292.

Lenhart, A., Madden, M., & Hitlin, P. (2005). *Teens and technology: Youth are leading the transition to a fully wired and mobile nation*. Washington, DC: Pew Internet and American Life Project. Retrieved July 6, 2009, from http://www.pewinternet.org/~media/FilesReports/2005/PIP_Teens_Tech_July2005web.pdf

Mason, K. L. (2008). Cyber bullying: A preliminary assessment for school personnel. *Psychology in the Schools, 45*, 65–91.

Mazerolle, P. (1998). Gender, general strain, and delinquency: An empirical examination. *Justice Quarterly, 15*, 65–91.

Mazerolle, P., & Piquero, A. (1997). Violent responses to strain: An examination of conditioning influences. *Violence and Victims, 12*, 323–343.

Moon, B., Morash, M., McCluskey, C. P., & Hwang, H. (2009). A comprehensive test of general strain theory: Key strains, situational and trait-based negative emotions, conditioning factors, and delinquency. *Journal of Research in Crime and Delinquency, 46*, 182–212.

Paternoster, R., Brame, R., Mazerolle, P., & Piquero, A. (1998). Using the correct statistical test for the equality of regression coefficients. *Criminology, 36*, 859–866.

Paternoster, R., & Mazerolle, P. (1994). General strain theory and delinquency: A replication and extension. *Journal of Research in Crime and Delinquency, 31*, 235–263.

Patton, G. C., Hemphill, S. A., Beyers, J. M., Bond, L., Tumbourou, J. W., McMorris, B. J., et al. (2007). Pubertal stage and deliberate self-harm in adolescents. *Journal of the American Academy of Child and Adolescent Psychiatry, 46*, 508–514.

Piquero, N. L., & Sealock, M. D. (2000). Generalizing general strain theory: An examination of an offending population. *Justice Quarterly, 17*, 449–484.

Piquero, N. L., & Sealock, M. D. (2004). Gender and general strain theory: A preliminary test of Broidy and Agnew's gender/GST hypotheses. *Justice Quarterly, 21*, 125–158.

Sharp, S. F., Terling-Watt, T. L., Atkins, L. A., Gilliam, J. T., & Sanders, A. (2001). Purging behavior in a sample of college females: A research note on general strain theory and female deviance. *Deviant Behavior, 22*, 171–188.

Vajani, M., Annest, J. L., Crosby, A. E., Alexander, J., & Millet, L. M. (2007). Nonfatal and fatal self-harm injuries among children aged 10–14 years—United States and Oregon, 2001–2003. *Suicide and Life-Threatening Behavior, 37*, 493–506.

Wang, J., Iannotti, R. J., & Nansel, T. R. (2009). School bullying among adolescents in the United States: Physical, verbal, relational, and cyber. *Journal of Adolescent Health, 45*, 368–375.

Warr, M. (1996). Organization and instigation in delinquent groups. *Criminology, 34*, 11–37.

White, N. A., & Loeber, R. (2008). Bullying and special education as predictors of serious delinquency. *Journal of Research in Crime and Delinquency, 45*, 380–397.

Ybarra, M., & Mitchell, K. (2004). Online aggressor/targets, aggressors, and targets: A comparison of associated youth characteristics. *Journal of Child Psychology and Psychiatry, 45*, 1308–1316.

DISCUSSION QUESTIONS

1. The effects of cyberbullying on male self-harm and suicidal ideation were significantly greater than the effects of cyberbullying on female self-harm and suicidal ideation. How can you explain this finding?

2. Why do you think that being bullied has an effect on delinquency?

3. What are the policy implications of this paper? How can schools and parents use these findings?

4. How can the measure of cyberbullying be improved or expanded for future research? Given the findings, do you think that cyberbullying is more problematic than traditional bullying? Why or why not?

◈

Introduction to Reading 2

Although colleges are generally safe places, there are still risks associated with college life. Wilcox and colleagues (2007) study college women's actual sexual assault, sexual coercion, physical assault, and stalking victimization experiences along with their assessments of campus risk, worry about crime, and precautionary behaviors. They do so through data collected from a telephone survey of 1,010 female undergraduate and graduate students enrolled at a Southeastern university. The sample was 84.2% White, with a median age of 21 years. For people who face a "real" threat of being victimized, you would think that they would (1) assess the campus as being the most risky, (2) express the highest levels of worry, and (3) exercise the most precautionary behaviors. The interrelationships between victimization, assessment of risk, worry about crime, and precautionary behavior are investigated—along with how the perpetrator may structure these relationships. How campuses report victimization is discussed in light of the researchers' findings.

A Multidimensional Examination of Campus Safety

Victimization, Perceptions of Danger, Worry About Crime, and Precautionary Behavior Among College Women in the Post-Clery Era

Pamela Wilcox, Carol E. Jordan, and Adam J. Pritchard

Leading fear-of-crime researcher Mark Warr (2000) suggests,

> Fear is a natural and commonplace emotion. Under many circumstances, it is a beneficial, even life-saving emotion. Under the wrong circumstances, it is an emotion that can unnecessarily constrain behavior, restrict freedom and personal opportunity and threaten the foundation of communities. (p. 482)

Warr implies that "wrong circumstances" for fear of crime are those in which there is a disconnect between fear levels and objective risk levels. "Fear, then, is not intrinsically bad. It is when fear is out of proportion to objective risk that it becomes dysfunctional" (p. 455).[1] Warr's delineation of fear as an experience with crime often very distinct from objective experience is an important one. Furthermore, his highlighting of both the positive and negative functions of fear (vis-à-vis objective risk) is important, though what constitutes "dysfunctionally out of proportion" has not yet been clearly defined (at either end of the spectrum) in the literature. After all, it makes intuitive sense to expect and even desire a somewhat higher level of fear than actual victimization within society, with some people's fear undoubtedly serving to reduce their future victimization risk, thus highlighting a very functional "feedback loop" or reciprocity between objective and subjective risk (Cook, 1986). At the same time, fear levels dramatically

higher than actual risk and/or experience level are unhealthy anxieties that can lead to unnecessarily restrictive behaviors. Further complicating the issue is the possibility that some people with high objective risks may have low fear levels (Warr, 1994), thus creating another type of dysfunctional fear—too little fear.

Although no threshold for a healthy or functional versus unhealthy or dysfunctional objective-to-subjective risk ratio has been established, it is clear that people experience crime in different ways—both objectively and subjectively—and that both types of experiences have important potential implications for well-being. Hence, knowledge about how people experience crime both objectively and subjectively is important information for those concerned with addressing crime and safety in a comprehensive, multidimensional fashion. To this end, during the past several decades scholars have studied the level of interconnectedness among victimization, perceptions of safety, individual victimization risk perception, emotional fear, and behavioral manifestations of fear (e.g., Baumer, 1985; Braungart, Braungart, & Hoyer, 1980; Ferraro, 1995; Fisher & Sloan, 2003; Garofalo, 1979; Lee & Ulmer, 2000; Skogan, 1987; Warr & Stafford, 1983; Wilcox Rountree, 1998; Wilcox Rountree & Land, 1996a, 1996b). Theoretical models such as the risk interpretation approach (Ferraro, 1995) or the general opportunity interpretation approach (Wilcox Rountree, 1998) have been put forth to explain the interrelationships among these experiences. These risk or opportunity interpretation models suggest that various individual and contextual

---

background characteristics, including previous victimization experiences, shape cognitive perceptions of perceived criminal opportunity and crime risk (i.e., perceived risk), which in turn increases emotional fear of crime and avoidance behavior (Ferraro, 1995; Wilcox Rountree, 1998). Extending this theoretical development, criminologists have attempted to address how interrelationships posited by such theoretical models might vary among different subgroups within the population and across different types of crime (Ferraro, 1995; Fisher & Sloan, 2003; Lane & Meeker, 2000; Wilcox Rountree, 1998). For instance, understanding women's relatively high levels of fear, despite relatively low levels of actual victimization and perceived risk, has been a source of important research. This research highlights not only women's uniquely elevated levels of fear (in comparison to men's fear) but their particularly elevated levels of fear of sexual assault specifically (e.g., Ferraro, 1995, 1996; Fisher & Sloan, 2003; May, 2001; Warr, 1984, 1985).

Unraveling women's fear of sexual assault vis-à-vis objective and perceived risk of sexual assault has received attention, but subjective perceptions and feelings about other crimes experienced by women, including those associated with physical assault and stalking, are less often addressed (but see Lane & Meeker, 2003; Tjaden & Thoennes, 2000). As such, we know little about how women's actual experiences with such other crimes relate to their subjective experiences. Furthermore, we know little about how an interrelationship among women's subjective and objective crime experiences might vary not only across type of crime but also across type of offender (e.g., stranger vs. acquaintance). Actual violence against women is largely intimate partner violence (Tjaden & Thoennes, 2000), yet that acquaintance violence takes place in a culture that touts the dangers of random, stranger-perpetrated violence (e.g., Best, 1999). As such, it is plausible that objective victimization experiences might be loosely coupled with subjective experiences regarding stranger-perpetrated crime, with fear being experienced well beyond the victim population. In contrast, objective and subjective experiences could be more closely linked for acquaintance-perpetrated crime, whereby women

who fear acquaintance-perpetrated crime are largely those who have experienced it.

In the present study, we address the above-mentioned gaps in the literature by (a) examining the interrelationships among objective and subjective crime experiences among women regarding three different types of crime—sexual assault or coercion, physical assault, and stalking—and (b) comparing the interrelationships among objective and subjective experiences across crimes involving strangers versus those involving acquaintances. We address these questions using survey data collected in spring 2004 from a sample of 1,010 college women.

The College Context

The population of college women is of particular interest in examining the interrelationships among objective crime experiences in that college women appear to be at greater objective risk of some forms of criminal victimization compared to similarly aged, noncollege counterparts (e.g., Fisher & Cullen, 2000; Shipman, 1994). In fact, the tradition of campuses downplaying crime was the focus of a concerted social movement in the late 1980s and 1990s, resulting in the passage of federal law designed to make the public reporting of campus crime mandatory, more extensive, and more accurate (e.g., Fisher, 1995; Hudge, 2000; Nicklin, 2000). The Campus Crime Disclosure Act (1998), later renamed the Clery Act (2000), amended the earlier Campus Security Act (1990) with new provisions mandating that schools report hate crimes in addition to those already required (index offenses), that schools include in their reports crime that occurs on property unowned by the college but contiguous to campus, and that schools receive punishment for noncompliance in crime reporting. If such legislative policies are successful in terms of their intent, today's college students should be well-informed about crime events on and around campus and should therefore objectively fear crimes most commonly experienced. However, campus crime reporting in the post-Clery era has been criticized for continued underreporting of crime for a variety of reasons. Numerous media and professional sources, for instance, have suggested that many campuses continue to underreport campus crime even in light of Clery because of jurisdictional confusion, organizational inefficiency, and concern with student (offender) confidentiality (Gregory, 2001; Hardy & Barrows, 2001; Kennedy, 2000; Leinwand, 2000; Megerson, 1992; Nicklin, 1999). Others have suggested that Clery-mandated crime reports underrepresent campus crime because they measure only crimes reported to police rather than victimization incidents, and they exclude categories with high incident rates such as larceny theft (Fisher, Hartman, Cullen, & Turner, 2002). With such limitations, subjective risk on the part of students might be dangerously deflated even in the post-Clery era. If so, fear of crime may be unique among college women because fear is lower as opposed to higher in relation to objective risk.

Moreover, the extent to which women have subjective perceptions of crime more closely versus more loosely coupled with objective risks may depend on crime characteristics such as offense type and victim-perpetrator relationship. Clery, in theory, should cover all three crimes examined herein, including those committed by strangers and acquaintances, however, there may be systematic bias in noncompliance whereby certain crimes (e.g., those committed by other students) are less often reported, less widely reported, or reported with fewer details. Furthermore, even if crimes committed by acquaintances, including other students, are reported by colleges and universities in the exact same manner as are crimes committed by strangers, these reports may be received differently on the part of college or university women. Fueled by media portrayals of random violence, women may react more strongly to reports of stranger-perpetrated violence than to acquaintance-perpetrated violence.

In summary, the present study cannot discern the extent to which colleges and universities comply with Clery or the manner in which campus women interpret reports that do surface at their institutions. Nonetheless, the emergence of Clery within a context in which the broader culture supports long-held biases toward magnifying stranger violence while at the same time most victimization of college women occurs at the hands of acquaintances (Fisher, Cullen, & Turner, 2000) provides an interesting confluence for exploring the interconnectedness of women's objective and subjective crime experiences and the extent to which these interconnections might vary across crime and perpetrator type.

Fear of Crime

Early work on fear of crime relied almost exclusively on either a General Social Survey (GSS) question—"Is there an area right around here (within a mile) where you would be afraid to walk alone at night?"—or one item from the National Crime Survey (NCS; "How safe do you feel or would you feel being out alone in your neighborhood at night?").[2] However, these measures were criticized on numerous grounds. First, as crime is not even mentioned in the questions, it was thought that the GSS and NCS measure tapped social concerns, or "urban unease," broader than fear of crime (e.g., DuBow, McCabe, & Kaplan, 1979; Garofalo & Laub, 1978). Further, Garofalo (1981) pointed out that the items measured anticipatory feelings ("How safe would you feel …"), which might be quite different from feelings one experiences when actually encountering the situation in question. Following on these criticisms, a good deal of work in the fear-of-crime literature in recent decades has focused on issues of conceptualization and operationalization. This work has led to recognition of a multidimensionality of fear of crime, with cognitive, emotional, and behavioral dimensions. Ferraro and LaGrange (1987), for instance, suggested that the NCS fear measure was more along the lines of a cognitive assessment of safety or judgment of risk than an emotionally or behaviorally based fear of crime because the question does not ask about actual feelings of worry or fear of crime or actual behavior in response to crime. Recent empirical work has supported the distinction highlighted by Ferraro and LaGrange between measures of cognitive perception of either general or personal risk (e.g., "How safe is your neighborhood from—?"; "How likely is it that—will happen to you?") and emotional fear (e.g., "How afraid/worried are you about—?"). Studies have found moderate correlations between the two types of measures (typically around .6), and although there are some similar covariates across risk and worry measures, extant research has also unearthed important etiological differences (e.g., Bennett & Flavin, 1994; Chiricos, Hogan, & Gertz, 1997; Ferraro, 1995; Fishman & Mesch, 1996; LaGrange & Ferraro, 1989; LaGrange, Ferraro, & Supancic, 1992; Lee & Ulmer, 2000; May & Dunaway, 2000; Mesch, 2000; Skogan, 1987; Warr, 1987; Wilcox Rountree & Land, 1996b). Previous research has also highlighted such distinctiveness with the behavioral dimension, showing that individual and contextual covariates vary somewhat across measures of risk perception, fear, and restricted or constrained routine activities (Ferraro, 1995; Wilcox Rountree & Land, 1996a).

Scholars have suggested further that fear of crime is not only multidimensional in the sense of having cognitive, emotional, and behavioral components but that it is also multidimensional in the sense of having crime-specific qualities. Previous studies have shown, for example, that the level and nature of fear vary depending on the crime category under consideration (Ferraro, 1995, 1996; Lee & Ulmer, 2000; May, 2001; Warr, 1984; Warr & Stafford, 1983; Wilcox Rountree, 1998). As such, fear of violence cannot be considered analogous to fear of property crime, and within the category of property crime fear of burglary may be quite different than fear of vandalism, for instance.

Rational Fear? Its Link to Objective Risk

Given multidimensionality in fear, how do different dimensions relate to actual crime risk? A great number of studies have addressed this issue, examining both the extent to which individual-level victimization experiences relate to fear and the extent to which rates of crime in the surrounding areas relate to fear. Unfortunately, most have addressed this issue without explicitly distinguishing and examining various dimensions of fear, thus confounding whether these effects might vary across cognitive, emotional, or

behavioral spheres (for exceptions, see Chiricos et al., 1997; Ferraro, 1995; Lee & Ulmer, 2000; Wilcox Rountree & Land, 1996b). Perhaps partly because of the failure to distinguish multiple dimensions of fear, findings regarding the effect of previous victimization on fear have been mixed with some studies supporting a positive relationship (e.g., Braungart et al., 1980; Garofalo, 1979; Skogan, 1987; Wilcox Rountree, 1998; Wilcox Rountree & Land, 1996b) and others questioning the strength of this relationship for at least some dimensions of fear (e.g., Baumer, 1985; Ferraro, 1995; Hindelang, Gottfredson, & Garofalo, 1978; Lane & Meeker, 2003; May, 2001; McGarrell, Giacomazzi, & Thurman, 1997). Still others have suggested that the link between victimization and fear is moderated by race, with the most pronounced effect of victimization on fear seen among Whites (Chiricos et al., 1997).

Studies assessing the link between actual risk and fear by estimating the effects of community rates of crime also reveal mixed results. Evidence exists of a positive effect of community crime on fear (Skogan & Maxfield, 1981), but other studies show weaker effects (Ferraro, 1995; Lee & Ulmer, 2000; Lewis & Maxfield, 1980; Lewis & Salem, 1986; Skogan, 1990; Taylor & Hale, 1986). Such effects have also been shown to be indirect, operating through local media coverage (Liska & Baccaglini, 1990), and effects of area crime have been shown to be conditional on a variety of other factors including individual race and age (Chiricos et al., 1997; Liska, Lawrence, & Sanchirico, 1982) and type of crime under consideration (Wilcox Rountree, 1998).

Extant research that differentiates between a cognitive component of fear assessing perceived risk of crime and an emotional component assessing worry or concern about crime has shown that the effects of personal victimization experiences and/or community rates of crime on emotional worry may operate through cognitive perceived risk (Chiricos et al., 1997; Ferraro, 1995; Lee & Ulmer, 2000) and perceived community incivilities (Wilcox Rountree, 1998). Finally, research differentiating a behavioral dimension of fear in terms of constrained behavior or safety precautions has been mixed, with some studies finding that victimization experiences and area rates of crime have little to do with safety precautions (Ferraro, 1995) and others providing evidence that victimization and area crime rates may have both direct effects on safety precautions and indirect effects through cognitive risk perceptions (Wilcox Rountree & Land, 1996a).

Campus Fear

Historically, most fear-of-crime scholarship has focused on the general adult population. However, coinciding with the heightened social awareness and political attention surrounding campus crime that began in the late 1980s, fear of crime on campus has also received increased attention in the past several decades. Still, scholarly empirical study of fear of crime among college students is scant, with important exceptions. Fisher and Nasar (1992, 1995; see also Nasar & Fisher, 1992, 1993), for instance, presented one of the first empirical studies of fear on campus with their work on the microlevel physical cues associated with fear in and around the Ohio State University's Werner Center for the Visual Arts. They found that certain aspects of the built and/or natural environment were associated with student fear, including "areas that were characterized by limited prospect, much concealment, and difficult escape" (Fisher & Nasar, 1995, p. 232; see also Day, 1999). Other single-campus surveys of fear have supported these spatial patterns and also highlighted temporal and sex-based differences in student levels of emotional concerns and/or cognitive risk perceptions. For instance, several single-campus studies have found that nighttime fear exceeds daytime fear among students (Fisher, Sloan, & Wilkins, 1995; McConnell, 1997). Furthermore, college women have reported higher levels of fear than do men, regardless of time of day (Fisher et al., 1995; McConnell, 1997) and across a variety of spatial domains, including campus jogging paths, campus parking lots, libraries, and so on (McConnell, 1997). Less research has been conducted on the behavioral dimension of fear among college women, though important work by Day (1994) suggested that much of the traditional campus safety initiatives aimed at reducing women's victimization and emotional fear thereof actually serve to, somewhat

ironically, further control and constrain college women's behavior.

In one of the only studies of campus fear using a national sample of college students, Fisher and Sloan (2003) examined daytime and nighttime emotional fear, along with cognitive personal risk perception, across college men and women for a variety of specific crimes (e.g., larceny theft, robbery, simple assault, aggravated assault, and rape). Women's personal risk perception and daytime emotional fear levels exceeded those of men for every crime except larceny theft. Women's nighttime emotional fear levels exceeded those of their male counterparts for all offenses examined. Furthermore, the link between perceived risk and fear, among women in particular, was not always obvious. For instance, among the five crime-specific risks or fears examined, women had the highest levels of perceived risk for larceny theft, yet that was the crime they feared the least at night. The crime that the college women feared most at night—rape—ranked third (out of five) among crime-specific risk perceptions. Although perceived risk was a significant positive predictor of nighttime fear of rape, the effects of various measures of objective crime experiences were less consistent. For instance, having been a victim of off-campus sexual or nonsexual assault was positively related to nighttime fear of rape, but on-campus sexual or nonsexual assault was not significant. The effect of lifetime experience with rape or sexual assault victimization specifically was also not significant, and the overall student sexual violence victimization rate was actually negatively related to nighttime fear of rape among college women.

In conclusion, the findings of Fisher and Sloan (2003) do not lend clear support for the notion of a close coupling among fear and perceived risk for college women, particularly when considering the crime of rape. Despite the importance of such findings, more work is needed to further unpack college subjective crime experiences. It would seem particularly important to extend the work of Fisher and Sloan by examining the links between objective and subjective crime experiences for other offenses affecting college women. Furthermore, it would seem important to examine whether the links between victimization and multiple dimensions of subjective crime experiences (i.e., cognitive, emotional, and behavioral) vary not only across different crimes but across different victim-offender relationships.

The Present Study

The present study therefore extends previous research by examining different crime-related experiences including actual victimization, perceptions of campus danger, emotional worry or concern among college women, and safety or precautionary behavior. We focus on experiences regarding three crime categories in particular: stalking, physical assault victimization, and sexual victimization. Furthermore, we examine these crime experiences with sensitivity to the possibility of domain-specificity in the sense of victimization and fear, and the interrelationships thereof, perhaps being conditional on victim-offender relationship (i.e., stranger crime vs. acquaintance crime).

Data

Data for the present study were collected from a sample of 1,010 women surveyed by telephone at a state university in the southeastern United States (referred to anonymously as *State U* hereafter). State U is a state-supported public university that includes an urban campus and large medical center complex. State U is located in an urban area with a population of 260,512. The University is unique among land-grant universities in that all its 16 colleges are located on one campus, resulting in a student population of more than 25,000. State U offers 88 certified degree programs that lead to bachelor's degrees and master's degrees in 93 fields and PhDs and other doctoral degrees in 60 programs.

Telephone interviews were conducted by specially trained interviewers contracted through Schulman, Ronca & Bucuvalas, Inc., using computer-assisted telephone interviewing procedures. These procedures allowed the survey instrument to incorporate a complex skip pattern, ensuring that participants were asked only relevant questions regarding the specific details of experiences with campus victimization. A university-provided list of 7,875 phone numbers for current female students was used to generate a random sample of 1,010 female students. The overall cooperation rate for valid student contacts was 83.5%. The interviews were conducted between April 1, 2004, and May 4, 2004, with each completed interview lasting an average of 17.5 minutes.

Consenting telephone contacts were included as participants if they reported that they were older than 18 and students at the university. Following this initial screening, participants were first asked questions regarding overall fear of crime on campus, then questioned regarding fear of specific types of victimization (stalking, sexual abuse or attack, physical abuse or attack) by either a known offender or a stranger, following the recommendation of Fisher and Sloan (2003). Next, participants were asked to provide basic background information; this section also included questions about current relationship status and alcohol or drug use. Finally, following the lead of recent studies of violence against women (Fisher & Cullen, 2000; Fisher et al., 2000; Tjaden & Thoennes, 2000), participants' previous victimization experiences were screened using 34 yes-no questions about specific types of events (see the appendix). Participants identified by these screening questions as victims of stalking, physical assault, or sexual victimization (including assault or rape, sexual coercion, and unwanted sexual contact) were then asked general follow-up questions regarding the circumstances (victim-offender relationship, occurrence before or during college, concurrence with other forms of victimization, etc.). Finally, detailed follow-up questions were asked only for the most recent events occurring while enrolled at State U.

Measures of Variables

As indicated above, numerous survey screening questions were utilized to determine whether respondents were classified as having ever been victims of stalking, physical assault, or sexual victimization (see the appendix). If respondents indicated that any of the events comprising these types of victimization had occurred, they were asked whether the event had occurred before college, while in college at someplace other than State U, or while enrolled at State U. For those where stalking, physical assault and sexual assault, coercion, or contact victimization occurred while at State U, they were asked to reveal the relationship of the offender characterizing the most recent incident. Based on this series of screening and follow-up questions, we constructed three categories for classifying respondents according to each of three types of victimization: (a) respondent's most recent State U stalking or physical assault or sexual victimization experience was stranger perpetrated, (b) respondent's most recent State U stalking or physical assault or sexual victimization experience was acquaintance perpetrated, or (c) respondent had not experienced stalking or physical assault or sexual victimization since being enrolled at State U.

We compare these various categories of campus victimization with measures of subjective crime experiences spanning cognitive, emotional, and behavioral dimensions. First, we measure cognitive assessment of campus danger with a single survey item asking respondents how safe their campus is from crime. Responses ranged from 1 (*very safe*) to 4 (*very unsafe*). We measure emotional worry about crime with six separate crime- and perpetrator-specific survey questions tapping level of worry (1 = *not really worried*, 4 = *very worried*) the respondent experiences regarding (a) personally being stalked by a stranger, (b) personally being stalked by an acquaintance, (c) personally being physically abused or attacked by a stranger, (d) personally being physically abused or attacked by an acquaintance, (e) personally being sexually abused or attacked by a stranger, and (f) personally being sexually abused or attacked by an acquaintance.[3]

We also measure crime experiences in the form of crime-related behavioral adjustments. For instance, we utilize a dichotomous variable to indicate (1 = *yes*, 0 = *no*) whether respondents have something they carry or keep at their home to protect themselves from crime. We also include a measure of avoidance behavior, utilizing a single survey item asking respondents how often they avoid places on or around State U's campus out of concern for personal safety. Responses ranged from 1 (*never*) to 4 (*always*).

We note that our measures of actual victimization versus various subjective perceptions of crime are somewhat different in terms of crime specificity and domain specificity. Victimizations include those that happened while enrolled at State U, either on or off campus. Assessment of danger refers to campus specifically and is a general and global rather than crime-specific and personal measure. Measures of emotional worry are crime specific but, like victimization, are not campus specific. Neither behavioral measure is crime specific, but one is campus specific (avoidance of campus), whereas the other (carrying or owning something for protection) is not. Given these fluctuations, we realize our examination of the interrelationships between these various crime-related experiences calls for qualification in that it cannot isolate linkages between campus victimization and campus-based danger, worry, or protective action. We nonetheless believe that examining college women's victimization (both on and off campus) vis-à-vis various dimensions of fear, whether or not the fear is specific to college or campus, is a valuable exercise in beginning to understand these linkages. We believe that women's subjective experiences with crime in terms of perceptions of the campus environment, worry about crime (both on and off campus), and precautionary behavior (both on and off campus) should be viewed in relation to crime victimization experienced while in college, whether on or off campus. How safe women perceive their campus to be, how strained women feel in terms of emotional concern about crime, and what sorts of avoidance and general precautionary behaviors they engage in all have important implications for general well-being of college women and should therefore be of interest to administrators, faculty, student affairs professionals, and parents.

In addition to these measures of various crime experiences, we also include in some of our subsequent analyses several control variables. We control for class standing through a series of dichotomous variables indicating (1 = *yes*, 0 = *no*) whether the respondent is a freshman, sophomore, junior, or senior, with the category of *graduate student* being the reference group in multivariate analyses. We control for respondent's race with a dichotomous variable indicating whether the respondent is non-White ($1 = yes, 0 = no$). We also control for whether the respondent is currently in a romantic relationship in the form of a spouse, a cohabiting partner, or a noncohabiting girlfriend or boyfriend ($1 = yes, 0 = no$). Household structure is controlled with a variable indicating the number of adults with whom the respondent lives.

Results

In examining the extent to which various dimensions of fear of crime among college women correspond with objective crime experiences in terms of personal victimization, we first discuss the extent of victimization experienced by sampled women. Overall, 35.6% of respondents had experienced stalking, physical assault, and/or sexual victimization while enrolled at State U. More specifically, 24.9% of participants experienced one type of victimization, 8.7% experienced two of the three types, and 1.4% experienced all three types. When examining victimization by perpetrator type within crime category, as depicted in Table 1, interesting differences

Table 1 Victimization Among Sample Respondents, While Students at State U

Victimization	%
Stalking victimization, stranger	7.0
Stalking victimization, acquaintance	10.2
No stalking victimization	82.8
Physical victimization, stranger	1.3
Physical victimization, acquaintance	8.6
No physical victimization	90.1
Sexual victimization, stranger	4.1
Sexual victimization, acquaintance	15.5
No sexual victimization	80.4

clearly surface. Table 1 reveals that sexual victimization by an acquaintance is the most prevalent form of victimization queried, with 15.5% of respondents indicating such experiences.[4] In contrast, physical assault by a stranger is the least prevalent, with 1.3% of respondents experiencing such victimization. For each of the three crime categories examined, acquaintance-perpetrated victimization is substantially more common than is stranger-perpetrated victimization. Fortunately, however, the modal category for each crime type is no victimization, with an overwhelming majority (more than 80.0%) indicating no such experiences.

Victimization and Assessment of Campus Danger

Our examination of the distribution of cognitive assessment of campus danger among respondents implies that college women may perceive the campus to be safer than their college victimization experience would suggest. According to this distribution, 15.5% of respondents indicated that they thought campus was unsafe—either somewhat or very unsafe—from crime. Interestingly, this percentage is quite a bit lower than the 35.6% who had actually experienced victimization while enrolled at State U.[5]

To begin to examine the extent to which assessment of campus danger relates to different types of victimization experiences, we plotted the mean levels of risk among the stranger-perpetrated, acquaintance-perpetrated, and no victimization categories for each of the three crimes under study. The results are shown in Figure 1. The data suggest that for stalking and physical assault, those experiencing stranger-perpetrated victimization actually report somewhat higher mean levels of campus danger than do acquaintance-perpetrated victims. Those not experiencing victimization, as expected, report the lowest levels of campus danger.

Comparisons of means using ANOVA indicate that these means are significantly different for stalking ($F = 10.221, p < .001$) and physical assault ($F = 7.938, p < .001$). However, in neither case is the contrast between stranger- and acquaintance-perpetrated victimization

Figure 1 Risk Perception Levels by Victimization Experiences

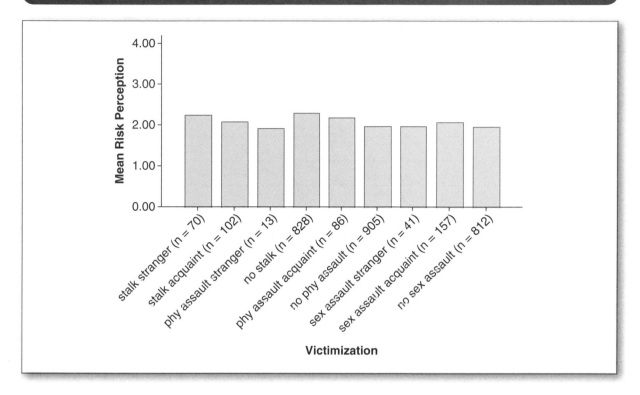

significant. For stalking, the only significant contrast is between stranger-perpetrated stalking victimization and no stalking victimization ($p < .001$). Keeping in mind the very small number of stranger-perpetrated physical assaults ($n = 13$), the only statistically significant contrast among physical assault categories was between acquaintance-perpetrated physical assault and no physical assault victimization ($p = .002$). Differences in mean campus danger for the different categories of sexual assault were not significant ($F = 2.138, p = .118$).

Finally, we consider the effect of victimization experiences on cognitive assessment of campus danger within a multivariate logistic regression model, predicting the odds of perceiving the campus to be unsafe (somewhat or very unsafe) as opposed to safe (somewhat or very safe). These results are presented in Table 2. Results indicate that assessing the campus as dangerous

is positively and significantly related to stranger-perpetrated stalking victimization. In comparison to those experiencing no stalking victimization (the reference category), victims of stranger-perpetrated stalking had 2.65 times greater odds of perceiving the campus to be unsafe. In contrast, it was physical assault by an acquaintance that significantly increased the odds of an unsafe assessment (by 78%) in comparison to no physical assault victimization. Odds of assessing the campus as dangerous were not significantly different for either victims of stranger or acquaintance sexual assaults in comparison to nonvictims of sexual assault.

Victimization and Worry About Crime

We next turn to an examination of the relationship between actual crime experiences and an emotional

Table 2   Logistic Regression Coefficients, Standard Errors, and Odds Ratios for Assessment of Campus Danger

Variable	Coefficient	SE	Odds Ratio
Stalking stranger	1.00*	.29	2.65
Stalking acquaintance	.50	.27	1.65
Physical victim stranger	.88	.65	2.41
Physical victim acquaintance	.60*	.28	1.78
Sex victim stranger	−.35	.50	.71
Sex victim acquaintance	.25	.24	1.28
Constant	−2.16		
Chi-square ($df = 13$)	34.50*		

NOTE: Model controls for class standing, race (non-White), whether in a romantic relationship, and number of adults living in same household. All control variables were nonsignificant except a dummy variable for freshman class standing. Freshman students had significantly higher odds of assessing State U's campus as dangerous in comparison to the graduate student reference category.

*$p < .05$.

dimension of fear of crime. For these purposes, we employ six crime- and offender-specific measures of worry about crime: worry about stranger-perpetrated and acquaintance-perpetrated stalking, physical assault, and sexual assault. Table 3 presents the percentage of respondents who indicated being worried—either somewhat or very worried—about each of these six offenses in comparison to the percentage of respondents who had actually experienced similar sorts of victimization while a student at State U. As Table 3 indicates, for all three stranger-perpetrated crimes in question, the percentage who are victimized is far less than the percentage who are fearful of such stranger-perpetrated crimes. The exact opposite pattern holds

Table 3   Percentage of Respondents Victimized Versus Somewhat or Very Worried by Victimization Type

Victimization	% Victimized	% Worried	Difference
Stalking stranger	7.00	21.83	−14.83
Stalking acquaintance	10.20	8.35	1.85
Physical victim stranger	1.29	38.42	−37.13
Physical victim acquaintance	8.57	6.83	1.74
Sex victim stranger	4.06	41.87	−37.81
Sex victim acquaintance	15.54	10.11	5.43

for the acquaintance-perpetrated victimizations and fears, where the percentage victimized exceeds the percentage fearful in all three instances, though the differences for acquaintance-perpetrated victimization versus worry were much smaller than the differences regarding stranger crimes.

Such patterns suggest a weak coupling of victimization and emotional fear, especially among stranger-perpetrated crimes. Indeed, bivariate correlations between victimization and fear measures are weak, ranging from −.15 to .15. Furthermore, results from multivariate logistic regression analysis of each crime-specific worry, shown in Table 4, support the idea that victimization experiences are generally weakly related to worry about crime, controlling for possible confounding effects. For each type of worry, two models are presented in Table 4—one without cognitive assessment of campus danger controlled and a second model with campus danger included. All of the coefficients for stalking and stranger-perpetrated victimization shown in Table 4 are in relation to the omitted *no stalking or physical assault or sexual victimization* category. All of the coefficients for campus danger are in relation to perceiving the campus to be somewhat or very safe.

Hence, regarding the first model estimating worry about stranger-perpetrated stalking, previous stranger stalking victims have 88% greater odds of worry than do nonvictims of stalking. However, that is the only significant victimization effect, and that effect disappears once assessment of campus danger is controlled. Campus danger is significantly related to worry about stranger stalking, with those who perceive State U as unsafe having 2.27 times higher odds of worry about stranger stalking than those who perceive State U as safe. Combining findings from Table 2 and Table 4, it appears as if assessment of campus danger may mediate the effects of stranger stalking victimization on worry about stalking. Those who have been stalked by strangers while enrolled at State U perceive their campus as unsafe and, in turn, worry about future stranger stalking. When estimating worry about stalking by acquaintances, only acquaintance-stalking victimization is significantly related to worry about acquaintance

stalking. Odds of acquaintance-stalking worry are more than 250% greater for previous victims of acquaintance stalking in comparison to nonvictims of stalking. Furthermore, this effect is not diminished on controlling for assessment of campus danger. Campus danger, in fact, has a nonsignificant effect.

In our estimation of worry about physical assault, the only type of victimization significantly related to worry about stranger-perpetrated physical assault was stranger-stalking victimization. Victims of stranger stalking had significantly higher odds of worry about physical assault than did nonvictims, even after controlling for assessment of campus danger. In contrast, in estimating worry about acquaintance-perpetrated physical assault, stranger-perpetrated physical assault victimization was the only significant effect, and it disappeared on controlling for campus danger.

Worry about stranger-perpetrated sexual assault appears unrelated to sexual assault victimization experiences, by acquaintance or stranger. The only significant effect revealed in the stranger sexual assault worry model was for acquaintance-stalking victimization, and this effect disappeared on controlling for assessment of campus danger. Furthermore, the model chi-square for estimation of stranger sexual assault worry without campus danger included is nonsignificant, suggesting that the covariates do not significantly improve the estimation in comparison to a null (intercept only) model. In contrast, worry about sexual assault by an acquaintance is linked to similar victimization experiences; acquaintance sexual assault victims had 1.75 times greater odds of worry than did nonvictims of sexual assault. Furthermore, those who had been physically assaulted by an acquaintance had just more than twice the odds of worry about acquaintance-perpetrated sexual assault than did nonvictims of physical assault. However, both of these effects became nonsignificant once we controlled for assessment of campus danger.

In sum, Table 4 indicates that there appears to be an overall fairly loose coupling between victimization experiences and worries about crime among college women. Each type of worry was related to some sort of

Table 4 Logistic Regression Coefficients, Standard Errors (in Parentheses), and Odds Ratios for Crime-Specific Worry ("Somewhat or Very Worried" Versus "Not Worried or Just a Little Worried")

	Stalking Worry				Physical Assault Worry				Sexual Assault Worry			
	Stranger		Acquaint		Stranger		Acquaint		Stranger		Acquaint	
Stalking Stranger	.63*	.50	.60	.57	.69*	.55*	−.01	−.13	.34	.16	.40	.23
	(.28)	(.29)	(.42)	(.43)	(.27)	(.27)	(.51)	(.53)	(.26)	(.27)	(.38)	(.39)
	1.88	1.65	1.82	1.77	1.99	1.74	.99	.88	1.40	1.17	1.49	1.26
Stalking Acquaintance	.13	.06	1.28*	1.27*	.38	.32	.57	.52	.48*	.42	.48	.38
	(.27)	(.27)	(.31)	(.31)	(.22)	(.23)	(.38)	(.39)	(.22)	(.23)	(.33)	(.34)
	1.14	1.06	3.60	3.55	1.46	1.38	1.78	1.69	1.61	1.52	1.62	1.47
Physical Victim Stranger	.57	.44	1.13	1.10	.81	.71	1.46*	1.37	.30	.14	.63	.45
	(.64)	(.66)	(.75)	(.75)	(.61)	(.63)	(.72)	(.73)	(.59)	(.62)	(.82)	(.85)
	1.77	1.55	3.10	3.01	2.25	2.04	4.29	3.92	1.35	1.15	1.88	1.56
Physical Victim Acquaintance	.17	.08	.18	.17	.36	.28	.59	.53	.02	−.09	.70*	.62
	(.28)	(.28)	(.40)	(.40)	(.24)	(.25)	(.41)	(.42)	(.24)	(.25)	(.33)	(.33)
	1.18	1.08	1.20	1.18	1.43	1.32	1.81	1.69	1.02	.91	2.02	1.85
Sex Victim Stranger	−.82	−.80	−1.34	−1.33	−.63	−.60	−.41	−.41	−.51	−.48	−1.76	−1.75
	(.50)	(.50)	(1.04)	(1.04)	(.37)	(.38)	(.77)	(.78)	(.35)	(.36)	(1.04)	(1.05)
	.44	.45	.26	.26	.53	.55	.66	.66	.60	.62	.17	.17
Sex Victim Acquaintance	.26	.23	.48	.47	.04	−.00	.55	.51	−.00	−.05	.56*	.51
	(.22)	(.22)	(.30)	(.31)	(.19)	(.20)	(.35)	(.35)	(.19)	(.20)	(.28)	(.28)
	1.30	1.26	1.61	1.60	1.04	.99	1.73	1.66	.99	.96	1.75	1.67
Campus Danger	—	.82*	—	.17	—	.99*	—	.60	—	1.17*	—	.96*
	—	(.20)	—	(.32)	—	(.19)	—	(.32)	—	(.19)	—	(.26)
	—	2.27	—	1.18	—	2.69	—	1.82	—	3.22	—	2.62
Constant	−1.94	−2.03	−2.54	−2.56	−.92	−1.04	−3.20	−3.26	−.55	−.68	−2.91	−3.04
Model Chi-Square ($df = 13$)	32.93*	48.77 *	39.35*	39.62*	28.28*	56.44*	34.65*	37.87*	17.80	56.71*	44.66*	57.73*

NOTE: Models control for class standing, race (non-White), whether in a romantic relationship, and number of adults living in same household. All control variables were nonsignificant except for race (non-White positive significant in all models). Sophomore was also positive and significant in the stalking models.

*$p < .05$.

previous victimization experience, but most victimization measures in each model were nonsignificant. Furthermore, in the exceptional instances where victimization does appear related to worry, the two are not always linked in a crime-specific and offender-specific way. Instead, in some instances, crossover effects occur, whereby previous victimization experiences of one type relate to worry about another type of crime. Thus, the effects of victimization on worry were not general but instead appeared very isolated, with no clear pattern emerging regarding which specific victimization experiences mattered more. Furthermore, most of the isolated victimization experiences that did emerge in initial models disappeared in secondary models that controlled for assessment of campus danger. Assessment of campus danger, in stranger-perpetrated crime in particular, and model fits for these worries tended to improve substantially on inclusions of campus danger. Interestingly, stranger-perpetrated sexual assaults appear to be feared more than other offenses (see Table 3), but the model estimating worry about stranger sexual assault (without campus danger controlled) had the lowest model chi-square among all models presented in Table 4. The second and third lowest chi-square values were associated with the other two models of stranger worries—worry about stranger-perpetrated physical assault and worry about stranger-perpetrated stalking (without campus danger controlled). Once assessment of campus danger was added to these models, however, their model fits were the three highest among all models presented in Table 4. Hence, results from Table 4 support what Table 3 also revealed—worry about stranger-perpetrated crimes and victimization experiences appear especially weakly linked. However, although worry about stranger-perpetrated crime appears weakly linked to previous victimization experiences, it appears strongly linked to cognitive perceptions of the campus environment as unsafe.

Victimization and Precautionary Behavior

The final dimension of fear of crime that we examine vis-à-vis victimization experiences is a behavioral dimension, including measures of self-protective behavior and avoidance behavior. In terms of self-protection, nearly 47.0% of students reported carrying or having in their home something to protect themselves from crime. Among that 47.0%, the most commonly named items were (a) mace or other spray (66.2%), (b) a cellular phone (12.1%), (c) a knife or sharp object (11.5%), (d) a gun (10.0%), and (e) a keychain or keys (9.4%). Figure 2 depicts the percentages that carry or own something for protection across different victimization categories. Although nonvictims are less likely than either stranger- or acquaintance-perpetrated victims to carry or own something for protection across all three offense types, the differences between stranger victims and acquaintance victims are inconsistent. Among stalking and physical assault victims, those experiencing victimization by acquaintances are more likely to carry, whereas victims of stranger sexual assault are more likely to carry than are acquaintance-perpetrated sexual assault victims. Differences among the three victim subgroups were only significant, however, in the case of sexual assault ($F = 3.615, p = .027$).

As another form of precautionary behavior, 81.5% of student respondents indicated avoiding campus (or areas right around campus) at least sometimes out of concern about crime; 35.8% engaged in such avoidance usually or always. Figure 3 depicts the mean level of avoidance across the different subgroups of victims for each of the three types of crime considered here; post hoc analyses of the three different contrasts involved in each three crime categories revealed no significant differences. Thus, as was the case for emotional worry, victimization status appears to have little to do with crime-related precautionary behavior in the form of either carrying or owning something for protection or campus avoidance.

Examining these relationships within multivariate models reveals similar findings. In Table 5, we present logistic regression models of protective carrying or owning and campus avoidance. As in Table 4, we present two models for each dependent variable—one without assessment of campus danger controlled and another model with campus danger included. Across all of the models for precautionary behavior there is only one

Figure 2   Percentages of Sampled Women Who Carry or Own Something for Protection by Victimization

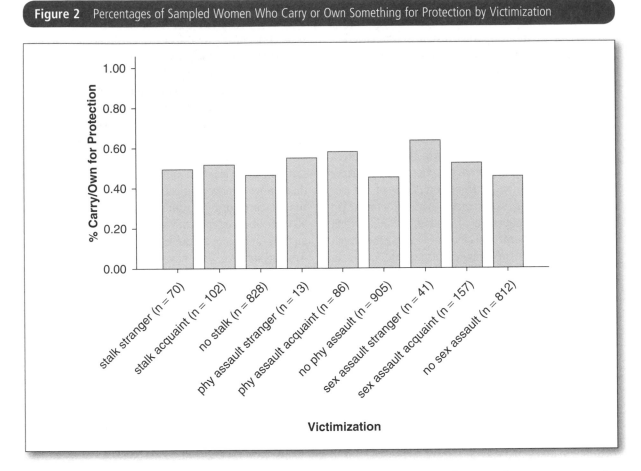

significant effect of victimization. Previous sexual victimization by a stranger more than doubled the odds of carrying or owning something for protection in comparison to sexual assault nonvictimization. Furthermore, this effect remained after controlling for assessment of campus danger. Aside from that important exception, however, Table 5 indicates that nonvictims in our sample differed little from victims (regardless of perpetrator type) in terms of precautionary behavior. Cognitive assessment of campus danger was significantly and positively related to campus avoidance, but it was unrelated to carrying or owning something for protection. The overall model fits for campus avoidance were, however, nonsignificant (with or without campus danger included); the model fit for carrying or owning something for protection was significant. Overall, then,

cognitive assessment of campus danger appears to play a less important role in understanding precautionary behavior than it does in understanding worry about stranger-perpetrated crimes, for instance (see Table 4).

Discussions and Conclusions

Extant research in the fear-of-crime tradition has been largely void of examination of the linkages among various objective and subjective crime experiences with particular attention to (a) multiple dimensions of women's subjective crime experiences and (b) perpetrator-specific experiences. In one of the only other studies estimating a relationship between victimization and

Figure 3 Mean Level of Campus Avoidance (Out of Concern for Personal Safety) by Victimization

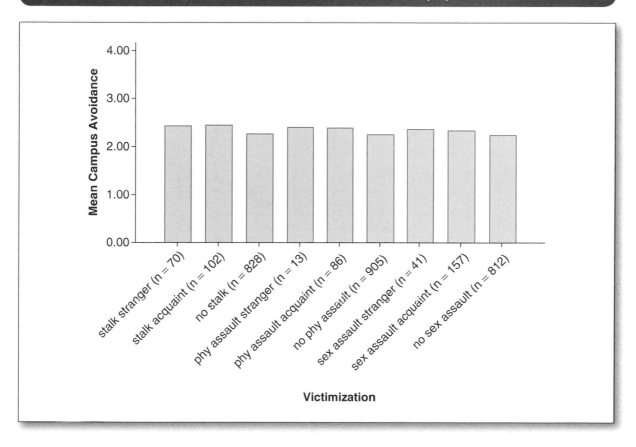

fear among college women, a weak linkage was found (Fisher & Sloan, 2003). The present study supports that weak linkage, with some important caveats to add.

Overall, we found that victimization was weakly related to multiple subjective crime experiences, especially crime- and perpetrator-specific measures of worry, safety precautions, and avoidance behavior. This overall weak relationship was qualitatively different, however, depending on the type of crime and/or fear under consideration. Victimization appears especially weakly related to worry about stranger-perpetrated crime, for instance, because the prevalence of worry among college women in our sample was much greater than the prevalence of victimization by strangers. Furthermore, individual victimization was a nonsignificant predictor of worry about stranger-perpetrated crimes. In addition,

the women in our sample engaged in precautionary behavior and avoidance behavior at rates in substantial excess of stranger victimization rates. In contrast, victimization rates by acquaintances tended to exceed rates of worry about acquaintance-perpetrated crimes among college women in our sample.

At first blush, therefore, it appears as if college women's worries are not entirely well placed in the sense that they appear to be most worried about stranger-perpetrated crime, whereas they are less worried about the acquaintance-perpetrated crime for which they experience higher objective risk. However, such conclusions are confounded by the fact that there is undoubtedly reciprocity between victimization experiences and various fear experiences, including worry. There may be a large disconnect between levels of precautionary

Table 5 Logistic Regression Coefficients, Standard Errors, and Odds Ratios for Precautionary Behavior

| Variable | Carry or Own Protective Device | | | | | | Campus Avoidance | | | | | |
| | Model 1 | | | Model 2 | | | Model 1 | | | Model 2 | | |
	C	SE	OR	C	SE	OR	C	SE	OR	C	SE	OR
Stalking stranger	−0.02	0.27	0.98	−0.07	0.27	0.94	0.32	0.27	1.38	0.24	0.27	1.28
Stalking acquaintance	0.08	0.22	1.08	0.06	0.22	1.06	0.39	0.22	1.47	0.35	0.23	1.42
Physical victim stranger	0.08	0.60	1.08	0.04	0.59	1.04	−0.17	0.63	0.84	−0.26	0.63	0.77
Physical victim acquaintance	0.39	0.24	1.47	0.36	0.24	1.43	0.32	0.24	1.38	0.28	0.25	1.32
Sex victim stranger	0.79*	0.34	2.20	0.80*	0.34	2.24	0.30	0.33	1.34	0.32	0.34	1.37
Sex victim acquaintance	0.13	0.19	1.14	0.12	0.19	1.12	0.14	0.19	1.15	0.12	0.20	1.13
Campus danger	—	—	—	0.29	0.19	1.33	—	—	—	0.51*	0.19	1.65
Constant	−0.41			−0.44			−0.84			−0.89		
Chi-square	27.58*			29.95*			13.52			20.59		

NOTE: C is coefficient, SE is standard error, OR is odds ratio. Models control for class standing, race (non-White), whether in a romantic relationship, and number of adults living in same household. All control variables were nonsignificant except for non-White (negative) and freshman and senior standing (positive) in the carry or own model.

*$p < .05$.

behavior, worry, and victimization, for instance, because the victim- and/or fear-related precautions are successful. If there is a feedback loop as part of a risk or opportunity interpretation model whereby fear and precautions lessen victimization (Cook, 1986), then an effect of victimization on fear or precautions may be difficult to unearth in a cross-sectional model. Our findings of a loose coupling between objective and subjective experiences is similar to several other studies (e.g., Baumer, 1985; Fisher & Sloan, 2003; Lane & Meeker, 2003; May, 2001), and scholars have suggested one important reason for such findings is unmeasured reciprocity (Cook, 1986; Skogan, 1987). Other scholars have suggested that personal victimization is not as important in shaping worry about crime and precautionary behavior as is "indirect victimization" (victimization of others within one's social network; e.g., Ferraro, 1995), local media exposure (Chiricos, Padgett, & Gertz, 2000), or personal, crime-specific risk perception

(e.g., Ferraro, 1995; Lee & Ulmer, 2000), which has been found to mediate the effects of victimization on emotional worry, making those effects almost entirely indirect. The data for our study, unfortunately, did not allow us to compare the effects of direct versus indirect victimization experiences or to control for media exposure or personal crime-specific risk perception. Our models, however, did show robust effects of generalized perceived campus danger in models of worry about stranger-perpetrated crimes, suggesting that women's perceived risk around campus affects this sort of worry much more than does direct victimization. Some of the effects of this general perception may be, more specifically, because of unmeasured indicators of risk and criminal opportunity such as friends' victimization, media or news exposure, or personal perceived risk.

Thus, although our findings hint at the loose coupling of objective and subjective crime experiences among college women, especially regarding stranger crime, we think, it is essential for future research to examine such relationships, while measuring both direct and indirect victimization, exposure to news or information about crime, and personal crime-specific risk perception, within nonrecursive models (e.g., Liska, Sanchirico, & Reed, 1988). In this way, the potential bidirectionality can be appropriately considered, and the source of perceived campus danger can be more clearly determined. Aside from their cross-sectional nature, findings from our study are further limited in that they focus on women from one large, public, Southeastern university. Despite the sample's selection from one university, sample characteristics are reasonably close to those revealed in previous national samples of college women, with our sample being somewhat older because of inclusion of a higher percentage of graduate students.

For instance, the national sample of college women analyzed by Fisher et al. (2000) was 80.0% White, 90.0% full-time, and 86.0% undergraduate, with a mean age of 21.54. In comparison, our sample was 84.0% White, 89.5% full-time, and 73.9% undergraduate, with a mean age of 23.5. Hence, although we do not claim that our findings are necessarily generalizable to all college women given the small scope of our sampling frame, we also do not think our sample is too unlike average college or university women, especially those at large, state research universities with a strong emphasis on graduate programs. As such, our findings should be useful to many universities across the country. However, future work should attempt to address whether there are relation-specific or even campus-specific effects regarding victim-fear relationships, possibly through the use of multistage sampling and multilevel modeling.

Despite its limitations, we feel that this study provides an important step toward delineating the relationship between victimization and multiple dimensions of fear among college women. Given the implications for quality-of-life indicators among college women (e.g., perceptions of campus, campus avoidance behavior, emotional distress), college administrators, faculty, and staff need to be aware of the subjective crime that many women experience

above and beyond any direct crime experiences, and they need to know whether these multidimensional indicators of fear exceed actual rates of victimization or, in contrast, whether students are less fearful than their actual risks would suggest. Our work has implications, therefore, for understanding crime experiences among college women in a post-Clery era. As noted earlier, crime reporting by campuses, as mandated by the Clery Act, has been criticized for under-representing actual campus crime because of organizational issues, methodological issues, and confidentiality concerns. These criticisms imply that much crime that actually takes place on college campuses is not getting reported in the Clery-mandated statistics, leading some to suggest that Clery represents much more symbolic as opposed to substantive reform (Fisher et al., 2002). Our findings hint that Clery-mandated statistics may be flawed not only in terms of how much crime they report but in the nature of that crime as well. Our findings suggest that students may not be getting accurate information about the specific characteristics of the crimes for which they are most at risk, as stranger-perpetrated worries predominated and are closely linked to unsafe perceptions of campus despite acquaintance-perpetrated victimization being more common. As such, it may not be sufficient simply to make crime statistics public to make people more appropriately aware of their risks. Rather, a more detailed discussion of actual risks is warranted. Nonspecific news of crime around campus, without more targeted education initiatives, may only serve to heighten worries regarding crime that students are least likely to experience—violence perpetrated by strangers—while also increasing both perceptions of campus danger and campus avoidance.

There is strong reason to believe that such changes can occur on college campuses without changing the Clery Act per se. In fact, research on the effectiveness of Clery has shown that few students actually pay attention to the published crime statistics that are formally mandated by Clery, but more students (especially women) do pay attention to other programs and information put forth by colleges that, although not formally mandated by the Clery Act, are probably a by-product of the awareness it has created (e.g., Gregory & Janosik, 2002; Janosik, 2001). Therefore, if more colleges incorporate into their informal (i.e., not Clery mandated) crime-related education and programming domain- and perpetrator-specific information about victimization risk, student fear in terms of perception of campus danger, worry about crime, and precautionary and avoidance behavior could become more congruent with actual risk. With such change, the Clery Act's effect, albeit indirect, could shift from symbolic to substantive.

✂ Appendix

Screening Items for Stalking, Physical Assault, and Sexual Assault

Stalking: Has anyone, male or female . . .

Sent you unsolicited letters?

Made unsolicited phone calls to you?

Stood outside your home, school, or workplace?

Showed up at places you were even though he or she had no business being there?

Left unwanted items for you to find?

Tried to communicate with you in other ways against your will?

Vandalized your property to destroy something you loved?

Physical assault: Has anyone . . .

Thrown something at you that could hurt you?

Pushed, grabbed, or shoved you?

Pulled your hair?

Slapped or hit you?

Kicked or bitten you?

Choked or attempted to drown you?

Hit you with some object?

Beat you up?

Threatened you with a gun?

Threatened you with a knife or other weapon besides a gun?

Used a gun on you?

Used a knife or other weapon on you besides a gun?

Sexual assault: Has anyone, male or female . . .

Made you have sexual intercourse by using force or threatening to harm you or someone close to you (by intercourse, I mean putting penis into vagina)?

Made you have oral sex by force or threat of harm (by oral sex, I mean did someone's mouth or tongue make contact with your vagina or anus or did your mouth or tongue make contact with someone else's genitals or anus)?

Made you have anal sex by force or threat of harm (by anal sex, I mean putting a penis in your anus or rectum)?

Used force or threat of force to sexually penetrate you with a foreign object?

Attempted or threatened but not succeeded in making you take part in any of the unwanted sexual experiences that I have just asked you about?

Not counting the above experiences has anyone . . .

Engaged you in any unwanted or uninvited touching of a sexual nature such as forced kissing, touching of private parts, grabbing, fondling, even if it was over your clothes?

Attempted or threatened but not succeeded in unwanted or uninvited touching of a sexual nature?

Made or tried to make you have sexual intercourse or sexual contact when you did not want to by making threats of nonphysical punishment such as lowering a grade, being demoted or fired from a job, damaging your reputation, or being excluded from a group?

Made or tried to make you have sexual intercourse or sexual contact when you did not want to by simply being overwhelmed by someone's verbal pressure or pestering?

Made or tried to make you have sexual intercourse or sexual contact when you did not want to by encouraging or pressuring you to use drugs?

Made or tried to make you have sexual intercourse or sexual contact when you did not want to by giving you drugs?

Not counting any incidents we have already discussed, have you experienced any other type of unwanted or uninvited sexual contact?

◤ Notes

1. Scholars such as Warr (2000) often cite the discrepancy between American levels of victimization and fear as evidence of possible dysfunction regarding fear. Recent data from the National Crime Victimization Survey, for instance, suggest that the overall rate of personal victimization per 1,000 persons age 12 and older in the United States is 22.6, suggesting that just about 2% of the population is at risk, objectively speaking (Catalano, 2004). Yet four decades' worth of time-series data from the National Opinion Research Center has consistently revealed that approximately 40% of Americans are fearful, as indicated by affirmative responses to the GSS question, "Is there any area right around here—within a mile—where you would be afraid to walk alone at night?" (see http://webapp.icpsr.umich.edu/GSS).

2. Since that early work, the National Crime Survey has been renamed. It is now the National Crime Victimization Survey.

3. Note that the term *worry* as opposed to *afraid* is used to measure emotional fear in our study. Our measure was based on methodological work by Ferraro and LaGrange (1987) and Ferraro (1995). Ferraro (1995) suggests specially that "measures of fear of crime should tap the emotional state of fear or worry," should contain "explicit reference to the type of crime [and] . . . should be aimed at assessing the phenomena in the subject's everyday life—not hypothetical or purposefully avoided situations" (p. 27). Our measure meets these criteria, and similar measures have been used in previous recent studies (Lane & Meeker, 2000; Miethe, 1992). Lane and Meeker (2000) suggest, in fact, that *worry* is often language used in lay conversations to convey the emotion of "being afraid."

4. Because our measures of worry about crime did not distinguish between rape and other sexual assault or abuse, we combine

these various sexual victimizations for the analyses presented herein. However, the reader should note that sexual coercion and/or unwanted sexual contact were far more common than was rape. In all, 3.5% had experienced sexual coercion or contact by a stranger.

5. It seems intuitive to suggest that some of the disparity between victimization levels and perceived campus danger levels is because of the fact that the survey asked about campus safety, whereas victimization included acts occurring either on or off campus. However, this logic is contradicted by the fact that when students were asked where on campus they were most fearful, they listed places both on and off campus. Based on these data, therefore, we conclude that when students assess the safety of campus or their fear levels on campus, they conceptualize *campus* as the greater campus vicinity, thus making the spatial referent between the victimization questions and the safety of fear questions more similar than they appear on the surface.

◤ References

Baumer, T. L. (1985). Testing a general model of fear of crime: Data from a national sample. *Journal of Research in Crime and Delinquency, 22*, 239–255.

Bennett, R. R., & Flavin, J. M. (1994). Determinates of fear of crime: The effects of cultural setting. *Justice Quarterly, 11*, 357–381.

Best, J. (1999). *Random violence: How we talk about new crime and new victims.* Berkeley: University of California Press.

Braungart, M. M., Braungart, R. S., & Hoyer, W. J. (1980). Age, sex, and social factors in fear of crime. *Sociological Focus, 13*, 55–66.

Campus Crime Disclosure Act, 20 U.S.C.A. 1092(f) (1998).

Campus Security Act, 20 U.S.C. 1092 (1990).

Catalano, S. (2004). *Criminal victimization, 2003* (NCJ 205445). Washington, DC: U.S. Department of Justice, Office of Justice Programs, Bureau of Justice Statistics.

Chiricos, T., Hogan, M., & Gertz, M. (1997). Racial composition of neighborhood and fear of crime. *Criminology, 38*, 755–785.

Chiricos, T., Padgett, K., & Gertz, M. (2000). Fear, TV news, and the reality of crime. *Criminology, 38*, 755–785.

Clery Act, 20 U.S.C. 1092(f) (2000).

Cook, P. J. (1986). The demand and supply of criminal opportunities. In M. Tonry & N. Morris (Eds.), *Crime and justice: An annual review of research* (Vol. 7, pp. 1–27). Chicago: University of Chicago Press.

Day, K. (1994). Conceptualizing women's fear of sexual assault on campus: A review of causes and recommendations for change. *Environment and Behavior, 26*, 742–765.

Day, K. (1999). Strangers in the night? Women's fear of sexual assault on urban college campuses. *Journal of Architectural and Planning Research, 16*, 289–312.

DuBow, F., McCabe, E., & Kaplan, G. (1979). Reactions to crime: *A critical review of the literature.* Washington, DC: National Institute of Law Enforcement and Criminal Justice, Government Printing Office.

Ferraro, K. F. (1995). *Fear of crime: Interpreting victimization risk.* Albany: State University of New York Press.

Ferraro, K. F. (1996). Women's fear of victimization: Shadow of sexual assault? *Social Forces, 75,* 667–690.

Ferraro, K. F., & LaGrange, R. (1987). The measurement of fear of crime. *Sociological Inquiry, 57,* 70–101.

Fisher, B. S. (1995). Crime and fear on campus. *Annals of the American Academy of Political and Social Science, 539,* 85–101.

Fisher, B. S., & Cullen, F. T. (2000). Measuring the sexual victimization of women: Evolution, current controversies, and future research. In D. Duffer (Ed.), *Measurement and analysis of crime and justice* (pp. 317–450). Washington, DC: U.S. Department of Justice.

Fisher, B. S., Cullen, F. T., & Turner, M. G. (2000). *The sexual victimization of college women* (NCJ 182369). Washington, DC: Office of Justice Programs.

Fisher, B. S., Hartman, J. L., Cullen, F. T., & Turner, M. G. (2002). Making campuses safer for students: The Clery Act as a symbolic legal reform. *Stetson Law Review, 32,* 61–89.

Fisher, B. S., & Nasar, J. L. (1992). Fear of crime in relation to three exterior site features: Prospect, refuge, and escape. *Environment and Behavior, 24,* 35–65.

Fisher, B. S., & Nasar, J. L. (1995). Fear spots in relation to microlevel physical cues: Exploring the overlooked. *Journal of Research in Crime and Delinquency, 32,* 214–239.

Fisher, B. S., & Sloan, J. J., III. (2003). Unraveling the fear of victimization among college women: Is the "shadow of sexual assault hypothesis" supported? *Justice Quarterly, 20,* 633–659.

Fisher B. S., Sloan, J. J., & Wilkins, D. L. (1995). Fear and perceived risk of victimization in an urban university setting. In B. S. Fisher & J. J. Sloan (Eds.), *Campus crime: Legal, social, and policy perspectives* (pp. 179–209). Springfield, IL: Charles C Thomas.

Fishman, G., & Mesch, G. S. (1996). Fear of crime in Israel: A multidimensional approach. *Social Science Quarterly, 77,* 76–89.

Garofalo, J. (1979). Victimization and the fear of crime. *Journal of Research in Crime and Delinquency, 16,* 80–97.

Garofalo, J. (1981). The fear of crime: Causes and consequences. *Journal of Criminal Law and Criminology, 72,* 839–857.

Garofalo, J., & Laub, J. (1978). The fear of crime: Broadening our perspective. *Victimology: An International Journal, 3,* 242–253.

Gregory, D. E. (2001). Crime on campus: Compliance, liability and safety. *Campus Law Enforcement Journal, 31,* 27–28.

Gregory, D. E., & Janosik, S. (2002). The Clery Act: How effective is it? Perceptions from the field—The current state of research and recommendations for improvement. *Stetson Law Review, 32,* 7–59.

Hardy, T., & Barrows, M. (2001). Campus rapes: Underreporting by schools boasting safe records. *IRE Journal, 24*(4), 16–17.

Hindelang, M., Gottfredson, M., & Garofalo, J. (1978). *Victims of personal crime.* Cambridge, MA: Ballinger.

Hudge, M. (2000). The campus security act. *Journal of Security Administration, 23*(2), 25–27.

Janosik, S. M. (2001). The impact of the Campus Crime Awareness Act of 1998 on student decision making. *NASPA Journal, 38,* 348–350.

Kennedy, M. (2000). Safety by the numbers. *American School and University, 73,* 34b–34f.

LaGrange, R. L., & Ferraro, K. F. (1989). Assessing age and gender differences in perceived risk and fear of crime. *Criminology, 27,* 697–719.

LaGrange, R. L., Ferraro, K. F., & Supancic, M. (1992). Perceived risk and fear of crime: Role of social and physical incivilities. *Journal of Research in Crime and Delinquency, 29,* 311–334.

Lane, J., & Meeker, J. W. (2000). Subcultural diversity and the fear of crime and gangs. *Crime & Delinquency, 46,* 497–521.

Lane, J., & Meeker, J. W. (2003). Women's and men's fear of gang crimes: Sexual and nonsexual assault as perceptually contemporaneous offenses. *Justice Quarterly, 20,* 337–371.

Lee, M. S., & Ulmer, J. T. (2000). Fear of crime among Korean Americans in Chicago communities. *Criminology, 38,* 1173–1206.

Leinwand, D. (2000, October 4). Campus crime underreported. *USA Today,* p. A1.

Lewis, D. A., & Maxfield, M. G. (1980). Fear in the neighborhoods: An investigation of the impact of crime. *Journal of Research in Crime and Delinquency, 17,* 160–189.

Lewis, D. A., & Salem, G. (1986). *Fear of crime: Incivility and the production of a social problem.* New Brunswick, NJ: Transaction Books.

Liska, A. E., & Baccaglini, W. (1990). Feeling safe by comparison: Crime in the newspapers. *Social Problems, 37,* 360–374.

Liska, A. E., Lawrence, J. J., & Sanchirico, A. (1982). Fear of crime as a social fact. *Social Forces, 60,* 760–770.

Liska, A. E., Sanchirico, A., & Reed, M. D. (1988). Fear of crime and constrained behavior: Specifying and estimating a reciprocal effects model. *Social Forces, 66,* 827–837.

May, D. C. (2001). The effect of fear of sexual victimization on adolescent fear of crime. *Sociological Spectrum, 21,* 141–174.

May, D. C., & Dunaway, G. R. (2000). Predictors of fear of crime among adolescents. *Sociological Spectrum, 20,* 149–168.

McConnell, E. H. (1997). Fear of crime on campus: A study of a southern university. *Journal of Security Administration, 20,* 22–46.

McGarrell, E. F., Giacomazzi, A., & Thurman, Q. C. (1997). Neighborhood disorder, integration, and fear of crime. *Justice Quarterly, 14,* 479–500.

Megerson, J. S. (1992). Crime Awareness and Campus Security Act of 1990: A paper tiger. *Campus Law Enforcement Journal, 22*(1), 24–27.

Mesch, G. (2000). Perceptions of risk, lifestyle activities, and fear of crime. *Deviant Behavior: An Interdisciplinary Journal, 21,* 47–62.

Miethe, T. D. (1992). *Testing theories of criminology and victimization in Seattle, 1960–1990.* Ann Arbor, MI: Inter-University Consortium for Political and Social Research.

Nasar, J. L., & Fisher, B. S. (1992). Design for vulnerability. *Sociology and Social Research, 76,* 48–58.

Nasar, J. L., & Fisher, B. S. (1993). Hot spots of crime and fear: A multimethod study. *Journal of Environmental Psychology, 13,* 187–206.

Nicklin, J. E. (1999). Colleges differ widely on how they tally incidents under crime-reporting law, fails to end debate over accuracy of statistics. *Chronicle of Higher Education, 45*(38), A41.

Nicklin, J. E. (2000). Shift in crime-reporting law fails to end debate over accuracy of statistics. *Chronicle of Higher Education, 46*(40), A50.

Shipman, M. (1994). Perceptions of campus police: News gathering and access to public records. *Newspaper Research Journal, 15*(2), 1–11.

Skogan, W. G. (1987). The impact of victimization on fear. *Crime & Delinquency, 33*, 135–154.

Skogan, W. G. (1990). *Disorder and decline.* New York: Free Press.

Skogan, W. G., & Maxfield, M. G. (1981). *Coping with crime: Individual and neighborhood reactions.* Beverly Hills, CA: Sage.

Taylor, R. B., & Hale, M. (1986). Testing alternative models of fear of crime. *Journal of Criminal Law and Criminology, 77,* 151–189.

Tjaden, R. B., & Thoennes, N. (2000). *Full report of the prevalence, incidence, and consequences of violence against women: Findings from the National Violence Against Women Survey* (NCJ 183781). Washington, DC: National Institute of Justice.

Warr, M. (1984). Fear of victimization: Why are women and the elderly more afraid? *Social Science Quarterly, 65,* 681–702.

Warr, M. (1985). Fear of rape among urban women. *Social Problems, 32,* 238–250.

Warr, M. (1987). Fear of victimization and sensitivity to risk. *Journal of Quantitative Criminology, 3,* 29–46.

Warr, M. (1994). Public perceptions and reactions to violent offending and victimization. In A. J. Reiss & J. A. Roth (Eds.), *Understanding and preventing violence: Consequence and control* (Vol. 4, pp. 1–66). Washington, DC: National Academy Press.

Warr, M. (2000). Fear of crime in the United States: Avenues for research and policy. In D. Duffer (Ed.), *Measurement and analysis of crime and justice: Criminal justice 2000* (pp. 451–489). Washington, DC: U.S. Department of Justice.

Warr, M., & Stafford, M. (1983). Fear of victimization: A look at the proximate causes. *Social Forces, 61,* 1033–1043.

Wilcox Rountree, P. (1998). A reexamination of the crime-fear linkage. *Journal of Research in Crime and Delinquency, 35,* 341–372.

Wilcox Rountree, P., & Land, K. C. (1996a). Burglary victimization, perceptions of crime risk, and routine activities: A multilevel analysis across Seattle neighborhoods and census tracts. *Journal of Research in Crime and Delinquency, 33,* 147–180.

Wilcox Rountree, P., & Land, K. C. (1996b). Perceived risk versus fear of crime: Empirical evidence of conceptually distance reactions in survey data. *Social Forces, 74,* 1353–1376.

DISCUSSION QUESTIONS

1. Why do you think that women, both college women and noncollege women, express greater levels of fear of crime than do men, even though women are generally less at risk for victimization?

2. A greater percentage of women experienced some type of victimization while enrolled at State U than indicated that they thought the campus was unsafe. How can you explain this finding?

3. One finding in the paper is that individual victimization was a nonsignificant predictor of worry about crime, once perception of campus danger was controlled for. What can explain this finding?

4. How can campuses address college women's victimization experiences and their fear of victimization? How do mandatory reporting laws such as the Clery Act shape college women's perceptions about the dangers they face?

◆

Introduction to Reading 3

Many students who major in criminal justice, psychology, and sociology will find themselves working in state government. For this reason, this article by Lord (1998), in which she surveyed 1,477 persons employed in state government in North Carolina, is included. Respondents were asked about their workplace violence experiences so their individual characteristics, job classifications, and perpetrators could be examined. This information can help inform managers how better to keep workplaces safe and identify and reduce risk to their employees.

Characteristics of Violence in State Government

Vivian B. Lord

Much has been publicized recently concerning the apparent increase in violent acts in workplaces all over the United States. According to a study by the National Institute for Occupational Safety and Health (1992), 750 people have been murdered at work each year since 1980. Northwestern National Life Insurance (1993) estimates that 2.2 million Americans are attacked each year and another 6.3 million workers are threatened while at their place of employment. The National Safe Workplace Institute (Kinney & Johnson, 1993) estimates that the cost of a workplace violence incident ranges up to $250,000, with a total of $4.2 billion annually. Less documented is the damage to worker morale and productivity that results from violent incidents or threats. Nonlethal forms of violence, including harassment and intimidation, can be directly linked to worker burnout, lower productivity, and increased health costs (Baron, 1993; Labig, 1995; Mantell & Albrecht, 1994).

Three areas of violence research that are relevant to the workplace are examined in the current study. The first area focuses on the individual aggressor, the second examines job classifications that appear to be above average risk for violent confrontations, and the third emphasizes the need to investigate the acts of workplace violence against women by domestic partners.

Continual efforts to identify personality traits of individual aggressors who will be violent in the workplace are attempted, but as noted by Labig (1995), there is little research to support specific personality profiles. A model developed by Monahan (as cited in Labig, 1995) focuses on situational aspects that interact with an individual's characteristics rather than solely on the personality characteristics themselves. He outlines a cycle with four parts:

- The individual confronts an incident that is experienced as stressful.
- The person reacts to the event with certain kinds of thoughts to which he or she is prone by his/her personality.
- These thoughts lead to emotional reactions.
- These reactions in turn determine the behavior that the individual will use to respond to the situation.

The cycle continues as other people in the individual's environment respond to his or her behavior. The responses of those people around the individual can either de-escalate or increase the likelihood of violence. Workplace violence is not a sudden act. Some experts believe that violent people give frequent and repeated warnings of their stress and possibilities of becoming violent.

Monahan's model emphasizes the importance of identifying potentially volatile situations before they reach the point of violence. This emphasis broadens the definition of workplace violence to include verbal forms of violence, as well as physical. Other experts in the area of workplace violence encourage employers to include verbal forms of violence, such as threats, harassment, and intimidation, in their violence prevention policies (Baron, 1993; Labig, 1995; Mantell & Albrecht, 1994).

Recent studies of jobs that appear to be high risk have found the following factors to be associated with workplace violence:

- Exchanging money;
- Working alone at night, and during early morning hours;
- Having money, valued items, jewelry, or other items that are easily exchanged for cash;

SOURCE: Lord (1998). Reprinted with permission.

- Performing public safety functions in the community;
- Working with patients, clients, or customers known to have or suspected of having a history of violence; and
- Working with employees with a history of assaults or who exhibit belligerent, intimidating, or threatening behavior toward others. (U.S. Department of Health and Human Services, 1993)

Identifying jobs with these high-risk factors highlights not only the features of the assailant but also the assailant's relationship with the victim. Although the first four factors might relate to strangers, the fifth factor is a closer relationship of client or patient, and the sixth is a peer or subordinate relationship. Lynch (1987) concluded the risk of workplace victimization was less a function of the occupation and more of the specific tasks and individual performers. Acts of violence from strangers in cases where money is exchanged, employees are working at vulnerable places or times, or employees are responsible for valuable materials are primarily acts of instrumental aggression used to commit an act of robbery. In contrast to the stranger relationship, clients, patients, and even subordinates or coworkers who have used violent behavior in the past will often continue using violence for a number of reasons including a means to get their needs met and reduce frustration. The need to consider the relationship between victim and assailant is also emphasized in the study by Northwestern National Life Insurance (1993). Its results conclude that 44% of violent acts are committed by customers or clients, 24% by strangers, 20% by coworkers, and 7% supervisors. Also, as discovered by Mayhew and Elliott in their international study (1989), managers in a number of different professions are at risk for verbal and physical violence. In all professions, managers are responsible for disciplinary action, terminations, selections, and promotions; all are actions that could cause anger and frustration.

Domestic violence has come to the attention of employers as more women work outside of the home. According to the U.S. Department of Justice (1994), more than 13,000 nonfatal acts of work site violence were committed against women by domestic partners in 1993. The workplace has become an easy place for significant others to locate and confront their partners.

To further examine these three relevant areas, the following questions are examined in this study: (a) What is the scope of workplace violence in one state government? (b) Is there a significant relationship between specific personal and job-related characteristics of victims and risk of workplace violence? (c) What is the relationship between victims and perpetrators? (d) What types of violence are employed? and (e) Is there a significant correlation between the victim/perpetrator relationship and the type of violence employed?

Sample

In August 1994, a 4% stratified random sample of all full-time employees of North Carolina state government agencies and universities was selected ($n = 3,500$). The population was stratified by department with every 23rd employee systematically selected. Usable questionnaires were returned by 42% ($n = 1,477$) of the sample. Based on an examination of characteristics of all state employees (North Carolina Office of State Personnel, 1995), the groups of respondents are similar to all employees in terms of gender, ethnicity, and job classification. As noted in Table 1, sample respondents are slightly overrepresented by females, Whites, and persons in professional and administrative positions.

Variables

Independent Variables. Five independent variables represent demographics and job-related indices of victimization in the workplace. These are age, sex, ethnicity, job classification, and supervision experience. The relationship of the perpetrator to the victim is divided into five categories: supervisor, fellow employee, family member, stranger, and customer/client.

Dependent Variables. Victims were directly asked if they had been subjected to workplace violence. As noted earlier, workplace violence is defined as any action in the workplace including harassment, threats, physical

Table 1 Comparison of Sample and Population

Characteristics	Population		Sample	
	Number	Percentage	Number	Percentage
Total	87,500	100.0	1,477	100.0
Sex				
Male	64,137	73.3	732	49.6
Female	23,450	26.8	725	49.2
Ethnicity				
Non-White	22,923	28.4	286	19.4
White	57,743	71.6	1,158	78.5
Job classification				
Maintenance	5,849	7.0	153	10.4
Protective service	10,857	13.5	191	12.9
Clerical	16,231	20.0	271	18.3
Skilled craft	11,310	14.0	98	6.6
Technician	12,378	15.3	188	12.7
Paraprofessional	21,812	27.0	57	3.9
Professional	298	4.0	639	43.3
Administrative	1,929	2.4	292	9.8

SOURCE: North Carolina Personnel Information Management Service (1995).

NOTE: Respondents may have checked more than one response.

attacks, or intentional property damage. The current study classifies violence into seven categories: name calling or obscenities, threats of physical harm, pushing, hitting with hands, kicking, hitting with an instrument, and sexual assault. The association of the type of violent experience to the perpetrator's relationship to the victim and victim characteristics was analyzed.

Analysis

To identify which individual and work characteristics of victims were associated with the risk of workplace violence, logistic regression was employed. Multiple regression analysis was used to determine the relationship between types of violence employed and the victims' individual characteristics, job classifications, and connection to their perpetrators. A p value of .05 was used for significance testing.

Results

Characteristics of Workplace Victimization

Distributions for respondent age, gender, ethnicity, job classification, and supervisory experience are provided

Table 2   Characteristics of Survey Respondents

Characteristics (Code Values)	Ever Been Victimized at Work?					
	Total		Yes (1)		No (0)	
	Number	Percentage	Number	Percentage	Number	Percentage
Total number of respondents	1,477	100	318	22.3	1,107	77.7
Age						
Less than 30 (1)	220	15.0	40	12.6	170	15.4
31–39 (2)	367	25.0	91	28.6	264	29.0
40–55 (3)	700	47.4	150	47.0	531	48.0
Over 55 (4)	171	11.6	34	11.0	132	12.0
Sex						
Male (0)	732	49.6	179	57.0	534	48.6
Female (1)	725	49.2	135	43.0	564	51.4
Ethnicity						
Non-White[a] (0)	286	19.4	58	18.2	217	19.6
White (1)	1,158	78.5	256	80.5	867	78.4
Job classification						
Maintenance (1)	153	10.4	27	8.5	121	10.9
Protective service (2)	191	12.9	90	28.3	97	8.8
Clerical (3)	271	18.3	39	12.3	224	20.2
Skilled craft (4)	98	6.6	25	7.9	71	6.4
Technician (5)	188	12.7	38	11.9	143	12.9
Paraprofessional (6)	57	3.9	13	4.1	42	3.8
Professional (7)	639	43.3	160	50.3	464	41.9
Administrative (8)	292	19.8	60	18.9	225	20.3
Supervisory responsibility						
No (0)	904	61.3	183	57.5	690	62.3
Yes (1)	551	37.4	133	41.8	403	36.4

NOTE: Respondents may have chosen more than one job classification.

a. Includes 20 respondents classified as Native American, 11 as Hispanic/Latino, and 8 as Asian American. The remaining 247 are African American.

in Table 2. Almost half of the respondents (47.4%) were between 40 and 55 years old. This age category correspondingly holds the largest number of individuals reporting victimization. Males had a slightly larger victimization response than females, but Whites and non-Whites reported similar percentages of victimization.

Protective service employees, who include state detention officers and state law enforcement, reported the greatest number of individuals victimized at 32.3%, but skilled craft employees and professionals also had fairly substantial percentages of victimization (26.0% and 25.6%, respectively). Perpetrators of violence against protective service employees could include inmates and suspects of crime. Professionals include employees who work with mentally ill patients. Although contact with inmates, suspects, and patients may explain to some extent violence that is attached to the job, skilled craft employees and many employees even in protective services or in the professional areas are not readily exposed to violent clients. Many of the violent acts must originate with internal sources. Slightly more of the victims have supervisory responsibilities (24.8%) than those without supervisory responsibilities (21.0%). This slightly

greater percentage of violence toward supervisors provides only slight support for Mayhew and Elliott's findings (1989) of greater violence against individuals who have managerial responsibilities.

Logistic regression was used to estimate the likelihood of victimization based on the five independent predictors, which include the five demographic and job-related indices (see Table 3).

The overall model is significant, but job classification is the only significant variable. When compared to employees scoring low on job class, persons in job categories with greater responsibility are more likely to be victims. These jobs include paraprofessional, professional, and administrative classifications.

Characteristics of Workplace Violence

In addition to the demographic and job classification indices, the perpetrator/victim relationship was considered in examining the type of workplace violence experienced by the survey respondents who answered this item affirmatively ($n = 318$). Distributions for type of violence experienced and perpetrator/victim relationship are provided in Table 4. The severity of workplace

Table 3   Logistic Regression Coefficients Predicting Victims of Workplace Violence by Personal and Work Characteristics

Independent Variables	Coefficient	Standard Error	Wald Statistic	Significance
Age	0.652	.0687	.8993	.3430
Gender	−.1392	.0891	2.4385	.1184
Ethnicity	−.0676	.0651	1.08	.2988
Supervision responsibility	−.0720	.0821	.7706	.3800
Job classification	.0483	.0232	4.318*	.0377
Constant	−2.9953	.2923	104.98	.0000
Model chi-square	80.82**			
n	1,416			
df	6			

$*p < .05. **p < .001.$

Table 4 Characteristics of Workplace Victimization

Characteristics (Code Values)	Number	Percentage
Total workplace victims	318	100
Type of violence experienced[a]		
Name calling/obscenities		
No (0)	58	16.8
Yes (1)	288	83.2
Threats of physical violence		
No (0)	131	37.9
Yes (1)	215	62.1
Pushing		
No (0)	228	66.3
Yes (1)	116	33.7
Hitting with hands		
No (0)	244	71.1
Yes (1)	99	28.9
Kicking		
No (0)	269	78.4
Yes (1)	74	21.6
Hitting with an instrument		
No (0)	298	87.1
Yes (1)	44	12.9
Rape/sexual assault		
No (0)	327	95.9
Yes (1)	14	4.1
Perpetrator/victim relationship[a]		
Supervisor		
No (0)	272	77.1
Yes (1)	81	22.9
Fellow employee		
No (0)	235	69.5
Yes (1)	103	30.5
Family member		
No (0)	325	97.6
Yes (1)	8	2.4
Stranger		
No (0)	253	74.2
Yes (1)	88	25.8
Customer/client		
No (0)	140	40.8
Yes (1)	203	59.2

a. Respondents could report more than one victimization.

violence decreases as the number of responses of violence increases such that name calling/obscenities is the response with the highest percentage (83.2%) and sexual assault has the lowest (4.1%). The division between verbal and physical assaults is fairly large, with name calling/obscenities and threats of physical violence clustering above 50% (83.2% and 62.1%, respectively), but pushing, hitting with hands, kicking, hitting with an instrument, and sexual assault grouped substantially below 50% of the responses (33.7%, 28.9%, 21.6%, 12.9%, and 4.1%, respectively). The customer/client overwhelmingly is reported as the role of the perpetrator (59.2%), although fellow employees and supervisors also are responsible for a substantial portion of the complaints (30.5% and 22.9%, respectively).

Multiple regressions were used to estimate the relationship between types of violence and nine independent variables. Four of the independent variables are victim characteristics of age, gender, ethnicity, and job classification. The other five variables are the victim/perpetrator relationships of supervisor, employee, family, stranger, and customer/client. The overall model is significant, with 20% of the variance of the type of violence explained ($R^2 = .203$). Victim characteristics and the relationship between the victim and perpetrator do significantly predict the severity of violence.

The independent variable customer/client provides the most explanation within the model (Table 5). Evidently, if the perpetrator is a customer or client, the probability of more severe violence against the

Table 5   Multiple Regression Coefficients Predicting Characteristics of Workplace Violence by Perpetrator/Victim Relationships and Personal and Job Characteristics

Characteristic	B	Standard Error	Beta	t	Sign. t
Age	.126	.098	.073	1.29	.197
Gender	−.167	.097	−.099	−1.72	.086
Ethnicity	−.1325	.108	−.065	−1.22	.222
Job classification	.011	.032	.018	.34	.731
Relationship					
Supervisor	−.197	.249	−.045	−.79	.430
Employee	−.04	.219	−.01	−.18	.856
Family	−.022	.601	−.002	−.04	.971
Stranger	.333	.219	.081	1.52	.129
Customer/client	1.45	.223	.392*	6.48	.000
Constant	1.45	.387		3.74	.000
$N = 322$; $R^2 = .203$; $df = 10$.					

NOTE: $*F = 7.93$. Sign. t = Significance of t.

employee increases ($t = 6.48, p < .001$). Lynch's findings (1987) of risk associated with public accessibility are supported with this result of customers/clients being more likely to perpetrate more severe violent acts.

Discussion

Characteristics of Workplace Victimization

Individual Victim

Individual characteristics of victims do not appear to differ significantly from employees who have not been victimized. Differences in job classifications provide the only meaningful difference. The age of victims is concentrated around 40 to 55, but this is also the age of the largest percentage of respondents and is representative of the overall population of state employees.

As noted earlier, the U.S. Department of Justice (1994) reported more than 13,000 nonfatal acts of work-site violence were committed against women by domestic partners in 1993. Although the current study discovers a substantial number of women who report victimization, it is not a significantly different number than men who reported victimization. In fact, fewer women than men reported victimization, although there were approximately equal numbers of male and female respondents, which also approximates the state employee population. The current study does not support a greater number of women than men exposed to violence in the workplace, but respondents also did not report many incidents of violence attributed to domestic violence. Violence appears to be more directed toward the employee's job responsibilities than his or her individual characteristics.

Victims by Job Classification

The current study does support the three risk factors associated with workplace violence reported by the National Institute for Occupational Safety and Health (1992) and Mayhew and Elliott (1989). As noted earlier, these factors included (a) public safety functions; (b) work with patients, clients, or customers having a history of violence; and (c) work with employees who exhibit belligerent or threatening behavior toward others.

Those individuals employed in protective services, and who are responsible for enforcement, detention, and other functions that often deal directly with violent situations, reported the highest percentage of victimization. Some violence is to be expected from suspects seeking to escape or inmates with a history of violence; therefore, the need to provide the necessary protection for employees becomes crucial. However, much of the violence originates with supervisors and coworkers.

Professional and administrative classifications who work with patients and clients or manage employees also reported high percentages of victimization. Although individuals who are mentally ill are no more likely to be violent than the average population, those who have been diagnosed with a major mental disorder such as schizophrenia or mania/bipolar disorder are more likely to react aggressively (Monahan, 1992). Feldmann and Johnson (1994) also found individuals who are diagnosed with borderline disorders, antisocial personality, drug or alcohol abuse, or bipolar disorder may respond poorly to stress and react with anger to perceived threats to their self-esteem.

The current research found only a slightly greater amount of violence against supervisors than nonsupervisors. However, managers often are involved in high-risk responsibilities. Mediation, conflict resolution, disciplinary action, and termination are among the many responsibilities that professionals and managers are expected to fulfill. The majority of these duties are likely to produce anger and stress, particularly if the individual feels threatened. Although the results of the current study cannot confirm the relationship of these responsibilities to the experiences with personal violence, the literature supports the volatility of these functions (Baron, 1993; Cawood, 1991; Manigan, 1994).

Relationship of the Victim to the Perpetrator

The findings of this current study in general support the findings by the Northwestern National Life Insurance study (1993). Similar to Northwestern, the largest percentage of violent responses identified the perpetrator

to be a customer or client. However, the current study's responses included much higher percentages of perpetrators as supervisors and fellow employees. Bandura (1986) notes that most assaultive actions of aggressors produce rewarding outcomes for them. He further notes the difficulties in training individuals in alternative means of solving problems; aggression procures immediate results, and alternative methods often take more time. Supervisors and fellow employees may use violence because it works with few consequences to them.

Although the literature attributes a portion of workplace violence to domestic problems invading the work site (Crisis Management International, 1995; Kinney & Johnson, 1993; U.S. Department of Justice, 1994), few respondents reported violence resulting from such problems. According to Major John Massey with the State Capital Police (personal communication, October 16, 1994), domestic disputes do follow employees to work, but the employees often are too embarrassed to report incidents of violence, and their supervisors support their reluctance to report such incidents. Even though the results of the current study are confidential, with no possible way of attaching names to surveys, victims may still be reluctant to report their own domestic violence.

Types of Violence

An indirect relationship was found between the severity of violence experienced and the percentage of responses; verbal abuse is experienced with greater frequency than physical abuse. There also is a distinct demarcation in the frequency in responses of verbal and physical incidents. As noted by Megargee (1976), a major inhibitor of violence is physical contact. A majority of aggressive individuals are comfortable using verbal aggression but will advance to physical aggression only if highly provoked. Violent individuals often will not cross the boundary into physical violence unless the provocation is major in their eyes. Additional research (Karsort, 1996) identifies the invasion of personal space, particularly the use of physical contact, as a major risk factor for future violence.

This finding does not minimize the seriousness of the physical assaults that were reported. In an attempt to explore variables that might predict the severity of violence, a model including the victim's demographic characteristics, job classification, and victim/assailant relationship does explain significant variability in the severity of the violent incident. A closer examination of the variables distinguishes the customer/client relationship to be the most meaningful variable in the model. Megargee (1976) describes an internal conflict within an individual between the motivation to use force and a variety of different inhibitors. One inhibitor is the personalizing of the victim; individuals are less likely to hurt somebody they know and with whom they work or live. A customer/client may find it less inhibiting to use violence if he or she is able to depersonalize the state employee as a bureaucrat.

Conclusion

This study raises more questions than it answers. Probably the only question it does answer conclusively is that violence does exist in state government, at least in one state. If the sample represents North Carolina's state government, 22% of its employees have been subjected to violent behavior that ranges from obscenities to physical and sexual assaults by supervisors, other employees, and customers. Although the current study does not support other findings that report significant workplace violence between domestic partners it does support the research citing the specific high-risk job classifications of public safety and managerial areas.

One of the purposes for the survey is to aid in the development of intervention strategies. It is important to identify employees who are particularly vulnerable to violence at the workplace. This study does provide some information that will be useful in targeting specific groups. For example, workplace violence prevention training for managers and public safety employees might be helpful.

North Carolina held a statewide conference on violence in the workplace for top administrators in the state government soon after the survey was conducted. The North Carolina Office of State Personnel also implemented policies and procedures that define workplace violence, reporting and investigating procedures, and a vehicle to train departmental crisis teams.

Future plans are being considered to survey state employees again to discern differences in responses.

Much more research is needed to understand the relationship between victims and perpetrators of workplace violence, the reasons that violent acts are used, and how agency administrators might intervene. This general study provides evidence that violence exists in state government and that it varies along different groups of people. It will take smaller, more precise studies to etch out the reasons for these differences.

References

Bandura, A. (1986). The social learning perspective. In J. Toch (Ed.), *Psychology of crime and criminal justice.* Prospect Heights, IL: Waveland Press.

Baron, S. A. (1993). *Violence in the workplace: A prevention and management guide for businesses.* Ventura: Pathfinder Publishing of California.

Cawood, J. (1991). On the edge: Assessing the violent employee. *Security Management,* September, 130–136.

Crisis Management International, Inc. (1995). *Workplace violence: Prevention, response, and employer liability.* Training material prepared for the North Carolina State Government.

Feldmann, T., & Johnson, P. (1994, August). *Violence in the workplace: A preliminary report on workplace violence database.* Paper presented at the annual meeting of the American Bar Association, New Orleans, LA.

Karsort, M. (1996). *A safe termination model for managers.* Presentation given at the Violence in the Workplace: An Agenda for Action Conference, Pennsylvania State University, March 25–27.

Kinney, J. A., & Johnson, D. L. (1993). *Breaking point: The workplace violence epidemic and what to do about it.* Chicago: National Safe Workplace Institute.

Labig, C. (1995). *Preventing violence in the workplace.* New York: American Management Association.

Lynch, J. P. (1987). Routine activity and victimization at work. *Journal of Quantitative Criminology, 3,* 283–300.

Manigan, C. (1994, April). The graveyard shift. *Public Management,* pp. 10–15.

Mantell, M., & Albrecht, S. (1994). *Ticking bombs: Defusing violence in the workplace.* Burr Ridge, IL: Irwin.

Mayhew, P., & Elliott, D. (1989). *The 1988 British Crime Survey.* London: Her Majesty's Stationery Office.

Megargee, E. (1976). The prediction of dangerous behavior. *Criminal Justice and Behavior, 3*(1), 3–21.

Monahan, J. (1992). Mental disorder and violent behavior. *American Psychologist, 47*(4), 511–521.

National Institute for Occupational Safety and Health. (1992). *Homicide in U.S. workplaces: A strategy for prevention and research.* Washington, DC: Government Printing Office.

North Carolina Office of State Personnel. (1995). *Policies and procedures for North Carolina state employees* (Section 9). Raleigh, NC: Author.

North Carolina Personnel Information Management Service. (1995). *1995 North Carolina position and employee listing* (internal report). Raleigh, NC: Author.

Northwestern National Life Insurance. (1993, October). *Fear and violence in the workplace.* Unpublished manuscript.

U.S. Department of Health and Human Services, Centers for Disease Control and Prevention. (1993). *Alert: Request for assistance in preventing homicides in the workplace.* Washington, DC: Government Printing Office.

U.S. Department of Justice. (1994). *Uniform crime reports: Crime in the United States.* Washington, DC: Government Printing Office.

DISCUSSION QUESTIONS

1. What do the study's findings suggest about preventing workplace violence in terms of the types of jobs that are most risky?

2. Do you think the findings would be different if the sample was taken from workers not in government jobs? Why or why not?

3. Given that public safety jobs are at high risk of nonfatal violence, does it give you an understanding of the stress these employees are under? How might this stress impact their daily work?

4. As the economy changes and people's view of the government shifts over time, would you expect that the extent of workplace violence would similarly shift? How so? Explain.

Introduction to Reading 4

One type of victimization that can occur at work is sexual harassment. Although this type of victimization is not discussed in this section, an article dealing with the topic is included since it is a type of victimization that can occur to both men and women and can have serious consequences for victims. This review piece covers the antecedents (individual and organizational) and responses to sexual harassment.

Sexual Harassment at Work

A Decade (Plus) of Progress

Anne M. O'Leary-Kelly, Lynn Bowes-Sperry, Collette Arens Bates, and Emily R. Lean

In a 1995 review of the sexual harassment (SH) literature, Lengnick-Hall identified limitations in research to date and suggested "what we don't know about sexual harassment far exceeds what we do know" (p. 841). Although it may always be true that existing knowledge is more limited than potential knowledge, it is worth exploring the degree to which researchers have closed this knowledge gap since the 1995 review. To what degree has there been progress in building knowledge around this consequential workplace phenomenon?

SOURCE: O'Leary-Kelly et al. (2009). Reprinted with permission of the Southern Management Association

⊠ Definition and Perceptions of SH

Definitions of SH as a Construct

Lengnick-Hall (1995:842) noted more than a decade ago that "construct confusion" had created many problems for SH research. At the time of his work in 1995, there was only two definitions of SH, legal and psychological. The legal definition (then and now) entails two types of SH: quid pro quo (QPQ) and hostile work environment (HWE). QPQ SH entails threats to make employment-related decisions (e.g., hiring, promotion, termination) on the basis of target compliance with requests for sexual favors, whereas HWE SH involves sex-related conduct that "unreasonably interfer[es] with an individual's work performance" or creates "an intimidating, hostile, or offensive working environment" (29 C.F.R. § 1604.11[a] [3]). In accordance with the legal definition, an individual is considered to have experienced SH if the sex-related behavior meets the requirements of either QPQ SH or HWE SH. Alternatively, an individual is considered to have experienced SH if he or she *feels* harassed (whether or not the sex-related behavior is illegal) under the psychological definition.

Current Definitions

As of 2008, there are four definitions of SH. In addition to the *legal* and *psychological* definitions described above, the construct of SH has been defined from *behavioral* and *sex-based* perspectives. The four perspectives differ in the extent to which they define SH as a subjective and/or objective phenomenon. The psychological and sex-based perspectives define SH subjectively. In accordance with the psychological perspective, SH is "unwanted sex-related behavior at work that is appraised by the recipient as offensive, exceeding her resources, or threatening her well-being" (Fitzgerald, Swan, & Magley, 1997:15). The sex-based perspective (Berdahl, 2007a: 644) defines *sex-based harassment* (SBH) as " behavior that derogates, demeans, or humiliates an individual based on that individual's sex" including "seemingly sex-neutral acts, such as repeated provocation, silencing,

exclusion, or sabotage, that are experienced by an individual because of sex."[1]

As described previously, the legal QPQ perspective is objective; if the sex-related behavior meets the provisions of the law, then it constitutes SH. The behavioral perspective also defines SH objectively. From a behavioral perspective, specific sex-related behaviors are considered SH whether or not they cause psychological discomfort to targets or are illegal (Bowes-Sperry & Tata, 1999). Fitzgerald and her colleagues (Fitzgerald, Gelfand, & Drasgow, 1995; Fitzgerald, Magley, Drasgow, & Waldo, 1999) argued that SH is a stable behavior construct consisting of three primary dimensions: gender harassment, which consists of sexual hostility (explicitly sexual verbal and nonverbal behaviors) and sexist hostility (insulting verbal and nonverbal behavior that are *not* sexual but are based on gender), unwanted sexual attention (unwelcome, offensive interest of a sexual nature), and sexual coercion (requests for sexual cooperation in return for job benefits).[2]

Finally, the legal HWE perspective includes both subjective and objective elements. The subjective element is that the plaintiff must prove that *he* or *she* was adversely affected by sex-related behavior; the objective element is that the plaintiff must prove a "reasonable person" would be affected in a similar way (Bowes-Sperry & Tata, 1999).

⊠ Theoretical Advances in SH Research

At the beginning of our review period, there were several theoretical explanations for SH. These included the sex-role spillover explanation (e.g., Gutek & Morasch, 1982), which suggested that SH results from the inappropriate carryover of sex-based expectations into work; the contact hypothesis (e.g., Gutek, Cohen, & Konrad, 1990), in which SH occurs because of the sexualized environment created by contact or interactions between men and women at work; and the power and dominance explanation (e.g., Cleveland & Kerst, 1993; Farley, 1978; MacKinnon, 1979), which suggests that SH occurs because of unequal power across men and women in society and the workplace. Recently, several additional theoretical explanations have emerged.

Organizational Influences on SH

At the time of the last review, Fitzgerald, Hulin, and Dragow (1994) had just presented a model that outlined the antecedents and consequences of workplace SH. This model, which regarded SH as a work stressor, suggested two organizational environment factors as direct antecedents to SH: job context (i.e., gender domination by men in a work group is associated with more frequent SH). The model predicted numerous negative outcomes from the experience of SH, including job-related, psychological, and health detriments. These relationships were expected to be moderated by the target's personal vulnerability and response style. As we will see, this model has had great influence on recent empirical research.

Harasser's Decision to Initiate SH

There is recent theoretical guidance on the question of why harassers choose SH actions. In an article that positioned SH as a form of aggressive work behavior, O'Leary-Kelly, Paetzold, and Griffin (2000) suggested a model in which potential harassers are viewed as decision makers pursuing valued goals. From this perspective, SH, like other forms of aggression, can serve a variety of actor goals, including emotional (desire to rid themselves of negative affect), retribution (desire to punish others for a perceived injustice), and self-presentational (desire to establish a desired social image). SH is viewed as a goal-directed behavior that is chosen when it is believed to have a high probability of success and low probability of punishment.

It also has been suggested that harassers are influenced by moral intensity perceptions (O'Leary-Kelly & Bowes-Sperry, 2001). According to Jones (1991), moral intensity is a multidimensional construct that assesses the degree of moral imperative inherent in an issue and influences progression through the stages of the ethical decision-making process. O'Leary-Kelly and Bowes-Sperry (2001) argued that there is much inherent to the SH phenomenon that discourages actors from regarding it as a high moral intensity issue. For example, moral intensity is lower when there is low social consensus regarding the act (as is the case for some types of SH), low proximity between the parties (targets and

harassers often are dissimilar in factors such as gender and job level), and low probability and magnitude of consequences (actors tend to underestimate the magnitude of harm done because targets often suffer in silence). Furthermore, they propose that if potential harassers do not recognize SH as an ethical issue, they will be more likely to engage in sexually harassing behavior; therefore, perceiving SH as low in moral intensity is expected to result in increased SH.

Target Responses to SH

Knapp, Faley, Ekeberg, and Dubois (1997) developed a typology of target responses based on theory from the whistle-blowing (Near & Miceli, 1985) and stress and coping (Lazarus & Folkman, 1984) literatures. They proposed that target responses to SH vary in terms of the focus of response (self vs. initiator) and the mode of response (self vs. supported), resulting in four response strategies: *avoidance/denial* (e.g., interpreting behavior as a joke), *social coping* (e.g., discussing the behavior with friends), *confrontation/negotiation* (e.g., asking the harasser to stop), and *advocacy seeking* (e.g., filing a formal report). Target decisions among these actions are proposed to be influenced directly by characteristics of the reporting process, target expectations regarding the outcomes of various responses, severity of SH experienced, and the target's level of psychological distress. Furthermore, these predictors are expected to be affected by targets' individual characteristics and power as well as characteristics of the workgroup, organization, and legal and economic environment. Tests of this typology are generally supportive (Malamut & Offermann, 2001; although for an exception, see Magley, 2002), including research establishing cross-cultural generalizability (Wasti & Cortina, 2002).

The O'Leary-Kelly, Paetzold, and Griffin, (2000) model also provides theoretical insights into target responses to SH. This model suggested that target perceptions of actor culpability are dependent on attribution judgments about actor intentionality and justifiability and on the foreseeability of negative target outcomes. Furthermore, target perceptions about the likelihood of future SH, which influence the target's chosen response, depend on attributions regarding the stability of actor behavior. Target's emotional and behavioral responses are expected to depend on their attribution judgments, their own goals (emotional, retributive, self-presentational), and their beliefs about the likelihood that various responses will facilitate goal success.

Broad Theoretical Approaches

Three recent articles took a broader approach to theory development (i.e., the focus was not on just targets or just harassers or just the organization). Berdahl's (2007a) theory of harassment based on sex extends the view of SH as goal-directed behavior and locates SH within the broader harassment literature. There are three central tenets of this theory. First, the "primary motive underlying *all* harassments is a desire to protect one's social status when it seems threatened" (2007a: 641, italics added). Second, the existence of gender hierarchy (at the societal level) renders sex a useful basis on which to harass. Third, distinctions are made within sexes as well as between them. Berdahl proposes that SH is influenced by both contextual and personal factors. Contextual factors included gender hierarchy at the organizational level as well as the existence and type of threat to social identity (i.e., threats that emphasize gender distinctions versus those that challenge them). Personal factors include the actor's sex, sexist attitudes, and gender role conformity. This theory also provides insight into regrets. Given that identity threat motivates actors to harass, individuals who threaten the gender identity of an actor (e.g., by confounding distinctions between the sexes) are likely to become targets of SH. This suggests that the most likely form of SH is men harassing women, especially women who challenge men's status; men will also harass other men who threaten their status and when women harass they will primarily target other women, particularly those who represent a status threat. Although we do not focus on legal theory in our review, it is worth noting that similar arguments have been made in legal journals. Franke (1997) argued that SH be regarded as a form of sex discrimination, not because men initiate it against women but because it is a tool for enforcing traditional gender norms, one that can be used against both women and men.

Another broad approach (O'Leary-Kelly, Tiedt, & Bowes-Sperry, 2004) uses accountability theory to explain SH, with accountability defined as the "perceived need to justify or defend a decision or action to some audience(s) which has potential reward and sanction power and where such rewards and sanctions are perceived as contingent on accountability conditions" (Frink & Klimoski 1998: 9). Accountability theory suggests conditions that limit the accountability harassers feel for their actions (e.g., fragmentation of responsibility, competing role expectations, reactance to new imposed standards on previously accepted behavior). Although no formal model was presented, theoretical principles explained why targets of SH often choose passive rather than direct or active responses (e.g., lack of clarity in prescriptions for behavior, identity implications). Finally, accountability theory provided insights into observers' inaction after witnessing SH (lack of connection between the event and the observer identity, ambiguity in role expectations).

In another broad theory piece, DeCoster, Estes, and Mueller (1999) applied the routine activities perspective from the criminology literature (Cohen & Felson, 1979; Hindelang, Gottfredson, & Garofalo, 1978) to explain SH victimization at work. They suggested that some individuals are more prone to victimization because their daily activities bring them in direct contact with predators (Cohen & Felson, 1979; DeCoster et al., 1999). More specifically, this suggests that three conditions are important to victimization: (a) a motivated harasser, (b) a suitable target (i.e., proximity to predators, material attractiveness to predators), and (c) the absence of guardians who can prevent the SH incident. As with the representation of harassers in O'Leary-Kelly, Paetzold, and Griffin (2000), this perspective assumes a rational harasser who strives to minimize costs and maximize outcomes. These theoretical propositions are tested, and findings indicate that guardianship (i.e., supportive supervisors, supportive work group cultures, work-group solidarity), target proximity (i.e., working in a male-dominated job or a highly populated job location), and target attractiveness (i.e., female targets being educated or having organization tenure, which are depicted as evidence of a power threat to male employees; being single) are predictive of the SH of women. These effects were additive but did not interact to predict victimization as the routine activities model would predict.

⊠ Antecedents to SH

A great deal of research since 1995 examines the conditions that prompt SH. As a framework for this discussion, we regard the harasser as a motivated actor who is driven by individual predispositions and who reacts to situational triggers. Therefore we discuss personal- or harasser-related antecedents and situational antecedents (including characteristics of the target and of the organizational climate).

Personal- or Harasser-Related Antecedents

Extensive research indicates that although harassers are most likely to be male (e.g., Gutek, 1985; Martindale, 1990; U.S. Merit Systems Protection Board [USMSPB], 1981, 1988, 1995), most men are not harassers. Many researchers have noted the lack of research attention given to harassers (e.g., Lucero, Middleton, Finch, & Valentine, 2003; O'Leary-Kelly, Paetzold, and Griffin, 2000), with a likely reason being the difficulty of gaining access to samples of adequate size. According to Pryor, Giedd, and Williams (1995), the first USMSPB (1981) survey attempted to examine characteristics of sexual harassers but was unable to do so because so few people responded affirmatively when asked if they had ever been accused of "sexually bothering" someone at work. However, three harasser-based antecedents have received some research attention: the likelihood to sexually harass, the position or role of the harasser, and the harasser's goals or motives.

Harasser Likelihood to Sexually Harass

Much of what we know about harassers comes from research conducted using Pryor's (1987) Likelihood to Sexually Harass (LSH) scale, which "measures a readiness to use social power for sexually exploitive purposes" (Pryor, Lavite, & Stoller, 1993: 74). Most empirical research on LSH (both that before and after 1995) has

focused on developing a personality profile of men (for exceptions, see Isbell, Swedish, & Gazan, 2005; Luthar & Luthar, 2008) who are likely to become sexual harassers. This research suggests that high LSH men are more likely than low LSH men to (a) be prone to sexual violence, that is they express a likelihood to rape, hold adversarial sexual beliefs, and accept rape myths (e.g., Bargh, Raymond, Pryor, & Strack, 1995; Begany & Milburn, 2002; Pryor, 1987), (b) connectively link the concept of social dominance with sexuality (Pryor & Stoller, 1994), (c) differentiate themselves from women, that is they prefer traditional male sex-role stereotypes, rate themselves as less feminine, and are lower in empathy, which is a stereotypically feminine characteristic (Driscoll, Kelly, & Henderson, 1998; Pryor, 1987), (d) have negative and hostile attitudes toward women (Begany & Milburn, 2002; Driscoll et al., 1998), and (e) have personalities that are high in authoritarianism, low in honesty humility and low in self-monitoring (Dall'Ara & Maas, 1999; K. Lee, Gizzarone, & Ashton, 2003; Pryor, 1987). There also is evidence that in certain situations, high (vs. low) LSH men are more likely to initiate unwanted sexual attention (Fitzgerald et al., 1994; Pryor, 1987; Pryor et al., 1993). More recent research has extended the predictive validity of LSH to other types of sexually harassing behaviors. The research of Maas and colleagues (Dall'Ara & Maas, 1999; Maas, Cadinu, Guarnieri, & Grasselli, 2003) established the validity of LSH for prediction of sexual hostility, which is a form of gender harassment (Fitzgerald et al., 1999). Their research indicates that high (vs. low) LSH men are more likely to send pornographic material to a female interaction partner and that this likelihood increases in response to gender identity threats. LSH also predicts sex-based (i.e., not sexual) harassment, such as asking sexist questions of a female during an interview (Rudman & Borgida, 1995), rating a female's performance or competency on task as low (Driscoll et al., 1998), spending less time with a female in a subordinate position (Murphy, Driscoll, & Kelly, 1999), and providing less feedback regarding the performance of a female they have been asked to evaluate (Murphy et al., 1999).

Research on observer perception of potential harassers provides evidence of the validity of the LSH measure. Participants who watched videotaped interactions between a man and woman were able to differentiate between high and low LSH men. More specifically, observers' ratings of men's LSH were positively related to men's self-reported LSH (Driscoll et al., 1998). Furthermore when asked, "What would it be like to have this man as your employer?" observers were more likely to provide negative (vs. positive) evaluations for men high (vs. low) in LSH (Craig, Kelly, & Driscoll, 2001).

Positions or Role of the Harasser

Although most SH research has focused on individuals within an organization, Gettman and Gelfand (2007) argued that clients and customers are also potential sources of SH, particularly in the service sector. They developed a theoretical model of client sexual harassment (CSH) based on the organizational model of Fitzgerald, Drasgow, Hulin, Gelfand, and Magley (1997), which they tested in two field studies. Their results indicated that client power (assessed using perceptions of target and organization dependency on the client) was significantly, strongly, and positively related to CSH.

Harasser Goals and Motives

Recent research has considered SH as goal-directed behavior chosen by an actor for a specific purpose (O'Leary-Kelly, Paetzold, & Griffin, 2000). Although empirical research that examines specific harasser motives is very limited, there is accumulating evidence that sexual harassers are motivated by social identity concerns (i.e., they initiate sexually harassing behavior with the goal of establishing or protecting a specific social identity). Maas and her colleagues (Dall'Ara & Mass, 1999; Maas et al., 2003) examined various aspects of identity threat using a "computer harassment paradigm" in which male participants interacted virtually with (fictitious) females to complete a task. Their results indicated that male participants exposed to gender identity threats (e.g., interacting with a woman espousing feminist values, having their masculinity questioned) were more likely than those with no exposure to engage in sexually hostile behavior. In addition, gender identity threat also predicted intentions to engage in sexually coercive behavior in future situations unrelated to the computer task. However, consistent with the notion that only some men enact SH, individual difference factors such as LSH, gender notification, and social

dominance orientation influenced the extent to which gender identity threats prompted SH. Finally, Berdahl's (2007b) finding that women with more masculine (as opposed to less masculine) personalities and occupations are more likely to be targets of SH implies that harassers are motivated to punish gender role violators.

Another aspect of the goal-directed harasser model (O'Leary-Kelly, Paetzold, & Griffin, 2000) is that harasser behaviors are enacted in accordance with target responses to initial SH. Recent empirical research by Lucero and her colleagues provides support for this perspective. Lucero et al. (2003) and Lucero, Allen, and Middleton (2006) used published arbitration cases to examine data on individuals who had been disciplined by their employer for SH. Their work suggested that these harassers could be distinguished by the nature of their behavior; some harassers had a more sexual repertoire of behaviors, whereas others had a more aggressive repertoire (Lucero et al., 2003). Furthermore, these repertoires remained consistent over time for the majority of harassers; when change did occur, it tended to entail the addition of a new type of SH behavior rather than the replacement of one behavior type with another (Lucero, Allen, & Middleton, 2006). Although this research focuses on harasser conduct, we can make inferences about the harasser motives that are the basis for this conduct. For example, it is reasonable to assume that individuals who initiate gender harassment (aggressive sex-based behavior) are not motivated by sexual desire.

Situational Antecedents

Recent research also has explored characteristics of the environment that are encountered by sexual harassers. Here we discuss research regarding characteristics of the target and of the organizational environment.

Target-Related Antecedents

Characteristics of the targets themselves may be associated with the occurrence of SH. Much research indicates that women are more likely than men to be targets (e.g., Berdahl & Moore, 2006; Gutek, 1985; Martindale, 1990; USMSPB, 1981, 1988, 1995). A provocative recent study (Berdahl, 2007b), however, suggested this finding may be, in part, a methodological artifact. Here, gender was relevant to being a target of SH only in male-dominated organizations, which have been the focus of most prior SH research.

As discussed previously, Berdahl (2007a) proposed that sexual harassers are motivated to punish individuals who violate gender-role norms. Berdahl (2007b: 425) investigated the effects of "personality gender" (i.e., the extent to which one's personality exhibits stereotypically masculine and feminine traits) and "occupational gender" (i.e., male- or female-dominated occupation) on becoming an SH target. Her results indicated that women who had more (vs. less) masculine personalities experienced more SH and those who occupied traditionally masculine (vs. feminine) jobs experienced more SH. This work suggests that SH is targeted at "uppity" women who step out of place by assuming characteristics considered more desirable for men (Berdahl, 2007b).

A recent large-scale study (Jackson & Newman, 2004), using USMSPB's survey of federal workers, examined the interplay of gender and other predictors of SH. For women, but not men, education and pay grade were positively associated with SH experiences. Furthermore, there was a stronger association between job status (blue collar, white collar) and SH for women than for men, with blue-collar women experiencing very high levels of SH; these findings appear consistent with the "uppity women" prediction just mentioned. It also was noteworthy that age had differential effects such that SH dropped off considerably for women as they aged, but this effect was less pronounced for men.

Ethnicity has also been examined as a potential target-related antecedent. Berdahl and Moore (2006) explored the effects of ethnicity and sex on various forms of SH. They found that being an ethnic minority in a workgroup was positively related to traditional forms of SH (e.g., gender harassment) and that ethnicity predicted "not-man-enough harassment" (e.g., not meeting masculine ideals, being too much like a woman). Similarly, Gettman and Gelfand (2007) found that non-White employees (in both professional and nonprofessional occupations) experienced more SH at the hand of clients and customers than did White employees.

Another study (Harned, Ormerod, Palmieri, Collinsworth, & Reed, 2002) examined target power as an antecedent to various types of SH (sexist hostility, sexual hostility, unwanted sexual attention, sexual coercion) and sexual assault. In a large-scale study of female members of the military, the authors examined negative conduct experienced at the hands of personnel employed in the military workplace. Two target-related antecedents were examined: organizational power (measured via pay grade and years of active duty service) and sociocultural power (measured via age, education, race/ethnicity, marital status). Their findings suggested that 4% of servicewomen reported an attempted or actual rape by colleagues within the past 12 months, and they found that both forms of power predicted SH and sexual assault (with lower power being associated with increased likelihood of SH and sexual assault).

Organizational Antecedents

The introduction of the organizational tolerance for sexual harassment (OTSH) construct by Hulin, Fitzgerald, and Drasgow (1996) prompted significant research on organizational antecedents to SH. OTSH is based on the Fitzgerald et al. (1994) model that outlined the organizational antecedents and consequences of work SH. OTSH, which reflects one dimension of an organization's overall climate, reflects respondents' perceptions of the contingencies between SH behavior and consequences, for targets and harassers, within their organizational context. In organizations characterized by strong OTSH perceptions, employees believe that reporting of SH is risky, that complainants are unlikely to be taken seriously, and that there would be few consequences for perpetrators (Hulin et al., 1996). There is now a well-established measure of OTSH (the OTSH Inventory), which has been used in multiple studies to date (e.g., Fitzgerald et al., 1997; Hulin et al., 1996; Wasti, Bergman, Glomb, & Drasgow, 2000). The inventory asks respondents to review six scenarios (depicting gender harassment, unwanted sexual attention, and sexual coercion; Fitzgerald et al., 1995; Gelfand, Fitzgerald, & Drasgow, 1995) and to rate their perceptions of the likely outcomes (i.e., risk to the complainant, degree to which complaint is taken seriously, consequences for behaviors initiated by a supervisor or coworker).

A recent meta-analysis demonstrates the importance of organizational factors as predictors of SH. Willness, Steel, and Lee (2007), in a meta-analytic review of 41 studies and almost 70,000 respondents, examined two organizational antecedents: organizational climate perceptions (e.g., OTSH) and job-gender context (e.g., proportion of women in workgroup, compositions of workgroup). Results indicated a significant and robust relationship between organizational climate and SH (weighted mean correlation corrected for reliability was equal to .364). Job-gender context also emerged as a significant predictor of SH experiences, but the effect size was smaller (corrected correlation = −.192). Results of a moderator analysis indicated these effects are strong for military samples.

Research using the OTSH Inventory demonstrates that the inventory predicts respondents' reports of SH (Fitzgerald et al., 1997; Hulin et al., 1996), making it a useful diagnostic tool for managers who want to anticipate hostile climates. Although we discuss SH consequences in detail later, it should be mentioned here that OTSH is directly associated with well-being-related variables such as work satisfaction, job withdrawal, life satisfaction, psychological well-being, anxiety and depression, physical health, and health satisfaction (e.g., Fitzgerald et al., 1997; Hulin et al., 1996) and that these effects occur for both male and female employees. Perhaps more surprising, one study (Hulin et al., 1996) demonstrated that OTSH explains more variance in well-being outcomes than does the direct experience of SH. Essentially, a high-OTSH climate is one in which employees perceive that they face considerable risk if they report SH (because of the normative nature of SH and because of the negative individual outcomes just mentioned). Oddly, this means that in those workplaces that are most poisoned, reporting of SH is least likely, suggesting a very negative spiral.

If, as suggested by the Willness et al. (2007) meta-analysis, an SH-tolerant organizational climate has negative effects, it is important to ask which aspects of the environment create this negative climate. Unfortunately, we found few studies that have explored this issue. In one notable exception, Williams, Fitzgerald, and Drasgow (1999: 306) examined three climate aspects in a military setting: organizational policies (formal written guidelines for behavior), organizational procedures ("formal or informal steps for filing grievances, investigating complaints, and enforcing penalties"), and various organizational practices (actual organizational actions around SH). Results suggest that one practice (implementation) was associated with SH reports (by both male and female employees). A second study examining climate factors (Amick & Sorenson, 2004) found that OTSH perceptions were strongest when respondents believed that coworkers held traditional attitudes toward women but that respondent job type (traditional or nontraditional) and workgroup gender mix were not significant predictors of OTSH.

Another outgrowth of the focus on organizational climate in recent research is the interesting question of whether it is reasonable to expect one SH climate within a work environment. In the initial work on OTSH (Hulin et al., 1996), it was noted that male and female employees held significantly different perceptions of OTSH, with women reporting higher levels. A qualitative study (Rogers & Henson, 1997) also provides support for the idea of multiple climates. Here, they found that temporary clerical workers operated in a more sexualized climate than did permanent workers, with temporary workers experiencing more SH and having less power to obtain remedy.

Two recent studies are interesting because they broaden the focus of either the organizational antecedent variables or the SH criterion variable. In the former, Mueller, DeCoster, and Estes (2001: 417) examined the relationship between general organizational conditions and SH, with general organizational conditions defined as those indicative of "modern methods of organizational control," including social integration, structural differentiation, decentralization, and formalization and legitimacy. These features of the general work environment were expected to be associated with lower levels of SH because they encourage coworkers to protect one another, they recognize professional behavior as necessary to organizational mobility, and they empower individuals to protect themselves from SH. In general, results supported these predictions.

The second, which broadened the focus of the SH criterion variable, examined SH experiences in the context of the climate for workplace civility (Lim & Cortina, 2005). In two studies of women in the federal court system, results provided support for the co-occurrence of SH and workplace incivility, in that almost all women who experienced SH also experienced incivility. These studies highlight that SH occurs within a broader context of mistreatment and disrespect and raise interesting questions about whether the same actors initiate both forms of negative conduct and whether the same organizational conditions might contribute to both. These results also emphasize the cumulative nature of multiple victimization, in that women who experienced both forms of mistreatment reported lower levels of organizational and psychological well-being.

✉ Responses to SH

Although most research on responses relates to the question of how SH targets respond or cope, there also is recent research examining the responses of SH observers and of the employer (organization).

Target Responses

Types of Target Responses

Target responses to SH have cognitive (e.g., labeling behavior as SH, discussed in detail earlier), emotional, and behavioral dimensions. Research indicates that pervasiveness and type of SH influence targets' cognitive or emotional responses (e.g., subjective appraisals of distress), which in turn influence behavioral responses such as confronting the harasser or seeking social support (Langhout et al., 2005; Malamut & Offermann, 2001). Furthermore, cognitive or emotional responses have been found to mediate relationships between other antecedents and target responses. For example, the target's appraisal of distress has been found to mediate the impact of occupational status, organizational climate, frequency and duration of SH, and power differentials between target and harasser on a variety of target responses (Bergman, Langhout, Palmieri, Cortina, & Fitzgerald, 2002; Langhout et al., 2005; Malamut & Offermann, 2001).

Research indicates that although targets engage in multiple strategies when responding to SH (e.g., Cortina & Wasti, 2005; Fitzgerald et al., 1995; Gruber, 1989;

Magley, 2002; Wasti & Cortina, 2002), some responses are more common than others. For example, both early and recent research has found that although many targets engage in avoidance responses, few ever formally report their experiences (e.g., Cochran, Frazier, & Olson, 1997; Culbertson, Rosenfeld, Booth-Kewley, & Magnusson, 1992; Malamut & Offermann, 2001; Martindale, 1990; USMSPB, 1981, 1995; Wasti & Cortina, 2002). Moreover, these multiple responses have been found to form specific *coping profiles* over time (Cortina & Wasti, 2005). In addition to knowledge regarding the frequency with which various responses are used, empirical research has identified numerous predictors of target responses. We use Knapp et al.'s (1997) theoretical model (described earlier) to structure our discussion of these predictors.

Personal or Target Antecedents

Knapp et al. (1997) proposed that personal characteristics of targets (e.g., age and gender) influence their behavioral responses to SH. Several demographic characteristics of targets have been found to influence their responses. Cortina and Wasti (2005) found that age was positively associated with a "detached" coping profile in which targets exhibit an absence of coping with the situation. Malamut and Offermann (2001) found that women were more likely than men to use social coping, advocacy seeking, and confrontation, yet women and men were equally likely to engage in avoidance and denial responses. Some studies have found effects of personal target characteristics on the coping response of reporting; target reporting has been found to be positively related to target education level and previous SH experience (Perry, Kulik, & Schmidtke, 1997) and negatively related to occupational status (Malamut & Offermann, 2001). Finally, cultural affiliation has been found to predict target responses and coping profiles; targets from collectivistic cultures are more likely than those from individualistic cultures to engage in avoidance, denial, and negotiating responses (Cortina & Wasti, 2005; Wasti & Cortina, 2002). Furthermore, perhaps because of cultural factors, Hispanic women who experience SH seek support from friends and family more than from formal organizational support mechanisms

(Cortina, 2004). Research indicates that personality characteristics also influence target responses. For example, target assertiveness has been found to be positively related to confronting the harasser (Adams-Roy & Barling, 1998), and conflict avoidance has been found to be negatively related to intentions to engage in such confrontation (Goldberg, 2007).

Organizational Antecedents

Empirical research on targets' responses indicates strong support for organizational antecedents. Various measures of perceived organizational climate have been found to influence target coping responses. For example, Malamut and Offermann (2001) found that targets' perceptions of OTSH were positively related to their use of avoidance denial, social coping, and advocacy seeking but not confrontation. Cortina and Wasti (2005) found that women who perceived higher levels of social support from organizational leaders (which is one aspect of climate) were more likely to fit the "detached" coping profile. Offermann and Malamut's (2002) results demonstrated that women who believed that military leaders (at multiple levels) made genuine efforts to end SH reported stronger reporting freedom, greater satisfactions with the reporting process, and more positive attitudes. It is noteworthy that these effects occurred after controlling for the effects of general work climate (having policies and procedures against SH). Cortina's (2004) study of Hispanic women found that these women sought more support from all sources (personal and organizational) when the harasser had high (vs. low) organizational power. Finally, research indicates that OTSH influences target reporting responses; Bergman et al. (2002) found an indirect effect (through SH history and frequency) of OTSH and Welsh and Gruber (1999) found a direct effect of OTSH.

Behavior-Based Antecedents

Research also indicates that SH severity is related to target responses. For example, Munson, Hulin, and Drasgow (2000) found that targets were more likely to use external coping strategies when SH severity was high than when it was low. Research on target reporting of SH (which is a type of external coping strategy) found that target reporting was more likely when harassers were supervisors, when there were multiple harassers, and when the type of behavior was sexual coercion (Lee, Heilmann, & Near, 2004; Welsh & Gruber, 1999). Bergman et al. (2002) found an indirect effect (through cognitive appraisal) of SH severity on reporting. Malamut and Offermann (2001) hypothesized that targets would use a full spectrum of strategies for severe SH (which they assessed in terms of SH type and SH frequency or duration). They found that as frequency or duration of SH increased, targets increased their use of avoidance denial, social coping, and advocacy seeking, but there was no effect on the use of confrontation. They also found that sexual coercion (the more severe type of SH) was positively related to the use of social coping and confrontation, but there was no effect on the use of avoidance denial or advocacy seeking. Finally, Cortina and Wasti (2005) found that SH severity was one of the strongest determinants of target responses; more specifically, as SH severity increased, so did the number of coping behaviors used by targets.

Rospenda, Richman, Ehmke, and Zlatoper (2005: 96) examined the influence of different patterns of SH and generalized work harassment (GWH; defined as "negative workplace interactions that affect the terms, conditions, or employment decisions related to an individual's job, or create a hostile, intimidating, or offensive working environment, but which are not based on legally protected social status characteristics") on use of services as a coping mechanism for SH and GWH. This study, which spanned a multiyear period, demonstrated that different patterns of harassment experiences are not equivalent. For both SH and GWH, they found that those who experienced intermittent (on and off harassment over time) or chronic (harassment that continues across time periods) harassment were most likely to seek professional services. Contrary to expectations, those experiencing SH remission (cessation of harassment) also reported increased service use.

Effectiveness of Target Responses

Knapp et al. (1997) proposed that avoidance or denial responses are the least effective (in that they do not stop SH) and most costly to organizations (in that they result in decreased productivity and turnover). They recommended that targets engage in advocacy-seeking behavior such as reporting SH to others within the organization, especially those with formal authority to take action. Because reporting SH is often ineffective and at times harmful to the target (Hessen-McInnis & Fitzgerald, 1998; Magley et al., 1999; Stockdale, 1998), Bergman et al. (2002) examined the reasonableness of reporting to determine if reporting was more effective under certain conditions than others. Their results suggested that it is not the act of reporting SH per se that determined consequences for targets but rather the organization's responses to such reporting.

Organizational Responses to SH

Although organizational policies, procedures and practices were discussed previously as organizational antecedents, they can also be conceptualized as organizational responses to the law. If a charge of SH is filed against an organization, the existence of SH awareness training can help establish an affirmative defense by demonstrating that it exercised reasonable care to prevent sexually harassing behavior. On the other hand, failure to provide training has resulted in employer liability for SH in U.S. federal courts (Zugelder, Champagne, & Maurer, 2006). Thus, organizational responses to SH can be characterized as either preventive (e.g., training) or corrective (e.g., disciplining or counseling harassers).

Although an important topic, there has not been much recent research examining organizational responses to SH. In a noteworthy exception, Bergman et al. (2002) investigated corrective organizational responses following a formal report of SH. More specifically, they examined antecedents and consequences of organizational remedies (e.g., disciplining the harasser), organizational retaliation (e.g., transferring targets who report SH against their will), and organizational minimization (e.g., encouraging targets to drop their complaints).

They found that all three organizational responses influenced the effectiveness of reporting (measured as targets' satisfaction with the reporting process). It is not surprising that organizational responses of retaliation and minimization were negatively related to targets' satisfaction

whereas providing organizational remedies was positively related. Furthermore, they found that targets' satisfaction with the reporting process was positively related to job-related, psychological, and health outcomes. Another exception is the study of women in the federal circuit court system by Miner-Rubino and Cortina (2007). These authors found that when organizations were unresponsive to an environment characterized by hostility toward women, female employees reported decreased levels of organizational commitment. With regard to preventive or proactive organizational responses, SH awareness training has been found to increase the likelihood of respondents labeling sex-related behavior as SH (Antecol & Cobb-Clark, 2003; Wilkerson, 1999).

It is important to note that theoretical models propose reciprocal influences between organizational and target responses. For example, Knapp et al. (1997) proposed that factors associated with the reporting process (e.g., failure to resolve previous SH complaints to the targets' satisfaction, extent to which procedures are understood) influence targets' responses to SH. Little empirical research, however, has addressed these predictions.

Observer Responses to SH

Raver and Gelfand (2005), in a study that found that the level of ambient SH (i.e., the general level of SH in the work group; Glomb et al., 1997) within a team was positively related to team conflict, argued that this conflict could result from observers adopting coping strategies such as confronting or refusing to speak to the harasser. Bowes-Sperry and colleagues conceptualized such actions as forms of observer intervention, which are one form of observer coping. Bowes-Sperry and Powell (1999) found support for an ethical decision-making model of observers' intentions to intervene in hypothetical scenarios of SH; both individual (i.e., ethical ideology) and situational (e.g., severity of SH) factors influenced observer intentions such that they were more likely to intervene if they recognized the incident as an ethical issue. As mentioned earlier, Bowes-Sperry and O'Leary-Kelly (2005) contributed a typology of observer intervention behaviors in SH; however, this typology has not been empirically tested.

Consequences of SH

SH has broad and negative consequences that affect SH targets, observers, and the organization as a whole. In this section, we highlight recent research that examines each of these forms of consequence.

Consequences for Targets of SH

The Willness et al. (2007) meta-analysis mentioned earlier examined not only antecedents to SH but also consequences experienced by SH targets.[3] Results demonstrated that most consequence variables proved to be significantly correlated with SH experience. SH experience was consistently associated with lower job satisfaction, regardless of how this latter construct was measured (individual facets, global measure), with effect sizes (weighted mean correlation corrected for reliability) ranging from $r_c = -.241$ to $r_c = -.316$. As predicted by the researchers, satisfaction with coworkers and supervisors (interpersonal work dimensions) was more negatively affected by the experience of SH than was work satisfaction. There was also a significant negative relationship between SH experience and organizational commitment ($r_c = -.249$), suggesting that this experience has a negative impact on attitudes toward the employer. Again, moderator analyses indicated that individuals in military contexts demonstrated distinct results; in military samples, there was a stronger relationship between the experience of SH and job satisfaction.

The findings related to psychological and health outcomes also demonstrate the highly negative effects of SH experience. There were significant relationships with mental health ($r_c = -.273$), physical health ($r_c = -.247$), and PTSD ($r_c = .247$). The relationship with life satisfaction was also significant but more limited ($r_c = -.119$), a finding that might be expected given the range of issues that compose life satisfaction.

Research not included in this meta-analysis also demonstrated the negative effects of SH. For example, Woodzicka and LaFrance (2005), who studied women exposed to mild gender harassment (e.g., asking a sexist question) during a job interview, found that SH had negative consequences for their performance. Participants in the harassing condition used significantly more diluted language; repeated words more frequently, exhibited more false starts, and were judged as having lower quality answers than those who were not harassed.

Recent work on CSH by Gettman and Gelfand (2007) demonstrated that the negative effects of SH are similar whether the harassment is initiated by organizational members or clients. Their cross-sectional study of professional women indicated that CSH is negatively related to job satisfaction and health satisfaction and positively related to psychological distress. Their study of CSH among nonprofessional food service workers indicates that CSH predicts job satisfaction, even after controlling for SH by organizational members.

Earlier, we described a study that examined a specific, and severe, form of SH—sexual assault (Harned et al., 2002). This cross-sectional study also examined job-related affect (e.g., supervisor, coworker, and work satisfaction) and psychological health (psychological well-being, health satisfaction). In general, the findings indicated that SH was most strongly related to job-related factors whereas sexual assault was most strongly related to health outcomes.

One series of studies (Rospenda, 2002; Rospenda et al., 2006; Rospenda, Richman, Ehmke, & Zlatoper, 2005) examined the effects of SH in context of a broader harassment construct, generalized work harassment (GWH). These studies posed SH as a form of WPA that is expected to be associated with other forms of aggression such as assault. In a study of university employees surveyed at multiple points in time, results suggested that both SH and GWH were associated with high levels of self-reported illness, injury, and assault. It is interesting that men (vs. women) suffered greater illness, injury, and assault as a result of SH experience.

The focus on SH climate in recent research seems to have sparked an interest in other related aspects of climate. Specifically, two studies have explored the effects of a generally misogynistic and hostile context on the well-being of employees who are not specifically targeted by hostile conduct. Miner-Rubino and Cortina (2004) examined, among federal circuit court employees, the well-being-related effects of working in an environment that includes incivility directed at women. This cross-sectional research suggested that women-directed incivility (actions akin to hostile environment SH) was associated with lower health satisfaction (but not lower work satisfaction) for both female and male employees. A second study (Miner-Rubino & Cortina, 2007) examined the vicarious effects of two aspects of climate: observed hostility toward women (a construct composed of both observation of workplace incivility directed at women and observation of SH behavior toward female employees) and perceptions of the organization's unresponsiveness to SH (measured with a scale other than the OTSH Inventory). This study supported the notion that employees, both male and female, who work in environments permeated with negative attitudes and behavior toward women experience negative effects on their psychological well-being and job satisfaction. Furthermore, organizational unresponsiveness to this negative environment is also associated with negative outcomes such as decreased job satisfaction and organizational commitment.

Several other cross-sectional studies suggested a more complicated picture in terms of the effects of SH on outcomes. In a sample of military employees, Murry, Sivasubramaniam, and Jacques (2001) found some evidence for supervisory support and a strong supervisory exchange relationship in mitigating negative effects of SH on attitudinal outcomes. Furthermore, a series of interesting studies raised the question of whether self-labeling is necessary for negative effects of SH to occur. Based on the stress and coping literature, it is reasonable to expect that labeling is a cognitive mediator between SH and negative outcomes of SH. However, two studies (Magley et al., 1999; Munson, Miner, & Hulin, 2001) demonstrated that regardless of labeling there were negative effects (e.g., on organizational commitment, group cohesion, emotions) from SH experiences (even mild SH) for both women and men.

Finally, one study addressed the question of *why* gender harassment has negative effects on target well-being (Parker & Griffin, 2002). In a cross-sectional study of police officers, there was support for a model

predicting that the negative effects of gender harassment on female officers' psychological distress were mediated by overperformance demands (i.e., the belief that one needs to constantly prove oneself to gain acceptance in the work group).

Consequences for Organizations

The Willness et al. (2007) meta-analysis mentioned earlier also examined the effects of SH on organization-related outcomes. They found negative relationships between SH and both work withdrawal ($r_c = .299$) and job withdrawal ($r_c = .161$), suggesting that targets of SH may respond with missed work and work distraction as well as intentions to leave the organization. Findings also indicated a significant negative correlation with workgroup productivity ($r_c = .221$), suggesting that SH has disruptive effects on employees' abilities to work effectively.

A recent longitudinal study (Sims, Drasgow, & Fitzgerald, 2005) of 11,000 military servicemen examined turnover behavior (leaving the military for reasons other than conclusion of term of duty, death, retirement, or transfer to an officer training program). Specifically, this research found direct effects of SH on turnover, not mediated effects as suggested by the Fitzgerald et al. (1997) model. They argued that the experience of SH can trigger an avoidance response (Magley, 2002) and a flight response (Mayes & Ganster, 1988), suggesting the possibility of a direct effect. This view of turnover, in which both mediated and direct effects are possible, is consistent with recent turnover models (T. W. Lee & Mitchell, 1994; T. W. Lee, Mitchell, Wise, & Fireman, 1996).

Several previously mentioned studies also demonstrated negative effects of SH on organizational outcomes. The study of CSH (Gettman & Gelfand, 2001) found that CSH was associated with lowered target organizational commitment, higher turnover intentions, and greater withdrawal from clients. In the Miner-Rubino and Cortino (2004) study of male and female federal circuit court employees, there was a significant interaction between observation of women-directed incivility and workgroup gender ratio on observer work withdrawal such that observed incivility had little effect on work withdrawal in female-skewed work units, but there was a positive relationship in male-skewed work units. This study demonstrated that an environment characterized by misogynistic behaviors has damaging effects for the organization, particularly when women already are underrepresented in the workplace.

In one of the few SH studies that has moved beyond the individual-level of analysis, Raver and Gelfand (2005) examined the effect of ambient SH on team-level process (conflict, cohesion, and OCB) and outcome (financial performance) variables. They found that overall ambient SH in the team was positively related to relationship conflict and task conflict. Furthermore, the impact of ambient SH on team processes and outcomes depended on the type or types of ambient SH experienced within the team. More specifically, ambient sexual hostility predicted team process and outcome variables, whereas ambient unwanted sexual attention predicted only team process variables and ambient sexist hostility predicted neither.

Discussion

Summary

After reviewing this body of research, we are encouraged by the significant progress that has been made toward understanding the nature of workplace SH since the last major literature review. In this section, we highlight a few of the reasons for our optimism. First, there has been significant progress on the theoretical front. We now have useful models that address harasser decisions and motives, target responses, and observer sense making and behavior. These theories are being used to frame research inquiries and empirical research testing these models is beginning to appear. These are positive trends that bring focus to this broad and diverse literature.

Second, in recent years researchers have adopted a broader focus in their studies of SH. For example, there is consideration of a broader range of potential harassers (e.g., clients), a broader range of conduct (e.g., sexual assault), a broader range of harasser motives (e.g., identity threat), and a broader range of

interested parties (with the focus on SH bystanders or observers). There is also an interesting trend toward situating SH within a broader realm of organizational misbehavior, including incivility and GWH. These trends are beneficial for two reasons. First, they extend our knowledge base about SH phenomenon. Second, they encourage us to recognize SH as an event situated in organizational life—that is, an event motivated by a range of factors in the organizational environment, an event witnessed by organizational members, and an event that occurs in conjunction with other organizational events. This contextualization of SH within the organizational environment can richen our research questions and results.

We suspect that this trend toward examining SH as "organizationally situated" resulted from the increased research attention given to organizational climates. We argued earlier that there has been a noticeable shift in the focus of SH research in the past decade toward an emphasis on organizational climate as a facilitator or inhibitor of SH. These years have brought well-tested models that identify climate-related antecedents and consequences and an often-used measure of the perceived organizational climate for SH. Although the vast majority of tests of these models examine climate at the individual level, some researchers are beginning to move to the team or group level to assess climate, a trend we hope will continue.

Finally, it is important to recognize that SH researchers are doing an effective job of cumulating research results. Meta-analyses have been used effectively to aggregate across studies examining similar research questions, such as gender effects in the labeling of SH and SH antecedents and consequences. This is important so that research can proceed effectively (i.e., so research on the same issues does not continue ad infinitum). However, it is important to note that meta-analyses do not correct for limitations in the data themselves, and one key issue in regard to data on SH phenomena is their cross-sectional nature. Most research, even that which proposes antecedents and consequences of SH, is not longitudinal. Although SH theory may provide some justification for posing certain variables as antecedents and others as consequences, the question of causality is largely unestablished (with a few notable exceptions such as Glomb et al., 1999). Because many reverse causality predictions are quite reasonable (e.g., instead of job satisfaction and organizational commitment being consequences of SH perceptions, perhaps highly satisfied and committed employees are less likely to perceive conduct as SH), this is an important limitation.

Conclusion

In the preface to her 1979 book introducing SH as a legal construct, Catherine MacKinnon (1979: xii) stated, "To date there are no 'systematic' studies of sexual harassment in the social-scientific sense." Put in that context, there has been remarkable progress toward understanding SH as a workplace phenomenon in a relatively short amount of time. Our review demonstrates that the SH literature continues to mature, with the emergence of stronger theory, new meta-analytic reviews of key findings, an enhanced focus on organizational contexts, and stronger integration with other workplace conduct. As the next decade of SH research unfolds, we hope for similar research progress. More important, we hope that the next decade brings evidence that research efforts are having an impact on the ability of work organizations to eliminate this harmful work-related conduct.

Notes

1. Berdahl (2007a) uses the term *sex-based harassment* (SBH) rather than *sexual harassment* because many of the behaviors examined by SH researchers are not actually sexual in nature (e.g., sexist hostility forms of gender harassment such as referring to women as bitches).

2. It should be noted that there is debate regarding this claim. See Gutek et al. (2004) for an extensive critique of the Sexual Experiences Questionnaire (Fitzgerald et al., 1995), which is the most common measure used in the behavioral perspective.

3. It should be noted that variables in this meta-analysis are identified as *antecedents* or *consequences* based on predictions made in SH theory, not based on causal analyses or longitudinal research designs. Although some studies included here were longitudinal (e.g., Glomb, Munson, Hulin, Bergman, & Drasgow, 1999), many were not.

References

Adams-Roy, J., & Barling, J. 1998. Predicting the decision to confront or report sexual harassment. *Journal of Organizational Behavior,* 19: 329–336.

Amick, N. J., & Sorenson, R. C. 2004. Factors influencing women's perceptions of a sexually hostile workplace. *Journal of Emotional Abuse: Interventions, Research & Theories of Psychological Mistreatment, Trauma & Nonphysical Aggression,* 4: 49–69.

Antecol, H., & Cobb-Clark, D. 2003. Does sexual harassment training change attitudes? A view from the federal level. *Social Science Quarterly,* 84: 826–842.

Bargh, J. A., Raymond, P., Pryor, J. B., & Strack, F. 1995. Attractiveness of the underlying: An automatic power-sex association and its consequences for sexual harassment and aggression. *Journal of Personality and Social Psychology,* 68: 768–781.

Begany, J. J., & Milburn, M. A. 2002. Psychological predictors of sexual harassment: Authoritarianism, hostile sexism, and rape myths. *Psychology of Men and Masculinity,* 3: 119–126.

Berdahl. J. L. 2007a. Harassment based on sex: Protecting social status in the context of gender hierarchy. *Academy of Management Review,* 32: 641–658.

Berdahl, J. L. 2007b. The sexual harassment of uppity women. *Journal of Applied Psychology,* 92: 425–437.

Berdahl, J. L., & Moore, C. 2006. Workplace harassment: Double jeopardy for minority women. *Journal of Applied Psychology,* 91(2): 426–436.

Bergman, M. E., Langhout, R. D., Palmieri, P. A., Cortina, L. M., & Fitzgerald, L. F. 2002. The (un)reasonableness of reporting: Antecedents and consequences of reporting sexual harassment. *Journal of Applied Psychology,* 87: 230–242.

Bowes-Sperry, L., & O' Leary-Kelly, A. M. 2005. To act or not to act: The dilemma faced by observers of sexual harassment. *Academy of Management Review,* 30(2): 288–306.

Bowes-Sperry, L., & Powell, G. N. 1999. Observers' reactions to social-sexual behavior at work: An ethical decision making perspective. *Journal of Management,* 25: 779–802.

Bowes-Sperry, L., & Tata, J. 1999. A multiperspective framework of sexual harassment: Reviewing two decades of research. In G. N. Powell (Ed.), *Handbook of gender and work*: 263–280. Thousand Oaks, CA: Sage.

Cleveland, J. N., & Kerst, M. E. 1993. Sexual harassment and perceptions of power: An under-articulated relationship. *Journal of Vocational Behavior,* 42: 46–67.

Cochran, C. C., Frazier, P. A., & Olson, A. M. 1997. Predictors of responses to unwanted sexual harassment. *Psychology of Women Quarterly,* 44: 588–608.

Cohen, L. E., & Felson, M. 1979. Social change and crime rate trends: A routine activity approach. *American Sociological Review,* 44: 588–608.

Cortina, L. 2004. Hispanic perspectives on sexual harassment and social support. *Personality and Social Psychology Bulletin,* 30(5): 570–584.

Cortina, L. M., & Wasti, S. A. 2005. Profiles in coping: Responses to sexual harassment across persons, organizations, and cultures. *Journal of Applied Psychology,* 90(1): 182–192.

Craig, T. Y., Kelly, J. R., & Driscoll, D. 2001. Participant perceptions of potential employers. *Sex Roles,* 44: 389–400.

Culbertson, A. L., Rosenfeld, P., Booth-Kewley, S., & Magnusson, P. 1992. *Assessment of sexual harassment in the Navy: Results of the 1989 Navy-wide survey* (Tech. Rep. 92-11). San Diego, CA: Naval Personnel Research and Development Center.

Dall'Ara, E., & Maas, A. 1999. Studying sexual harassment in the laboratory: Are egalitarian women at higher risk? *Sex Roles,* 41: 681–704.

DeCoster, S., Estes, S. B., & Mueller, C. W. 1999. Routine activities and sexual harassment in the workplace. *Work and Occupations,* 26: 21–49.

Driscoll, D. M., Kelly, J. R., & Henderson, W. L. 1998. Can perceivers identify likelihood to sexually harass? *Sex Roles,* 38(7–8): 557–588.

Farley, L. 1978. *Sexual shakedown: The sexual harassment of women on the job.* New York: Warner Books.

Fitzgerald, L. F., Drasgow, F., Hulin, C. L., Gelfand, M. J., & Magley, V. J. 1997. Antecedents and consequences of sexual harassment in organizations: A test of an integrated model. *Journal of Applied Psychology,* 82: 578–589.

Fitzgerald, L. F., Gelfand, M. J., & Drasgow, F. 1995. Measuring sexual harassment: Theoretical and psychometric advances. *Basic and Applied Social Psychology,* 17: 425–445.

Fitzgerald, L. F., Hulin, C. L., & Drasgow, F. 1994. The antecedents and consequences of sexual harassment in organizations: An integrated model. In G. P. Keita & J. J. Hurrell (Eds.), *Job stress in a changing workforce: Investigating gender, diversity, and family issues:* 55–73. Washington, DC: American Psychological Association.

Fitzgerald, L. F., Magley, V. J., Drasgow, F., & Waldo, C. R. 1999. Measuring sexual harassment in the military: The Sexual Experiences Questionnaire (SEQ-DoD). *Military Psychology,* 11(3): 243–263.

Fitzgerald, L. F., Swan, S., & Magley, V. J. 1997. But was it really sexual harassment? Legal, behavioral, and psychological definitions of the workplace victimization of women. In W. O'Donohue (Ed.), *Sexual harassment: Theory, research, and treatment:* 5–28. Boston: Allyn & Bacon.

Franke, K. M. 1997. What's wrong with sexual harassment? *Stanford Law Review,* 49(4): 691–772.

Frink, D. D., & Klimoski, R. J. 1998. Toward a theory of accountability in organizations and human resources management. In G. R. Ferris (Ed.), *Research in personnel and human resources management:* 1–51. Stamford, CT: JAI.

Gelfand, M. J., Fitzgerald, L. F., & Drasgow, F. 1995. The structure of sexual harassment: A confirmatory analysis across cultures and settings. *Journal of Vocational Behavior,* 47: 164–177.

Gettman, H. J., & Gelfand, M. J. 2007. When the customer shouldn't be king: Antecedents and consequences of sexual harassment by clients and customers. *Journal of Applied Psychology,* 92(3): 757–770.

Glomb, T. M., Munson, L. J., Hulin, C. L., Bergman, M. E., & Drasgow, F. 1999. Structural equation models of sexual harassment: Longitudinal explorations and cross-sectional generalizations. *Journal of Applied Psychology,* 84: 14–28.

Glomb, T. M., Richman, W. L., Hulin, C. L., Drasgow, F., Schneider, K. T., & Fitzgerald, L. F. 1997. Ambient sexual harassment: An integrated model of antecedents and consequences. *Organizational Behavior and Human Decision Processes,* 71(3): 309–328.

Goldberg, C. B. 2007. The impact of training and conflict avoidance on responses to sexual harassment. *Psychology of Women Quarterly,* 31: 62–72.

Gruber, J. E. 1989. How women handle sexual harassment: A literature review. *Sociology and Social Research,* 74: 3–9.

Gutek, B. 1985. *Sex and the workplace.* San Francisco: Jossey-Bass.

Gutek, B. A., Cohen, A. G., & Konrad, A. M. 1990. Predicting social-sexual behavior at work: A contact hypothesis. *Academy of Management Journal,* 33: 560–577.

Gutek, B. A., & Morasch, B. 1982. Sex-ratios, sex-role spillover, and sexual harassment of women at work. *Journal of Social Issues,* 38: 55–74.

Gutek, B. A., Murphy, R. O., & Douma, B. 2004. A review and critique of the Sexual Experiences Questionnaire (SEQ). *Law and Human Behavior,* 28: 457–482.

Harned, M. S., Ormerod, A. J., Palmieri, P. A., Collinsworth, L. L., & Reed, M. 2002. Sexual assault and other types of sexual harassment by workplace personnel: A comparison of antecedents and consequences. *Journal of Occupational Health Psychology,* 7: 174–188.

Hessen-McInnis, M. S., & Fitzgerald, L. R. 1998. Sexual harassment: A preliminary test of an integrative model. *Journal of Applied Social Psychology,* 27(10): 877–901.

Hindelang, M. J., Gottfredson, M., & Garofalo, J. 1978. *Victims of personal crime: An empirical foundation for a theory of personal victimization.* Cambridge, MA: Ballinger.

Hulin, C. L., Fitzgerald, L. F., & Drasgow, F. 1996. Organizational influences on sexual harassment. In M. S. Stockdale (Ed.), *Sexual harassment in the workplace,* vol. 5: 127–150. Thousand Oaks, CA: Sage.

Isbell, L. M., Swedish, K., & Gazan, D. B. 2005. Who says it's sexual harassment? The effects of gender and likelihood to sexually harass on legal judgments of sexual harassment. *Journal of Social Psychology,* 35: 745–772.

Jackson, R. A., & Newman, M. A. 2004. Sexual harassment in the federal workplace revisited: Influences on sexual harassment by gender. *Public Administration Review,* 64: 705–717.

Jones, T. M. 1991. Ethical decision making by individuals in organizations: An issue-contingent model. *Academy of Management Review,* 16: 366–395.

Knapp, D. E., Faley, R. H., Ekeberg, W. C., & Dubois, C. L. Z. 1997. Determinants of target responses to sexual harassment: A conceptual framework. *Academy of Management Review,* 22: 687–729.

Langhout, R., Bergman, M., Cortina, L., Fitzgerald, L., Drasgow, F., & Williams, J. 2005. Sexual harassment severity: Assessing situational and personal determinants and outcomes. *Journal of Applied Social Psychology,* 35(5): 975–1007.

Lazarus, R. S., & Folkman, S. 1984. *Stress, appraisal and coping.* New York: Springer.

Lee, J., Heilmann, S., & Near, J. 2004. Blowing the whistle on sexual harassment: Test of a model of predictors and outcomes. *Human Relations*, 297–322.

Lee, K., Gizzarone, M., & Ashton, M. C. 2003. Personality and the likelihood to sexually harass. *Sex Roles*, 49: 59–69.

Lee, T. W., & Mitchell, T. R. 1994. An alternative approach: The unfolding model of voluntary employee turnover. *Academy of Management Review*, 19: 51–89.

Lee, T. W., Mitchell, T. R., Wise, L., & Fireman, S. 1996. An unfolding model of voluntary employee turnover. *Academy of Management Review*, 39: 5–36.

Lengnick-Hall, M. L. 1995. Sexual harassment research: A methodological critique. *Personnel Psychology*, 48: 841–864.

Lim, S., & Cortina, L. M. 2005. Interpersonal mistreatment in the workplace: The interface and impact of general incivility and sexual harassment. *Journal of Applied Psychology*, 90: 483–496.

Lucero, M., Allen, R., & Middleton, K. 2006. Sexual harassers: Behaviors, motives, and change over time. *Sex Roles*, 55: 331–343.

Lucero, M. A., Middleton, K., Finch, W., & Valentine, S. 2003. An empirical investigation of sexual harassers: Toward a perpetrator typology. *Human Relations*, 56(12): 1561–1483.

Luthar, H. K., & Luthar, V. K. 2008. Likelihood to sexually harass: A comparison among American, Indian, and Chinese students. *International Journal of Cross Cultural Management*, 8: 59–77.

Maas, A., Cadinu, M., Guarnieri, G., & Grasselli, A. 2003. Sexual harassment under identity threat: The computer harassment paradigm. *Journal of Personality and Social Psychology*, 85(5): 853–870.

MacKinnon, C. A. 1979. *Sexual harassment of working women.* New Haven, CT: Yale University Press.

Magley, V. J. 2002. Coping with sexual harassment: Reconceptualizing women's resistance. *Journal of Personality and Social Psychology*, 83: 930–946.

Magley, V., Hulin, C., Fitzgerald, L., & DeNardo, M. 1999. Outcomes of self-labeling sexual harassment. *Journal of Applied Psychology*, 84(3): 390–402.

Malamut, A. B., & Offermann, L. R. 2001. Coping with sexual harassment: Personal, environmental, and cognitive determinants. *Journal of Applied Psychology*, 86: 1152–1166.

Martindale, M. 1990. *Sexual harassment in the military: 1988.* Washington, D.C.: Manpower Data Center, Department of Defense.

Mayes, B. T., & Ganster, D. C. 1988. Exit and voice: A test of hypotheses based on fight/flight responses to job stress. *Journal of Organizational Behavior*, 9: 199–216.

Miner-Rubino, K., & Cortina, L. M. 2004. Working in a context of hostility toward women: Implications for employees' well-being. *Journal of Occupational Health Psychology*, 9: 107–122.

Miner-Rubino, K., & Cortina, L. M. 2007. Beyond targets: Consequences of vicarious exposure to misogyny at work. *Journal of Applied Psychology*, 92: 1254–1269.

Mueller, C. W., DeCoster, S., & Estes, S. B. 2001. Sexual harassment in the workplace: Unanticipated consequences of modern social control in organizations. *Work and Occupations*, 28(4): 411–446.

Munson, L. J., Hulin, C., & Drasgow, F. 2000. Longitudinal analysis of dispositional influences and sexual harassment: Effect on job and psychology outcomes. *Personnel Psychology*, 53: 21–46

Munson, L. J., Miner, A. G., & Hulin, C. 2001. Labeling sexual harassment in the military: An extension and replication. *Journal of Applied Psychology*, 86(2): 293–303.

Murphy, J. D., Driscoll, D. M., & Kelly, J. R. 1999. Differences in the nonverbal behavior of men who vary in the likelihood to sexually harass. *Journal of Social Behavior & Personality*, 14: 133–129.

Murry, W. D., Sivasubramaniam, N., & Jacques, P. H. 2001. Supervisory support, social exchange relationship, and sexual harassment consequences: A test of competing models. *Leadership Quarterly*, 12: 1–29.

Near, J. P., & Miceli, M. P. 1985. Organizational dissidence: The case of whistle-blowing. *Journal of Business Ethics*, 4: 1–16.

Offermann, L. R., & Malamut, A. B. 2002. When leaders harass: The impact of target perceptions of organizational leadership and climate on harassment reporting and outcomes. *Journal of Applied Psychology*, 87: 885–893.

O'Leary-Kelly, A. M., & Bowes-Sperry, L. 2001. Sexual harassment as unethical behavior: The role of moral intensity. *Human Resource Management Review*, 11: 73–92.

O'Leary-Kelly, A. M., Paetzold, R. L., & Griffin, R. W 2000. Sexual harassment as aggressive behavior: An actor-based perspective. *Academy of Management Review*, 25: 372–388.

O'Leary-Kelly, A. M., Tiedt, P., & Bowes-Sperry, L. 2004. Answering accountability questions in sexual harassment: Insights regarding harasser, targets, and observers. *Human Resource Management Review*, 14: 85–106.

Parker, S. K., Griffin, M. A. 2002. What is so bad about a little name-calling? Negative consequences of gender harassment for overperformance demands and distress. *Journal of Occupational Health Psychology*, 7(3): 195–210.

Perry, E. L., Kulik, C. T., & Schmidtke, J. M. 1997. Blowing the whistle: Determinants of reponses to sexual harassment. *Basic and Applied Social Psychology*, 19(4): 457–482.

Pryor, J. B. 1987. Sexual harassment proclivities in men. *Sex Roles*, 17: 267–290.

Pryor, J. B., Giedd, J. L., & Williams, K. B. 1995. A social psychological model for predicting sexual harassment. *Journal of Social Issues*, 51(1): 69–84.

Pryor, J. B., Lavite, C. M., & Stoller, L. M. 1993. A social psychological analysis of sexual harassment: The person/situation interaction. *Journal of Vocational Behavior*, 42: 68–83.

Pryor, J. B., & Stoller, L. M. 1994. Sexual cognition processes in men high in the likelihood to sexually harass. *Personality and Social Psychology Bulletin*, 20: 163–169.

Raver, J. L., & Gelfand, M. J. 2005. Beyond the individual victim: Linking sexual harassment team processes and team performance. *Academy of Management Journal*, 48(4): 387–400.

Rogers, J. K., & Henson, K. D. 1997. "Hey, why don't you wear a shorter skirt?" Structural vulnerability and the organization of sexual harassment in temporary clerical employment. *Gender & Society,* 11(2): 215–237.

Rospenda, K. 2002. Workplace harassment, services utilization, and drinking outcomes. *Journal of Occupational Health Psychology,* 7(2): 141–155.

Rospenda, K. M., Richman, J., Ehmke, J., & Zlatoper, K. 2005. Is workplace harassment hazardous to your health? *Journal of Business & Psychology,* 20: 95–110.

Rudman, L. A., & Borgida, E. 1995. The afterglow of construct accessibility: The behavior consequences of priming men to view women as sex objects. *Journal of Experimental Social Psychology,* 31: 493–517.

Sims, C. S., Drasgow, F., & Fitzgerald, L. R. 2005. The effects of sexual harassment on turnover in the military: Time dependent modeling. *Journal of Applied Psychology,* 90(6): 1141–1152.

Stockdale, M. S. 1998. The direct and moderating influences of sexual-harassment pervasiveness, coping strategies, and gender on work-related outcomes. *Psychology of Women Quarterly,* 22(4): 521–535.

U.S. Merit Systems Protection Board. 1981. *Sexual harassment in the federal workplace. Is it a problem?* Washington, DC: Government Printing Office.

U.S. Merit Systems Protection Board. 1988. *Sexual harassment in the federal workplace. Is it a problem?* Washington, DC: Government Printing Office.

U.S. Merit Systems Protection Board. 1995. *Sexual harassment in the federal workplace: Trends, progress and continuing challenges.* Washington, DC: Government Printing Office.

Wasti, S. A., Bergman, M. E., Glomb, T. M., & Drasgow, F. 2000. Test of the cross-cultural generalizability of a model of sexual harassment. *Journal of Applied Psychology,* 85(5): 766–778.

Wasti, S. A., & Cortina, L. M. 2002. Coping in context: Sociocultural determinants of responses to sexual harassment. *Journal of Personality and Social Psychology,* 83: 394–405.

Welsh, S., & Gruber, J. E. 1999. Not taking it anymore: Women who report or file complaints of sexual harassment. *Canadian Review of Sociology and Anthropology,* 36: 559–583.

Wilkerson, J. 1999. The impact of job level and prior training on sexual harassment labeling and remedy choice. *Journal of Applied Social Psychology,* 29(8): 1605–1623.

Williams, J. H., Fitzgerald, L. F., & Drasgow, F. 1999. The effects of organizational practices on sexual harassment and individual outcomes in the military. *Military Psychology,* 11(3): 303–328.

Willness, C. R., Steel, P., & Lee, K. 2007. A meta-analysis of the antecedents and consequences of workplace sexual harassment. *Personnel Psychology,* 60(1): 127–162.

Woodzicka, J. A., & LaFrance, M. 2005. The effects of subtle sexual harassment on women's performance in a job interview. *Sex Roles,* 51(1–2): 67–77.

Zugelder, M. T., Champagne, P. J., & Maurer, S. D. 2006. An affirmative defense of sexual harassment by managers and supervisors: Analyzing employer liability and protecting employee rights in the United States. *Employee Responsibilities & Rights Journal,* 18: 111–122.

DISCUSSION QUESTIONS

1. What are the risk factors for sexual harassment?

2. What are the similarities between sexual harassment and other types of victimization at work, if any? Think of who it is likely to happen to, its consequences, its risk factors, etc.

3. How is coping structured by individual and organizational factors? Given the fact that coping differs across individuals and contexts, what does this suggest about how we should respond to victims?

4. What else would you like to know about sexual harassment victimization?

5. What are the similarities between sexual harassment and bullying?

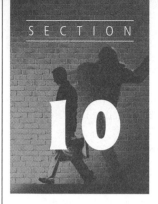

10

Property and Identity Theft Victimization

So far, we mainly have discussed violent or person victimizations. These are the types of victimization that commonly make the news—murders, rapes, assaults, and child abuse, to name a few. We tend to think about these when we hear the word *victimization*. But these are not the most common types of victimization. Instead, the most common type is property victimization. In fact, it is theft! Although this is not the type of victimization most commonly shown on the evening news, it is what will most likely happen to you if you are unfortunate enough to become a crime victim. There are, of course, other types of property victimization, and we will discuss those as well. And in today's world, criminals are acting in innovative ways. No longer does an offender need to be in your presence to victimize you. Indeed, the Internet has opened up numerous ways for you to have your money taken, your privacy violated, and even your identity stolen. In this section, we will discuss these two types of victimization that generally receive less attention in the field of victimology but nonetheless impact numerous persons each year.

Property Victimization

Victimization is generally discussed in two categories: personal victimization and property victimization. Personal victimization, as measured in the National Crime Victimization Survey (NCVS), includes simple assault, aggravated assault, rape and sexual assault, and robbery. Property victimization as measured in the NCVS includes theft, motor vehicle theft, and household burglary. There are other types of property victimization, but we will limit our discussion to these three types.

Theft

Theft, also called larceny-theft, is defined by the Federal Bureau of Investigation (FBI, 2010) as the "unlawful taking, carrying, leading, or riding away of property from the possession or constructive possession of another." To be classified as a theft, the item cannot be taken by force, violence, or fraud. As such, if items are taken via embezzlement, confidence games, check fraud, or forgery, it is not considered larceny-theft (FBI, 2010). If these types of action are excluded, what, then, would be considered theft? Stealing a bicycle, shoplifting, pick-pocketing, or stealing any item or property, so long as it is not achieved through force, violence, or fraud, as noted, is theft (FBI, 2010).

Extent of Theft

Theft is the most common crime victimization experienced by Americans. This is supported by statistics from the Uniform Crime Reports (UCR) and those from the NCVS. According to the UCR, in 2009, there were more than 6 million larceny-thefts, resulting in a rate of 2,060.9 per 100,000 persons in the United States (FBI, 2010). Larceny-thefts composed almost 68% of all property crimes that occurred in 2009 (FBI, 2010). The extent of theft is also captured in the NCVS. For 2009, the NCVS estimates that there were 2 times the number (11,709,830) of thefts (not including attempted and completed purse snatching and pocket picking) compared with the UCR data (Truman & Rand, 2010). Theft composed 58% of all victimizations in the NCVS (Bureau of Justice Statistics, 2007). Why do you think the estimates from the NCVS are so much higher for theft compared with those from the UCR?

Characteristics of Theft

The UCR and the NCVS also describe the characteristics of theft in the United States. According to the UCR, the average larceny costs the victim $864 in lost property, which amounts to a $5.5 billion total loss for all larcenies in 2009 (FBI, 2010). According to the NCVS, losses for larceny are somewhat lower, with a mean dollar loss of $403 (Bureau of Justice Statistics, 2007). You are probably not surprised to learn that most thefts go unreported to police. In fact, only about 30% of all thefts were reported to the police in 2009 (Truman & Rand, 2010). The low reporting rates probably contribute to the low recovery rates of stolen property. In more than 86% of larceny victimizations reported in the NCVS, the victim noted that none of his or her property was recovered (Bureau of Justice Statistics, 2007).

You may wonder what items people most commonly steal. More than one third of larcenies reported to the police (remember that the UCR includes only those crimes the police know about) were thefts of motor vehicle parts, accessories, and contents (FBI, 2010). When persons were asked in the NCVS what was stolen, almost 36% indicated that personal effects such as portable electronics, photographic equipment, jewelry, and clothing were taken (Bureau of Justice Statistics, 2007).

Who Are Theft Victims?

Although virtually any possession can be stolen, theft rates do differ according to household characteristics measured in the NCVS. Rates of theft are highest for households headed by persons who are White, while Hispanics have higher theft victimization rates than do non-Hispanics (Bureau of Justice Statistics, 2007). Households headed by younger people (12–19 years of age) report higher rates of theft victimization than households headed by older persons (Bureau of Justice Statistics, 2007). Households that have a total income less than $7,500 annually have the highest theft rates (Bureau of Justice Statistics, 2007). Also, as the number of people in the household increases, so does the risk of theft. In fact, households with six or more people were 3 times more likely to experience a theft than were single-person households (Truman & Rand, 2010).

Risk Factors for Theft Victimization

Theft is so common, at least in terms of victimization, that it may seem as though risk is ubiquitous and theft amounts to little more than having your property in the wrong place at the wrong time. Remember, though, that researchers have linked victimization risk to certain qualities and characteristics, often due to a person's lifestyle and routine activities. Theft victimization also has been explained using this perspective. In fact, Cohen and Felson, in their routine activities theory, linked increases in property crime to the production developments that made goods smaller, more durable, and easier to move. At the individual level, exposure to delinquent or criminal others increases the chances that a person will become the victim of theft. In a study of college students, Mustaine and Tewksbury (1998) found that smoking marijuana increased the risk of minor theft victimization, while threatening another with a gun or threatening without a weapon increased the risk of minor and major theft victimizations. Living in a neighborhood that had too much crime or was too noisy also positively impacted minor victimization risk (Mustaine & Tewksbury, 1998). In addition, having capable guardianship decreased the chances of theft. College students who had a dog in their residence and who lived in rural areas were less likely to experience minor larceny victimization (Mustaine & Tewksbury, 1998). Installing extra locks also decreased the chances of major larceny victimization for college students (Mustaine & Tewksbury, 1998).

College students may be deemed suitable targets for theft as well. Students who eat out frequently found themselves at a greater risk of minor theft victimization than did others (Mustaine & Tewksbury, 1998), likely because these students were exposing themselves to many people and displaying their wallets and money. Students who left their homes often to study were also more likely to be theft victims than were others, likely because they were exposing their personal items to would-be offenders (Mustaine & Tewksbury, 1998). Beyond this study of routine activities theory and theft among college students, the research on the reasons why some people become victims of theft has not been fully explored. Can you think of other theories or explanations that could explain why certain people have their possessions stolen?

Motor Vehicle Theft

Anyone who has a car has probably thought about the possibility of having it stolen, but **motor vehicle theft** can also occur with other self-propelled vehicles that run on land, such as sport utility vehicles, trucks, buses, motorcycles, motor scooters, all-terrain vehicles, and snowmobiles (FBI, 2010). When we discuss motor vehicle theft, it covers the theft or attempted theft of the vehicle itself. If a car is broken into, for example, and the contents taken, it is a larceny-theft, not a motor vehicle theft. When auto theft involves the taking of a car by an offender who is armed and the vehicle is occupied, it is called **carjacking.** In this section, we limit our discussion to the nonviolent victimization of motor vehicle theft.

Extent of Motor Vehicle Theft Victimization

In 2009, 794,616 motor vehicle thefts were reported to the police—this number of reported thefts equates to a motor vehicle theft rate of 258.8 per 100,000 persons in the United States (FBI, 2010). For motor vehicle theft, the number of reports to the police, as shown in the UCR, is not too different from the estimate in the NCVS. In 2009, there were 735,770 motor vehicle thefts according to the NCVS. Why are motor vehicle theft estimates so close for the UCR and NCVS while estimates for theft, as you read above, are so divergent? The answer is probably tied to the nature of the victimization. If your car were stolen, don't you think you would be quite likely to report it to the police? You would want to get the police on the case so your car could be found. You also need a police report for insurance purposes. These reasons likely contribute to the similar estimates.

Characteristics of Motor Vehicle Theft Victimization

Along with knowing the extent to which motor vehicles are stolen, we also have an idea of the characteristics of motor vehicle thefts that occur in the United States. The costs associated with motor vehicle theft are high.

According to the UCR, each stolen vehicle costs, on average, $6,505 (FBI, 2010), and the average is $6,286 according to the NCVS (Bureau of Justice Statistics, 2007). For the nation as a whole, motor vehicle theft costs almost $5.2 billion annually (FBI, 2010). However, financial costs are not the only ones that victims incur when their motor vehicles are stolen. More than one fifth of victims report missing at least 1 day of work due to motor vehicle theft (Bureau of Justice Statistics, 2007).

Table 10.1	Top 10 Most Frequently Stolen Passenger Vehicles in 2009

Rank	Year/Make/Model
1	1994 Honda Accord
2	1995 Honda Civic
3	1991 Toyota Camry
4	1997 Ford F-150 Pickup
5	2004 Dodge Ram Pickup
6	2000 Dodge Caravan
7	1994 Chevrolet Pickup (full size)
8	1994 Acura Integra
9	2002 Ford Explorer
10	2009 Toyota Corolla

SOURCE: Insurance Information Institute (2011). Reprinted by permission of the National Insurance Crime Bureau.

▲ Photo 10.1   The Cadillac Escalade had the highest theft-loss rate for passenger vehicles that are 1 to 3 years old in the United States.

Photo credit: © John Lamm/Transtock/Corbis

As previously mentioned, motor vehicle theft is widely reported to the police (more than 80%). About half the victims of motor vehicle theft report that all their property (in this case, their vehicle) was recovered (Bureau of Justice Statistics, 2007). Motor vehicle theft is most likely to occur at night, near the victim's home (Bureau of Justice Statistics, 2007), and in urban areas (FBI, 2010). Almost half of motor vehicle thefts occur when the victims are sleeping (Bureau of Justice Statistics, 2007).

As you may imagine, though, not all vehicles are equally likely to be stolen. To determine the cars most likely to be stolen, the National Insurance Crime Bureau (NICB) compiles a list using National Crime Information Center data based on police reports (Insurance Information Institute, 2011). See Table 10.1 for a list of the cars that were most stolen in 2009. After you examine the list, consider that the costs of these cars when stolen may not be that much compared with the costs of more expensive cars. When insurance claims rates and the size of insurance payments are considered together, the Highway Loss Data Institute has identified that the 2009 Cadillac Escalade has the highest theft-loss rate for passenger vehicles that are 1 to 3 years old in the United States, followed by the Ford F-250, the Infiniti G37, the Dodge Charger V8, and the Chevrolet Corvette Z06 V8 (Inside Line, 2010). When the number of these vehicles that are stolen, along with the costs associated with their loss, are totaled and compared with the number of each make and model insured, their theft costs the insurance industry the most. These values take into account the value of the contents of the vehicles as well, so it does not include just the cost of the vehicle.

Along with certain types of cars being prime targets, certain cities have higher-than-average rates of motor vehicle theft. In 2010, the Fresno, California, Metropolitan Statistical Area had the highest motor vehicle theft rate per capita in the United States, according to data compiled by the NICB (Insurance Information Institute, 2011). All the top 10 cities for highest motor vehicle theft rates are in the Western part of the United States. See Table 10.2 for the list.

Who Are Motor Vehicle Theft Victims?

Anyone can have their car stolen—do you know someone who has had this happen? Does this person have the common

Table 10.2 Top 10 U.S. Metropolitan Areas With Highest Motor Vehicle Theft Rates in 2010

Rank	Metropolitan Statistical Area	Vehicles Stolen	Rate
1	Fresno, CA	7,559	812.4
2	Modesto, CA	3,878	753.81
3	Bakersfield, CA	5,623	669.70
4	Spokane, WA	2,673	586.35
5	Vallejo-Fairfield, CA	2,392	578.69
6	Sacramento-Arden-Arcade-Roseville, CA	11,881	552.83
7	Stockton, CA	3,779	551.43
8	Visalia-Porterville, CA	2,409	544.80
9	San Francisco-Oakland-Fremont, CA	22,617	521.68
10	Yakima, WA	1,266	520.49

SOURCE: Insurance Information Institute (2011). Reprinted by permission of the National Insurance Crime Bureau.

characteristics of the "typical" car theft victim? Read below to answer this question. Households headed by persons between the ages of 12 and 19 and the ages of 20 and 34 have the highest rates of motor vehicle theft (Truman & Rand, 2010). Households headed by persons who are Black or by persons of two or more races have higher rates of motor vehicle theft than do others (Truman & Rand, 2010). Households that rent rather than own their home, have total incomes of $7,500 per year or less, and have more people in them (single-person households have the lowest rates) experience higher rates of motor vehicle theft than do other households (Truman & Rand, 2010).

Risk Factors for Motor Vehicle Theft Victimization

Routine activities theory provides an excellent underpinning for understanding why some motor vehicles are stolen while others remain in the possession of their rightful owners. Remember, the three constructs relevant to this theoretical perspective are motivated offenders, suitable targets, and lack of capable guardianship. Motivated offenders are assumed to be present. What makes a vehicle a suitable target, though? Vehicles that are easy to steal are going to be deemed more suitable than those that are more difficult to steal. Vehicles with keys in the ignition are definitely suitable targets! Similarly, vehicles left unlocked are more suitable than locked vehicles. Cars parked on streets with poor lighting are suitable targets, while those parked in locked garages are not suitable targets. Also consider why a person would be compelled to steal a car. One reason is to sell the parts. A suitable target, then, would be a vehicle whose parts are easy to sell. A suitable vehicle would be a popular model, in that there would be a resale market for its parts. Common vehicles would be easier to sell parts from than more rare models that hardly anyone drives. In addition, vehicles that do not have Vehicle Identification Numbers (VINs) etched in the windows are more suitable than vehicles with these numbers. Thieves will not want to replace the windows after stealing the vehicle.

Capable guardianship is also an important concept to consider. Guardianship for vehicles can be physical—those things that physically provide protection from theft. For cars, this can be locks, steering wheel column locks, car alarms, ignition cut-off systems, and electronic tracking devices. It is a bit more difficult to provide social guardianship for a vehicle, as people are bound to leave their cars unattended at some point. That is why it is imperative that vehicles are left locked at all times, with personal belongings and electronics stowed out of sight. To read about how environmental factors contribute to the incidence of motor vehicle theft and repeated motor vehicle theft, read Marissa P. Levy and Christine Tartaro's (2010) article in this section.

Response to Motor Vehicle Theft

As previously discussed, motor vehicle theft has one of the highest reporting rates of all crimes. Even so, only 12% of motor vehicle thefts are cleared by police (FBI, 2010). When a case is cleared by police, it means the police either made an arrest in the case or cleared the case exceptionally. To clear a case exceptionally, the police agency must know who the offender is, be able to support an arrest, and know where the offender is, but be unable to make an arrest and have the charges filed (FBI, 2010). These low clearance rates suggest that many motor vehicle thieves are not being apprehended. In addition to police actions, other efforts have been undertaken to reduce motor vehicle theft victimization.

Several key pieces of federal legislation have been passed to deal with motor vehicle theft. The **Motor Vehicle Theft Law Enforcement Act** (1984; Pub. L. 98-547) required manufacturers to stamp identification numbers on parts in high-theft major passenger car lines (the impetus for the VIN) to make it easier to trace parts that are stolen from vehicles. Penalties for altering or removing these numbers were also increased. Investigators believe that requiring the marking of vehicle parts aids in the arrest of individuals who steal parts or vehicles (Finn, 2000). This act also made interstate trafficking in stolen vehicles part of the federal racketeering statutes (Insurance Information Institute, 2011). In 1992, the **Anti-Car Theft Act** (Pub. L. 105-119) made carjacking a federal crime and provided funding to link all state motor vehicle departments. This funding was created in hopes that states would be able to share title, registration, and salvage information to make it more difficult for stolen vehicles to be easily titled and registered (SmartMotorist.com, n.d.). This led to the development of the National Motor Vehicle Title Information System. It also expanded the parts-marking identification number requirement to include more than just high-theft passenger car line parts. Vans and utility vehicles that have higher-than-average theft rates and half of passenger cars, vans, and SUVs with lower-than-average theft rates were also required to have their major parts (engines, transmissions, and 12 other major parts) marked (SmartMotorist.com, n.d.). As part of the Violent Crime Control and Law Enforcement Act, 2 years later, the **Motor Vehicle Theft Prevention Act of 1994** authorized the development of a national, voluntary motor vehicle theft prevention program. Those people who wished to participate could place on their vehicles a decal alerting police that their car is not normally operated between the hours of 1 A.M. and 5 A.M. (Federal Grants Wire, 2011). In 1996, the ability of states and the federal government to determine if a car was stolen was strengthened by the **Anti-Car Theft Improvements Act of 1996**. This act served to upgrade state motor vehicle department databases that contain title information (SmartMotorist.com, n.d.).

Federal legislation is not the only way to combat motor vehicle theft victimization. In many states, anti-car theft groups have formed that comprise law enforcement, insurers, and consumers. These groups work together to promote awareness of motor vehicle theft and to pass state and local legislation to reduce motor vehicle theft (Insurance Information Institute, 2011). Many states attach fees to car registration or car insurance policies that go toward funding such groups. For example, Michigan has a program called Help Eliminate Auto Theft (HEAT), funded through auto insurance policies ($1 from each policy), that has created a hotline for people to report thefts and "chop shop" operations. In 25 years of operation, the program has aided in the recovery of more than 4,200 vehicles (Insurance Information Institute, 2011).

Other ways to reduce motor vehicle theft are centered on target-hardening approaches. **Target hardening** is the process whereby a target is made more difficult for an offender to attack. This approach is likely to be successful in reducing motor vehicle theft. In fact, it has been argued that "the most promising preventive approach is through the manufacture of more secure vehicles" (Clarke & Harris, 1992, p. 1). Locking car doors and installing security devices can be quite beneficial in reducing motor vehicle theft. In addition, installing a steering wheel column lock, such as The Club, can reduce the likelihood that a motor vehicle will be stolen. In the event that your car is stolen, you can increase your chances of recovering it by having an electronic tracking device. One such device, called LoJack, uses a radio transmitter. If your car is stolen, the police can track your car via the radio signal to locate it. LoJack (2011) reports a 90% recovery rate for stolen vehicles equipped with this system. Electronic immobilizers can also be used that disable the ignition, starter, and fuel systems until the system is activated by a computer-coded key (Linden & Chaturvedi, 2005). Research conducted in Canada and Australia indicates that these electronic immobilization systems are quite successful—theft rates for vehicles that have these systems are low (Potter & Thomas, 2001; Tabachneck, Norup, Thomason, & Motlagh, 2000).

Household Burglary

Perhaps one of the scariest things you could think of is to be alone, in the privacy of your own home, when an intruder comes in with a weapon—stealing whatever he or she wants and threatening you. If this were to occur, you would be the victim of a household or residential burglary. The FBI (2010) defines **burglary** as "the unlawful entry of a structure to commit a felony or theft." To understand this definition, you must understand what constitutes a structure. The FBI's definition recognizes structures to include houses, apartments, house trailers, houseboats when used as a permanent dwelling, barns, offices, railroad cars, vessels, and stables. Also important to note, entry into the structure does not have to occur forcefully. Entry can be achieved either via (1) forceful entry or (2) unlawful entry without force. Forceful entry includes such things as breaking a window or forcing a door open, while unlawful entry without force would occur if an offender used an unlocked or opened door to enter the structure. Attempted forceful entry of a structure can also constitute a burglary. For the purposes of this section, we will focus only on residential or household burglary—those burglaries of structures where persons live, which made up 73% of all burglaries reported to the police in 2009 (FBI, 2010).

You may also have heard the term **home invasion** used to refer to household burglary. This term is most commonly used to refer to a specific type of residential burglary that occurs when someone is home and the offender intends to harm, use force, or use violence against the residents, as in the example presented above. Most typically, a home invasion involves forceful entry. The term *home invasion* has been incorporated into some state laws, where such an offense is a specific crime. For example, in Florida, a home invasion robbery occurs when a person enters a dwelling with the intent to commit a robbery of the persons inside and does, in fact, do so (FSS 812.135). For household burglary to occur more generally, however, intent on the part of the offender to harm the occupants or rob them is not needed, just that the offender enters the dwelling to commit a felony or theft.

Extent of Household Burglary

According to the UCR, in 2009 there were 2,199,125 burglaries, 1,596,008 of which were burglaries of dwellings (FBI, 2010). Of all property crimes during that year, almost 24% were burglaries (including nonresidential burglary). Most household burglary occurred during the day, perhaps because this is the time when people are at work and their homes are left unguarded. We will return to this point again. On average, each residential burglary resulted in a loss of $2,163. Remember, however, from our discussion in Section 2, that the UCR includes only information about crimes that are known to law enforcement—typically because the crime is reported to the police. About half of all crimes go unreported. As such, the UCR likely underrepresents the true amount of crime victimization that occurs in the United States. Also, the UCR provides only limited information about the offense, criminal, and victim. To find out about victimizations that are reported and not reported to the police and to understand the patterns and characteristics of victims and victimization events, we often use the NCVS. So, let's discuss residential burglary using statistics from the NCVS.

According to the NCVS, in 2009, there were 3,134,920 household burglaries, which does not include household burglaries in which someone became a victim of a violent crime or was threatened (Truman & Rand, 2010). According to NCVS classification rules, those incidents would be classified as personal victimizations according to what occurred (e.g., rape/sexual assault, simple assault). If a household burglary occurred and no one was home or no one was hurt or threatened, it would be classified as a property victimization. It is interesting to examine both types of incidents, however, and the Bureau of Justice Statistics has recently published a report analyzing victimization that occurs during household burglary (Catalano, 2010). When taking into account all household burglaries, it is estimated that 3.7 million households are burglarized each year (Catalano, 2010).

Characteristics of Household Burglary

On average, between 2003 and 2007, most household burglaries were achieved through unlawful entry through an unlocked or open door or window rather than forceful entry (Catalano, 2010). When forceful entry was used or attempted, most typically, a door was damaged or removed or a window pane was damaged or removed (Catalano, 2010). In about 28% of household burglaries, someone was present during the incident, and 7% of households that were burglarized had someone injured as a result (Catalano, 2010). Most typically, the type of violence that occurred was a simple assault (15%), followed by a robbery (7%), and rape/sexual assault (3%) (Catalano, 2010). For those burglaries that occurred when no one was home, one fourth of victims said they were at work at the time and slightly less than one fourth said they were participating in a leisure activity away from home (Catalano, 2010). About 40% of household burglaries occurred between 6 A.M. and 6 P.M. (Catalano, 2010). A larger percentage of burglaries occurred during this time period when no one was home, as compared with burglaries that occurred when someone was present (Catalano, 2010). Even though burglary might seem to be one crime that would be perpetrated by a stranger, almost two thirds (65%) of household burglaries that involved violence were perpetrated by someone known to the victim (Catalano, 2010).

What Households Are Burglarized?

If a burglar is intent on entering a house, which house does he or she choose? Is it random or more systematic? Well, it appears that some houses are at greater risk than others for burglary. Households composed of married couples without children had the lowest rates of burglary when someone was home and when someone was not home, on average, between 2003 and 2007 (Catalano, 2010). Households composed of single males had the highest rates of burglary when a household member was not present, while households composed of single females had the highest rates of burglary when a household member was present. When examining households with children, single-male-headed households had the highest burglary rates when a household member was not present, and single-female-headed households had the highest burglary rates when someone was present (Catalano, 2010).

Race, income, age, and type of housing unit are also risk factors for burglary. Households that have an American Indian or Alaska Native head of household had the highest rates of burglary when no one was home, whereas when at least one person was present, American Indian, Alaska Native, or households headed by persons of more than one race were equally likely to be burglarized. Lower-income households and those headed by younger persons (aged 12–19) had higher rates of burglary than did higher-income households and those headed by older persons

(Catalano, 2010). Finally, burglary rates were highest for households residing in rental properties. When a household member was not present, households residing in hotel, motel, or rooming houses faced the greatest risk of burglary, while households living in mobile homes faced the greatest risk of burglary when a household member was present (Catalano, 2010).

Risk Factors for Household Burglary

As with most types of victimization, researchers have attempted to identify why some households are more likely to be burglarized compared with others. This line of inquiry has mainly focused on the decision making of burglars—why they select certain homes to break into but leave others alone. This body of research has concluded that, at least to some degree, household burglars choose suitable targets that lack capable guardians (Cromwell, Olson, & Avary, 1991; Tunnell, 1992; Wright & Decker, 1994). In other words, routine activities theory is a viable explanation for burglary victimization. Remember from Section 2 that routine activities theory suggests victimization is likely to occur when motivated offenders, suitable targets, and lack of capable guardianship coalesce in time and space. Let's consider household burglary from this perspective.

Motivated offenders do not need to be explained, according to routing activities theory (Cohen & Felson, 1979). Given opportunity, then, motivated offenders will burgle homes. The selection of homes is what is interesting. What makes a home a suitable target? Ethnographic research conducted via interviews on burglars shows that they tend to target middle- or high-income neighborhoods (Rengert & Wasilchick, 1985), and quantitative research predicting burglary indicates that expensive homes are most prone to burglary (Fishman, Hakim, & Shachmurove, 1998)—even though statistics from the NCVS (that do not control for other factors) indicate that lower-income households are more prone to household burglary (Catalano, 2010). A target also becomes suitable to a burglar when it lies on a familiar route that is also along major arterial routes (Rengert & Wasilchick, 1985).

Households that lack capable guardians are also most likely to be burgled. For instance, homes are most likely to be burgled during the day, when people are at work, and most burglaries occur when no one is home (Catalano, 2010). Because burglars seek out unoccupied homes, research shows that one way to reduce the chances of having your home burglarized is by making it appear that someone is home. Ways to do this are by keeping a car in your driveway, keeping lights on outside and inside, and keeping a television or radio on even if you are not home (Fishman et al., 1998).

In addition, houses located on quiet residential streets are more prone to burglary than others (Fishman et al., 1998), whereas households in high-density areas are less likely to be burglarized (Catalano, 2010). Homes on quiet streets are less likely to have watchful eyes, and burglars may feel that they can successfully get away with their crime. Although it is difficult to change the location of your home, simply having a burglar alarm reduces the chances that your house will be burglarized (Fishman et al., 1998).

⬚ Identify Theft

As technology advances, the means by which a person's property and money are taken also change. Especially with the widespread use of computers and the ready access people have to them, it is easier today than ever for people to access other people's money illegally. With key pieces of identifying information, criminals can assume your identity and either access your existing accounts or open new accounts in your name. Collectively, these actions are known as identify theft, and this section describes its occurrence, its effects, and the system's response.

Generally, **identify theft** is defined as the use of personal identifying information by a person to commit some type of fraud (Federal Deposit Insurance Corporation [FDIC], 2004). According to the FDIC (2004), it is one of the fastest-growing types of consumer fraud. Identify theft became a federal offense in 1998 when the **Identify Theft and Assumption Deterrence Act** was passed. It made it a crime for anyone to

> knowingly transfer or use, without lawful authority, a means of identification of another person with the intent to commit, or to aid or abet, any unlawful activity that constitutes a violation of Federal law, or that constitutes a felony under any applicable State or local law. (FDIC, 2004, pp. 4–5)

Identity theft covers taking over an existing account (**account hijacking**), creating new accounts, or creating a synthetic identity to obtain services or benefits fraudulently (FDIC, 2004).

What are the common ways identity theft occurs? Computers have allowed criminals easy access to your information, but they do not necessarily need computers to steal your personal information. One way they can steal your information is by **shoulder surfing**. Shoulder surfing involves criminals watching while you input numbers such as a telephone calling card number or credit card number. They also can listen while you say your credit card number over the telephone, such as, while you are placing an order or booking a trip (U.S. Department of Justice, n.d.). Account numbers can be stolen via **skimming** devices placed on automated teller machines (ATMs) that record a person's debit account number, password, account information, and other data that can then be used to imprint a new card and withdraw funds (Berg, 2009).

Your physical mail is also a valuable resource for offenders wanting to steal your identity. Criminals may **dumpster dive,** going through garbage cans or communal dumpsters or bins in hopes of finding documents with account numbers on them, checks, or bank and credit statements (U.S. Department of Justice, n.d.). Criminals may find in your trash or in your mailbox preapproved credit card offers that they can send off in your name. You may not even know that there are open credit lines in your name that others are using.

The Internet is also widely used for identity theft. One common way of executing account hijacking is through **phishing.** When phishing occurs, deceptive e-mails or fake websites are used to get people to provide usernames, passwords, and sometimes account numbers to perpetrators who then are able to use this information to access accounts (FDIC, 2004). For example, an e-mail is sent to a person indicating a problem with his account and he is asked to log in to his account through a hyperlink included in the e-mail. Unfortunately, the hyperlink takes the person to a spoofed website. If the person does, indeed, go to the site through the hyperlink and enter his username and password, the information is then collected and the account can be hijacked (FDIC, 2004). You may wonder how often phishing occurs. In 2006, more than 109 million adults in the United States reported receiving a phishing e-mail (Gartner, 2006). Between 5% (FDIC, 2004) and 22% (Gartner, 2006) of the people who receive such phishing e-mails respond to them (FDIC, 2004). Account hijacking can also occur when people hack into financial institutions or service providers and access customers' information (FDIC, 2004). Accounts also may be hijacked using software, called **spyware,** that collects information from unsuspecting persons' computers. This software is surreptitiously loaded when a person opens an e-mail attachment or clicks on a pop-up advertisement. The software then tracks keystrokes and information

▲ **Photo 10.2**   This ATM has a mini-camera installed so that pin codes can be stolen.

Photo credit: ©Fredrik Von Erichsen/dpa/Corbis

such as usernames, passwords, and account information and sends it to the defrauder (Berg, 2009). It may not take sophisticated software, however, to hijack an account. It is estimated that 65% to 70% of identity theft is committed by employees who have access to sensitive personal data (as cited in FDIC, 2004). See the box for an example of a real case of identity theft.

EXAMPLE OF IDENTITY THEFT CASE

30-year-old Anthony Eugene Vaughn was arrested for at least 1,000 counts of second-degree identity theft and two counts of first-degree identity theft. More than 1,000 people had their driver's licenses, credit cards, and Social Security numbers stolen by this Thurston County criminal, who is said to have committed the largest identity theft this country has ever seen! Detectives believe that other thieves, who aided in opening several fraudulent bank accounts in victims' names, accompanied Vaughn. Stolen ATM cards and credit cards were traced back to illegal purchases. More than 40 boxes of evidence were seized, which included driver's licenses, credit cards, credit card swipers, and Social Security cards/numbers. This case has been traced back to a motor vehicle rummage on the Capitol Campus. The vehicle, which belonged to a state employee, held thousands of people's names and Social Security numbers. Further, Vaughn was identified as a suspect in the burglary of a safe belonging to a construction company in Mason County. A detective found that Vaughn used his home phone number to open a fraudulent account using the stolen documents from the Mason County burglary.

SOURCE: Pawloski (2011). Reprinted with permission.

Extent of Identify Theft Victimization

It is quite difficult to know exactly how much identify theft victimization occurs each year in the United States. It is possible that you could have your identify stolen and be completely unaware—or if you did become aware, it could be weeks, months, or even years after it initially occurred. Even if you were aware your identity had been stolen, you may not notify the police. And the UCR does not neatly measure identify theft; it is not one of the eight index crimes that the FBI includes in its yearly report. To begin to understand the extent and characteristics of identity theft, an identity theft victimization supplement to the NCVS was collected for the first time in 2008. Unlike the NCVS, only persons over the age of 16 participated in the survey, and persons were asked about experiences of identity theft they experienced in the previous 2 years (Langton & Planty, 2010). Persons completing the **Identity Theft Supplement** (ITS) were also people participating in the NCVS. The final sample in the ITS included 56,480 persons.

The ITS collected data on the use or attempted use of existing accounts (account hijacking), use of personal information to open a new account, or misuse of personal information for other fraudulent purposes (Langton & Planty, 2010). Existing accounts that a person may hijack could include a credit card account, savings account, mortgage account, loan, or checking account. Persons may want to use personal information to obtain a job, for medical care, when interacting with law enforcement, to rent an apartment, or for government benefits (Langton & Planty, 2010). Based on findings from the 2008 ITS, it was estimated that 11.7 million people were the victims of identity theft between 2006 and 2008 (Langton & Planty, 2010). Thus, 1 in 20 Americans in a 2-year period was a victim of this type of victimization. Most commonly, victims reported that they had an existing credit card account misused or attempted to be misused (Langton & Planty, 2010). More than half (53%) of identity theft victims reported this type of victimization (Langton & Planty, 2010). Next most common was the use of an existing bank account. About 1.7 million persons reported that someone had fraudulently used their information to open a new account, and about 618,000 people said that their personal information had been used to commit other crimes—to receive medical care, to receive government benefits, or to be given to law enforcement (Langton & Planty, 2010). Recurring identify theft victimization was also common. About 16% of victims indicated that they had experienced more than one type of identity theft during the 2-year period (Langton & Planty, 2010).

There are other ways to assess the extent to which people are victimized by identity theft. One of these ways is to look to agencies where people report their victimizations. The Federal Trade Commission (FTC) publishes a report of complaints it and other agencies (The Internet Crime Complaint Center, the Better Business Bureau, Canada's PhoneBusters, United States Inspection Service, Identity Theft Assistance Center, and the National Fraud Information Center, among others) receive regarding fraud and identity theft. In this report, the Consumer Sentinel Network Databook, identity theft was the number one complaint received by the FTC in 2009 (FTC, 2010). There were 278,078 identity theft complaints made to the FTC. Of these complaints, 23% were classified as "other"—victims were uncertain what type of identity theft occurred or it fit into a miscellaneous category. Of those victims who knew what type of identity theft occurred, 17% indicated it was credit card fraud and 16% were the victims of government documents to receive benefits fraud (FTC, 2010).

Who Is Victimized by Identify Theft?

Identity theft can happen to anyone, but results from the ITS show to whom identify theft is most likely to occur. Females make up the majority of identity theft victims. Fifty-three percent of victims were female (Langton & Planty, 2010). Remember that young people (ages 12–24 and 25–34) are the typical victims of most property and personal victimizations (Truman & Rand, 2010). Younger persons are more likely to be victims of identity theft than are older persons. Six percent of persons between the ages of 16 and 24, and 5.9% of persons between the ages of 25 and 34 reported identity theft, compared with 3.7% of persons 65 or older (Langton & Planty, 2010). Reports to the FTC also indicate that younger persons are more often victims of identity theft than are older persons. Almost one third of complaints filed with the FTC were by persons 29 years old or younger (FTC, 2010). Differences across race and ethnic groups were also found in the ITS. Persons of more than one race were more likely than others (10.4% of persons) to report experiencing identity theft (Langton & Planty, 2010). A greater percentage of White, non-Hispanics experienced identity theft than did Hispanics and Black, non-Hispanics. There also was a difference in the types of identity theft experienced. A greater percentage of White respondents compared with Black respondents reported experiencing identity theft that involved theft of an existing account (Langton & Planty, 2010). Income is also relevant for understanding who is likely to be victimized by identity theft. As household income increases, so too does a person's likelihood of being victimized. Persons living in households with incomes of $75,000 or more were more likely than others to be identity theft victims (Langton & Planty, 2010). Also, similar to what was found with race, persons living in households with incomes of $75,000 or more were victims of theft of existing accounts more often than were persons living in households with lower incomes (Langton & Planty, 2010). Locales also differed in their rates of reports of identity theft. The states with the five highest rates of identity theft reports for 2009 are, in order, (1) Florida, (2) Arizona, (3) Texas, (4) California, and (5) Nevada. For an example of a state's 2009 occurrences of identity theft, see Table 10.3, which shows the extent of reports in Arizona for 2009.

Characteristics of Identity Theft Victimizations

One of the interesting things about identity theft victimization is that people may be unaware that they have been victimized. Others may not have any idea as to how the offender was able to victimize them. Not knowing how the offender stole your identity can be especially problematic for prevention! In the ITS, victims were asked if they knew how the identity theft had occurred. About 40% of victims indicated that they knew how their identity had been stolen. Of those victims, about 1 in 3 believed that their identity had been stolen during a purchase or transaction, and about 1 in 5 indicated that they thought their identity had been stolen from a wallet or checkbook. About 14% thought that their personal information had been stolen from personnel or other files at their office, and 8% thought that friends or family had accessed their personal information (Langton & Planty, 2010).

What do victims of identity theft do after realizing that their identities have been stolen? According to findings from the ITS, only 17% of identity theft victims notified a law enforcement agency about the incident (Langton & Planty, 2010). Victims who had offenders try to open a new account in their name (28%) and who had offenders misuse their personal information (26%) were more likely than victims whose offenders tried to use an existing account (13%) to report the incident to the police. In the ITS, victims were asked why they did not report to the police (Langton & Planty, 2010). Almost half of those victims who did not report the incident to the police said they handled it another way. About 20% said they did not report because they did not lose any money. Others (15%) indicated that they did not know the police could help, suggesting that they were not sure their experience was in fact criminal or the police would be able to do anything in response. A small percentage (7%) said they were embarrassed, afraid, or thought that reporting would be inconvenient.

If victims do not tell the police, are they telling others? Results from the ITS indicate that they often do disclose the incident. Sixty-eight percent of victims contacted a credit card company or bank to report that their account or personal information had been misused, about 15% notified a credit bureau, 7% contacted a credit monitoring service, 3% told a government consumer affairs agency or similar consumer protection agency (e.g., Better Business Bureau), 4% notified an agency that issues identification documentation, and 1% reported the incident to the FTC (Langton & Planty, 2010).

Consequences of Identity Theft

Reporting is not the only consequence victims face after being victimized by identity theft. One consequence is the financial cost that being a victim of identity theft carries. The majority of victims incurred some financial cost associated with their victimization. These costs can be direct or indirect. **Direct costs** are monies and the value of goods and services taken. **Indirect costs** consist of things such as legal bills, phone calls, bounced checks, and notary fees (Langton & Planty, 2010). The ITS estimates that identity theft costs from 2007 to 2009 totaled $17.3 billion (Langton & Planty, 2010). Of those victims who suffered a loss, the average was $2,400, and 23% suffered an out-of-pocket cost that was not reimbursed (Langton & Planty, 2010). In addition to financial costs, 27% of victims noted that they had to spend more than a month trying to clear up problems

Table 10.3 Reports of Identity Theft in Arizona, 2009 (complaints *n* = 7,875)

Rank	Identity Theft Type	Complaints	Percentage
1	Employment-related fraud	2,328	30
2	Government documents or benefits fraud	1,200	15
3	Credit card fraud	942	12
4	Phone or utilities	928	12
5	Bank fraud	657	8
6	Loan fraud	331	4
	Other	1,742	22
	Attempted	354	4

SOURCE: FTC (2010, p. 20).

resulting from having their identities stolen. Most victims, however, were able to resolve their problems in a day or less (Langton & Planty, 2010).

Finally, identity theft may take an emotional toll on victims, similar to other victimizations. About one in five victims of identity theft in the ITS indicated that they experienced severe emotional stress, 6% reported they had significant problems with relationships between family members or friends as a result of the incident, and 3% reported that they experienced significant problems at work or school as a result of being victimized (Langton & Planty, 2010). Although fewer identity theft victims reported problems with relationships at work and school and having severe emotional distress as compared with violent crime victims, it should be evident that identity theft does carry consequences for many victims.

Risk Factors for Identity Theft Victimization

Routine activities theory helps us understand who is vulnerable to identity theft. Motivated offenders are, by definition, present. It is when motivated offenders come into contact with suitable targets in time and space without capable guardianship that identity theft is likely to occur. Suitable targets are likely to be those who are easily victimized. Why would a person be an easy target for identity theft? Consider the mechanisms by which offenders steal a person's identity. Leaving personal identification intact in the trash, making phone calls in public places to discuss personal information, and using public Internet connections to access personal accounts all make a person a suitable target. A novice to the Internet or a person who does not understand the intricacies of how secure Internet sites work will also be a more suitable target than a savvier Internet user. Notably, even persons who are aware of the dangers are still likely at risk. What is interesting about identity theft achieved via the Internet is that potential victims will not or cannot curtail their Internet usage, even if they fear victimization (Cox, Johnson, & Richards, 2009). Capable guardianship, then, is a particularly important concept for identity theft, especially when considering online accounts, usernames, and passwords (Cox et al., 2009). Guardianship often comes in the form of warning e-mails, anti-virus software, and firewalls (Cox et al., 2009). But simply having bank accounts, credit card accounts, and a Social Security number puts you at risk of having this personal information stolen! For this reason, as discussed in a later section, everyone should take precautions to protect these pieces of information. Before reading about these precautions, turn to the article in this section by Travis C. Pratt, Kristy Holtfreter, and Michael D. Reisig (2010). In this article, they discuss another form of victimization that can occur online—Internet fraud—and how online activities structure risk of victimization.

Responses to Identity Theft Victimization

As noted in the first part of this section, the Identity Theft and Assumption Deterrence Act of 1998 (Pub. L. 105-318) introduced identity theft into the federal criminal code. It made it a crime for anyone to

> knowingly transfer or use, without lawful authority, a means of identification of another person with the intent to commit, or to aid or abet, any unlawful activity that constitutes a violation of Federal law, or that constitutes a felony under any applicable State or local law. (Pub. L. 105-318)

The **New Fair and Accurate Credit Transactions Act of 2003** (Pub. L. 108-159) enacted several provisions to reduce identity theft and also to help identity theft victims. This act

- mandated that the three major credit reporting bureaus provide a free copy of credit reports to consumers once every 12 months;

- created the National Fraud Alert System, which allows persons who reasonably suspect that they are or will become a victim of identity theft, or who are on active duty in the military, to place an alert on their credit file;
- provided that credit card receipts must have shorter account numbers and developed rules for financial institutions and creditors for the disposal of sensitive credit report information;
- noted that if a person establishes that she or he has been a victim of identity theft, credit reporting agencies are required to stop reporting the fraudulent account information;
- established that creditors and businesses must provide copies of records of fraudulent accounts or transactions related to them so that victims can attempt to prove their victimization; and
- provided for victims to report accounts affected by identity theft directly to creditors (FTC, 2004).

The **CAN-SPAM Act of 2003** also addressed identity theft. This act made it a federal crime to send spam e-mail under certain circumstances (Pub. L. 108-187). For the specifics of this act, see the boxed item below. In 2004, the **Identity Theft Penalty Enhancement Act** (Pub. L. 108-275) was passed and created the crime of aggravated identity theft, which added 2 years to the sentence of anyone convicted of unlawful identity theft during and in relation to specified felony violations. It also created an additional 5-year prison term enhancement for unlawful identity theft related to terrorist acts. In addition to these federal laws, each of the 50 states and the District of Columbia all have laws addressing identity theft. See the boxed item on the next page for the law in Illinois.

CAN-SPAM ACT OF 2003

IN GENERAL—Whoever, in or affecting interstate or foreign commerce, knowingly—

(1) accesses a protected computer without authorization, and intentionally initiates the transmission of multiple commercial electronic mail messages from or through such computer,

(2) uses a protected computer to relay or retransmit multiple commercial electronic mail messages, with the intent to deceive or mislead recipients, or any Internet access service, as to the origin of such messages,

(3) materially falsifies header information in multiple commercial electronic mail messages and intentionally initiates the transmission of such messages,

(4) registers, using information that materially falsifies the identity of the actual registrant, for five or more electronic mail accounts or online user accounts or two or more domain names, and intentionally initiates the transmission of multiple commercial electronic mail messages from any combination of such accounts or domain names, or

(5) falsely represents oneself to be the registrant or the legitimate successor in interest to the registrant of 5 or more Internet Protocol addresses, and intentionally initiates the transmission of multiple commercial electronic mail messages from such addresses, or conspires to do so, shall be punished as provided in subsection (b).

SOURCE: Federal Trade Commission (2004).

IDENTITY THEFT LAW, ILLINOIS

A person commits the offense of identity theft when he or she knowingly uses any personal identifying information or personal identification document of another person to fraudulently obtain credit, money, goods, services, or other property. Penalties for identity theft depend on the value of the theft.

If the value of the credit, money, goods, services, or other property is less than $300, the crime is a Class 4 felony, punishable by one to three years in prison and a fine up to $25,000. Subsequent offenses are upgraded to a Class 3 felony, punishable by two to five years in prison, as is the penalty for people who have previously been convicted of certain crimes, including burglary, theft, or fraud. If the amount is between $300 and $2,000, it is a Class 3 felony; between $2,000 and $10,000 is a Class 2 felony (punishable by two to seven years in prison); between $10,000 and $100,000 is a Class 1 felony (punishable by four to fifteen years in prison); and over $100,000 is a Class X felony (punishable by six to thirty years in prison).

Penalties are increased one step (with the exception of a Class X felony) if the victim of the offense is an active duty member of the Armed Services or Reserve Forces of the United States or of the Illinois National Guard serving in a foreign country.

SOURCE: National Conference of State Legislatures (2010). Reprinted by permission of the National Conference of State Legislatures.

Not only can laws reduce the occurrence of identity theft, but there are many things individuals can do to reduce their chances of having their identities stolen. When going out of town, you should have your mail held at the post office or have a trusted friend or family member remove it from your mailbox. Do not discuss your personal information (address, telephone number, Social Security number) at an open telephone booth or on a phone in public near people who can hear you and possibly intercept your information (U.S. Department of Justice, n.d.). Do not give out your personal information to people who *call you* on the telephone, even if they say they are from your bank or work for a business with whom you have an account. If they do, in fact, work with one of these businesses, they should have your information. If you *call them*, however, you should expect to be asked to verify your identity (U.S. Department of Justice, n.d.). Your personal checks should display the least amount of personal information possible—do not put your Social Security number or your phone number on your checks. Phone calls notifying you of exciting prizes you have won that require you to give your Social Security number, credit card number, or bank account number may be fraudulent. Ask for these offers to be sent to you in writing (U.S. Department of Justice, n.d.). Shred all documents that display your account numbers or personal information before discarding them.

As you can tell, as technology advances, the ways in which your property and goods can be stolen also expand. It will take new prevention strategies to reduce your chances of being victimized. Nonetheless, the connections between the types of property victimization and the factors that place individuals and homes at risk are evident. Although property crimes may not always make the news, these types of offenses are the ones you are most likely to experience.

SUMMARY

- Theft made up more than half the reported cases of property crime in the National Crime Victimization Survey (NCVS) and 68% of all property crimes in the Uniform Crime Reports (UCR) in 2009.
- The characteristics of your household contribute to your chances of property crime victimization—persons living in households headed by persons who are White, young, and with a total income less than $7,500 have the highest theft victimization rates, according to the NCVS.

- There are characteristics that make certain vehicles' and cities' motor vehicle theft rates exceed the rates of others.
- Within households, there are even certain races of people who are more likely to be burglarized and/or experience some form of property victimization.
- In discussing motor vehicle theft, routine activities theory provides that some vehicles are more suitable targets than others.
- Motor vehicle theft is one of the most reported crimes to the police.
- In response to the fact that many motor vehicle thieves are not apprehended, federal legislation has been passed to reduce, eliminate, and protect potential victims and those who already have been victimized.
- Target hardening is the process whereby targets are made more difficult for offenders to attack.
- Household burglary is not always achieved through forceful entry; it can also be accomplished through unlawful entry.
- Most household burglary is committed during the day, when no one is home, via unlawful entry, and by someone known to the victim.
- Routine activities theory can be used to explain why some households are burglarized and others are not. Those houses that are well guarded are less likely to be burglarized than others. Also, houses that are deemed suitable targets are more likely to be burglarized.
- The advancement of technology has made it easier for criminals to steal other people's identities.
- Common methods of identity theft include shoulder surfing, skimming, dumpster diving, phishing, and using spyware.
- The UCR does not neatly measure identity theft, but both the Federal Trade Commission and the Identity Theft Supplement to the NCVS measure the extent to which people's identities are stolen.
- It is estimated that 6% of persons between the ages of 25 and 34 had their identities stolen between 2006 and 2008.
- A person's income and race are indicators of the likelihood that their identities will be stolen. A person with more income is a more suitable target for criminals.
- As with other forms of victimization, identity theft can cause more than just financial loss. Victims of identity theft often suffer emotional stress as well.
- To reduce your chances of becoming a victim of identity theft, certain precautions should be taken, such as destroying old mail and credit information rather than simply tossing it in the garbage.
- Though many people do not report their crimes to the police, the NCVS indicates that they contact other entities, such as a credit bureau, to inform them of their victimization.
- The fact that laws for identity theft have been outlined in the federal criminal code signifies the increasing seriousness of this issue, which continues to plague many people. The enhancement of severity of punishment is now addressed in the CAN-SPAM Act of 2003.

DISCUSSION QUESTIONS

1. What about the top 10 most frequently stolen passenger vehicles makes them the most likely to be stolen?

2. Why are theft rates so high in the top 10 U.S. metropolitan areas for motor vehicle theft? Why are motor vehicle theft rates so high in the Western United States?

3. Why do the states with the five highest rates of identity theft reports for 2009 have such high rates?

4. What are some things one can do in order to make his or her house less likely to be burglarized?

5. What makes a person a more likely target for identity theft?

KEY TERMS

theft

motor vehicle theft

carjacking

Motor Vehicle Theft Law Enforcement
Act (1984)

Anti-Car Theft Act (1992)

Motor Vehicle Theft Prevention
Act of 1994

Anti-Car Theft Improvements
Act of 1996

target hardening

burglary

home invasion

identity theft

Identity Theft and Assumption
Deterrence Act (1998)

account hijacking

shoulder surfing

skimming

dumpster dive

phishing

spyware

Identity Theft Supplement

direct costs

indirect costs

New Fair and Accurate Credit
Transactions Act of 2003

CAN-SPAM Act of 2003

Identity Theft Penalty Enhancement
Act (2004)

INTERNET RESOURCES

"Auto Theft" (http://www.iii.org/media/hottopics/insurance/test4/)

The Insurance Information Institute website contains statistics from the Federal Bureau of Investigation's Uniform Crime Reports about the types of cars that are most likely to be stolen and in which cities/states cars are most likely to be stolen, as well as the car models least likely to be stolen and the cities/states in which cars are least likely to be stolen. The website also provides this information as it pertains to motorcycle theft. A historical background information section addresses anti-theft laws as well.

Edmunds Inside Line (http://www.insideline.com/)

This website is a great source for information on theft complaints for particular popular motor vehicles. The site not only lists the newest vehicles, but it also lists the issues/complaints that buyers can expect. Recent news articles regarding motor vehicle theft, including new technologies to prevent motor vehicle theft, are also highlighted.

"Preventing Auto Theft" (http://www.geico.com/information/safety/auto/preventing-auto-theft/)

This Geico webpage provides several tips to prevent you from becoming the next victim of auto theft. There is a breakdown of what types of auto thieves exist most commonly, as well as the various types of anti-theft systems available and what those systems do.

Fighting Back Against Identity Theft (http://www.ftc.gov/bcp/edu/microsites/idtheft/)

The Federal Trade Commission's identity theft website provides introductory information about identity theft for consumers, businesses, law enforcement, and persons in the military. It also includes ways to report identity theft victimization. Data about identity theft and ways to prevent identity theft are also discussed.

"Burglary of Single-Family Houses" (http://www.popcenter.org/problems/burglary_home/)

The Center for Problem-Oriented Policing provides a webpage dedicated to the topic of burglary. This webguide provides information about burglary, the types of housing units that are at risk, risk factors for burglary, and the types of goods typically stolen. There is also a discussion about burglars.

Introduction to Reading 1

Single and repeat auto theft is examined in Levy and Tartaro's (2010) paper. They examine auto theft in Atlantic City, New Jersey, using the Watchers, Activity Nodes, Location, Lighting, and Security indices (W.A.L.L.S.). Their paper is important in that it identifies characteristics about certain places that contribute to the risk of repeat auto theft. In this way, they identify characteristics—not of the automobile but, rather, of the place—that can be targeted for intervention.

Auto Theft

A Site-Survey and Analysis of Environmental Crime Factors in Atlantic City, NJ

Marissa P. Levy and Christine Tartaro

Literature Review

Site-level auto theft research is conducted in order to obtain a better understanding of the physical environment surrounding the locations targeted by offenders. Some research has compared locations with crime to those without crime to determine the environmental differences that may have led to victimization in one location but not the other (Bichler-Robertson and Potchak, 2002). In order to conduct this type of research, three general criminological foundations must be considered: Routine Activities Approach (Cohen and Felson, 1979), Rational Choice Perspective (Cornish and Clarke, 1986) and Situational Crime Prevention (Jacobs, 1961; Jeffery, 1971; Brantingham and Brantingham, 1981).

Drawing from the work of Hawley (1950) and other human ecologists, the Routine Activities Approach emphasized the context in which offenders choose to commit crimes. Criminal events are routine activities socially organized in time and space. These routine activities influence the circumstances in which offenders carry out criminal acts. In particular, Cohen and Felson (1979) and Felson and Cohen (1980) originally hypothesize that the dispersion of activities away from households and families would increase the opportunity for crime and thus generate higher crime rates.

Changes in routine activity patterns can alter crime rates by affecting the convergence in space and time of the three elements of a crime: motivated offenders, suitable targets and the absence of a capable guardian (Cohen and Felson, 1979). A lack of any one of these elements is enough to prevent the completion of a crime. If the number of motivated offenders or suitable targets were to remain the same, changes in routine activities could modify the likelihood of their convergence in space and time, creating more opportunities for crimes to occur. The idea that a crime could be pinpointed to a specific spot (a house within a neighborhood), instead of an entire neighborhood, dramatically changes the way criminologists think about crime.

The Rational Choice Perspective focuses more on the offender's decision-making process. This utilitarian

SOURCE: Levy and Tartaro (2010). Reprinted with permission of Palgrave-MacMillan.

cognitive process precedes and informs an offender's decision of whether to commit a particular crime at a particular time and place (Cornish and Clarke, 1986). This idea originates from Bentham (1948) and suggests that if the net result of an act is anticipated to produce a positive consequence, then the individual engages in the act. If the anticipated consequence of the criminal act is negative, then the individual refrains from such action. Both Bentham (1948) and Cornish and Clarke (1986) suggest that there is more than strictly a physical component to the commission of a crime.

During the course of the decision-making process, the offenders may consider such things as potential gains (monetary gains as well as those which must be liquidated), potential costs (punishment for getting caught, jail time, conviction), the best time of day to conduct the crime, the risk associated with certain locations (with regard to apprehension, confrontation, etc.) and then finally make a decision based on the net gain or loss associated with all of these factors (Cornish and Clarke, 1986). Rational Choice Perspective is applied in this study when researchers attempt to capture cues in the environment that make the target an attractive one (e.g. a situation where the offender is less likely to be caught and a greater reward or payoff is expected) or an unattractive one (e.g. a crime in which a greater risk of being caught and/or a smaller reward or payoff exists). Information that the offender possesses regarding surveillance, accessibility, lighting and security of the location plays a role in the decision to commit the crime. Each small factor that is considered has an effect—whether consciously or subconsciously–on the overall decision-making process of the offender, according to the Rational Choice Perspective.

Finally, Situational Crime Prevention identifies the physical components and structures that can facilitate a criminal act. Researchers have experimented with manipulating the environment in order to prevent or eliminate certain types of crime. Jeffery (1971) believes that crime can be prevented by creating invisible boundaries that people recognize to be areas that are off-limits, even if they are not physical barriers preventing access and egress.

Jacobs (1961, 1993) also suggests that spaces, especially public spaces, play a role in crime prevention.

Her suggestions are based on the residential environment; by keeping block segments short and personable—clean, friendly and full of children playing—Jacobs believes that they will remain crime free. She also advocates the needs for *secondary diversity* in primary use areas. Areas that are designed for only one purpose will be unoccupied by local residents when they are not able to use those resources (e.g. a playground in winter). This invites outsiders to use these vacated areas for criminal purposes (Jacobs, 1961). Jacobs identifies three places that are harmful unless the locations are controlled: parking lots, trucking depots and gas stations.

W.A.L.L.S.

Routine Activity Approach, Rational Choice Perspective and Situational Crime Prevention provide a framework from which other environmental criminologists can draw. These concepts, part drawn from theory and part learned from practice, aid researchers in identification and categorization of crime-specific and location-specific crime prevention measures. In this research project five situational factors are tested for significance, Watchers, Activity Nodes, Location, Lighting and Security (W.A.L.L.S.). These five factors have been identified and empirically tested in prior auto theft research projects (Levy, 2006) and will be tested for significance herein.

The "Watcher" variable represents the presence or absence of capable guardians of the auto theft location. "Activity Nodes" refers to the locations of crime generators and attractors near the auto theft location. "Location" refers to the environmental design of the location used to conceal criminal acts. "Lighting" is defined as the quality and quantity of lighting near auto theft locations. "Security" refers to the environmental cues meant to deter offenders.

Watchers

Watchers are individuals who exert a presence in the community and provide informal surveillance in these locations, including residents, local storeowners, consumers and pedestrians. They can be the same people every day or different people each day depending on the individual's daily activity patterns and the land use in the area. Watchers are an important part of the landscape of crime, since they can deter offenders simply by their presence and are rarely required to do any more than go about their daily routine in order to be effective. Watchers have been referred to as "capable guardians" by Cohen and Felson (1979) and Felson and Cohen (1980), "handlers" by Felson (1986) and Jochelson (1997) and "place managers" by Eck (1994) and Eck and Wartell (1999); each scholar has researched their importance.

With regard to auto theft research, Smith (1996) found that the presence of informal surveillance by pedestrians may be an important factor in the reduction of auto thefts and related crimes. Smith (1996) also posited that the presence of uniformed officers was one of the best crime prevention methods. These uniformed officers provide a sense of formal guardianship over the area, as suggested by Cohen and Felson (1979), while the presence of pedestrians enhances informal surveillance.

Watchers have been identified as an important environmental characteristic for crimes other than auto theft. Cohen and Felson (1979) discuss the importance of guardianship in any type of property crime by arguing that locations that are absent capable guardians may suffer increased victimization and higher crime rates than other locations. Felson and Clarke (1998) acknowledge that while capable guardians "cannot be everywhere . . . nor can the most likely offenders be everywhere" (p. 14). In some circumstances fear may deter guardians from shopping or visiting an area. Patrons may take cues from the environment (more bars on the windows or shutters on the doors) and interpret them as signs of crime or violence. This may cause them to question their own safety and to shop elsewhere (Fisher, 1991). The end result is fewer eyes on the street—less informal surveillance.

Activity Nodes

Activity nodes are locations with multiple functions that draw heavy use for both legal and illegal activities at similar times. Activity nodes, especially those open late (Automatic Teller Machines [ATMs], payphones, gas stations, bars, bus stops, convenience stores, casinos, etc.) bring people who are looking to commit a crime and those who are simply going about their day (or evening) into the same location at the same time. This puts individuals engaging in legal activities at greater risk for victimization than those who are not present at these locations.

Fleming (1999) found that some teenagers use shopping malls as a "hunting ground" for auto theft. These teens would use the public atmosphere of the malls as a cover for their illegal auto theft activities. Fleming noted that malls provided a "large pool of easy-to-steal vehicles" (1999, p. 76) and the ability for thieves to act as if they were engaging in a routine activity. Malls and other activity nodes, such as schools, provide additional awareness space (Brantingham and Brantingham, 1993) for juvenile auto thieves. Light *et al.* (1993) found that most of their sample of auto thieves became involved in auto-related crime in their early to mid-teens, when over two-thirds were still officially in school. Roncek and Faggiani (1985) and Roncek and Lobosco (1983) also discuss the relationship between the location of neighborhood schools and property crime.

Some common activity nodes that have been connected to criminal events are: ATMs (Scott, 2002), bars (Cavan, 1966; Graham *et al.*, 1980; Roncek and Bell, 1981; Rossmo and Fisher, 1993; Homel and Clark, 1994; Rossmo, 1995; Block and Block, 1995), casinos, convenience stores and gas stations (Crow and Bull, 1975; Jeffery *et al.*, 1987), hotels (Cook *et al.*, 1993; Huang *et al.*, 1998), payphones, shopping areas (Engstad, 1975; Gibbs and Shelly, 1982) and transportation hubs (Levine and Wachs, 1986; Levy, 1994; Loukaitou-Sideris, 1999).

Locations

The characteristics of the location help the offender to determine if it is a "good" target to select. Offenders will use landscape and design features to determine if their crime will be concealed and if the layout of the area will impede their escape after the crime is committed. Jeffery (1971) noted that there is a decision model that offenders use during the actual commission of a criminal act. "The decision to commit a crime involves the past experience of the subject, the immediate opportunity for a crime, plus the chances of apprehension or injury" (Jeffery, 1971, p. 251). The last part of the decision, the likelihood of apprehension, is discussed with regard to the layout of the location to be targeted. All other things being equal, offenders will choose locations that provide an easy getaway and enhance their chance of escaping apprehension. Nichols (1980) suggests that after an offender decides to commit a crime she/he engages in a search pattern of potential sites. This is conducted by using a "mental map" (Nichols, 1980, p. 156) of known locations to help the offender evaluate his or her options.

With regard to auto theft, Webb *et al.* (1992) found that in addition to lighting and surveillance, the parking lot layout was an important factor in the decision to steal a car. Parking lots have better attendant visibility which contributes to a lower risk of theft from cars as well as auto theft (Webb *et al.*, 1992). Parking garages that are partially or fully enclosed and elevated above ground offer less natural surveillance (Smith, 1996). This problem is enhanced by sloping ramps and multiple floors that further decrease informal surveillance inside the garage. High-risk facilities require access control and well-maintained grounds and surveillance (Smith, 1996). If parking garages have these features, they are deemed to be safer. For residential parking, Clarke and Mayhew found that "parking in a domestic garage at night is safer by a factor of 20 than parking in a driveway or other private place, and safer by a factor of 50 than parking in a street near home" (1994, p. 91). This indicates an increased risk for victimization for those who park on the street instead of in a residential garage.

Location is an important contribution to the W.A.L.L.S. characteristics because it measures a part of the offender's decision-making process. Burglars consider the street position of the building (Buck *et al.*, 1993), adjacency to freeways and open escape routes (Smith, 1996; Graham, 2001), search patterns (Nichols, 1980), physical barriers (Wright and Logie, 1988; Michael *et al.*, 2001) and symbolic barriers (Brown, 1985; Perkins *et al.*, 1992; Brown and Bentley, 1993; Kuo and Sullivan, 2001) when choosing a location in which

to offend. Auto thieves consider these and other site-level factors when deciding when and where to offend.

Lighting

The presence and quality of lighting is an important site-level factor considered before the commission of a crime. Both the amount of lighting that is properly working and the intensity of that lighting is important in the overall lighting of streets at night. Poor lighting can provide cover for offenders. Just as offenders use hedges and walls to prevent their detection, dark alleys and dimly lit streets hinder residents and other watchers from their involvement in informal surveillance. "Lighting is often installed with little thought to its functional ability to illuminate specific areas" (Bopp, 1982, p. 79). Furthermore, Bopp (1982) found that the level of lighting as well as its placement is critical.

Smith (1996) suggested that it is important for parking lots to have proper, bright, directed lighting to increase the levels of surveillance. Since parking garages are either partially or fully enclosed, cutting off natural light from entering and prohibiting natural surveillance from the street, artificial lighting is even more important (Smith, 1996; Poyner, 1997). Garages that can minimize ramps and utilize flat surfaces for parking areas will maximize the positive effects of light and enhance natural surveillance from outside of the garage (Painter, 1994). White stain on the ceiling can reflect light and increase uniformity in the parking garage (Painter, 1994; La Vigne, 1997).

Security

Webb *et al.* (1992) noted that the layout of the parking structure is a factor considered in auto theft. Parking structures with manned exits have the lowest risk for both auto theft and theft from autos (Webb *et al.*, 1992). Manned exits not only provide increased surveillance, but bars or mechanical arms on the exits provide a physical barrier to prevent cars from leaving the structures. Light *et al.* (1993) found that 35 percent of interviewed auto thieves indicated that they would be deterred by alarms. Fleming *et al.* (1994) found that nearly 75 percent of offenders would be deterred by an alarm or other security device.

Summary

The aforementioned research has provided evidence of a link between environmental characteristics and offenders' decision-making processes for property crimes. More work is needed, however, to develop a deeper understanding of target selection and the important situational factors contributing to crime. In an effort to provide more information, some researchers have moved toward a place-intensive method of studying crime (Bichler-Robertson and Johnson, 2001; Boba, 2001; Boesch, 2001; Woods, 2001; Bichler-Robertson and Potchak, 2002; Levy, 2006). This research seeks to uncover the specific factors present at locations in which crimes occur by conducting site-level research of both repeat and single victimization locations.

The following hypotheses are tested in this study:

1. Location with poor *surveillance/guardianship* will have more auto theft than those with good surveillance.

2. Autos parked in locations with *activity nodes* in the area will have more auto theft than those parked in areas without activity nodes.

3. Locations that have landscape and design features that *provide cover for offenders* will have more auto theft than those without these features.

4. Locations in areas with *poor lighting* will have more auto theft than those with good lighting.

5. Locations that have *less security and cues indicating security* will have more auto theft than those with more security or more security cues.

⊠ Methodology

Site Selection

The Atlantic City Police Department provided the researchers with data on auto thefts from 2004 to 2005. There were 283 auto thefts reported to the police department during those 2 years (167 in 2004 and 116 in 2005). The police provided information on the location of the vehicle at the time of the theft. The researchers were interested in identifying repeat victimization locations as well as single victimization locations. Repeat victimization locations are defined as any street segment on which more than one auto theft occurred during the study period. These locations were matched to comparison sites with only one reported auto theft during the 2-year study period (Table 1). The matching criteria are proximity to repeat victimization location, zoning (commercial or residential) and size/volume of the parking structure. Matching based on proximity in this study is defined by a single victimization location on the same street segment or adjoining block as a repeat victimization location. Cases are also matched on zoning in order to ensure that each single victimization auto theft in the commercial area is matched to a repeat auto theft also in a commercial area. The same logic was used for residentially zoned locations. For the residential area, the researchers identified 18 repeat victimization locations that were

Table 1 Sample of Auto Thefts in Atlantic City, NJ 2004–2005

	F	%
Residential auto thefts		
Single victimization locations	18	50
Repeat victimization locations	18	50
Parking		
Street parking	20	55.6
Garage/driveway parking	10	27.8
Lot parking	6	16.7
Additional street parking is available	33	92.2
Commercial auto thefts		
Single victimization locations	16	50
Repeat victimization locations	16	50
Parking		
Street parking	20	64.5
Garage parking	4	12.9
Lot parking	7	22.6

matched to an additional 18 single-theft locations. In the commercial areas, there are 16 repeat theft locations matched with 16 nearby single-theft locations. The single-theft locations were coded as 1, while the repeat theft locations were coded as 2.

In 2006, two researchers visited each selected theft location twice—once during the day and once after sunset. Daytime observations took approximately 3–5 min per site. One researcher filled out the site survey instrument while the other took photos of the block with a GPS camera. The primary purpose of the GPS camera is to record the longitude and latitude of the auto theft location so that it could be mapped in Arc View for future analyses. The photos are also used to double-check discrepancies in the data collected with the survey instrument and to demonstrate interesting observations in the environment. The researchers were careful to note any changes in the environment that may have taken place between the time of the crime and data collection. One area of Atlantic City was purposefully excluded from this analysis due to a recent redevelopment initiative.

The majority of the residential parking at the auto theft sites consists of street parking (56 percent), with the rest of the parking coded as garage or driveway (28 percent) or lot parking (17 percent) (Table 1). Nearly all locations (92 percent) offer street parking in addition to garage, driveway or lot spots. As with the residential areas, most of the theft locations in the commercial area consist of street parking (65 percent). Four incidents (13 percent) took place in parking garages, and seven (23 percent) were in parking lots.

Residential Area Variables

Residential Watchers

Information about the residential watchers variable is presented in Table 2. Several dichotomous variables were used to capture information about informal surveillance at the auto theft sites. The observation instrument included several measures of watchers that may provide informal surveillance. These data were used to create a Watchers Index. This index included information about whether people inside the dwelling near the auto theft location are able to see parked vehicles from inside the house/apartment (yes = 1, no = 0), whether

pedestrians can see where the car was located (yes = 1, no = 0), whether landscaping obstructs sight-lines, thereby making it more difficult to see the car (yes = 0, no = 1) and whether the dwelling is set back from the road (yes = 0, no = 1). This index was weighted to reflect the potential relative impact of each variable. The issue of whether pedestrians can see the car—and likely the car thief—was deemed the most important variable in this index, so a "yes" answer to this question was given four points. The next important variable was whether people inside the house or apartment near where the vehicle was stolen could see the vehicle, so an affirmative answer was given three points. If the landscaping was not obstructing the view of the automobile, two points were added. Finally, if the dwelling was *not* set back from the road, one point would be added to the index. The possible scores on the Watchers Index ranged from 0 to 10, with higher numbers indicating greater pedestrian and residential visibility of the vehicle location. The mean score was 4.32 (s.d. = 1.92).

Two additional indices were created to measure informal surveillance. The researchers measured the amount of pedestrian and vehicle traffic during day and night observations. For pedestrian traffic, the researchers coded light traffic as "1," moderate traffic as "2" and heavy traffic as "3." Light pedestrian traffic was defined as 0–5 people on that street segment during the 3-min observation period, moderate was 6–10 people, and heavy was at least 11 people in that area. Automobile traffic was measured in a similar fashion, with coding options ranging from 1 = light to 3 = heavy. Observations of vehicle traffic also took place during a 3-min period, and 0–5 non-parked vehicles were considered light, 6–10 were moderate and 11 + were heavy. The daytime pedestrian and vehicle traffic values were added together to create an index that had scores that varied from 2 to 6 (mean = 3.03; s.d. = 1.36). The same was done for nighttime pedestrian and vehicle traffic (mean = 2.86; s.d. = 0.93).

Residential Activity Nodes

The researchers identified five activity nodes in the residential areas: ATMs or payphones in sight of the auto theft locations, gas stations in sight, bars in sight, bus stop in sight and schools within two blocks. For the Activity Nodes Index, the aforementioned locations were all weighted evenly to reflect their potential influence in drawing people to that particular area. Locations with an ATM/payphone in sight, gas station in sight, bar in sight, bus stop in sight or school within two blocks received one point for each activity node present. The mean for the index was 1.28 (s.d. = 1.523).

Residential Location

The Location Index for the residential sites included six variables, and all were weighted in accordance with their potential impact on providing an attractive or unattractive location to commit auto theft. Theft sites with a design-related edge were given four points, sites with a converged alley or access in the back of the residence were given three points, two-way streets were given three points, sites on roads with a

four-way intersection at the end were given three points, locations with only street parking were given three points, locations with lot parking were given two points, a location with no residents or pedestrians hanging out during the observation period was given two points and sites with an apartment as the residence were given one point. Scores for Location Index range from 0 to 16, with a high score indicating a more attractive location for an auto theft to occur. The mean for this index was 9.72 (s.d. = 3.75).

Residential Lighting and Security

The quality of lighting at the theft sites was assessed by two researchers during the evening observations. Lighting was rated either as good "2," average "1" or poor "0." The mean for the quality of the lighting

Table 2 Descriptive Statistics for Residential Location in Atlantic City, NJ 2004–2005

	F (yes)	% (yes)	Mean	s.d.
Watcher variables				
People in dwelling can see car (yes = 1; no = 0)	14	38.9	—	—
Pedestrians can see car location (yes = 1; no = 0)	8	22.2	—	—
Landscaping obstructs sight of car (yes = 0; no = 1)	9	25.0	—	—
Dwelling set back from road (yes = 0; no = 1)	10	27.8	—	—
Watcher Index[a]	—	—	4.32	1.92
Pedestrian and vehicle traffic during the day[b]	—	—	3.03	1.36
Pedestrian and vehicle traffic during the night[b]	—	—	2.86	0.93
Activity nodes variables				
ATM or pay phone in sight	14	38.9	—	—
Gas station in sight	2	5.6	—	—
Bar in sight	9	25.0	—	—
Bus stop in sight	14	38.9	—	—
Schools	7	18.9	—	—
Activity Nodes Index[c]	—	—	1.28	1.523

(Continued)

Table 2 (Continued)

	F (yes)	% (yes)	Mean	s.d.
Location variables				
Design-related edge (yes = 1; no = 0)	9	25.0		
Covered alley/access in back of residence (yes = 1; no = 0)	6	16.7		
Type of residence				
Home/Duplex = 0	21	58.3		
Apartment = 1	15	41.7		
Type of street				
1-way = 0	16	44.4		
2-way = 1	20	55.6		
Type of road				
T (dead end) = 0	13	36.1		
+ (thru street) = 1	23	63.9		
Street parking	20	55.5		
Lot parking	6	16.7		
People hanging out	14	38.9		
Location index[d]			9.72	3.75
Lighting and security variables				
Quality of lighting: good = 2, avr = 1, poor = 0			0.64	0.68
Visible sign of security system = 0, no sign = 1			0.78	0.42

[a]Watchers Index = (pedestrian can see car = 4 points) + (people in dwelling can see car = 3 points) + (landscaping is NOT obstructing dweller's view of car = 2 points) + (dwelling is NOT set back from the road = 1 point).

[b]Traffic = (light = 1; moderate = 2; heavy = 3) Daytime and nighttime traffic totals = pedestrian traffic + vehicle traffic.

[c]Activity Nodes Index = (every ATM or payphone in sight = 1) + (bar in sight = 1) + (bus stop in sight = 1) + (gas station in sight = 1) + (location with a school within 2 blocks = 1).

[d]Location Index = (design-related edge = 4) + (covered alley/access in the back = 3) + (type of street = 3) + (type of road = 3) + (street parking = 3) + (lot parking = 2) + (absence of people hanging out = 2) + (type of residence = 1).

is 0.64 (s.d. = 0.68). Researchers took note of any signs of a security system at the dwelling where the theft was reported. Locations with some sign of a security system were coded "1," while those without were coded "0." The mean for this variable was 0.78 (s.d. = 0.42).

Commercial Area Variables

Commercial Watchers

The Watchers Index consisted of four variables (Table 3). Locations with businesses that provided a view of the vehicle from the front of the building were given four points, locations where pedestrians can see the vehicle from the front without an obstructed view were given three points, those with businesses that provided a view of the vehicle location from the back of the building were given two points and then one point was added to the index for sites where pedestrians can see the theft location from the back of the store. The highest possible score on the index was 10. The Watchers Index mean was 3.5 (s.d. = 4.32).

The pedestrian and vehicle traffic measures were coded in the same manner as they were in the residential area (1 = light traffic, 2 = moderate and 3 = heavy). Scores for pedestrian and vehicle traffic during the day were added, resulting in a mean of 3.87 (s.d. = 1.45). The mean score for evening foot and vehicle traffic was 4.45 (s.d. = 1.43). The heavier evening traffic reflects the fact that many of the commercial locations were near the part of the city with the casinos, shops and restaurants.

Commercial Activity Nodes

The commercial area included numerous activity nodes. This research was conducted in Atlantic City, and the largest attractions in the area are the casinos. Any theft location within two blocks of a casino or the major concert hall was given two points towards the Activity Nodes Index. Other activity nodes that were weighted with two points were the ATMs/payphones and convenience stores, due to their 24-hr a day access. Activity nodes that were weighted with one point were gas stations and schools since they are not always accessible. The mean for the commercial Activity Nodes Index was 1.71 (s.d. = 1.13).

Commercial Location

Seven variables were included in the commercial area Location Index. Areas with a design-related edge were given four points, four points for the theft taking

Table 3   Descriptive Statistics for Commercial Locations in Atlantic City, NJ 2004–2005

	F (yes)	% (yes)	Mean	s.d.
Watchers variables				
Business with view of the garage/lot/parking space from front	19	61.3		
Business with view of the garage/lot/parking space from back	18	58.1		
Pedestrians can see G/L/S from front w/o blocked view	22	71.0		
Pedestrians can see G/L/S from back w/o blocked view	18	58.1		
Watchers Index[a]			3.5	4.32
Pedestrian and vehicle traffic during the day[b]			3.87	1.45
Pedestrian and vehicle traffic during the night[b]			4.45	1.43
Activity nodes variables				
ATM or payphone in sight	21	67.7		
Gas station in sight	6	19.4		
Casinos or concert venue w/in 2 blocks	7	22.6		
Convenience store	15	48.4		
School	15	48.4		
Activity Nodes Index[c]	4	12.9	1.71	1.13

(Continued)

Table 3 (Continued)

	F (yes)	% (yes)	Mean	s.d.
Location variables				
Design-related edge (yes = 1; no = 0)	8	25.8		
Covered alley/access in back of business (yes = 1; no = 0)	9	29.0		
Enclosed garage/lot (yes = 0; no = 1)	6	19.4		
Type of parking				
Garage/lot = 0	11	35.5		
Single spots = 1	20	64.5		
Type of street				
1-way = 0	9	29.0		
2-way = 1	22	71.0		
Pedestrian access on foot (yes = 1; no = 0)	24	77.4		
People hanging out on block after 8pm or closing (yes = 0; no = 1)	27	87.1		
Location Index[d]			10.87	4.00
Lighting and security variables				
Visible security sign/stickers	7	22.6		
Security guard at exit or location	5	16.1		
Security cameras	8	25.8		
Security Index[e]			3.13	1.45
Lighting at night (Good = 2, Average = 1, Poor = 0)			1.39	0.62

[a]Watchers Index = (business with a view from the front = 4) + (pedestrians can see from the front of the business = 3) + (business with a view from the back = 2) + (pedestrians with a view from the back = 1).

[b]Traffic = (light = 1; moderate = 2; heavy = 3) Daytime and nighttime traffic totals = pedestrian traffic + vehicle traffic.

[c]Activity Node Index = (ATM or payphone in sight = 2) + (convenience store in sight = 2) + (casino within 2 blocks = 2) + (gas station in sight = 1) + (school within 2 blocks = 1).

[d]Location Index = (presence of a design-related edge = 4) + (enclosed lot or garage = 4) + (presence of a covered alley in the back of the building = 3) + (street parking = 3) + (two-way street = 3) + (pedestrian has foot access to spot = 2) + (absence of people hanging out at night = 1).

[e]Security Index = (no security signs/stickers = 2) + (no security cameras = 1) + (no security guards = 1).

place in a garage or enclosed lot, three points for the presence of a covered alley in the back of the building, three points for street parking, three points for a theft on a two-way street, two points if the lot or garage had easy access for pedestrians on foot and one point if there were no pedestrians hanging out in the area after dark. The highest possible score for the Location Index was 17. The mean was 10.87 (s.d. = 4.00).

Commercial Lighting and Security

The Security Index included information from three variables. Businesses where the thefts occurred received two points if they had no visible signs of a security system, one point if they lacked security cameras in the parking area and one point if the parking area lacked security guards. A higher score on the Security Index indicates low levels of security. The mean score for the Security Index was 3.23 (s.d. = 1.45). The measures for quality of lighting at night ranged from 0 = poor to 2 = good. The mean was 1.39 (s.d. = 0.62).

⬕ Analysis

Residential Locations—Bivariate and Multivariate Analyses

The results of the bivariate analysis for the residential locations are displayed in Table 4. The Watchers Index related to type of theft in that a higher score on this index was associated with repeat victimization locations ($r = 0.526$, $P < 0.01$). The presence of activity nodes also had a positive correlation with the type of location ($r = 0.471$, $P < 0.01$), meaning that sites with more activity nodes are more likely to be repeat victimization locations. There was also a correlation between lighting and the dependent variable ($r = -0.342$, $P < 0.05$), with single victimization locations being more likely to have better lighting. Finally, lack of visible signs of a security system on the dwelling was associated with repeat victimization locations rather than single-theft areas ($r = 0.431$, $P < 0.01$). Daytime car and pedestrian traffic, nighttime car and pedestrian traffic and the Location Index were not significantly related to the type of auto theft location.

The logistic regression model for residential auto theft is displayed in Table 5.

While several independent variables exhibited significant associations with the location of the thefts in the bivariate analysis, none were significant when included together in the logistic regression model.

Table 4 Bivariate Correlations for Residential Locations, Atlantic City, NJ 2004–2005

	(1)	(2)	(3)	(4)	(5)	(6)	(7)	(8)
Type of theft (1) (1 = single victimization; 2 = repeat victimization)	1							
Watchers Index (2)	0.526**	1						
Daytime car and pedestrian traffic (3)	0.268	0.198	1					
Nighttime car and pedestrian traffic (4)	−0.083	0.057	0.048	1				
Presence of Activity Nodes (5)	0.471*	0.451*	0.368*	−0.113	1			
Lighting (6) (good = 2; average = 1; poor = 0)	−0.342*	−0.236	−0.204	0.459**	−0.285	1		
Location Index (7)	0.245	0.084	−0.105	0.079	0.189	−0.040	1	
Sign of security system (8) (Yes = 0, No = 1)	0.431**	0.319	0.111	0.065	−0.277	0.287	0.008	1

**Correlation is significant at the 0.01 level (two-tailed).

*Correlation is significant at the 0.05 level (two-tailed).

Table 5 Multivariate Logistic Regression for Residential Location, Atlantic City, NJ 2004–2005

	B	S.E.	Sig.	Exp(B)
Watchers Index	0.729	0.446	0.103	2.072
Daytime car and pedestrian traffic	0.060	0.440	0.892	1.062
Nighttime car and pedestrian traffic	−0.359	0.613	0.558	0.698
Activity Nodes Index	0.417	0.461	0.366	1.517
Lighting (Poor = 0 → Good = 2)	0.495	0.913	0.587	1.641
Location Index	0.182	0.157	0.248	1.199
Signs of a security system (No = 0; Yes = 1)	−1.487	1.236	0.229	0.226
Constant	−4.508	3.289	0.170	0.011
$\chi^2 = 19.921$, d.f. = 7, $P = 0.006$	−2 log Likelihood = 29.28	Nagelkerke $R^2 = 0.580$		

Commercial Locations–Bivariate and Multivariate Analyses

Unlike the residential locations, the Watchers Index was not associated with the amount of victimization in the commercial locations (Table 6). The presence of activity nodes, however, was significant, with a higher score on the Activity Nodes Index being associated more with repeat victimization locations than single victimization locations ($r = 0.528$, $P < 0.01$). The Location Index is also significant ($r = 0.392$, $P < 0.05$), with more criminogenic locations associated with repeat victimization locations. Car and pedestrian traffic, quality of lighting and security indicators were not correlated with the type of theft location.

The results of the logistic regression model for commercial auto theft locations are displayed in Table 7. Only two variables remained statistically significant at the 0.05 level, the Activity Nodes Index and the Location Index. Locations that had more activity nodes were more likely to suffer repeat victimization ($B = 1.628$, $P = 0.05$). In commercial areas, repeat victimization locations were 5.10 times more likely to have criminogenic activity nodes.

Locations that received higher scores for being more attractive to potential car thieves, due to their design features, were more likely to be among the repeat victimization locations ($B = 0.479$, $P < 0.05$). In commercial areas, repeat victimization locations were 1.614 times more likely to have poor environmental designs than single victimization locations. The overall multivariate model is significant ($\chi^2 = 23.58$, d.f. = 7, $P = 0.001$) and explains a substantial amount of variance in the dependent variable (Nagelkerke $R^2 = 0.745$).

✂ Discussion

Watchers, Activity Nodes, Location, Lighting and Security have been measured and tested to determine their relationship to auto theft and many other types of crime. The ways in which these variables are measured, in this and other studies, may certainly have an impact on whether or not they achieve statistical significance. Bichler-Robertson and Potchak (2002) used a blended approach to crime analysis by incorporating community-level GIS analysis and a site-level survey to study commercial burglary. Indices were created for each of the S.T.E.A.L. variables: surveillability, target hardening,

Table 6   Bivariate Correlations for Commercial Locations, Atlantic City, NJ 2004–2005

	(1)	(2)	(3)	(4)	(5)	(6)	(7)	(8)
Type of theft (1) (1 = single victimization; 2 = repeat victimization)	1							
Watchers Index (2)	0.169	1						
Daytime car and pedestrian traffic (3)	0.308	−0.003	1					
Nighttime car and pedestrian traffic (4)	0.307	0.188	0.700**	1				
Presence of Activity Nodes (5)	0.528*	−0.097	0.280	0.145	1			
Lighting (6) (good = 2; average = 1; poor = 0)	−0.152	0.021	−0.020	−0.016	-0.264	1		
Location Index (7)	0.392*	0.091	−0.095	0.069	−0.104	−0.155	1	
Security Index (8) (higher score = less security)	0.326	−0.226	0.008	0.259	0.327	0.244	0.028	1

*Correlation is significant at the 0.05 level (two-tailed).

**Correlation is significant at the 0.01 level (two-tailed).

Table 7   Multivariate Logistic Regression for Commercial Locations, Atlantic City, NJ 2004–2005

	B	s.e.	Sig.	Exp(B)
Watchers Index	0.219	0.240	0.361	1.245
Daytime car and pedestrian traffic	1.125	0.825	0.172	3.081
Nighttime car and pedestrian traffic	−0.446	0.756	0.555	0.640
Activity Nodes Index	**1.628**	**0.831**	**0.050**	**5.095**
Lighting (Poor = 0 → Good = 2)	0.277	1.187	0.816	1.319
Location Index	**0.479**	**0.234**	**0.041**	**1.614**
Security Index (higher score = less security)	0.417	0.757	0.581	1.518
Constant	−12.934	5.071	0.011	0.000
$\chi^2 = 23.58$, d.f. = 7, $P = 0.001$	−2 log Likelihood = 16.312	Nagelkerke $R^2 = 0.745$		

Significant variables are given in bold.

edge effects, accessibility and liquidation potential in order to determine their impact on burglars' target selection preferences.

Levy (2006) created an index for each of the W.A.L.L.S. variables and used an Independent Samples *t*-test to determine if there were significant environmental differences between single and repeat victimization auto theft locations. In both residential and commercial areas one of the most widely studied environmental factors, activity nodes, was not found to have a statistically significant impact on frequency of victimization. Despite this, much of the crime prevention and pattern analysis literature has indicated that activity nodes present a multitude of opportunities for offenders. Roncek and Faggiani (1985) and Roncek and Lobosco (1983) found that in residential areas, houses with schools located within one block experienced an increase in property crime compared to locations that did not have a school on the block.

Research on commercial locations has identified similar patterns. Rossmo (1995) studied bars and the effects of simultaneous closing times. He found that when bars close simultaneously and dump patrons on the streets in drunk and agitated states, there is an increased chance these patrons will commit crime. Convenience stores were also found to contribute to robberies if they carried large amounts of cash and were situated in areas that had poor natural surveillance (Hunter and Jeffery, 1992).

In the current study, the number of activity nodes and the characteristics of the location of the vehicle were significant factors in predicting whether the site would be a single or repeat victimization location. Commercial areas with ATMs and payphones, schools, convenience stores, gas stations, or casinos or concert venues were also locations where more than one car was likely to be stolen during the 2-year study period. The activity nodes bring both offenders and non-offenders into these areas every day. It is during offenders' routine daily activities that they notice a particular site to be an attractive area to commit an auto theft. For juveniles, who are often perpetrators of auto theft, locations near schools, convenience stores, and in Atlantic City, near the casinos and boardwalk, are places where

they frequently visit. These areas are familiar, but not too familiar, to them. The locations are not in their immediate neighborhoods (since they live in residential areas), but they are close enough to travel there on foot or bicycle and then leave with a car. For auto theft perpetrators who are non-Atlantic City residents, it is not surprising that the activity nodes in the commercial areas are better able to predict repeat auto theft locations than in the residential areas. Visitors to the city typically gravitate to the boardwalk and a block or two off of the boardwalk; these areas are almost exclusively commercial. Public transportation in the form of buses and trains are available for people to travel to the city, with both the train station and bus terminal located in the commercial sections of the city. The commercial areas have plenty of cars along the streets, in lots, and in garages. Some of the locations are attractive, while others are not.

Brantingham and Brantingham (1993) may help to explain the significance of location in the current research. They discussed activity nodes by noting the importance of not only the nodes themselves but also the paths used to get to them and the edges that they create. These paths and edges mentioned by Brantingham and Brantingham (1993) were measured in the location variable. Type of street (one-way or two-way) and type of road (dead-end or thru street) were measured as dichotomous variables and included in the location index representing paths to the location. Edges were also measured dichotomously and included in the location index by determining if the location caused a design-related edge and if the location had covered alley access in the back.

Our finding that, for the commercial areas, the repeat victimization locations were more likely to have poor environmental designs and more criminogenic activity nodes than the single victimization locations echoes findings in other property crime research. Bichler-Robertson and Potchak (2002) found that when measuring surveillability, target-hardening, edge effects, accessibility, liquidation potential and awareness space, only the variables in the accessibility index (covered access in the back of business, alley that leads to the back of the premise, and buildings with more than two doors) were significantly more likely be located in hot commercial burglary zones. Camp (1968) found that the physical layout of the back of a location was an important target selection factor for bank robbers. Graham (2001) and Webb *et al.* (1992) also found that thieves considered escape routes during target selection processes for convenience store robberies and auto thefts, respectively.

The statistical significance of the location index in commercial areas, but not residential areas, is interesting. Perhaps the inconsistent results are due to the location of the research study. Prior auto theft research conducted in Lexington, Kentucky (Levy, 2006) found that in both residential and commercial locations, the environmental design of the location was significantly related to frequency of victimization. In residential areas of Lexington only 10.3 percent of cars in the sample were parked on the street (compared to 55.5 percent in Atlantic City) and 35.2 percent were parked in a lot (compared to only 16.7 percent in Atlantic City). Owing to the small sample size of this study ($N = 36$ residential locations) and the larger sample size in Lexington ($N = 68$ residential locations), it is possible that with a larger, more varied sample, the location index may be statistically related to victimization in Atlantic City, NJ.

With regard to the implications that this study has for crime prevention efforts, it is clearly important to examine the areas with many activity nodes and attractive locations for auto thieves. A good question to ask is how can the repeat victimization areas be made less attractive to potential offenders? There must be a reason why either the same offender has chosen to return to that site or that a second offender has approached that area and found it attractive.

The authors have been working with the Special Improvement District in Atlantic City to implement situational crime prevention techniques on the city streets. The first phase of this revitalization included new street lamps and flower boxes that served to improve the look and feel of the streets in Atlantic City. The second phase included a crime prevention audit of local stores on the main drag, Atlantic Avenue, in Atlantic City. The local stores on Atlantic Avenue account for the majority of the activity nodes (besides casinos) in Atlantic City. The audit revealed several crime prevention tactics were needed but also that many storeowners felt very disconnected from the others on the block. These feelings prevented them from sharing experiences and barred them from caring about the neighborhood. It was our sense that this disconnect reinforced the criminals' ability to successfully steal from many stores many times. The first action taken as a result of the audit included the development and implementation for redesign of the window displays and store aisles to create better visibility both in and outside of the store. The second phase was to foster a sense of neighborhood responsibility, encouraging the storeowners to work together and act as a community. The storeowners have since spent more time cleaning up the outside of their stores and sidewalks. Perhaps this behavior will create an environment in which auto thieves will feel more vulnerable.

Limitations and Suggestions for Future Research

The current study indicates a number of advances in the field of environmental criminology. However, there are three main suggestions for future research. First, since there has been limited research conducted on auto theft using site-surveys, the measurement for many of the variables has been adapted for research studies on other property crime. As Cornish and Clarke (1986) suggest, crime prevention techniques should be crime-specific. In this spirit we suggest that future research aim is to study auto theft in another location using a similar site-survey and measurement for the W.A.L.L.S. variables.

Second, continued research should be conducted in Atlantic City. By focusing on increasing the sample size and expanding the data collection study period, a longitudinal design could help to increase the validity of the findings as well as demonstrate and track changes in the environment. As the city continues to undergo a period of redevelopment and revitalization, it is imperative to determine how this will impact auto theft victimization, especially with regard to the environmental design of the high victimization areas. This longitudinal design would also serve as a replication of the W.A.L.L.S. indices, perhaps leading to re-specification of some W.A.L.L.S. variables.

Finally, the blending of a qualitative and quantitative design should be studied in other locations and in regard to other types of property crime. Bichler-Robertson and Potchak (2002) have used this "blended approach" to study commercial burglary. A blending of qualitative and quantitative methods is essential for site-level data collection that is inherently qualitative (when it seeks to capture the thought process of offenders as they are selecting targets) and quantitative (when it seeks to determine if the road is one-way or two-way and if the house has a garage or no garage). By combining these methodologies researchers can more clearly delineate the environmental characteristics that encourage crime from those that help to prevent it.

References

Bentham, J. (1948) *An Introduction to the Principles of Morals and Legislation.* New York: Kegan Paul.

Bichler-Robertson, G. and Johnson, M. (2001) *A Blended Approach to Examining Environmental Cues: Using GIS to Explore Environmental Factors Associated with Commercial Burglary.* Paper Presented at the Annual Meeting of the Western Society of Criminology, Portland, OR.

Bichler-Robertson, G. and Potchak, M. (2002) Testing the Importance of Target Selection Factors Associated with Commercial Burglary Using the Blended Approach. *Security Journal.* Vol. 15, No. 4, 41–61.

Block, R. L. and Block C. R. (1995) Space, Place and Crime: Hot Spot Areas and Hot Places of Liquor-Related Crime. In Eck, J. and Weisburd, D. (eds) *Crime and Place.* Monsey, NY: Criminal Justice Press, pp. 145–183.

Boba, R. (2001) Introduction. In Bair, S., Boba, R., Fritz., N., Helms, D. and Hick, S. (eds) *Advanced Crime Mapping Topics: Results of the First Invitational Advanced Crime Mapping Topics Symposium.* Denver, CO: National Law Enforcement and Corrections Technology Center.

Boesch Jr., W. (2001) In Bair, S., Boba, R., Fritz., N., Helms, D. and Hick, S. (eds) *Advanced Crime Mapping Topics: Results of the First Invitational Advanced Crime Mapping Topics Symposium.* Denver, CO: National Law Enforcement and Corrections Technology Center.

Bopp, W. J. (1982) The Effect of Lighting on Crime and Personal Injuries. In Schultz, D. O. and Service, J. G. (eds) *Security Litigations and Related Matters.* Springfield, IL: Charles C. Thomas Publishing, pp. 75–84.

Brantingham, P. J. and Brantingham, P. L. (1981) *Environmental Criminology.* Prospect Heights, IL: Waveland Press.

Brantingham, P. L. and Brantingham, P. J. (1993) Nodes, Paths and Edges: Considerations on the Complexity of Crime and Physical Environment. *Journal of Environmental Psychology.* Vol. 13, No. 3, pp. 3–23.

Brown, B. (1985) Residential Territories: Cues to Burglar Vulnerability. *Journal of Architectural Planning and Research.* Vol. 2, pp. 321–243.

Brown, B. and Bentley, D. L. (1993) Residential Burglars Judge Risk: The Role of Territoriality. *Journal of Environmental Psychology.* Vol. 13, pp. 51–61.

Buck, A., Hakim, S. and Regert, G. (1993) Burglar Alarms and Choice Behavior of Burglars: A Suburban Phenomenon. *Journal of Criminal Justice.* Vol. 13, pp. 497–507.

Camp, G. M. (1968) *Nothing to Lose: A Study of Bank Robbery in America.* Doctoral Dissertation, Yale University, New Haven, CT.

Cavan, S. (1966) *Liquor License: An Ethnography of Behavior.* Chicago, IL: Aldine Publishing.

Clarke, R. and Mayhew, P. (1994) Parking Patterns and Car Theft Risks: Policy Relevant Findings from the British Crime Survey. In Clarke, R. V. (ed.) *Crime Prevention Studies.* Vol. 4, Monsey, NY: Criminal Justice Press.

Cohen, L. E. and Felson, M. (1979) Social Change and Crime Rate Trends: A Routine Activity Approach. *American Sociological Review.* Vol. 44, No. 4, pp. 588–608.

Cook Jr., W. J., Merlo, A. V. and McHugh, T. (1993) Business Travel and Criminal Victimization. *Security Journal.* Vol. 4, No. 4, pp. 177–184.

Cornish, D. B. and Clarke, R. V. (1986) *The Reasoning Criminal.* New York: Verlag.

Crow, W. and Bull, J. L. (1975) *Robbery Deterrence: An Applied Behavioral Science Demonstration.* La Jolla, CA: Western Behavioral Sciences Institute.

Eck, J. E. (1994) *Drug Markets and Drug Places: A Case-Control Study of the Spatial Structure of Illicit Drug Dealing.* Dissertation, University of Michigan College Park.

Eck, J. and Wartell, J. (1999) *Reducing Crime and Drug Dealing by Improving Place Management: A Randomized Experiment.* Washington, DC: National Institute of Justice.

Engstad, P. A. (1975) Environmental Opportunities and the Ecology of Crime. In Silverman, R. A. and Teevan, J. J. (eds) *Crime in Canadian Society.* Toronto: Butterworth's.

Felson, M. (1986) Linking Criminal Choices, Routine Activities, Informal Control, and Criminal Outcomes. In Cornish, D. B. and Clarke, R. V. (eds) *The Reasoning Criminal: Rational Choice Perspectives on Offending.* New York, NY: Springer-Velar, pp. 119–128.

Felson, M. and Clarke, R. V. (1998) *Opportunity Makes the Thief: Practical Theory for Crime Prevention.* Police Research Series, Paper #98. London, UK: Research, Development and Statistics Directorate, U.K. Home Office.

Felson, M. and Cohen, L. E. (1980) Human Ecology and Crime: A Routine Activity Approach. *Human Ecology.* Vol. 8, No. 4, pp. 389–406.

Fisher, B. (1991) Neighborhood Business Proprietors' Reactions to Crime [in Minnesota]. *Journal of Security Administration,* Vol. 14, No. 2, pp. 23–54.

Fleming, Z. (1999) The Thrill of It All: Youthful Offenders and Auto Theft. In Cromwell, P. (ed.) *Their Own Words: Criminals on Crime* 2nd ed. Los Angeles: Roxbury, pp. 71–79.

Fleming, Z., Brantingham, P. L. and Brantingham, P. J. (1994) Exploring Auto Theft in British Columbia. In Clarke, R. V. (ed.) *Crime Prevention Studies*. Vol. 3, Monsey, NJ: Criminal Justice Press, pp. 47–90.

Gibbs, J. J. and Shelly, P. L. (1982) Life in the Fast Lane: A Retrospective View by Commercial Thieves. *Journal of Research in Crime and Delinquency*. Vol. 19, No. 2, pp. 299–330.

Graham, D. (2001) *Preventing Armed Convenience Store Robbery: A Fusion of Environmental and Social Strategies*. Unpublished document.

Graham, K., La Rocque, L., Yetman, R., Ross, T. J. and Guistra, E. (1980) Aggression and Barroom Environments. *Journal of Studies on Alcohol*. Vol. 41, No. 3, pp. 277–292.

Hawley, A. (1950) *Human Ecology: A Theory of Community Structure*. New York: Ronald.

Homel, R. and Clark, J. (1994) The Prediction and Prevention of Violence in Pubs and Clubs. In Clarke, R. V. (ed.) *Crime Prevention Studies*. Vol. 3, Monsey, NY: *Criminal Justice Press*, pp. 1–45.

Huang, W. S., Kwag, M. and Streib, G. (1998) Exploring the Relationship Between Hotel Characteristics and Crime. *FIU Hospitality Review*. Vol. 16, No. 1, pp. 81–93.

Hunter, R. D. and Jeffery, C. R. (1992) Preventing Convenience Store Robbery Through Environmental Design. In Clarke, R. V. (ed.) *Situational Crime Prevention: Successful Case Studies*. Albany, NY: Harrow and Heston.

Jacobs, J. (1961) *The Death and Life of Great American Cities: The Failure of Town Planning*. London: Peregrine Books.

Jacobs, J. (1993) *The Death and Life of Great American Cities*. New York: Random House Inc.

Jeffery, C. R. (1971) *Crime Prevention Through Environmental Design*. Beverly Hills, CA: Sage.

Jeffery, C. R., Hunter, R. D. and Griswold, J. (1987) Crime Prevention and Computer Analysis of Convenience Store Robberies in Tallahassee, Florida. *Security Systems*. August, pp. 33–40.

Jochelson, R. (1997) *Crime and Place: An Analysis of Assaults and Robberies in Inner Sydney*. Sydney, Australia: New South Wales Bureau of Crime Statistics and Research, Attorney General's Department.

Kuo, F. E. and Sullivan, W. (2001) Environment and Crime in the Inner City: Does Vegetation Reduce Crime? *Environment and Behavior*. Vol. 33, No. 3, pp. 343–367.

La Vigne, N. G. (1997) *Visibility and Vigilance: Metro's Situational Approach to Preventing Subway Crime*. Washington, DC: U.S. Department of Justice, Office of Justice Programs, National Institute of Justice, Research in Brief.

Levine, N. and Wachs, M. (1986) Bus Crime in Los Angeles: I—Measuring the Incidence. *Transportation Research A*. Vol. 20, No. 4, pp. 273–284.

Levy, M. P. (2006) *Place-Based Crime Prevention: Using Opportunity Structures and Environmental Characteristics to Estimate Crime*. Dissertation Abstracts International (UMI No. 3223500).

Levy, N. J. (1994) *Crime in New York City's Subways: A Study and Analysis of the Issues with Recommendations to Enhance Safety and the Public's Perceptions of Safety within the Subway System*. Albany. NY: NYS Senate Transportation Committee, Senate Transportation Committee/Legislative Commission on Critical Transportation Choices.

Light, R., Nee, C. and Ingham, H. (1993) *Car Theft: The Offender's Perspective*. London: HMSO Books.

Loukaitou-Sideris, A. (1999) Hot Spots of Bus Stop Crime. *Journal of the American Planning Association*. Vol. 65, No. 4, pp. 395–411.

Michael, S. E., Hull, R. B. and Zahm, D. (2001) Environmental Factors Influencing Auto Burglary: A Case Study. *Environment and Behavior*. Vol. 33, No. 3, pp. 368–388.

Nichols, W. W. (1980) Mental Maps, Social Characteristics, and Criminal Mobility. In Georges-Abeyie, D. E. and Harries, K. (eds) *Crime: A Spatial Perspective*. New York: Columbia University Press.

Painter, K. A. (1994) The Impact of Street Lighting on Crime, Fear, and Pedestrian Street Use. *Security Journal*. Vol. 5, pp. 116–124.

Perkins, D. D., Meeks, J. W. and Taylor, R. B. (1992) The Physical Environment of Street Blocks and Resident Perceptions of Crime and Disorder: Implications for Theory and Measurement. *Journal of Environmental Psychology*. Vol. 12, pp. 21–34.

Poyner, B. (1997) Situational Crime Prevention in Two Parking Facilities. In Clarke, R. V. (ed.) *Situational Crime Prevention: Successful Case Studies*. Monsey, NY: Willow Tree Press, pp. 157–166.

Roncek, D. W. and Bell, R. (1981) Bars, Blocks, and Crimes. *Journal of Environmental Systems*. Vol. 11, No. 1, pp. 35–47.

Roncek, D. W. and Faggiani, D. (1985) High Schools and Crime: A Replication. *Sociological Quarterly*. Vol. 26, No. 4, pp. 491–505.

Roncek, D. W. and Lobosco, A. (1983) The Effect of High School in Their Neighborhoods. *Social Science Quarterly*. Vol. 64, No. 3, pp. 599–613.

Rossmo, D. K. (1995) Overview: Multivariate Spatial Profiles as a Tool in Crime Investigation. In Block, C. R., Daboub, M. and Fregly, S. (eds) *Crime Analysis Through Computer Mapping*. Washington, DC: Police Executive Research Forum, pp. 65–98.

Rossmo, D. K. and Fisher, D. K. (1993) Problem-Oriented Policing: A Cooperative Approach in Mount Pleasant, Vancouver. *RCMP Gazette*. Vol. 55, No. 1, pp. 1–9.

Scott, M. S. (2002) *Robbery at Automated Teller Machines*. Problem-Oriented Guides for Police Series. No. 8. U.S. Washington, DC: U.S. Department of Justice, Office of Community Oriented Policing Services (COPS).

Smith, M. S. (1996) *Crime Prevention through Environment Design in Parking Facilities*. Washington, DC: U.S. Department of Justice, Office of Justice Programs, National Institute of Justice, Research in Brief.

Webb, B., Brown, B. and Bennett, K. (1992) *Preventing Car Crime in Car Parks*. Crime Prevention Unite Series: Paper No. 34. London: Home Office Police Department.

Woods, M. (2001) Site Mapping Technology and Crime Mapping. In Bair, S., Boba, R., Fritz, N., Helms, D. and Hick, S. (eds) *Advanced Crime Mapping Topics: Results of the First Invitational Advanced Crime Mapping Topics Symposium*. Denver, CO: National Law Enforcement and Corrections Technology Center.

Wright, R. and Logie, R. (1988) How Young Burglars Choose Targets. *The Howard Journal*. Vol. 27, No. 2, pp. 92–105.

DISCUSSION QUESTIONS

1. What do the findings of the paper suggest about what can be done to prevent repeat auto theft?

2. Would you expect that victimization rates for other types of offenses would be similarly high in the same areas? Why or why not?

3. How good are the W.A.L.L.S. indices as measures to predict auto theft? Explain. What else could be used to predict single and/or repeat auto theft?

4. Do you think these results would be generalizable beyond Atlantic City, New Jersey? Why or why not?

◈

Introduction to Reading 2

Much attention has been given to the fact that houses that are burglarized once are at increased risk of being burgled again. To address this risk, households can adopt security and other target-hardening measures that make it more difficult for the home to be burglarized. Hirschfield et al. (2010) evaluate the effectiveness of a widespread target-hardening program in the city of Liverpool, North West England, that was in place between July 2005 and December 2007. Their evaluation is unique in that it studied 1,739 properties to examine burglary and target hardening and determine whether target hardening was being employed in the most effective manner.

Linking Burglary and Target Hardening at the Property Level

New Insights Into Victimization and Burglary Protection

Alex Hirschfield, Andrew Newton, and Michelle Rogerson

Background

Research has identified patterns in the distribution of domestic burglary both at the area level and at the level of individual properties. Routine activities theory (Cohen & Felson, 1979) explains the role of the physical and social environment in the generation of spatial concentrations of burglary. For example, the interrelated variables of demographic composition, housing tenure, residential turnover, and levels of guardianship have been linked to spatial variations in the volume and location of crime (Bottoms & Wiles, 1988). These

SOURCE: Hirschfield et al. (2010). Reprinted with permission.

operate in conjunction with physical aspects of the environment such as street layout patterns and transport routes (Brantingham & Brantingham, 1993; Groff & La Vigne, 2001).

At the household level, the most significant finding has been the identification of repeat victimization and the discovery that prior victimization is the single best predictor of future victimization (Farrell, 2005; Pease, 1998). Repeat victimization is particularly prominent in high-crime areas (Trickett, Osborn, Seymour, & Pease, 1992). Offenders use available cues to discriminate between potential burglary targets. Research informed by rational choice theory (Cornish & Clarke, 1986) highlights offenders' use of cues that signal the likely value of goods within the property, the lack of occupancy, poor security, and low levels of natural surveillance (Coupe & Blake, 2006; Nee & Taylor, 2000; Wright, Logie, & Decker, 1995). Repeat victimization occurs partly because the factors that "flag" a property as a suitable target remain consistent over time and partly because the increased knowledge an offender gains during victimization "boosts" a property's status as a suitable target for the future (Tseloni & Pease, 2003). Victimization may also boost the vulnerability of neighboring properties for future burglary giving rise to "near-repeat" burglaries (Townsley, Homel, & Chaseling, 2003).

It is clear that a substantial and growing evidence base exists regarding the burglary problem, its manifestation, and spatial patterning. The evidence highlights the potential to reduce burglary through situational crime prevention and the manipulation of the risk, effort, and rewards as perceived by the would-be burglar (Clarke, 1997), and it advocates the prioritization of properties that have either already been burgled or that are in close proximity to victimized properties (Anderson, Chenery, & Pease, 1995; Chenery, Holt, & Pease, 1997). There is however a dearth of consistent and timely information on the manifestation of crime prevention policy, on the patterning of crime prevention activities, and more specifically, how these relate to crime and crime theories (Hirschfield & Newton, 2008). This article posits that patterns of crime prevention policy have been less thoroughly researched and questions whether our existing knowledge of crime patterns is sufficient to inform the delivery of crime prevention policy.

Lessons on the effectiveness of crime prevention measures have been derived from evaluation exercises that have sought to establish how far observed reductions in burglary can be attributed to prevention measures (Hamilton-Smith & Kent, 2005). These studies bring together information on burglary patterns and changes in these over time with "policy data" on burglary reduction interventions (Hirschfield, 2004). The evaluation of the Reducing Burglary Initiative in England and Wales indicated a significant positive relationship between the intensity of burglary reduction measures measured at the area level (i.e., the scale and timing of action on the ground) and the number of burglaries prevented (Bowers, Johnson, & Hirschfield, 2004b). This is indicative of an approach whereby information on burglary and that on crime prevention activity are effectively two data sets covering the same time periods and areas but not linked at the level of the individual property. This reflects a common dilemma in this field of research, namely, the absence of data that directly links the "problem," in this case details about burglary at a given property with the "treatment," namely, the target hardening installed at the property in question.

This article focuses on the analyses that can be conducted and new insights that can be gained when information on domestic burglary and target-hardening activity is linked at the level of the individual property. It draws on a recent policy evaluation that examined the impact of target hardening on domestic burglary in the City of Liverpool, North West England, between July 2005 and December 2007. *Target hardening* is a term used to describe the process of increasing the security of a property to make it more difficult to burgle, thereby increasing the effort needed by the offender to gain entry to a property. The intended outcome is ultimately to deter the offender from burgling an individual property. Target hardening has been employed internationally and has been widely cited as an effective strategy for burglary reduction (Hamilton-Smith & Kent, 2005; Hirschfield, 2004; Millie & Hough, 2004).

During the period analyzed, the community-safety partnership in Liverpool, Liverpool Citysafe,[1] installed target hardening at 1,739 properties, at a total cost of £911,715. Target hardening comprised the fitting of new doors and fitting of security chains to door and window locks, installation of alarms, the fitting of movement detection lighting, and fitting of security chains to doors.[2] Funding for target hardening was derived from several different funding streams but predominantly through Liverpool Citysafe's community-safety budget, the Housing Market Renewal Initiative (HMRI), and the Neighbourhood Renewal Initiative.[3] Each funding stream carried its own set of objectives for which target hardening was intended to meet. For example, one of the objectives of the HMRI funding was to encourage residents to remain in the area while regeneration took place around them. Some of the target hardening installed to prevent crime was not aimed primarily at burglary reduction but at domestic violence and criminal damage. To ensure coordination, all target hardening, regardless of funding source, was conducted by Liverpool Citysafe installers. Target hardening was installed at no cost to residents. An unknown number of residents declined the offer of target hardening (these details were not recorded). The majority of target hardening was rolled out across Liverpool one area at a time, the locations of which reflected the geographical boundaries of the variant funding streams (Newton, Rogerson, & Hirschfield, 2008).

An innovative feature of the evaluation of target hardening in Liverpool was the creation of longitudinal data on burglary and publicly funded target hardening for individual properties in Liverpool. This paved the way for a number of analyses that simply would not be possible using data just on domestic burglary or target-hardening records that had not been linked to burglary at the address level. For example, the burglary status of each target-hardened property (e.g., not burgled, burgled once, burgled twice or more) and the target-hardening status of each burgled property (e.g., target hardened, not target hardened) could be identified and linked. This could be broken down further by the timing and characteristics of the burglary and the nature, timing, and expenditure of each target-hardening episode. The extent to which all of this varied by location (e.g., within or outside of crime hot spots) could also be explored.

This article focuses on the opportunities that linked data provide for more in-depth and insightful analyses of the crime/intervention relationship. Linking burglary and target hardening can result in having the best of both worlds. All of the flexibility and advantages that come with disaggregate crime data remain, including the ability to identify repeat victimization, the opportunity to reveal spatial and temporal patterns in crime, and the option to aggregate burglary data to higher spatial levels (e.g., census zones, regeneration areas, different communities). In addition, there is the added value of being able to identify repeat episodes of target hardening, temporal and spatial patterns in target hardening, and to relate all of this to burglary at both the address and area level.

Data and Method

A range of data sets were obtained to undertake this research and to address the questions outlined above. Police force analysts provided data extracted from their crime information system on recorded domestic burglary (including burglary attempts). The data contained information on the date, time, location and modus operandi of each offence. Details were provided for 14,262 burglaries reported to Merseyside police during the 3-year period from January 2005 to December 2007.

Data on target hardening were provided by the local community-safety partnership, Liverpool Citysafe. The data set was essentially an administrative database used to record and monitor the 1,739 target-hardening installations conducted in the period July 2005 to December 2007. The target-hardening data included type of measures installed (for example, locks, bolts, and new doors), the amount spent, and the date of intervention. Additional information on administrative boundaries (e.g., census output areas, regeneration zones) and sociodemographic indicators was also made available.

The burglary records and the target-hardening data were joined together using the residential address of each property. The process was facilitated by the use of a geographic information system (GIS) and the

National Land Property Gazetteer, which contains all the addresses in the United Kingdom. Once the records were linked, the extent to which properties were target hardened on more than one occasion could be identified and this compared with the number of burglaries over a given time period. The time interval between burglaries could be calculated and any differences in this between target-hardened and non-target-hardened properties established. The extent to which these findings are consistent across different communities in Liverpool or vary by the intensity of crime hot spots could also be explored. The focus of this article is on the use of these data sets to examine the research questions outlined earlier.

Limitations to Data

There were a number of limitations to this research that must be borne in mind when considering the results. The analysis is based on secondary analysis of existing administrative data sets. Police data exclude burglaries not reported to the police, estimated at around 35% of all burglaries in England and Wales (Walker, Kershaw, & Nicholas, 2009). The target-hardening data only included publicly funded installations. The British Crime Survey estimates that between 96% and 98% of homes in England and Wales have some form of security device installed (e.g., deadlocks, window locks, security lights, or burglar alarms; Walker et al., 2009). It is clear prior to the target-hardening program, the majority of homes in Liverpool would have had some level of security. However, it was not possible to estimate the number of properties with existing target hardening or the level, adequacy, or appropriateness of that security. It is safe to assume that at least some of the "unprotected" properties would have had some degree of security.

The research was constrained by the use of a 2.5-year monitoring period. Both burglaries and target-hardening activity occurring before or indeed after this time period were not included in the analysis. Thus properties that were identified as having no prior burglaries may have been burgled before the monitoring period and having no prior burglaries may have been burgled before the monitoring period and indeed

thereafter. The truncation of data should be borne in mind in the consideration of the property profile categories (PPCs) described below.

Finally, the analysis reported here is concerned with the relationships between target hardening and burglary. However, burglary reduction was only one of the objectives underpinning the target hardening of properties in Liverpool. Thus, the analysis did not take into account the relationships between target hardening and other factors such as reductions in other types of crime, improvements in residents' quality of life, impacts on neighborhood satisfaction, and housing demand.

Construction of Property Profiles

Data on the presence or absence of burglary and/or (publicly funded) hardening at each property and the timing of these "events" were used to construct a series of PPCs. These were mutually exclusive groups defined by the trajectory of events affecting each property during the 2.5-year monitoring period and effectively classified properties according to their vulnerability to burglary, receipt of target hardening, and subsequent outcomes. The following PPCs were identified:

1. neither burgled nor target hardened;

2. burgled and never target hardened;

3. never burgled, target hardened, no subsequent burglary;

4. prior burglary, then target hardened, no subsequent burglary;

5. prior burglary, then target hardened, subsequently burgled; and

6. no prior burglary, target hardened, then subsequently burgled.

The first category accounted for the vast majority of properties in Liverpool (94%). Properties in this group were not included in the analysis as they did not appear in either the burglary or target-hardening data sets. The analyses concentrated on Categories 2 through 6.

In all categories other than Categories 2 and 3, both burglary and target-hardening episodes had occurred at the properties in question albeit in a different order. Category 2 identified properties that were vulnerable but never protected; Category 3 defined properties that had been protected but never victimized either prior to target hardening or subsequently (i.e., up to the end of the monitoring period). Of particular interest from a policy evaluation point of view was Category 4. This group provided some indication of the fate of properties with a burglary history that had been protected through target hardening. Categories 5 and 6, on the other hand, identified properties where target-hardening measures had not resulted in immunity to future victimization.

Interestingly, because the linked records contained data on the type of target hardening implemented and expenditure levels, the modus operandi of the burglary, and the dates of each burglary and target-hardening "event," it was possible to explore variations in these variables between each PPC. Thus, differences between PPCs in repeat victimization, the duration of intervals between burglaries, and concentrations of both burglary and target hardening within crime hot spots and other zones could all be explored. Initial analyses involved calculating the number and proportion of properties in each category and identifying levels of repeat victimization.

A GIS was used to generate a map of burglary hot spots on which the distribution of properties in each PPC could be plotted.[4] Thus, it was possible to visualize the distribution of properties in relation to hot spots generally and by the intensity of crime (i.e., burglary) within them. The GIS was also used to count the number of properties found in hot spots with different intensities of burglary. Thus, the extent to which properties in each PPC fell into the most intense high-crime areas could be established.

Results

The findings from the analyses reported here focus mainly on properties occupying each of the mutually exclusive PPCs. The number and percentage of properties falling into each PPC is displayed in two pie charts,

Figures 1 and 2, below. The first pie chart (Figure 1) expresses the number of properties in each PPC as a percentage of all monitored properties (i.e., either target hardened, burgled, or both). The second pie chart (Figure 2) shows the number and percentage of target-hardened properties occupying each PPC.

The first PPC (not shown in Figures 1 or 2) contains properties that were neither burgled nor target hardened during the monitoring period. This category

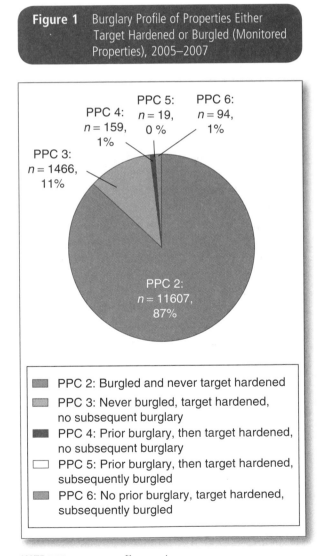

Figure 1 Burglary Profile of Properties Either Target Hardened or Burgled (Monitored Properties), 2005–2007

PPC 4: *n* = 159, 1%
PPC 5: *n* = 19, 0 %
PPC 6: *n* = 94, 1%
PPC 3: *n* = 1466, 11%
PPC 2: *n* = 11607, 87%

PPC 2: Burgled and never target hardened

PPC 3: Never burgled, target hardened, no subsequent burglary

PPC 4: Prior burglary, then target hardened, no subsequent burglary

PPC 5: Prior burglary, then target hardened, subsequently burgled

PPC 6: No prior burglary, target hardened, subsequently burgled

NOTE: PPC = property profile categories

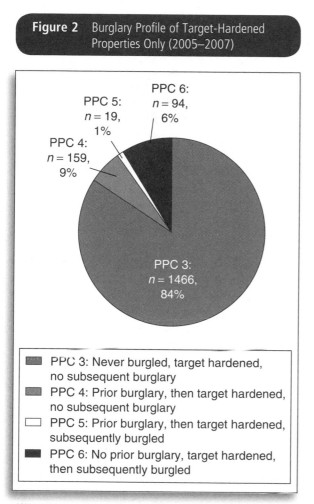

Figure 2 Burglary Profile of Target-Hardened Properties Only (2005–2007)

PPC 6:
$n = 94$,
6%

PPC 5:
$n = 19$,
1%

PPC 4:
$n = 159$,
9%

PPC 3:
$n = 1466$,
84%

▨ PPC 3: Never burgled, target hardened, no subsequent burglary

▨ PPC 4: Prior burglary, then target hardened, no subsequent burglary

☐ PPC 5: Prior burglary, then target hardened, subsequently burgled

■ PPC 6: No prior burglary, target hardened, then subsequently burgled

NOTE: PPC = property profile categories

represents the overwhelming majority (94%) of Liverpool's properties and highlights the fact that although domestic burglary is classified as a "high-volume crime," it still affects only a small proportion of the total housing stock. Inevitably, an unknown number of properties in this major category would have experienced burglaries that were not reported to the police. A further unknown number of properties will have been protected by privately funded security measures although, regrettably, data were not available on these.

The second largest category (PPC2, Figure 1) includes those properties that were burgled during the monitoring period but did not receive target hardening.

These 11,607 properties constituted 6% of all Liverpool properties and 87% of all monitored properties. Thus, the overwhelming majority of properties that had been victimized had not received publicly funded target hardening.

The question then arises as to which properties did receive target-hardening measures. Target-hardened properties were divided into four groups (PPCs 3 through 6, shown in Figure 2). Most of them (1,466 or 84% of the total) fell into PPC3 (namely, properties that were target hardened although they had not been burgled prior to the intervention nor subsequently, that is, up to the end of the monitoring period). PPCs 4 and 5 represent the 10% of monitored properties (178) that were burgled prior to target hardening. The 159 properties in PPC4 were free of burglary subsequent to target hardening. This sequence accords most closely with the desired outcomes of the target hardening program. Although representing just 9% of all properties target hardened, these 159 properties actually accounted for 60% of target-hardened and burgled properties indicating some degree of success for target hardening.

The two remaining PPCs identify properties that were burgled after they were target hardened. Neither sequence accord to the desired program outcomes. The larger of the groups, with 94 properties (6% of all those target hardened), was PPC6, namely, properties without burglary prior to target hardening that were subsequently victimized. The fate of these properties is counterintuitive, a reversal of the anticipated outcomes of situational crime prevention. Clearly, in this case, target hardening appears not to have been effective in preventing a burglary. Although this might have been the case, an alternative proposition, namely, that target hardening raised the risk of burglary is less tenable. The remaining unsatisfactory scenario is PPC5: properties that were burgled, target hardened and then burgled again. This was a much smaller group comprising just 19 properties.

Repeat Victimization

Future burglary risk increases with the number of prior burglaries experienced; this emphasizes the need to direct burglary prevention toward prior victims and repeat victims (Ashton, Brown, Senior, & Pease, 1998; Everson & Pease, 2001; Farrell, 2005). Targeting of repeatedly victimized properties was lower than would be expected. Of the 1,663 homes that experienced two or more burglaries over the analysis period, only 82 (5%) were target hardened. Table 1 provides a breakdown of the level of repeat victimization in PPC2, those 11,607 properties that experienced burglary but had not been protected with target hardening during the monitoring period. Within this group, there were 1,581 properties that had been repeatedly victimized with over 300 burgled three or more times. The research literature suggests that these would be the most appropriate candidates for protection through target hardening and yet, as shown in Table 2, only a small fraction of target-hardening resources were directed at such properties.

The prior burglary histories of the 1,738 target-hardened properties are provided in Table 2. Of these, just under 89.8% (1,560) had no prior burglary. It is possible that some of the latter may have been burgled just prior to the monitoring period and/or may have been targeted due to their close proximity to other burglaries. Most of the target hardening covering burgled properties went to those burgled just once. Only 2% (3) of target-hardened properties had experienced repeat victimization. Therefore, while 87% of burgled properties had not been target hardened, just under

Table 2 The Burglary Status of Target-Hardened Properties Prior to Target Hardening

	Target-Hardened Properties (n)	Target-Hardened Properties (%)
No prior burglary	1,560	89.8
One prior burglary	141	8.1
Two or more prior burglaries	37	2.1
Total	1,738	100.0

90% of target-hardened properties had not been burgled previously. It is somewhat surprising that such a small proportion of target hardening was directed toward properties with a prior history of burglary given that this group is the most vulnerable to subsequent burglary. Importantly, these statistics would not have emerged without linking burglary addresses to those of target-hardened properties.

Clearly, information beyond the mere sequence of events is needed to interpret the different scenarios represented by the PPCs. Additional insights can be gained by examining the location of the properties (low- vs. high-crime areas), the proportion of attempted versus successful burglaries, changes in offender modus operandi, and differences in outcome by type of target hardening. Other factors such as the responsible use of target hardening and how well maintained the equipment is over time, although important, were beyond the remit of this research.

Some of these factors are examined briefly, below. However, the main point to be emphasized here is that linking burglary and target-hardening data at the individual address level generates valuable insights not only about the efficacy of resource allocation decisions but also about vulnerability, risk and outcomes for individual properties.

High-Crime Areas

The inclusion of a range of geographical codes for each property afforded the opportunity to explore

Table 1 Repeat Burglaries in Group 2: Burgled Never Target Hardened (2005–2007)

Number of Burglaries	Number of Properties	Percentage	Cumulative Percentage
1	10,026	86.4	86.4
2	1,218	10.5	96.9
3	226	1.9	98.8
4	77	0.7	99.5
5 plus*	60	0.5	100.0
Total	11,607	100.0	100.0

*15 of these properties contained 'multiple addresses'

relationships between victimization, target hardening, and the spatial concentration both of burglary and target-hardening activity. Of particular interest, in this context, is the relationship between victimization, target hardening, and outcomes at the property level represented by the PPCs and geographical variations in the concentration of crime.

Disaggregate burglary data were used to define burglary hot spots of varying intensity within Liverpool. Although hot spots can change in location and intensity over time, the hot spots identified remained relatively stable over the period. A hot-spot intensity score was assigned to each monitored property based on its proximity to the hot-spot areas, and properties were assigned into quintiles according to the level of burglary risk at their location. The concentration of properties in burglary hot spots by PPC for the entire monitoring period (July 2005 to December 2007) is shown in Table 3.

Burgled properties overall were categorized into five equal groups from lowest risk of burglary to highest risk of burglary based on burglary hot spots. For properties "burgled and target hardened" during the monitoring period (irrespective of the order of events), almost two thirds of these were situated in the two highest area risk categories for burglary. Forty percent of those "burgled, target hardened, and then burgled again" (PPC5) were in the most intense burglary hot spots, and 33% of those "not burgled, target hardened, and subsequently burgled" (PPC6) were also in this highest risk areas. However, a far larger proportion (86%) of the properties that had been target hardened despite not having been burgled (PPC3) were also located in hot spots with the highest burglary risk (i.e., Quintiles 4 and 5). This possibly accounts for why they were target hardened even though they had not been victimized.

The message emanating from this analysis is that there was a clear area-based dimension underpinning the allocation of target hardening. This was strongest for properties target hardened without any prior or subsequent burglary (PPC3) followed by those previously burgled, target hardened, and then free from burglary (PPC4). In both cases, properties were in the most intense burglary hot spots where, arguably, the risks of being targeted by offenders were greatest.

Table 3   Percentage of Properties by Profile and Burglary Hot-Spot Risk (2005–2007)

Property Profile Category		Hot Spots Lowest risk ⟵⟶ highest risk					N
		1	2	3	4	5	
All properties burgled		20.0	20.0	20.0	20.0	20.0	15,088
2	Burgled and never target hardened	21.5	21.0	20.2	19.4	18.0	11,540
3	Never burgled, target hardened, no subsequent burglary	2.2	3.5	8.1	32.1	54.1	1,466
4	Prior burglary, then target hardened, no subsequent burglary	6.3	7.5	14.5	32.1	39.6	159
5	Prior burglary, then target hardened, subsequent burglary	10.0	15.0	20.0	15.0	40.0	20
6	No prior burglary, target hardened, then subsequently burgled	7.4	16.0	20.2	23.4	33.0	94
All properties burgled and target hardened		7.0	11.0	16.8	27.8	37.4	273

Viewed in isolation from this spatial context, the decision to concentrate so much of the target hardening on nonvictimized properties might be regarded as questionable and unjust. However, with the knowledge that 86% of the properties were, in fact, located in Liverpool's highest crime hot spots, the decision makes more sense. A closer analysis of the spatial distribution of target hardening, reported in an earlier article, revealed a strong concentration of activity in Liverpool's Neighbourhood Renewal Areas and other regeneration zones (Newton et al., 2008). This reflected the fact that target hardening was seen as a means of boosting neighborhood satisfaction and as a way of strengthening housing demand in deprived communities as well as a set of interventions for reducing burglary. The allocation of target hardening has focused attention on high-crime areas rather than individual high-risk properties. The following section considers the impact that this patterning of crime prevention activity has had on the effectiveness of the intervention.

The Cost Effectiveness of Target Hardening

The effectiveness of target hardening was assessed by comparing the incidence of burglary in target-hardened properties before and after installation with the incidence of burglary for comparable periods in non-target-hardened home. Estimates of "expected burglaries" for the period following target hardening would change in the same way as in the non-target-hardened properties (for more details on this methodology, see Bowers, Johnson, & Hirschfield, 2004a; Johnson et al., 2004). The availability of individual property-level data offers the potential to refine this method considerably by enabling anticipated changed in crime to be applied specifically to those properties that have been target hardened rather than to all properties within an area, many of which would not have been protected.

Over the monitoring period, the total burglaries observed in the target-hardened properties were 163 and were fewer than the number of burglaries expected based on the change in the non-target-hardened properties. This suggests a small net reduction in burglary.

However, given the £917,761 investment in target hardening and the estimated value of the prevented burglaries of £527,128 (Duborg, Hamed, & Thorns, 2005), the cost benefit ratio for the initiative over the period was 0.57, meaning that for every £1 invested in target hardening only 57p was returned in benefits. At less than one, this ratio indicates that the target-hardening initiative had failed to break even. Importantly, analysis of burglary change and cost effectiveness was repeated for each of the 30 electoral wards within Liverpool revealing notable variations. The cost benefits of target hardening varied dramatically across the city from 61.2 to −14.9 (with a mean of 15). Initial investigations into the factors underlying this variation have pointed to the role of targeting. As shown in Table 4, the greatest cost benefits returns were found neither in the wards with the highest rate of, or spend on, target hardening nor in the wards with the highest level of burglary, but in those wards in which the greatest proportion of target-hardening activity was directed toward properties with a prior history of burglary. This strong positive correlation was highly significant. This analysis, albeit preliminary, suggests that while across Liverpool there was a close alignment between areas of crime prevention activity and high burglary, this does not guarantee the targeting of those properties at greatest risk, and it is at the individual level that targeting appears to be most important to the effectiveness of the intervention.

Further Insights Into Burglary and Target Hardening

Further light can be shed on both the nature of the burglary event and the performance of target hardening by scrutinizing offender modus operandi, the balance between successful and attempted burglaries, and the impact of different interventions across the PPC. It is possible that burglaries following a target-hardening event could either be unsuccessful attempts (demonstrating that target hardening has been successful) or committed through insecure doors or windows (because target hardening may not have been operational at the time of the burglary). However, these explanations were not borne out by the analysis.

Table 4 Correlations Between Target-Hardening Activity and Returned Cost Benefit Ratios Measured at the Electoral Ward Level (2005–2007; $n = 30$)

Target Hardening	M	SD	Correlation[a] With Cost Benefit Ratio	
Total number of properties target hardened	57.9	102.3	−.279	ns
Target-hardening rate per 1,000 properties	2.5	4.3	−.183	ns
Cost of target hardening per 1,000 properties (£)	3.8	8.2	−.379	*
Burglary rate per 1,000 properties	23.8	6.8	−.224	ns
% target-hardened properties with prior burglary	21.3	18.8	+.681	**

NOTE: ns = correlation is not significant

a. Spearman rho.

$*p < .05$ (two-tailed). $**p < .01$ (two-tailed).

Although 15% (25) of burglaries occurring subsequent to target hardening (PPCs 5 and 6) were attempted burglaries, this is equivalent to the overall proportion of attempted burglaries across Liverpool. Therefore, the proportion of attempted burglaries did not appear to be higher for target-hardened properties that were subsequently victimized.

The proportion of distraction burglaries among PPCs 5 and 6 (less than 1%) was equivalent to that for the rest of Liverpool. This negates the possibility that burglaries following target hardening occurred as a result of a tactical shift to distraction burglary. Of the 162 burglaries occurring at properties that had previously been target hardened, (PPCs 5 and 6) 20% (32) were committed via insecure doors or windows, which is only marginally lower than the proportion of insecure burglaries across Liverpool (22%).

There were few discernable difference in the level or variety of target-hardening measures received between target-hardened properties that were subsequently burgled and those that were not. Properties in the four PPCs that were target hardened received an average of two to three different types of products. No significant association was found between the number of target-hardening product types and the occurrence of a repeat burglary. The same was true with respect to

the total number of burglaries following interventions. However, as shown in Table 5 below, those properties suffering a burglary following target hardening (PPCs 5 and 6) were less likely to have received passive infra-red (PIR) lighting than properties not suffering a subsequent burglary (PPCs 3 and 4). However, the numbers in these categories were small and the difference was not statistically significant.

Linking Burglary and Target Hardening: Policy Challenges

By linking the analysis of crime and crime prevention patterns, this article has identified that despite a close alignment between crime and crime prevention activity at an area level, only a minority of the properties at highest risk of burglary were provided with target hardening. No victimized properties appear to have been target hardened due to their proximity to intense burglary hot spots; however, the success of this strategy will necessarily be limited if the victimized and repeat-victimized properties contributing to high-burglary rates are not themselves protected. This argument is supported by the evidence that cost effectiveness was improved when the intervention was targeted at the property level. Further investigation is needed to

Table 5 Profile Status by Proportion of Properties Receiving PIR Lighting (2005–2007)

Property Profile Category		Receiving PIR Lighting (n)	Receiving PIR Lighting (%)
Target hardened, no subsequent burglary	PPC3: Never burgled, target hardened, no subsequent burglary	1,065	73
	PPC4: Prior burglary, then target hardened, no subsequent burglary	84	53
Target hardened, subsequently burgled	PPC5: prior burglary, then target hardened, subsequently burgled	5	25
	PPC6: No prior burglary, target hardened, then subsequently burgled	22	23

NOTE: PIR= Passive Infra-Red Sensor lighting that comes on when motion is detected; PPC = property profile categories.

understand the dynamics of targeting and the competing benefits of property versus area-based targeting.

Findings from the evaluation suggested possible reasons why practitioners favored area-based targeting. First, area-based implementation is more straightforward to plan, timetable, and coordinate proving logistically more straightforward and more economical to implement. Second, to target at-risk properties, they must first be identified and located, a more complex task than the identification of high-crime areas. The profiles presented in this article are the culmination of considerable data processing and cross-referencing involving the joining of target-hardening and burglary data and their integration with sophisticated spatial analysis software. The availability of data and the ability of practitioners to make the best use of information to inform their targeting decisions may be a factor in the targeting strategies adopted. The effectiveness of property-level targeting will hinge on the extent and nature of data sharing, data quality, and the availability of skills and software. Finally, burglary reduction was only one of the policy objectives for which target hardening was implemented. The various policy objectives were not necessarily in competition; to the contrary, there is an apparent synergy to the policy aims of crime reduction, neighborhood regeneration, and neighborhood satisfaction. However, attempting to combine these priorities may have diluted the impact on burglary reduction.

Conclusions

Research on burglary and target hardening is not generally conducted at the level of the individual property and certainly not for large numbers of properties spanning a city the size of Liverpool. Thus, individual properties have not been the primary units of analysis for exploring relationships between burglary and target hardening. This article has demonstrated not only that analysis at the level of the individual property is possible but also that by doing so new insights can be gained about burglary risk, the presence or absence of protective measures, and burglary outcomes over selected time periods. Thus, it has been possible to generate new data about the burglary status of target-hardened properties and about the target-hardening status of burgled properties. This has been explored by the frequency of victimization and according to the extent to which properties were located in burglary hot spots differentiated by the intensity of crime found within them.

Some of the results were unexpected. For example, a large number of properties that were repeatedly burgled were not protected by target hardening; conversely, many of those that were protected by target hardening had not been burgled, and many of the nonvictimized properties that were target hardened were located in the most intense crime hot spots where the risks of victimization were greatest. Thus, although target hardening predominantly took place in high-crime areas, this did not guarantee the protection of the highest risk properties within these areas. Of concern is the finding that the strategies for allocating target hardening may have diluted the cost effectiveness of the program in reducing burglary. None of this would have been apparent without linked data on burglary and target hardening for individual households.

There is considerable scope to broaden the range of analyses reported in this article. Geographically referenced, individual-level property data pave the way for even more ambitious analyses. For example, the interval between burglaries for individual properties can be considered in the same way as periods of desistance from crime of individual offenders. This raises the prospect of applying survival analysis used in recidivism studies to evaluate the impact of intensive supervision and other interventions on desistence from crime to compare intervals between burglaries for protected and nonprotected properties. The results from experimental research, that has already been undertaken in this area (Bowers, Lab, & Johnson, 2008), is encouraging and suggests that this might be a viable approach for evaluating the impact of target hardening in both preventing and delaying victimization.

This article has focused on the geographical alignment of the risk of crime and crime prevention activities. However, risk may not be the only concern of policy makers. Millie (2008) and Ekblom and Sidebottom (2008) have highlighted the need to consider the risk, vulnerability, and impact of crime separately. This raises questions regarding how to prioritize target hardening based on the likely risk and potential consequences of burglary, should properties with the highest risk of burglary be targeted or those residents deemed to be more vulnerable to the consequence of burglary, or some combination of or all of these factors. It is apparent that analysis of the crime patterns in isolation from the policy environment can result in the prescription of crime prevention solutions that are not practical, desirable, or effective in practice. Moving forward requires the integration of crime theory with an improved understanding of the practicalities of delivery and the wider policy goals of crime prevention.

✉ Notes

1. In England and Wales, the Crime and Disorder Act 1998 and the Police Reform Act 2002 place a duty on specific agencies to form partnerships within the community and to tackle crime, disorder, and misuse of drugs. This partnership in Liverpool is known as Citysafe.

2. Target hardening is used to refer to all measures covered by the Liverpool Citysafe "Target-Hardening Strategy." The authors acknowledge that some of these improvements, for example lighting, may not be viewed as strictly target-hardening strategies as they do not improve the actual physical toughness of a property.

3. The Housing Market Renewal Initiative is a national program aiming to rebuild housing markets and communities in areas where demand for housing is weak and which have seen a significant decline in population, dereliction, poor services, and poor social conditions as a result. The neighborhood Renewal Fund is a nonprescriptive grant, which has been made available to England's most deprived areas.

4. The hot-spot analysis technique employed was Kernel Density Estimation (KDE) interpolation, using CrimeStat 3 (http://www.icpsr.umich.edu/CRIMESTAT/). See Newton, Rogerson, and Hirschfield (2008) for further discussion of the methodology used.

✉ References

Anderson, D., Chenery, S., & Pease, K. (1995). *Biting back: Tackling repeat burglary and car crime* (Crime Detection and Prevention Series No. 58). London: Home Office.

Ashton, J., Brown, B., Senior, B., & Pease, K. (1998). Repeat victimization: Offender accounts. *International Journal of Risk, Security and Crime Prevention, 3,* 269–279.

Bottoms, A., & Wiles, P. (1988). Crime and housing policy: A framework for crime prevention analysis. In T. Hope & M. Shaw (Eds.), *Communities and crime reduction* (pp. 84–98). London: HMSO.

Bowers, K. J., Johnson, S. D., & Hirschfield, A. (2004a). Closing off opportunities for crime: An evaluation of alley-gating. *European Journal on Criminal Policy and Research, 10,* 283–308.

Bowers, K. J., Johnson, S. D., & Hirschfield, A. (2004b). The measurement of crime prevention intensity and its impact on level of crime. *British Journal of Criminology, 44,* 419–440.

Bowers, K. J., Lab, S. P., & Johnson, S. D. (2008). Evaluating crime prevention: The potential of using individual-level matched control designs. *Policing, 2,* 218–225.

Brantingham, P., & Brantingham, P. (1993). Environmental routine and situation: Towards a pattern theory of crime. *Advances in Criminological Theory, 5,* 259–294.

Chenery, S., Holt, J., & Pease, K. (1997). *Biting back II: Reducing repeat victimization in Huddersfield* (Crime and Prevention Series No. 82). London: Home Office.

Clarke, R. V. (1997). *Situational crime prevention: Successful case studies.* Monsey, NY: Criminal Justice Press.

Cohen, L. E., & Felson, M. (1979). Social change and crime rate trends: A routine activities approach. *American Sociological Review, 44,* 588–608.

Cornish, D. B., & Clarke, R. V. (1986). *The reasoning criminal.* New York: Springer Verlang.

Coupe, R. T., & Blake, L. (2006). Daylight and darkness targeting strategies and the risks of being seen at residential burglaries. *Criminology, 44,* 431–463.

Duborg, R., Hamed, J., & Thorns, J. (2005). *The economic and social costs of crime against individuals.* London: Home Office.

Ekblom, P., & Sidebottom, A. (2008). What do you mean, "Is it secure?" Redesigning language to be fit for the task of assessing the security of domestic and personal electronic goods. *European Journal of Criminal Policy Research, 14,* 61–87.

Everson, S., & Pease, K. (2001). Crime against the same person and place: Detection opportunity and offender targeting. In G. Farrell & K. Pease (Eds.), *Repeat victimization* (pp. 199–220). Monsey, NY: Criminal Justice Press.

Farrell, G. (2005). Progress and prospects in the prevention of repeat victimization. In N. Tilley (Ed.), *Handbook of crime prevention and community safety* (pp. 143–170). Collumpton, UK: Willan.

Groff, E., & La Vigne, N. (2001). Mapping an opportunity surface of residential burglary. *Journal of Research in Crime and Delinquency, 38,* 257–278.

Hamilton-Smith, N., & Kent, A. (2005). Preventing domestic burglary. In N. Tilley (Ed.), *Handbook of crime prevention and community safety* (pp. 415–457). Collumpton, UK: Willan.

Hirschfield, A. (2004). *The impact of the Reducing Burglary Initiative in the north of England* (Home Office Online Report 40/04). London: Home Office.

Hirschfield, A., & Newton, A. (2008). The crime-crime prevention relationship: A Manchester case study. *Built Environment, 34,* 104–120.

Johnson, S. D., Bowers, K. J., Jordan, P., Mallender, J., Davidson, N., & Hirschfield, A. (2004). Estimating crime reduction outcomes: How many crimes were prevented? *Evaluation, 10,* 327–348.

Millie, A. (2008). Vulnerability and risk: Some lessons from the U.K. Reducing Burglary Initiative. *Police Practice & Research, 9,* 183–198.

Millie, A., & Hough, M. (2004). *Assessing the impact of the Reducing Burglary Initiative in Southern England and Wales.* Home Office Online Report 42/24, London: Home Office.

Nee, C., & Taylor, M. (2000). Residential burglary in the Republic of Ireland: A situational perspective. *Howard Journal of Criminal Justice, 27,* 105–116.

Newton, A. D., Rogerson, M., & Hirschfield, A. (2008). Relating target hardening to burglary risk: Experiences from Liverpool. *Papers from the British Criminology Conference: An Online Journal Published by the British Society of Criminology, 8,* 153–174.

Pease, K. (1998). *Repeat victimization: Taking stock* (Crime Prevention & Detection Paper 90). London: Home Office.

Townsley, M., Homel, R., & Chaseling, J. (2003). Infectious burglaries: A test of the near repeat hypothesis. *British Journal of Criminology, 43,* 615–633.

Trickett, A., Osborn, D. R., Seymour, J., & Pease, K. (1992). What is different about high-crime areas? *British Journal of Criminology, 32,* 18–89.

Tseloni, A. M., & Pease, K. (2003). Repeat personal victimization: "Flags" or "boosts"? *British Journal of Criminology, 43,* 196–212.

Walker, A., Kershaw, C., & Nicholas, S. (2009). *Crime in England and Wales 2008/9.* London: Home Office.

Wright, R. T., Logie, R. H., & Decker, S. H. (1995). Criminal expertise and offender decision making: An experimental study of target selection process in residential burglary. *Journal of Research in Crime and Delinquency, 32,* 39–53.

DISCUSSION QUESTIONS

1. One of the findings was that targeting properties that had prior burglary risk would be most effective in burglary intervention. How important is this finding for victimization reduction programming overall?

2. What do these findings suggest about the theoretical explanations for repeat victimization: the flag and boost explanations identified by Hirschfield et al. that are similar to risk heterogeneity and state dependence?

3. What are your impressions of target hardening as a prevention strategy for burglary? What else can be done to prevent burglary victimization? Is target hardening enough? Why or why not?

Introduction to Reading 3

People can also be victimized online by being defrauded. Specifically, consumers can be targeted via online auctions, through e-mail solicitations, purchases, and fraudulent investment deals, among others. Pratt and colleagues (2010) seek to investigate whether personal characteristics or routine activities theory explains why some people become targets of online fraud. To do so, telephone surveys of 922 adults in the state of Florida were completed from December 21, 2004, to January 25, 2005. They discuss how their findings can inform crime prevention.

Routine Online Activity and Internet Fraud Targeting

Extending the Generality of Routine Activity Theory

Travis C. Pratt, Kristy Holtfreter, and Michael D. Reisig

Consumer fraud is a form of economic crime involving the "intentional deception or attempted deception of a victim with the promise of goods, services, or other benefits that are nonexistent, unnecessary, were never intended to be provided, or were grossly misrepresented" (Titus 2001:57). The Federal Trade Commission (FTC) reports that nearly one third of American adults believe that they are targeted by fraudsters annually, with 11.2 percent of the population experiencing victimization. According to the FTC, the median financial loss resulting from fraud victimization is approximately $220 per victim (Anderson 2004). The estimated annual financial loss resulting from consumer fraud is quite high—approximately $680 million (FTC 2006). Some evidence suggests that the number of victims and costs of fraud to society exceed the figures of those victimized by serious street crime (Moore and Mills 1990; Titus 2001). Although consumer fraud scams take many forms, fraud perpetrators are increasingly using the Internet to attract suitable targets (Grabowsky and Smith 1998; Newman

and Clarke 2003; Wall 2005, 2007a; Wilson et al. 2006; Yar 2006).

The creation and expansion of the Internet has given offenders an unlimited range of time to make contact with potential victims (Newman and Clarke 2003; Wall 2007a; Yar 2006). The Federal Bureau of Investigation (2001) defines Internet fraud as:

> any fraudulent scheme in which one or more components of the Internet, such as web sites, chat rooms, and e-mail, play a significant role in offering nonexistent goods or services to consumers, communicating false or fraudulent representations about the schemes to consumers, or transmitting victims' funds, access devices, or other items of value to the control of the scheme's perpetrators.

Government and media accounts suggest that victim reports of Internet fraud are increasing (Internet Crime Complaint Center 2006), yet Internet use continues to rise, and consumers now perform many of their daily activities online (e.g., banking, communication, and shopping; see Fletcher 2007).

Despite the prevalence of Internet fraud targeting and victimization, theoretically informed empirical research in criminology has emerged only recently. Much of the empirical literature has focused on the sociodemographic characteristics of fraud targets/victims (see e.g., the discussion by Holtfreter, Reisig, and Pratt 2008). These issues have been empirically examined by researchers from other academic disciplines (e.g., marketing, advertising, and information technology). Because of their different substantive interests, these researchers may have little or no understanding of theoretical perspectives regarding criminal victimization in general, let alone offense-specific explanations of targeting and victimizations. As a result, such research ultimately offers little insight into the causal mechanisms that underlie fraud targeting and victimization (i.e., *why* consumers with certain sociodemographic characteristics are more likely to be targeted for victimization). As a response to this problem, a growing body of criminological research has focused on Internet-related victimization (e.g., credit card fraud, cyber bullying, and identify theft, among others; Grabowsky and Smith 1998; Levi 1998; McNally and Newman 2008; Newman and Clarke 2003; Wall 2005, 2007a; Yar 2006). While our understanding of the Internet victimization context has increased, additional insight into offense-specific explanations of targeting would improve crime prevention efforts (Newman and Clarke 2003).

To that end, the present study extends the literature on criminal victimization to the study of Internet fraud targeting. In particular, Cohen and Felson's (1979) routine activity theory predicts that aggregate changes in legitimate opportunity structures, coupled with the lack of capable guardianship, will increase the convergence in time and space of motivated offenders and suitable targets. Consistent with this argument, Newman and Clarke (2003) noted that changes in society, and subsequent changes in crime patterns, can be attributed to technological advances. Clarke (1999:1) identified common attributes of "hot products" most frequently targeted by thieves, noting that they are "concealable, removable, available, valuable, enjoyable, and disposable ("C.R.A.V.E.D.")." Applying this model to cyberspace, Newman and Clarke argued that the main target of Internet crime is information.[1] This theoretical model is not only beneficial for understanding the causal mechanisms underlying fraud targeting and victimization, but also for developing crime prevention strategies (Clarke 1995). In an Internet fraud context, routine activity theory and situational crime prevention have important implications for crime control policy in that potential targets can be educated about changing their online routines to reduce the chances of fraud targeting. Situational crime prevention also provides guidance for e-commerce merchants and Internet service providers to strengthen targets and reduce criminal opportunities. With these theoretical perspectives in mind, the current study has two objectives. First, we investigate the extent to which sociodemographic characteristics explain routine online activities reflecting exposure (e.g., hours spent online and purchasing from Internet Web sites). Second, we assess the influences that such routine online activities have, in turn, on the likelihood of Internet fraud targeting.

✉ Fraud Targeting and Fraud Victimization

Internet Use and Internet Fraud

An estimated 22.3 million people in the United States access the Internet (Nielsen Online 2008). They do so for a variety of reasons: to exchange e-mail, obtain news, entertain themselves, and manage their finances. Online shopping is another common form of Internet activity. Indeed, a recent survey found that 66 percent of Internet-connected adults reportedly made online purchases (Horrigan 2008). In 2007 American consumers spent $157.4 billion online, and sales are projected to rise to $316 billion in 2010. Online shopping has become so prevalent that this form of spending now accounts for 12 percent of total retail sales (U.S. Census Bureau 2008).

Despite the growth in e-commerce, studies show that a majority of online consumers believe the risks associated with Internet use are far from benign. For example, a recent IBM-sponsored survey found that approximately 70 percent of Internet users reported that they believe they are more likely to be the victim of a cybercrime than a physical crime (Criminal Law Reporter 2006). What is more, a large number of Internet users perceive the risks associated with using their credit card to make an online purchase to outweigh the benefits (i.e., convenience; Bhatnagar, Misra, and Rao 2000; Horrigan 2008).

Chief among consumers' concerns about online victimization is theft, especially credit card theft—a fear that may not be unfounded (Reisig, Pratt, and Holtfreter 2009). As stated previously, nearly one third of the adult population in the United States reports that they have been targeted by perpetrators of consumer fraud, with nearly one in nine people reporting having been victimized in the past year (Anderson 2004). Extrapolating this percentage to the 221 million Internet users would translate into nearly 25 million fraud victims—roughly 18 times the number of murder, rape, armed robbery, and aggravated assault victims combined (U.S. Department of Justice 2008). Given the financial importance of fraud victimization and its relevance to such a large portion of the online population, it is important to understand the nature of this form of criminal victimization.

The Importance of Fraud Targeting

Contact between the victim and the offender is a necessary but not sufficient condition for fraud victimization since the methods offenders use to target consumers vary considerably. For example, consumers can be targeted through print media sources (e.g., mail, newspapers, magazine advertisements, and catalogs), telemarketing, and television. Research also shows that consumers are targeted via the Internet (Anderson 2004; Holtfreter, Reisig, and Blomberg 2006). Consumer protection legislation at both federal and state levels has addressed fraud victimization via telephone, mail, and television. In comparison, however, the Internet

remains largely unregulated (Anderson 2004; Burns, Whitworth, and Thompson 2004; Fletcher 2007). Despite the existence of consumer protection statutes, the multijurisdictional context of Internet fraud makes prosecution of perpetrators difficult (Levi and Wall 2004; Newman and Clarke 2003; Wall 2007b; Yar 2006).

Equally troubling is the concentration on victimization as a primary outcome in the bulk of the prior empirical fraud research; an important yet neglected area of research is target selection (Fattah 1993). As others have acknowledged, crime victims do not constitute a representative sample or an unbiased cross-section of the general population (Budd 1999; Lauritsen 2001). By the time a fraud attempt is successful, victims have lost money, perpetrators are long gone, and it is difficult for legal authorities to initiate an investigation (Grazioli and Jarvenpaa 2001; Pontell, Brown, and Tosouni 2008). From a crime prevention perspective, research that identifies the circumstances of elevated risk for those yet to be victimized (i.e., targets) is critical (Clarke 1999; Felson 1986; Johnson et al. 2007).

Research in other victimization contexts suggests that there is a common underlying process of targeting (Dugan and Apel 2005). In particular, offenders have preferences for the types of targets that they consider suitable. Since demographic preferences (e.g., age and gender) cannot be directly observed online, cyber criminals target consumers during the course of their regular Internet routines. Insight into how these routines, in turn, influence fraud targeting is not only theoretically important, but also policy relevant in that established predictors of targeting will assist legal authorities in developing strategies to equip consumers and businesses to defend themselves against fraud attempts.

Routine Activity Theory and Fraud Targeting

The diversity of methods used by fraud perpetrators to make contact with potential victims has made the prediction of fraud targeting a difficult undertaking. Despite theoretical advances, much of the empirical literature in criminology has focused solely on identifying sociodemographic correlates of fraud targeting and victimization (e.g., age, gender, marital status, and socioeconomic status). As such, with regard to routine online activities, some consistent sociodemographic differences between online shoppers and nonshoppers have been observed. Online shoppers tend to be relatively well educated and occupy higher income brackets (Farag et al. 2006; Ratchford, Talukadar, and Lee 2001; Soopramanien and Robertson 2007; Stranahan and Kosiel 2007; Swinyard and Smith 2003). Several studies also show that males are more likely to make purchases online (Chang and Samuel 2004; Farag et al. 2006; Korgaonkar and Wolin 2002; Soopramanien and Robertson 2007). Research findings regarding the differential use of online shopping across age and racial/ethnic groups are far less conclusive (Chang and Samuel 2004; Sorce, Perotti, and Widrick 2005; Stranahan and Kosiel 2007). In addition, studies that focus on time spent online, generally speaking, reveal differential patterns of Internet usage across social groups. For example, research shows that higher income and better educated Internet users spend more time online (Korgaonkar and Wolin 1999), as do men (Fallows 2005; Nielsen Online 2008).

Although this research is certainly useful for marketing purposes, such a focus is problematic since empirical evidence suggests that fraud perpetrators target consumers from all backgrounds (Holtfreter et al. 2006) and that fraud victimization cannot be consistently predicted with only sociodemographic variables (Holtfreter et al. 2008; Titus, Heinzelmann, and Boyle 1995). A theoretical understanding of fraud targeting/victimization is therefore necessary if we are to fully understand the nature of crime in this context. In this study we therefore draw upon theoretical discussions regarding criminal victimization in general, as well as the fraud-specific literature rooted in routine activity theory and situational crime prevention (Newman and Clarke 2003).

Cohen and Felson (1979) argued that structural changes in aggregate routines can influence the convergence in time and space of motivated offenders and suitable targets in the absence of capable guardians. The theory, which assumes a constant supply of motivated

offenders, focuses on the behaviors, activities, and situational contexts that place would-be targets at risk for victimization. In the study of street crime victimization, routine activity theory has informed public policy, particularly with the development of situational crime prevention strategies intended to increase guardianship (Clarke 1995). Early applications of this model focused on revealing "profiles" of likely victims of violence, such as young, minority males. Taking this logic a step further, Clarke (1999) broadened our understanding of nonviolent crime targeting on "hot" products (i.e., those most likely to be selected by thieves).

Recent studies of street crime have focused on theoretically based indicators of routines and lifestyles that may increase victimization risk for certain individuals. In the street crime context, research by Stewart, Scheck, and colleagues has revealed that deviant lifestyles involving various illegal behaviors (e.g., prostitution and drug use) increase exposure to motivated offenders, thereby elevating the odds of violent victimization (see e.g., Schreck and Fisher 2004; Schreck, Stewart, and Fisher 2006; Schreck, Wright, and Miller 2002; Stewart, Elifson, and Sterk 2004). The applicability of traditional indicators of routine activities (e.g., deviant lifestyles) has been called into question in the consumer fraud context, however, since individuals are typically targeted by fraudsters during the course of routine, nondeviant behaviors and day-to-day lifestyle choices (Holtfreter et al. 2008). Put simply, the types of routine behaviors that could be considered "risky" shift when considering fraud targeting (i.e., from deviant to nondeviant behaviors), and such behaviors may be specific to the context under consideration (e.g., cyberspace).

Accordingly, Cohen and Felson (1979:591) noted long ago that "technological advances designed for legitimate purposes" can influence the nature of criminal victimization. Among these potential changes, Cohen and Felson (1979:599) pointed to "changes in the sales of consumer goods" as a contributing factor to increased criminal opportunities. The penetration of the Internet into consumer lifestyles represents a key structural change that is relevant to a routine activity explanation of fraud targeting. Approximately 88 percent of American households have online access (Parks

Associates 2008). This is a considerable increase from U.S. census estimates of 50 percent in 2001 and 18 percent in 1997 (Day, Janus, and Davis 2005). Evidence suggests that consumers are increasingly willing to buy online, but Internet use (e.g., purchasing from Web sites and time spent online) varies across consumers from different walks of life. Direct-to-consumer marketing communication channels such as the Internet provide cost- and time-efficient options, but may create unguarded exposure to fraudsters (Lin 2006). As Cohen and Felson (1979) predicted, fraudsters feed upon the legal activities (e.g., modes of commerce reflecting unguarded exposure) of everyday life to target potential victims (Langenderfer and Shimp 2001; Shover, Coffey, and Hobbs 2003; Shover, Coffey, and Sanders 2004).

As Newman and Clarke (2003:78) argue, the Internet, and online shopping in particular, presents multiple opportunities for fraud targeting and victimization. In fact, while tangible items (e.g., electronics) were once "hot products," the transformations associated with e-commerce crime, particularly information storage and transmission, has created very lucrative targets. Indeed, because retailer databases containing attractive consumer information (e.g., names, addresses, passwords, credit/debit card and/or bank account details) fit the characteristics Clarke (1999) refers to as C.R.A.V.E.D., they have become a coveted target for cyber fraudsters (Newman and Clarke 2003). Although routine activity theory predicts that more time spent away from home will increase victimization risk in other contexts (e.g., burglary of an unguarded residence), this proposition is less applicable to Internet fraud targeting, given that consumers often engage in routine online activities (e.g., shopping on Web sites) while in the safe confines of their own homes. What this means is that in the fraud victimization context, staying home and spending time (and money) online exposes potential targets for victimization.[2]

⊠ Current Focus

The current study addresses two issues. First, guided by prior marketing and consumer research, we investigate the extent to which sociodemographic characteristics explain routine online activities, such as purchasing goods from an Internet Web site. Second, we assess whether routine online activities significantly increase the odds of Internet fraud targeting, net of consumer attributes (e.g., age, race, and income). Routine activity theory suggests that online activities reflect greater unguarded exposure to potential motivated offenders. Because we are able to directly measure online activities, we hypothesize that the effects of consumer attributes on Internet fraud targeting are, comparatively speaking, far less salient. We carry out our objectives using data from a telephone survey of adult consumers conducted in the state of Florida.

Dependent Variable

Internet Fraud Targeting

A series of questions were used to determine whether participants were targeted by fraud perpetrators via the Internet. First, survey respondents were asked whether "there ever was a time you felt you were the subject of consumer fraud attempt?" Next, participants were asked when the attempt took place. Because memory decay bias was a concern, cases that occurred within one year of the interview were selected. Participants who reported that they had been targeted within the past year were read a list of 13 items describing common fraud schemes (e.g., "an investment deal that turned out to be phony" and "agreed to buy a product or service for a certain price but was later charged a lot more"). An open-ended response was administered to respondents who did not select an item from the list. Open-ended responses were carefully screened to ensure that these experiences actually reflected consumer fraud (see Titus 2001). Using these 16 survey items, we determined that 15.18 percent of participants were recent targets of consumer fraud. Finally, respondents were asked how they first learned about the fraud attempt. Three of the options included in the survey were Internet related: (1) through an Internet auction site, (2) from an Internet Web site (other than an auction site), and (3) from an email. Targeting experiences that initially took place via these three online avenues were selected and included in the outcome measure.

Approximately 3 percent of the sample reported that they were targeted by perpetrators of consumer fraud via the Internet during the year leading up to the survey. Internet fraud targeting is a binary variable (1 = yes, 0 = no).

Independent Variables

Routine Online Activity

Two key variables of interest were used to measure respondents' routine online activity. For the first variable, hours spent online, participants were first administered a screening question. Specifically, respondents were asked whether they "ever go online to access the Internet or World Wide Web or send and receive email?" If they responded "yes," they were then asked "how many hours each week would you say you spend on the Internet?" The mean number of hours spent online each week was 6.51 ($SD = 11.24$), comparable to the 8 hours per week reported in a national survey (Harris Poll 2005). The second variable, Internet Web site purchase, captured whether participants had purchased

something from an Internet Web site during the year prior to the survey (1 = yes, 0 = no). Nearly 45 percent of respondents indicated that they had recently made such a purchase.

Personal Characteristics

Nine sociodemographic variables were included in the multivariate analyses: age (respondent's age in years), education (1 = some grade school, 2 = some high school, 3 = high school graduate [or equivalency], 4 = technical or vocational school, 5 = some college, 6 = college graduate, and 7 = graduate/professional degree), income (1 = less than \$20,000, 2 = \$20,000 to \$40,000, 3 = \$40,001 to \$60,000, 4 = \$60,001 to \$80,000, 5 = \$80,001 to \$100,000, and 6 = more than \$100,000), retired (1 = yes), homeowner (1 = yes), male (1 = yes), married (1 = yes), black (1 = yes), and Latino (1 = yes). Descriptive statistics for the variables used in the current study are provided in Table 1.

Turning to our first research objective, we examined whether consumers' personal attributes were associated with routine online activity. Table 2 presents two models,

Table 1 Descriptive Statistics ($N = 922$)

Variables	M	SD
Internet fraud targeting (1 = targeted online in the past year, 0 = otherwise)	0.03	0.16
Hours spent online (number of hours spent online each week)	6.51	11.24
Internet Web site purchase (1 = purchased something online in the past year, 0 = otherwise)	0.45	0.50
Age (respondents' age in years)	57.38	18.43
Education (1 = some grade school to 7 = graduate or professional degree)	4.63	1.56
Income (1 = less than \$20,000 to 6 = \$100,000 or more)	3.13	1.51
Retired (1 = retired respondent, 0 = otherwise)	0.43	0.50
Homeowner (1 = respondent owns home, 0 = otherwise)	0.83	0.37
Male (1 = male respondent, 0 = female respondent)	0.45	0.50
Married (1 = married respondent, 0 = otherwise)	0.59	0.49
Black (1 = African-American respondent, 0 = otherwise)	0.09	0.29
Latino (1 = Latino respondent, 0 = otherwise)	0.09	0.29

one for the number of hours survey respondents spent online and a second model for whether participants recently purchased goods from an Internet Web site. The model χ^2 statistics indicate that both models fit the data well. The hours spent online negative binomial model is featured on the left-hand side of Table 2. The dispersion parameter (α) is 2.33, and the likelihood ratio test of alpha is 7432.64 ($p < .001$). These findings indicate that the data are consistent with the model. Looking at the regression estimates, we found that five sociodemographic variables were significant ($p < .05$) correlates of self-reported time spent on the Internet each week. For example, younger respondents spent significantly more time online relative to older participants. The exponentiated coefficient shows that each unit increase in age corresponds to a 3 percent decrease in time online relative to whites, Asians, and other racial minorities who make up the omitted category. In

comparison to other racial groups, blacks spend approximately 39 percent less time online each week. Males, homeowners, and participants with higher levels of formal education, however, spend significantly more time online.

The second model in Table 2 provides the effects of sociodemographic characteristics on whether purchases were made from an Internet Web site. Three variables—age, black, and retired—are inversely related to the outcome measure in the logistic regression equation. For example, the exponentiated coefficient indicates that being retired reduced the odds of making an online purchase by 37 percent. A similar odds ratio is observed for African American participants. The effects of income, married, education, and homeownership are positive and statistically significant. For education, a one-unit increase corresponds to a 41 percent increase in the odds that participants

Table 2 Routine Online Activity Regression Models ($N = 922$)

Variables	Hours Spent Online[a]				Internet Web Site Purchase[b]			
	b	SE	exp(b)	z Test	b	SE	exp(b)	z Test
Constant	1.93	.32	—	5.94**	−.94	.40	—	−2.36*
Age	−0.03	.01	0.97	−5.76**	−.04	.01	0.96	−5.94**
Education	0.16	.04	1.17	3.82**	.34	.06	1.41	5.91**
Income	0.05	.05	1.05	1.09	.25	.06	1.29	4.18**
Retired	−0.03	.17	0.97	−0.19	−.46	.23	0.63	−2.02*
Homeowner	0.63	.19	1.87	3.30**	.45	.22	1.57	2.03*
Male	0.28	.12	1.32	2.37*	.07	.16	1.08	0.47
Married	0.07	.13	1.07	0.51	.55	.17	1.74	3.27**
Black	−0.50	.20	0.61	−2.53*	−.74	.26	0.48	−2.83**
Latino	−0.05	.19	0.95	−0.26	.04	.28	1.04	0.15
	Model $\chi^2 = 116.67$** McFadden's $R^2 = .03$				Model $\chi^2 = 187.25$** McFadden's $R^2 = .21$			

NOTE: Entries are unstandardized coefficients (b) and robust standard errors.

a. Negative binomial regression model (likelihood ratio of alpha $= 7432.64, p < .001$).

b. Logistic regression model.

*$p < .05$. **$p < .01$. (two-tailed test)

make online purchases. In sum, findings show that six of the nine sociodemographic variables included in the logistic regression model significantly predicted whether respondents purchased goods from an Internet Web site during the 12 months prior to the administration of the survey. Overall, the findings in Table 2 are highly consistent with prior marketing and consumer research in that different sociodemographic characteristics shape routine online activity among adult consumers (Chang and Samuel 2004; Farag et al. 2006; Korgaonkar and Wolin 2002; Ratchford et al. 2001; Soopramanien and Robertson 2007).

To determine whether routine online activity and personal attributes are empirically linked to Internet fraud targeting, three logistic regression models are presented in Table 3. Model 1 features the nine sociodemographic variables used in Table 2. Only two variables—age and education—are associated with Internet fraud targeting. Younger and more educated individuals are significantly more likely to be targets of consumer fraud via the Internet. In model 2, Internet fraud targeting was regressed onto the two routine online activity variables. Note that we apply a square-root transformation to the hours spent online variable to reduce the

Table 3 Internet Fraud Targeting Logistic Regression Models ($N = 922$)

Variables	\multicolumn{12}{c}{Internet Fraud Targeting}
	\multicolumn{4}{c}{Model 1}	\multicolumn{4}{c}{Model 2}	\multicolumn{4}{c}{Model 3}
	b	SE	exp(b)	z Test	b	SE	exp(b)	z Test	b	SE	exp(b)	z Test
Constant	−3.94	1.31	—	−3.01**	−5.21	0.54	—	−9.73**	−4.19	1.48	—	−3.31**
Hours spent online[a]	—	—	—		0.21	0.10	1.24	2.19*	0.19	0.10	1.21	1.86[†]
Internet Web site purchase	—	—	—		1.56	0.58	4.77	2.71**	1.36	0.66	3.90	2.06*
Age	−0.04	0.02	0.96	−2.13*	—	—	—		−0.03	0.02	0.97	−1.53
Education	0.33	0.15	1.39	2.21*	—	—	—		0.19	0.15	1.20	1.26
Income	0.02	0.17	1.02	0.11	—	—	—		−0.05	0.18	0.95	−0.30
Retired	0.37	0.84	1.44	0.44	—	—	—		0.54	0.75	1.71	0.72
Homeowner	0.16	0.63	1.18	0.26	—	—	—		0.02	0.63	1.02	0.04
Male	0.56	0.42	1.75	1.33	—	—	—		0.51	0.44	1.66	1.16
Married	0.20	0.43	1.23	0.47	—	—	—		0.17	0.45	1.19	0.38
Black	0.26	0.69	1.30	0.38	—	—	—		0.49	0.70	1.62	0.69
Latino	−0.49	0.70	0.61	−0.70	—	—	—		−0.59	0.73	0.56	−0.80
	\multicolumn{4}{l}{Model $\chi^2 = 45.05$** McFadden's $R^2 = .06$}	\multicolumn{4}{l}{Model $\chi^2 = 17.23$** McFadden's $R^2 = .09$}	\multicolumn{4}{l}{Model $\chi^2 = 51.27$** McFadden's $R^2 = .11$}

NOTE: Entries are unstandardized coefficients (b) and robust standard errors.

a. Square root transformation.

*$p < .05$. **$p < .01$. (two-tailed test) $^{†}p < .05$ (one-tailed test)

skew that resulted from a few individuals who reported that they spend a considerable amount of time online each week ($M = 2.27$, $SD = 1.54$).[3] When compared to model 1, model 2 provides a good fit to the data as indicated by McFadden's R^2. Both of the covariates included in the logistic regression equation were statistically significant. Making a purchase from a Web site has a substantial effect on Internet fraud targeting. More specifically, buying something from a Web site increases the odds of Internet fraud targeting by 377 percent. Thus far, the results from the first two models in Table 3 indicate that routine online activity is more important in explaining Internet fraud targeting relative to consumer attributes.

A saturated model that included both personal characteristics and routine online activity variables is featured on the right-hand side of Table 3. The effects of age and education failed to reach statistical significance ($p = .13$ and $.21$, respectively) when the two routine online activity variables were included in the model, suggesting that the effect of education and age on Internet fraud targeting is fully explained by the number of hours consumers spend online and whether they make purchases from Internet Web sites. The effect of making online purchases on Internet fraud targeting persisted in model 3. Under this model specification, those who make purchases from Web sites increase the odds that they will be targeted by cyberfraudsters by 290 percent. The effect of hours spent online on Internet fraud targeting was also positive and statistically significant ($p < .05$, one-tailed test).

⚲ Discussion

Internet fraud, like victimization for many other types of crime, affects a large portion of the population. Furthermore, it is reasonable to expect fraud victimization rates to continue to increase as opportunities for online activity (both legal and illegal) expand and as a greater portion of the population feels comfortable performing all manner of tasks online. Indeed, rates of Internet use in general have been increasing steadily over the years, and online economic activity has increased accordingly (U.S. Census Bureau 2008; U.S.

Department of Commerce 2008). Given the capacity of Internet fraud to affect so many lives, the purpose of the present study was to gain a better theoretical and empirical understanding of this form of criminal victimization—in particular, to uncover the key factors that predict fraud targeting online. To that end, the work presented here leads to four conclusions.

First, the previous empirical work on fraud targeting, which has relied almost exclusively on correlating respondents' sociodemographic characteristics with various dimensions of experiences with fraudsters (see Holtfreter et al. 2008), has failed to fully capture the nature of this form of criminal victimization. In short, when it comes to understanding why certain people are targeted by fraud perpetrators, it is simply not enough to focus on demographics while ignoring criminologically informed indicators of the routine online activities that put people at risk for victimization. To be sure, the effects of all of our measured sociodemographic characteristics on fraud targeting were fully mediated by the variables measured in respondents' routine online activities. Assumptions that rely on inferences about the relationships between demographic characteristics and fraud targeting are therefore fundamentally misplaced. Instead, understanding the problem of fraud targeting requires an appreciation of how online exposure shapes the opportunity structure for victimization in this context.

Second, and relatedly, routine activity theory and situational crime prevention provide an excellent framework for informing the study of this area of crime targeting. The routine activity perspective (Cohen and Felson 1979; Cohen, Kluegel, and Land 1981; see also Hindelang, Gottfredson, and Garafolo 1978) has been a staple of the criminal victimization literature in general for nearly three decades now (Pratt and Cullen 2005). It has been applied most often to the explanation of violent victimization (Belknap 1987; Dugan and Apel 2005; Dugan, Nagin, and Rosenfeld 2003; Gottfredson 1981; Jensen and Brownfield 1986; Kennedy and Forde 1990; Messner and Tardiff 1987; Miethe and Meier 1994; Mitchell and Finkelhor 2001; Rodgers and Roberts 1995; Sampson 1987; Sampson and Lauritsen 1990; Sampson and Wooldredge 1987). Scholars have also integrated routine activity and self-control theories into recent assessments of why those who are more likely to offend are also more likely to be victims of crime (Holtfreter et al. 2010; Piquero et al. 2005; Schreck et al. 2002, 2006; Schreck and Fisher 2004; Stewart et al. 2004). In addition to the recent empirical work of Holtfreter et al. (2008), we can now extend the "generality" of the routine activity framework to explain Internet fraud targeting as well (see also Newman and Clarke 2003).

We should also note that we examined only one form of online targeting for criminal victimization in a sample of adults (i.e., fraud). Whether routine activities put consumers at risk for other types of fraud targeting (e.g., identity theft) is an empirical question in need of further scrutiny. The expansion of Internet use—particularly among younger citizens—has resulted in new opportunities for additional problems such as child exploitation (Ferraro, Casey, and McGrath 2004; Finkelhor, Mitchell, and Wolak 2000; Wortley and Smallbone 2006). This issue has caught the attention of the media, the public, and policymakers alike, and legislation is currently being produced in an effort to combat these types of crime (Gelber 2006). Coupled with existing theoretical work (see e.g., Newman and Clarke 2003; Wall 2007a, 2007b; Willison 2008; Yar 2006), the analyses presented here should help to guide future empirical studies in this area as scholars continue to investigate the kinds of online routines that put children at risk.

Third, the policy implications of the work presented here are consistent with Dugan and Apel's (2005:700) discussion of "target-initiated exposure reduction" methods that citizens can use to combat attempted fraud by motivated offenders. Thus, crime control policies could be developed with an eye toward educating citizens about using various safeguards when shopping online (e.g., secure servers and virus protection software). Just as we can teach potential sexual assault targets to change their daily routines, we can educate potential fraud targets about altering their online activities, minimizing exposure to the criminal opportunity structure. This approach highlights the need to think about the prevention of crime in general,

and online victimization in particular, in ways that move beyond a strict focus on the criminal justice system (see e.g., the discussion by Pratt 2009). Instead, parents, schools, and employers will each be critical to any efforts at educating citizens on how to reduce their exposure to online risks. Recent empirical research (Reisig et al. 2009) has found that Internet users who perceive higher risk of having their credit card stolen online reduce their potential exposure through behavioral adaptations (i.e., by spending less time online and making fewer Internet purchases), a finding that lends support to target-initiated exposure reduction in an online context.

Given that the Internet is a public place with a wide criminal opportunity structure where personal, face-to-face contact is not needed in fraud attempts, educating consumers about online risk is just one piece of the puzzle. Motivated offenders will continue to select targets that are "C.R.A.V.E.D." (See Clarke 1997, 1999), suggesting that more extensive opportunity reduction strategies beyond those directed at modifying individual-level consumer behaviors are warranted. For example, Newman and Clarke's (2003) application of Clarke's (1999) opportunity-reducing techniques to e-commerce crime suggests key strategies relevant to target hardening of information systems maintained by online retailers, employers, and Internet service providers (Berg 2008; Newman and Clarke 2003; Willison 2008). What is more, techniques such as increasing risks and reducing rewards may carry the added strength of what has been termed the "diffusion of crime control benefits" (Clarke and Weisburd 1994:167; Hesseling 1995; Weisburd et al. 2006). In short, methods designed to limit opportunity for identity theft may also reduce other types of fraud victimization (Newman 2008).

Finally, it is important to note that many of the correlates of fraud targeting found in our study appear to differ from those typically associated with targeting for street crime. In particular, where street/violent crime targeting and victimization is typically associated with younger minority males, those targeted for Internet fraud tended to be younger and better educated. Nevertheless, these differences obscure a more important finding: that both violent crime and Internet fraud forms of victimization still share the same underlying causal mechanism associated with the routine activities of potential crime targets. This point highlights most clearly the need to apply a firm theoretical understanding of the nature of fraud victimization to empirical evaluation. The purpose of our study was to accomplish this task by extending routine activity theory into this domain of victimization. In doing so, we have provided another piece of evidence indicating that routine activity theory is truly a "general" framework for understanding criminal victimization. Future empirical research might expand on our efforts by also empirically examining situational characteristics (e.g., opportunity measures of perceived risks and rewards) that may be more germane to situational crime prevention.

⊠ Notes

1. Viewed in this way, in the original routine activity perspective—which emphasized the conversion in physical space of offenders and victims—the notion of "space" can be extended to crimes that do not necessarily involve "direct-contact" predatory victimization.

2. A key distinction is that the concept of "place" in offline crime typically refers to a small area, such as a street (see Eck and Weisberg 1995). In a street crime context, offenders may select victims based on situational characteristics of the place (e.g., proximity to a suitable target). In comparison, the Internet is a virtual place, not bound by the geographical and spatial constraints inherent in street crime (Wall 2007a; Yar 2006). Online, the problem of physical proximity to suitable targets is neutralized (Newman and Clarke 2003). This unique context therefore provides a vast array of criminal opportunities for motivated offenders.

3. Prior to reexpressing the distribution, a constant (+1) was added to the scores to eliminate zero values.

⊠ References

Anderson, Keith B. 2004. *Consumer Fraud in the United States: An FTC Survey.* Washington, DC: Federal Trade Commission.

Belknap, Joanne. 1987. "Routine Activity Theory and the Risk of Rape: Analyzing Ten Years of National Crime Survey Data." *Criminal Justice Policy Review* 2:337–56.

Berg, Sara. 2008. "Preventing Identity Theft through Information Technology." Pp. 151–67 in *Perspectives on Identify Theft*, edited by M. M. McNally and G. R. Newman. Monsey, NY: Criminal Justice Press.

Bhatnagar, Amit, Sanjog Misra, and H. Raghav Rao. 2000. "On Risk, Convenience, and Internet Shopping Behavior: Why Some Consumers Are Online Shoppers While Others Are Not." *Communications of the Association for Computing Machinery* 33:98–105.

Budd, Tracey. 1999. *Burglary of Domestic Dwellings: Findings from the British Crime Survey.* London: Home Office Statistical Bulletin.

Burns, Ronald G., Keith H. Y. Whitworth, and Carol Y. Thompson. 2004. "Assessing Law Enforcement Preparedness to Address Internet Fraud." *Journal of Criminal Justice* 32:477–93.

Chang, Joshua and Nicholas Samuel. 2004. "Internet Shopper Demographics and Buying Behavior in Australia." *Journal of American Academy of Business* 5:171–76.

Clarke, Ronald V. 1995. "Situational Crime Prevention." Pp. 91–150 in *Building a Safer Society: Strategic Approaches to Crime Prevention*, edited by M. Tonry and D. P. Farrington. Chicago: University of Chicago Press.

Clarke, Ronald V. 1997. *Situational Crime Prevention: Successful Case Studies.* 2nd ed. Albany, NY: Harrow & Heston.

Clarke, Ronald V. 1999. *Hot Products: Understanding, Anticipating, and Reducing Demand for Stolen Goods* (Police Research Series, Paper 112). London: Home Office.

Clarke, Ronald V. and David Weisburd. 1994. "Diffusion of Crime Control Benefits: Observations on the Reverse of Displacement." Pp. 165–83 in *Crime Prevention Studies.* Vol. 2, edited by R. V. Clarke. Monsey, NY: Criminal Justice Press.

Cohen, Lawrence E. and Marcus Felson. 1979. "Social Change and Crime Rate Trends: A Routine Activity Approach." *American Sociological Review* 52:170–83.

Cohen, Lawrence E., James R. Kluegel, and Kenneth C. Land. 1981. "Social Inequality and Predatory Criminal Victimization: An Exposition and Test of a Formal Theory." *American Sociological Review* 46:505–24.

Criminal Law Reporter. 2006. "More Americans Fear Cybercrime than Physical Crime, According to IBM Study. *Criminal Law Reporter* 78:725.

Day, Jennifer C., Alex Janus, and Jessica Davis. 2005. *Computer and Internet Use in the United States.* Washington, DC: U.S. Census Bureau.

Dugan, Laura and Robert Apel. 2005. "The Differential Risk of Retaliation by Relational Distance: A More General Model of Violent Victimization." *Criminology* 43:697–728.

Dugan, Laura, Daniel S. Nagin, and Richard Rosenfeld. 2003. "Exposure Reduction or Retaliation? The Effects of Domestic Violence Resources on Intimate-Partner Homicide." *Law and Society Review* 37:169–98.

Eck, John E. and David Weisburd. 1995. "Crime Places in Crime Theory." Pp. 1–33 in *Crime and Place, Crime Prevention Studies.* Vol. 4, edited by J. E. Eck and D. Weisburd. Monsey, NY: Criminal Justice Press.

Fallows, Deborah. 2005. *How Women and Men Use the Internet.* Washington, DC; Pew Internet and American Life Project.

Farag, Sendg, Jesse Weltevreden, Ton Van Rietbergen, Marty Dijst, and Frank Van Ort. 2006. "E-Shopping in the Netherlands: Does

Geography Matter?" *Environment and Planning B: Planning and Design* 33:59–74.

Fattah, Ezzat. 1993. "The Rational Choice/Opportunity Perspectives as a Vehicle for Integrating Criminological and Victimological Theories." Pp. 225–58 in *Routine Activity and Rational Choice,* edited by R. Clarke and M. Felson. New York: Transaction.

Federal Bureau of Investigation. 2001. Press release: Internet fraud investigation "operation cyber loss." Retrieved http://www.fbi.gov/Pressre1/pewaaew101/ifcc052301.htm.

Federal Trade Commission. 2006. *Consumer Fraud and Identity Theft Complaint Data.* Washington, DC: Author.

Felson, Marcus. 1986. "Linking Criminal Choices, Routine Activities, Informal Social Control, and Criminal Outcomes." Pp. 119–28 in *The Reasoning Criminal: Rational Choice Perspective on Offending,* edited by D. B. Cornish and R. V. Clarke. New York/Berlin: Springer-Verlag.

Ferraro, Monique Mattei, Eoghan Casey, and Michael McGrath. 2004. *Investigating Child Exploitation and Pornography.* San Diego, CA: Academic Press.

Finkelhor, David, Kimberly J. Mitchell, and Janis Wolak. 2000. *Online Victimization: A Report on the Nation's Youth.* Washington, DC: Office of Juvenile Justice and Delinquency Prevention.

Fletcher, Nigel. 2007. "Challenges for Regulating Fraud in Cyberspace." *Journal of Financial Crime* 14:190–207.

Gelber, Alexandra. 2006. "Federal Jurisdiction in Child Pornography Cases." *United States Attorney's Bulletin* 54(7): 3–7.

Gottfredson, Michael R. 1981. "On the Etiology of Criminal Victimization." *Journal of Criminal Law and Criminology* 72:714–26.

Grabowsky, Peter N. and Russell G. Smith. 1998. *Crime in the Digital Age: Controlling Telecommunications and Cyberspace Illegalities.* New Brunswick, NJ: Transaction Publishers.

Grazioli, Stefano and Sirkka L. Jarvenpaa. 2001. "Deceived: Under Target Online." *Communications of the ACM* 46:196–205.

Harris Poll. 2005. "Almost Three-Quarters of all American Adults Go Online. Harris Interactive Poll." Retrieved March 5, 2010 (http://www.Harrisinteractive.com/harris_poll/indes.asp?PID=569).

Hesseling, Rene B. P. 1995. "Displacement: A Review of the Literature." Pp.197–230 in *Crime Prevention studies.* Vol. 3, edited by R. V. Clarke. Monsey, NY: Criminal Justice Press.

Hindelang, Michael J., Michael R. Gottfredson, and James Garafolo. 1978. *Victims of Personal Crime.* Cambridge, MA: Ballinger.

Holtfreter, Kristy, Michael D. Reisig, and Thomas G. Blomberg. 2006. "Consumer Fraud Victimization in Florida: An Empirical Study." *St. Thomas Law Review* 18:761–89.

Holtfreter, Kristy, Michael D. Reisig, and Travis C. Pratt. 2008. "Low Self-Control, Routine Activities, and Fraud Victimization." *Criminology* 46:189–220.

Holtfreter, Kristy, Michael D. Reisig, Nicole Leeper Piquero, and Alex R. Piquero. 2010. "Low Self-Control and Fraud: Offending, Victimization, and Their Overlap." *Criminal Justice and Behavior* 37:188–203.

Horrigan, John B. (2008). *Online Shopping.* Washington, DC: Pew Internet & American Life Project.

Internet Crime Complaint Center. 2006. *Internet Crime Report.* Washington, DC: The National White Collar Crime Center and the Federal Bureau of Investigation.

Jensen, Gary F. and David Brownfield. 1986. "Gender, Lifestyles, and Victimization: Beyond Routine Activity." *Violence and Victims* 1:85–99.

Johnson, Shane D., Wim Bernasco, Kate Bowers, Henk Elffers, Jerry Ratcliffe, George Rengert, and Michael Townsley. 2007. "Space-time Patterns of Risk: A Cross-national Assessment of Residential Burglary Victimization." *Journal of Quantitative Criminology* 23:201–19.

Kennedy, Leslie W. and David R. Forde. 1990. "Routine Activities and Crime: An Analysis of Victimization in Canada." *Criminology* 28:137–52.

Korgaonkar, Pradeep K. and Lori D. Wolin. 1999. "A Multivariate Analysis of Web Usage." *Journal of Advertising Research* 39:53–68.

Korgaonkar, Pradeep and Lori D. Wolin. 2002. "Web Usage, Advertising, and Shopping: Relationship Patterns." *Internet Research* 12:191–204.

Langenderfer, Jeff and Terrance A. Shimp. 2001. "Consumer Vulnerability to Scams, Swindles, and Fraud: A New Theory of Visceral Influences on Persuasion." *Psychology and Marketing* 18:763–83.

Lauritsen, Janet L. 2001. "The Social Ecology of Violent Victimization: Individual and Contextual Effects in the NCVS." *Journal of Quantitative Criminology* 17:3–32.

Levi, Michael. 1998. "Organizing Plastic Fraud: Enterprise Criminals and the Side-stepping of Fraud Prevention." *The Howard Journal* 37:423–38.

Levi, Michael and David S. Wall. 2004. "Technologies, Security, and Privacy in the Post-9/11 European Information Society." *Journal of Law and Society* 31:194–220.

Lin, Carolyn A. 2006. "Interactive Media Technology and Electronic Shopping." Pp. 203–22 in *Communication Technology and Social Change: Theory and Implications,* edited by C. A. Lin and D. J. Atkin. New York: Routledge.

McNally, Megan M. and Graeme R. Newman. 2008. *Perspectives on Identity Theft. Crime Prevention Studies.* Vol. 23. Monsey, NY: Criminal Justice Press.

Messner, Steven F. and Kenneth Tardiff. 1987. "The Social Ecology of Urban Homicide: An Application of the 'Routine Activities' Approach." *Criminology* 23:241–67.

Miethe, Terance and Robert F. Meier. 1994. *Crime and Its Social Context: Toward an Integrated Theory of Offenders, Victims, and Situations.* Albany: State University of New York Press.

Mitchell, Kimberly J. and David Finkelhor. 2001. "Risk of Crime Victimization Among Youth Exposed to Domestic Violence." *Journal of Interpersonal Violence* 26:944–64.

Moore, Elizabeth and Michael Mills. 1990. "The Neglected Victims and Unexamined Costs of White-Collar Crime." *Crime and Delinquency* 36:408–18.

Newman, Graeme R. 2008. "Identity Theft and Opportunity." Pp. 9–31 in *Perspectives on Identity Theft. Crime Prevention Studies.* Vol. 23, edited by M. M. McNally and G. R. Newman. Monsey, NY: Criminal Justice Press.

Newman, Graeme R. and Ronald V. Clarke. 2003. *Superhighway Robbery: Preventing E-Commerce Crime.* Devon, UK: Willan Publishing.

Nielsen Online. 2008. *Nielson Online Reports Topline U.S. data for March 2008.* New York: Nielsen Company.

Parks Associates. 2008. *Only One in Five Households Has Never Used E-Mail.* Dallas, TX: Parks Associates.

Piquero, Alex R., John MacDonald, Adam Dobrin, Leah E. Daigle, and Francis T. Cullen. 2005. "Self-control, Violent Offending, and Homicide Victimization: Assessing the General Theory of Crime." *Journal of Quantitative Criminology* 21:55–71.

Pontell, Henry N., Gregory C. Brown and Anastasia Tosouni. 2008. "Stolen Identities: A Victim Survey." Pp. 57–85 in *Perspectives on Identity Theft. Crime Prevention Studies.* Vol. 23, edited by M. M. McNally and G. R. Newman. Monsey, NY: Criminal Justice Press.

Pratt, Travis C. 2009. *Addicted to Incarceration: Corrections Policy and the Politics of Misinformation in the United States.* Thousand Oaks, CA: Sage.

Pratt, Travis C. and Francis T. Cullen. 2005. "Assessing Macro-level Predictors and Theories of Crime: A Meta-analysis." *Crime and Justice: A Review of Research* 32:373–450.

Ratchford, Brian T., Debabrata Talukdar, and Myung-Soo Lee. 2001. "A Model of Consumer Choice of the Internet as an Information Source." *International Journal of Electronic Commerce* 5:7–21.

Reisig, Michael D., Travis C. Pratt, and Kristy Holtfreter. 2009. "Perceived Risk of Internet Theft Victimization: Examining the Effects of Social Vulnerability and Impulsivity." *Criminal Justice & Behavior* 36:369–84.

Rodgers, Karen and Georgia Roberts. 1995. "Women's Non-spousal Multiple Victimization: A Test of the Routine Activities Theory." *Canadian Journal of Criminology* 37:363–91.

Sampson, Robert J. 1987. "Personal Violence by Strangers: An Extension and Test of the Opportunity Model of Predatory Victimization." *Journal of Criminal Law and Criminology* 78:327–56.

Sampson, Robert J. and Janet L. Lauritsen. 1990. "Deviant Lifestyles, Proximity to Crime, and the Offender-victim Link in Personal Violence." *Journal of Research in Crime and Delinquency* 27:110–39.

Sampson, Robert J. and John D. Wooldredge. 1987. "Linking the Micro- and Macro-level Dimensions of Lifestyle-Routine Activity and Opportunity Models of Predatory Victimization." *Journal of Quantitative Criminology* 3:371–93.

Schreck, Christopher J. and Bonnie S. Fisher. 2004. "Specifying the Influence of Family and Peers on Violent Victimization: Extending Routine Activities and Lifestyles Theories." *Journal of Interpersonal Violence* 19:1021–41.

Schreck, Christopher J., Eric A. Stewart, and Bonnie S. Fisher. 2006. "Self-control, Victimization, and the Influence on Risky Lifestyles: A Longitudinal Analysis Using Panel Data." *Journal of Quantitative Criminology* 22:319–40.

Schreck, Christopher J., Richard A. Wright, and J. Mitchell Miller. 2002. "A Study of Individual and Situational Antecedents of Violent Victimization." *Justice Quarterly* 19:159–80.

Shover, Neal, Glenn S. Coffey, and Dick Hobbs. 2003. "Crime on the Line: Telemarketing Fraud and the Changing Nature of Professional Crime." *British Journal of Criminology* 43:489–505.

Shover, Neal, Glenn S. Coffey, and Clinton R. Sanders. 2004. "Dialing for Dollars: Opportunities, Justifications, and Telemarketing Fraud." *Qualitative Sociology* 27:59–75.

Soopramanien, Didier and Alastair Robertson. 2007. "Adoption and Usage of Online Shopping: An Empirical Analysis of the Characteristics of 'Buyers,' 'Browsers' and 'Non-Internet shoppers.'" *Journal of Retailing and Consumer Services* 14:73–82.

Sorce, Patricia, Victor Perotti, and Stanley Widrick. 2005. "Attitude and Age Differences in Online Buying." *International Journal of Retail and Distribution Management* 33:122–32.

Stewart, Eric A., Kirk W. Elifson, and Claire E. Sterk. 2004. "Integrating the General Theory of Crime into an Explanation of Violent Victimization among Female Offenders." *Justice Quarterly* 21:159–81.

Stranahan, Harriet and Dorota Kosiel. 2007. "E-Tail Spending Patterns and the Importance of Online Store Familiarity." *Internet Research* 4:421–34.

Swinyard, William R. and Scott Smith. 2003. "Why People (Don't) Shop Online: A Lifestyle Study of the Internet Consumer." *Psychology and Marketing* 20:567–97.

Titus, Richard, 2001. "Personal Fraud and Its Victims." Pp. 57–74 in *Crimes of Privilege: Readings in White-Collar Crime,* edited by N. Shover and J. P. Wright. Oxford, UK: Oxford University Press.

Titus, Richard, Fred Heinzelmann, and John Boyle. 1995. "Victimization of Persons by Fraud." *Crime and Delinquency* 41:54–72.

U.S. Census Bureau. 2008. *Projected Online Retail Sales.* Washington, DC: U.S. Census Bureau, Housing and Household Economic Statistics Division.

U.S. Department of Commerce. 2008. *U.S. Census Bureau News: Quarterly Retail E-commerce Sales, 4th Quarter 2007.* Washington, DC: U.S. Department of Commerce.

U.S. Department of Justice. 2008. *Crime in the United States, 2006.* Washington, DC: U.S. Department of Justice.

Wall, David S. 2005. "The Internet as a Conduit for Criminals." Pp. 77–98 in *Information Technology and the Criminal Justice System,* edited by A. Pattavina. Thousand Oaks, CA: Sage.

Wall, David S. 2007a. *Cybercrime: The Transformation of Crime in the Information Age.* Cambridge, UK: Polity.

Wall, David S. 2007b. "Policing Cybercrimes: Situating the Public in Networks of Security within Cyberspace." *Police Practice and Research* 8:183–205.

Weisburd, David, Laura A. Wyckoff, Justin Ready, John E. Eck, Joshua C. Hinkle, and Frank Gajweski. 2006. "Does Crime Just Move Around the Corner? A Controlled Study of Spatial Displacement and Diffusion of Crime Control Benefits." *Criminology* 44:549–92.

Willison, Robert. 2008. "Applying Situational Crime Prevention to the Information Systems Security Context." Pp. 169–92 in *Perspectives on Identity Theft. Crime Prevention Studies*. Vol. 23, edited by M. M. McNally and G. R. Newman. Monsey, NY: Criminal Justice Press.

Wilson, Debbie, Alison Patterson, Gemma Powell, and Rachelle Hembury. 2006. "Fraud and Technology Crimes: Findings from the 2003/04 British Crime Survey, the 2004 Offending, Crime and Justice Survey and Administrative Sources. Online Report 9/06."

London: Research, Development and Statistics Directorate, Home Office. Retrieved March 5, 2010 (http://www.homeoffice.gov.uk/rds/offending-survey.html).

Wortley, Richard and Stephen Smallbone. 2006. "Applying Situational Principles to Sexual Offenses against Children." Pp. 7–35 in *Situational Prevention of Child Sexual Abuse, Crime Prevention Studies*. Vol. 19, edited by R. Wortley and S. Smallbone. Monsey, NY: Criminal Justice Press.

Yar, Majid. 2006. *Cybercrime and Society*. Thousand Oaks, CA: Sage.

DISCUSSION QUESTIONS

1. What other theories may explain why people become victimized by Internet fraud? Explain how these theories could apply.

2. What is situational crime prevention? What can people do to protect themselves from Internet fraud? What can the government and the criminal justice system do to protect people from Internet fraud victimization?

3. What makes Internet fraud victimization a unique victimization? Consider the context in which it occurs, the fear of victimization that people experience, the victim-offender relationship, detection, etc.

4. What personal characteristics are related to hours spent online and making Internet website purchases? Why are these characteristics related to these Internet-based behaviors? How do these behaviors impact risk of Internet fraud victimization?

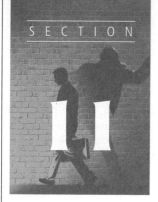

Contemporary Issues in Victimology

Victims of Hate Crimes, Human Trafficking, and Terrorism

C indy was living in rural China and going to school when she was offered the opportunity to work at a restaurant in Ghana that friends of her neighbor and her husband had opened. Cindy dropped out of school and went with the couple to Ghana. Instead of working in a restaurant, however, Cindy fell victim to a Chinese sex trafficking ring. She was taken to live in a brothel, and her passport and return ticket were taken from her. She was forced to engage in commercial prostitution and was beaten anytime she refused. The money she made from prostitution was taken from her by her traffickers, who told her that she owed money for her travel expenses and accommodations (U.S. Department of State, 2010).

What happened to Cindy is, unfortunately, probably more common than we realize. It is known as human trafficking—a contemporary issue in the field of victimology. It is a type of victimization in which a victim is targeted because of what he or she can provide in terms of services, work, or sex. Other types of victimization that we are becoming aware of involve people being targeted because of an offender's hate or bias. Terrorism also often creates victims because of the desire to harm a particular group or person. This section deals with these emerging issues in victimology.

⊠ Victims of Hate Crimes

Some victims are targeted because of their qualities—not because of what they can provide in terms of services, work, or sex, but because the offender wants to attack them out of hate or bias. Qualities that an offender may target are sexual orientation, race, and religion. When victims are targeted under these circumstances, they are classified as **hate crime** or **bias crime** victims.

What Is Hate Crime Victimization?

Beginning in 1990 with the passage of the **Hate Crime Statistics Act** (28 U.S.C. 534), the attorney general is required to collect data on hate crimes perpetrated based on race, religion, sexual orientation, or ethnicity. This requirement was expanded via the Violent Crime Control and Law Enforcement Act of 1994 to include hate crimes based on disability. It was not until 2009, when Congress passed the **Matthew Shepard and James Byrd, Jr. Hate Crime Prevention Act,** that crimes motivated by bias against gender and gender identity and crimes committed by and against juveniles were included in the hate crimes data collection efforts. To meet these mandates, the attorney general has charged the Federal Bureau of Investigation (FBI) with data collection (Harlow, 2005). In doing so, the FBI acquires data from law enforcement agencies about the crimes of murder, nonnegligent manslaughter, forcible rape, aggravated assault, simple assault, intimidation, arson, destruction of property, damage of property, and vandalism of property that are hate or bias motivated. Currently, the FBI defines hate crimes as any of the above mentioned offenses or offenses that law enforcement agencies currently report on through their Uniform Crime Reports (UCR) or National Incident-Based Reporting System data collection programs that are perpetrated based on race, religion, sexual orientation, ethnicity, disability, gender, and gender identity (FBI, 2010a). How do the police know, though, if a crime should be considered a hate crime? Obviously, a crime that is reported to the police is supposed to be included in the statistics they report to the FBI through the UCR data collection. It is up to the police agency, however, to determine if the incident will be classified as a hate crime. Law enforcement, then, has to use their judgment to classify incidents—it is not enough that a victim is part of a protected class (i.e., a racial or ethnic minority). To determine hate crime status, law enforcement use victim and witness information as well as information from the investigation. Information they consider may include the following (Nolan, McDevitt, Cronin, & Farrell, 2004):

- Hate speech the offender used
- Hate symbols left by the offender (e.g., a swastika painted on a synagogue)
- Timing of incident to occur on day of significance for victim's or suspect's group
- Previous history of hate crime perpetration by the suspect or suspect is a known member of organized hate group
- Previous history of hate crime incidents at or around location of current incident
- Absence of other obvious motives of victimization

▲ **Photo 11.1**   On the night of October 6–7, 1998, Matthew Shepard was beaten, pistol-whipped, robbed, and tied to a tree by two men in Laramie, Wyoming. He died on October 12, 1998 from his injuries. At trial, it was revealed that he was targeted due to his sexual orientation. In 2009, the Matthew Shepard and James Byrd, Jr. Hate Crime Prevention Act was passed.

Photo credit: Gina Van Hoof. Reprinted with permission of the Matthew Shepard Foundation.

We will discuss the extent of hate crime victimization based on the data collected from the FBI below, but before we do, a discussion about how hate crime is defined in the National Crime Victimization Survey (NCVS) is warranted.

As you know, the NCVS asks individuals about their victimization experiences during the previous 6 months. Beginning in the July 2000 data collection, persons were asked in the incident report if they suspected that they were victimized because of one or more of their personal characteristics: race, ethnicity, religion, sexual orientation, and disability (Harlow, 2005). That question served as a hate crime **screen question**. If they answered affirmatively, they were also asked if they had any evidence that the incident was a crime of hate, bigotry, or prejudice. Evidence would include the offender used "derogatory language, the offender left hate symbols, or the police confirmed that a hate crime had taken place" (Harlow, 2005, p. 2). If they answered yes to a hate crime screen question and yes to a question regarding evidence of hate, bigotry, or prejudice motivation, then the incident was classified as a hate crime victimization.

Extent of Hate Crime Victimization

According to the statistics compiled by the FBI, in 2009, 8,336 people were the victims of hate crimes in the United States. Similar to other victimizations, there are differences when comparing data from the UCR with data from the NCVS. Keep in mind that UCR statistics reflect official data compiled by the FBI based on law enforcement agencies, who must make an official determination that a crime they are made aware of is, in fact, a hate crime before report-ing it to the FBI as such. The NCVS data rely on the victim's assessment of the incident being motivated by bias. One of the key differences is the extent of hate crime victimizations in the NCVS as compared with the UCR. Between 2000 and 2003, there was an annual average of 210,000 hate crime victimizations in the Unites States according to the NCVS (Harlow, 2005). Although the NCVS measures both victimizations reported to the police and those not reported, victims indicated that about 92,000 of these incidents were, in fact, reported to the police.

Who Are Hate Crime Victims?

Individual Characteristics

One of the interesting findings from the NCVS survey data is that rates of hate crime victimization are similar across demographic groups. In other words, hate crime victimization rates were not different across gender, race, and ethnic groups. This is contrary to what is found for victimization more generally, for which there are differences in victimization rates for males and females, for Whites and Blacks, and for Hispanics and non-Hispanics. Also, hate crime victimization rates do not vary according to educational attainment. There are, however, some characteristics associated with an increased hate crime victimization risk. Persons who are young (17 or younger, 18–20, and 21–29) have higher hate crime victimization rates than do older persons. In addition, persons who are separated or divorced or who have never married are more likely to be victims of a hate crime than those who are widowed or married. Where you live also impacts risk. Persons living in urban areas face a greater chance of hate crime victim-ization than those living in rural or suburban areas. Income level is the final characteristic that differentiates risk. Persons living in households with a total income less than $25,000 have the greatest hate crime victimization rates (Harlow, 2005).

Type of Hate Crime Victimization Experienced

Of the victimizations reported in the UCR, the most common motivation for the offense was the victim's race. In fact, almost half (49%) of all hate crime victims were targeted for this characteristic. The next most common motivation was the victim's religious beliefs (19%), followed by sexual orientation (18%), ethnicity/national origin (13%), and disability (1%) (FBI, 2010b).

Within these categories, the FBI's data further delineate the characteristics of victims for which they are targeted. In terms of racially motivated bias crimes, an overwhelming majority of victims, 72%, were targeted because of an offender's anti-Black bias. Seventeen percent of victims were targeted because of anti-White bias on the part of the offender. See Table 11.1 for more details regarding the percentage of victims targeted. Of the

Table 11.1 Reasons Hate Crime Victims Reported Being Victimized, UCR 2009	
Type of Bias	**% Victims Experienced**
Offender's racial bias (*n* = 4,057)	
Anti-Black bias	71.5
Anti-White bias	16.5
Anti-Asian/Pacific Islander bias	3.7
Anti-American Indian/Alaskan Native bias	2.1
Bias against group of individuals in which more than one race was represented	6.2
Offender's religious bias (*n* = 1,575)	
Anti-Jewish bias	71.9
Anti-Islamic bias	8.4
Anti-Catholic bias	3.7
Anti-Protestant bias	2.7
Anti-atheist/agnostic bias	0.7
Bias against other religions	8.3
Bias against groups of individuals of varying religions	4.3
Offender's sexual-orientation bias (*n* = 1,482)	
Anti-male homosexual bias	55.1
Anti-homosexual bias	26.4
Anti-female homosexual bias	15.3
Anti-bisexual bias	1.8
Anti-heterosexual bias	1.4
Offender's ethnicity/national origin bias (*n* = 1,109)	
Anti-Hispanic bias	62.4
Bias against other ethnicities/national origins	37.6
Offender's disability bias (*n* = 99)	
Anti-mental disability bias	74.7
Anti-physical disability bias	25.2

SOURCE: FBI (2010b).

victims targeted for anti-religious reasons, most were targeted because of anti-Jewish bias. In fact, 72% of victims of anti-religious hate crime were victimized for this reason. As seen in Table 11.1, other victims are targeted for anti-Islamic bias, anti-Catholic bias, anti-Protestant bias, anti-atheist/agnostic bias, and for biases against other religions. Victims targeted due to sexual-orientation bias were most commonly victimized as a result of anti-male homosexual bias (55%), with more than one quarter of victims being targeted because of an offender's anti-homosexual bias.

Table 11.1 displays other common reasons victims were targeted. Victims of hate crimes were also targeted because of an offender's bias toward their ethnicity or national origin. Most typically, these crimes occurred because of an anti-Hispanic bias—62% of victims who were targeted because of ethnicity or national origin bias. The remaining victims in this category were targeted because of other ethnicities or national origins. Finally, offenders may target victims due to a bias against persons with disabilities. Most commonly, victims are targeted due to an anti–mental disability bias. See Table 11.1 for the percentage of victims who were targeted due to disability bias (FBI, 2010b).

Similar to what was found in the data collected by the FBI, NCVS data indicate that race was the most common motivation (55% of incidents) for hate crime victimizations (Harlow, 2005). The second most common motivation victims reported for why they were targeted was their association with persons with certain characteristics. In 29% of incidents, ethnicity was identified as being the reason a victim was targeted (Harlow, 2005). In 18% of incidents, a victim's sexual orientation provoked the victimization (Harlow, 2005). Religion was mentioned in about one in eight hate crime victimizations, and about one in nine hate crime victimizations were motivated by disability (Harlow, 2005).

Special Case: Sexual-Orientation-Bias-Motivated Hate Crime Victimization

Although the FBI's UCR data collection and NCVS data both include information about sexual-orientation-based victimization experiences, other efforts have been made to identity the extent to which persons are victimized due to their sexual orientation or gender identity as well as the effects this specific type of hate crime victimization has on victims. One of the major sources of information on this type of hate crime victimization is the National Coalition of Anti-Violence Programs (NCAVP, 2010), which collects data from a network of some 38 antiviolence organizations that "monitor, respond to, and work to end hate and domestic violence, HIV-related violence, pick-up crimes, rape, sexual assault, and other forms of violence that affect LGBTQ [lesbian, gay, bisexual, transgendered, and queer] communities" (p. 1). Individuals who contact these antiviolence organizations and report victimizations will then be included in the NCAVP report of **anti-LGBTQ hate violence**. Based on this report, in 2009, there were 2,181 victims of anti-LGBTQ hate violence (NCVAP, 2010). There were 22 murders associated with anti-LGBTQ hate violence in this same year according to the NCAVP (2010) report. Notice that both of these estimates are higher than what was found when examining hate crime statistics in the UCR. Other research using survey data estimates that about 20% of gay, lesbian, and bisexual individuals have experienced a victimization based on their sexual orientation during their adult life (Herek, 2009). Read about the extent and types of victimization experienced by gay, lesbian, and bisexual individuals in the article in this section by Gregory Herek (2009). Also, see box for an example of a real victim of an anti-LGBTQ attack.

TARA'S STORY OF EXPERIENCING ANTI-LGBTQ VICTIMIZATION

Tara, 24, White, queer, non-transgender woman

I was going to the community center like I did every week for work meetings, and this man that always stared at me and yelled that he would "turn me straight" was there. After my meeting I was walking to my car and the man followed me and raped me. The whole time he was talking about making me a straight girl like I should be. I called NCAVP after my nightmares about that night started coming back and I would get panic attacks when I had to go to work. They gave me support and got me connected with a counselor that has really helped me to work on healing from this.

SOURCE: NCAVP (2010).

Characteristics of Hate Crime Victimizations

Of the hate crime victimizations known to the police that occurred in 2009, 62% were crimes against the person (FBI, 2010b). Of these, 45% were crimes of intimidation, 35% were simple assaults, and 19% were aggravated assaults (FBI, 2010b). Only eight murders and nine rape hate crime offenses were reported to the police (FBI, 2010b). Of the crimes against property, 83% were acts of destruction, damage, or vandalism (FBI, 2010b).

Findings from the NCVS also show that hate crimes were most commonly associated with violent victimizations. Rape/sexual assault, robbery, and assault together accounted for 84%, on average, of the hate crime victimizations occurring between 2000 and 2003 (Harlow, 2005). Hate crime victimizations were more likely to be violent victimizations than were non-hate crime victimizations (Harlow, 2005). More than half of all hate crime victimizations were simple assaults; however, 19% were aggravated assaults, 5% were robberies, and 4% were rapes/sexual assaults (Harlow, 2005). Theft was the most common property hate crime victimization experienced (Harlow, 2005).

The largest percentage of hate crime victimizations occurred during the daytime, specifically between noon and 6 P.M. (Harlow, 2005). About 4 in 10 of all hate crime victimizations occurred during this time period (Harlow, 2005). Slightly more than one quarter of hate crime victimizations happened between 6 P.M. and midnight (Harlow, 2005).

Most hate crime victimizations occurred in public areas such as at a commercial establishment, a parking area, or on the street (Harlow, 2005). Slightly more than one in five of all hate crime victimizations took place within 5 miles of the victim's home (Harlow, 2005). According to the NCAVP report, anti-LGBTQ victimizations spiked during the month of October 2009, when the Matthew Shepard and James Byrd, Jr. Hate Crime Prevention Act was passed to include gender and gender identity in the federal hate crimes law (NCAVP, 2010).

Persons who indicated in the NCVS that they had experienced a hate crime victimization were asked if they told the police about the incident. About 44% of hate crime victimizations were reported to the police (Harlow, 2005). Research on hate crime victimization reporting indicates that minority victims are less likely than nonminority victims to report hate crime victimizations (Zaykowski, 2010). Reporting rates were about the same for violent hate crime victimizations and violent non–hate crime victimizations (Harlow, 2005). Overall reporting rates may be linked to the victim-offender relationship—more than half of violent hate crime victimizations were perpetrated by strangers (Harlow, 2005). Not all research on reporting has found that reporting is equally likely for hate crime victimizations and non–hate crime victimizations. When comparing hate crime victimizations based on sexual orientation to non–hate crime victimizations among individuals living in Sacramento, California, hate crimes were less likely to be reported to the police (Herek, Gillis, & Cogan, 1999). This lack of reporting may be due to the response that victims receive from the police. In the NCAVP (2010) report, the majority of victims who reported to the police indicated that the police response was something other than courteous. Eleven percent of victims who experienced anti-LGBTQ victimizations said the police were verbally abusive without slurs, 7% indicated they were verbally abusive with slurs, and 3% said the police were physically abusive (NCAVP, 2010).

Risk Factors for Hate Crime Victimization

Obviously, given what hate crime victimizations are, certain groups are at risk due to their characteristics. It is, in fact, personal characteristics or what an offender perceives to be personal characteristics—race, sexual orientation, religion, gender identity—that motivate hate crime victimizations. For example, a recent study found that gay men, lesbians, bisexuals, and people who have had a same-sex partner are 1.5 to 2 times more likely to experience violence as are people in the general population (Roberts, Austin, Corliss, Vandermorris, & Koenen, 2010). It also has been noted that for anti-LGBTQ hate crime victimizations, people may be targeted because offenders perceive that they are unlikely to notify the police due to the systemic discrimination they have faced (NCAVP, 2010).

Not all people who have characteristics covered in hate crime laws, however, are at risk of being victimized. Routine activities and lifestyles theory may be useful to understand why some people are victimized and others are not. According to this perspective, motivated offenders are ubiquitous and do not need to be explained (Cohen & Felson, 1979; Hindelang, Gottfredson, & Garafolo, 1978). Being a suitable target increases risk, though, as does a lack of capable guardianship. A person who has characteristics an offender "hates" is likely to activate an already motivated offender. For hate crimes, a motivated offender can simply see a person whom he or she perceives to have the characteristics the offender does not like. When this target lacks capable guardianship, a hate crime victimization is likely to occur.

In addition to individual risk factors, hate crime victimization may be a response to perceived threat—violations of territory or property, violations of what is sacred, and violations of status (Ehrlich, 1992). Following this threat explanation, economic competition by minorities has been proposed as a reason why racially and ethnically motivated hate crimes occur (Finn & McNeil, 1987). In addition, there may be community-level factors related to hate crime victimization. Lyons (2007), for instance, found that anti-Black hate crimes were more likely to occur in communities that were organized, while anti-White hate crimes occurred more frequently in disorganized communities. Accordingly, it may not be individual characteristics or routine activities per se that place individuals or groups at risk of hate crime victimization but, rather, underlying cultural or structural conditions.

Consequences of Hate Crime Victimization

We have discussed the innumerable consequences that come with many types of victimization, but the consequences of hate crime victimization deserve special consideration given that such victims are targeted because of characteristics they often cannot change. There is little doubt in victims' minds that their experience was born of hate. Think about how hard this must be to digest and live with afterward—it may result in a person seeing the world as particularly hostile and unsafe. Further, such a victimization may be particularly difficult to forgive (West & Wiley-Cordone, 1999). Because of hate motivation, the community also faces unique consequences.

Consequences for Individuals

Hate crime victimization likely comes with a heavy price for victims. Remember that hate crime victimizations are more likely than non–hate crime victimizations to be violent (Harlow, 2005). In addition, hate crime victims report a greater percentage of victimizations that resulted in injury than do non–hate crime victims (Harlow, 2005). In short, there are often physical consequences associated with hate crime victimization.

Along with these physical consequences, victims of hate crimes also commonly experience psychological consequences. As noted, victims are often unlikely to be capable of changing the characteristic for which they were targeted. As such, they may fear revictimization and feel that there is little they can do to prevent a subsequent attack. Fear may not be all that victims experience. Hate crime victimization has been linked to distress symptoms such as depression, stress, and anger (Roberts et al., 2010). For example, lesbians, gay men, and bisexuals who have experienced a bias crime victimization reported higher levels of fear of crime and perceived vulnerability and less belief in the benevolence of people compared with victims of non-bias crimes and nonvictims (Herek et al., 1999). For hate crime victimizations based on sexual orientation, another consequence is posttraumatic stress disorder (PTSD). In fact, gay men, lesbians, bisexuals, and persons who have had same sex-partners who have been violently victimized are twice as likely to experience PTSD in response to the experience as are people in the general population (Roberts et al., 2010). This heightened response to hate crime victimization for gay, lesbian, and bisexual individuals may be due to the fact that their sexual orientation is an important part of their self-concept and also the reason they were victimized (Herek et al., 1999).

Consequences for the Community

When a hate crime victimization occurs, especially if it is made public, others in the community who share characteristics of the victim may become fearful that they will also become a target. This fear may cause individuals to change their behaviors and may limit their activities (West & Wiley-Cordone, 1999). To the extent that people identify with the victim, they may also experience **secondary victimization** whereby they suffer similar psychological consequences to those of the direct victim (West & Wiley-Cordone, 1999). The community may find that people are less likely to visit as tourists or to move there, and real estate values may plummet after publicized hate crime incidents (West & Wiley-Cordone, 1999).

Responses to Hate Crime Victimization

Legislation

It was not until the 1980s that hate crime victimization received widespread formal attention from the criminal justice system and the media. As special-interest groups began documenting victimizations that they perceived to be motivated by hate or bias, a movement toward recognizing hate crimes as a specialized category of crime took hold. In part, this movement resulted in the passage of legislation that attached enhanced penalties to crimes motivated by hate and that mandated the collection of data.

The first federal law specifically addressing hate crimes, as previously mentioned, was the Hate Crime Statistics Act (Pub. L. 101-275), which mandated the collection of data on hate crimes perpetrated based on race, religion, sexual orientation, or ethnicity. As noted, this requirement was expanded to include hate crimes based on disability via the Violent Crime Control and Law Enforcement Act of 1994 (Pub. L. No. 103-322). This law also directed the U.S. Sentencing Commission to develop or amend federal sentencing guidelines to provide enhancement of at least three offense levels for those offenses determined by judges beyond a reasonable doubt to be hate crimes. But how were hate crimes treated in federal court before 1990? There was not specific federal legislation that dealt with hate crimes. Rather, Title 18, U.S.C., Section 245 was used to hold people accountable for willfully injuring, intimidating, or interfering with, or attempting to do so by force or threat of force, people because of race, color, religion, or national origin during the course of federally protected activities—such as when attending school, participating as a juror, participating in an activity or program administered by the state or local government, or as a patron of any public accommodation. Although it served to protect certain classes of people, if a federally protected activity was not being engaged in at the time of the offense, the federal government could not intervene unless another federal interest was involved. This limit was removed with the passage of the Matthew Shepard and James Byrd, Jr. Hate Crime Prevention Act. This created a federal hate crimes law that eliminated the government's need to prove that victims were participating in a federally protected activity. Instead, crimes in which the federal government would otherwise have jurisdiction are included (e.g., the offender traveled with the victim across state lines). There are other federal laws in place that serve to prevent bias-motivated crimes, such as the **Church Arson Prevention Act of 1996,** which prohibits damage to religious property because of the religious, racial, or ethnic characteristics of the property or the obstruction of or attempt to obstruct by force or threat of force a person exercising religious beliefs.

In addition to these pieces of federal legislation, states are also free to enact hate crime legislation to criminalize behavior that occurs within their borders. As recently as mid-year 2009, all but four states had hate crime legislation that included crimes based on individual characteristics (National Center for Crime Victims, 2008). See the following box for an example of California's hate crime law provisions.

CALIFORNIA'S HATE CRIME LAW PROVISIONS, CAL PEN CODE § 422.6

Who Is Protected?

Any person who is injured or threatened or has property damaged because of actual or perceived "race, color, religion, ancestry, national origin, disability, gender, or sexual orientation."

What Is the Penalty?

Persons who commit hate crime(s) as described above shall be punished by imprisonment for no more than one year or may be subject to a fine for no more than $5,000 and or both imprisonment and the fine. In addition, the court may order that community service be completed, not to exceed 400 hours, that is performed over a period of no more than 350 days (with consideration of his or her employment or school attendance).

SOURCE: FindLaw (2011).

Criminal Justice System Response

The criminal justice system is also set up to address this special type of crime victimization. As mentioned, the police are required to record and report data on hate crimes that occur. Research on compliance with the Hate Crime Statistics Act is somewhat sobering. King (2007) reported that police compliance with the requirements is less likely in jurisdictions that have greater populations of Black residents. This is particularly salient when considered in conjunction with the fact that racially motivated hate crimes are more likely to occur in these same areas. He further found that in the South, the correlation between the size of the Black population and police compliance with the reporting requirements of the Hate Crime Statistics Act was negative—in areas with high Black populations, adherence was low (King, 2007). Why do you think police are less likely to meet the requirements of federal reporting in these areas? What are the implications for these communities?

Also, most states have some type of hate crime legislation that specifically designates crimes motivated by hate as a special type of crime deserving of a special penalty. Remember, though, that even without hate crime laws, criminal behavior is criminal behavior and a hate crime statute need not be in place for police to make an arrest when there is reason to do so. To address hate crimes, some police departments have specialty units (Levin & Amster, 2007). Boston was the first city, in 1978, to create a special unit whose mission was to respond to community disorders, including hate- or bias-motivated crimes (Levin & Amster, 2007). New York City founded a similar unit in 1980 (Levin & Amster, 2007). Many other jurisdictions today have similar units designed to address hate crimes, which have produced positive impacts on hate crime (Levin & Amster, 2007).

Once a crime is identified as being bias motivated, the prosecutor makes the decision whether to charge it as a hate crime. Some research suggests that prosecutors use their discretion, as they do when making charging decisions in other cases, when making this important decision. In fact, research shows that prosecutors are likely to charge bias crimes only when they are clearly bias motivated and when evidence overtly shows this bias (Bell, 2002; Phillips, 2009). Those cases in which bias is but one motive are unlikely to be charged as bias crimes—only "clean" cases, in which bias is the singular motive, are likely to be charged as bias crimes by a prosecutor (Maroney, 1998). Thus, when you consider that police must initially determine whether a crime is a hate crime (and only after the victim has decided to report) and then the prosecutor must determine whether to charge it as a hate crime—but only after the first step has been taken by the police—it should make sense that few hate crime victimizations actually result in formal charges in the criminal justice system.

Even if a case is not formally charged, hate crime victims can receive assistance from other sources. Victims of hate crimes can access traditional services designed for crime victims. That is, they are eligible to receive victim compensation so long as they meet eligibility requirements, they can receive services from Victim-Witness Assistance Programs, they can agree to participate in victim-offender mediation or reconciliation programs, and they can exercise any rights granted to them in the jurisdiction in which they were victimized. Along with these traditional services, they can also seek assistance from special interest organizations and groups (e.g., Kansas City Anti-Violence Project).

⊠ Victims of Human Trafficking

Human trafficking has become a hotbed issue domestically and internationally, particularly in the past two decades. Interestingly, although media and governmental attention have grown dramatically over this time period, human trafficking is not a new victimization; it has its roots in slavery. Throughout history, we see examples of various forms of slavery, and America was not immune to the practice. Even though laws were passed in the United States to abolish slavery in 1863, these laws did not specifically address the issue of recruiting people for prostitution in another country and did not end the selling or forcing of African Americans into labor to pay debts (Logan, Walker, & Hunt, 2009). In other words, although race-based slavery was condemned, other forms of enslavement and trafficking were under way and were not similarly abolished (Logan et al., 2009).

What Is Human Trafficking?

Trafficking was not a federal crime until 2000, although the behaviors that are now prohibited may have been criminalized in other statutes. The Trafficking Victims Protection Act of 2000 (Pub. L. 106-386) defines human trafficking as

(1) sex trafficking in which a commercial sex act is induced by force, fraud, or coercion, or in which the person induced to perform such act has not attained 18 years of age; or (2) the recruitment, harboring, transportation, provision, or obtaining of a person for labor or services, through the use of force, fraud, or coercion for the purpose of subjection to involuntary servitude, peonage, debt bondage, or slavery.

This is not the only definition of trafficking. The United Nations Protocol to Prevent, Suppress, and Punish Trafficking in Persons defines trafficking as

the recruitment, transportation, transfer, harbouring or receipt of persons, by means of the threat or use of force or other means of coercion, of abduction, of fraud, of deception, of the abuse of power or of a position of vulnerability or of the giving or receiving of payments or benefits to achieve the consent of a person having control over another person, for the purpose of exploitation. Exploitation shall include, at a minimum, the exploitation of the prostitution of others or other forms of sexual exploitation, forced labour or services, slavery or practices similar to slavery, servitude or the removal of organs (Article 2 Statement of Purpose).

The type of human trafficking related to sexual exploitation is known as **sex trafficking**. The type of human trafficking used to exploit someone for labor is known as **labor trafficking**. Persons do not have to be transported across international borders to be trafficked. Trafficking can be done within a country's borders. When this occurs, it is known as **domestic human trafficking**. When persons are transported into another country, it is called **transnational human trafficking**. Take note, however, that the actual transport of a person is not necessary for an incident to be considered human trafficking—you can become a victim of trafficking within your own home. If you were held against your will in your home and were made to perform sex acts on persons who came to the house, you would be a victim of sex trafficking.

Let's look at the different types of trafficking a little more closely. Sex trafficking occurs when a person is coerced, forced, or deceived into prostitution. If the initial decision to participate in prostitution was consensual but the person is forced or coerced into continuing, then he or she is a victim of sex trafficking (U.S. Department of State, 2010). In this way, an initial consensual decision to engage in prostitution does not preclude a person from being considered a sex trafficking victim if that person wants to quit but is forced to continue. Persons can be forced to work in sex clubs, dance in strip clubs, work in massage parlors, or be otherwise enslaved or forced into other types of commercial sex work (e.g., street prostitution, pornography). Age is also a consideration when determining sex trafficking status. In the United States, if the victim is under the age of 18, there does not have to be force, fraud, or coercion in order for sex trafficking to occur. The victim simply being under the age of 18 and being recruited, harbored, transported, or obtained for the purpose of a commercial sex act is "enough" for sex trafficking to occur (Shared Hope International, 2010). Sex trafficking can also occur within debt bondage, another form of human trafficking discussed next.

Labor trafficking can be further divided into several types—bonded labor, forced labor, and involuntary domestic servitude. **Bonded labor,** also referred to as **debt bondage,** is the most common method of enslaving victims (U.S. Department of Health and Human Services [DHHS], 2011b). This type of labor trafficking occurs when a person is enslaved to work off a debt he or she owes when the terms or conditions of the debt were not previously known or properly defined or when the services the victim provides are not calculated in a reasonable fashion and "counted" toward the debt owed (DHHS, 2011b). The second type of labor trafficking is **forced labor,** which occurs when victims are forced to work under the threat of violence or punishment, their freedom is restricted, and some level of ownership over them is exercised (DHHS, 2011b). The third type of labor trafficking is **involuntary domestic servitude**. This type of trafficking involves victims who are forced to work as domestic workers. Often, these victims work in isolation on private property, which makes their situations difficult to detect since authorities may not be able to enter and inspect these locations (U.S. Department of State, 2010). Think back to Cindy's story at the beginning of this section. What type of human trafficking did she experience? How do you know?

Extent of Human Trafficking

As you may imagine, it is quite difficult to know the true extent to which people are trafficked within the United States and internationally. Generally, it is widely accepted that most human trafficking goes undetected, so reliance on official data sources most likely grossly underestimates the extent of human trafficking. The reasons that most human trafficking remains undetected are many. First, human trafficking is, by its nature, underground and hidden (Schauer & Wheaton, 2006). It is a type of victimization that occurs out of sight of most people and, therefore, is difficult to detect. Second, many countries do not treat trafficking seriously and in some countries law enforcement and other officials actively participate in human trafficking. Third, most victims do not report their victimizations to authorities. Many victims are unable to report and others are unwilling to report. Victims are often taught to fear law enforcement and other groups that would be able to help them (U.S. Department of State, 2010). Other victims receive threats of further harm if they report their victimizations to authorities. Fourth, victims are often a mobile population—they are frequently moved from one location to the next—and, thus, counting them is difficult at best (Schauer & Wheaton, 2006). Fifth, the definition of what constitutes human trafficking is not clear in every jurisdiction, despite formal definitions put forth by the United Nations (Gozdziak & Collett, 2005). Likewise, not all countries collect data about the number of human trafficking victims within their borders.

Despite these hindrances, there have been efforts to determine the extent to which humans are trafficked. The U.S. Department of State (2010) publishes the Trafficking in Persons Report (TIP Report), which provides information about trafficking throughout the world using information from U.S. embassies, government officials, nongovernmental and international organizations, reports, research trips to regions included in the report, and information submitted to the State Department's e-mail address for the TIP Report. From this report, it is estimated that 12.3 million persons are in forced labor, bonded labor, and forced prostitution in the world and as many as 2 million children are in the global commercial sex trade (U.S. Department of State, 2010). It is estimated that between 600,000 and 4 million persons each year are trafficked worldwide (McCabe & Manian, 2010). In 2009 alone, 49,105 victims were identified in the TIP Report. Although much of the attention surrounding human trafficking is on sex trafficking, according to the TIP Report, there are more people trafficked for the purposes of forced labor than for commercial sex (U.S. Department of State, 2010). In fact, it is estimated that for every one person forced into prostitution, there are nine people forced into labor (U.S. Department of State, 2010).

Domestic trafficking within the United States also has been investigated. As part of the Trafficking of Victims Protection Reauthorization Act of 2005 (Pub. L. 109-164), biennial reporting on human trafficking is now required. In response to this requirement, the U.S. Department of Justice has supported the creation of the Human Trafficking Reporting System, which provides information on investigations opened by 38 federally funded human-trafficking task forces in the United States (Kyckelhahn, Beck, & Cohen, 2009). Based on this data collection, from January 1, 2007, to September 30, 2008, there were 1,229 investigations opened by these 38 agencies, 112 of which were confirmed to be human trafficking (123 had a decision pending and another 708 had no data about their confirmation) (Kyckelhahn et al., 2009). The vast majority of these incidents (83%) were sex trafficking incidents.

Who Is Trafficked?

For the same reasons it is difficult to know the true extent to which human trafficking occurs, it is also difficult to know who the "typical" victim of trafficking is. As a result, the description of who is most likely to be a human trafficking victim is likely only as reliable as the data available. Nevertheless, it is estimated that females make up 56% of all victims of human trafficking worldwide (U.S. Department of State, 2010). In the United States, females make up 94% of confirmed human trafficking incidents reported by task forces (Kyckelhahn et al., 2009). Females make up a larger percentage of victims of sex trafficking (almost 100%) than of labor trafficking (61%) in the United States, as reported by task forces (Kyckelhahn et al., 2009). In terms of who is trafficked in the United States, data indicate that the majority are immigrants (Logan, Walker, & Hunt, 2009). Worldwide, about half of all victims of trafficking are under the age of 18 (McCabe & Manian, 2010). Trafficking appears to be more prevalent in some areas of the world than in others. In Asia and the Pacific, the prevalence of trafficking is estimated to be 3 persons per 1,000 inhabitants, compared with the worldwide estimate of 1.8 per 1,000 inhabitants (U.S. Department of State, 2010).

Russia supplies most of the women and girls for sex trafficking (Schauer & Wheaton, 2006). This means that Russia is a popular **source country** for sex trafficking. Other former Soviet bloc countries are also popular source countries for the sex trafficking industry (Schauer & Wheaton, 2006). The countries that receive victims of human trafficking are known as **destination countries**. Germany is the top destination country for sex trafficking of women and girls (Schauer & Wheaton, 2006). The United States is the second most popular destination country for sex trafficking, with most victims coming from Asia, Mexico, and former Soviet bloc countries (Schauer & Wheaton, 2006). Other popular destination countries are Italy, the Netherlands, Japan, Greece, India, Thailand, and Australia (Schauer & Wheaton, 2006).

Anyone can also become a victim of labor trafficking. Victims can be children or adults and women or men. Even so, women and children are more likely to become victims of labor trafficking than are others (DHHS, 2011b). Victims may work in legitimate jobs such as domestic, factory, or construction work, while others may be involved in illegal activities such as distributing drugs (DHHS, 2011b).

Risk Factors for Human Trafficking

Individual Risk Factors

One of the most pervasive factors placing persons at risk of being trafficked is living in extreme poverty. People living in some countries face little opportunity for gainful employment, and women in these countries especially face economic and social oppression (Schauer & Wheaton, 2006). Persons living in extreme poverty and/or those who are oppressed are vulnerable to accepting promises of work in other countries—work that sometimes turns into labor trafficking when they are forced into debt bondage upon arrival in the destination country. This desire for work may be the impetus for females willingly going with traffickers to other countries only to find that they owe their traffickers exorbitant amounts of money for falsified documents used to enter the country. They may also realize that the only work "available" to them is sex work (Miller, Decker, Silverman, & Raj, 2007).

Country-Level Risk Factors

Not only are there factors that may place individuals at risk for being trafficked, either by making them suitable targets or by pushing them into choices that lead them into trafficking victimization, but characteristics about countries also can make trafficking more or less likely. Countries characterized by high levels of civil unrest and violence are more likely to have trafficking networks operating within them (Logan et al., 2009). Similarly, countries that provide little opportunity for social mobility and few economic opportunities are likely to have more victims of trafficking (DHHS, 2011b). In addition, some countries have greater levels of acceptance of trafficking and their governments do little to address this problem, even sometimes working to facilitate trafficking and not punishing traffickers (Logan et al., 2009). In countries where government officials and law enforcement are corrupt and easily bought off, traffickers are likely to pay off these officials. For example, in Bosnia and Herzegovina, victims have reported police officers actually participating in their transport and their traffickers giving "something" to the Border Police (Rathgeber, 2002). The role of women in many countries also contributes to their likelihood of being trafficked. Women who live in cultures that objectify and stigmatize them while at the same time barring them from legitimate employment opportunities may be more likely to be victims of trafficking than those who live in cultures where women are treated with more respect and allowed entry into legitimate employment (Schauer & Wheaton, 2006).

Consequences for Victims of Human Trafficking

There are obvious physical health consequences for persons who have been trafficked. They are often harmed physically while being held against their will. In addition, victims who are trafficked for labor are often forced to work long hours in deplorable conditions, which can take a toll on their bodies. These physical effects can be both immediate and long-lasting. Victims of labor trafficking may be exposed to dangerous working conditions and, as a result, may suffer health problems such as back pain, hearing loss, cardiovascular and respiratory problems, and limb amputations (DHHS, 2011b).

Health consequences also have been identified for victims of sex trafficking. In a study of women who had received services in Europe for sexual exploitation, 90% of the victims reported some type of sexual violence while they were trafficked and 76% indicated they had been physically abused during their trafficking experience (Zimmerman et al., 2008). In addition to experiencing violence, the majority of these 192 women and adolescent girls reported experiencing headaches, dizzy spells, back pain, memory difficulty, stomach pain, pelvic pain, and gynecological symptoms during the previous 2 weeks (Zimmerman et al., 2008). Other research on victims of sex trafficking has noted forced or coerced use of drugs or alcohol (Zimmerman et al., 2003).

In addition to these physical health consequences, the mental health outcomes of trafficking victims have been examined. Reported psychological effects of labor trafficking include shame, anxiety disorders, PTSD, phobias, panic attacks, and depression (DHHS, 2011b). Women and girls trafficked for sex exploitation reported high levels of depression, anxiety, hostility (Zimmerman et al., 2003, 2008), PTSD, suicidal ideation, and suicide attempts (Zimmerman et al., 2003).

Response to Human Trafficking Victims

International Response

In 2000, the United Nations adopted the Protocol to Prevent, Suppress, and Punish Trafficking in Persons, Especially Women and Children (called the **Palermo Protocol**). This international protocol called for the criminalization of human trafficking and outlined how governments should respond to the problem of trafficking. Specifically, the Palermo Protocol identified that governments should include elements of prevention, criminal prosecution, and victim protection—the "3P" paradigm of governmental response (U.S. Department of State, 2010). Despite the Palermo Protocol being adopted more than 10 years ago, there are still 62 countries that have not convicted a trafficker under laws in compliance with the Palermo Protocol (U.S. Department of State, 2010). In addition, in 2009 only 29 countries were rated in Tier 1 of the TIP Report, indicating that they are in full compliance with the Trafficking Victims Protection Act's minimum standards (U.S. Department of State, 2010).

Victims who have traveled to foreign countries also have special needs tied to their immigration status. According to the Palermo Protocol, governments should not simply deport human trafficking victims if they are found to be in countries illegally. Despite this mandate, 104 countries do not have laws, policies, or regulations to prevent such deportation (U.S. Department of State, 2010). So how do countries deal with victims of human trafficking who are in foreign countries? One response is **repatriation** of foreign victims, although this should be done only if it serves the victims' best interests. It is possible that returning victims to their places of citizenship will put them back in the same context and conditions that led to their being trafficked originally, and some face violence or death upon return to their home countries (U.S. Department of State, 2010).

United States Governmental/Criminal Justice Response

The United States has recognized that trafficked persons who are non-U.S. citizens have special needs regarding immigration and citizenship. To address this issue, victims can become **certified** so they can receive the same benefits and services from federal and/or state programs generally provided to refugees (DHHS, 2011a). To become certified, victims of human trafficking must meet three criteria: (1) be a victim of a severe form of trafficking according to the Trafficking Victims Protection Act of 2000; (2) be willing to assist in the investigation and prosecution of trafficking cases (or be unable to because of physical or psychological trauma); and (3) have completed a bona fide application for a T visa or received Continued Presence from the Department of Homeland Security to be able to contribute to the prosecution of human traffickers. A **T visa** allows victims of human trafficking to become temporary residents of the United States. After 3 years, persons with T visas may be eligible for permanent resident status (DHHS, 2011a).

One way in which trafficking has been addressed in the United States is through the adoption of formal law enforcement task forces and investigative entities. For example, an FBI initiative to disrupt human trafficking in the United States and abroad began in 2004. The FBI works with more than 71 human trafficking task force working groups throughout the United States. Since 2004 the FBI has doubled the number of open human trafficking investigations and quadrupled the number of prosecutions and convictions (FBI, n.d.). Research on trafficking task forces indicates that areas that have these task forces have higher detection rates and are more likely to succeed in prosecuting traffickers (Farrell, McDevitt, & Fahy, 2008). The FBI also employs victim specialists who work with victims of human trafficking to assist them with legal needs as well as other services such as child care, immigration issues, employment, education, and job training (FBI, n.d.).

Often, law enforcement officers must arrest victims for the crimes in which they are involved, such as prostitution (Shared Hope International, 2010). Even if doing so keeps the victim safe and frees him or her from his or her captor, this criminalization of the victim's behavior can have serious consequences since it brings the victim into the formal criminal justice system. For example, arresting a victim may actually make him or her ineligible for some victim services, such as victim compensation (Shared Hope International, 2010). To remedy this situation, some states have passed laws that make minors immune to prosecution for prostitution—for example, there is a presumption that minors are coerced into committing prostitution by another person (Shared Hope International, 2010). Victims, as defined by U.S. law, do have the right not to be held in detention or to be charged with crimes in relation to the trafficking offense, so long as they are willing to cooperate with the criminal investigation and prosecution of their trafficking offenders (Logan et al., 2009). If the victim is unwilling to assist in the formal criminal justice process, then she or her may not be protected from charges or deportation (Logan et al., 2009).

Victim Services

Along with the growing response to the problem of human trafficking, numerous resources for victims have been developed and instituted. Victims of trafficking often experience a higher level of distress than other victims, since when they seek assistance or are discovered they often have few, if any, resources of their own (Logan et al., 2009). Despite this need, victims of trafficking have fewer resources available to them than do victims of other crimes (Logan et al., 2009).

One option that may be available for victims is to be taken to a protective shelter. However, there is not enough space for all victims to use this option. For example, in the United States, there are fewer than 100 beds in facilities that specialize in treating victims of sex trafficking who are under the age of 18 (Shared Hope International, 2010). Reports from other countries also show a shortage of open beds in safe houses. For example, in Bosnia and Herzegovina, women victims have slept in chairs at police stations while waiting to be questioned instead of being taken to a safe house for the night (Rathgeber, 2002). It is also important to consider that female victims should be interviewed by female police officers who are specially trained in how to respond to victims of trafficking (Rathgeber, 2002). In the United States and abroad, victims should also be offered support counselors or advocates if they testify in court (Rathgeber, 2002). As human trafficking continues to receive attention as a serious problem throughout the world, the resources available to victims will, hopefully, continue to expand to meet their unique needs.

⊠ Victims of Terrorism

Perhaps no victimization event in the past 20 years has impacted people in the United States more profoundly than the terrorist attacks that took place on September 11, 2011, in which almost 3,000 people lost their lives when planes were hijacked and crashed into the World Trade Center, into the Pentagon, and in Shanksville, Pennsylvania (National Commission on Terrorist Attacks upon the United States, 2004). An event so cataclysmic and profound, fortunately, does not happen very often. But people are impacted by terrorism throughout the world in less publicized incidents. This section discusses the extent to which terrorism victimization occurs, to whom it occurs, the impact it has, and how victims of terrorism can be assisted.

▲ **Photo 11.2** The aftermath of the World Trade Center attacks on 9/11.

Photo credit: © Bernd Obermann/Corbis

Extent of Terrorism Victimization

Outside Lahore High Court in GPO Chowk, Lahore, Punjab, Pakistan on January 10, 2008, a suicide bomber detonated an improvised explosive device that was strapped to his body as he approached a group of riot police. In this single incident, 17 police officers were killed, along with 8 civilians. Eighty other people were wounded and at least six vehicles were destroyed (National Counterterrorism Center [NCTC], 2009). Despite its severity, you probably did not hear about it, given that it was but one of the many terrorist attacks against noncombatants that occurred during that year.

It is difficult to know to what extent people are victimized by terrorism throughout the world. Not all incidents are recorded or reported—many may occur in remote areas of the world. The most comprehensive data available on terrorism incidents in the world is reported by the National Counterterrorism Center (NCTC), which is required by the U.S. State Department to compile statistical information on terrorism incidents—the number of individuals killed, injured, or kidnapped by terrorist groups during the preceding calendar year "as reported in open source media" (NCTC, 2009, p. 1). According to this report, about 11,000 terrorist attacks occurred worldwide in 2009. These attacks resulted in more than 58,000 persons being victimized—indicating that multiple people were harmed in many of the incidents (NCTC, 2010). U.S. citizens were relatively safe from terrorism in 2009—9 citizens were killed, 14 were injured, and 4 were kidnapped worldwide (NCTC, 2010).

Who Are Victims of Terrorism?

Although there were 11,000 terrorist attacks, almost 48,000 persons were victimized, indicating that multiple people were harmed in many of the attacks (NCTC, 2010). Of these 48,000 persons, two thirds were civilians. Although it is difficult to know exactly who the victims are in all cases, data indicate that police officers made up 14% of all those who were injured or killed (NCTC, 2010). Other commonly targeted people are government officials, employees, and contractors—making up 5% of terrorist victims in total (NCTC, 2010). In terms of fatalities, more than half of all deaths attributed to terrorism were of civilians and 16% were of police officers (NCTC, 2010). Most of the victims of terrorism in 2009 were Muslims (NCTC, 2010). Sadly, children are disproportionately victimized by terrorism (NCTC, 2009).

Researchers examining data on terrorism collected from other sources have also attempted to identify the characteristics of persons who are most likely to be victims of terrorism. In a study of terrorism that occurred in Uruguay, Northern Ireland, Spain, Germany, Italy, and Cyprus, it was found that most victims were between the ages of 20 and 39 years, were males, and were members of security forces (Hewitt, 1988). Research on terrorism victims in Northern Ireland confirms this finding regarding males being the most likely victims (Fay, Morrissey, & Smyth, 1999). Other research examining Israeli civilians killed in Israel via acts of terror between September 1993 and the end of 2003 found that the majority of victims were male. Further, persons aged 17 to 24 composed 30% of fatalities attributed to terrorism in Israel even though they made up only 14% of the population (Feniger & Yuchtman-Yaar, 2010). In Israel, most of the fatalities were Jewish; this finding is contrary to the NCTC (2010) report, which found that, at least in 2009, Muslims were the largest group of victims.

Characteristics of Terrorism Victimizations

In 2009, most of the terrorism incidents were perpetrated via armed attacks, bombings, and kidnappings (NCTC, 2010). There was also indication that terrorists targeted first responders at attack sites (NCTC, 2010). Suicide attacks composed a small fraction of all terrorist activities recorded in the NCTC report (299 out of the 11,000 attacks). Of all attacks that resulted in injury (not just suicide attacks), terrorists most commonly used improvised explosive devices, explosives, and vehicle bombs (NCTC, 2010). Of the terrorist attacks that resulted in deaths, almost half were caused by bombs (NCTC, 2010). Armed attacks caused the second most deaths (NCTC, 2010).

Terrorist attacks impact persons throughout the world, but some areas were disproportionately hit. Almost 44% of the attacks occurred in South Asia, including 6,270 fatalities attributed to terrorism, which was 42% of all fatalities worldwide (NCTC, 2010). The second area hit hardest by terrorism was the Near East, with almost 30% of all terrorist attacks occurring there (NCTC, 2010). Iraq alone accounted for 75% of the terrorist attacks that occurred in the Near East during 2009 and 3,654 deaths (NCTC, 2010). Iraq was the country with the most attacks and deaths due to terrorism in 2009 (NCTC, 2010). In comparison, there were 444 attacks and 377 deaths in the Western Hemisphere due to terrorism; 323 of these deaths occurred in Colombia (NCTC, 2010). See Figure 11.1 for the breakdown of deaths resulting from terrorism in 2009.

Interestingly, when analysis of terrorism data is expanded across time from 1970 to 2006, a somewhat different picture emerges. Latin America was the region that experienced the greatest percentage of fatal and nonfatal attacks, followed by Western Europe (LaFree, Morris, & Dugan, 2010). Moreover, data show that terrorism grew during the 1970s, with a peak in 1992 and another in 2006 (Dugan, LaFree, Cragin, & Kasupski, 2008).

Risk Factors for Terrorism Victimization

One of the scariest things about terrorism is its unpredictability. Although some terrorism is targeted at specific groups or individuals, other terrorist activities are targeted more generally at "enemy" groups (Feniger &

Figure 11.1 Deaths From Terrorism by Country, 2009

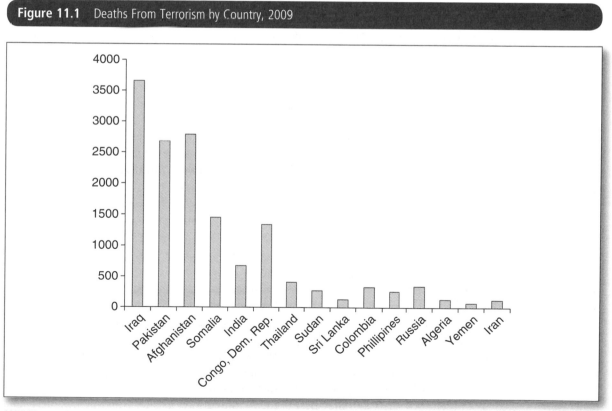

SOURCE: NCTC (2010).

Yuchtman-Yaar, 2010). This type of terrorism appears to be indiscriminate, targeting anonymous individuals who belong or appear to belong to a group, but the precise identities of the individuals do not matter (Feniger & Yuchtman-Yaar, 2010). Obviously, some people are more at risk of being victimized than others, given that they belong to what has been identified as an "enemy" group. In addition, the amount of time that people spend in public spaces has been linked to terrorism victimization risk, particularly for indiscriminate forms of terrorism (Feniger & Yuchtman-Yaar, 2010).

These factors make sense when considered from a routine activities theory perspective. Persons who are suitable targets (i.e., deemed part of the "enemy" group) and who lack capable guardianship are at risk of being victims of terrorism when they come into contact in time and space with motivated offenders. Similarly, places that are suitable targets and are without capable guardianship are more likely to be targeted. As noted in the NCTC (2010) report, the United States has implemented strategies to track motivated offenders, harden targets, and create capable guardianship since 9/11; since then, the United States has not suffered a major attack, although other factors such as political forces are also noted as likely playing a role in terrorism activity.

Consequences of Terrorism on Victims

The loss of human life is inarguably the most tangible consequence of terrorism; however, terrorism has far-reaching impacts on victims and, as you will see, on those who are exposed to terrorism but not directly victimized. Along with physical injury, victims of terrorism are likely to be psychologically impacted by their experience. Basic assumptions about the world may change for victims. They may no longer believe that the world is a safe place where good things happen to good people; instead, they may realize that evil things can and do happen even to good people (Gonzalez, Schofield, & Gillis, 2001). In addition to this impact on people's worldview, survivors may suffer from PTSD, anxiety disorders (North et al., 1999), major depression, panic disorder, and agoraphobia (Gabriel et al., 2007).

Following the attacks in Mumbai, India, in November 2008 that killed 164 people and injured at least 308 people, acute stress disorder was assessed in victims admitted to one of the public hospitals. Of the 74 victims, 30% were found to be suffering from acute stress disorder (Balasinorwala, 2009). Survivors of the 1995 Oklahoma City terrorist bombing also experienced negative consequences postdisaster. In a study of 50 of the survivors, 22% experienced bombing-related PTSD (Tucker et al., 2010). Persons who were injured in the March 11, 2004 terrorist attacks in Madrid had prevalence rates of PTSD in the 2 months after the attacks that were 40 times higher than expected given previous rates of PTSD in the Spanish adult population before the attacks (Gabriel et al., 2007).

Acts of terrorism also affect people who are not directly victimized. People may hear about the terrorism in the media, or they may know someone who was harmed. Think back to 9/11—it was difficult to turn the television on in the days and weeks that followed without seeing a news story or update about the events. This exposure, while informative, also could have been damaging (Slone & Shoshani, 2008). Indeed, research shows that persons not directly involved in terrorist attacks still may experience psychological trauma afterward. Research on Americans following the 9/11 terrorist attacks showed that between 3 and 5 days after the event, 44% of people surveyed experienced substantial stress reactions (e.g., repeated disturbed memories, thoughts, or dreams about what happened; difficulty concentrating) (Schuster et al., 2001). Further research investigating the impacts of 9/11 showed that almost 6% of persons surveyed reported PTSD symptoms 6 months after the attacks (Silver, Holman, McIntosh, Poulin, & Gil-Rivas, 2002). Terrorism also has indirect impacts on children. In a study that screened children after the Oklahoma City bombing, it was found that 34% of middle and high school students were worried about their own safety and the safety of their family 2 months later (Gurwitch, Pfefferbaum, & Leftwich, 2002).

Persons who lose loved ones in terrorist attacks may be hit especially hard. A common response is the desire to take revenge on the attacker, although actually taking action along these lines is quite rare (Miller, 2004). More common is a feeling of vulnerability and fear that may lead to isolation, changing daily routines, installing alarms, refusing to be alone or to go out at night, or carrying weapons for safety (Miller, 2004). They may have "survivor's guilt" or feel as though they should have somehow foreseen the attack or kept their loved one safe (Miller, 2004). They may suffer from a loss of appetite, difficulty sleeping, gastrointestinal problems, cardiovascular disease, anxiety, and depression (Miller, 2004).

Responses to Victims of Terrorism

Persons who are victimized by terrorism may access many of the same services available to other types of crime victims, but they may not view themselves in the same light as someone who is the victim of a more traditional crime. As you may imagine, they may not turn to the formal criminal justice system for assistance as might a person who has experienced, for example, a household robbery. Nonetheless, there is awareness that victims of terrorism have special needs and rights that deserve attention.

In the United States, by definition, terrorist acts are federal crimes (Reno, Marcus, Leary, & Turman, 2000). As such, the federal government has developed numerous resources for victims of terrorism. One set of resources centers on legislation. Following the Iran hostage crisis, two pieces of legislation were passed to provide a remedy to the victims. The **Hostage Relief Act of 1980** (Pub. L. 96-449) provided victims, their spouses, and dependent children benefits such as compensation for medical costs, deferral of taxes and penalties, and reimbursement for educational and training costs. It did not, however, provide money to compensate for wages lost during captivity. To remedy this, then-President Reagan signed into law the **Victims of Terrorism Compensation Act** (title VIII of Pub. L. 99-399).

Other pieces of legislation have been passed in response to specific terrorism acts, such as the **Aviation Security Improvement Act** (Pub. L. 101-604); however, the most grand-sweeping pieces of legislation dealing with terrorism were passed in response to the 9/11 attacks. The **Victims of Terrorism Tax Relief Act (2001)** (Pub. L. 107-134) mandates that qualifying payments made to the families of victims of "qualified disasters" for disaster-related expenses are not taxable. Further, death benefits would not be counted as taxable income if the death was a result of the Oklahoma City bombing, the attacks on 9/11, or anthrax-related attacks between September 11, 2011, and January 1, 2002. Crime victim compensation was also extended to victims of 9/11 and their families through the **Air Transportation Safety and System Stabilization Act (2001)** (Pub. L. 107-42). This act established a compensation fund for victims who were killed or physically harmed; they, their spouses, and/or their dependents may receive benefits from the fund. Awards vary for individuals and families (e.g., compensation based on loss of future earnings), and only economic losses that victims could be compensated for in their state in a tort claim are covered (Levin, 2002). In addition, a presumptive amount of noneconomic damages was set—$250,000 for each deceased person, plus $100,000 for a widowed spouse and for each dependent child (Levin, 2002). On average, each family was predicted to be awarded $1.85 million (Levin, 2002). Research on the victims of 9/11 and their families shows that a total of $8.7 billion in benefits has been provided to civilians killed or seriously injured in the 9/11 attacks and their families (Dixon & Stern, 2004). Of this, about $6 million was funded through the Victims Compensation Fund, with awards ranging from $250,000 to $7.1 million (Dixon & Stern, 2004). Other persons also received compensation, such as emergency responders and businesses, and insurance companies and charities also provided compensation to victims. Figure 11.2 shows the amounts that different groups of victims received in quantified benefit payments from different sources.

Figure 11.2   9/11 Quantified Benefits by Victim Group for Those Killed or Seriously Injured at the World Trade Center, the Pentagon, and the Pennsylvania Plane Crash Site

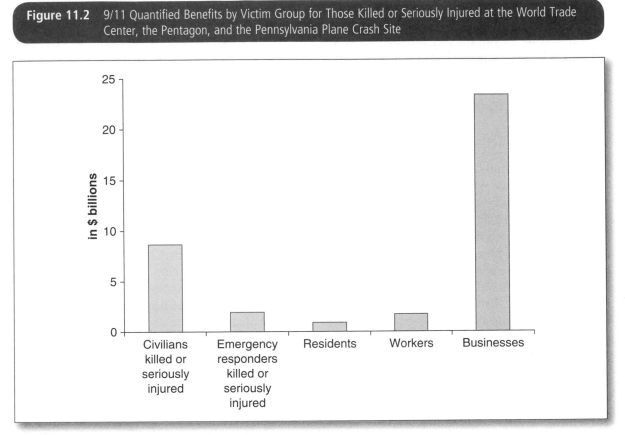

SOURCE: RAND Corporation (2004). Reprinted with permission.

Beyond financial compensation, the Office for Victims of Crime has identified several recommendations for services and policies for victims of terrorism in the United States. Specifically, this agency recommends that

- unnecessary delays in death notification should be avoided;
- release of victim remains and death notification should be handled in a sensitive manner;
- directly after a terrorism event, a "compassion center" should be established where victims and their families can go for information, crisis counseling, and privacy;
- mental health assessment and services should be provided;
- victim compensation and other services' applications should be streamlined—services should be made available in a timely manner, paperwork should be minimized, and agencies should coordinate services;
- states should establish emergency funds that can be used quickly in cases of terrorism;
- citizens who are victimized outside the U.S. should be eligible for compensation and services;
- federal personnel who work with victims of terrorism should receive training on victims' rights laws and services;
- the FBI should keep victims informed of the state of the investigation and events in the case, and identification of victims and access to victim contact information must be established and maintained;

- federal agencies should have a team that can be mobilized on-site to provide support to terrorism victims;
- prosecuting offices need to keep victims apprised of their cases and services while providing support;
- victims should be provided with a media liaison through the U.S. Department of Justice to help with media requests so they can avoid added trauma; and
- federal agencies that employ persons who may be targeted by terrorism should provide information about procedures for responding to employees and their families in the event of victimization (Reno et al., 2000).

Victims of terrorism in other countries are also afforded rights and protections. The Council of Europe's Guidelines on the Protection of Victims of Terrorist Acts (2005, as cited in Kilchling, n.d.) outlines that timely compensation should be given to victims of terrorism and their close family for direct physical and psychological harm. Further, the European Union has provided that victims of terrorism and their family members be protected during criminal proceedings and be provided restitution (Kilchling, n.d.). Victims of terrorism in Israel are also afforded access to compensation for property damage and bodily injuries. Family members of persons killed by terrorism can also be compensated (Kilchling, n.d.).

Although there are other contemporary issues in the field of victimology, human trafficking, hate crime victimization, and terrorism victimization are emerging areas of study that impact not just the United States but the global community as well. As these areas garner further attention, developing methodologies to determine the true extent to which each type of victimization occurs should be at the forefront. Identifying victims is critical to developing policies and implementing services to assist them.

SUMMARY

- The two major types of human trafficking are sex trafficking and labor trafficking. Human trafficking can be domestic or transnational.
- Much like other forms of sexual victimization, sex trafficking considers the age of the victim. In the United States, persons under 18 do not have to be forced or coerced for their victimization to be considered sex trafficking.
- Labor trafficking includes bonded labor and debt bondage, which are the most common forms of trafficking.
- Depending on the cultural gender role of women, one country may be more susceptible to human trafficking than another.
- Government officials sometimes participate in trafficking, and many countries do not treat trafficking seriously; trafficking is often difficult to detect, and estimates of its extent likely underestimate the problem.
- The Trafficking in Persons Report provides estimates of the extent of human trafficking in the United States.
- Females are more likely to be victims of human trafficking than are men. Poverty is one of the most salient risk factors for human trafficking.
- Countries characterized by high levels of civil unrest and violence are more likely to have trafficking networks operating within them. Countries that do not have opportunities for upward social mobility are also more likely to have human trafficking problems.
- Victims of human trafficking experience not only physical health consequences but also mental health problems such as shame, anxiety disorders, posttraumatic stress disorder, phobias, panic attacks, and depression.
- As an international response to human trafficking, the "3P" paradigm—prevention, criminal prosecution, and victim protection—was developed.
- A critical issue with foreign victims of human trafficking is that returning them to their legal home countries could put them at risk for revictimization.
- Victims who meet certain criteria can become certified and obtain a T visa allowing them temporary resident status in the United States.
- Some states have laws of protection for victims who are arrested for the crimes they commit while being trafficked.

- There is a lack of adequate space in protective shelters for victims of human trafficking (e.g., the United States has fewer than 100 beds for victims under age 18 in facilities that specialize in treating human trafficking).
- When victims are targeted out of hate or bias related to race, perceived race, religion, or sexual orientation, it is considered a hate crime.
- Generally, there are differences across gender, race, and ethnic groups for types of persons who are victimized, but this is not the case with hate crime victims.
- Hate crimes are sometimes committed due to the offender perceiving a threat from the victim.
- Because victims of hate crimes usually cannot change the characteristic for which they were targeted, there is a real fear of revictimization.
- If and when a hate crime is made public to the community, there may be some communal consequence, such as tourists deciding not to visit.
- The criminal justice system, specifically prosecutors, use discretion when determining whether the crime committed was a hate crime.
- About 11,000 terrorist attacks occurred worldwide in 2009.
- In 2009, most terrorism victims were male, civilian, and Muslim.
- In 2009, most of the incidents of terrorism were perpetrated via armed attacks, bombings, and kidnappings.
- Obviously, some individuals are more at risk of terrorism victimization than others, given that they belong to a group identified as an "enemy." In addition, the amount of time people spend in public spaces has been linked to terrorism victimization risk, particularly for indiscriminate forms of terrorism.
- Along with physical injury, victims of terrorism are likely to be psychologically impacted by their experiences. Victims may experience posttraumatic stress disorder, depression, anxiety, agoraphobia, and panic disorders.
- The Air Transportation Safety and System Stabilization Act (2001) made it possible for victims of terrorism to be compensated for their losses.
- Victims of terrorism in other countries are also afforded rights and protections.

DISCUSSION QUESTIONS

1. What are some country-level factors that place women at risk of experiencing human trafficking? How do women's roles in their countries protect them or make them more likely to be trafficked?

2. What are some ways the government protects minors who are victims of human trafficking?

3. As it pertains to reporting hate crimes, why do you think the estimates from the Uniform Crime Reports are so different from those of the National Coalition of Anti-Violence Programs?

4. Why do you think some research shows minority victims as more likely to report their hate crimes than nonminority victims?

5. Could it be argued that being a member of security forces puts one at greater risk for terrorism victimization? Why or why not?

KEY TERMS

hate crime

bias crime

Hate Crime Statistics Act (1990)

Matthew Shepard and James Byrd, Jr. Hate Crime Prevention Act

screen question

anti-LGBTQ hate violence

secondary victimization

Church Arson Prevention Act of 1996

sex trafficking

labor trafficking

domestic human trafficking

transnational human trafficking

bonded labor

debt bondage

forced labor

involuntary domestic servitude

source country

destination countries

Palermo Protocol

repatriation

certified

T visa

Hostage Relief Act of 1980

Victims of Terrorism Compensation Act

Aviation Security Improvement Act

Victims of Terrorism Tax Relief Act (2001)

Air Transportation Safety and System Stabilization Act (2001)

INTERNET RESOURCES

U.S. Department of State (http://www.state.gov/)

This website provides detailed information about our country's security as well as the security protocols of other countries with which the Unites States currently allies. It also includes information on counterterrorism, along with country reports on terrorism activities and responses. Go here for information about disputes, terrorism, and other international affairs that impact our security.

"Terrorism" (http://www.fema.gov/hazard/terrorism/index.shtm)

The Federal Emergency Management Agency website provides general information about terrorism, explosions, threats, and homeland security. There are links to emergency contacts for victims. The section on disaster survivors would be most helpful for those who have experienced terrorism victimization. It is clear from this website that the U.S. government is involved in homeland security and ensuring the public is protected.

Humantrafficking.org (http://www.humantrafficking.org/countries/united_states_of_america)

This website provides information about what the government is currently doing to stop human trafficking. Several awareness campaigns are listed with detailed information about the prevention of human trafficking. Also provided are links to other countries that deal with human trafficking and information on what projects they currently use to combat this form of victimization.

"Hate Crime Statistics 2009: Victims" (http://www2.fbi.gov/ucr/hc2009/victims.html)

The U.S. Department of Justice uses the Uniform Crime Reports to generate current statistics on the victims of hate crime (i.e., race, age, religious belief, sexual preference, etc.). The types of hate crime that occurred in 2009, as well as the statistical tables that depict those crimes, are available on this website. The types of offenders that typically commit hate crimes are also detailed.

"Hate Crimes in the United States" (http://www.civilrights.org/hatecrimes/united-states/)

The Leadership Conference on Civil and Human Rights website provides detailed information about hate crimes in the United States. You can find information about the extent of hate crimes in general, but the website also provides statistics and descriptions about hate crimes against specific groups, such as the homeless, immigrants, children, and specific religious adherents. Links to other resources are also provided.

Introduction to Reading 1

Herek's (2009) study is important in that from it, national estimates of the extent of violence and stigma-related experiences of gay, lesbian, and bisexual adults were produced, giving us an idea about the true extent of this type of hate crime victimization. In addition, the differential experiences of stigma-related events are shown for gay, lesbian, and bisexual men and women. These findings were produced from data on 662 self-identified gay, lesbian, or bisexual persons aged 18 and over. Individuals in the study were part of a larger existing panel of data, the Knowledge Networks panel, a group of individuals who have agreed to participate in online surveys. For the current study, a sample of people completed an online survey in which they were asked about their experiences of eight different forms of stigma they had encountered since the age of 18.

Hate Crimes and Stigma-Related Experiences Among Sexual Minority Adults in the United States

Prevalence Estimates From a National Probability Sample

Gregory M. Herek

In 1989, the National Institute of Mental Health convened an expert panel on antigay violence to review existing knowledge and identify research needs. The panel named *collection of prevalence data* as a top research priority and urged that such data be obtained from probability samples when possible (Herek & Berrill, 1990). Since then, data collected by the Federal Bureau of Investigation (e.g., 2005), the National Crime Victimization Survey (Harlow, 2005), and the National Coalition of Anti-Violence Programs (e.g., 2005) have shown that criminal enactments of sexual stigma are widespread (Herek & Sims, 2007). However, prevalence data on criminal victimization among lesbians, gay men, and bisexuals are still fragmentary and derived almost entirely from convenience samples.

Berrill (1992) compiled data from 24 published and unpublished studies conducted between 1977 and 1991 by academic researchers and community-based organizations, all but 1 of them using convenience samples of gay men, lesbians, and bisexuals. Across studies, a median of 9% of respondents reported having been the target of an aggravated assault (i.e., assault with a weapon) because of their sexual orientation; 17% reported simple physical assault (i.e., without a weapon); 19% reported vandalism of their personal property; 44% had been threatened with violence; 33% had been spat on; and 80% had been verbally harassed (Berrill, 1992). Most of the studies did not report data separately by respondents' gender or sexual orientation.

More recently, in a study of 2,259 lesbian, gay, and bisexual adults in the greater Sacramento, California, area, 28% of gay men, 19% of lesbians, 27% of bisexual men, and 15% of bisexual women reported having experienced some type of criminal victimization since

SOURCE: Herek (2009). Reprinted with permission.

age 16 because of their sexual orientation ($n = 898, 980$, 191, and 190, respectively; Herek, Gillis, & Cogan, 1999). This includes respondents who reported experiencing a simple or aggravated assault (13% of gay men, 7% of lesbians, 11% of bisexual men, and 5% of bisexual women) or a sexual assault based on their sexual orientation (4% of gay men, 3% of lesbians, 7% of bisexual men, and 4% of bisexual women).

Other research has focused on particular age groups in sexual minority communities. In a sample of 1,248 young gay and bisexual men ($M = 23$ years, range = 18–27 years) recruited in three Southwestern U.S. cities, 5% reported they had experienced physical violence because of their sexual orientation during the previous 6 months (Huebner, Rebchook, & Kegeles, 2004). In a sample of 194 lesbian, gay, and bisexual youths (age range = 15–21 yrs) recruited from service agencies across the United States, 9% reported at least one aggravated assault based on their sexual orientation, 18% had experienced a simple assault, 22% had been sexually assaulted, and 44% had been threatened with attack (Pilkington & D'Augelli, 1995). In a study of sexual minority youths recruited through community-based organizations in New York City and its suburbs, D'Augelli and his colleagues found that 11% reported physical violence based on their sexual orientation, 9% reported sexual violence, and 78% reported verbal threats or harassment (D'Augelli, Grossman, & Starks, 2006). At the other end of the age continuum, D'Augelli and Grossman (2001) document the lifetime occurrence of hate crime victimization among older (> 59 years) lesbian, gay, and bisexual adults recruited from across the United States ($n = 416$). In that sample, 16% had been physically attacked at some time in their life, 7% had been sexually assaulted, 11% reported having had objects thrown at them, and 29% had been threatened with violence.

It is difficult to use these studies to derive an estimate of the population prevalence of hate crime victimization against U.S. sexual minorities because of variations in how they categorized crimes, the time frames within which they assessed victimization, and how they reported their data (e.g., some studies reported findings separately for men and women, or

homosexuals and bisexuals, whereas others did not). Moreover, because nearly all of the surveys used convenience samples, the extent to which their results describe the entire U.S. gay, lesbian, and bisexual population cannot be determined.

Prevalence data collected in three studies with probability samples further confirm that hate crime victimization is widespread. In a 1989 *San Francisco Examiner* national telephone survey, 5% of gay men ($n = 287$) and 10% of lesbians ($n = 113$) reported having been physically abused or assaulted in the previous year because they were gay (Results of Poll, 1989). In a 2000 Kaiser Family Foundation (KFF) survey of 405 lesbian, gay, and bisexual adults residing in major U.S. population centers, 32% of respondents said they had been targeted for violence against their person or property because of their sexual orientation (Kaiser Family Foundation, 2001). In a probability sample of 912 Latino men who have sex with men, recruited from social venues in New York, Miami, and Los Angeles, 10% reported they had experienced violence as an adult because of their sexual orientation or femininity (Diaz, Ayala, Bein, Henne, & Marin, 2001). These pioneering surveys yielded valuable data but are nevertheless limited in the generalizability of their findings. Neither the *San Francisco Examiner* poll nor the KFF poll were published in a peer-reviewed journal. Few published details are available about the methodology of the *Examiner* survey, making it particularly difficult to evaluate. The KFF poll sampled respondents only in 15 U.S. cities, which may limit its generalizability, and the survey of Latino men focused on specific venues in only three cities.

The present article addresses a gap in current knowledge by reporting data on the prevalence of anti-gay violence and related experiences in a national probability sample of sexual minority adults. Violence against individuals because of their presumed sexual orientation is conceptualized here as a manifestation of *sexual stigma,* that is, society's negative regard for any nonheterosexual behavior, identity, relationship, or community (Herek, 2004, 2008). Sexual stigma is a cultural belief system through which homosexuality is denigrated, discredited, and socially constructed as invalid relative to heterosexuality (Herek, 2008; Herek, Chopp, & Strohl, 2007). As with other forms of stigma, sexual stigma is expressed through society's institutions (e.g., through discriminatory laws and policies) and by its individual members. Individual enactments of stigma can range from personal ostracism to criminal attacks against people perceived to be homosexual or bisexual.

Sexual stigma has important consequences for sexual minority individuals. Whereas being the target of any violent crime can have negative psychological effects, victims of antigay violence are at heightened risk for psychological distress (Herek et al., 1999; Mills et al., 2004; see also McDevitt, Balboni, Garcia, & Gu, 2001). Hate crimes may have especially negative psychological sequelae because they attack a core aspect of the victim's personal identity and community membership, components of the self that are particularly important to sexual minority individuals because of the stresses created by sexual stigma (Garnets, Herek, & Levy, 1990; Herek et al., 1999; Herek et al., 2007). In addition to stigma's direct effects, sexual minority individuals' awareness of its extent and their expectancies about when it will be enacted create a subjective sense of threat. This *felt stigma* (e.g., Scambler & Hopkins, 1986) can motivate them to engage in a variety of proactive behaviors aimed at shielding themselves from enacted stigma. Such strategies (e.g., concealing their sexual orientation) can protect them from directly experiencing enacted stigma but also restrict their opportunities for having normal social interactions and receiving social support (Herek, 2008).

✉ Measures

The present data are based on a subset of questions from questions from a larger questionnaire. Only the relevant variables are discussed here.

Enacted Stigma

Eight questions were posed about how often respondents had experienced different forms of enacted stigma since age 18 because someone perceived them to be lesbian or bisexual (female respondents) or gay or bisexual (male respondents). The response options were *never, once, twice,* and *three or more times.* The questions assessed experiences of enacted stigma within three general categories: criminal victimization, harassment and threats, and discrimination.

To assess criminal victimization, respondents were asked how often they had experienced a crime against their person ("You were hit, beaten, physically attacked, or sexually assaulted") or property ("You were robbed, or your property was stolen, vandalized, or purposely damaged") or an attempted crime ("Someone *tried* to attack you, rob you, or damage your property, but they didn't succeed") based on their sexual orientation. To assess harassment and threats, respondents were asked about their experiences with antigay threats ("Someone threatened you with violence") and harassment ("Someone verbally insulted or abused you" and "Someone threw an object at you"). To assess discrimination, respondents were asked about their experience with sexual orientation discrimination in employment ("You were fired from your job or denied a job or promotion") and housing ("You were prevented from moving into a house or apartment by a landlord or realtor").

Felt Stigma

Felt stigma was assessed with three statements, each accompanied by a 5-point Likert-type response scale ranging from *strongly agree* to *strongly disagree.* Utilizing Web software capabilities, the item wording was customized by inserting respondents' preferred term for characterizing their sexual orientation (gay, lesbian, bisexual, queer, etc.); each respondent had selected this term earlier in the questionnaire. The three statements (as worded for respondents who indicated gay was their preferred self-identifying term) were (a) "Most people where I live think less of a person who is gay"; (b) "Most employers where I live will hire openly gay people if they are qualified for the job"; (c) "Most people where I live would not want someone who is openly gay to take care of their children."

Demographic Data

The survey included a question about the respondents' specific sexual orientation (bisexual or homosexual). Other demographic data—including respondents' gender, age, race and ethnicity, and highest educational level completed—had been collected previously by Knowledge Networks. Design weights were calculated and assigned to each case to adjust for the fact that cases had unequal probability of selection.

⧖ Results

Sample Characteristics

The final sample consisted of 311 women (152 lesbians, 159 bisexuals) and 351 men (241 gay men, 110 bisexuals). When design weights were applied, the weighted sample was 14.6% lesbian (95% confidence interval [CI] = 11.6%–18.2%), 34.8% gay male (CI = 28.9%–41.2%), 23.7% bisexual female (CI = 18.8%–29.3%), and 26.9% bisexual male (CI = 19.1%–36.4%). Unless otherwise indicated, the weighted data are used hereafter.

The respondents' mean age was 39 years (CI= 37.2–40.8). Gay men were significantly older (M = 45.3) than lesbians (M = 40.1), bisexual men (M = 36.6), and bisexual women (M = 31.8 years). In addition, lesbians were significantly older than bisexual women (for all statistically significant differences reported here, $p <$ .05). The sample was 65.4% non-Hispanic White, 15.6% non-Hispanic Black, and 12.5% Hispanic, with the remaining 6.5% from other races or of mixed race or ethnicity. Compared with lesbians and bisexual women, significantly fewer bisexual men were non-Hispanic White (43.0%, compared with 74.4% of lesbians and 77.5% of bisexual women). Bisexual men were also substantially less likely than gay men (70.5%) to be non-Hispanic White, but the difference was not statistically significant. Bisexual men were more likely than other respondents to be Hispanic (20.6%) or non-Hispanic Black (28.6%), but the differences were not statistically significant. Most respondents had earned a bachelor's degree (32.9%) or attended some college (31.4%). Only 7.3% did not have a high school diploma

or equivalent. Compared with gay men and lesbians, bisexual men were significantly less likely to have a bachelor's degree: 15.9% of bisexual men had a degree, compared with 46.4% of gay men and 40.9% of lesbians. A more detailed demographic description of the sample will be reported elsewhere.

Experiences With Enacted Stigma

As shown in the last column of Table 1, 13.1% of the sample reported having experienced violence against their person based on their sexual orientation at least once during their adult life, and 14.9% had experienced a property crime. Approximately 1 in 5 reported experiencing one or both types of crime, and this proportion increased to about 25% when attempted crimes were included. Overall, 12.5% of respondents reported having objects thrown at them because of their sexual orientation; 23.4% had been threatened with violence, and 49.2% had experienced verbal abuse. More than 1 respondent in 10 (11.2%) reported having experienced housing or employment discrimination because of her or his sexual orientation.

Across the sexual orientation groups, gay men reported the highest levels of enacted stigma. They were significantly more likely than others to report experiences of antigay violence (24.9%) and antigay property crimes (28.1%). More than one third of gay men (37.6%) reported experiencing one or both types of crimes, compared with 12.5% of lesbians, 10.7% of bisexual men, and 12.7% of bisexual women. As indicated by the overlapping CIs in Table 2, differences among the latter groups were too small to be considered reliable. Gay men were significantly more likely than bisexual men to report having had objects thrown at them (21.1% vs. 5.6%). They were significantly more likely than lesbians and bisexual women to have been threatened with violence (35.4% of gay men vs. 17.3% of lesbians and 14.1% of bisexual women), and significantly more likely than bisexual women to report verbal abuses because of their sexual orientation (63.0% vs. 34.3%). Employment and housing discrimination were significantly more likely among gay men and lesbians (reported by 17.7% and 16.3%, respectively) than among bisexual men and women (3.7% and 6.8%, respectively).

Table 1 Proportion of Respondents Who Reported *Ever* Experiencing Each Category of Enacted Stigma

Type of Enacted Stigma	Group (Unweighted *N*)				
	Gay Men (*n* = 241)	Lesbians (*n* = 152)	Bisexual Men (*n* =110)	Bisexual Women (*n* = 159)	Total (*n* = 662)
Violence	24.9%	7.1%	6.9%	6.7%	13.1%
	17.3–34.5	3.7–13.1	3.1–14.5	3.3–13.0	9.7–17.6
Property crime	28.1%	10.2%	7.7%	6.3%	14.9%
	20.9–36.6	6.0–16.8	3.8–15.1	2.0–17.8	11.4–19.3
Violence or property crime	37.6%	12.5%	10.7%	12.7%	20.9%
	29.1–46.9	7.9–19.3	5.6–19.5	6.6–23.1	16.5–26.1
Attempted crime	21.5%	8.3%	16.3%	5.6%	14.4%
	14.7–30.3	3.9–16.9	6.3–36.0	2.5–12.0	10.1–20.2
Violence, property crime, or attempted crime	39.0%	15.4%	20.1%	14.6%	24.8%
	30.4–48.3	9.5–24.0	9.0–38.9	8.0–25.1	19.4–31.0
Objects thrown	21.1%	14.6%	5.6%	6.8%	12.5%
	14.4–29.8	8.9–23.0	2.4–12.5	3.6–12.5	9.4–16.6
Threatened with violence	35.4%	17.3%	19.0%	14.1%	23.4%
	27.2–44.6	11.0–26.2	8.2–38.1	7.8–24.2	18.2–29.5
Verbal abuse	63.0%	54.5%	41.4%	34.3%	49.2%
	53.8–71.4	44.4–64.3	22.9–62.7	24.8–45.1	42.0–56.4
Job or housing discrimination	17.7%	16.3%	3.7%	6.8%	11.2%
	12.1–25.0	10.3–24.7	1.4–9.6	3.9–11.7	8.5–14.6

NOTE: Table reports parameter estimates and 95% confidence intervals for proportion of respondents experiencing each form of enacted stigma at least once, using weighted data.

As noted above, the groups differed significantly in age, race and ethnicity, and educational level, which might account for the observed difference on enacted stigma. To test this hypothesis, separate logistic regression equations were computed for each type of enacted stigma (violence, objects thrown, etc.) as well as for the combined categories of criminal victimization (comprising violence, property crime, and attempted crime) and harassment (comprising threats, objects thrown, and verbal abuse). In each equation, the dependent variable was dichotomized (*ever experiencing* that form

of enacted stigma vs. *never experiencing* it). The independent variables were age, education, race and ethnicity (coded as non-Hispanic White vs. Hispanic or non-White), and sexual orientation (coded as gay male vs. other groups). In the equations for the combined categories of criminal victimization and harassment, the odds ratios (ORs) for sexual orientation were statistically significant, indicating gay men were significantly more likely to report experiencing both categories of enacted stigma, even when demographic differences are controlled. (For criminal victimization,

OR = 2.45, 95% CI = 1.36–4.41; for harassment, OR = 2.10, CI = 1.23–3.60.) The same pattern was observed for all of the individual forms of enacted stigma except attempted crimes (for which none of the variables yielded a statistically significant OR).

A similar procedure was followed to assess whether lesbians' and gay men's greater likelihood of experiencing employment or housing discrimination was explained by group differences in age, education, and race. In this equation, sexual orientation was coded as homosexual versus bisexual. The OR for sexual orientation (3.29, CI = 1.58–6.85) was significant, indicating that the differences in discrimination were not due to demographic differences.

Felt Stigma

As shown in Table 2, substantial minorities of respondents expressed some degree of felt stigma. More than one third (34.6%) agreed (*strongly* or *somewhat*) that most people where they live think less of a sexual minority individual; 25.5% disagreed that most employers will hire qualified sexual minority individuals; and 40.6% agreed that most people would not want a sexual minority individual to care for their children. Overall, a majority of respondents (54.7%, CI = 47.1–62.1%) gave at least one response indicative of felt stigma. As indicated by the overlapping CIs in Table 2, there were few significant differences among the sexual orientation groups in their responses to the individual felt stigma items. The only exception is that gay men were significantly less likely than lesbians to perceive sexual stigma in hiring: 54.8% agreed that most employers in their area will hire a qualified sexual minority person, compared with 32.3% of lesbians who agreed with this statement.

Responses to the three statements were summed (with hiring question responses reversed) and divided by the number of items to create a felt stigma scale ($\alpha = .71$, using unweighted data). Scores could range from 1 (low felt stigma) to 5 (high felt stigma). As shown in the first row of Table 2, gay men scored the lowest on the scale and lesbians the highest (Ms = 2.79 and 3.11, respectively), but differences among the sexual orientation groups were not statistically significant.

Association of Felt Stigma With Enacted Stigma

Felt stigma scores were higher among respondents who reported having experienced person or property crimes or attempted crimes based on their sexual orientation compared to those who had not (Ms = 3.2 vs. 2.8, respectively), those who had experienced verbal threats or harassment compared to those who had not (Ms = 3.0 vs. 2.8, respectively), and those who had experienced employment or housing discrimination compared to those who had not (Ms = 3.2 and 2.8, respectively). These differences were not statistically significant.

However, significant differences were observed on these variables within some sexual orientation groups. Lesbians who said they had experienced crimes or attempted crimes scored significantly higher on felt stigma compared with those who had not (Ms = 3.6 vs. 3.0, respectively). The difference was also significant among bisexual men (Ms = 3.6 vs. 2.6, respectively), but not gay men (Ms = 2.9 vs. 2.8) or bisexual women (Ms = 3.7 vs. 2.9). Felt stigma was significantly higher among lesbians who reported having experienced employment or housing discrimination (Ms = 3.7 vs. 3.0 among those who had not). The difference was also significant for bisexual men (Ms = 3.9 vs. 2.8), but not for gay men (Ms = 2.9 vs. 2.8) or bisexual women (Ms = 3.2 vs. 2.9).

◤ Discussion

The present study yields the most reliable estimates to data of the prevalence of antigay victimization in the United States. The data indicate that approximately 20% of the U.S. sexual minority population has experienced a crime against person or property since age 18 based on their sexual orientation. With attempted crimes included, the proportion increases to roughly 25%. Harassment is considerably more widespread, with about half of sexual minority adults reporting verbal abuse at some time in their adult life as a consequence of their sexual orientation. More than 1 sexual minority adult in 10 has experienced housing or employment discrimination because of her or his sexual orientation.

Table 2 Felt Stigma Scale Scores and Responses to Individual Items

Item		Group (Unweighted N)				
		Gay Men (*n* = 241)	Lesbians (*n* = 152)	Bisexual Men (*n* = 110)	Bisexual Women (*n* = 159)	Total (*n* = 662)
Scale score (mean)		2.79	3.11	2.83	2.96	2.89
		(2.65–2.94)	(2.92–3.29)	(2.31–3.34)	(2.64–3.27)	(2.72–3.06)
"Most people where I live think less of a person who is [L/G/B]."	Agree	33.2%	44.0%	28.6%	37.7%	34.6%
		(25.3–42.1)	(34.3–54.1)	(16.5–44.9)	(27.7–48.9)	(28.9–40.9)
	In middle	23.0%	24.6%	24.8%	19.1%	22.8%
		(15.6–32.6)	(16.4–35.2)	(10.2–48.8)	(12.1–28.8)	(16.9–30.0)
	Disagree	43.8%	31.4%	46.6%	43.2%	42.6%
		(35.1–52.9)	(23.0–41.1)	(26.6–67.7)	(31.8–55.3)	(35.3–50.2)
"Most employers where I live will hire openly [L/G/B] people if they are qualified for the job."	Agree	54.8%	32.3%	49.1%	35.4%	45.4%
		(45.5–63.7)	(24.3–41.5)	(29.0–69.6)	(24.6–48.0)	(38.1–52.9)
	In middle	27.5%	40.6%	19.6%	35.2%	29.1%
		(19.3–37.6)	(30.8–51.2)	(10.4–33.8)	(25.7–46.1)	(23.7–35.2)
	Disagree	17.7%	27.1%	31.3%	29.4%	25.5%
		(12.7–24.1)	(19.0–37.1)	(16.9–50.6)	(20.1–40.9)	(20.2–31.6)
"Most people where I live would not want someone who is openly [L/G/B] to take care of their children."	Agree	41.2%	44.1%	39.1%	39.1%	40.6%
		(32.5–50.6)	(34.5–54.2)	(22.5–58.6)	(28.9–50.4)	(34.0–47.5)
	In middle	32.1%	30.3%	20.8%	30.6%	28.4%
		(24.0–41.5)	(21.3–40.9)	(10.9–35.9)	(21.5–41.5)	(23.1–34.4)
	Disagree	26.7%	25.6%	40.1%	30.3%	31.0%
		(20.0–34.6)	(17.9–35.2)	(20.1–64.2)	(19.6–43.6)	(23.5–39.7)

NOTE: Table reports weighted percentage of respondents (with 95% confidence intervals) in each response category, and weighted mean scores (with 95% confidence intervals) for combined felt stigma scale. *Strongly agree* and *somewhat agree* responses are combined into *agree* category; *strongly disagree* and *disagree somewhat* responses are combined into *disagree* category. In each item, the respondent's preferred label for her or his own sexual orientation (e.g., "gay") was substituted for [L/G/B].

The likelihood of experiencing victimization clearly is not uniform among sexual minorities. Gay men are at the greatest risk for person and property crimes. Approximately 38% of gay men in the present sample reported they had experienced one or both types of criminal victimization. Gay men also were more likely than lesbians and bisexuals to be harassed because of their sexual orientation. This pattern is consistent with previous findings that sexual minority men are at greater risk for antigay victimization than are sexual minority women (D'Augelli & Grossman, 2001; Herek et al., 1999). Several factors may account for this greater risk. Men are more likely than women to be victims of violent crime in general, especially crimes committed by strangers (Catalano & Bureau of Justice Statistics, 2005). Most such crimes are perpetrated by heterosexually identified men, who tend to hold more hostile attitudes toward sexual minority males than toward sexual minority females (Herek, 2002a, 2002b). In addition, gay men may be more visible targets than sexual minority women and bisexual men because, for example, they may be more likely to frequent gay-oriented venues and the public spaces around them.

Lesbians and gay men were significantly more likely than bisexuals to report discrimination based on their sexual orientation. This pattern probably cannot be attributed to attitudinal differences among the agents of discrimination because heterosexuals' attitudes toward bisexuals tend to be somewhat more negative than their attitudes toward gay men and lesbians (Herek, 2002b). Instead, homosexual adults' greater visibility probably makes then more vulnerable to discrimination in workplace and housing settings, compared with bisexuals. Additional data collected in the present study suggest that bisexual men and women are less likely than gay men and lesbians to disclose their sexual orientation to others in a variety of social contexts, including the workplace. In addition, to the extent that homosexual adults are more likely than bisexuals to cohabit with a same-sex partner (because many coupled bisexuals have a different-sex partner), the former are probably more readily labeled as gay by landlords and realtors and thus are more subject to discrimination.

As noted above, comparisons of prevalence estimates across previous studies are problematic because of differences in sampling strategies, question wording, time frames, and data-reporting conventions. Estimates of lifetime victimization from the present data are higher than those reported in some published studies (e.g., D'Augelli & Grossman, 2001; Herek et al., 1999) but lower than in others (Kaiser Family Foundation, 2001). The difference between the current study and the KFF survey may be due to the fact that the latter was conducted only with urban residents, who are more likely than nonurban residents to experience crime of all sorts (Herek & Sims, 2007). Differences between the present study and those conducted by D'Augelli and Grossman (2001) and Herek et al. (1999) may result from the latter's use of convenience samples whose representativeness cannot be determined; the current data probably provide a more accurate estimate

of the extent of victimization experiences within the sexual minority population.

The current study's estimates of the extent of felt stigma in the sexual minority population are another unique contribution. About 55% of respondents manifested some degree of felt stigma. It tended to be higher among respondents who had experienced enactments of stigma. For criminal victimization and employment discrimination, this pattern was mainly due to significant differences among lesbians and bisexual men who had experienced enacted stigma versus those who had not. Because the data reported here are retrospective and cross-sectional, the causal and temporal direction of these relationships, if any, cannot be determined. Experiencing enacted stigma is likely to increase an individual's subjective sense of vulnerability related to her and his sexual orientation (Herek et al., 1999), which could result in a positive correlation between enacted and felt stigma. However, other explanations for the pattern are also plausible. For example, persons with high levels of felt stigma may have a heightened sensitivity to the occurrence of stigma enactments and consequently may be more likely than others to attribute ambiguous incidents to stigma. Previous research on the cognitive strategies used by sexual minority crime victims for assessing their attackers' motives, however, suggests that relatively unambiguous cues (e.g., antigay verbal abuse) often accompany hate crimes based on sexual orientation (Herek, Cogan, & Gillis, 2002).

Even in the absence of additional surveys with probability samples, however, the present data have important policy implications. They demonstrate that the experience of violence and property crime is disturbingly widespread among sexual minority adults, especially gay men. Thus, they highlight the ongoing need for criminal justice programs to prevent and deter such crimes, and the need for victim services that will help to alleviate the physical, economic, social, and psychological consequences of such crimes (e.g., Herek et al., 1999; Herek & Sims, 2007).

In addition, the psychological toll of antigay hate crimes and harassment should be considered by mental health professionals and by researchers conducting studies of psychological distress and well-being in this population. Some research, for example, suggests that individuals who have engaged in homosexual behavior may be at greater risk than exclusively heterosexual adults for some forms of psychological distress (Herek & Garnets, 2007). Most studies, however, have not assessed how experiences of victimization and harassment might explain this pattern (for an exception, see Mays & Cochran, 2001). Given the association between antigay victimization and heightened psychological distress (Herek et al., 1999) and the present study's finding that such victimization has been experienced by roughly one in eight lesbians and bisexuals, and nearly 4 gay men in 10, it seems likely that the associations observed in past research between sexual orientation and psychological problems are attributable, at least in part, to such victimization. Moreover, the fact that more than half of the respondents in the present study experienced some degree of felt stigma related to their sexual orientation further highlights the extent to which sexual minorities are subjected to stressors that heterosexuals do not experience (e.g., Meyer, 2003).

Ever since the Hate Crimes Statistic Act became law in 1990, marking the federal government's first official recognition of the problem of violence and crime against people because of their sexual orientation, researchers have attempted to document the extent and prevalence of antigay victimization. The present study makes an important contribution to this effort. Such data can assist law enforcement agencies, service providers, and sexual minority communities in alleviating and preventing the problems created by sexual stigma.

References

Berrill, K. (1992). Antigay violence and victimization in the United States: An overview. In G. Herek & K. Berrill (Eds.), *Hate crimes: Confronting violence against lesbians and gay men* (pp. 19–45). Thousand Oaks, CA: Sage.

Catalano, S., & Bureau of Justice Statistics. (2005). *Criminal victimization, 2004*. Retrieved February 19, 2006, from http://www.ojp.usdoj.gov/bjs/abstract/cv04.htm

D'Augelli, A., & Grossman, A. (2001). Disclosure of sexual orientation, victimization and mental health among lesbian, gay, and bisexual older adults. *Journal of Interpersonal Violence, 16,* 1008–1027.

D'Augelli, A., Grossman, A., & Starks, M. (2006). Childhood gender atypicality, victimization, and PTSD among lesbian, gay, and bisexual youth. *Journal of Interpersonal Violence, 21,* 1462–1482.

Diaz, R., Ayala, G., Bein, E., Henne, J., & Marin, B. (2001). The impact of homophobia, poverty, and racism on the mental health of gay and bisexual Latino men: Findings from 3 U.S. cities. *American Journal of Public Health, 91,* 927–932.

Federal Bureau of Investigation. (2005). *Hate crime statistics 2004.* Washington, DC: U.S. Department of Justice.

Garnets, L., Herek, G., & Levy, B. (1990). Violence and victimization of lesbians and gay men: Mental health consequences. *Journal of Interpersonal Violence, 5,* 366–383.

Harlow, C. (2005). *Hate crimes reported by victims and police.* Washington, DC: U.S. Department of Justice. Retrieved July, 4, 2006, from http://www.ojp.usdoj.gov/bjs/pub/pdf/hcrvp.pdf

Herek, G. (2002a). Gender gaps in public opinion about lesbians and gay men. *Public Opinion Quarterly, 66,* 40–66.

Herek, G. (2002b). Heterosexuals' attitudes toward bisexual men and women in the United States. *Journal of Sex Research, 39,* 264–274.

Herek, G. (2004). Beyond "homophobia": Thinking about sexual stigma and prejudice in the twenty-first century. *Sexuality Research and Social Policy, 1*(2), 6–24.

Herek, G. M. (2008). Understanding sexual stigma and sexual prejudice in the United States: A conceptual framework. In D. Hope (Ed.), *Contemporary perspectives on lesbian, gay and bisexual identities: The 54th Nebraska Symposium on Motivation.* New York: Springer.

Herek, G., & Berrill, K. (1990). Anti-gay violence and mental health: Setting an agenda for research. *Journal of Interpersonal Violence, 5,* 414–423.

Herek, G., Chopp, R., & Strohl, D. (2007). Sexual stigma: Putting sexual minority health issues in context. In I. Meyer & M. Northridge (Eds.), *The health of sexual minorities: Public health perspectives on lesbian, gay, bisexual, and transgender populations* (pp. 171–208). New York: Springer.

Herek, G., Cogan, J., & Gillis, J. (2002). Victim experience in hate crimes based on sexual orientation. *Journal of Social Issues, 58,* 319–339.

Herek, G., & Garnets, L. (2007). Sexual orientation and mental health. *Annual Review of Clinical Psychology, 3,* 353–375.

Herek, G., Gillis, J., & Cogan, J. (1999). Psychological sequelae of hate crime victimization among lesbian, gay, and bisexual adults. *Journal of Consulting and Clinical Psychology, 67,* 945–951.

Herek, G., & Sims, C. (2007). Sexual orientation and violent victimization: Hate crimes and intimate partner violence among gay and bisexual males in the United States. In R. J. Wolitski, R. Stall, & R. O. Valdiserri (Eds.), *Unequal opportunity: Health disparities among gay and bisexual men in the United States* (pp. 35–71). New York: Oxford University Press.

Huebner, D., Rebchook, G., & Kegeles, S. (2004). Experiences of harassment, discrimination, and physical violence among young gay and bisexual men. *American Journal of Public Health, 94,* 1200–1203.

Kaiser Family Foundation. (2001). *Inside-out: A report on the experiences of lesbians, gays, and bisexuals in America and the public's view on issues and politics related to sexual orientation.* Menlo Park, CA: Author. Retrieved November 14, 2001, from http://www.kff.org

Mays, V. M., & Cochran, S. D. (2001). Mental health correlates of perceived discrimination among lesbian, gay, and bisexual adults in the United States. *American Journal of Public Health, 91,* 1869–1876.

McDevitt, J., Balboni, J., Garcia, L., & Gu, J. (2001). Consequences for victims: A comparison of bias- and non-bias-motivated assaults. *American Behavioral Scientist, 45,* 697–713.

Meyer, I. (2003). Prejudice, social stress, and mental health in lesbian, gay, and bisexual populations: Conceptual issues and research evidence. *Psychological Bulletin, 129,* 674–697.

Mills, T., Paul, J., Stall, R., Pollack, L., Canchola, J., Chang, Y., Moskowitz, J., & Catania, J. (2004). Distress and depression in men who have sex with men: The urban men's health study. *American Journal of Psychiatry, 161,* 278–285.

National Coalition of Anti-Violence Programs. (2005). *Anti-lesbian, gay, bisexual, and transgender violence in 2004.* New York: Author. Retrieved May 1, 2006, from http://www.ncavp.org/

Pilkington, N. W., & D'Augelli, A. R. (1995). Victimization of lesbian, gay, and bisexual youth in community settings. *Journal of Community Psychology, 23,* 34–56.

Results of poll. (1989, June 6). *San Francisco Examiner,* p. A-19.

Scambler, G., & Hopkins, A. (1986). Being epileptic: Coming to terms with stigma. *Sociology of Health and Illness, 8,* 26–43.

DISCUSSION QUESTIONS

1. Describe the person who is most at risk of experiencing stigma-related victimization based on the findings presented in this article. Why do think this is the case?

2. Why do lesbians and bisexual men who have experienced crimes have higher levels of felt stigma compared with those who have not experienced crimes?

3. Do you think the effects of felt stigma are similar to the effects of enacted stigma? Why or why not?

4. How can these findings be used to impact policy or programs to reduce victimization?

◈

Introduction to Reading 2

Human trafficking is a problem that impacts countries worldwide. In this article, Lebov (2009) presents research based on interviews with 28 persons working in agencies that assist victims of human trafficking. These people discussed their experiences with the 79 victims who had contact with their agencies in Scotland between April 2007 and March 2008. In part, Lebov's research aims to detail a picture of the problem of and response to human trafficking in Scotland. In this way, the piece provides an overview of victims of human trafficking and how the system responds to this type of victimization in Scotland.

Human Trafficking in Scotland, 2007/08

Korin Lebov

The phenomenon of human trafficking has received increased attention in recent years from policy makers, academics and non-governmental organizations (NGOs). The policy focus in the UK has culminated in the signing of the Council of Europe Convention on Action against Trafficking in Human

SOURCE: Lebov (2009). Reprinted with permission.

Beings in March 2007 and the creation of a joint action plan in 2007 by the UK and Scottish Governments with the purpose of adhering to provisions contained in the 2000 Palermo Protocol to Prevent, Suppress and Punish Trafficking in Persons, especially Women and Children, by April 2009. However, little empirical research on this subject has been carried out in the UK (Marie and Skidmore 2007; CEOP 2007; Dowling et al. 2007; Skrivánková 2006; IOM 2005; Dickson 2004; Zimmerman et al. 2003; Kelly and Regan 2000) and none to date has focused specifically on the trafficking of adults in Scotland.

This article describes a small-scale research project carried out between September 2007 and April 2008 as part of work within the Scottish Government to improve the evidence base around human trafficking in Scotland. The objectives of the research were to construct a more detailed national picture of the extent and nature of human trafficking in Scotland and to examine issues and challenges for policing and victim care in Scottish context.

The most commonly used definition of human trafficking is the broad one set out in the Palermo Protocol to Prevent, Suppress and Punish Trafficking in Persons, especially Women and Children, which supplements the United Nations Convention against Transnational Organized Crime. It defines human trafficking as follows.

'Trafficking in human beings' shall mean the recruitment, transportation, transfer, harbouring or receipt of persons, by means of the threat or use of force or other forms of coercion, of abduction, of fraud, of deception, of the abuse of power or of a position of vulnerability or of the giving or receiving of payments or benefits to achieve the consent of a person having control over another person, for the purpose of exploitation. Exploitation shall include, at a minimum, the exploitation of the prostitution of others or other forms of sexual exploitation, forced labour or services, slavery or practices similar to slavery, servitude or the removal of organs. (United Nations 2000: Article 3(a))

It is not possible to present a full academic literature review within this paper. However, it is useful to broadly situate the current study within a brief overview of that which is most relevant. Although it is a topic of relatively recent academic and policy interest, there is now a growing body of international literature relating to human trafficking, which spans the overlapping discourses of migration (Agustin 2006a, 2006b and 2005; Chapkis 2003; Gulcur and IIkkaracan 2002; Kelly 2002; Mai 2001), prostitution and sex work (Bindel and Kelly 2003; Aghatise 2004), labour exploitation (Agustin 2004; Dowling et al. 2007; Hughes et al. 1999), and gender studies (Doezema 2000; Pickup 1998; Pyle 2001).

Much of the literature acknowledges the difficulties in establishing the scale and nature of human trafficking (EUROPOL 2008; Kelly and Regan 2000:7). Any estimations of scale provided are wide-ranging. For example, the 2009 US State Department's Trafficking in Persons report suggests that there are at least 12.3 million adults and children in forced labour, bonded labour and commercial sexual servitude worldwide at any given time (US State Department 2009). Kelly and Regan estimated that the scale of trafficking in women into and within the UK lay within the range of 142 and 1,420 women per year (2000:22). Another Home Office (2003) report estimated that 4,000 women had been brought into the UK for the purposes of prostitution, but it is unclear how many of these women had been trafficked (see also Viuhko in this issue). During the Pentameter police operation,[1] 84 women were seized in raids across the UK, but it is not clear how many of them had been trafficked. Although there have been various newspaper reports estimating the numbers of trafficked victims in Scotland (see e.g. *The Herald*, 1 February 2009), these estimations have not been based on any empirical research and are essentially crude estimates.

There are also few consistent accounts of the nature or experiences of human trafficking in the UK (see, for example, IOM 2005; Zimmerman et al. 2003; Kelly and Regan 2000). This is probably related both to the dearth of empirical data in this field and the multiplicity of its victims' experiences. Indeed, some of the literature (Agustin 2006a; Doezema 2000) focuses specifically on the challenges of diagnosing human trafficking as distinct from the much larger phenomena of migration, labour exploitation and sex work, and it becomes clear that to arrive at any understanding of human trafficking one must also understand the other phenomena with which it seems inextricably linked.

Human trafficking occurs in Scotland in the context of a society that has a small but growing migrant population. Official immigration figures doubled between 2001/2 and 2006/7. In 2001/2, 18,357 people identified themselves as migrating to Scotland from overseas, whereas in 2006/7 this figure stood at 37,800 (General Register Office for Scotland 2009). These have recently joined the European Union, but also foreign nationals from other countries who come to the UK, usually seeking to improve their standard of living (see Migration Policy Institute 2007).

The sex trade, which is most is prevalent in the major Scottish cities of Glasgow, Edinburgh, Stirling, Aberdeen and Dundee, has increasingly moved 'indoors' in recent years (i.e. from street location to saunas, massage parlours and private residences).[2] This is likely to be the result of increased policing and lower levels of tolerance for street prostitution and new legislation introduced in late 2006, which criminalizes those who purchase sex. This movement may also be related to the changing demographics of workers. The industry has traditionally been occupied by individuals from Scotland and other parts of the UK, but in recent years this has changed and it is now estimated that approximately 50 percent of indoor sex workers in Glasgow—the largest urban centre for sex work—are from outside the UK. In the year or so prior to the research there was also a notable movement within the indoor sex industry—from saunas and massage parlours to private houses and flats.[3]

⚑ The UK and Scottish Policy Context

In March 2007, the Home Secretary signed the Council of Europe Convention on Action against Trafficking in Human Beings, 2005 (ECAT) on behalf of the UK. At the same time, the Home Office and the Scottish Executive (now the Scottish government) jointly published The UK Action Plan on Tackling Human Trafficking (2007), which outlined the initial measures needed to implement the Convention. Scotland has a significant role to play in the development of this work, which includes the provision of support for victims of trafficking and the arrest and prosecution of those individuals responsible.

Two associated recent developments are particularly noteworthy. In November 2006, the UK Human Trafficking Centre was established in Sheffield, England. This is a multiagency centre, designed as a hub of expertise to support law enforcement agencies across the UK in dealing with cases of human trafficking. Launched in early 2006, operation Pentameter was the first UK-wide police operation to focus specifically on tackling human trafficking for the purposes of sexual exploitation. The operation was carried out over three months and involved all UK police forces, along with the UK Border Agency (UKBA), the Serious Organised Crime Agency (SOCA), the Scottish Crime and Drug Enforcement Agency (SCDEA) and others. The original operation was followed by Pentameter 2 in 2007, a much longer operation in which the focus was widened to include other forms of trafficking in adults and children. Although the research report here does not represent the results of Pentameter 2, the timescales of the two did overlap and it appears that increased police activity generated larger amounts of data for analysis (ACPOS 2008).

In Scotland, the only non-law enforcement agency focusing specifically on victims of trafficking is the TARA (Trafficking Awareness Raising Alliance) Project, which provides services to female adult victims who have been trafficked into sexual exploitation. It liaises with the police, housing departments, health professionals, agencies that work with women in the sex trade and other relevant NGOs. Other agencies, including the Scottish Refugee Council, the International Organization for Migration (IOM) and the Women and Children's Department of the Legal Services Agency, also provide specialist support to victims at one or more stages of the process. Barnardo's and Save the Children provide services to children and young people who have been trafficked or exploited, as do local authority social work services.

Methodology

This research was exploratory in nature and relied primarily on an analysis of semi-structured interviews ($n = 28$) that I conducted between October 2007 and February 2008. A review of the relevant international literature (both academic and government reports) was also carried out, along with documentary analysis of training materials and progress reports relating to individual organizations to gain a richer and more objective understanding of human trafficking in Scotland. Basic content analysis was used when reviewing these materials.[4] Given the limited resources as well as the sensitivities associated with victimization in this area, it was not possible to interview victims of human trafficking. The Appendix summarizes the types and number of professionals interviewed within each organization.

The review of the relevant international literature and other documentary materials was used to construct a framework of themes to be explored during the interviews. Given the range of different organizations included in the primary data collection, and the wide and relatively unexplored nature of the topic, it was deemed most appropriate to use a semi-structured interview schedule. All interviews lasted between 90 and 120 minutes.

In most cases interviews were carried out face-to-face in a private room within the interviewee's workplace. In a small number of cases ($n = 2$) where distance or timescales prevented face-to-face meetings, interviews were conducted by telephone. In line with the preferences of some interviewees, interviews were not audio-recorded. Instead, I took extensive notes and transcripts of interviews were provided to the interviewee, who added to and/or amended them where necessary. Data relating to individual organizations were anonymized and marked only as 'police' or 'victim care.' All the data were systematically coded and manually analysed by content using a schema I created comprising variables associated with the various themes explored (e.g. age, ethnicity, location and travel route).

All of the findings presented relate to various forms of trafficking among adults. For the purposes of this research, individuals were counted as 'victims' of human trafficking if they identified themselves as such, and if the relevant organization interviewed had reasonable grounds to consider that one or more elements of the Palermo Protocol definition applied to them. Individuals were counted as 'suspected victims' if they identified themselves as such, or if the organization interviewed considered that one or more elements of the Protocol definition were relevant. Although this is a rather crude measure, it is arguably more accurate than previous estimates discussed earlier because it is based on empirical research. For ease of reference, all 'victims' and 'suspected victims' are referred to as 'victims' here.

Limitations of the Research

First, because the analysis was based on historical and secondary data relating to incidents that could not be verified for their validity, and only on cases that had come to the attention of statutory and non-statutory agencies, the results must be treated with a degree of caution.[5] Given that this is a relatively new area of policing and victim care and a very new area of empirical research, attempting to estimate the extent of unreported incidents would not be methodologically advisable. As documented earlier in this paper, previous estimates made in the UK (see Kelly and Regan 2000) have been wide ranging and largely unsupported by empirical evidence.

Secondly, although every effort was made to conduct systematic analysis of the data, as is often characteristic of historical data, the various data sources are incomplete in some areas.

Thirdly, as documented in the findings section, victims of trafficking are often difficult to identify. In some cases, the conclusions drawn may reflect characteristics of migrant sex workers as much as those of trafficking victims. Fourthly, the requirement of anonymity, especially relating to the characteristics of individual victims, produces limitations, including gaps and possible duplication. In some cases, two or more agencies may have encountered the same individuals. Although efforts were made to avoid 'double counting,' some of these may not have been successful owing to the incapacity to identify individual victims.

⚐ Findings

The National Picture

The picture of human trafficking in Scotland is a complex one with numerous gaps. A large part of what we currently know about this crime comes from police intelligence and operational activity. However, victim care services and NGOs can also provide us with information about those affected. Notwithstanding the limitations of the data, between March 2008 and April 2009, 79 victims of human trafficking came into contact with participating agencies. Of these cases, 52 (66 percent) involved women and 21 (27 percent) failed to identify the gender of the victim. Of the three main agencies, the police dealt with the greatest proportion of cases ($n = 35$; 44 percent) followed by TARA ($n = 23$; 29 percent) and then NGOs ($n = 21$; 27 percent).

The majority of cases involved females who were believed to have been trafficked into sexual exploitation ($n = 50$; 63 percent)[6] and a smaller proportion represent groups of both males and females who appeared to have been trafficked for exploitation in other industries ($n = 28$; 35 percent). It is difficult to tell whether this pattern is consistent with international experiences owing to a lack of empirical evidence, especially that involving trafficked males and/or trafficking for labour exploitation (Skrivánková 2006). For a high proportion of victims who came to the attention of the police and NGOs, gender was unknown ($n = 21$; 27 percent). This is because in several cases involving groups of males and females the gender of individuals was not recorded before they left. The apparent difference in scope between sexual and labour exploitation may be accounted for, in part, by real factors, such as the influx in recent years of relatively cheap labour from the newer EU countries. However, the picture of trafficking for labour exploitation may be distorted by the nature of some of these situations of exploitation; for instance, domestic servitude, which may be even more easily concealed than brothels operating in private residences, may be occurring but not visible to agencies (Department of Justice, Equality and Law Reform, Republic of Ireland 2009). In many cases the age of victims is unknown. Where age was known, it spanned a wide range from individuals in their late teens to those in their early forties.

Nature of Experiences

Analysis of the interview data showed that the nature of individuals' experiences varied widely. Many of those who had been recovered by police in operations during Pentameter 2 presented as illegal migrants whose travel to, and work in, the UK had been facilitated by an unidentified third party.[7] In most of these cases, individuals claimed that they knew what work they would be doing and were content to be doing it. This supports previous research that suggests that 'the majority of those persons identified as having been brought illegally into a country to work in exploitative conditions have given their initial consent to this process' (Munro 2006:328). However, some reported having been deceived and/or coerced into the work they were doing. This element alone distinguishes such cases as being human trafficking rather than smuggling or facilitated illegal immigration. Several victims who made contact with agencies interviewed as part of this research also claimed that an 'agent' had taken their identity documents from them upon arrival in the UK. Many had been debt bonded or charged large fees for the arrangement of work permits and travel, and females recovered in brothels often cited the repayment of this debt as their main reason for remaining in the sex industry. Some appeared to be frightened about what would happen to them if they did not repay this debt. These characteristics of human trafficking echo those documented in research elsewhere (Zimmerman et al. 2003). Most refused victim support services or accommodation and fled either before being interviewed or shortly after an initial interview.

A somewhat different picture was presented by victims encountered by victim care services and other victim-focused agencies in Scotland—TARA, the Scottish Refugee Council and the Legal Service Agency. Several of the victims they met had also actively sought to travel abroad to work and earn money but had been trafficked into situations of exploitation. This supports findings from previous research carried out in other jurisdictions (IOM 2005). However, others had fled some form of persecution (usually gender based) in their country of origin, and had been introduced to individuals who offered to help them with travel, accommodation or employment but then trafficked them into exploitation. Few of the (mainly) women and girls encountered by these agencies had been paid for prostitution. The purpose of this prostitution was usually solely to repay a discretionary debt imposed by the traffickers for travel or accommodation. Similarly, only a small number of these individuals had freedom of movement once they were trafficked and most felt under threat in their countries of origin, either because traffickers had knowledge of where their families lived or because of the likelihood that they would face discrimination from the local community. Such constraints on freedom and implicit coercion echo those illustrated in previous research by Agustin (2006a), Gulcur and Ilkkaracan (2002), and Coomaraswamy (1998).

The differences between the police and victim services in the pictures of trafficking victims drawn within the 20 interviews may represent real differences in the experiences of exploitation. Indeed, the client groups appeared to comprise different individuals: Very few of those encountered by the police were referred to victim services, and vice versa. However, it is likely that some of these differences may be accounted for by victims' perceptions of the agencies they initially encounter and the amount of time taken to build trust before disclosing their experiences (Refugee Council 2008). The interview data showed that many victims and suspected victims appeared to mistrust and/or fear any official authorities, particularly the police. These findings are supported by evidence that most victims who have had very traumatic experiences may not be capable of reflecting on or disclosing these experiences for some time due to posttraumatic stress disorder (Zimmerman et al. 2003).

Location

In cases of trafficking for sexual exploitation, all victims were recovered in private flats or houses that were being used to operate brothels. This may have been partially because saunas and massage parlours were receiving increased attention from both the police and

the media owing to Pentameter 2. It was acknowledged by interviewees, however, that such establishments remain places of risk, and agencies continued to monitor intelligence from those working with women they consider to be exploited.

In the small number of cases of labour exploitation encountered ($n = 15$; 19 percent), victims were found to be working in restaurants or takeaways and/or living in private multi-occupancy dwellings. Disregarding the very small numbers, this finding resonates with Skrivánková's observation that 'exploitation and forced labour mostly occur in industries that depend on casual and temporary labour' (2006:15)

Ethnicity

In just over a quarter of cases ($n = 21$) the country or region of origin of the victims and suspected victims was unknown. Where the country of origin was known, most originated from Asian ($n = 27$) and African ($n = 21$) countries. Three of the interviewees considered that the expansion of the EU in recent years had marked a general movement of the trafficking business from East European countries to other continents. Although East European migrants remain at risk of exploitation in various ways, it is no longer as profitable for traffickers to facilitate their travel, employment and accommodation because most can now reside and work freely in the UK. Within the Asian group, most identified themselves as Chinese.[8] Others were identified as Malaysian, Thai, Pakistani and Vietnamese. The remaining group of victims comprised smaller numbers of Romanians, Slovakians, Nigerians and Brazilians.

However, the picture perceived by other agencies that have come into contact with victims is somewhat different. Of those being supported by the TARA Project during the fieldwork period, 3 were Nigerian, 1 was Somali, 1 was Pakistani, 1 was Ugandan, 1 was Lithuanian, 1 was Gambian and 1 was Russian. Of the 10 victims of trafficking who had been clients of the Legal Services Agency since April 2007, 6 were Nigerian, 3 were from Eastern Africa and only 1 was Asian. Similarly, clients of the Scottish Refugee Council who had disclosed being victims of trafficking were mainly from West African countries, in particular Nigeria, Sierra Leone and Ghana.[9] Some were also from Pakistan but none were Chinese. In cases where traffickers, agents, pimps, madams and other facilitators were identified, most were from the same countries or regions of origin as their respective victims. Again, the largest proportion presented, or was reported, as being Chinese. In a small number of cases, individuals from Scotland or other parts of the UK had been linked to some element of trafficking activity.

Routes

Little is known about the precise routes of human trafficking into and through the UK. However, the literature that does refer to routes (see, for example, Home Office and Scottish Executive 2007) identifies the UK primarily as a destination country. This is supported by evidence from the current study. From the available data, the main trafficking route into Scotland appeared to be from London, where individuals travelled directly from source countries or via other European countries. They were usually accompanied, and/or were met at an airport, by a male or female linked to the trafficking organization. From interviews with professionals it seemed that most victims entered the UK using counterfeit documentation. Most were no longer in possession of such documentation by the time they came to the attention of the police or other agencies. In some cases victims reported that they had been forced to hand over their papers to an agent. Another significant route appeared to be from Belfast (Northern Ireland) to Stranraer (Western Scotland) via the Republic of Ireland (usually Dublin) and other European countries.

Links With Other Types of Organized Crime

International evidence suggests that organized criminals do not tend to specialize in one crime type, but rather they are increasingly flexible in terms of the activities they pursue and the other individuals or groups they work with (see Wright 2006).[10] It was clear that several of the women whose travel to the UK was facilitated had initially approached individuals or groups whom they knew to be involved in other forms of organized crime. Most notably these included Chinese 'snakehead' groups. In some of these cases, police also uncovered other evidence of multiple links with such individuals active in the UK and abroad from material found at the scene, such as mobile phones and financial documents, and from existing entries on the Scottish Intelligence Database.

Interviews highlighted the close relationship between human trafficking and other forms of organized crime. One human trafficking operation revealed links with various individuals who were known by English police forces to be involved in the distribution of Class A drugs. An individual who employed illegal migrants in his restaurants was known to be involved in various other forms of criminality, including serious fraud, money laundering and importation of contraband cigarettes. Various males connected to a brothel that was the focus of a human trafficking operation were found to be involved in criminality in other parts of Scotland and the UK, including the cultivation of cannabis and other drug offences, the distribution of counterfeit DVDs and credit card fraud. An individual involved in human trafficking and pimping of a female from various premises was known to have been involved in the supply of cocaine.

Policy and Practice Implications for Policing and Victim Care

In this section I examine the main implications of the key findings for policy and practice.

Focus of Pentameter 2

Although the European Convention definition encompasses child trafficking and human trafficking for labour exploitation, as well as human trafficking for sexual exploitation, and it was initially understood that Pentameter 2 would tackle all of these forms of exploitation, the overwhelming focus of intelligence gathering and operational work in the UK appeared to remain on sexual exploitation among adults. The findings in this study suggest that this was true of Scotland.

Priority of Human Trafficking

Human trafficking has become a high priority for Scottish police forces and there is a centrally driven will to include human trafficking as core police business, which has very high-level support. In practice, however, data from interviews with police personnel showed that action on human trafficking was still considered to be an addition to everyday police work. Forces set their own intelligence and operational priorities within individual control strategies and these strategies relate to their particular local public concerns.

Actionable Intelligence

It was clear from interview responses that there was generally a lack of actionable intelligence to support suspicions of human trafficking and applications for warrants. Police can make routine visits to saunas, massage parlours and private residences that may be functioning as brothels, but they cannot force entry without a warrant and there needs to be justification for this. Usually the level of intelligence required for a warrant is gleaned by surveillance, first-hand observation or reliable third-party sources. According to police personnel interviewed, these require intense and costly resourcing.

Issues Relating to the Identification of Victims

Normally, victims and witnesses represent the primary source of evidence relating to any crime. However, in about half of the cases included in this study suspected 'victims' did not identify themselves as such. Where victims did self-identify, they often left wherever they were staying during, or immediately after, initial debriefing. Mistrust of the police was thought to be one reason for this. Another reason may have been fear of deportation, regardless of their immigration status. If they were debt bonded they may have feared what would happen to them or their families if they did not pay off the rest of their debt and/or if they agreed to act

as witnesses of a crime. There is some evidence from the Poppy Project and other specialized victim care service operating in England and Wales, that offering a holistic service to victims of trafficking, as opposed to simply providing accommodation, may serve to address some of these issues (Dickson 2004). However, another emerging factor appeared to be that some individuals simply did not want to go back to the life they had left behind before they came to the UK and that they would remain in situations of exploitation unless viable economic alternatives were presented to them. This finding echoes research that emphasizes the rational decision-making that characterizes individuals who come to be known in western countries as victims of trafficking (Agustin 2005).

Unknown Communities

In Scotland, the migrant population has grown substantially in recent years (Scottish Government 2009). For example, there are now in excess of 77 nationalities residing in Glasgow, and even in much smaller cities and towns substantial proportions of the population are made up of new and often transient ethnic minority communities. With language barriers and mistrust of the authorities, especially police, some communities are in effect 'closed' to regular policing and other mainstream service provision. Where there is willingness to report serious crime to the police there are often problems with interpretation, in terms of both a lack of appropriate interpreters and the reliability of those available.

Multi-Agency Services

Although UK and Scottish policies describe a victim-centred approach to human trafficking, the competing priorities of various statutory and non-statutory agencies can be problematic. Although there is a good working relationship between the police, TARA and UKBA, in practice it is the police who carry out operations and the extent and nature of the involvement of other agencies is usually determined by their initial judgement. An organization such as TARA may be best placed to properly identify victims of human trafficking and gain their trust, but two law enforcement officials interviewed expressed fears that involving non-statutory agencies could jeopardize the security of operations.

Prosecutions

Between 2003 and 2008 in England and Wales there were 84 successful human trafficking prosecutions resulting in some of the most severe sentences in Europe. In Scotland, although there have been none for human trafficking to date. Data gathered during the course of this research showed that this is likely to result from a combination of factors including: an unclear intelligence picture; low levels of awareness among the public; an absence of witnesses; difficulties with translation during the debriefing of witnesses; further training needs among police and prosecution professionals; and some difficulties in obtaining warrants, including a perceived tendency for Sheriffs to favour the familiar language of brothel-keeping instead of newer legislation relating to human trafficking. All of these factors contribute to a lack of evidence that might support a successful prosecution for human trafficking.

Appendix

The type and number of professionals interviewed within each organization:

Organization	Profession	Number Interviewed
Strathclyde Police	Detective Inspector	2
Lothian and Borders Police	Detective Chief Inspector, Detective Sergeant, Detective Constables	4 (across 3 interviews)
Fife Constabulary	Detective Inspector	2
Northern Constabulary	Detective Chief Inspector	1
Dumfries and Galloway	Detective Inspector	1
Central Scotland Police	Detective Inspector	1
Tayside Police	Detective Inspector	1
Grampian Police	Detective Inspector	1
Scottish Crime and Drug Enforcement Agency	Deputy Principal Analyst, Senior Analyst, Detective Chief Inspector	3 (across 2 interviews)
Her Majesty's Revenue and Customs	Senior Taxation Official	1
UK Borders Agency	Senior Immigration Officials	4
TARA Project	Counter Trafficking Development Officer	1
Scottish Refugee Council	Asylum Support Team Leader	1
Scot-Pep (Scottish Prostitution Education Project)	Coordinator	1
Women and Children's Department of the Legal Services Agency	Legal Executive	1
International Organization for Migration	Director and Senior Migration Advisor	2
Save the Children	Programme Coordinator	1
Barnardo's	Director of Policy	1
Amnesty International Scotland	Researcher	1
Child Exploitation and Online Protection	Research Officer	1

◺ Notes

1. Pentameter and Pentameter 2 were UK-wide police operations aimed at rescuing and protecting victims of trafficking and identifying and bringing to justice those involved in committing trafficking offences. The operations also had significant intelligence-gathering and awareness-raising aspects. More information can be found at http://www.pentmeter.police.uk/ (accessed 28 September 2009).

2. Informal interview with anonymous sources at Strathclyde Police on 25 September 2007.

3. Informal interview with anonymous sources at Strathclyde Police on 25 September 2007.

4. The following were included in this documentary analysis: quarterly progress reports prepared by TARA for the Scottish government; a human trafficking pack prepared for Senior Investigating Officer training; and a draft memorandum of understanding between TARA, Strathclyde Police and UKBA.

5. By secondary data, I mean data collected from professionals working with victims and suspected victims, rather than data collected from victims themselves.

6. This is misleading because TARA, the only victim care agency specifically for victims of trafficking, works only with females.

7. Because the observations were gleaned from the interviews, the responding agency would not, or was unable to, provide specific numbers.

8. Some of these individuals may actually originate from other Asian countries, but China does not accept the return of individuals from other countries unless it can be proven that they are Chinese. This may in part explain why many Asians claim to be Chinese when they are from other Asian countries.

9. The Scottish Refugee Council supported six trafficking victims between April 2007 and March 2008.

10. There are various definitions of organized crime in existence (see United Nations 2000; EU Directorate General on Freedom, Security and Justice at http://ec.europa.eu/justice-home/doc-centre/crime/trafficking/wai/doc-crime-human-trafficking-en.htm, accessed 28 September 2009) but they all suggest that it involves serious crime carried out by more than two associated people collaborating over a prolonged period of time. Serious crime refers to conduct constituting an offence punishable by a maximum deprivation of liberty of at least four years or a more serious penalty.

◺ References

ACPOS [Association of Chief Police Officers in Scotland] (2008). *Operation Pentameter 2 exposes human trafficking*. News release, 2 July. URL (accessed 9 September 2009): http://www.acpos.police.uk/Documents/News%20Releases/ACPOS%20%Pentameter%202.pdf.

Aghatise, E. (2004). Trafficking for prostitution in Italy: Possible effects of government proposals for legalization of brothels in Italy. *Violence against Women 10*, 1126–55.

Agustin, L. A. (2004). A migrant world of services. *Social Politics 10*, 377–96.

Agustin, L. A. (2005). Migrants in the mistress's house: Other voices in the 'trafficking' debate. *Social Politics 12*, 96–117.

Agustin, L. A. (2006a). The conundrum of women's agency: Migration and the sex industry. In R. Campbell and M. O'Neill (eds.) *Sex work now*. Devon: Willan Publishing.

Agustin, L.A. (2006b). The disappearing of a migration category: Migrants who sell sex. *Journal of Ethnic and Migration Studies 32*, 29–47.

Bindel, J. and Kelly, L. A. (2003) *Critical examination of responses to prostitution in four countries: Victoria, Australia; Ireland; the Netherlands; and Sweden*. Child and Women Abuse Studies Unit of London Metropolitan University.

CEOP [Child Exploitation and Online Protection] (2007). *A scoping project on child trafficking in the UK*. London: CEOP, URL (accessed 9 September 2009): http://polis.osce.org/library/f/2973/1500/GOV-GBR-RPT-2973-EN-A%20Project%200n%20Child%20Trafficking%20in%20the %20UK.pdf.

Chapkis, W. (2003) Trafficking, migration and the law. *Gender and Society 17*, 923–37.

Coomaraswamy, R. (1998). *Report of the special rapporteur on violence against women: Its causes and consequences*. Paper presented at the Commission on Human Rights, Fifty-fourth session, 30 March, Geneva.

Department of Justice, Equality and Law Reform, Republic of Ireland (2009). About blue blindfold. URL (accessed 9 September 2009): http://www.blueblindfold.gov.ie./.

Dickson, S. (2004). *When women are trafficked: Quantifying the gendered experience of trafficking in the UK*. The POPPY Project.

Doezema, J. (2000). Loose women or lost women: The re-emergence of the myth of white slavery in contemporary discourses of trafficking in women. *Gender Studies 18*, 23–50.

Dowling, S., Moreton, K. and Wright, L. (2007). *Trafficking for the purposes of labour exploitation: A literature review*. Home Office Online report 10/07. URL (accessed 9 September 2009): http://www.homeoffice.gov.uk/pdfs07/rdsolr1007.pdf.

EUROPOL (2008). *Trafficking human beings in the European Union: A Europol perspective*. URL (Accessed 9 September 2009): http://www.europol.europa.eu/publications/Serious-Crime-Overviews/Trafficking-in-human-beings-2008.pdf.

General Register Office for Scotland (2009). Migration statistics. 26 February. URL (accessed 9 September 2009): http://www.gro-scotland.gov.uk/statistics/migration/index.html.

Gulcur, L. and IIkkaracan, P. (2002). The 'Natasha' experience: Migrant sex workers from the former Soviet Union and Eastern Europe in Turkey. *Women's Studies International Forum 25*, 411–21.

Home Office (2003). *Crime reduction toolkits: People trafficking*. London: Home Office.

Home Office and Scottish Executive (2007). *UK action plan on tackling human trafficking*. URL (accessed 9 September 2009): http://www.homeoffice.gov.uk/documents/human-traffickaction-plan?view-Binary.

Hughes, D., Sporcic, L., Mendelsohn, N. and Chirgwin, V. (1999) *The factbook on global sexual exploitation.* Coalition Against Trafficking in Women.

IOM [International Organization for Migration] (2005). Data and research on human trafficking. *International Migration, 43.*

Kelly, E. (2002). *Journeys of jeopardy: A review of research on trafficking in women and children in Europe.* Geneva: International Organization for Migration.

Kelly, L. and Regan, L. (2000). *Stopping traffic: Exploring the extent of, and responses to, trafficking in women for sexual exploitation in the UK.* London: Research, Development and Statistics Directorate, Home Office.

Mai, N. (2001). Transforming traditions: A critical analysis of the trafficking and exploitation of young Albanian girls in Italy. In R. King (ed.) *Mediterranean Passage: Migration and new cultural encounters in Southern Europe,* 258–78. Liverpool: Liverpool University Press.

Marie, A. and Skidmore, P. (2007). *A summary report mapping the scale of internal trafficking in the UK based on a survey of Barnardo's anti-sexual exploitation and missing services.* London: Barnardo's Research and Development and London Metropolitan University.

Migration Policy Institute (2007). *Europe's disappearing internal borders.* Fact Sheet No. 20, December. URL (accessed 9 September 2009): http://www.migrationpolicy.org/pubs/FS20-SchengenDisappearingBorders-121807.pdf.

Munro, V. E. (2006). Stopping traffic? A comparative study of responses to the trafficking in women for prostitution. *British Journal of Criminology 46,* 318–33.

Pickup, G. (1998). Deconstructing trafficking in women: The example of Russia. *Journal of International Studies 27,* 995–1021.

Pyle, J. L. (2001). Sex, maids, and export processing: Risks and reasons for gendered global production networks. *International Journal of Politics, Culture and Society 15,* 55–76.

Refugee Council (2008). *Refugee Council response to the Home Affairs Committee Inquiry into Human Trafficking.* London: Refugee Council, February. URL (accessed 9 September 2009): http://www.refugeecouncil.org.uk/policy/responses/2008/hic-trafficking.htm.

Skrivánková, K. (2006). *Trafficking for forced labour: UK country report.* Anti-Slavery International.

United Nations (2000). *Protocol to Prevent, Suppress and Punish Trafficking in Persons, especially Women and Children, supplementing the United Nations Convention Against Transnational Organized Crime.* URL (accessed 9 September 2009): http://www.uncjin.org/Documents/Conventions/dcatoc/final-documents-2/convention-%20traff-eng.pdf.

US State Department (2009). *Trafficking in Persons Report.* Washington, DC: US State Department. URL (accessed 9 September 2009): http://www.state.gov/g/tip/rls/tiprpt/2009/.

Wright, A. (2006). *Organised Crime.* Devon: Willan Publishing.

Zimmerman, C. et al. (2003). *The health risks and consequences of trafficking in women and adolescents. Findings from a European study.* London: London School of Hygiene and Tropical Medicine. URL (accessed 9 September 2009): http://www.oas.org/atip/global%20Report/Zimmerman%20TIP%20HEALTH.pdf.

DISCUSSION QUESTIONS

1. What is the extent and nature of human trafficking in Scotland?

2. What barriers are in place that prevent victims of trafficking from seeking help from authorities or social service agencies?

3. What are the limitations of relying on data from agencies to make conclusions about human trafficking in a country?

4. What similarities and differences exist between human trafficking in the United States and in Scotland?

5. What can be done to better serve victims of human trafficking?

Introduction to Reading 3

Using a sample of 532 inhabitants of New York City and Washington, D.C., who completed a telephone survey in 2006, Nellis's (2009) research examined the predictors of fear of terrorism, use of avoidance behaviors, and information-seeking behaviors. The author provides a gendered analysis that examines whether males and females differ in their levels of fear, avoidance behaviors, and information-seeking behaviors. This is similar to what previous research has found regarding females having elevated levels of fear of crime even though they are less likely than males to experience most types of victimization.

Gender Differences in Fear of Terrorism

Ashley Marie Nellis

Research on fear of crime has examined the role of gender through various lenses and frequently finds that women are more fearful than men (Ferraro, 1995; Fisher & Sloan, 2003; Madriz, 1997; Warr, 1984, 2000). Many attribute this difference to women's heightened fear of sexual assault (Ferraro, 1995) or physical assault (Lane & Meeker, 2003b). Gender differences are certainly not unique to crime reactions: Research in related disciplines finds that women respond differently to stressful events than men do (Chu, Seery, Ence, Holman, & Silver, 2006). The present research examines possible gender differences in the ways that individuals cope with terrorism-related information. Coping is operationalized with measures of fear, perceived risk, avoidance behaviors, and information-seeking behaviors that were collected in 2006 through telephone surveys of 532 inhabitants of New York and Washington, D.C. The fear of crime research offers a unique foundation with which to examine public reactions to terrorism-related information. The present study, which uses terrorism as the measure of violent victimization, allows testing of the relationship between gender and the most recent type of victimization fear, terrorism.

The present study aims to further clarify the understanding of gender differences in fear of victimization in three ways. First, it measures possible behavioral responses to fear in addition to traditionally used emotional responses, though conclusions about temporal ordering of fear and behaviors are limited by the cross-sectional nature of the data. Second, this study aims to better understand the fear victimization puzzle by specifying a perpetrator—a terrorist, in this case—as advocated in studies on fear of gang-related crime (Lane, 2002; Lane & Meeker, 2003a, 2003b, 2005). Finally, it examines responses to a form of violent victimization in a context that does not include the opportunity of sexual assault, thereby allowing the test of gender differences without the prospect of this interacting with fear to amplify the women's fear response.

In addition to gender, which is discussed in great detail in the following section, several factors have been found to facilitate victimization fear, and these are included in the present study as well. These factors include indirect victimization exposure to television news, income, race, and age. Though political ideology is one commonly considered correlate of crime fear, it is expected that political orientation can facilitate terrorism fear simply because the terrorism discussion has been so politically charged in this country over the past 8 years. Therefore, a measure of party identification is included as a possible predictor of terrorism fear as well.

Terrorism is likely to remain on the public radar for some time, though worries have eased somewhat since the most recent domestic terrorist attack on September 11, 2001 (Saad, 2004). Measurements of fear of terrorism were originally asked in response to the bombing of the Alfred P. Murrah Building in 1995 in Oklahoma City. The main terrorism-related question that has been used since this time to measure fear of terrorism is: "How worried are you that you or someone in your family will become a victim of terrorism—very worried, somewhat worried, not too worried, or not worried at all?" Data show that just after the Oklahoma City bombings 42% of Americans were somewhat or very worried that they or a family member could be a victim of a terrorist attack similar to the one in Oklahoma City (Jones, 2000). However, the last poll taken before September 11, 2001, showed that only 24% of Americans were very or somewhat worried (Saad, 2004). Research conducted on the day of the September 11, 2001, attacks indicated that 58% of the respondents were somewhat (35%) or very (23%) worried about being victimized by terrorism. And

SOURCE: Nellis (2009). Reprinted with permission.

beginning on October 2001, the percentages began to decline, fluctuating between 28% and 49% between late 2001 and 2004 (Saad, 2004). A Gallup poll taken in June 2005 showed that only 8% of Americans were very worried about terrorism, and more than 60% were not too worried or not worried at all (The Gallup Organization, 2005).

Similar to the present study, measurements of precautionary behavior have been made since September 11, 2001, as well. Preliminary Gallup poll data collected during the 3 months following September 11, 2001, show clear behavioral responses to the attacks: 40% of respondents handled their mail with more caution; 35% reported seeking information about bioterrorism; and 10% noted voluntary changes in their travel plans, avoidance of crowded places, stockpiling of goods, and/or purchasing of a weapon. More than 33% reported changing or planning to change their behavior as a result of September 11, 2001.

In general, fear of crime studies focus on emotional and psychological based definitions of fear, but some research has examined protective and precautionary actions taken as correlates of fear and found an association (Warr, 1984, 1994, 2000). In his national assessment of predictors of fear, Ferraro (1995) discovered that fear and precautionary behaviors were rare but distinct outcomes of a perceived victimization risk. Dubow, McCabe, and Kaplan (1979) speculated that avoiding dangerous places (i.e., spatial avoidance) was another indicator of fear of victimization. Finally in their study, Smith and Uchida (1988) determined that perceived risk of crime was a significant predictor of purchasing a weapon for protection.

✄ Evidence of Gendered Perceptions of Fear of Crime

More than 25 years of research shows that gender is the best predictor of fear of crime, often more than 2 times as strong as other predictors (Ferraro, 1995). This is sometimes referred to as the paradox of fear, given that women are also less likely to be victimized than their male counterparts.

Several reasons for women's elevated fears have been explained in the research literature. Hale (1996) reports two general explanations. First, both official and unofficial victimization data fail to capture the full nature of women's victimization. This may be because crimes that are more commonly carried out against women (e.g., rape and domestic violence) are more likely to go unreported. A second reason may be that women are more commonly subjected to a subcriminal type of victimization categorized by "'hey honey' harassments" that produce fear but are not directly tied to crime (Stanko, 1988, 1995). Scholars argue that if these types of victimization were included in overall victimization rates, the fear of crime would not appear to be disproportionate to the rates of victimization for men and women (Sacco, 1990; Tulloch, 2000).

Another possible explanation for higher fear of crime among women is that women feel more physically and socially vulnerable, which translates into higher perceptions of crime risk (Warr, 1994; Warr & Stafford, 1983). Skogan and Maxfield (1981) suggested that physical vulnerability refers to the openness to attack, powerlessness to resist attack, and exposure to traumatic physical and emotional consequences if attacked. If attacked, women are more physically vulnerable because they tend to be physically less able to defend themselves. Therefore, their fears about crime are the result of a pronounced focus on the seriousness of the outcome rather than a reduced concentration on the likelihood of its occurrence (Warr & Stafford, 1983). Social vulnerability refers to the concept that some groups are more vulnerable to victimization due to their social standing. Social status—being female and vulnerable, in this case—may cause women to feel more susceptible to victimization.

Socialization differences between men and women may contribute to disparities in fear of crime. Gordon and Riger (1989) asserted that women's fears of crime are tied up in their subordinate position to men in society, contributing to their sense of vulnerability. This suggests that their subordinate position in society and subsequent lack of power leads to fear of crime through an increased sense of vulnerability to victimization. Evidence for the vulnerability explanation of women's unexpectedly high level of fear about crime may also be found by looking at typical images of victims (Madriz, 1997). Sacco (1990), however, considers socialization to be an indication of spuriousness. He explains that women are socialized to be cautious and reserved, whereas men are socialized to be risk takers. Therefore, socialization determines fear of crime levels of both women and men. In the present study socialization differences by gender may emerge in terrorism responses if women feel more vulnerable to attack because they feel more vulnerable generally.

Others express the opinion that men are socialized not to express fear and that suspicions of greater fear among women may in reality reflect a suppressed expression of fear among men (Goodey, 1997; Sutton & Farrall, 2005). As an alternative explanation, Warr and Stafford find that fear is a multiplicative function of perceived risk and perceived seriousness. Because of women's higher sensitivity to risk, they have higher levels of fear of crime (Warr, 1994; Warr & Stafford, 1983). For the purpose of studying terrorism fears, if men are less accurate in their assessment of terrorism risk as suggested by Smith and Tortensson (1997), they may report being less afraid.

Finally, Ferraro (1995) clarified our understanding of women's fear through his pioneering research allowing for impact of offense-specific fears rather than crime up to the point. With respect to fear of crime among women this proved to be especially enlightening. The introduction of offense-specific sectioning about crimes revealed that women's fear of crime could more accurately be defined as a heightened fear of sexual assault (Ferraro, 1995). This is explained with the reasoning that women are more inclined to add a master-offense crime of sexual assault to their calculations and thus their responses about crime fears (Ferraro, 1996). Some suggest that women's fears may be higher than men's fears even for nonpersonal crimes like burglary because they could be perceived contemporaneously with a burglary (Warr, 1984). Researchers commonly refer to this as the shadow effect of fear of crime.

Although the impact of a possible sexual assault is not tested in the present study, it is nevertheless important to examine whether gender differences persist with victimization fears in hypothetical instances where the threat of sexual assault is not an issue.

The present study examines the possibility that gender differences in actions taken will emerge just as differences in fear of terrorism will emerge. Specifically, women will be more likely to engage in avoidance and information-seeking behaviors than men. Women will also be more worried about another terrorist attack than men.

Data Collection and Method

A telephone survey, designed to test the impact of media consumption on perceived risk and fear of terrorism, was conducted in March and April of 2006. Surveys were conducted by 12 trained telephone interviewers at American University. A total of 532 interviews were obtained from adults (18 and older) living in the metropolitan statistical areas (MAS) of New York City and Washington, D.C. These areas were selected to achieve adequate resonance of terrorism-related events, as these areas were the sites of the terrorist attacks on September 11, 2001. Results should be interpreted with this in mind; generalizability to the rest of the country cannot be assumed. In addition, because the data are cross-sectional, a causal model that examines the strength of fear as a predictor of actions taken is not undertaken. Therefore, this study does not determine the temporal order of relationship; rather, it seeks to identify associations among theoretically relevant variables.

A random sample from these two metropolitan statistical areas, with sampling techniques used to mirror the sampling frame according to race, income, age, and gender, was provided by Survey Sampling, Inc. Within-household randomization was obtained utilizing the "last birthday method." Interviewers asked to speak with someone in the household, 18 years of age or older, who most recently had a birthday. This helps to reduce the age and gender bias often inherent in telephone survey research (Bourque & Fielder, 2003). Interviews took approximately 15 min and each subject was asked questions about worry about and perceived risk of terrorism, actions taken in responses to changes in government-issued alerts about terrorism risk, exposure to the news media in general, exposure to terrorism-related news, and additional questions about related topics. Demographic data were also obtained.

The response rate for this survey was fairly low (41%). Though appropriate checks were conducted to ensure that the sample obtained was similar in demographic composition to the population from which it was drawn, it is unknown whether those who participated in this survey were different in their assessments of fear, their engagement in avoidance or information-seeking behaviors, media exposure, or other relevant variables. Findings must be considered with this potential caveat in mind.

Low response rates are a universal drawback to survey research design and are studied in their own right. One recent study examined differences in responses among two separately administered random-digit dial telephone surveys to examine whether varying response rates (60.6% vs. 36.0%) produced significantly different results in the topics discussed: media attention, political engagement, social trust and connectedness, and political attitudes. Though some demographic differences were observed, changes in nonresponse rates did not result in significant

differences in the survey topics (Keeter, Miller, Kohum, Groves, & Presser, 2000).

Table 1 displays basic descriptive information about the sample. A larger percentage of women (56%) than men (43%) participated in the study. One reason this is not surprising is that women are often more likely to answer the phone than men are when both are present in the residence (Groves & Kahn, 1979; Lavrakas, 1987). Other research finds that men are more likely to refuse participation in a telephone survey than women are (Smith, 1983). The final sample size is 547 due to missing gender data from 5 respondents.

To determine the racial and ethnic information from interviewees, individuals were asked, "What is your race or ethnicity" in an open-ended form and later coded according to standard U.S. Census categories. Responses were recorded for the purpose of this study into White (73%) and non-White (27%). Age was determined by asking respondents what year they were born and subsequently recoding their answers into their current age. The mean age was 50 years ($SD = 15.80$) with participants' age ranging from 18 to 89 years.

Respondents' incomes were roughly similar to those in the population from which they were drawn.

Table 1   Description of the Sample

Variable	Coding	Frequency	Percentage
Sex	0 = male	231	43.8
	1 = female	296	56.2
Age	Number (range 18–89 years)	—	—
	Mean	50	N/A
	SD	15.8	N/A
Race	0 = non-White	136	27.0
	1 = White	367	73.0
Income	1 = less than $20,000	28	6.1
	2 = $21,000–49,000	79	17.2
	3 = $50,000–100,000	168	36.7
	4 = more than $100,000	183	40.0

It is important to remember that these two sampled areas comprise an affluent segment of the nation's population. Exact income levels were not asked of respondents in the present study. Rather, they were asked which of the following categories best describes their total household income, before taxes. In the previous year (2005): less than $20,000, between $21,000 and $49,000, between $50,000 and $100,000, more than $100,000. In all, 12% ($n = 64$) of respondents declined to answer this question. The mean and median household income category among those who responded is $50,000–$100,000.

Dependent Variables

Three dependent variables are used to measure fear. The first, fear of terrorism, taps the emotional construct for fear of terrorism and asks respondents to report how worried they are of becoming a victim in a terrorist attack on a 10-point scale (from 1 = *no fear* to 10 = *great fear*). Specifically, respondents were asked, "Using a scale of 1 to 10, where 1 equals *not worried at all*, and 10 equals *extremely worried*, please tell me how much or how little you worry about the following events happening: I will be the victim in a terrorist attack." The mean score was 3.82 ($SD = 5.24$).

Though some researchers object to using questions with the word *worry* to measure fear, it has generally been established that the term *worry* is a satisfactory measure of fear (Lane & Meeker, 2000; Warr & Ellison, 2000).

The second and third dependent variables tap behavioral precautions one could take in association with fear and are guided by the expectation that fear is connected to activities to reduce it. The first behavioral measure, avoidance behaviors scale, is an additive scale of 8 avoidance behaviors items ($\alpha = .83, n = 429$). The mean score was 2.11 ($SD = 2.31$) with a possible range from 0 (*no behaviors taken)* to 8 (*eight behaviors taken)*. The second measure, information-seeking scale, is a 3-item additive scale ($\alpha = .55, n = 511$). Though the Cronbach's alpha is a rather low reliability score, this scale was retained for theoretical reasons. The mean score was 1.47 ($SD = 0.99$) with a possible range from 0 (*no information-seeking actions taken*) to 3 (*three information-seeking*

actions taken; see Appendix A for scale construction and descriptive statistics for these two scales).

The introductory statement for this section of the survey asked respondents to think of times immediately following changes in the U.S. Terror Advisory System (TAS), more commonly known as the terror-alert system that has five colors to indicate risk of terrorism. Before offering their responses, interviewers read the respondents the following statement:

> Occasionally, the TAS's terror-alert level is elevated to indicate a higher risk of terrorism. When you hear that the terror alert has been raised, please tell me if you have ever taken any of the following actions because of your concern about terrorism. Please answer "yes" or "no."

Therefore, individuals may have engaged in avoidance and information-seeking behaviors, but to tie their activities to terrorism-related events rather than for other reasons, the survey asked respondents to limit their activities to these periods just following changes in the TAS. It is entirely possible that, though not determined in this study, individuals engaged in some or all of the mentioned activities out of terrorism fear at some point or another.

Independent Variables

The current study follows the research that clearly distinguishes fear measures from perceived risk measures (Ferraro, 1995) and includes a measure of perceived risk of terrorism. Specifically, survey respondents were asked five questions pertaining to different types of terrorist attacks. These items were then summed to create the personal perceived risk scale that ranged from 1 to 46 (M = 17.31, $SD = 12.81$). The Cronbach's alpha was .91 ($n = 515$; see Appendix A for scale construction and descriptive statistics for this scale).

Research also suggests that the media figures prominently in fear of crime (see Eschholz, 1997; Hale, 1996 for a review). Coverage of terrorism and threats of terrorism have been frequent over the past 6 years. In addition, terrorism news is presented in ways similar to crime news coverage in many ways (Althea, 2002). Consequently, media exposure is included as an independent variable. Specifically, respondents were asked, "In the past two weeks, how many hours of news did you watch on television?" Responses for local news, national news, and cable news were summed to a total TV news score, a continuous variable. The average response was 16 hr of TV news in the past 2 weeks, or just over an hour a day, but responses ranged from 0 to 168 hours ($SD = 18.78$).

Some research suggests that vicarious victimization—that is, knowing someone who was victimized by crime—may facilitate fear (Covington & Taylor, 1991; Skogan & Maxfield, 1981). For this reason, a measure of indirect fear is included here as well. Specifically, respondents were asked, "Do you know anyone who has been the victim of a terrorist attack, either here or abroad?" Responses were coded dichotomously (0 = no, 1 = yes). Slightly fewer than half, or 43.8% ($n = 532$), knew someone who had been a victim of a terrorist event. This high figure is not surprising when considering the sample locations of New York and Washington, D.C. In addition, the survey deliberately allowed respondents to self-define what it meant to be a victim of terrorism. For some this might include death or physical injury and for others victimization might include loss of a business as a result of the event.

Political conservatism was measured using the following question: "In terms of your political opinions, would you consider yourself to be (a) very conservative, (b) somewhat conservative, (c) mixed, (d) somewhat liberal, or (e) very liberal?" Responses were treated as an interval-level scale from 1 (*very liberal*) to 5 (*very conservative*). The mean score was 2.96 ($SD = 1.16$).

Control Variables

Previous research suggests that race (Chiricos, McEntire, & Gertz, 2001; Chiricos, Padgett, & Gertz, 2001) and age (Ferraro & LaGrange, 1992; Holloway & Jefferson, 1997; LaGrange & Ferraro, 1989) are frequently associated with fear of crime. To a lesser extent, income has been found to be associated with fear of crime as well (Baumer, 1985; Garofalo & Laub, 1978;

Will & McGrath, 1995). Therefore, the estimated models controlled for self-reported age, race, and income.

Findings

This study utilized ordinary least squares (OLS) regression to explore the multivariate relationship between gender and fear, and Poisson regression to explore the relationship between behaviors undertaken and gender. This analysis explores between-gender differences by estimating separate models by gender, and then examining the regression coefficients between models for women versus models for men. Fear of terrorism is a continuous variable, so OLS is an appropriate statistical technique. The avoidance behavior scale and the information-seeking scale are more appropriately estimated with a Poisson model, because these are essentially count data (Cohen, Cohen, West, & Aiken, 2003). Before multivariate models were run, all appropriate diagnostic tests were conducted.

Table 2 displays results from disaggregating variable means by gender and also tests for differences with an independent sample t test. Results show that women expressed higher levels of worry than men on a scale of 1 to 10 (M = 3.13 for men vs. M = 4.39 for women), and these differences were statistically significant, $t(519) = -5.76, p < .001$.

The results presented in Table 2 also demonstrate that women and men differ on the two scales measuring behavior. In terms of information-seeking behaviors, women were more likely to seek information from outside sources such as the local government, the news, or friends and family (M = 1.61 for women, M = 1.30 for men); these differences were statistically significant, $t(506) = -3.59, p < .001$. Women were also more likely to report engaging in avoidance behaviors (M = 2.59 for women, M = 1.55 for men). These differences were statistically significant as well $t(484) = -5.07, p < .001$. Women also watched significantly more TV news and reported greater levels of perceived terrorism risk.

Bivariate correlations among variables in the models were also run separately on men and women. For men, fear of terrorism was associated with minority status, indirect victimization, and perceived risk of terrorism.

Table 2   Means by Gender, Independent Samples *t* Test for Different Means

Variable	Men(0)		Women(1)			
	Mean	SD	Mean	SD	Range	t Value
Fear of terrorism	3.13	2.289	4.39	2.634	1–10	−5.756**
Information-seeking behaviors	1.30	0.983	1.61	0.967	0–3	−3.593**
Avoidance behaviors	1.55	2.032	2.59	2.434	0–8	−5.067**
Age	49	16.248	51	15.420	18–86	−1.230
White/non-White	0.74	0.438	0.72	0.450	0–1	0.610
Income	3.26	0.862	2.98	0.910	1–4	3.276**
Political Conservatism	3.12	1.149	2.82	1.146	1–5	2.932**
Indirect victim	0.46	0.499	0.43	0.495	0–1	0.761
TV news	14.22	17.624	18.01	19.522	0–168	−2.285*
Perceived risk	13.83	11.412	20.31	13.168	1–46	−5.877**

NOTE: *t* value was determined through independent sample *t* test.

*$p < .05.$ **$p < .01.$

Indirect victimization, TV news exposure, and personal perceived risk were significantly associated with avoiding places or events. Specifically, younger men, men from minority groups, and men with lower incomes were more likely to participate in avoidance behaviors than older, White, and higher-income men. Minority group membership, more frequent TV news consumption, indirect victimization, and heightened perceived risk of terrorism were associated with information seeking among men. Politically conservative men also reported statistically significantly more information-seeking behaviors.

For women, many of the same patterns emerge. Younger women report more fear and engage in more avoidance and information-seeking behaviors. The same is true for minority women. Like men, women's enhanced sense of perceived risk is also strongly associated with greater fear and engagement in more avoidance and information-seeking behaviors. Greater amounts of watching TV news was positively related to scores on both of the behavior scales, but not with fear. Finally, politically conservative women were more likely to avoid places or events and to seek information,

but political conservatism was not statistically related to the other two dependent variables.

Next, independent and control variables were entered into various multivariate models to examine their influence on fear of terrorism. As shown in Table 3 risk perception is the only highly significant predictor of terrorism fear that emerges among women, though age is marginally significant. Younger women and those with a sense of greater personal perceived risk have the greatest amounts of terrorism fear, controlling for the remaining relevant factors.

TV news exposure is not significantly associated with greater amounts of terrorism fear, which is unexpected, considering that terrorism stories often dominate the news. It is possible that TV news exposure was more predictive of fear closer to the time of the attacks, but that women have become somewhat desensitized to terrorism news over time; 5 years had passed since a domestic terrorist attack had occurred. Overall, these predictors explained 35.6% of the variance in terrorism fear among women.

Table 4 presents the regression results of the relationship between these seven individual characteristics and two categories of precautionary behaviors, avoidance and information seeking, among women. Both models were statistically significant. Among the individual predictors, political conservatism, frequency of television news exposure, and perceived risk were associated with engaging in statistically significantly more avoidance behaviors; however, perceived risk of terrorism was the only significant predictor of information-seeking behaviors.

As reported in Table 5, statistically significant predictors of terrorism fear among men included minority group membership, indirect victimization experience, and a heightened sense of terrorism risk. These independent variables taken together explained 28.7% of the variance in terrorism fear.

In terms of avoidance behaviors, minorities, those who were more politically conservative, those who watched more TV news, and those reporting heightened perceptions of risk reported greater likelihood to avoid places or events, net of the other factors. However, only age (younger men) and minority status influenced information seeking among men. (See Table 6.)

Perceived risk of terrorism was unassociated with information seeking for men, though it had

Table 3 OLS Model of Terrorism Fear on Individual-Level Predictors Among Women

Variable	Fear of Terrorism	
	Unstandardized Beta (SE)	Standardized Beta
Age	−0.02+ (0.01)	−.11
White/non-White	−0.27 (0.35)	−.05
Income	0.11 (0.17)	.04
Political conservatism	0.02 (0.12)	.01
Indirect victimization	0.27 (0.30)	.05
TV news	0.00 (0.01)	.03
Perceived risk	0.11*** (0.01)	.54
Constant	−2.72 (0.89)	—
R²	0.36	—
Adjusted R²	0.34	—
N	229	—

***p <.01.

Table 4 Poisson Regression Model of Precautionary Behaviors on Individual-Level Predictors Among Women

Variable	Avoidance Behavior Scale		Information-Seeking Scale	
	Beta	SE	Beta	SE
Age	0.00	0.00	−0.01	0.00
White/non-White	−0.10	0.10	0.02	0.12
Income	−0.00	0.05	−0.05	0.06
Political conservatism	0.16***	0.04	0.02	0.05
Indirect victimization	0.04	0.09	−0.02	0.11
TV news	0.01***	0.00	0.00	0.00
Perceived risk	0.03***	0.00	0.01***	0.00
Constant	−0.15	0.26	0.53	0.31
Log likelihood	−446.62	—	−316.33	—
Likelihood ratio test	120.82***	—	18.42**	—
McFadden's adjusted R²	0.12	—	0.03	—
N	209	—	224	—

**p < .05. ***p < .01.

Table 5 OLS Modes of Terrorism Fear on Individual-Level Predictors Among Men

Variable	Fear of Terrorism	
	Unstandardized Beta (SE)	Standardized Beta
Age	−0.01 (0.01)	−.07
White/non-White	−0.57** (0.34)	−.12
Income	0.10 (0.17)	.04
Political conservatism	−0.03 (0.11)	−.02
Indirect victimization	0.52** (0.27)	.13
TV news	−0.01 (0.01)	−.04
Perceived risk	0.09*** (0.01)	.47
Constant	2.23 (0.75)	—
R^2	0.29	—
Adjusted R^2	0.26	—
N	192	—

$**p < .05. ***p < .01.$

been a strong predictor for information seeking among women (see Table 6). This lends preliminary support to the perspective that fear and risk are distinct constructs that sometimes fail intuitive understanding. Furthermore, these findings suggest that perceived risk is associated with greater fear among men, but not with attempts to seek out information. Some argue that risk of a catastrophic event is very difficult to personalize, whereas fear is less so, and this may thus account for the disparity (Sunstein, 2003). This difference may also speak to the claim referenced earlier that men are more inclined to access their risk inaccurately.

Finally, perceived risk emerged as a significant predictor in the model of terrorism fear and the model of avoidance behaviors for both men and women, signaling the need to determine whether the coefficients were significantly different by conducting the appropriate z test (Paternoster, Brume, Mazerolle, & Piquero, 1998). The results of the equivalent parameters test indicated results for men and women are statistically significantly different from each other ($z = 4.84, p < .01$).

Table 6 Poisson Regression Model of Precautionary Behaviors on Individual-Level Predictors Among Men

Variable	Avoidance Behavior Scale		Information-Seeking Scale	
	Beta	SE	Beta	SE
Age	−0.00	0.00	−0.01**	0.00
White/non-White	−0.34**	0.14	−0.310	0.15
Income	0.09	0.08	−0.00	0.08
Political conservatism	0.19***	0.06	0.02	0.06
Indirect victimization	0.04	0.12	0.14	0.13
TV news	0.01***	0.00	0.00	0.01
Perceived risk	0.02***	0.01	0.01	0.01
Constant	−0.48	0.35	0.64	0.35
Log likelihood	−352.06	—	−255.07	—
Likelihood ratio test	53.40***	—	17.18***	—
McFadden's adjusted R^2	7.05	—	3.25	—
N	185	—	190	—

$**p < .05. ***p < .01.$

Discussion

Overall, these results are consistent with findings elsewhere in that they uncover gender-specific differences in terrorism fear. Women reported greater amounts of fear of terrorism than men and these differences were statistically significant. Women also reported statistically significantly greater likelihood to seek information than men and also engaged in many more avoidance behaviors. These findings provide evidence in support of the vulnerability perspective. Women perceived a much greater risk of terrorism than men did as well, and this result was statistically significant.

The gender difference in fear and behavioral responses to terrorism-related information calls into question the previously explained "shadow of sexual assault" premise that fear of crime for women is more accurately a fear of rape or sexual assault. However, if sexual assault was the only motivator for women's elevated fears of victimization, gender effects would not have been expected to emerge in the present study. Terrorism, equally likely to affect women and men, is a type of violent victimization that excludes the possibility of sexual assault. The observation that women are more afraid of terrorism suggests that something else may be co-occurring with terrorism fear for women.

A more reasonable explanation is that, collectively, women have a greater sense of physical and social vulnerability for the reasons expressed earlier, and this transfers to their feelings about terrorism as well. In the alternative, women may be more willing to admit to being fearful and taking actions to protect themselves than men are because of the social expectations imposed on men not to express fear, as suggested by Hurwitz and Smithey (1998), and Sutton and Farrall (2005).

Others claim that reports of fear are not exclusively about fear of physical harm but also about the apparent unpredictability and lack of control that accompany events like crime or, in this case, terrorism (Lane & Meeker, 2003). These additional concerns could be especially poignant for terrorism due to the nearly excusive reliance on the government for risk estimates and other pertinent information about terrorist threats.

This study's findings also demonstrated that women's responses to terrorism-related information, as well as their proclivity to engage in avoidance and information-seeking behaviors, is different from men's. Bivariate correlations for men revealed that minorities, younger, and low-income men reported greater amounts of fear. Each of these characteristics can also be identified as an expression of social vulnerability in the American culture. Greater physical vulnerability is expressed through statistically significantly greater risk perception among men. The multivariate models showed that status as a minority is a consistent predictor across the three models, controlling for other variables; for the model predicting information seeking, this was the strongest predictor.

For women, age, minority status, and risk perception were correlated with each of the dependent variables at the bivariate level. Results showed a strong, positive relationship between risk perception and fear, avoidance behaviors, and information-seeking behaviors. The multivariate regressions showed that perceived risk is a strong predictor of fear, avoidance behaviors, and information seeking in most cases.

This study sought to replicate what is known from the fear of crime research to determine whether gender patterns persist for fear of terrorism. Without exception, gender appears as a significant contributor to fear, with women reporting greater fear than men. The reality is that women are at no greater or lesser danger of terrorist attack than men are, but explanations from the fear of crime literature help to understand why women are more fearful and take more precautions than men do.

This research sought to explore fear among those who had some familiarity with terrorism, so the cities of New York and Washington, D.C., were selected deliberately because of the September 11, 2001 attacks that took place in these locations. It is reasonable to expect that findings would be different in a location where terrorist attacks had not taken place because terrorism would not resonate with these residents as much. Future research could replicate this study in other locales to see what impact, if any, proximity to terrorist attacks has on fear and precautionary behaviors.

Conclusion

Fear is a powerful motivator for individuals' behaviors and attitudes. Several decades of crime research have focused on the various factors that facilitate fear of crime. Researchers have consistently shown that gender is among the strongest predictors of fear of crime. More recently, fear of terrorism has been studied to determine the extent to which individuals are internalizing the messages they hear from the government, the media, and informal social networks. Only 7 years ago, America experienced the deadliest terrorist attack in its history, so there is ample justification for some worry about terrorism. The present study explored the possibility that women report greater fear of terrorism than men and engage in more behaviors that they believe protect them from terrorism and/or inform them about terrorism threats.

Fear of terrorism can have a variety of consequences such as negative economic impacts from reduced air travel and tourism; individual psychosocial problems such as depression, anxiety, and insomnia (Bleich, Gelkopf, & Solomon, 2003); and public support for antiterrorism policies that encroach on civil liberties (Huddy, Feldman, Taber, & Lahav, 2005; Nellis, 2007). The federal government, which has dedicated multiple agencies to thwarting future terrorist attacks, often promotes antiterrorism policies in the alleged interest of reducing the public fear of terrorism. It is, therefore, essential to accurately gauge public fear levels through comprehensive research dedicated to better understanding levels of fear and its correlates.

✖ Appendix A: Scale Construction and Descriptive Statistics

Avoidance Behaviors and Information-Seeking Behaviors

"Occasionally, the U.S. Advisory system's 'terror alert' level is elevated to indicate a higher risk of terrorism. When you hear that the terror alert has been raised, please tell me if you have ever taken any of the following actions because of your concern about terrorism. Please answer 'yes' (yes = 1) or 'no' (no = 0)."

Descriptive Statistics of Avoidance Behavior Scale Items, by Gender

	Men		Women		
Behavior	Yes	No	Yes	No	t value
Used more caution in daily routine	83	144	141	142	−3.019**
Avoided public transportation	51	173	113	163	−4.378**
Avoided air travel	47	178	95	186	−3.241**
Bought protective gear	27	200	39	243	−0.645
Avoided certain cities that have been the target of terrorist attacks	57	168	137	142	−5.609**
Avoided tall buildings	29	198	53	227	−1.874
Avoided crowds	42	184	77	203	−2.359*
Avoided landmarks	27	200	62	220	−3.000**

$*p < .10. **p < .15.$

Descriptive Statistics for Items in Information-Seeking Behavior Scale

	Men		Women		
Behavior	Yes	No	Yes	No	t Value
Sought information through the news	161	65	225	57	−2.248*
Sought information from my local government	44	183	77	206	−2.069*
Sought information from family and friends	89	138	154	129	−3.451**

$*p < .10. **p < .05.$

"Apart from the issue of worries about terrorism, it is useful to know people's idea of risk of certain events occurring to them or someone they know. Please tell me the risk you think there is of the following incidents occurring, on a scale of 1 to 10, where 1 equals *very unlikely* and 10 equals *extremely likely* to occur."

Personal Perceived Risk to Terrorism

	Men	Women		
Behavior	**M (SD)**	**M(SD)**	**Range**	**t value**
I could be on plane that is hijacked	3.71 (2.64)	5.10 (3.07)	1–10	−5.475**
I could be on a subway or bus that is bombed	4.29 (2.69)	5.67 (3.19)	1–10	−5.254**
I could be in a tall building during a terrorist attack	3.31 (2.63)	4.48 (3.10)	1–10	−4.580**
I could be the victim in a suicide bombing	3.26 (2.67)	4.66 (3.10)	1–10	−5.472**
I could witness a suicide bombing	3.35 (2.81)	4.38 (3.08)	1–10	−3.946**

**p < .05.

References

Althea, D. (2002).*Creating fear: News and the construction of crisis.* New York: Walter de Gruyter, Inc.

Baumer, T. L. (1985). Testing a general model of fear of crime: Data from a national sample. *Journal of Research in Crime and Delinquency, 22,* 239–255.

Bleich, A., Gelkopf, M., & Solomon, Z. (2003). Exposure to terrorism, stress-related mental health symptoms, and coping behaviors among a nationally representative sample in Israel. *JAMA, 290,* 612–620.

Bourque, L., & Fielder, E. P. (2003). *The survey kit 2: How to conduct telephone surveys* (2nd ed.). Thousand Oaks, CA: Sage.

Chiricos, T., McEntire, R., & Gertz, M. (2001). Perceived Racial and Ethnic Composition of Neighborhood and Perceived Risk of Crime. *Social Problems, 48*(3), 322–340.

Chiricos, T., Padgett, K., & Gertz, M. (2001). Fear, TV news, and the reality of crime. *Criminology, 38,* 755–786.

Chu, T. Q., Seery, M. D., Ence, W. A., Holman, E. A., & Silver, R. (2006). Ethnicity and gender in the face of a terrorist attack: A national longitudinal study of immediate responses and outcomes two years after September 11. *Basic and Applied Social Psychology, 28,* 291–301.

Cohen, J., Cohen, P., West, S. G., & Aiken, L. S. (2003). *Applied multiple regression/correlation analysis for the behavioral sciences* (3rd ed.). Mahwah, NJ: Lawrence Erlbaum.

Covington, J., & Taylor, R. (1991). Fear of crime in residential neighborhoods: Implications of between and within-neighborhood sources for current models. *The Sociological Quarterly, 32,*321–349.

Dubow, F., McCabe, E., & Kaplan, G. (1979). *Reactions to crime: A critical review of the literature.* Washington, DC: U.S. Department of Justice.

Eschholz, S. (1997).The media and fear of crime: A survey of the research. *Journal of Law and Public Policy, 9*(1), 37–59.

Ferraro, K. (1995). *Fear of crime: Interpreting victimization risk.* Albany: State University of New York Press.

Ferraro, K. (1996). Women's fear of victimization: Shadow of sexual assault? *Social Forces, 75,* 667–690.

Ferraro, K., & LaGrange, R. L. (1992). Are older people most afraid of crime? Reconsidering age differences in fear of victimization. *Journal of Gerontology, 47,* 233–244.

Fisher, B. S., & Sloan, J. J. (2003). Unraveling the fear of victimization among college women: Is the "shadow of sexual assault" hypothesis supported? *Justice Quarterly, 20,* 633–659.

Garofalo, J., & Laub, J. (1978). The fear of crime: Broadening our perspective. *Victimology, 3,* 242–253.

Goodey, J. (1997). Boys don't cry: Masculinities, fear of crime, and fearlessness. *British Journal of Criminology, 37,* 401–418.

Gordon, M., & Riger, S. (1989). *The female fear: The social cost of rape.* New York: Free Press.

Groves, R., & Kahn, R. (1979). *Surveys by telephone: A national comparison with personal interviews.* San Diego, CA: Academic Press.

Hale, C. (1996). Fear of crime: A review of the literature. *International Review of Victimology, 4,* 79–150.

Holloway, W., & Jefferson, T. (1997). The risk society in an age of anxiety: Situating fear of crime. *British Journal of Sociology, 48*(2), 255–266.

Huddy, L., Feldman, S., Taber, C., & Lahav, G. (2005). Threat, anxiety, and support of antiterrorism policies. *American Journal of Political Science, 49,* 593–608.

Hurwitz, J., & Smithey, S. (1998). Gender differences on crime and punishment. *Political Research Quarterly, 51,* 89–115.

Jones, J. (2000). *Americans less concerned about terrorist attacks five years after Oklahoma City.* Washington, DC: The Gallup Organization.

Keeter, S., Miller, C., Kohut, A., Groves, R. M., & Presser, S. (2000). Consequences of reducing nonresponse in a national telephone survey. *Public Opinion Quarterly, 64,* 125–148.

LaGrange, R. L., & Ferraro, K. (1989). Assessing age and gender differences in perceived risk and fear of crime. *Criminology, 27*(4), 697–720.

Lane, J. (2002). Fear of gang crime: A qualitative examination of the four perspectives. *Journal of Research in Crime and Delinquency, 39,* 437–471.

Lane, J., & Meeker, J. (2000). Subcultural diversity and the fear of crime and gangs. *Crime & Delinquency, 46,* 497–521.

Lane, J., & Meeker, J. (2003a). Ethnicity, information sources, and fear of crime. *Deviant Behavior, 24*(1), 1–26.

Lane, J., & Meeker, J. (2003b). Women's and men's fear of gang crimes: Sexual and nonsexual assault as perceptually contemporaneous offenses. *Justice Quarterly, 20,* 337–371.

Lane, J., & Meeker, J. (2005). Theories and fear of gang crime among Whites and Latinos: A replication and extension of prior research. *Journal of Criminal Justice, 33,* 627–641.

Lavrakas, P. (1987). *Telephone survey methods: Sampling, selection, and supervision.* Thousand Oaks, CA: Sage.

Madriz, E. I. (1997). Images of criminals and victims: A study on women's fear and social control. *Gender & Society, 11,* 342–356.

Nellis, A. M. (2007). *How does the American public cope with terrorism-related information?* Dissertation submitted to American University, Washington, DC.

Paternoster, R., Brume, R., Mazerolle, P., & Piquero, A. (1998). Using the correct statistical test for the equality of regression coefficients. *Criminology, 36,* 859–866.

Saad, L. (2004, September 10). *Three years after 9/11, most Americans carrying on normally.* Washington, DC: The Gallup Organization.

Sacco, V. F. (1990). Gender, fear, and victimization: A preliminary application of power-control theory. *Sociological Spectrum, 10,* 485–506.

Skogan, W., & Maxfield, M. G. (1981). *Coping with crime: Individual and neighborhood reactions.* Beverly Hills, CA: Sage.

Smith, T. W. (1983). The hidden 25 percent: An analysis of nonresponse on the 1980 General Social Survey. *Public Opinion Quarterly, 47,* 386–404.

Smith, W. R., & Tortensson, M. (1997). Gender differences in risk perception and neutralizing the fear of crime: Towards resolving the paradoxes. *British Journal of Criminology. 37,* 608–634.

Smith, D. A., & Uchida, C. D. (1988). The social organization of self-help: A study of defensive weapon ownership. *American Sociological Review, 53,* 94–102.

Stanko, E. A. (1988). Hidden violence against women. In M. Maguire & J. Pointing (Eds.), *Victims: A New Deal?* New York: Open University Press.

Stanko, E. (1995). Women, crime and fear. *Annals of the American Academy of Political Science, 539,* 46–58.

Sunstein, C. (2003). Terrorism and probability neglect. *The Journal of Risk and Uncertainty, 26,* 121–136.

Sutton, R. M., & Farrall, S. (2005). Gender, socially desirable responding and the fear of crime. *British Journal of Criminology, 45,* 212–224.

Tulloch, M. (2000). The meaning of age differences in the fear of crime. *British Journal of Criminology, 40,* 451–467.

The Gallup Organization. (2005). Gallup Brain Search: "Terrorism" [Database of survey items]. Retrieved December 15, 2008 from http://brain.gallup.com/

Warr, M. (1984). Fear of victimization: Why are women and the elderly more afraid? *Social Science Quarterly, 65,* 681–702.

Warr, M. (1994). Public perceptions and reactions to violent offending victimization. In A. J. Reiss, Jr. & J. A. Roth (Eds.), *Understanding and preventing violence: Consequences and control* (Vol. 4, pp. 1–66). Washington, DC: National Research Council.

Warr, M. (2000). *Fear of crime in the United States: Avenues for research and policy.* Washington, DC: National Institute of Justice.

Warr, M., & Ellison, C. G. (2000). Rethinking social reactions to crime: Personal and altruistic fear in family households. *American Journal of Sociology, 106,* 551.

Warr, M., & Stafford, M. (1983). Fear of victimization: A look at the proximate causes. *Social Forces, 61,* 1033–1043.

Will, J. A., & McGrath, J. H. (1995). Crime, neighborhood perceptions, and the underclass: The relationship between fear of crime and class position. *Journal of Criminal Justice, 23,* 163–176.

DISCUSSION QUESTIONS

1. Why are women more fearful of terrorism than are men? Why do they use more avoidance behaviors and seek more information from outside sources than do men?

2. Why does watching TV news increase use of avoidance behaviors among women and men?

3. What individual factors are related to the outcome measures for females and males? What differences exist in these factors and why?

4. Only one variable predicts information-seeking behavior for females and males; why do you think the other variables were not predictive of information seeking? What variables not included in the model may predict why people would seek information from outside sources?

5. Are there beneficial effects of fear? Explain.

Glossary

abandonment: When an elderly person is left by the caregiver in charge of him or her

account hijacking: Taking over someone else's existing account without consent

acute battering phase: The second phase in the cycle of violence, in which the abuser engages in major and often serious physically assaultive behavior

age-graded theory of adult social bonds: Proposes that marriage and employment can help one desist from criminal behavior

Air Transportation Safety and System Stabilization Act (2001): Compensation for victims of 9/11 who were killed or injured and their families

Anti-Car Theft Act (1992): Made carjacking a federal offense and provided funding to link motor vehicle databases

Anti-Car Theft Improvements Act of 1996: Upgraded state motor vehicle departments' databases to help identify stolen cars

anti-LGBTQ hate violence: Violence committed against people because of their sexual orientation or identity

Antiterrorism and Effective Death Penalty Act (1996): Required restitution for violent crimes and increased funds available to victims of terrorism

anxiety: An affective disorder or state often experienced as irrational and excessive fear and worry, which may be coupled with feelings of tension and restlessness, vigilance, irritability, and difficulty concentrating

Aviation Security Improvement Act: Legislation passed in response to terrorism

avoidance/numbing symptoms: Regular avoidance of stimuli associated with the traumatic event and numbness of response

baseless allegations: There is not enough evidence to prove an assault occurred

behavioral self-blame: When a person believes she or he did something to cause victimization

behavioral strategies: Actions workers can take to reduce their chances of victimization

Benjamin Mendelsohn: "Father of victimology"; coined the term *victimology* in the mid-1940s

bias crime: A crime committed against a person because of characteristics such as race, religion, or sexual orientation

bonded labor: Person is enslaved to work off a debt when the conditions of the debt were not previously known

bounding: Giving a time frame to reference in order to aid recall

bullying: Intentional infliction of injury repeatedly over time by a more powerful perpetrator over a less powerful victim

burglary: Entering a structure unlawfully to commit a felony or theft

CAN-SPAM Act of 2003: Made sending spam e-mail a crime in certain circumstances

capable guardianship: Means by which a person or target can be effectively guarded to prevent a victimization from occurring

carjacking: The taking of an occupied vehicle by an armed offender

certified: Allows victims of human trafficking access to the same services usually given to refugees

characterological self-blame: Person ascribes blame to a nonmodifiable source, such as one's character

Child Abuse Prevention and Treatment Act (1974): Provided definitions for child abuse and neglect and established mandatory reporting laws

child sexual abuse: Unwanted or forced sexual contact with a child, engaging children in sex work, or exposing children to sexually explicit material

Child Victims' Bill of Rights (1990): Gave victims' rights to children who were victims and witnesses

Church Arson Prevention Act of 1996: Prevents damaging religious buildings and preventing people from practicing their religious beliefs through threats or force civil litigation: Victims may sue their offenders in civil court to recoup costs and to compensate for emotional harm

civil rights movement: Advocated against racism and discrimination, noting that all Americans have rights that are protected by the U.S. Constitution

classification: Inmates are assessed and placed in appropriate institutions based on needs and characteristics

Code of Hammurabi: Early Babylonian code that emphasized the restoration of equity between the offender and the victim

coerced sexual contact: The offender uses psychological or emotional coercion to touch, grope, rub, pet, lick, or suck the breasts, lips, or genitals of the victim

cohabitation: Couples live together but are not married

Conflict Tactics Scale (CTS): Measurement tool used to gauge levels and use of various conflict tactics in intimate relationships

conjugal visits: Visits on prison grounds that allow married inmates to spend private time with their spouses

control deficit: When the amount of control a person exercises is outweighed by the control he or she is subject to

control ratio: Control surplus and control deficit considered together

control surplus: When the control one has exceeds the amount of control one is subject to

control-balance theory: The amount of control one possesses over others and the amount of control to which one is subject; the ratio of control influences the risk of engaging in deviant behavior

costs of crime: Mental, physical, and monetary loss that victims of crime incur

Crime Control Act (1990): Created a federal bill of rights for victims

Crime Victims with Disabilities Act of 1998: Required the National Crime Victimization Survey to collect information on people with disabilities

criminal intent incidents: Offender has no real relationship to the business where the crime occurs

customer/client incidents: A person with a legitimate reason for being at a business becomes violent while at that business

cyberbullying: Bullying over the Internet, by cell phone, or through another form of digital technology

cycle of violence: A common pattern of abuse that involves different phases: tension building phase, acute battering phase, and honeymoon phase; first developed by Lenore Walker in 1979

dark figure of crime: Unreported crime that occurs in prisons or jails

debt bondage: Person is enslaved to work off a debt when the conditions of the debt were not previously known

deinstitutionalization: Closing of institutions for mentally disordered people

deliberate indifference: Prison officials know inmates are at risk but do nothing to help

delinquent peers: People involved in delinquency with whom a person spends time; having such peers increases one's likelihood of victimization

dependency theory: As the dependence of the elderly increases, their rates of abuse and neglect increase

dependency-stress model: Theory that disabled children are more dependent on caregivers, which causes more stress that can lead to abuse

depression: A mood disorder characterized by sleep disturbances, changes in eating habits, feelings of guilt and worthlessness, and irritability. These symptoms interfere with a person's everyday life.

destination countries: Countries that receive victims of human trafficking

developmental disability: Serious, chronic impairment of major activities such as self-care, language, learning, mobility, and capacity for independent living

direct bullying: Physical and verbal actions performed in the presence of the victim

direct costs: Monies and the value of goods and services taken as a result of identity theft

direct property losses: When victims possessions are taken or damaged

diversion: Offender not formally charged if she or he completes required programs

domestic human trafficking: Trafficking that occurs within a country's borders

drug or alcohol facilitated rape: Victim is given drugs or alcohol without his or her knowledge or consent and then raped while under the influence

dumpster dive: Going through trash to find papers with personal identifying information

economic costs: Financial costs associated with victimization

Elder Justice Act: Provided funding for Adult Protective Services and for the prevention and detection of elder abuse

emotional abuse: Behavior such as yelling at or verbally degrading a partner or child; can also take the form of belittling, shaming, humiliating, ignoring, rejecting, or limiting physical contact

emotional or psychological elder abuse: Causing emotional pain by ridiculing, demeaning, humiliating, etc., an elderly person

environmental design: Making the workplace more secure and less of a target for crime

false allegations: Report of a sexual assault that did not happen family or community group conferencing: Victim, offender, family, friends, and supporters talk about the impact and consequences of a crime

family structure: Household style or shape

Federal Crime Victims with Disabilities Awareness Act of 1998: Required that statistics on disabled victims of crime be collected through the National Crime Victimization Survey

Federal Victim Witness Protection Act (1982): Developed and implemented guidelines for how officials respond to victims and witnesses

financial exploitation: Illegal or improper use of an elderly person's property, assets, or funds

forced labor: Victims are held and forced to work under threat of violence or punishment

forceful physical strategies: Physical resistance by the victim against the offender, such as shoving, biting, hitting, etc.

forceful verbal strategy: Verbal attempts to scare the offender or attract the attention of others

forcible rape: Offender uses or threatens to use force to achieve penetration

gene x environment interaction: Genes interact with environmental features to shape behavior

general theory of crime: Proposes that a person with low self-control will engage in crime if given the opportunity

Gun-Free Schools Act: Mandates that schools receiving federal funding must suspend for at least 1 year any student who brings a gun to school

Hans von Hentig: Developed a victim typology based on characteristics of the victim that increase risk of victimization

hate crime: A crime committed against a person because of characteristics such as race, religion, or sexual orientation

Hate Crime Statistics Act (1990): Requires that the attorney general collect statistics on hate crimes

hierarchy rule: If more than one Part I offense occurs in the same incident report, only the most serious offense will be counted in the reporting process

home invasion: Burglary of a residence in which the offender uses force against the residents

homicide survivors: People whose loved ones have been murdered

honeymoon phase: The third phase in the cycle of violence, in which the abuser is calm and loving and most probably begging his partner for forgiveness

Hostage Relief Act of 1980: Provided funding to help hostage victims, their spouses, and their children

hot spots: Areas that are crime-prone

hyperarousal: Persistent arousal symptomology; for example, not being able to sleep, being hypervigilant, and having problems concentrating

identity theft: Using another's personal identifying information to commit fraud

Identity Theft and Assumption Deterrence Act (1998): Made identity theft a federal offense

Identity Theft Penalty Enhancement Act (2004): Created the crime of aggravated identity theft for identity theft associated with certain felonies

Identity Theft Supplement: Supplement to the National Crime Victimization Survey in which data on extent of account hijacking, use and misuse of personal information was collected

incapacitated rape: Rape occurs when victim is unable to consent because of being unconscious, drugged, or otherwise incapacitated

incident report: Detailed questions about a victimization experience

indirect bullying: Subtle actions such as isolating, excluding, and making obscene gestures; often called social bullying

indirect costs: Legal bills, bounced checks, and other costs associated with identity theft

injury: A negative health outcome of intimate partner violence

intimate partner: A husband or wife, an ex-husband or ex-wife, a boyfriend or girlfriend, or a dating partner

intimate terrorism: Severe, persistent, and frequent abuse within intimate relationships that tends to escalate over time

intrusive recollection: Reexperiencing trauma through recurring or intrusive recollections or nightmares, feeling as though the event were recurring, and/or intense psychological distress when exposed to cues that symbolize or resemble a component of the traumatic event

involuntary domestic servitude: Victims are forced to work as domestic workers

Jeanne Clery Disclosure of Campus Security Policy and Campus Crime Statistics Act: Requires schools to publish an annual crime report, report warnings of threats, and protect the rights of victims and offenders in cases handled on campus

Justice for All Act (2004): Enforced victims' rights and provided funds to test the backlog of rape kits

labor trafficking: Human trafficking with the goal of exploiting someone for labor

learned helplessness: Victims believe they are unable to change the situation and stop trying to resist

lex talionis: An eye for an eye

life-course perspective: Examines the development of and desistance from offending and other behaviors over time

Long-Term Care Ombudsman Program: Receives complaints about elderly mistreatment in long-term care facilities

lost productivity: Being unable to work, go to school, or complete everyday tasks because of being victimized

mandatory arrest policies: Require arrest by police officers when there is probable cause that a crime was committed and enough evidence exists for an arrest

mandatory reporting law: Requires certain professionals, such as doctors, to report suspected cases of child abuse

Marvin Wolfgang: Used Philadelphia homicide data to conduct the first empirical investigation of victim precipitation

Matthew Shepard and James Byrd, Jr. Hate Crime Prevention Act: Included gender-based hate crimes and hate crimes by and against juveniles in data collection

medical care costs: Costs associated with treating victims of crime

Menachem Amir: Studied victim provocation in rapes

mental health care costs: Psychiatric care required as a result of being victimized

Minneapolis Domestic Violence Experiment: Conducted in 1984 by Lawrence Sherman and Richard Berk to examine the deterrent effect of arrest on domestic violence perpetrators

misdemeanor: A crime that usually is less serious than a felony and carries a maximum penalty of a year in jail

motivated offenders: People who will commit crime if given an opportunity

motor vehicle theft: The unlawful taking of another's vehicle

Motor Vehicle Theft Law Enforcement Act (1984): Required manufacturers to put identification numbers on car parts to make them easier to trace when stolen

Motor Vehicle Theft Prevention Act of 1994: Developed a voluntary, national motor vehicle theft prevention program

National Child Abuse and Neglect Data System (NCANDS): Annual analysis of data on child abuse reports submitted to child protective services

National College Women Sexual Victimization Study (NCWSV): A nationally representative study of sexual victimization among college women, conducted in 1997

National Crime Survey: First ever government-sponsored victimization survey; relied on victims to recall their own victimization experiences

National Crime Victimization Survey (NCVS): National survey of households that is used to generate annual estimates of victimization in the United States

National Elder Abuse Incidence Study: Measured the incidence of mistreatment of people over age 60 in domestic settings

National Incidence Study (NIS): Report on cases investigated by child protective services and cases identified by professionals in the community

National Inmate Survey: Self-report survey on inmates' rates of victimization

National Social Life, Health, and Aging Project: Self-report survey of elderly people about their experiences of abuse

National Study of Drug or Alcohol Facilitated, Incapacitated, and Forcible Rape: A nation study of three types of rape

National Survey of Youth in Custody: Survey of youth in state and private correctional facilities

National Violence Against Women Survey (NVAWS): Telephone survey of 8,000 men and 8,000 women about violence they have experienced

neglect: When a child's basic needs are not met; also, when someone with the responsibility of caring for an elderly person fails to fulfill their caretaking obligations

neighborhood context: Features of neighborhoods that impact risk for victimization

New Fair and Accurate Credit Transactions Act of 2003: Enacted provisions to prevent identity theft and help victims

no-drop prosecution: The victim is not able to drop the charges against the offender and the prosecutor's discretion in deciding to charge is curtailed

noncontact sexual abuse: Forms of sexual victimization that do not involve touching or penetration; includes verbal and visual abuse

nonforceful physical strategies: Nonforceful physical attempts to stop an assault, such as trying to escape the attack by running away

nonforceful verbal strategy: Nonaggressive attempts to get the offender to stop, such as talking to or pleading with the offender

notification: The right of victims to be kept apprised of key events in their cases

Occupational Safety and Health Administration (OSHA): Federal agency in the U.S. Department of Labor that provides guidelines for workplace health and safety

Older Americans Act of 1965: Protected the rights of and provided funding for the elderly

organizational and administrative controls: Strategies that administrators and agencies can implement to reduce the risk of workplace victimization in their organizations

Palermo Protocol: Criminalized human trafficking and established procedures for government response

parity hypothesis: The idea that, to stop an incident, the victim's level of self-protection should match the offender's level of attack participation and consultation: Rights given to victims to encourage participation in the criminal justice system; also provide victims the right to discuss their cases with the prosecutor and/or judge before key decisions are made

patriarchy: A form of social organization in which the man is dominant and is allowed to control women and children

peacemaking circles: Gathering of victim, offender, community members, and sometimes criminal justice officials to promote healing

permissive arrest policies: Policies that do not mandate or presume that an arrest will be made by law enforcement when warranted; allow police to use their discretion

personal relationship incidents: Perpetrator has a relationship with the victim and targets him or her while at work

phishing: Use of fake websites and e-mails to trick people into providing personal information

physical abuse: Injury or physical harm of another person, such as a child

physical bullying: Hitting, punching, shoving, or other physical forms of violence

physical elder abuse: Nonaccidental harm of an elderly person causing pain, injury, or impairment

physical injury: Physical harm suffered that may include bruises, soreness, scratches, cuts, broken bones, contracted diseases, and stab or gunshot wounds

physical violence: Includes hitting, slapping, kicking, punching, choking, and throwing objects at another person

posttraumatic stress disorder: Psychiatric anxiety disorder caused by experiencing traumatic events such as war, violence, etc.

power: A person's ability to impose his or her will on another person

power of attorney abuse: Misusing access to an elderly person's money

presumptive arrest policies: Arrest policies that presume an arrest will be made when probable cause exists to do so

principle of homogamy: People who share characteristics of offenders are more at risk of victimization, given that they are more likely to come into contact with offenders

Prison Rape Elimination Act of 2003: Requires the Bureau of Justice Statistics to analyze incidence and effects of prison rape

pro-arrest policies: Require arrests in specific situations in which certain criteria are met

protective custody: Secure inmate housing in which inmates are separated from others

protective order: Order secured to keep one person away from another

rape: Nonconsensual contact between the penis and the vulva or anus, or penetration of the vulva or anus, or contact between the mouth and penis, vulva, or anus, or penetration of another person's genital or anal opening with a finger, hand, or object, accompanied by force or threat of force

rape shield laws: Prohibit the defense from using the victim's sexual history in court

recurring victimization: When a person or place is victimized more than once in any way

repatriation: Returning trafficking victims to their native countries

repeat victimization: When a person is victimized more than once in the same way

reporting: Disclosing the victimization to the police

residential mobility: The percentage of persons 5 years and older living in a different house from 5 years before

resistance strategy: Something the victim does to try to stop or prevent an attack

restitution: Money or services paid to victims of crimes by the offenders

restorative justice: A movement recognizing that crime is a harm caused not just to the state but to the victim and his or her community. It seeks to use all entities in response to crime and allows for input from the offender, the victim, and community members harmed by the offense in making a determination of how to repair the harm caused by the offender.

retribution: A criminal is punished because he or she deserved it, and the punishment is equal to the harm caused

revictimization: When a person is victimized more than once over the course of the life span

right to a speedy trial: Victims' interests are considered when judges rule on postponement of trial dates

right to protection: Safety measures provided to victims

risk heterogeneity: Characteristics about a person that, if left unchanged, place him or her at greater risk of being victimized repeatedly

risky lifestyles: Engaging in risky behaviors that expose people to situations likely to increase their victimization risk

routine activities and lifestyle theory: A person's routine activities and lifestyle place him or her at risk of being victimized. Risk is highest when motivated offenders, lack of capable guardianship, and suitable targets coalesce in time and space.

safe haven laws: Allow mothers or caregivers in crisis to leave their babies at designated locations anonymously without risk of punishment

school victimization: Victimization of students on school grounds, buildings, buses, or at school events or functions screen questions: Used to cue respondents or jog their memories as to whether they experienced any of seven types of criminal victimization in the previous 6 months

screening: Questions asked by a medical professional to determine if a victimization occurred, identifying victims in order to provide them service referrals

secondary victimization: When people identify with the victim and suffer psychological consequences similar to the victim's

self-blame: Victims believe they are responsible for their own victimization

self-esteem: Beliefs and emotions about a person's own self-worth or value

self-neglect: When an elderly person fails to care for or adequately protect him- or herself

self-protective action: Something the victim does to try to stop or prevent an attack; generally classified into one of four types: forceful physical, nonforceful physical, forceful verbal, or nonforceful verbal

self-worth: A person's own perception of his or her worth or value

sentencing circle: Gathering of victim, offender, community members, and sometimes criminal justice officials to determine the offender's sentence

sex crime units: Special units within police departments trained to examine victims and collect evidence in sexual assault cases

sex trafficking: Human trafficking for the purpose of sexual exploitation

sexual assault nurse examiner (SANE): Registered nurses specially trained in examining victims of sexual assault and collecting forensic evidence

sexual assault response team (SART): Coordinate medical and criminal justice responses to rape and sexual assault victims

sexual coercion: Offender manipulates victim to engage in unwanted sex

sexual elder abuse: Sexual contact with an elderly person without his or her consent

Sexual Experiences Survey (SES): Widespread measurement tool developed by Koss that measures rape, sexual coercion, and sexual contact

sexual victimization: Encompasses any type of victimization involving sexual behavior perpetrated against an individual

sexual violence: Includes unwanted sexual contact, sexual coercion, and rape

shaken baby syndrome: Brain hemorrhages, skull fractures, and retinal hemorrhages caused by shaking

shoulder surfing: When offenders get information such as credit card numbers by watching or listening to someone enter them

situational couple violence: When conflict gets out of control and results in violence; also called common couple violence

skimming: Using devices to extract personal information from ATM machines

social distance: Feelings people have about others with whom they spend time

social interactionist perspective: Proposes that distressed individuals behave aggressively, which then elicits an aggressive response from others

social learning: People learn behavior by observing others engaging in it and by having their own behavior reinforced

source country: A country that supplies trafficking victims

spousal or marital privilege laws: Provide an exception that victims may not have to testify against their abusers if they are legally married to them

spyware: Software that surreptitiously collects information from unsuspecting people's computers

state dependence: The way a victim and offender respond to an incidence of victimization effects their likelihood of being involved in future victimization

statutory rape: When a person below the age of consent has sex

Stephen Schafer: Argued that victims have a functional responsibility not to provoke others into victimizing or harming them and that they also should actively attempt to prevent that from occurring

stressor: A traumatic event

structural density: The percentage of units in structures of five or more units

subintentional homicide: The victim facilitates her or his own death by using poor judgment, placing him- or herself at risk, living a risky lifestyle, or using alcohol or drugs

suitable targets: Victims chosen by offenders based on their attractiveness in the situation/crime

symptomology: Exhibiting symptoms of a mental disorder

system costs: Costs paid by society in response to victimization (e.g., law enforcement, insurance costs)

T visa: Allows victims of human trafficking to become temporary citizens of the United States

target hardening: Making it more difficult for an offender to attack a certain target

tension building phase: The first phase in the cycle of violence, in which positive and charming behavior on the part of the abuser lasts until pressures and more serious events generate tension

theft: The taking of another person's property

theory of low self-control: Proposes that people who have low self-control are more likely to be involved in crime and delinquency transnational human trafficking: When victims are transported from one country to another

unfounded: Assault is determined to be untrue or a case lacks evidence to move forward

Uniform Crime Reports (UCR): Annual reports of the amount of crime reported to or known by the police in a year

unwanted sexual contact: Person is touched in an erogenous zone, but it does not involve attempted or completed penetration

unwanted sexual contact with force: Person is touched in an erogenous zone, but not penetrated, through use or threat of force

verbal abuse: Offender makes offensive sexual comments or noises to victim

verbal bullying: Direct name calling and threating

vicarious victimization: The effect one person's victimization has on others

victim compensation: The right of victims to have monies that they lost due to victimization repaid to them by the state

victim facilitation: When a victim unintentionally makes it easier for an offender to commit a crime

victim impact statement: Statement made to the court by the victim or his/her family about the harm caused and the desired sentence for the offender

victim precipitation: The extent to which a victim is responsible for his or her own victimization

victim provocation: When a person does something that incites another person to commit an illegal act

victimization theory: Generally, a set of testable propositions designed to explain why a person is victimized

victim-offender mediation programs: Sessions led by a third party in which the victim and offender meet face-to-face to come to a mutually satisfactory agreement as to what should happen to the offender—often through the development of a restitution plan

victimology: The scientific study of victims and victimization

Victims of Crime Act (1984): Created the Office for Victims of Crime and provided funds for victim compensation

Victims of Terrorism Compensation Act: Provided money to compensate for wages lost during captivity

Victims of Terrorism Tax Relief Act (2001): Payments made to victims of qualifying disasters and death benefits for those lost in terrorist acts such as 9/11 cannot be taxed

victims' rights: Rights given to victims to enhance their privacy, protection, and participation

Victims' Rights and Restitution Act (1990): Guaranteed victims the right to restitution

Victims' Rights Clarification Act (1997): Allowed victims to make impact statements and attend their offenders' trials

victims' rights movement: Movement centered on giving victims a voice in the criminal justice system and providing them rights

victim/witness assistance programs (VWAPs): Provide aid to victims during the investigation and criminal justice process

violence: The intentional physical harm of another person

Violence Against Women Act (1994): Gave money to programs for prevention and treatment of female victims

Violence Against Women Act (2000): Provided funding for rape prevention and education and domestic violence victims, and included Internet stalking as a crime

Violent Crime Control and Law Enforcement Act (1994): Increased funds for victim compensation and created the national sex offender registry

visual abuse: Victim is forced to view sexual acts, pictures, and/or videos

women's movement: Recognized the need for female victims of crime to receive special attention and help due to the fact that victimizations such as sexual assault and domestic violence are byproducts of sexism, traditional sex roles, emphasis on traditional family values, and the economic subjugation of women

worker-on-worker incidents: When a current or former employee attacks another employee

workplace victimization: Victimization that occurs while a person is on duty at work

zero-tolerance policies: Specific, established punishments for students involved in fighting, violence, or bringing weapons to school

References

Abbey, A. (2002). Alcohol-related sexual assault: A common problem among college students. *Journal of Studies on Alcohol, 14*, 118–128.

Administration on Aging. (2008). *Investments in change: Enhancing the health and independence of older adults.* Washington, DC: U.S. Department of Health and Human Services, Administration on Aging. Retrieved from http://www.aoa.gov/AoARoot/Program_Results/docs/2008/AOA_2008_AnnualReport.pdf

Afifi, T. O., & Brownridge, D. A. (2008). Physical abuse of children born to adolescent mothers: The continuation of the relationship into adult motherhood and the role of identity. In T. I. Richardson & M. V. Williams (Eds.), *Child Abuse and Violence* (pp. 19–42). Hauppauge, NY: Nova Science.

Allen, M. (2007). *Lesbian, gay, bisexual and trans (LGBT) communities and domestic violence: Information and resources.* Harrisburg, PA: National Resource Center on Domestic Violence. Retrieved from http://www.vawnet.org/Assoc_Files_VAWnet/NRC_LGBTDV-Full.pdf

Allen, N. H. (1980). *Homicide: Perspectives on prevention.* New York: Human Sciences Press.

American Psychiatric Association. (2000). *Diagnostic and statistical manual of mental disorders* (4th ed., text revision). Washington, DC: Author.

Amick-McMullan, A., Kilpatrick, D. G., & Veronen, L. J. (1989). Family survivors of homicide victims: A behavioral analysis. *The Behavior Therapist, 12*, 75–79.

Angel, C. (2005). *Crime victims meet their offenders: Testing the impact of restorative justice on victims' post-traumatic stress symptoms.* PhD dissertation, University of Pennsylvania.

Appelbaum, P., Robbins, P., & Monahan, J. (2000). Violence and delusions: Data from the MacArthur Violence Risk Assessment Study. *American Journal of Psychiatry, 157*, 566–572.

Archer, J. (1999). Assessment of the reliability of the Conflict Tactics Scale: A meta-analytic review. *Journal of Interpersonal Violence, 14*, 1263–1289.

Archer, J. (2000). Sex differences in aggression between heterosexual partners: A meta-analytic review. *Psychological Bulletin, 126*, 651–680.

Arseneault, L., Moffit, T. E., Caspi, A., Taylor, P. J., & Silva, P. A. (2000). Mental disorders and violence in the total birth cohort: Results from the Dunedin study. *Archives of General Psychiatry, 57*, 979–986.

Arseneault, L., Walsh, E., Trzesniewski, K., Newcombe, R., Caspi, A., & Moffitt, T. E. (2006). Bullying victimization uniquely contributes to adjustment problems in young children: A nationally representative cohort study. *Pediatrics, 118*, 130–138.

Bachman, R. (1994). Violence and theft in the workplace. In *Crime Data Brief: National Crime Victimization Survey* (p. 1). Washington, DC: Bureau of Justice Statistics.

Bachman, R., & Meloy, M. L. (2008). The epidemiology of violence against the elderly: Implications for primary and secondary prevention. *Journal of Contemporary Criminal Justice, 24*(2), 186–197.

Badgley, R. F., Allard, H. A., McCormick, N., Proudfoot, P. M., Fortin, D., Ogilvie, D., et al. (1984). Sexual Offences Against Children. Catalogue no. J2–50/1984E. Ottowa: Department of Supply and Services.

Baladerian, N. J. (1991). Sexual abuse of people with developmental disabilities. *Sexuality and Disability, 9*(4), 323–335.

Balasinorwala, V. P. (2009). Acute stress disorder in victims after terror attacks in Mumbai, India. *British Journal of Psychiatry, 195*, 462.

Bandes, S. (1999). Victim standing. *Utah Law Review, 331.*

Banyard V. L., Plante, E. G., & Moynihan, M. M. (2007). *Rape prevention through bystander education: Final report to NIJ for grant 2002-WG-BX-0009.* Retrieved February 28, 2007, from www.ncjrs.org/pdffiles1/nij/grants/208701.pdf

Barnett, O. W., & LaViolette, A. D. (1993). *It could happen to anyone: Why battered women stay.* Newbury Park, CA: Sage.

Barton, G., & Vevea, B. (2010, July 10). 2 court systems didn't make link to Milwaukee assault suspect: Restraining orders not reviewed for criminal conduct. *Milwaukee Journal Sentinel.* Retrieved from http://www.jsonline.com/news/milwaukee/98177914.html

Baskin, D. R., Sommers, I., & Steadman, H. J. (1991). Assessing the impact of psychiatric impairment on prison violence. *Journal of Criminal Justice, 19*, 271–280.

Baum, K., & Klaus, P. (2005). *Violent victimization of college students, 1995–2002.* (NCJ Report No. 206836). Washington, DC: Bureau of Justice Statistics.

Baum, K., Catalano, S., Rand, M., & Rose, K. (2009). *Stalking victimization in the United States.* Washington, DC: U.S. Department of Justice, Bureau of Justice Statistics.

Baum, R., & Moore, K. (2002). *Lesbian, gay, bisexual and transgender domestic violence in 2001.* New York: National Coalition of Anti-Violence Programs.

Beaver, K. M., Wright, J. P., DeLisi, M., Daigle, L. E., Swatt, M. L., & Gibson, C. L. (2007). Evidence of a gene X environment interaction in the creation of victimization: Results from a longitudinal sample of adolescents. *International Journal of Offender Therapy and Comparative Criminology, 51*(6), 620–645.

Beck, A. J., & Harrison, P. M. (2010). *Sexual victimization in prisons and jails reported by inmates, 2008–09*. Washington, DC: Bureau of Justice Statistics, U.S. Department of Justice.

Beck, A. J., Harrison, P. M., & Guerino, P. (2010). *Special report: Sexual victimization in juvenile facilities reported by youth, 2008–09*. Washington, DC: Bureau of Justice Statistics, U.S. Department of Justice. Retrieved from http://bjs.ojp.usdoj.gov/content/pub/pdf/svjfry09.pdf

Beitchman, J. H., Zucker, K. J., Hood, J. E., da Costa, G. A., Akman, D., & Cassavia, E. (1992). A review of the long-term effects of child sexual abuse. *Child Abuse and Neglect, 16*, 101–118.

Bell, J. (2002). *Policing hatred: Law enforcement, civil rights, and hate crime.* New York: New York University Press.

Bennett Cattaneo, L., & Goodman, L. (2003). Victim-reported risk factors for continued abusive behavior: Assessing the dangerousness of arrested batterers. *Journal of Community Psychology, 31*, 349–369.

Bennett Cattaneo, L., & Goodman, L. A. (2005). Risk factors for reabuse in intimate partner violence: A cross-disciplinary critical review. *Trauma, Violence and Abuse: A Review Journal, 6*, 141–175.

Bennett, J. (2010, October 4). From lockers to lockup: School bullying in the digital age can have tragic consequences. But should it be a crime? *Newsweek.* Retrieved from http://www.newsweek.com/2010/10/04/phoebe-prince-should-bullying-be-a-crime.html?GT1=43002

Bennett, K. J. (2003). Legal and social issues surrounding closed circuit television of child victims and witnesses. *Journal of Aggression, Maltreatment and Trauma, 8*, 233–271.

Benson, M. L., Fox, G. L., DeMaris, A., & Van Wyk, J. (2003). Neighborhood disadvantage, individual economic distress and violence against women in intimate relationships. *Journal of Quantitative Criminology, 19*(3), 207–235.

Berg, S. (2009). Identity theft causes, correlates, and factors: A content analysis. In F. Schmalleger & M. Pittaro (Eds.), *Crimes of the Internet* (pp. 225–250). Upper Saddle River, NJ: Pearson-Prentice Hall.

Berios, D. C., & Grady, D. (1991). Domestic violence: Risk factors and outcomes. *Western Journal of Medicine, 155*, 133–135.

Bettencourt, B. A., & Miller, N. (1996). Gender differences in aggression as a function of provocation: A meta-analysis. *Psychological Bulletin, 119*(3), 422–447.

Blaauw, E., Winkel, F. W., Arensman, E., Sheridan, L., & Freeve, A. (2002). The toll of stalking: The relationship features of stalking and psychopathology of victims. *Journal of Interpersonal Violence, 17*, 50–63.

Blumenthal, J. A. (2009). Affective forecasting and capital sentencing: Reducing the effect of victim impact statements. *American Criminal Law Review, 46*, 107–126.

Bond, L., Carlin, J. B., Thomas, L., Ruin, K., & Patton, G. (2001). Does bullying cause emotional problems? A prospective study of young teenagers. *British Medical Journal, 323*, 480–484.

Bond, P. G., & Webb, J. R. (2010). Child abuse, neglect, and maltreatment. In B. S. Fisher & S. P. Lab (Eds.), *Encyclopedia of victimology and crime prevention* (Vol. 1, pp. 75–83). Los Angeles: Sage.

Bonderman, J. (2001). *Working with victims of gun violence.* Washington, DC: United States Department of Justice Office for Victims of Crime.

Bonnie, R. J., & Wallace, R. B. (2003). *Elder mistreatment: Abuse, neglect and exploitation in aging America.* Washington, DC: National Academies Press.

Bookwala, J., Frieze, I., Smith, C., & Ryan, K. (1992). Predictors of dating violence: A multivariate analysis. *Violence and Victims, 7*, 297–311.

Breitenbecher, K. H. (2001). Sexual assault on college campuses: Is an ounce of prevention enough? *Applied and Preventative Psychology, 9*(1), 23–52.

Browne, A., & Finkelhor, D. (1986). The impact of child sexual abuse: A review of the research. *Psychological Bulletin, 99*, 66–77.

Browning, S., & Erickson, P. (2009). Neighborhood disadvantage, alcohol use, and violent victimization. *Youth Violence and Juvenile Justice, 7*(4), 331–349.

Brownridge, D. A. (2006). Partner violence against women with disabilities: Prevalence, risk, and explanations. *Violence Against Women, 12*, 805–822.

Bui, H. I. (2001). In the adopted land: Abused immigrant women and the criminal justice system. Westport, CT: Praeger.

Bune, K. L. (2007, February 4). *Marital privilege law sends wrong message.* Retrieved from http://www.officer.com/article/10250193/marital-privilege-law-sends-wrong-message

Bureau of Justice Statistics. (2006a). Criminal victimization in the United States: Statistical tables. *Criminal Victimization in the United States.* Washington, DC: U.S. Department of Justice.

Bureau of Justice Statistics. (2006b). *National Crime Victimization Survey.* Washington, DC: United States Department of Justice.

Bureau of Justice Statistics. (2007). *National Crime Victimization Survey (NCVS).* Washington, DC: United States Department of Justice.

Bureau of Justice Statistics. (2008). *National Crime Victimization Survey, 2005.* Ann Arbor, MI: Inter-University Consortium for Political and Social Research. Retrieved from http://www.icpsr.umich.edu/icpsrweb/NACJD/studies/22746

Bureau of Justice Statistics. (2008). *National Crime Victimization Survey.* Washington, DC: United States Department of Justice.

Bureau of Justice Statistics. (2010, February). *Criminal victimization in the United States, 2007: Statistical tables.* Washington, DC: U.S. Department of Justice, Bureau of Justice Statistics.

Bureau of Labor Statistics. (2010). *Fact sheet: Workplace shootings.* Washington, DC: Bureau of Labor Statistics, U.S. Department of Labor. Retrieved from http://www.bls.gov/iif/oshwc/cfoi/osar0014.htm

Bureau of Labor Statistics. (2011, August 25). *National Census of Fatal Occupational Injuries in 2010 (preliminary results).* Washington, DC: Bureau of Labor Statistics, U.S. Department of Labor. Retrieved from http://www.bls.gov/news.release/pdf/cfoi.pdf

Buschur, C. (2010). Expert testimony. In Fisher, B. S., & Lab, S. P. (Eds.), *Encyclopedia of victimology and crime prevention* (Vol. 1, pp. 364–365). Thousand Oaks, CA: Sage.

Buzawa, E. (2007). Victims of domestic violence. In R. C. Davis, A. J. Lurigio, & S. Herman (Eds.), *Victims of crime* (3rd ed., pp. 55–74). Thousand Oaks, CA: Sage.

Buzawa, E. S., & Buzawa, C. G. (1993). Opening the doors: The changing police response to domestic violence. In R. G. Dunham & G. P. Alpert (Eds.), *Critical issues in policing* (pp. 551–567). Prospect Heights, IL: Waveland.

Campbell, R., Patterson, D., & Lichty, L. F. (2005). The effectiveness of sexual assault nurse examiner (SANE) programs: A review of the psychological, medical, legal, and community outcomes. *Trauma, Violence, and Abuse, 6*(4), 313–329.

Cantwell, H. B. (1999). The neglect of child neglect. In M. E. Helfer, R. S. Kempe, & R. D. Krugman (Eds.), *The Battered Child* (pp. 347–373). Chicago: University of Chicago Press.

Carbone-Lopez, K., & Kruttschnitt, C. (2010). Risky relationships? Assortative mating and women's experiences of intimate partner violence. *Crime and Delinquency, 56,* 358–384.

Carter, S. D., & Bath, C. (2007). The evolution and components of the Jeanne Clery Act: Implications for higher education. In B. S. Fisher & J. J. Sloan (Eds.), *Campus crime: Legal, social and policy perspectives* (2nd ed., pp. 27–44). Springfield, IL: Charles C. Thomas.

Cass, A. I. (2007). Routine activities and sexual assault: An analysis of individual and school level factors. *Violence and Victims, 22*(3), 350–366.

Catalano, S. (2010). Victimization during household burglary. *National Crime Victimization Survey,* 1–12.

Catalano, S., Smith, E., Snyder, H., & Rand, M. (2009). *Female victims of violence.* Washington, DC: U.S. Department of Justice, Bureau of Justice Statistics. Retrieved from http://bjs.ojp.usdoj.gov/content/pub/pdf/fvv.pdf

Centers for Disease Control and Prevention. (2003). *Costs of intimate partner violence against women.* Atlanta, GA: Centers for Disease Control and Prevention, National Center for Injury Prevention and Control.

Centers for Disease Control and Prevention. (2004, May 21). Youth risk behavior surveillance, United States, 2003. *Surveillance Summaries, 53* (no. SS-02). Atlanta, GA: Author.

Centers for Disease Control and Prevention. (2006). *HIV statistics and surveillance.* Atlanta, GA: Author.

Centers for Disease Control and Prevention. (2010). Developmental disabilities. Washington, DC: Department of Health and Human Services, Centers for Disease Control and Prevention. Retrieved from http://www.cdc.gov/ncbddd/dd/

Centers for Disease Control and Prevention. (2010). Intimate partner violence prevention: Consequences. Retrieved from http://www.cdc.gov/ViolencePrevention/intimatepartnerviolence/consequences.html

Centers for Disease Control and Prevention. (2010, June 4). Youth risk behavior surveillance: United States, 2009. *Surveillance Summaries, 59.*

Centre for Justice and Reconciliation. (2008). *What is restorative justice?* (Briefing paper). Retrieved from http://www.pfi.org/cjr

Chalk, R., & King, P. (1998). *Violence in families: Assessing prevention and treatment programs.* Washington, DC: National Academy Press.

Chandek, M. S., & Porter, C. O. L. H. (1998). The efficacy of expectancy disconfirmation in explaining crime victim satisfaction with the police. *Police Quarterly, 1,* 21–40.

Child Welfare Information Gateway. (2008a). *Long-term consequences of child abuse and neglect.* Retrieved from http://www.childwelfare.gov/pubs/factsheets/long_term_consequences.cfm

Child Welfare Information Gateway. (2008b). *What is child abuse and neglect?* Retrieved from http://www.childwelfare.gov/pubs/factsheets/whatiscan.cfm

Child Welfare Information Gateway. (2009). *Parental substance use and the child welfare system.* Retrieved from http://www.childwelfare.gov/pubs/factsheets/parentalsubabuse.cfm

Child Welfare Information Gateway. (2010a). *Infant safe haven laws: Summary of state laws.* Retrieved from http://www.childwelfare.gov/systemwide/laws_policies/statutes/safehaven.cfm

Child Welfare Information Gateway. (2010b). *Mandatory reporters of child abuse and neglect: Summary of state laws.* Retrieved from http://www.childwelfare.gov/systemwide/laws_policies/statutes/manda.cfm

Chon, S. (2010, February 24). *Carr v. United States (08-1301).* Ithaca, NY: Legal Information Institute, Cornell University Law School. Retrieved from http://topics.law.cornell.edu/supct/cert/08-1301

Chonco, N. R. (1989). Sexual assaults among male inmates: A descriptive study. *The Prison Journal, 69,* 72–82.

Clarke, R. V. G. (1980). "Situational" crime prevention: Theory and practice. *British Journal of Criminology, 20*(2), 136–147.

Clarke, R. V. G. (1982). Crime prevention through environmental management and design. In J. Gunn & D. P. Farrington (Eds.), *Abnormal offenders, delinquency, and the criminal justice system* (pp. 213–230). Chichester: John Wiley.

Clarke, R. V., & Harris, P. M. (1992). *Auto theft and its prevention.* In M. Tonry (Ed.), *Crime and justice: A review of research* (Vol. 16, pp. 1–52). Chicago: University of Chicago Press.

Classen, C., Pales, O. G., & Aggarwa, R. (2005). Sexual revictimization: A review of the empirical literature. *Trauma, Violence and Abuse, 6*(2), 103–129.

Clay-Warner, J. (2002). Avoiding rape: The effects of situational factors on rape. *Violence and Victims, 17*(6), 691–705.

Clery, H., & Clery, C. (2008). *What Jeanne didn't know.* Retrieved from http://www.securityoncampus.net/index.php?option=com_content&view=category&layout=blog&id=34&Itemid=53

Clodfelter, T. A., Turner, M. G., Hartman, J. L., & Kuhns, J. B. (2010). Sexual harassment victimization during emerging adulthood: A test of routine activities theory and a general theory of crime. *Crime & Delinquency, 56,* 455–481.

CNN Wire Staff. (2011). Swiss social worker admits to 114 sexual assaults on disabled. *CNN World.* Retrieved from http://articles.cnn.com/2011-02-01/world/switzerland.sex.charges_1_sexual-assaults-social-worker-information-on-similar-cases?_s=PM:WORLD

Cohen, L. E., & Felson, M. (1979). On estimating the social costs of national economic policy: A critical examination of the Brenner Study. *Social Indicators Research, 6,* 251–259.

Cohen, L. E., & Felson, M. (1979). Social change and crime rate trends: A routine activities approach. *American Sociological Review, 44*(4), 588–608.

Coker, A. L., Smith, P. H., Bethea, L., King, M. R., & McKeown, R. E. (2000). Physical health consequences of physical and psychological intimate partner violence. *Archives of Family Medicine, 9,* 1–7.

Coker, A. L., Smith, P. H., Thompson, M. P., McKeown, R. E., Bethea, L., & Davis, K. E. (2002). Social support protects against the negative effects of partner violence on mental health. *Journal of Women's Health & Gender-Based Medicine, 5,* 465–476.

Conron, K. J., Beardslee, W., Koenen, K. C., Buka, S. L., & Gortmaker, S. L. (2009). A longitudinal study of maternal depression and child maltreatment in a national sample of families investigated by child protective services. *Archives of Pediatric and Adolescent Medicine, 163*(10), 922–930.

Cose, E. (1994, August 8). Truths about spouse abuse. *Newsweek,* p. 49.

Cox, R. W., Johnson, T. A., & Richards, G. E. (2009). Routine activity theory and Internet crime. In F. Schmalleger & M. Pittaro (Eds.), *Crimes of the Internet* (pp. 302–316). Upper Saddle River, NJ: Pearson-Prentice Hall.

Cromwell, P. F., Olson, J. N., & Avary, D. W. (1991). *Breaking and entering: An ethnographic analysis of burglary.* Newbury Park, CA: Sage.

Cross, T. P., Walsh, W. A., Simone, M., & Jones, L. M. (2003). Prosecution of child abuse: A meta-analysis of rates of criminal justice decisions. *Trauma, Violence and Abuse, 4*(4), 323–340.

Daigle, L. E., & Fisher, B. S. (2010). Rape. In B. S. Fisher & S. P. Lab (Eds.), *Encyclopedia of victimology and crime prevention* (Vol. 2, pp. 708–715). Los Angeles: Sage.

Daigle, L. E., Beaver, K. M., & Hartman, J. L. (2008). A life-course approach to the study of victimization and offending behaviors. *Victims and Offenders, 3*(4), 365–390.

Daigle, L. E., Fisher, B. S., & Cullen, F. T. (2008). The violent and sexual victimization of college women. *Journal of Interpersonal Violence, 23*(9), 1296–1313.

Daigle, L. E., Fisher, B. S., & Stewart, M. (2009). The effectiveness of sexual victimization prevention among college students: A summary of "what works." *Victims and Offenders, 4,* 398–404.

Davies, K. (2010). Victim assistance programs, United States. In B. S. Fisher & S. P. Lab (Eds.), *Encyclopedia of victimology and crime prevention* (Vol. 2, pp. 968–969). Thousand Oaks, CA: Sage.

Davis, R. C., & Mulford, C. (2008). Victim rights and new remedies: Finally getting victims their due. *Journal of Contemporary Criminal Justice, 24*(2), 198–208.

Davis, R. C., Henley, M., & Smith, B. (1990). *Victim impact statements: Their effects on court outcomes and victim satisfaction.* New York: New York City Victim Service Agency.

Davis, R. C., Smith, B. E., & Davies, H. J. (2001). Effects of no-drop prosecution of domestic violence upon conviction rates. *Justice Research and Policy, 3,* 1–13.

Dawkins, J. L. (1996). Bullying, physical disability and the pediatric patient. *Developmental Medicine and Child Neurology, 38,* 603–612.

Dawson, M. & Dinovitzer, R. (2001). Victim cooperation and the prosecution of domestic violence in a specialized court. *Justice Quarterly, 18,* 593–622.

Deess, P. (1999). *Victims' rights: Notification, consultation, participation, services, compensation, and remedies in the criminal justice process.* New York: Vera Institute of Justice.

Deinstitutionalization. (2011). *Encyclopedia of mental disorders.* Retrieved on February 12, 2011, from http://www.minddisorders.com/Br-Del/Deinstitutionalization.html

Devoe, J. F., Bauer, L., & Hill, M. R. (2010, July). *Student victimization in U.S. schools: Results from the 2007 School Crime Supplement to the National Crime Victimization Survey.* Retrieved from nces.ed.gov/pubs2010/2010319.pdf

Diedrich, J., & Fauber, J. (2006, November 14). Gunshot costs echo through economy: From hospitals to jails, price of violence adds up quickly. *Milwaukee Wisconsin Journal Sentinel.* Retrieved February 7, 2010, from http://www.jsonline.com/news/milwaukee/29205944.html.

Dixon, L., & Stern, R. K. (2004). *Compensating the victims of 9/11* (Research brief). RAND. Retrieved from http://www.rand.org/pubs/research_briefs/RB9087.html

Dobash, R. E., & Dobash, R. P. (1979). *Violence against wives: A case against the patriarchy.* New York: Free Press.

Dobash, R. E., & Dobash, R. P. (1992). *Women, violence and social change.* London: Routledge.

Dragiewicz, M. (2010). Conflict Tactics Scale (CTS/CTS2). In B. S. Fisher & S. P. Lab (Eds.), *Encyclopedia of victimology and crime prevention* (Vol. 1, pp. 136–139). Los Angeles: Sage.

Dryden-Edwards, R. (2007). Anxiety. *Emedicinehealth.* Retrieved from http://www.emedicinehealth.com/anxiety/article_em.htm#Anxiety%20Overview

Dugan, L., LaFree, G., Cragin, K., & Kasupski, A. (2008). *Building and analyzing a comprehensive open source database on global terrorist events.* Washington, DC: U.S. Department of Justice. Retrieved from http://www.ncjrs.gov/pdffiles1/nij/grants/223287.pdf

Duhart, D. T. (2001). *National Crime Victimization Survey: Violence in the workplace, 1993–1999* (Special Report). Washington, DC: Bureau of Justice Statistics. Retrieved July 27, 2006, from http://www.ojp.usdoj.gov/bjs/pub/pdf/vw99.pdf

Dupont-Morales, T. (2009). Von Hentig's typologies. In J. K. Wilson (Ed.), *Praeger handbook of victimology* (pp. 308–309). Santa Barbara, CA: Praeger.

Egan, S. K., & Perry, D. G. (1998). Does low self-regard invite victimization? *Developmental Psychology, 34,* 299–309.

Ehrlich, H. J. (1992). The ecology of anti-gay violence. In G. M. Herek & K. T. Berrill (Eds.), *Hate crimes: Confronting violence against lesbians and gay men* (pp. 105–122). Thousand Oaks, CA: Sage.

Eigenberg, H. M. (1989). Male rape: An empirical examination of correctional officers' attitudes toward rape in prison. *The Prison Journal, 68*(1), 39–56.

Elias, R. (1984). Alienating the victim: Compensation and victim attitudes. *Journal of Social Issues, 40*(1), 103–116.

Erez, E., & Globokar, J. (2010). Victim impact statements. In B. S. Fisher & S. P. Lab (Eds.), *Encyclopedia of victimology and crime prevention* (Vol. 2, pp. 974–975). Los Angeles: Sage.

Erez, E., & Tontodonato, P. (1992). Victim participation in sentencing and satisfaction with justice. *Justice Quarterly, 9*(2), 393–417.

Erez, E., Roeger, L., & Morgan, F. (1994). *Victim impact statements in South Australia: An evaluation.* Adelaide: Office of Crime Statistics, South Australian Attorney General's Department.

Ericson, N. (2001, June). *Addressing the problem of juvenile bullying* (OJJDP Fact Sheet #27). Washington, DC: U.S. Department of Justice, Office of Justice Programs, Office of Juvenile Justice and Delinquency Prevention. Retrieved from http://www.ncjrs.gov/pdffiles1/ojjdp/fs200127.pdf

Esbensen, F. A., Huizinga, D., & Menard, S. (1999). Family context and criminal victimization in adolescence. *Youth and Society, 31*(2), 168–198.

Escobar, G. (2010). *Prosecuting intimate partner violence: Do victim services affect case outcomes?* Paper presented at the ASC Annual Meeting, San Francisco. Retrieved May 31, 2011, from http://www.allacademic.com/meta/p431247_index.html

Espelage, D. L., & Swearer, S. M. (2003). Research on school bullying and victimization: What have we learned and where do we go from here? *School Psychology Review, 32*(3), 365–383.

Estrich, S. (1988). *Real rape: How the legal system victimizes women who say no.* Cambridge, MA: Harvard University Press.

Family Violence Prevention Fund. (2010). *Intimate partner violence and Healthy People 2010 fact sheet.* San Francisco: Author. Retrieved from http://www.futureswithoutviolence.org/userfiles/file/Children_and_Families/ipv.pdf

Farmer v. Brennan (92–7247), 511 U.S. 825 (1994).

Farrell, A., McDevitt, J., & Fahy, S. (2008). *Understanding and improving law enforcement responses to human trafficking, final report.* Rockville, MD: National Institute of Justice.

Farrell, G., Phillips, C., & Pease, K. (1995). Like taking candy: Why does repeat victimization occur? *British Journal of Criminology, 35,* 384–399.

Fay, M., Morrissey, M., & Smythe, M. (1999). *Northern Ireland's troubles: The human costs.* Sterling, VA: Pluto.

Federal Bureau of Investigation. (2006). *Crime in the United States, 2006.* Retrieved from http://www.fbi.gov/about-us/cjis/ucr/crime-in-the-u.s/2006/

Federal Bureau of Investigation. (2009). *Crime in the United States, 2009*. Retrieved from http://www.fbi.gov/about-us/cjis/ucr/crime-in-the-u.s/2009/crime2009

Federal Bureau of Investigation. (2010, September). *Crime in the United States, 2009*. Retrieved from http://www2.fbi.gov/ucr/cius2009/index.html

Federal Bureau of Investigation. (2010a, November). *Hate crime statistics, 2009: Methodology*. Retrieved from http://www2.fbi.gov/ucr/hc2009/methodology.html

Federal Bureau of Investigation. (2010b, November). *Hate crime statistics, 2009: Victims*. Retrieved from http://www2.fbi.gov/ucr/hc2009/victims.html

Federal Bureau of Investigation. (n.d.). Human trafficking. Retrieved from http://www.fbi.gov/about-us/investigate/civilrights/human_trafficking

Federal Bureau of Investigation. (n.d.). *National Incident Based Reporting System (NIBRS): General information*. Retrieved from http://www2.fbi.gov/ucr/faqs.htm

Federal Deposit Insurance Corporation. (2004, December 14). *Putting an end to account-hijacking identity theft*. Retrieved from http://www.fdic.gov/consumers/consumer/idtheftstudy/identity_theft.pdf

Federal Grants Wire. (2011). Motor Vehicle Theft Prevention Act Program (16.597). Retrieved from http://www.federalgrantswire.com/motor-vehicle-theft-protection-act-program.html

Federal Trade Commission. (2004). *CAN-SPAM Act: A compliance guide for business*. Retrieved from http://business.ftc.gov/documents/bus61-can-spam-act-compliance-guide-business

Federal Trade Commission. (2010). *Consumer sentinel network data book: January–December 2009*. Retrieved from http://www.ftc.gov/sentinel/reports/sentinel-annual-reports/sentinel-cy2009.pdf

Fedus, D. B. (2010). Elder abuse, neglect, and maltreatment. In B. S. Fisher & S. P. Lab (Eds.), *Encyclopedia of victimology and victim prevention* (Vol. 1, pp. 348–350). Thousand Oaks, CA: Sage.

Felson, R. B. (2002). Reasons for reporting and not reporting domestic violence to the police. *Criminology, 40*, 617–648.

Felson, R. M. (1992). Routine activities and crime prevention: Armchair concepts and practical action. *Studies on Crime and Crime Prevention, 1*(1), 30–34.

Feniger, Y., & Yuchtman-Yaar, E. (2010). Risk groups in exposure to terror: The case of Israel's citizens. *Social Forces, 88*, 1451–1462.

FindLaw. (2011). CAL. PEN. CODE § 422.6: California Code—Section 422.6. Retrieved from http://codes.lp.findlaw.com/cacode/PEN/3/1/11.6/2/s422.6

Finkelhor, D., Turner, H. A., Ormrod, R. K., & Hamby, S. (2009). Violence, abuse and crime exposure in a national sample of children and youth. *Pediatrics, 124*(5), 1–14.

Finn, P. (2000). Labeling automobile parts to combat theft. *FBI Law Enforcement Bulletin, 69*, 10–14.

Finn, P., & McNeil, T. (1987). *The response of the criminal justice system to bias crimes*. Cambridge, MA: Abt Associations.

Fischer, K., & Rose, M. (1995). When "enough is enough": Battered women's decision making around court orders of protection. *Crime & Delinquency, 41*, 414–429.

Fisher, B. S., & Cullen, F. T. (2000). Measuring the sexual victimization of women: Evolution, current controversies, and future research. In D. Duffee (Ed.), *Criminal justice 2000 volumes: Vol. 4. Measurement and analysis of crime and justice* (pp. 317–390). Washington, DC: National Institute of Justice.

Fisher, B. S., Cullen, F. T., & Turner, M. G. (1998). *The extent and nature of sexual victimization among college women: A national-level analysis; Final report*. Washington, DC: U.S. Department of Justice, National Institute of Justice.

Fisher, B. S., Cullen, F. T., & Turner, M. G. (2000). *The sexual victimization of college women*. Washington, DC: U.S. Department of Justice, Bureau of Justice Statistics and National Institute of Justice.

Fisher, B. S., Daigle, L. E. & Cullen, F. T. (2010). *Unsafe in the Ivory Tower: The sexual victimization of college women*. Thousand Oaks, CA: Sage.

Fisher, B. S., Daigle, L. E., & Cullen, F. T. (2010). What distinguishes single from recurrent sexual victims? The role of lifestyle-routine activities and first-incident characteristics. *Justice Quarterly, 27*(1), 102–129.

Fisher, B. S., Daigle, L. E. Cullen, F. T., & Turner, M. G. (2003). Reporting sexual victimization to the police: Results from a national-level study of college women. *Criminal Justice and Behavior, 30*(6), 6–38.

Fisher, B. S., Daigle, L. E., Cullen, F. T., & Santana, S. A. (2007). Assessing the efficacy of the protective action-sexual victimization completion nexus. *Violence and Victims, 22*(1), 18–42.

Fisher, B. S., Karjane, H. M., Cullen, F. T., Blevins, K. R., Santana, S. A., & Daigle, L. E. (2007). Reporting sexual assault and the Clery Act: Situating findings from the National Campus Sexual Assault Policy Study within college women's experiences. In B. S. Fisher & J. J. Sloan (Eds.), *Campus crime: Legal, social and policy perspectives* (2nd ed., pp. 65–86). Springfield, IL: Charles C. Thomas.

Fisher, B. S., Sloan, J. J., Cullen, F. T., & Lu, C. (1998). Crime in the Ivory Tower: The level and sources of student victimization. *Criminology, 36*, 671–710.

Fishman, G., Hakim, S., & Shachmurove, Y. (1998). The use of household survey data: The probability of property crime victimization. *Journal of Economic and Social Measurement, 24*, 1–13.

Fliege, H., Lee, J., Grimm, A., & Klapp, B. F. (2009). Risk factors and correlates of deliberate self-harm behavior: A systematic review. *Journal of Psychosomatic Research, 66*, 477–493.

Focht-New, G., Clements, P. T., Barol, B., Faulkner, M. J., & Service, K. P. (2008). Persons with developmental disabilities exposed to interpersonal violence and crime: Strategies and guidance for assessment. *Perspectives in Psychiatric Care, 44*, 3–13.

Foshee, V. A., Bauman, K. E., & Linder, G. F. (1999). Family violence and perpetration of adolescent dating violence: Examining social learning and social control processes. *Journal of Marriage and the Family, 61*, 331–343.

Frye, V. (2001). Examining homicide's contribution to pregnancy-associated deaths. *Journal of the American Medical Association, 285*(11), 1510–1511.

Gabriel, R., Ferrando, L., Corton, E. S., Mingote, C., Garcia-Cambo, E., Liria, A. F., et al. (2007). Psychopathological consequences after a terrorist attack: An epidemiological study among victims, the general population, and police officers. *European Psychiatry, 22*, 339–346.

Garner, J., & Maxwell, C. (2009). Prosecution and conviction rates for intimate partner violence. *Criminal Justice Review, 34*, 44–79.

Gartner. (2006). Gartner says number of phishing e-mails sent to U.S. adults nearly doubles in just two years. Retrieved June 6, 2011, from http://www.gartner.com/it/page.jsp?id=498245

Garvin, M. (2010). Victims' rights movement, United States. In B. S. Fisher & S. P. Lab (Eds.), *Encyclopedia of victimology and crime prevention* (Vol. 2, pp. 1019–1020). Los Angeles: Sage.

Gerkin, P. M. (2009). Participation in victim-offender mediation: Lessons learned from observations. *Criminal Justice Review, 34,* 226–247.

Goldkamp, J. S., Weiland, D., Collins, M., & White, M. (1996). *Role of drug and alcohol abuse in domestic violence and its treatment: Dade County's domestic violence experiment, appendices to the final report.* Washington, DC: U.S. Department of Justice.

Goldman, R. (2010, March 29). Teens indicted after allegedly taunting girl who hanged herself. Retrieved from http://abcnews.go.com/Technology/TheLaw/teens-charged-bullying-massgirl-kill/story?id=10231357

Gómez, A. M. (2010, January 7). Testing the cycle of violence hypothesis: Child abuse and adolescent dating violence as predictors of intimate partner violence in young adulthood. *Youth & Society* (online first). Retrieved from http://yas.sagepub.com/content/early/2010/01/07/0044118X09358313.full.pdf+html

Gonzales, A. R., Schofield, R. B., & Gillis, J. W. (2001). *Responding to victims of terrorism and mass violence crimes: Coordination and collaboration between American Red Cross workers and crime victim service providers.* Washington, DC: U.S. Department of Justice, Office for Victims of Crime. Retrieved from http://www.ojp.usdoj.gov/ovc/publications/infores/redcross/ncj209681.pdf

Goodman, L. A., Salyers, M. P., Mueser, K. T., Rosenberg, S. D., Swartz, M., Essock, S. M., et al. (2001). Recent victimization in women and men with severe mental illness: Prevalence and correlates. *Journal of Traumatic Stress, 14,* 615–632.

Goodman, L., Thompson, K., Weinfurt, K., Corl, S., Acker, P., Mueser, K., et al. (1999). Reliability of reports of violent victimization and PTSD among men and women with SMI. *Journal of Traumatic Stress, 12,* 587–599.

Gottfredson, M. R., & Hirschi, T. (1990). *A general theory of crime.* Stanford, CA: Stanford University Press.

Gover, A., MacDonald, J., & Alpert, G. (2003). Combating domestic violence: Findings from an evaluation of a local domestic violence court. *Criminology & Public Policy, 3,* 109–132.

Gozdziak, E. M., & Collett, E. A. (2005). Research on human trafficking in North America: A review of literature. *International Migration, 43,* 1–2.

Greenfield, L., & Henneberg, M. (2000). *Alcohol, crime, and the criminal justice system.* Paper commissioned for the Alcohol Policy XII Conference, Washington, DC.

Grills, A. E., & Ollendick, T. H. (2002). Peer victimization, global self-worth, and anxiety in middle school children. *Journal of Clinical Child and Adolescent Psychology, 31*(1), 59–68.

Groff, E. R. (2007). Simulation for theory testing and experimentation: An example using routine activity theory and street robbery. *Journal of Quantitative Criminology, 23*(2), 75–103.

Grothaus, R. S. (1985). Abuse of women with disabilities. In S. E. Brown, D. Connors, & N. Stern (Eds.), *With the power of each breath: A disabled woman's anthology* (1st ed., pp. 124–132). Pittsburg: Cleiss.

Gundy-Yoder, A. V. (2010). Victims' rights legislation, federal, United States. In B. S. Fisher & S. P. Lab (Eds.), *Encyclopedia of victimology and crime prevention* (Vol. 2, pp. 1012–1013). Los Angeles: Sage.

Gurwitch, R. H., Pfefferbaum, B., & Leftwich, M. T. J. (2002). The impact of terrorism on children: Considerations for a new era. *Journal of Trauma Practice, 1,* 101–124.

Guterman, N. B., & Lee, Y. (2005). The role of fathers in risk for physical child abuse and neglect: Possible pathways and unanswered questions. *Child Maltreatment, 10,* 136–149.

Hard, S. (1986). *Sexual abuse of the developmentally disabled: A case study.* Paper presented at the National Conference of Executives of Associations for Retarded Citizens, Omaha, NE.

Harlow, C. W. (2005). *Hate crimes reported by victims and police* (Special report, NCJ 20991). Washington, DC: U.S. Department of Justice, Bureau of Justice Statistics.

Harrell, E., & Rand, M. (2010). Disabilities and victimization: NCVS statistical overview. Retrieved from http://ovc.ncjrs.gov/ncvrw2010/pdf/6_StatisticalOverviews.pdf

Hart, T. C. (2007). Violent victimization of college students: Findings from the National Crime Victimization Survey. In J. Sloan and B. Fisher (Eds.), *Campus crime: Legal, social, and policy perspectives* (2nd ed., pp. 129–146). Springfield, IL: C. C. Thomas.

Hart, T. C., & Rennison, C. (2003). *Reporting crime to the police, 1992–2000.* Washington, DC: United States Bureau of Justice Statistics.

Hartley, D., Biddle, E., & Jenkins, L. (2005). Societal cost of workplace homicides in the United States. *American Journal of Industrial Medicine, 47*(6), 518–527.

Hartman, J. L., & Alligood, K. (2010). Protection/restraining orders. In B. S. Fisher & S. P. Lab (Eds.), *Encyclopedia of victimology and crime prevention* (Vol. 2, pp. 688–689). Los Angeles: Sage.

Hartman, J. L., & Belknap, J. (2003). Beyond the gatekeepers: Court professionals' self-reported attitudes about experiences with misdemeanor domestic violence cases. *Criminal Justice and Behavior, 30,* 349–373.

Hay, C., Meldrum, R., & Mann, K. (2010). Traditional bullying, cyber bullying, and deviance: A general strain theory approach. *Journal of Contemporary Criminal Justice, 26,* 130–147.

Haynie, D. L., Nansel, T., Eitel, P., Crump, A. D., Saylor, K., Yu, K., et al. (2001). Bullies, victims, and bully/victims: Distinct groups of at-risk youth. *Journal of Early Adolescence, 21*(1), 29–49.

Headden, S. (1996, July 1). Guns, money & medicine. *U.S. News & World Report, 121*(1), 31–40.

Hensley, C., Koscheski, M., & Tewksbury, R. (2005). Examining the characteristics of sexual assault targets in maximum security prisons. *Journal of Interpersonal Violence, 40,* 667–697.

Hensley, C., Tewksbury, R., & Castle, T. (2003). Characteristics of prison sexual assault targets in male Oklahoma correctional facilities. *Journal of Interpersonal Violence, 18,* 595–606.

Herek, G. M. (2009). Hate crimes and stigma-related experiences among sexual minority adults in the United States: Prevalence estimates from a national probability sample. *Journal of Interpersonal Violence, 24,* 54–74.

Herek, G. M., Gillis, J. R., & Cogan, J. C. (1999). Psychological sequelae of hate crime victimization among lesbian, gay, and bisexual adults. *Journal of Consulting and Clinical Psychology, 67,* 945–951.

Herkov, M. J., & Biernat, M. (1997). Assessment of PTSD symptoms in a community exposed to serial murder. *Journal of Clinical Psychology, 53,* 809–815.

Herman, S. (2004, April 28). *Supporting and protecting victims: Making it happen.* Keynote address at the National Victims Conference, London. Retrieved from http://www.ncvc.org/NCVC/main.aspx?dbName=DocumentViewer&DocumentID=38044

Hewitt, C. (1988). The costs of terrorism: A cross-national study of six countries. *Studies in Conflict and Terrorism, 11*, 169–180.

Heyman, R. E., & Smith, A. (2002). Do child abuse and interparental violence lead to adulthood family violence? *Journal of Marriage and Family, 64,* 864–870.

Hiday, V. A. (1997). Understanding the connection between mental illness and violence. *International Journal of Law and Psychiatry, 20,* 399–417.

Hiday, V. A., Swartz, M. S., Swanson, J. W., Borum, R., & Wagner, H. R. (1999). Criminal victimization of persons with severe mental illness. *Psychiatric Services, 50,* 62–68.

Hiday, V. A., Swartz, M. S., Swanson, J. W., Borum, R., & Wagner, H. R. (2002). Impact of outpatient commitment on victimization of people with severe mental illness. *American Journal of Psychiatry, 159,* 1403–1411.

Hindelang, M. J., Gottfredson, M. R., & Garafolo, J. (1978). *Victims of personal crime: An empirical foundation for a theory of personal victimization.* Cambridge, MA: Ballinger.

Hirschel, D. (2008, July 25). *Domestic violence cases: What research shows about arrest and dual arrest rates.* Washington, DC: National Institute of Justice.

Hirschel, D., Buzawa, E., Pattavina, A., Faggiani, D., & Reuland, M. (2007, May). *Explaining the prevalence, context, and consequences of dual arrest in intimate partner cases.* Washington, DC: U.S. Department of Justice.

Hirschel, J., Dean, C., & Lumb, R. (1994). The relative contribution of domestic violence to assault and injury of police officers. *Justice Quarterly, 11,* 99–117.

Hirschfield, A., Newton, A., & Rogerson, M. (2010). Linking burglary and target hardening at the property level: New insights into victimization and burglary protection. *Criminal Justice Policy Review, 21,* 319–337.

Hodge, P. D. (1999). National law enforcement programs to prevent, detect, investigate and prosecute elder abuse and neglect in healthcare facilities. *Journal of Elder Abuse and Neglect, 9*(4), 23–41.

Holt, T. J., & Bossler, A. M. (2009). Examining the applicability of lifestyle-routine activities theory for cybercrime victimization. *Deviant Behavior, 30*(1), 1–25.

Howley, S., & Dorris, C. (2007). Legal rights for crime victims in the criminal justice system. In R. C. Davis, A. J. Lurigio, & S. Herman (Eds.), *Victims of crime* (3rd ed., pp. 299–314). Thousand Oaks, CA: Sage.

Howton, A. J. (2010). Sexual assault response team (SART). In B. S. Fisher & S. P. Lab (Eds.), *Encyclopedia of victimology and crime prevention* (Vol. 2, pp. 855–856). Los Angeles: Sage.

Hudson, C. G. (2005). Socioeconomic status and mental illness: Tests of the social causation and selection hypotheses. *American Journal of Orthopsychiatry, 75,* 3–18.

Inside Line. (2010, August 3). Cadillac Escalade tops auto theft list. Retrieved from http://www.insideline.com/cadillac/escalade/cadillac-escalade-tops-auto-theft-list.html

Insurance Information Institute. (2011, June). Auto theft. Retrieved from http://www.iii.org/media/hottopics/insurance/test4/

Irwin, J. (1980). *Prisons in turmoil.* Boston: Little, Brown.

Iyengar, R. (2007, August 7). The protection battered spouses don't need (Editorial). *New York Times.* Retrieved from http://www.nytimes.com/2007/08/07/opinion/07iyengar.html

Janoff-Bulman, R. (1979). Characterological versus behavioral self-blame: Inquiries into depression and rape. *Journal of Personality and Social Psychology, 37*(10), 1798–1809.

Jaquier, V. (2010). The role of the gay male and lesbian community. In B. S. Fisher & S. P. Lab (Eds.), *Encyclopedia of victimology and crime prevention* (Vol. 1, pp. 313–314). Los Angeles: Sage.

Jenkins, E. L. (2010). Workplace violence, United States. In B. S. Fisher & S. P. Lab (Eds.), *Encyclopedia of victimology and crime prevention* (Vol. 2, pp. 1078–1084). Los Angeles: Sage.

Jerin, R. A., Moriarty, L. J., & Gibson, M. A. (1996). Victim service or self-service: An analysis of prosecution-based victim-witness assistance programs and services. *Criminal Justice Policy Review, 7*(2), 142–154.

Jesperson, A., Lalumiere, M. L., & Seto, M. C. (2009). Sexual abuse history among adult sex offenders: A meta-analysis. *Child Abuse and Neglect, 33*(3), 179–192.

Johnson, I., & Sigler, R. (2000). Forced sexual intercourse among intimates. *Journal of Family Violence, 15,* 95–108.

Johnson, M. P. (2006). Conflict and control: Gender symmetry and asymmetry in domestic violence. *Violence Against Women, 12*(11), 1003–1018.

Johnson, M., & Kercher, G. (2009, January). *Personal victimization of college students.* Huntsville, TX: Crime Victims' Institute, Criminal Justice Center, Sam Houston State University. Retrieved from http://www.crimevictimsinstitute.org/documents/CSVictimizationFinal.pdf

Johnson, R. J., Rew, L., & Sternglanz, W. (2006). The relationship between childhood sexual abuse and sexual health practices of homeless adolescents. *Adolescence, 41*(162), 221–234.

Johnstone, G. (2002). *Restorative justice: Ideas, values, debates.* Portland, OR: Willan.

Jordan, C. (2004). Intimate partner violence and the justice system: An examination of the interface. *Journal of Interpersonal Violence, 19,* 1412–1434.

Justice Solutions. (2002, July). *Victim impact statement resource package.* Retrieved from http://www.justicesolutions.org/art_pub_victim_impact_resource.htm#sample_victim_impact

Kang, S., Magura, S., Laudet, A., & Whitney, S. (1999). Adverse effect of child abuse victimization among substance-using women in treatment. *Journal of Interpersonal Violence, 14*(5), 657–670.

Kanno, H., & Newhill, C. (2009). Social workers and battered women: The need to study client violence in the domestic field. *Journal of Aggression, Maltreatment and Trauma, 18,* 46–63.

Kaufman, M. T. (1998, April 18). Marvin E. Wolfgang, 73, dies; leading figure in criminology. *The New York Times.* Retrieved from http://www.nytimes.com/1998/04/18/us/marvin-e-wolfgang-73-dies-leading-figure-in-criminology.html

Keilitz, S., Hannaford, P., & Efkeman, H. (1997). *Civil protection orders: The benefits and limitations for victims of domestic violence.* Williamsburg, VA: National Center for State Courts. Retrieved from http://www.ncjrs.gov/pdffiles1/pr/172223.pdf

Kelley, B. T., Thornberry, T. P., & Smith, C. A. (1997). In the wake of child maltreatment. *Juvenile Justice Bulletin.* Washington, DC: Office of Juvenile Justice and Delinquency Prevention, U.S. Department of Justice.

Kennard, K. L. (1989). The victim's veto: A way to increase victim impact on criminal case dispositions. *California Law Review, 77,* 417.

Kerley, K. R., Hochstetler, A., & Copes, H. (2009). Self-control, prison victimization, and prison infractions. *Criminal Justice Review, 34,* 553–568.

Kerley, K. R., Xu, X., & Sirisunyaluck, B. (2008). Self-control, intimate partner abuse, and intimate partner victimization: Testing the general theory of crime in Thailand. *Deviant Behavior, 29*, 503–532.

Kessler, R. C., Sonnega, A., Bromet, E., Hughes, M., & Nelson, C. B. (1995). Posttraumatic stress disorder in the National Comorbidity Survey. *Archives of General Psychiatry, 52*(12), 1048–1060.

Kilchling, M. (n.d.). Victims of terrorism: An overview on international legislation on the support and compensation for victims of terrorists threats. Retrieved from http://www.mpicc.de

Kilpatrick, D. C., Edmunds, C., & Seymour, A. (1992). Rape in America: A report to the nation. In *The National Women's Study*. Washington, DC: National Institute of Drug Abuse, National Victim's Center, and National Crime Victims Research and Treatment Center at the Medical University of South Carolina.

Kilpatrick, D. G., & Acierno, R. (2003). Mental health needs of crime victims: Epidemiology and outcomes. *Journal of Traumatic Stress, 16*(2), 119–132.

Kilpatrick, D. G., & Tidwell, R. (1989). *Victims' rights and services in South Carolina: The dream, the law, the reality*. Charleston: Crime Victims Research and Treatment Center, Medical University of South Carolina.

Kilpatrick, D. G., Amick, A., & Resnick, H. S. (1990). *The impact of homicide on surviving family members*. Charleston: Crime Victims Research and Treatment Center, Medical University of South Carolina.

Kilpatrick, D. G., Resnick, H. S., Ruggiero, K., Conoscenti, L. M., & McCauley, J. (2007). *Drug facilitated, incapacitated, and forcible rape: A national study*. Washington, DC: U.S. Department of Justice.

King, R. D. (2007). The contest of minority group threat: Race, institutions and complying with hate crime law. *Law and Society, 41*, 36–41.

Klaus, P. A., & Maston, C. T. (2008). *Criminal victimization in the United States, 2006*. Washington, DC: Bureau of Justice Statistics. Retrieved from http://bjs.ojp.usdoj.gov/index.cfm?ty=pbdetail&iid=1094

Klein, L. (2010). Victim compensation. In B. S. Fisher & S. P. Lab (Eds.), *Encyclopedia of victimology and crime prevention* (Vol. 2, pp. 971–974). Los Angeles: Sage.

Knauer, S. (2002). *Recovering from sexual abuse, addictions, and compulsive behaviors: "Numb" survivors*. Binghamton, NY: Haworth Social Work Practice Press.

Knowles, G. J. (1999). Male prison rape: A search for causation and prevention. *Howard Journal of Criminal Justice, 38*, 267–282.

Kochenderfer, B. J., & Ladd, G. W. (1996). Peer victimization: Cause or consequence of school maladjustment? *Child Development, 67*, 1305–1317.

Koss, M. P. (1985). The hidden rape victim. *Psychology of Women Quarterly, 48*, 193–212.

Koss, M. P. (1988). Hidden rape. In A. W. Burgess (Ed.), *Rape and sexual assault* (pp. 3–25). New York: Garland.

Koss, M. P., Abey, A., Campbell, R., Cook, S., Norris, J., Testa, M., et al. (2007). Revising the SES: A collaborative process to improve assessment of sexual aggression and victimization. *Psychology of Women Quarterly, 31*(4), 357–370.

Koss, M. P., Gidycz, C. A., & Wisniewski, N. (1987). The scope of rape: Incidence and prevalence of sexual aggression and victimization in a national sample of higher education students. *Journal of Consulting and Clinical Psychology, 55*, 162–170.

Kotz, P. (2010, January 28). Phoebe Prince, 15, commits suicide after onslaught of cyber-bullying from fellow students. Retrieved from http://www.truecrimereport.com/2010/01/phoebe_prince_15_commits_suici.php

Krasnoff, M., & Moscati, R. (2002). Domestic violence screening and referral can be effective. *Annals of Emergency Medicine, 40*, 485–492.

Kyckelhahn, T., Beck, A. J., & Cohen, T. H. (2009). *Characteristics of suspected human trafficking incidents, 2007–2008*. Washington, DC: U.S. Department of Justice, Bureau of Justice Statistics.

LaFree, G., Morris, N. A., & Dugan, L. (2010). Cross-national patterns of terrorism: Comparing trajectories for total, attributed and fatal attacks, 1970–2006. *Global Perspectives, 50*, 622–649.

Lahm, K. F. (2009). Physical and property victimization behind bars: A multilevel examination. *International Journal of Offender Therapy and Comparative Criminology, 53*, 348–365.

Langton, L., & Planty, M. (2010). Victims of identity theft, 2008. *National Crime Victimization Survey Supplement*, pp. 1–19.

Laub, J. H., & Lauritsen, J. L. (1998). The interdependence of school violence with neighborhood and family conditions. In D. S. Elliott, B. A. Hamburg, & K. R. Williams (Eds.), *Violence in American schools* (pp. 127–158). New York: Cambridge University Press.

Laumann, E. I., Leitsch, S. A., & Waite, L. J. (2008). Elder mistreatment in the United States: Prevalence estimates from a nationally representative study. *Journal of Gerontology, Social Sciences, 63*(4), S248–S254.

Lauritsen, J. L., Laub, J. H., & Sampson, R. J. (1992). Conventional and delinquent activities: Implications for the prevention of violent victimization among adolescents. *Violence and Victims, 7*(2), 91–108.

Layman, M., Gidycz, C. A., & Lynn, S. J. (1996). Unacknowledged versus acknowledged rape victims: Situational factors and posttraumatic stress. *Journal of Abnormal Psychology, 105*, 124–131.

Lebov, K. (2009). *Human trafficking in Scotland, 2007/08*. Edinburgh: Queens Printers of Scotland.

Levin, A. (2005). People with mental illness more often crime victims. *Psychiatric News, 40*, 16.

Levin, B., & Amster, S. (2007). Making hate history: Hate crime and policing in America's most diverse city. *American Behavioral Scientist, 51*, 319–348.

Levin, R. (2002). *September 11 victim compensation fund: A model for compensating terrorism victims?* Retrieved from http://www.kentlaw.edu/honorsscholars/2002students/Levin.html

Levy, M. P., & Tartaro, C. (2010). Auto theft: A site-survey and analysis of environmental crime factors in Atlantic City, NJ. *Security Journal, 23*, 75–94.

Limber, S. P., & Small, M. A. (2003). State laws and policies to address bullying in U.S. schools. *School Psychology Review, 32*(3), 445–455.

Linden, R., & Chaturvedi, R. (2005). The need for comprehensive crime prevention planning: The case of motor vehicle theft. *Canadian Journal of Criminology and Criminal Justice, 47*, 251–270.

Link, B., Monahan, J., Stueve, A., & Cullen, F. T. (1999). Real in their consequences: A sociological approach to understanding the association between psychotic symptoms and violence. *American Sociological Review, 64*, 316–332.

Littel, K. (2001). Sexual assault nurse examiner programs: Improving the community response to sexual assault victims. *Office for Victims of Crime Bulletin, 4*, 1–19.

Little, L. (2002). Middle-class mothers' perceptions of peer and sibling victimization among children with Asperger's syndrome and nonverbal learning disorders. *Issues in Comprehensive Pediatric Nursing, 25*, 43–57.

Littleton, H. L., Axson, D., Breitkopf, C. R., & Berenson, A. (2006). Rape acknowledgment and postassault experiences: How acknowledgment status related to disclosure, coping, worldview, and reactions received from others. *Violence and Victims, 21*(6), 761–778.

Logan, T. K., Walker, R., & Hunt, G. (2009). Understanding human trafficking in the United States. *Trauma, Violence, & Abuse, 10*, 3–30.

LoJack. (2011). How it works. Retrieved from http://www.lojack.com/why/pages/how-lojack-works.aspx

Lonsway, K. A., & Fitzgerald, L. F. (1994). Rape myths: In review. *Psychology of Women Quarterly, 18*, 133–164.

Lonsway, K. A., Archambault, J., & Berkowitz, A. B. (2007). *False reports: Moving beyond the issue to successfully investigate and prosecute non-stranger sexual assault.* San Luis Obispo, CA: EVAW International.

Lord, V. B. (1998). Characteristics of violence in state government. *Journal of Interpersonal Violence, 13,* 489–503.

Lynch, D. R. (1997). The nature of occupational stress among public defenders. *Justice System Journal, 19*(1), 17–35.

Lyon, E., & Lane, S. (2009). *Meeting survivors' needs: A multi-state study of domestic violence shelter experiences.* National Resource Center on Domestic Violence and UConn School of Social Work.

Lyon, E., Lane, S., & Menard, A. (2008, October). Review of relevant literature. In *Meeting survivors' needs: A multi-state study of domestic violence shelter experiences* (pp. 22–26). Retrieved from http://new.vawnet.org/Assoc_Files_VAWnet/MeetingSurvivorsNeeds-FullReport.pdf

Lyons, C. J. (2007). Community (Dis)Organization and Racially Motivated Crime. *American Journal of Sociology, 113,* 815–863.

Maltz, W. (2001). *The sexual healing journey: A guide for survivors of sexual abuse.* New York: HarperCollins.

Marge, D. K. (2003). *A call to action: Ending crimes of violence against children and adults with disabilities; A report to the nation.* Syracuse: SUNY Upstate Medical University, Department of Physical Medicine and Rehabilitation.

Margolin, L. (1992). Child abuse by mothers' boyfriends: Why the overrepresentation? *Child Abuse & Neglect, 16*(4), 541–551.

Maroney, T. A. (1998). The struggle against hate crime: Movement at a crossroads. *NYU Law Review, 73,* 564–620.

Marshall, T. F. (1999). *Restorative justice: An overview.* London: Home Office, Research Development and Statistics Division.

Martin, G., Bergen, H. A., Richardson, A. S., Roeger, L., & Allison, S. (2004). Sexual abuse and suicidality: Gender differences in a large community sample of adolescents. *Child Abuse & Neglect, 28,* 491–503.

Martin, S. L., Ray, N., Sotres-Alvarez, D., Kupper, L. L., Moracco, K. E., Dickens, P. A., et al. (2006). Physical and sexual assault of women with disabilities. *Violence Against Women, 12,* 823–837.

Marx, B. P., Calhoun, K. S., Wilson, A. E., & Meyerson, L. A. (2001). Sexual revictimization prevention: An outcome evaluation. *Journal of Consulting and Clinical Psychology, 69,* 25–32.

Maryland Crime Victims' Resource Center. (2007). Retrieved from http://www.mdcrimevictims.org/

Maston, C. T. (2010, March 2). *Criminal victimization in the United States, 2007.* Washington, DC: Bureau of Justice Statistics. Retrieved from http://bjs.ojp.usdoj.gov/index.cfm?ty=pbdetail&iid=1743

Matthias, R. E., & Benjamin, A. E. (2003). Abuse and neglect of clients in agency-based and consumer-directed home care. *Health & Social Work, 28,* 174–184.

Maxwell, C., Garner, J., & Fagan, J. (2001). *The effects of arrest on intimate partner violence: New evidence from the spouse assault replication program.* Washington, DC: National Institute of Justice. Retrieved from http://www.ncjrs.gov/txtfiles1/nij/188199.txt

May, D. (2010). Victimization of school teachers/staff. In B. S. Fisher & S. P. Lab (Eds.), *Encyclopedia of victimology and crime prevention* (Vol. 2, pp. 833–835). Los Angeles: Sage.

McCabe, A., & Manian, S. (2010). *Sex trafficking: A global perspective.* Plymouth, UK: Lexington Books.

McCartney, J. R., & Campbell, V. A. (1998). Confirmed abuse cases in public residential facilities for persons with mental retardation: A multi-state study. *Mental Retardation, 36,* 465–473.

McClellan, D. S., Farabee, D., & Crouch, B. M. (1997). Early victimization, drug use, and criminality: A comparison of male and female prisoners. *Criminal Justice and Behavior, 24*(4), 455–476.

McCold, P., & Wachtel, B. (1998). *Restorative policing experiment: The Bethlehem Pennsylvania Police Family Group Conferencing Project.* Bethlehem, PA: Real Justice.

McCorkle, R. C. (1992). Personal precautions to violence in prison. *Criminal Justice and Behavior, 19,* 160–173.

McFarlane, J., Hughes, R. B., Nosek, M. A., Groff, J. Y., Swedlend, N., & Dolan Mullen, P. (2001). Abuse assessment screen-disability (AAS-D): Measuring frequency, type, and perpetrator of abuse toward women with physical disabilities. *Journal of Women's Health and Gender-Based Medicine, 10,* 861–866.

McGarrell, E., Olivares, K., Crawford, K., & Kroovand, N. (2000). *Returning justice to the community: The Indianapolis Juvenile Restorative Justice Experiment.* Indianapolis, IN: Hudson Institute.

McMullin, D., Wirth, R. J., & White, J. W. (2007). The impact of sexual victimization on personality: A longitudinal study. *Sex Roles, 56,* 403–414.

Menard, S. (2002, February). Short- and long-term consequences of adolescent victimization. *Youth Violence Research Bulletin.* Washington, DC: U.S. Department of Justice, Office of Juvenile Justice and Delinquency Prevention.

Mendelsohn, B. (1947, March). *New biopsychosocial horizons: Victimology.* Paper presented to the Psychiatric Society of Bucharest, Coltzea State Hospital, Hungary.

Meyers, T. W. (2002). Policing and sexual assault: Strategies for successful victim interviews. In L. J. Moriarty & M. L. Dantzker (Eds.), *Policing and victims* (pp. 57–73). Upper Saddle River, NJ: Pearson.

Miller, E., Decker, M. R., Silverman, J. G., & Raj, A. (2007). Migration, sexual exploitation and women's health: A case report from a community health center. *Violence Against Women, 13,* 486–497.

Miller, H. V., & Miller, J. M. (2010). School-based bullying prevention. In B. S. Fisher & S. P. Lab (Eds.), *Encyclopedia of victimology and crime prevention* (Vol. 2, pp. 817–819). Los Angeles: Sage.

Miller, L. (2004). Psychotherapeutic interventions for survivors of terrorism. *American Journal of Psychiatry, 58,* 1–16.

Miller, T. R., Cohen, M. A., & Wiersema, B. (1996). *Victim costs and consequences: A new look.* Washington, DC: United States National Institute of Justice.

Minnesota Department of Health. (1998). Tools about sexual violence: General prevention tool #25; Effects of sexual victimization. In *A place to start: A resource kit for preventing sexual violence.* St. Paul, MN: Author.

Morgan, P. (2010). Workplace violence training and education. In B. S. Fisher & S. P. Lab (Eds.), *Encyclopedia of victimology and crime prevention* (Vol. 2, pp. 1076–1078). Los Angeles: Sage.

Murphy, S. A., Tapper, V. J., Johnson, L. C., & Lohan, J. (2003). Suicide ideation among parents bereaved by the violent deaths of their children. *Issues in Mental Health Nursing, 24*(1), 5–25.

Mustaine, E. E. (2010). Stalking. In B. S. Fisher & S. P. Lab (Eds.), *Encyclopedia of victimology and crime prevention* (Vol. 2, pp. 900–904). Los Angeles: Sage.

Mustaine, E. E., & Tewksbury, R. (1998). Predicting risks of larceny theft victimization: A routine activity analysis using refined lifestyle measures. *Criminology, 36*(4), 829–857.

Mustaine, E. E., & Tewksbury, R. (1998). Victimization risks at leisure: A gender-specific analysis. *Violence and Victims, 13,* 3–21.

Mustaine, E. E., & Tewksbury, R. (1999). A routine activity theory explanation for women's stalking victimizations. *Violence Against Women, 5*(1), 43–62.

Mustaine, E. E., & Tewksbury, R. (2007). Collateral consequences and community reentry for registered sex offenders with child victims: Are the challenges even greater? *Journal of Offender Rehabilitation, 46*(1/2), 113–131.

Mustaine, E. E., & Tewksbury, R. (2007). The routine activities and criminal victimization of students: Lifestyle and related factors. In B. S. Fisher & J. J. Sloan (Eds.), *Campus crime: Legal, social, and policy perspectives* (pp. 147–166). Springfield, IL: Charles C. Thomas.

Nannini, A. (2006). Sexual assault patterns among women with and without disabilities seeking survivor services. *Women's Health Issues, 16,* 372–379.

Nansel, T. R., Overpeck, M., Pilla, R. S., Ruan, J., Simons-Morton, B., & Scheidt, P. (2001). Bullying behaviors among U.S. youth: Prevalence and association with psychosocial adjustment. *Journal of the American Medical Association, 285,* 2094–2100.

National Association of Crime Victim Compensation Boards. (2009). National conference to mark 25 years of VOCA grants to states. *Crime Victim Compensation Quarterly, 2,* 1. Retrieved from http://www .nacvcb.org/NACVCB/files/ccLibraryFiles/Filename/000000000025/ 20092.pdf

National Association of State Boards of Education. (2010, December 3). Bullying, harassment, and hazing. *State School Healthy Policy Database.* Retrieved from http://nasbe.org/healthy_schools/hs/state.php?state= Florida#Bullying, Harassment and Hazing

National Center for Prosecution of Child Abuse, National District Attorneys Association. (2010, July). *NDAA presence of support person for child witness compilation.* Retrieved from http://www.ndaa.org/pdf/Presence% 20of%20Support%20Persons%20for%20Child%20Witnesses%202010 .pdf

National Center for Victims of Crime. (1999). *Victim impact statements.* Retrieved from http://www.ncvc.org/ncvc/main.aspx?dbName=Docu mentViewer&DocumentID=32515

National Center for Victims of Crime. (2008). *Domestic violence: Why victims may stay.* Washington, DC: Author. Retrieved from http://www.ncvc .org/ncvc/main.aspx?dbName=DocumentViewer&Documen tID=32347

National Center for Victims of Crime. (2008). Hate crimes. Retrieved from http://www.ncvc.org/ncvc/main.aspx?dbName=DocumentViewer&D ocumentID=32356

National Center for Victims of Crime. (2008). Sexual assault. Retrieved from http://www.ncvc.org/ncvc/main.aspx?dbName=DocumentViewer&D ocumentID=32369

National Center for Victims of Crime. (2009). *About victims' rights.* Retrieved from http://www.victimlaw.info/victimlaw/pages/victimsRight.jsp

National Center for Victims of Crime. (n.d.). Stalking facts. Retrieved from http://www.ncvc.org/src/main.aspx?dbID=DB_statistics195

National Center on Domestic and Sexual Violence. (2011). Family justice centers. Retrieved from http://www.ncdsv.org/ncd_linksfamilyjustice .html

National Center on Elder Abuse. (2011). *Why should I care about elder abuse?* Retrieved from http://www.ncea.aoa.gov/ncearoot/Main_Site/pdf/ publication/NCEA_WhatIsAbuse-2010.pdf

National Coalition Against Domestic Violence. (2007). *Domestic violence facts.* Retrieved from http://www.ncadv.org/files/DomesticViolenceFac tSheet(National).pdf

National Coalition of Anti-Violence Programs. (2010). *Hate violence against the lesbian, gay, bisexual, transgender and queer communities in the United States in 2009.* New York: Author. Retrieved from http://www .avp.org/documents/NCAVP2009HateViolenceReportforWeb_000.pdf

National Commission on Terrorist Attacks upon the United States. (2004, July 22). *The 9/11 Commission report.* Retrieved from http://www.9-11 commission.gov/report/911Report.pdf

National Conference of State Legislatures. (2010). Identity theft state statutes. Retrieved from http://www.ncsl.org/?tabid=12538

National Council on Disability. (2007). *Breaking the silence on crime victims with disabilities in the United States* (Joint statement by the National Council on Disability, the Association of University Centers, and the National Center for Victims of Crime). Retrieved from http://www.ilru .org/html/training/webcasts/handouts/2007/05-30-NCVC/abstract .html

National Counterterrorism Center. (2009, April 30). *2008 report on terrorism.* Washington, DC: Office of the Director of National Intelligence, National Counterterrorism Center. Retrieved from http://www.fas.org/ irp/threat/nctc2008.pdf

National Counterterrorism Center. (2010, April 30). *2009 report on terrorism.* Washington, DC: Office of the Director of National Intelligence, National Counterterrorism Center. Retrieved from http://www.nctc .gov/witsbanner/docs/2009_report_on_terrorism.pdf

National Court Appointed Special Advocates. (2010). *Evidence of effectiveness.* Retrieved from http://www.casaforchildren.org/site/c .mtJSJ7MPIsE/b.5332511/k.7D2A/Evidence_of_Effectiveness.htm

National Crime Victim Bar Association. (2007). *Civil justice for victims of crime.* Washington, DC: Author. Retrieved from http://www.victimbar .org/vb/AGP.Net/Components/documentViewer/Download.aspxnz? DocumentID=43749

National District Attorneys Association. (2008, November). *Legislation permitting the use of anatomical dolls in child abuse cases.* Retrieved from http://www.ndaa.org/pdf/statutes-anatomical-dolls-112008.pdf

National Institute for Occupational Safety and Health. (1995, May). *Preventing homicide in the workplace* (Publication No. 93-109). Washington, DC: Department of Health and Human Services.

National Institute on Drug Abuse. (1995). *NIDA notes* (Vol. 10). Bethesda, MD: Author. Retrieved from http://archives.drugabuse.gov/NIDA_ Notes/NN95index.html

National Institute on Drug Abuse. (1998). *NIDA notes* (Vol. 13, No. 5). Bethesda, MD: Author. Retrieved from http://archives.drugabuse.gov/NIDA_Notes/NN98index.html#Number5

National Research Council. (1996). *Understanding violence against women.* Washington, DC: National Academy Press.

National Victims' Constitutional Amendment Passage. (n.d.). Retrieved from http://www.nvcap.org/

Neeves, S. (2008). *An examination of power differentials and intimate partner violence in lesbian relationships.* Unpublished master's thesis, Virginia Polytechnic Institute and State University. Retrieved from http://scholar.lib.vt.edu/theses/available/etd-05012008-211729/unrestricted/NeevesThesisIPVLesbian.pdf

Nellis, A. M. (2009). Gender differences in fear of terrorism. *Journal of Contemporary Criminal Justice, 25,* 322–340.

Nettelbeck, T., & Wilson, C. (2002). Personal vulnerability to victimization of people with mental retardation. *Trauma, Violence, and Abuse, 3,* 289–306.

Nicholas, S., Povey, D., Walker, A., & Kershaw, C. (2005). *Crime in England and Wales 2004/2005.* United Kingdom: Great Britain Home Office Research Development and Statistics Directorate.

Nies, Y., James, S. D., & Netter, S. (2010, January 28). Mean girls: Cyberbullying blamed for teen suicides. *Good Morning America, ABC News.* Retrieved from http://abcnews.go.com/GMA/Parenting/girls-teen-suicide-calls-attention-cyberbullying/story?id=9685026

Nolan, J. J., McDevitt, S., Cronin, S., & Farrell, A. (2004). Learning to see hate crimes: A framework for understanding and clarifying ambiguities in bias crime classification. *Criminal Justice Studies: A Critical Journal of Crime, Law, and Society, 17,* 91–105.

Noonan, M. E. (2010a). *Deaths in custody reporting program: Mortality in local jails, 2000–2007* (Special report). Washington, DC: Bureau of Justice Statistics, U.S. Department of Justice.

Noonan, M. E. (2010b). *Deaths in custody reporting program: Mortality in state prisons, 2000–2007* (Special report). Washington, DC: Bureau of Justice Statistics, U.S. Department of Justice.

North, C. S., Nixon, S. J., Shariat, S., Mallonee, S., McMillen, J. C., Spitznagel, E. L., et al. (1999). Psychiatric disorders among survivors of the Oklahoma City bombing. *JAMA, 282,* 755–762.

Nosek, M. A. (1996). Sexual abuse of women with physical disabilities. In D. M. Krotoski, M. A. Nosek, & M. A. Turk (Eds.), *Women with physical disabilities: Achieving and maintaining health and well-being.* (pp. 153–173). Baltimore: Paul H. Brookes.

Nugent, W. R., & Paddock, J. B. (1995). The effect of victim-offender mediation on severity of response. *Conflict Resolution Quarterly, 12*(4), 353–367.

O'Leary-Kelly, A. M., Bowes-Sperry, L., Bates, C. A., & Lean, E. R. (2009). Sexual harassment at work: A decade (plus) of progress. *Journal of Management, 35,* 503–536.

Occupational Safety and Health Administration. (n.d.). Safety and health topics: Workplace violence. Retrieved from http://www.osha.gov/SLTC/workplaceviolence/

Office for Victims of Crime. (2010, April). *Crime Victims Fund: OVC fact sheet.* Retrieved from http://www.ojp.usdoj.gov/ovc/publications/factshts/cvf2010/fs_000319.html

Office of Justice Programs. (2002, January). *Enforcement of protective orders* (Legal Series, Bulletin #4). Washington, DC: U.S. Department of Justice. Retrieved from https://www.ncjrs.gov/ovc_archives/bulletins/legalseries/bulletin4/ncj189190.pdf

Okun, L. E. (1986). *Women abuse: Facts replacing myths.* Albany: SUNY Press.

Olweus Bullying Prevention Program. (2011). *State and federal bullying information.* Retrieved from http://www.olweus.org/public/bullying_laws.page

Olweus, D. (1991). Bully/victim problems among schoolchildren: Basic facts and effects of a school-based intervention program. In D. J. Pepler & K. H. Rubin (Eds.), *The development and treatment of childhood aggression* (pp. 411–448). Hillsdale, NJ: Erlbaum.

Olweus, D. (1993). *Bullying at school: What we know and what we can do.* New York: Blackwell.

Olweus, D. (1993). Victimization by peers: Antecedents and long-term outcomes. In K. H. Rubin & J. B. Asendorpf (Eds.), *Social withdrawal, inhibition, and shyness in childhood* (pp. 315–341). Hillsdale, NJ: Erlbaum.

Olweus, D. (2007). Bullies and victims at school: Are they the same pupils? *British Journal of Educational Psychology, 77,* 441–464.

Orel, N. A. (2010). Elder abuse, neglect, and maltreatment: Institutional. In B. S. Fisher & S. P. Lab (Eds.), *Encyclopedia of victimology and crime prevention* (Vol. 1, pp. 351–353). Thousand Oaks, CA: Sage.

Pagelow, M. D. (1997). *Battered women: A historical research review and some common myths.* Binghamton, NY: Haworth Maltreatment & Trauma Press.

Pawloski, J. (2011, January 8). Olympia-area man arrested in Thurston County's largest ever ID-theft case. *The Olympian.* Retrieved from http://www.theolympian.com/2011/01/07/1498356/id-theft-bust-largest-ever-in.html

Payne v. Tennessee, 501 U.S. 808 (1991).

Payne, B. J., & Gainey, R. R. (2009). *Family violence and criminal justice, 3rd edition: A life-course approach.* Providence, NJ: Matthew Bender.

Payne, B. K., & Cikovic, R. (1995). An empirical examination of the characteristics, consequences and causes of elder abuse in nursing homes. *Journal of Elder Abuse and Neglect, 7*(4), 61–74.

Pellegrini, A. D. (1998). Bullies and victims in school: A review and call for research. *Journal of Applied Developmental Psychology, 19,* 165–176.

Petersilia, J. R. (2001). Crime victims with developmental disabilities: A review essay. *Criminal Justice and Behavior, 28,* 655–694.

Petersilia, J. (2010). A retrospective view of corrections reform in the Schwarzenegger administration. *Federal Sentencing Reporter, 22,* 148–153.

Pfohl, S. J. (1977). The "discovery" of child abuse. *Sociological Problems, 24,* 310–323.

Phillips, N. (2009). The prosecution of hate crimes: The limitations of hate crime typologies. *Journal of Interpersonal Violence, 24,* 883–905.

Pillemer, K., & Finkelhor, D. (1988). The prevalence of elder abuse: A random sample survey. *The Gerontologist, 28*(1), 51–57.

Piquero, A. R., & Hickman, M. (2003). Extending Tittle's control-balance theory to account for victimization. *Criminal Justice and Behavior, 30*(3), 282–301.

Piquero, A., MacDonald, J., Dobrin, A., Daigle, L. E., & Cullen, F. T. (2005). Self-control, violent offending, and homicide victimization: Assessing the general theory of crime. *Journal of Quantitative Criminology, 21,* 55–71.

Pleck, E. (1987). *Domestic tyranny: The making of American social policy against family.* New York: Oxford University Press.

Potter, R., & Thomas, P. (2001). *Engine immobilisers: How effective are they?* Melbourne, Australia: National Motor Vehicle Theft Reduction Council.

Powers, L. E., Curry, M. A., Oschwald, M., & Maley, S. (2002). Barriers and strategies in addressing abuse: A survey of disabled women's experiences. *Journal of Rehabilitation, 68,* 4–14.

Pratt, T. C., Holtfreter, K., & Reisig, M. D. (2010). Routine online activity and Internet fraud targeting: Extending the generality of routine activity theory. *Journal of Research in Crime and Delinquency, 47,* 267–296.

Prevent Child Abuse America. (2000). *Current trends in child abuse reporting and fatalities: The 2000 fifty state survey.* Chicago: National Center on Child Abuse Prevention Research.

Ptacek, J. (1999). *Battered women in the courtroom: The power of judicial responses.* Boston: Northeastern University Press.

RAINN. (2009). *Kentucky mandatory reporting requirements regarding children.* Retrieved from http://www.rainn.org/pdf-files-and-other-documents/Public-Policy/Legal-resources/2009-Mandatory-Report/Kentucky09C.pdf

RAINN. (2009). *Ohio mandatory reporting requirements regarding elders/disabled.* Retrieved from http://www.rainn.org/files/reportingdatabase/Ohio/OhioElderlyMandatoryReporting.pdf

RAINN. (2009). The laws in your state. Retrieved from http://www.rainn.org/public-policy/laws-in-your-state

RAINN. (2011). Georgia. Retrieved from http://www.rainn.org/files/reportingdatabase/Georgia/GeorgiaHIV.pdf

RAINN. (2011a). *Alabama mandatory reporting requirements regarding elders/disabled.* Retrieved from http://www.rainn.org/files/reportingdatabase/Alabama/AlabamaElderlyMandatoryReporting.pdf

RAINN. (2011b). *The laws in your state.* Retrieved from http://www.rainn.org/public-policy/laws-in-your-state

RAND Corporation. (2004). Compensating the victims of 9/11. *Research Brief.* Retrieved from http://www.rand.org/pubs/research_briefs/RB9087/index1.html

Rand, M. (2008). *Criminal victimization, 2007.* Washington, DC: U.S. Department of Justice Bureau of Justice Statistics.

Rand, M. (2009). *Criminal victimization, 2008.* Washington, DC: U.S. Department of Justice, Bureau of Justice Statistics.

Rand, M. R. (1997). *Violence-related injuries treated in hospital emergency departments* (Special report). Washington, DC: Bureau of Justice Statistics.

Rand, M., & Rennison, C. (2004). *How much violence against women is there?* Washington, DC: National Criminal Justice Reference Service. Retrieved from http://www.ncjrs.gov/pdffiles1/nij/199702.pdf

Raphael, J. (2008). Book Review: Taylor Jr., S., & Johnson, K. C. (2007). "Until proven innocent: Political correctness and the shameful injustices of the Duke Lacrosse rape case." *Violence Against Women, 14,* 370–375.

Rathgeber, C. (2002). The victimization of women through human trafficking: An aftermath of war? *European Journal of Crime, Criminal Law and Criminal Justice, 10,* 152–163.

Rauma, D. (1984). Going for the gold: Prosecutorial decision making in cases of wife assault. *Social Science Research, 13,* 321–351.

Reaves, B. A. (2008, February). *Campus law enforcement, 2004–05* (Special Report NCJ 219374). Washington, DC: Bureau of Justice Statistics.

Redmond, L. (1989). *Surviving when someone you love was murdered: A professional's guide to group grief therapy for families and friends of murder victims.* Clearwater, FL: Psychological Consultation and Educational Services.

Reiss, A. (1980). Victim proneness in repeat victimization by type of crime. In S. Fienberg & A. Reiss (Eds.). *Indicators of Crime and Criminal Justice: Quantitative Studies* (pp. 41–53). Washington, DC: U.S. Department of Justice.

Rengert, G., & Wasilchick, J. (1985). *Suburban burglary: A time and place for everything.* Springfield, IL: Charles C. Thomas.

Rennison, C. M. (1999). *Criminal victimization in 1998: Changes 1997–1998 with trends 1993–1998.* Washington, DC: Bureau of Justice and Statistics.

Rennison, C. M. (2002). *Criminal victimization 2001: Changes 2000–2001 with trends 1993–2001.* Washington, DC: U.S. Government Printing Office.

Reno, J., Marcus, D., Leary, M. L., & Turman, K. M. (2000). *Responding to terrorism victims: Oklahoma City and beyond.* Washington, DC: U.S Department of Justice, Office of Justice Programs, Office for Victims of Crime.

Renzetti, C. M., & Miley, C. H. (1996). *Violence in gay and lesbian domestic partnerships.* New York: Haworth.

Resnick, H., Monnier, J., & Seals, B. (2002). Rape-related HIV risk concerns among recent rape victims. *Journal of Interpersonal Violence, 17*(7), 746–759.

Rice, K. J., & Smith, W. R. (2002). Socioecological models of automotive theft: Integrating routine activity and social disorganization approaches. *Journal of Research in Crime and Delinquency, 39*(3), 304–336.

Ricketts, M. (2010). School violence. In B. S. Fisher & S. P. Lab (Eds.), *Encyclopedia of victimology and crime prevention* (Vol. 2, pp. 835–840). Los Angeles: Sage.

Rigby, K. (1997). Bullying and suicide among children. *Beyond Bullying News, 2,* 3–4.

Rigby, K., & Slee, P. (1993). Dimensions of interpersonal relating among Australian school children: Implications for psychological well-being. *Journal of Social Psychology, 131,* 615–627.

Riggs, D. S., Caulfield, M. B., & Street, A. E. (2000). Risk for domestic violence: Factors associated with perpetration and victimization. *Journal of Clinical Psychology, 56,* 1289–1316.

Ritsche, D. F. (2006, September). *Sex crime legislation* (Informational Bulletin 06-3). Madison: State of Wisconsin Legislative Reference Bureau.

Robers, S., Zhang, J., Truman, J., & Snyder, T. D. (2010, November). *Indicators of school crime and safety: 2010* (NCES 2011-002/NCJ 230812). Washington, DC: National Center for Education Statistics, U.S. Department of Education, and Bureau of Justice Statistics, Office of Justice Programs, U.S. Department of Justice.

Roberts, A. L., Austin, S. B., Corliss, H. L., Vandermorris, M. D., & Koenen, K. C. (2010). Pervasive trauma exposure among U.S. sexual orientation minority adults and risk of posttraumatic stress disorder. *American Journal of Public Health, 100,* 2433–2441.

Robinson, L., de Benedictis, T., & Segal, J. (2011). *Elder abuse and neglect: Warning signs, risk factors, prevention, and help.* Retrieved from http://helpguide.org/mental/elder_abuse_physical_emotional_sexual_neglect.htm

Rosenbaum, M. E. (1986). The acquaintance process: Looking mainly backward. *Journal of Personality and Social Psychology, 51,* 1156–1166.

Rounds-Bryant, J., Kristiansen, P. L., Fairbank, J. A., & Hubbard, R. L. (1998). Substance use, mental disorders, abuse and crime: Gender comparisons among a national sample of adolescent drug treatment centers. *Journal of Child and Adolescent Substance Abuse, 7*(4), 19–34.

Ruback, R. B., Menard, K. S., Outlaw, M. C., & Shaffer, J. N. (1999). Normative advice to campus crime victims: Effects of gender, age, and alcohol. *Violence and Victims, 14*(4), 381–396.

Rycus, J., & Hughes, R. (1998). *Field guide to child welfare. Volume IV: Placement and permanence.* Washington, DC: Child Welfare League of America and Columbus, OH: Institute for Human Services.

Saisan, J., Smith, M., & Segal, J. (2011, June). *Child abuse and neglect: Recognizing and preventing child abuse.* Retrieved from http://www.helpguide.org/mental/child_abuse_physical_emotional_sexual_neglect.htm

Salasin, S. E., & Rich, R. F. (1993). Mental health policy for victims of violence: The case against women. In J. P. Wilson and B. Raphael (Eds.), *International handbook of traumatic stress syndromes* (pp. 947–955). New York: Plenum.

Sampson, R. J. (1985). Neighborhood and crime: The structural determinants of personal victimization. *Journal of Research in Crime and Delinquency, 22*(1), 7–40.

Sampson, R. J., & Laub, J. H. (1993). *Crime in the making: Pathways and turning points through life.* Cambridge, MA: Harvard University Press.

Sampson, R. J., Raudenbush, S. W., & Earls, F. (1997). Neighborhoods and violent crime: A multilevel study of collective efficacy. *Science, 277,* 918–924.

Schafer, S. (1968). The victim and his criminal: A study in functional responsibility. New York: Random House.

Schauer, E. J., & Wheaton, E. M. (2006). Sex trafficking in the United States: A literature review. *Criminal Justice Review, 31,* 146–169.

Schmidt, J., & Steury, E. (1989). Prosecutorial discretion in filing charges in domestic violence cases. *Criminology, 27,* 487–510.

Schreck, C. J. (1999). Criminal victimization and low self-control: An extension and test of a general theory of crime. *Justice Quarterly, 16*(3), 633–654.

Schreck, C. J., & Fisher, B. S. (2004). Specifying the influence of family and peers on violent victimization: Extending routine activities and lifestyles theories. *Journal of Interpersonal Violence, 19*(9), 1021–1041.

Schreck, C. J., Stewart, E. A., & Fisher, B. S. (2006). Self-control, victimization, and their influence on risky activities and delinquent friends: A longitudinal analysis using panel data. *Journal of Quantitative Criminology, 22,* 319–340.

Schuster, M. A., Stein, B. D., Jaycox, L. H., Collins, R. L., Marshall, G. N., Elliott, M. N., et al. (2001). A national survey of stress reactions after the September 11, 2001, terrorist attacks. *New England Journal of Medicine, 345,* 1507–1512.

Schwartz, M. D., & Pitts, V. L. (1995). Exploring a feminist routine activities approach to explaining sexual assault. *Justice Quarterly, 12*(1), 9–31.

Scott, K. D., Schafer, J., & Greenfield, T. K. (1999). The role of alcohol in physical assault perpetration and victimization. *Journal of Studies on Alcohol, 60,* 528–536.

Sedlak, A. J., Mettenburg, J., Basena, M., Petta, I., McPherson, K., Greene, A., et al. (2010). *Fourth national incidence study of child abuse and neglect (NIS-4): Report to Congress executive summary.* Washington, DC: U.S. Department of Health and Human Services, Administration for Children and Families.

Seligman, M. E. P. (1975). Helplessness: On depression, development, and death; A series of books in psychology. New York: W. H. Freeman.

Sexual assault laws of Alabama. (n.d.). Retrieved from http://www.ageofconsent.com/alabama.htm

Shared Hope International. (2010). *Protected innocence legislative framework: Methodology.* Retrieved from http://media.aclj.org/pdf/shared-hope-intl_protected-innocence-methodology.pdf

Shepherd, J. M. (2002). Reflections on a rape trial: The role of rape myths and jury selection in the outcome of a trial. *Affilia, 17,* 69–92.

Sherman, L. (1992). *Policing domestic violence: Experiments and dilemmas.* New York: Free Press.

Sherman, L. W., & Berk, R. A. (1984). The specific deterrent effects of arrest for domestic assault. *American Sociological Review, 49,* 261–272.

Sherman, L. W., Strang, H., Angel, C., Woods, D., Barnes, G., Bennett, S., et al. (2005). Effects of face-to-face restorative justice on victims of crime in four randomized, controlled trials. *Journal of Experimental Criminology, 1*(3), 367–395.

Siegel, R. S., La Greca, A. M., & Harrison, H. M. (2009). Peer victimization and social anxiety in adolescents: Prospective and reciprocal relationships. *Journal of Youth and Adolescence, 38*(8), 1096–1109.

Silver, E. (2002). Mental disorder and violent victimization: The mediating role of involvement in conflicted social relationships. *Criminology, 40,* 191–212.

Silver, E., Arseneault, L., Langley, J., Caspi, A., & Moffitt, T. (2005). Mental disorder and violent victimization in a total birth cohort. *American Journal of Public Health, 95,* 2015–2021.

Silver, R. C., Holman, E. A., McIntosh, D. N., Poulin, M., & Gil-Rivas, V. (2002). Nationwide longitudinal study of psychological responses to September 11. *JAMA, 288,* 1235–1244.

Silverman, J. G., Raj, A., Mucci, L. A., & Hathaway, J. E. (2001). Dating violence against adolescent girls and associated substance use, unhealthy weight control, sexual risk behavior, pregnancy and suicidality. *Journal of the American Medical Association, 286,* 572–579.

Sloan, J. J., & Shoemaker, J. (2007). State-level Cleary Act initiatives: Symbolic politics or substantive policy? In B. S. Fisher & J. J. Sloan (Eds.), *Campus crime: Legal, social, and policy perspectives* (pp. 102–121). Springfield, IL: Charles C. Thomas.

Slone, M., & Shoshani, A. (2008). Indirect victimization from terrorism: A proposed post-exposure intervention. *Journal of Mental Health Counseling, 30,* 255–266.

SmartMotorist.com. (n.d.). History of auto-theft legislation. Retrieved from http://www.smartmotorist.com/auto-security-systems/history-of-auto-theft-legislation.html

Smith, B. L., Sloan, J. J., & Ward, R. M. (1990). Public support for the victims' rights movement: Results of a statewide survey. *Crime and Delinquency, 36*(4), 488–502.

Smith, Gambrell, & Russell, LLP. (2005). Workplace violence: Recognizing risk factors and formulating prevention strategies. *Trust the Leaders, 13.* Retrieved from http://www.sgrlaw.com/resources/trust_the_leaders/leaders_issues/ttl13/869/

Smith, N. E., & Batiuk, M. E. (1989). Sexual victimization and inmate social interaction. *The Prison Journal, 69,* 29–38.

Smolak, L., & Murnen, S. K. (2002). A meta-analytic examination of the relationship between childhood sexual abuse and eating disorders. *International Journal of Eating Disorders, 31,* 136–150.

Sobsey, D. (1994). *Violence and abuse in the lives of people with disabilities: The end of silent acceptance?* Baltimore: Paul H. Brookes.

Sobsey, D., & Doe, T. (1991). Patterns of sexual abuse and assault. *Journal of Sexuality and Disability, 9,* 243–259.

Spohn, C., & Holleran, D. (2004). On the use of the total incarceration variable in sentencing research. *Criminology, 42*(1), 211–240.

Stark, E. (1984, May). The unspeakable family secret. *Psychology Today,* pp. 42–46.

State v. Johnson, No. C4-92-251, 1993, Minn. App. LEXIS 617 (Minn. App. June 9, 1993).

Stepakoff, S. (1998). Effects of sexual victimization on suicidal ideation and behavior in U.S. college women. *Suicide and Life-Threatening Behavior, 28*(1), 107–126.

Stets, J., & Straus, M. (1990). Gender differences in reporting marital violence and its medical and psychological consequences. In M. Straus & R. Gelles (Eds.), *Physical violence in American families* (pp. 227–244). New Brunswick, NJ: Transaction.

Stewart, E. A., Elifson, K. W., & Sterk, C. E. (2004). Integrating the general theory of crime into an explanation of violent victimization among female offenders. *Justice Quarterly, 21*, 159–180.

Stiegel, L. A. (2008). *Durable power of attorney abuse: It's a crime too; A national center on elder abuse fact sheet for consumers.* Retrieved from http://www.ncea.aoa.gov/ncearoot/main_site/pdf/publication/DurablePowerOfAttorneyAbuseFactSheet_Consumers.pdf

Stiegel, L., Klem, E., & Turner, J. (2007). *Neglect of older persons: An introduction to legal issues related to caregiver duty and liability.* Retrieved from http://www.ncea.aoa.gov/ncearoot/main_site/pdf/publication/NeglectOfOlderPersons.pdf

Stombler, M. (1994). "Buddies" or "slutties": The collective sexual reputation of fraternity little sisters. *Gender and Society, 8,* 297–323.

Storch, E. A. (2003). Reliability and factor structure of the Sport Anxiety Questionnaire in fifth- and sixth-grade children. *Psychological Reports, 93,* 160.

Strang, H. (2002). *Repair or revenge: Victims and restorative justice.* Oxford, UK: Clarendon.

Straus, M. A. (2007). Conflict tactics scales. In N. A. Jackson (Ed.), *Encyclopedia of violence* (pp. 190–197). New York: Routledge, Taylor & Francis Group.

Straus, M. A., Hamby, S. L., Boney-McCoy, S., & Sugarman, D. B. (1996). The revised Conflict Tactics Scale (CTS2): Development and preliminary psychometric data. *Journal of Family Issues, 17*(3), 283–316.

Sugarman, D. B., & Frankel, S. L. (1996). Patriarchal ideology and wife-assault: A meta-analytic review. *Journal of Family Violence, 11,* 13–40.

Sullivan, P. M., & Knutson, J. F. (2000). Maltreatment and disabilities: A population-based epidemiological study. *Child Abuse & Neglect, 24,* 1257–1273.

Sweeting, H., Young, R., West, P., & Der, G. (2006). Peer victimization and depression in early-mid adolescence: A longitudinal study. *British Journal of Educational Psychology, 76*(3), 577–594.

Sygnatur, E. F., & Toscano, G. A. (2000, Spring). Work-related homicides: The facts. *Compensation and Working Conditions,* 3–8.

Tabachneck, A., Norup, H., Thomason, S., & Motlagh, P. (2000). *VICC approved theft deterrent systems: A study into the impact of VICC approved theft deterrent systems on insurance theft claim frequency and loss cost.* Don Mills, Ontario: Vehicle Information Centre of Canada.

Tatara, T., Kuzmeskus, L. B., Duckhorn, E., & Bivens, L. (1998). *National Elder Abuse Incidence Study: Final report.* Washington, DC: Administration on Aging, U.S. Department of Health and Human Services. Retrieved from http://aoa.gov/AoA_Programs/Elder_Rights/Elder_Abuse/docs/ABuseReport_Full.pdf

Taylor, B. G., Davis, R. C., & Maxwell, C. D. (2001). The effects of a group batterer treatment program in Brooklyn. *Justice Quarterly, 18,* 170–201.

Taylor, T. J., Peterson, D., Esbenson, F., & Freng, A. (2007). Gang membership as a risk factor for adolescent violent victimization. *Journal of Research in Crime and Delinquency, 44*(4), 351–380.

Teasdale, B. (2009). Mental disorder and violent victimization. *Criminal Justice and Behavior, 36,* 513–535.

Teaster, P. B., Otto, J. M., Dugar, T. A., Mendiondo, M. S., Abner, E. L., & Cecil, K. A. (2006). *The 2004 survey of state adult protective services: Abuse of adults 60 years of age and older.* Retrieved from http://www.elderabusecenter.org/pdf/2-14-06%20FINAL%2060+REPORT.pdf

Teplin, L. A., McClelland, G. M., Abram, K. M., & Weiner, D. A. (2005). Crime victimization in adults with severe mental illness: Comparison with the National Crime Victimization Survey. *Archives of General Psychiatry, 62,* 911–921.

Tewksbury, R., & Mustaine, E. E. (2000). Routine activities and vandalism: A theoretical and empirical study. *Journal of Crime and Justice, 23,* 81–110.

Tewksbury, R., & Pedro, D. (2003). The role of alcohol in victimization. In L. J. Moriarty (Ed.), *Controversies in victimology* (pp. 25–42). Cincinnati, OH: Anderson.

The Gadsden Times. (2011, April 8). Gadsden mother convicted of aggravated child abuse of 22-month-old son. Retrieved from http://www.gadsdentimes.com/article/20110408/NEWS/110409788/1084/NEWS?Title=Gadsden mother-convicted-of-aggravated-child-abuse-of-22-month-old-son

The Smoking Gun. (2010, April 15). Ben Roethlisberger's bad play: Police reports detail NFL quarterback's unseemly night in Georgia. Retrieved from http://www.thesmokinggun.com/documents/crime/ben-rocthlisbergers-bad-play

Thompson, M. P., Kaslow, N. J., Price, A. W., Williams, K., & Kingree, J. B. (1998). Role of secondary stressors in the parental death–child distress relation. *Journal of Abnormal Child Psychology, 26*(5), 357–366.

Tittle, C. R. (1995). *Control balance: Toward a general theory of deviance.* Boulder, CO: Westview.

Tittle, C. R. (1997). Thoughts stimulated by Braithwaite's analysis of control balance theory. *Theoretical Criminology, 1*(1), 99–110.

Tjaden, P., & Thoennes, N. (1998, November). Prevalence, incidence, and consequences of violence against women: Findings from the National Violence Against Women Survey. *Research in Brief.* Washington, DC: U.S. Department of Justice, National Institute of Justice and U.S. Department of Health and Human Services, Centers for Disease Control and Prevention. Retrieved from https://www.ncjrs.gov/pdffiles/172837.pdf

Tjaden, P., & Thoennes, N. (2000). *Full report of the prevalence, incidence and consequences of violence against women.* Washington, DC: National Institute of Justice, Centers for Disease Control and Prevention.

Tjaden, P., & Thoennes, N. (2000). Prevalence and consequences of male-to-female and female-to-male intimate partner violence as measured by the National Violence Against Women Survey. *Violence Against Women, 6,* 142–161.

Tjaden, P., & Thoennes, N. (2006). *Extent, nature, and consequences of rape victimization: Findings from the National Violence Against Women Survey.* Rockville, MD: National Institute of Justice.

Tjaden, P., Thoennes, N., & Allison, C. J. (1999). Comparing violence over the life span in samples of same-sex and opposite cohabitants. *Violence and Victims, 14,* 413–425.

Tobolowsky, P. (1999). Victim participation in the criminal justice process: Fifteen years after the President's Task Force on Victims of Crime. *Criminal and Civil Confinement, 25*(21), 21–105.

Toch, H. (1977). *Police, prisons, and the problem of violence.* Washington, DC: U.S. Government Printing Office.

Truman, J. L., & Rand, M. R. (2010). *Criminal victimization, 2009.* Washington, DC: Bureau of Justice Statistics, Office of Justice Programs, U.S. Department of Justice. Retrieved from http://bjs.ojp.usdoj.gov/content/pub/pdf/cv09.pdf

Tucker, P., Pfefferbaum, B., North, C. S., Kent, A., Jeon-Slaughter, H., & Parker, D. E. (2010). Biological correlates of direct exposure to terrorism several years postdisaster. *Annals of Clinical Psychiatry, 22,* 186–195.

Tunnell, K. (1992). *Choosing crime: The criminal calculus of property offenders.* Chicago: Nelson-Hall.

Turner, H. A., Finkelhor, D., & Ormond, R. (2010). The effect of adolescent victimization on self-concept and depressive symptoms. *Child Maltreatment, 15*(1), 76–90.

Tynes, B., & Giang, M. (2009). P01-298 Online victimization, depression, and anxiety among adolescents in the U.S. *European Psychiatry, 24*(1), S686.

U.S. Department of Health and Human Services. (1996, July). *Violence in the workplace: Risk factors and prevention strategies* (DHHS [NIOSH] Publication No. 96-100). Retrieved from http://www.cdc.gov/niosh/violcont.html

U.S. Department of Health and Human Services. (1999, April). *Blending perspectives and building common ground: A report to Congress on substance abuse and child protection.* Washington, DC: U.S. Government Printing Office. Retrieved from http://aspe.hhs.gov/hsp/subabuse99/subabuse.htm

U.S. Department of Health and Human Services. (2010). *Child maltreatment, 2009.* Washington, DC: Author. Retrieved from www.acf.hhs.gov/programs/cb/pubs/cm09/cm09.pdf

U.S. Department of Health and Human Services. (2011a). *Certification for victims of trafficking fact sheet.* Retrieved from http://www.acf.hhs.gov/trafficking/about/cert_victims.pdf

U.S. Department of Health and Human Services. (2011b). *Labor trafficking fact sheet.* Retrieved from http://www.acf.hhs.gov/trafficking/about/fact_labor.pdf

U.S. Department of Justice. (n.d.). Identity theft and identity fraud. Retrieved from http://www.justice.gov/criminal/fraud/websites/idtheft.html

U.S. Department of State. (2010, June). *Trafficking in persons report* (10th ed.). Retrieved from http://www.state.gov/documents/organization/142979.pdf

U.S. National Library of Medicine. (2011). *Shaken baby syndrome.* Washington, DC: National Institutes of Health. Retrieved from http://www.nlm.nih.gov/medlineplus/ency/article/000004.htm

Ulicyn, G. R., White, G., Bradford, B., & Matthews, R. M. (1990). Consumer exploitation by attendants: How often does it happen and can anything be done about it? *Rehabilitation Counseling Bulletin, 33,* 240–246.

Ullman, S. E. (2007). A 10-year update of "review and critique of empirical studies of rape avoidance." *Criminal Justice and Behavior, 34,* 411–429.

Umbreit, M. S. (1994a). Crime victims confront their offenders: The impact of the Minneapolis Mediation Program. *Research on Social Work Practice, 4*(4), 436–447.

Umbreit, M. S. (1994b). *Victim meets offender: The impact of restorative justice and mediation.* Monsey, NY: Criminal Justice Press.

Umbreit, M. S. (2000). *Peacemaking and spirituality: A journey toward healing and strength.* Saint Paul: Center for Restorative Justice and Peacemaking, University of Minnesota. Retrieved from http://www.cehd.umn.edu/ssw/rjp/resources/Forgiveness/Peacemaking_and_Spirituality_Journey_Toward_Healing.pdf

Umbreit, M. S., Coates, R. B., & Kalanj, B. (1994). *Victim meets offender: The impact of restorative justice and mediation.* Monsey, NY: Willow Tree Press.

Umbreit, M., & Greenwood, J. (2000). *Guidelines for victim-sensitive victim offender mediation: Restorative justice through dialogue.* Washington, DC: Office for Victims of Crime, Office of Justice Programs.

Valenti-Hein, D., & Schwartz, L. (1995). *The sexual abuse interview for those with developmental disabilities.* Santa Barbara: CA: James Stanfield Company.

van Dijk, J. J. M., van Kesteren, J. N., & Smit, P. (2008). Background to the International Crime Victims Survey. In *Criminal victimisation in international perspective: Key findings from the 2004–2005 ICVS and EU ICS* (pp. 21–23). The Hague: Boom Legal. Retrieved from http://rechten.uvt.nl/icvs/pdffiles/ICVS2004_05.pdf

Verdugo, M. A., Bermejo, B. G., & Fuertes, J. (1995). Maltreatment of intellectually handicapped children and adolescents. *Child Abuse and Neglect, 19,* 205–215.

Victim-Offender Reconciliation Program Information and Resource Center. (2006). *About victim-offender mediation and reconciliation.* Retrieved from http://www.vorp.com/

Virginia Department of Corrections. (2010). Victim services. Retrieved August 30, 2011, from http://www.vadoc.state.va.us/victim

von Hentig, H. (1948). The criminal and his victim: Studies in the sociobiology of crime. Cambridge, MA: Yale University Press.

Waalen, J., Goodwin, M., Spitz, A., Petersen, R., & Saltzman, L. (2000). Screening for intimate partner violence by health care providers: Barriers and interventions. *American Journal of Preventative Medicine, 19,* 230–237.

Wacker, J. L., Parish, S. L., & Macy, R. J. (2008). Sexual assault and women with cognitive disabilities: Codifying discrimination in the states. *Journal of Disability Policy Studies, 19,* 86–94.

Walker, L. E. (1979). *The battered woman.* New York: Harper & Row.

Wallace, H. (2007). Victimology: Legal, psychological, and social perspectives (2nd ed.). Boston: Pearson.

Wallace, R. M. M. (1997). *International human rights: Text and materials.* London: Sweet & Maxwell.

Watts-English, T., Fortson, B. L., Gibler, N., Hooper, S. R., & De Bellis, D. (2006). The psychology of maltreatment in childhood. *Journal of Social Issues, 62*(4), 717–736.

Weinberg, M. K., & Tronick, E. Z. (1998). Emotional characteristics of infants associated with maternal depression and anxiety. *Pediatrics, 102,* 1298–1304.

Welsh, B. C., Loeber, R., Stevens, B. R., Stouthamer-Loeber, M., Cohen, M. A., & Farrington, D. P. (2008). Costs of juvenile crime in urban areas: A longitudinal perspective. *Youth Violence and Juvenile Justice, 6,* 3–27.

West, K., & Wiley-Cordone, J. (1999). Healing the hate: Innovations in hate crime prevention. *Illinois Council for the Prevention of Violence, 1,* 1–4.

Widom, C. S. (1989). Child abuse, neglect, and adult behavior. *American Journal of Orthopsychiatry, 59*(3), 355–367.

Widom, C. S. (1989). *The intergenerational transmission of violence.* New York: Harry Frank Guggenheim Foundation.

Widom, C. S. (2000). Understanding the consequences of child victimization. In R. M. Reece (Ed.), *Treatment of child abuse: Common ground for mental health, medical and legal practitioners* (pp. 339–361). Baltimore, MD: Johns Hopkins University Press.

Widom, C. S. (2000, January). Childhood victimization: Early adversity, later psychopathology. *National Institute of Justice Journal*, 2–9.

Wilcox, P., Jordan, C. E., & Pritchard, A. J. (2007). A multidimensional examination of campus safety: Victimization, perceptions of danger, worry about crime, and precautionary behavior among college women in the post-Clery era. *Crime & Delinquency, 53*, 219–254.

Wilcox, W. B., & Dew, J. (2008, January). *Protectors or perpetrators? Fathers, mothers, and child abuse and neglect* (Research brief 7). New York: Institute for American Values: Center for Marriage and Families.

Williams, K., Chambers, M., Logan, S., & Robinson, D. (1996). Association of common health symptoms with bullying in primary school children. *BMJ, 313*, 17–19.

Wilson, C., & Brewer, N. (1992). The incidence of criminal victimization of individuals with an intellectual disability. *Australian Psychologist, 27*, 114–117.

Wilson, C., Seaman, L., & Nettelbeck, T. (1996). Vulnerability to criminal exploitation: Influence of interpersonal competence differences among people with mental retardation. *Journal of Intellectual Disability Research, 40*, 10–19.

Wilson, M., & Daly, M. (1993). Spousal homicide risk and estrangement. *Violence and Victims, 8*, 3–16.

Wisconsin Coalition Against Sexual Assault. (2003). *People with disabilities and sexual assault: Information sheet series*. Madison, WI: Author. Retrieved from http://kyasap.brinkster.net/Portals/0/pdfs/Disabilitiesandsexualassault.pdf

Wisner, C., Gilmer, T., Saltman, L., & Zink, T. (1999). Intimate partner violence against women: Do victims cost health plans more? *Journal of Family Practice, 48*, 6439–6443.

Wolf, M., Holt, V., Kernic, M., & Rivara, F. (2000). Who gets protection orders for intimate partner violence? *American Journal of Preventative Medicine, 19*, 286–291.

Wolff, N., Blitz, C., & Shi, J. (2007). Rates of sexual victimization in prisoners with and without mental disorders. *Psychiatric Services, 58*, 1087–1094.

Wolff, N., Shi, J., & Siegel, J. A. (2009a). Patterns of victimization among male and female inmates: Evidence of an enduring legacy. *Violence and Victims, 24*, 469–484.

Wolff, N., Shi, J., & Siegel, J. A. (2009b). Understanding physical victimization inside prisons: Factors that predict risk. *Justice Quarterly, 26*, 445–475.

Wolfgang, M. E. (1957). Victim precipitated criminal homicide. *Journal of Criminal Law, Criminology, and Police Science, 48*(1), 1–11.

Wolfgang, M. E., & Ferracuti, F. (1967). *The Subculture of Violence: Towards an Integrated Theory in Criminology*. London: Tavistock Publications.

Wooldredge, J. D. (1998). Inmate lifestyles and opportunities for victimization. *Journal of Research in Crime and Delinquency, 35*, 480–502.

Wright, R. T., & Decker, S. (1994). *Burglars on the job: Streetlife and residential break-ins*. Boston: Northeastern University Press.

Yllo, K. A., & Straus, M. A. (1990). Patriarchy and violence against wives: The impact of structural and normative factors. In M. A. Straus & R. J. Gelles (Eds.), *Physical violence in American families* (pp. 383–389). New Brunswick, NJ: Transaction.

Young, M. A. (1989). Crime, violence, and terrorism. In R. Gist & B. Lubin (Eds.), *Psychological aspects of disaster* (pp. 61–85). New York: Wiley.

Young, M. E., Nosek, M. A., Howland, C., Chanpong, G., & Rintala, D. H. (1997). Prevalence of abuse of women with physical disabilities. *Archives of Physical Medicine & Rehabilitation, 78*(12, Suppl. 5), S34–S38.

Young, M., & Stein, J. (2004). The history of the crime victims' movement in the United States: A component of the Office for Victims of Crime oral history project. Washington, DC: United States Department of Justice.

Yun, I., Johnson, M., & Kercher, G. (2005). *Victim impact statements: What victims have to say*. Huntsville, TX: Sam Houston State University Crime Victims' Institute.

Zaykowski, H. (2010). Racial disparities in hate crime reporting. *Violence and Victims, 25*, 378–394.

Zielinski, D. (2009). Child maltreatment and adult socioeconomic well-being. *Child Abuse and Neglect, 33*(10), 666–678.

Zimmerman, C., Hossain, M., Yun, K., Gaidaiev, N., Techomarova, M., Cirrochi, R. A., et al. (2008). The health of trafficked women: A survey of women entering posttrafficking services in Europe. *American Journal of Public Health, 98*, 55–59.

Zimmerman, C., Yun, K., Watts, C., Trappolin, L., Treppete, M., Bimbi, F., et al. (2003). *The health risks and consequences of trafficking women and adolescents: Findings from a European study*. London: London School of Hygiene and Tropical Medicine.

Zink, T. M., Fisher, B. F., Regan, S. L., & Pabst, S. (2005). The prevalence and incidence of intimate partner violence in older women in primary care practices. *Journal of General Internal Medicine, 20*, 884–888.

Zubrick, S. R., Silburn, L., Gurrin, H., Teoh, C., Shepherd, J., Carlton, J., et al. (1997). *Western Australian Child Health Survey: Education, health and competence*. Perth: Australian Bureau of Statistics and the TVW Institute for Child Health Research.

Index

About the Editor

Leah E. Daigle is assistant professor in the Department of Criminal Justice in the Andrew Young School of Policy Studies at Georgia State University. She received her PhD in criminal justice from the University of Cincinnati in 2005. Her most recent research has centered on repeat sexual victimization of college women and responses women use during and after being sexually victimized. Her other research interests include the development and continuation of offending and victimization across the life course. She is coauthor of *Criminals in the Making: Criminality Across the Life Course* and *Unsafe in the Ivory Tower: The Sexual Victimization of College Women,* which was awarded the 2011 Outstanding Book Award by the Academy of Criminal Justice Sciences. She has also published numerous peer-reviewed articles that have appeared in outlets such as *Justice Quarterly, Journal of Quantitative Criminology, Journal of Interpersonal Violence,* and *Victims and Offenders.*